AMERICAN VACUUM SOCIETY CLASSICS

VACUUM SEALING TECHNIQUES

AMERICAN VACUUM SOCIETY CLASSICS

H. F. Dylla, Series Editor-in-Chief

Basic Data of Plasma Physics: The Fundamental Data on Electrical Discharges in Gases
Sanborn C. Brown

Field Emission and Field Ionization
Robert Gomer

Vacuum Technology and Space Simulation
David H. Holkeboer, Donald W. Jones, Frank Pagano, and Donald J. Santeler

The Physical Basis of Ultrahigh Vacuum
P. A. Redhead, J. P. Hobson, and E. V. Kornelsen

Handbook of Electron Tube and Vacuum Techniques
Fred Rosebury

Vacuum Sealing Techniques
A. Roth

Ionized Gases
A. von Engel

AMERICAN VACUUM SOCIETY CLASSICS

VACUUM SEALING TECHNIQUES

A. Roth

AIP PRESS

American Institute of Physics New York

AIP Press
American Institute of Physics
500 Sunnyside Boulevard
Woodbury, NY 11797-2999

Library of Congress Cataloging-in-Publication Data

Roth, A. (Alexander)
Vacuum sealing techniques /by A. Roth.
p. cm.—(American Vacum Society classics)
Includes bibliographical references and index.
ISBN 1-56396-259-4
1. Vacuum technology. 2. Sealing (Technology) I. Title. II. Series.
TJ940.R619 1993 93-27399
621.5'5–dc20 CIP

Series Preface

The science of producing and measuring high and ultrahigh vacuum environments has fundamental interest for basic research in addition to a wide variety of important practical applications. Basic research involving particle physics, atomic and molecular physics, plasma physics, physical chemistry, and surface science often involves careful production, control, and measurement of a vacuum environment in order to perform experiments. Practical applications of vacuum science and technology are found in many materials processing techniques used for microelectronic, photonic, and magnetic materials and in the simulation of space and rarefied gas environments.

As in most of modern science, the rapid development of the field has been accompanied by a parallel growth in the related technical literature including specialized journals, monographs, and textbooks. There exist early publications in vacuum science and technology which have attained the status of indispensable references among practitioners, lecturers, and students of the field. Many of these "classic" publications have gone out of print and are currently unavailable to newcomers to the field. The present series, commissioned by the American Vacuum Society, and published by AIP Press to celebrate the 40th anniversary of the Society in 1993, is entitled the "**American Vacuum Society Classics.**"

The American Vacuum Society Classics will reprint important books from the last four decades that continue to have significant impact on the modern development of the field. It is the goal of the American Vacuum Society Classics to reprint these books in a high quality and affordable format to ensure wide availability to the technical community, individual researchers, and students.

H. F. Dylla
Continuous Electron Beam Accelerator Facility
Newport News, VA
and
College of William and Mary
Williamsburg, VA

PREFACE

This book is intended to be a systematic answer to the many questions concerning "vacuum sealing" encountered by the author during his work as a designer, technologist and consultant.

Vacuum technology has expanded recently, extending to numerous new fields and progressing towards higher vacua. With this development, a technical literature of a considerable expanse has grown, dealing with particular solutions for specific problems or reviewing the general aspects of this technology. The aim of this book is not to duplicate those books dealing with the questions: "how to obtain a vacuum", "how to measure the vacuum" or "how to use the vacuum for a particular application", but to complement them by answering the question of "how to seal a vacuum system".

Any physicist, chemist, or engineer, who has built or used a vacuum system, knows that to insure and to maintain the vacuum tightness at the required level constitutes one of the major problems. Leak hunting, plugging, and repeated testing of the vacuum system after its construction is time consuming and expensive. It can be avoided only by careful design and construction of the constituent parts and adequate assembly of the vacuum system.

The number of solutions developed for the construction of vacuum seals is very large, as indicated by the fact that over a thousand references are listed in this book. The author has attempted to make a systematic and detailed classification of all these vacuum seals. The book follows this classification, and discusses the various seals, in two fundamental groups: (a) basic vacuum seals (Chapters 2 and 3); (b) specialized vacuum seals (Chapters 4 to 7). As these groups are then divided according to the constructional features of the seals, summarized cross references are included in order to allow the reader to find easily the description of the various seals meeting the particular requirements (e.g. ultra-high vacuum, cryogenic, etc.).

To the many authors cited as reference throughout the book is due the credit for their solutions, and to them this author acknowledges his indebtedness. It was attempted to give credit throughout the book to all the authors, publishers, institutions and industrial establishments, and omissions in the acknowledgments are unintentional.

The author wishes to express his thanks to the numerous authors, publishers, institutions and industrial establishments for their kind permission to reproduce material from their publications, and for their co-operation in supplying the relevant technical information.

The author is very grateful to Mr. S. Mandelzweig M.Sc. (Eng.), who kindly undertook the editing of the manuscript, and made many valuable suggestions, which were very useful in improving the text.

The author wishes to express his indebtedness to the Management and Staff of Pergamon Press Ltd. for their kind co-operation.

A. ROTH

CONTENTS

CHAPTER 1

THE VACUUM AND SURROUNDING SPACE

1.1 THE VACUUM

The Latin word *vacuum* means "empty". The term "vacuum" refers to a given space filled with gas at pressures below atmospheric (American Vacuum Society[26]).

11.1 Pressure and mean free path

According to the kinetic theory* a gas exerts pressure on the enclosing walls because of the elastic impact of molecules on these walls. Thus the *pressure* is the force applied by the gas per unit area of real wall surface or imaginary surface located inside the gas.

The *pressure units* used in vacuum measurements are expressed as *force per unit area* (e.g. Newtons per square metre, dynes per square centimetre, pounds per square inch, kilograms per square metre, etc.) or as *height of a column of liquid* balanced by the gas at the given pressure (e.g. centimetre or inch of water; inch, millimetre or micron of mercury, etc.).

The pressure unit recommended by the American Vacuum Society[26] to be used in vacuum technology is the *torr* (millimetre of mercury) defined as 1/760 of a standard atmosphere or 1,013,250/760 dyne/cm^2.

Factors for the conversion of the various pressure units used in vacuum techniques are listed in Appendix A.1.

In a gas the molecules travel in straight line until they collide with other molecules or with the walls of the vacuum system. The average distance travelled by a gas molecule between two successive collisions is known as its *mean free path.* The mean free path λ is inversely proportional to the number of molecules per unit volume, thus at a given temperature it is inversely proportional to the pressure:†

$$\lambda_{(\mathrm{cm})} = \frac{k}{p_{(\mu)}}$$

* Details on kinetic theory of gases see e.g. Dushman[314, 315], Heinze[533].

† For the effect of temperature on the mean free path see Dushman[314, 315].

TABLE 1.1. VALUES OF *k* (CM.MICRON) AT 20 °C

Gas	*k*	Gas	*k*
Xenon	3.00	Nitrogen	5.1
Water vapour	3.40	Air	5.1
Carbon dioxide	3.34	Oxygen	5.4
Chlorine	3.47	Mercury vapour	6.3
Krypton	4.05	Hydrogen	9.3
Argon	5.07	Neon	10.4
		Helium	14.6

Table 1.1 lists the values of the constant k for various gases.* These values are in fact the mean free paths (expressed in cm) at a pressure of 10^{-3} torr (1 micron).

11.2 Low, high and ultra-high vacuum

The lowest pressure to be obtained, measured and maintained in a vacuum chamber or system determines the main characteristics of the system, the pumps to be used, the gauges and the seals.

According to the lowest pressure and the corresponding mean free path *vacuum ranges* were established.

The *low vacuum* range includes pressures less than atmospheric and greater than 25 torr. The lower limit of this range was set at that pressure (about one inch of mercury) corresponding to approximately the vapour pressure of water at 25 °C.

Medium vacuum is considered to be in the range of 25 to 10^{-3} torr. Another classification divides the same range (760 to 10^{-3}) into *rough vacuum* (760 to 1 torr) and *fine vacuum* (1 to 10^{-3} torr).

In any case the low vacuum is the range of water vapours and the medium or fine vacuum the range of other vapours (mercury, hydrocarbons, etc.).

High vacuum is defined as the condition in a gas-filled space at pressures less than 10^{-3} torr. This range is used to be divided into *high, very high and ultra-high vacuum* (see Table 1.2). The high and very high vacuum range is limited to use materials having corresponding low vapour pressures at room temperature. This is the range of elastomer gasket seals (see Sec. 3.8). The ultra-high vacuum requires baking. This practically restricts the list of materials to a few metals, glasses and ceramics, and the gaskets to metallic ones

* For ions $\lambda_i = \sqrt{2}\cdot\lambda$, for electrons $\lambda_e = 4\sqrt{2}\cdot\lambda$; for λ of gas mixtures see Heinze[533].

(see Sec. 14.31 and Adam[3], Pirani[992], Munday[906], Alpert[18], Vanderslice[1255], Steinherz[1182], Kirchner[678], Schweizer[1113], Martinson[820], Trendelenburg[1236]).

11.3 Vacuum systems

A vacuum system is the assembly of the components used to obtain, measure and maintain the vacuum in a vessel, chamber or device. Any vacuum system is made up of a pump (or pumps), gauges and pipes connecting them together. The system contains also valves, motion seals, electric lead-throughs,

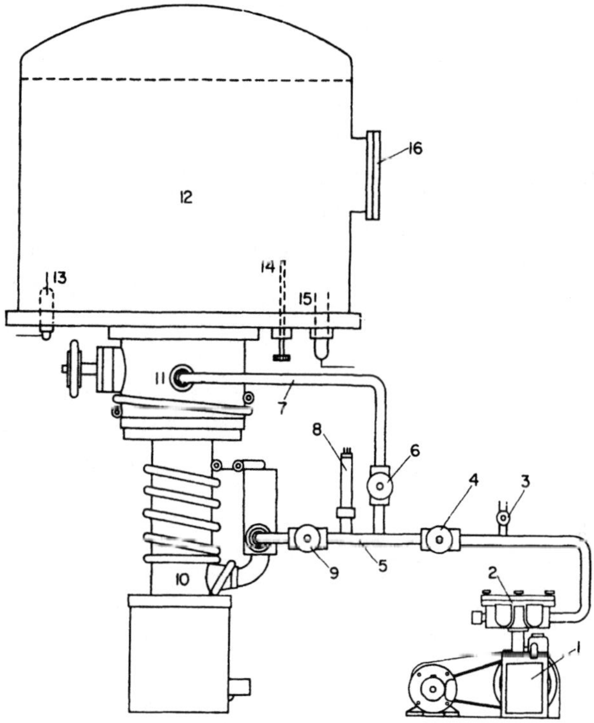

FIG. 1.1 Vacuum system. (1) rotary backing pump; (2) moisture trap with window; (3) air admittance valve; (4) throttling valve; (5) backing line; (6) roughing valve; (7) roughing line; (8) Pirani gauge; (9) backing valve; (10) diffusion pump; (11) baffle valve; (12) vacuum chamber; (13) electric lead-through; (14) shaft seal; (15) Penning gauge head; (16) window

etc. Figure 1.1 presents a typical vacuum system, showing the component parts and the usual way to connect the *roughing* and *backing* pumping lines.

This is a *dynamic or kinetic system* in which the required vacuum is obtained by the use of fast pumps in spite of the outgassing load and even the presence of small leaks (see Section 1.3). Obviously such systems require continuous pumping, and generally have dismountable seals (see Chapter 3). Dynamic

TABLE 1.2. VACUUM RANGES

American Vacuum Society[26], Diels[280], Dutton[318], Mönch[879], Dushman[314]

Pressure (torr)	Mean free path	Vacuum range		Pumps	Gauges	Materials	Seals*
760	0.06 μ	Rough	Low	Piston Rotary	Diaphragm Liquid level	Any	
25	2 μ	Rough / Fine	Medium	Rotary	McLeod Tesla coil Pirani gauge Alphatron	Waxes, glasses, ceramics metals and elastomers Porous and high vapour pressure materials excluded	Any elastomer seals (Section 3.8), waxed seals (Sections 3.1, 3.2), soldered seals (Section 3.5), greased ground seals (Section 3.6), liquid seals (Section 3.7)
1	50 μ	Fine	Medium				
10^{-3}	5 cm	High		Molecular Diffusion	McLeod Penning Ionization	Only glass, metal, ceramics and low vapour pressure elastomers	
10^{-6}	50 m	Very high		Molecular Diffusion Ion pump	Ionization Bayard-Alpert	Preferably clean bakeable materials (glass, copper, stainless steel, ceramics, Viton)	Elastomer (Section 3.8), greaseless ground seals (Section 3.6), glass–glass (Section 2.3), glass–metal (Section 2.4), guard vacuum (Sections 3.6 and 37.23), ethoxyline (Section 3.3), silver chloride (Section 3.4)
10^{-9}	50 km	Ultra-high		Ion pump Cryopump	Bayard-Alpert Magnetron	Only bakeable materials and with low permeability for gases	Welded and brazed (Section 2.2), glass–metal (Section 2.4), glass–ceramic and ceramic–metal (Section 2.5), metal gasket (Section 3.8)

* Bakeable seals see Table 1.6; cryogenic seals see Table 1.7.

systems are generally larger, but the chamber to be evacuated can be of any size from centimetres to metres. Large chambers are described e.g. by Kurie[723], Mark[804], Crawley[244], Ata[54], Prevot[1009].

In contrast to such systems, a *static* vacuum system is defined as one in which the outgassing and leaks are extremely small. After the required vacuum has been reached such systems can be sealed off, maintaining their vacuum for very long periods. These vacuum systems are generally based on permanent seals (see Chapter 2). The ultra-high vacuum systems are classified as static systems, due to their extreme tightness and cleanliness.

As the scope of the present book is limited to the vacuum sealing techniques, for the principles and techniques connected with pumps, gauges, traps etc., refer to: Barrington[88a], Brombacher[170], Buch[180a], Champeix[202], Davy[266], Dunoyer[311], Dushman[314, 315], Eschbach[346], Guthrie[491, 490a], Heinze[533], Holland-Merten[587], Inanananda[616], Jaeckel[623], Laporte[741], Loeb[779], Leblanc[751], Leck[752], Martin[818], Mönch[878], Morand[886], Nöller[936], Pirani[992], Pollard[997], Reiman[1045], Roth[1083], Spinks[1171a], Steinherz[1181a], Strong[1207], Turnbull[1240a], Wagner[1280], Yarwood[1339].

An incomplete list of the firms manufacturing vacuum equipment is: Alley[14a], Associated Electrical Ind.[52], Atlas-Werke[53a], Balzers[444], Beaudoin[101a], Comp. Gen. de Radiologie[224], Consolidated Vacuum Corp.[231], Deutsche-Vacuumapparate[278], Edwards[328], Eisler[335], Genevac[441], Granville-Phillips[469], Heraeus[542], Kammerer[652], Labor. de Basses Pressions[728], Leybold[762], Micafil[853a], National Research Corp.[912], N.G.N. Electrical[926], Officine Galileo[953], Pfeiffer[987], Pulsometer[1018], Seavom[1116a], Sogev[1162b], Ultek[1251a], Vactronic[1253b], Varian[1257a], Veeco[1262], Via-Vac[1266a], V.I.C.[1266b], Welch[1294].

1.2 INTERACTION BETWEEN THE SURROUNDING SPACE AND THE VACUUM

Vacuum technique is actually the attempt to isolate a given space from the surrounding and to evacuate its gas content. The surrounding materials exert their "opposition" to this attempt. This fact was once called "horror vacuum", now it is known as a combined effect of impedance, leak, permeation, outgassing, etc.

The degree of vacuum reached in a system corresponds to an equilibrium pressure between two opposing mechanisms: the gas flowing out due to the exhausting action of the pumps and the gas flowing in due to permeation, real and virtual leaks (see Section 1.3).

12.1 Gas flow, throughput and impedance

At any opening which connects two spaces with different pressures the gas flows from the high pressure side to that with lower pressure.

The quantity of gas flowing across a given opening is known as *throughput* if the flow occurs between two parts of a vacuum system, and as *leak rate* if the flow occurs from an external source into the system.

The throughput or leak rate is expressed generally in torr.litre per second, lusec (litre.micron per second) or micron.cubic feet per minute. Appendix A.3 lists conversion factors for these and other throughput units.

The throughput Q (or the leak rate) depends on the impedance Z or on the conductance C of the connecting opening and on the pressure difference between the two sides of the opening:

$$Q = \frac{P_1 - P_2}{Z} = (P_1 - P_2)C.$$

The impedance Z is the resistance of the opening to the gas flow. In vacuum calculations it is generally preferable to use the inverse of the impedance, i.e. the conductance C (Witty[1329]).

The value of the conductance depends on: the kind of the gas flowing through the opening, the regime of the gas flow, the dimensions and shape of the component, etc. The formulae to calculate the conductance in various cases are listed in Appendix A.5. For details on flow and conductance calculations and measurements we refer to Bureau[182], Sherriff[1131], Oatley[950], Stark[1179], Roth[1086], Delafosse[271], Florescu[386], Barrett[86], Dunoyer[310], Normand[939], Nerken[919], Santeler[1098], Moenich[870a], Dong[286], Eschbach[348], Lawrence[748], Geller[438].

The total conductance C of a number of components with individual conductances C_1, C_2, C_3 . . . is

$$\frac{1}{C} = \frac{1}{C_1} + \frac{1}{C_2} + \frac{1}{C_3} + \ldots$$

if the components are connected in series, and

$$C = C_1 + C_2 + C_3 + \ldots$$

if they are connected in parallel.

For seals, separating the vacuum space from the surroundings, the smallest possible conductances are required (Section 14.1). Internal seals connecting two parts of a vacuum system (valves) require the largest possible conductance when open, and small conductances when closed (Section 61.31).

12.2 Gas permeation

Gases have the possibility to flow through solids even if the openings present are not large enough to permit a regular flow. The passage of a gas into, through and out of a solid barrier having no holes large enough to permit more than a small fraction of the gas to pass through any one hole is known as *permeation* (American Vacuum Society[26]). The steady state rate of flow in these conditions is the *permeability coefficient* or simply the *permeability*. This is usually expressed in cubic centimetres of gas at STP flowing per second through a square centimetre of cross section, per millimetre of wall thickness and 1 torr of pressure drop across the barrier. Other units as well as conversion factors for permeability are listed in Appendix A.4.

The values for the permeability of various materials (expressed in the above mentioned units) are: for metals, less than 10^{-10}; for ceramics and glasses, from 10^{-12} to 10^{-13}; and for elastomers, from 10^{-10} to 10^{-12} (Jaeckel[624]). General features of the permeation of various gases through various solids are desribed in Section 21.12. For detailed discussions refer to Barrer[83–85], Norton[944], Klemenc[688], Frank[397], Hill[561], Collins[222].

The permeability can be lowered by using thicker walls or by coating the walls (Section 3.2). Eschbach[347] reports that the permeability of iron sheet for hydrogen (at 150 °C) was often lowered by a factor by coating it with enamel.

1.3 REAL AND VIRTUAL LEAKS

An ideal vacuum chamber should maintain forever the vacuum (pressure) reached at the moment of its separation from the pumps. Any real chamber presents a rise in pressure after being isolated from the pumping system. The pressure rise is produced by the gas which penetrates from outside into the chamber Q_L and/or evolves from the components (walls) inside the chamber Q_D. The rate of pressure rise in unit time dp/dt is given by:

$$\frac{dp}{dt} = \frac{Q}{V},$$

where Q is the total gas evolved in the chamber* (expressed in lusec or similar units, see Appendix A.3) and V is the volume of the chamber.

Q consists of two parts*

$$Q = Q_L + Q_D,$$

where Q_L is the real leak rate and Q_D the virtual leak rate.

The shape of the pressure rise curve shows whether the leak is real, virtual or a combination of them (see Fig. 1.2).

The *ultimate vacuum* which can be reached in a given vacuum vessel is

$$p_u = \frac{Q}{S_p}$$

* Q also includes the gas which penetrates through the walls by permeation (Sections 12.2 and 21.12), which is negligible in all cases except in ultra-high vacuum work or when by mistake the wall is porous (Sections 21.12 and 21.2).

where S_p is the pumping speed at the vessel. If S_0 is the pumping speed at the inlet of the pump, and C the conductance of the connecting parts between pump and vessel, then S_p results from

$$\frac{1}{S_p} = \frac{1}{C} + \frac{1}{S_0}.$$

13.1 Real leaks and tightness

The *real leak rate* is defined (American Vacuum Society[26]) as the quantity of gas (in p.V. units) flowing per unit time into the system from an *external source.* Obviously a perfectly tight vacuum system or chamber is one having a zero real leak rate, but to achieve this is as impossible as it is to reach zero pressure.

The leak rate is expressed in lusec, or torr.litre/sec, etc. (see Appendix A.3). Indirectly the leak rate of a given system or chamber is sometimes expressed as the pressure rise in a given time and for a specific volume or as the time required for a given quantity of gas to flow into the system. Table 1.3 compares these specifications. For example, a leak rate of 10^{-4} lusec in a vessel of 5 litres means that the pressure inside the vessel will rise (Table 1.3) by

TABLE 1.3. LEAK RATE SPECIFICATION

Leak rate (lusec)	Pressure rise in 1 litre volume	Time for 1 micron pressure rise/litre	Time for 1 cm³ STP gas inflow	Equivalent opening
1	1 μ/sec	1 sec	12.7 min	Rectangular slit with 1 cm width, 0.1 mm height and 1 cm depth
10^{-1}	6 μ/min	10 sec	2.1 hr	Rectangular slit with 1 cm width, 30 μ height and 1 cm depth
10^{-2}	36 μ/hr	1.66 min	21 hr	Capillary 1 cm long and 7 μ dia
10^{-3}	3.6 μ/hr	16.6 min	8.7 days	Capillary 1 cm long 4 μ dia
10^{-4}	8.6 μ/day (24 hr)	2.77 hr	87 days	Capillary 1 cm long 1.8 μ dia
10^{-5}	0.86 μ/day	27.7 hr	2.4 yr	Capillary 1 cm long 0.8 μ dia
10^{-6}	31 μ/yr	11.6 days	24 yr	Capillary 1 cm long 0.4 μ dia
10^{-7}	3 μ/yr	116 days	240 yr	Capillary 1 cm long 0.2 μ dia

8.6/5 = 1.7 micron/day, or that the time required for a pressure rise of one micron will be $2.77 \times 5 = 14$ hours. This vessel will need 87 days to receive 1 cm^3 (STP) of gas. The dimensions of a "hole" producing a leak rate of 10^{-4} lusec from the atmosphere into a vacuum are: 1 cm length and about 1.8 micron diameter (Ochert[951a]). The admissible leak rates of various seals are listed in Table 1.5. For leak rates of calibrated capillary leaks see Fig. 6.115.

13.2 Virtual leaks and degassing

The *virtual leaks* are defined (American Vacuum Society[26]) as the semblance of a leak in a vacuum system caused by slow release of sorbed or occluded gas. At present it is considered that the lower level of gas evolution from a square centimetre of the best materials, after a long period of pumping, is not less than 1×10^{-8} torr.litre per second or 1×10^{-5} lusec (Trendelenburg[1236]). The gas evolution from materials regularly used in building vacuum systems can be summarized (Jaeckel[624]) as being: 10^{-4} lusec/cm^2 for metals, 10^{-3} lusec/cm^2 for ceramics, and 10^{-2} lusec/cm^2 for elastomers.

The gas evolution rates of the materials to be used can be lowered if they are degassed in vacuum (Section 21.13) by baking, ion bombardment (Martin[812]) or other methods.*

The physically adsorbed gases can be removed by heating in vacuum to a few hundred degrees C, but to remove the chemisorbed or dissolved gases requires the system to be baked up to about 500 °C (Schweizer[1113]).

If an evacuated chamber is opened to atmosphere the pumping time needed to attain again a given vacuum in the vessel is short if the time that it was exposed is not too long (Morand[886]). Thus a degassed system should always be maintained under vacuum.

The degassing of the materials used in the construction of vacuum systems is discussed in Section 21.13. A graphical method for the evaluation of degassing and its influence on the design of the pumping systems is described by Santeler[1098]. For more detailed discussions on degassing refer to: Becker[102], Biram[128], Boettcher[144], Dayton[268], Ennos[344], Flecken[382], Garbe[425], Hobson[567], Jaeckel[624], Lawson[749], Markley[806], Pagano[961], Spence[1168], Stark[1179], Stoll[1200], Varady[1256], Turner[1241].

13.3 Leak detection

13.31 Pressure rise method. In approaching the problem of a supposed leak it is necessary to determine first whether or not a leak actually exists. Plotting of pressure vs. time curves (Fig. 1.2) will assist in determining the actual leak rate. First the system is exhausted to a stable minimum pressure. To minimize

* Groszkowsky[487] describes a method to obtain gas desorbtion by rubbing the surfaces in vacuum.

the effect of the vapours present in the system it is useful to employ also a trap with liquid nitrogen. When no further improvement in the pressure is evident during further pumping, the pumps are valved off from the system. The recorded pressure rise vs. time curve during a long period of isolation shows (Roth[1085], Delafosse[271], Korolev[712]) if the rise of pressure is produced only by real leaks (curve a straight line as curve 1, Fig. 1.2), only by virtual leaks (curve tends to a limiting maximum value as curve 2, Fig. 1.2) or by a combined effect of real and virtual leaks (curve 3, Fig. 1.2). The recording

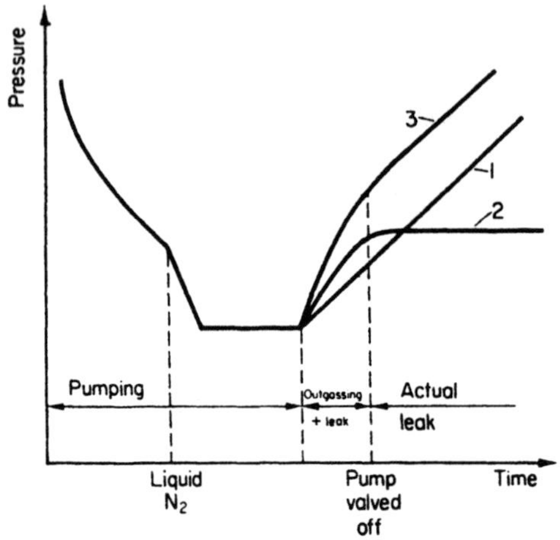

FIG. 1.2 Pressure vs. time, during the exhaust and after valving-off. Curve (1) only real leaks; curve (2) only virtual leaks; curve (3) real and virtual leaks together

should be continued (especially in cases 2 and 3) until the shape of the curve becomes evident. In curve 3 the first part of the curve shows the two effects together, while the slope of the straight part shows the leak rate (see Table 1.3) of the unit volume $\Delta p/\Delta t$. Thus the leak rate Q_L of the tested volume is

$$Q_L = V \frac{\Delta p}{\Delta t}.$$

The pressure rise method is the most direct but the slowest. The advantage of this method is that if the pressure rise is satisfactory the system is not only tight but also reasonably free of volatile impurities.

After testing the whole system, it is convenient to plot the pressure rise curve for isolated sections of the system in order to locate any leak. It is not necessary to relocate the vacuum gauge when doing this, but only to run the curve first for the section containing the gauge and then to open the valves one at a time. In this way the curve is recorded for increasingly larger portions of the system. The product between the various slopes and the respective volumes will show which section (or sections) contain the leak.

If quicker or more systematic tests are required a *hood* is placed over the component to be tested and both are evacuated. Air at atmospheric pressure is then admitted to the hood. If a leak is present this will produce a pressure rise in the evacuated component.

13.32 Test gas methods. In order to obtain higher sensitivity and to locate the exact position of the leaks a series of leak testing methods were developed, using various test gases (or liquids). In principle in all these tests the gas (or liquid) is spread over the outside of the suspected part and the presence of the gas is tested inside, or the the vacuum system is filled with the test gas and its presence detected outside. As a detailed description of the various techniques, cannot be given here they are just listed in Table 1.4.

It should be pointed out that the use of *liquids* (acetone, ether, benzene) has the danger of washing dirt or dust particles into the leak thus to plug it temporarily. Such a leak cannot be located (for the moment) but it re-opens later. For details on leak detection refer to e.g. Bicknell[122], Blears[134, 134a], Briggs[168], Cross[250], Doré[288], Dushman[314], Fleischmann[383], Guthrie[490], Jacobs[620], King[673b], Klopfer[691], Kronberger[721], Lineweaver[770a], Nivière[934], Ochert[951], Piatti[988], Pirani[992], Pollard[997], Steckelmacher[1180a], Stohr[1198], Turnbull[1240a], Ziock[1355].

Various leak detectors are available from most of the firms manufacturing vacuum equipment (e.g. see list at Section 11.3).

1.4 THE SEALS AND THEIR CLASSIFICATIONS

A vacuum system or even a vacuum chamber cannot be constructed as a single unit. One must use various components of various shapes and different materials and provide for the possibility to change the parts or to open and close the chambers. These various parts are joined together using various seals.

A *seal* can be defined as a means to prevent leakage through a joint, but the term "seal" is used as well to denote the sealed joint itself.

The classification of the seals used in vacuum techniques can be based on: the purpose of the seals, the requirements, the materials or the construction techniques used.

(a) According to the *purpose* the vacuum seals can be classified as:

1. Seals against gas penetration — purpose only tightness (Chapters 2 and 3).
2. Tight seals for the transmission of electric current (Chapter 4).
3. Tight seals for the transmission of motion (Chapter 5).
4. Tight seals for material transfer (Chapter 6).
5. Tight seals for radiation transmission (Chapter 7).

(b) According to the *requirements* the classification is:

1. Seals for low, high or ultra-high vacuum (Table 1.2).
2. Seals for high or low temperatures (Tables 1.6 and 1.7).
3. Rigid or flexible seals (Sections 51.3–51.7).
4. Seals resistant to chemical corrosion (Section 14.33).

(c) According to the *materials* joined together the seals can be classified as:

1. Metal to metal (Section 2.2).
2. Glass to glass (Section 2.3).
3. Glass to metal (Section 2.4).
4. Ceramic to glass (Section 25.21).
5. Ceramic to metal (Sections 25.22–25.25).
6. Wax or resin to glass or metal (Sections 3.1–3.4).
7. Elastomer to glass or metal (Section 3.8).

(d) According to the *degree of permanency* the seals are classified as:

1. Permanent seals (Chapter 2).
3. Semi-permanent seals (Sections 3.1–3.5).
4. Demountable seals (Sections 3.6–3.8).

(e) According to the *sealing technique* used there are:

1. Welded and fusion seals (Sections 22.1, 22.3, 2.3, 2.4).
2. Brazed or soldered seals (Sections 22.2, 22.3, 25.3, 3.5).
3. Wax and resin seals (Sections 3.1–3.4).
4. Ground and lapped seals (Section 3.6).
5. Liquid seals (Section 3.7).
8. Gasket seals (Section 3.8).

The choice of the seals to be used for each particular case is determined by the most important factor among those listed in (a) to (e). Obviously the ideal seal would correspond to the purpose, and fulfil the tightness (Section 14.1), the functional (Section 14.2) and the special (Section 14.3) requirements. In practice the design must be a compromise between requirements, possibilities and cost.

14.1. Vacuum tightness, the essential requirement

A common requirement and a permanent problem of all the vacuum seals is their *tightness*. Any vacuum seal must be tight but must not necessarily be hermetic.

A *hermetic seal* is designed to permit no detectable leak through it (on a very sensitive detector such as a helium mass spectrometer) while a *tight seal* is just free of leaks according to a given specification (American Vacuum Society[26]). The maximum leak rate (Section 1.3) to be specified in a particular case is determined either by the *ultimate vacuum* to be reached or by the permitted *pressure rise* (see Fig. 1.2).

In a dynamic system (Section 11.3) the specific leak rate q to be specified is determined* by the ultimate pressure p_u to be obtained, by the pumping speed S_p available and the length L of the seal:

$$q < \frac{p_u \cdot S_p}{L}.$$

The specific leak rate is the leak rate per unit length of seal.

In sealed-off vacuum vessels (or static vacuum systems) the specified leak rate q depends* on the admissible pressure rise per unit time dp/dt (see Table 1.3), the volume V of the vessel and the length L of the seal:

$$q < \frac{dp}{dt}\frac{V}{L}.$$

Table 1.5 lists some values quoted in the literature as the maximum specified leak rates and the specific leak rates for various seals respectively.

14.2 Functional requirements

Besides their function to prevent gas penetration, some vacuum seals must be capable of allowing the transmission of an electric current or of a motion, the transfer of material or the passage of radiation into the system. Simple vacuum systems may have just a few of such seals; on very large plants this number can extend to thousands (Camack[191]).

14.21 Electric current transmission. Very often it is necessary to transmit electric current from the atmospheric side into the vacuum chamber. For this purpose the use of electrically insulated and vacuum sealed conductors is required.

To fulfil the electrical, mechanical and vacuum requirements the seals used for electric current transmission are based on glass, ceramic or elastomers. Chapter 4 describes in detail the various such seals.

14.22 Motion transmission. In order to change the position of any part built in the vacuum chamber or system, motion must be transmitted from outside without opening the evacuated space.

A large number of techniques were developed to fulfil the various requirements of motion transmission into evacuated spaces. The techniques are based on mechanical transmission (as bellows, diaphragms, ground seals, shaft seals see Section 5.1), and on magnetic (Section 5.2) or electric transmission (Section 5.3).

14.23 Material transfer. In vacuum applications various materials need to be transferred from outside into the vacuum chamber or from the vacuum cham-

* The virtual leak (Section 1.3) is assumed to be negligible, i.e. the system is after a thorough degassing.

TABLE 1.4. LEAK DETECTION METHODS

Method	Test gas (liquid)	Test principle	Pressure range	Minimum detectable air leak (lusec)	Remarks
Flame wavering, hissing	air, nitrogen	The gas stream is heard by its hissing or is seen by the wavering of a flame	up to 3 atm	40	Draught free or quiet room required
Electric discharge	acetone, methanol, CO_2, hydrogen	Colour change of the discharge	–	10	Cross[250]
Wet outside surface	pressurized liquid	The inside is filled with liquid; observe the points where outside will be wet	up to 3 atm	4	Involves wetting the inside of the vessel and subsequent cleaning
Bubbles in liquid	air, nitrogen	The leak is indicated by a bubble appearing where the gas—pressurized inside—can pass to the outside	up to 3 atm	0.1	
Bubbles on soap film	air, nitrogen		up to 3 atm	0.04 *(8×10^{-3})	*If soap film maintained 5 mi (Guthrie[491])
Ammonia fumes	ammonia, CO_2, SO_2	Ammonia inside; detected outside with CO_2 or HCl or CO_2 resp. SO_2 inside; detected outside with ammonia	up to 3 atm	0.04	

Ammonia sensitive paper	ammonia	Ammonia inside; where leaks to outside produces black spots on wet ammonia sensitive paper rolled over the parts	about 2 atm	1×10^{-3} #(1×10^{-5})	#Detectable in about 30 hr (Leybold[762])
Single Pirani gauge	CO_2	The test gas changes the thermal conduction inside the gauge	1 to 1×10^{-3} torr	2×10^{-2}	
	hydrogen			1×10^{-2}	
	butane			5×10^{-3}	
Differential Pirani	CO_2	A pair of gauges, one sensitive to both air and test gas—the other connected via a trap is sensitive only to air	1 to 10^{-6} torr	1×10^{-3}	
	butane			5×10^{-4}	
Charcoal Pirani	hydrogen	Cooled charcoal trap is inserted in the gauge line to reduce the effect of pressure fluctuations	1 to 10^{-6} torr	4×10^{-4}	
Single ionization gauge	hydrogen	Test gas changes gauge reading	10^{-3} to 10^{-8} torr	5×10^{-3}	
	CO_2			1×10^{-3}	
	butane			1×10^{-4}	
Differential ionization gauge	CO_2	Pair of ionization gauges arranged as in diff. Pirani test	10^{-3} to 10^{-8} torr	3×10^{-6}	
	butane			5×10^{-7}	
Palladium barrier	hydrogen	The gauge (Pirani, ionization) separated from vacuum system by a Pd barrier, which when hot is permeable to H_2 only	as by Pirani or ionization	5×10^{-5}	Care not to poison the gauge by impurities

(Table 1.4 continued)

Method	Test gas (liquid)	Test principle	Pressure range	Minimum detectable air leak (lusec)	Remarks
Halogen detector	freon (CCl_2F_2) trichloroethylene tetrachlorcarbon	Based on the positive ion emission from hot Pt anode when exposed to traces of halides	2×10^{-1} to 7×10^{-2} torr (optimum range)	10^{-2} to 10^{-3}	Sudden loss of sensitivity after exposure to high concentration of halide vapours. Advisable to check frequently the sensitivity
Mass spectrometer (cold cathode)	helium	Based on separation of ions produced by the test gas from those formed by the residual gases	10^{-2} to 10^{-4} torr	10^{-5}	Can be used only at high vacuum. A separate leak testing pumping system is required in addition to the pumping system of the mass spectrometer
Mass spectrometer (hot cathode ion source)	hydrogen		5×10^{-4} to 10^{-8} torr	5×10^{-6}	
	argon			5×10^{-6}	
	helium			5×10^{-7}	
RF mass spectrometer	helium		10^{-5} to 10^{-10} torr	10^{-7}	
Omegatron	hydrogen, argon	Acceleration of ions	10^{-6} to 10^{-10} torr	4×10^{-8}	
Ion pump	hydrogen, air	Ionization	10^{-6} to 10^{-10} torr	1×10^{-7}	Joung[1341]

TABLE 1.5. ADMISSIBLE LEAK RATES

Seal (vacuum system)	Description	Leak rate	Reference
Glove box	General rule based on pressure rise of 0.05% per hr, and a volume of 250 litre	25 lusec	White[1306a]
Vacuum plant for metal casting	Total leak rate	1 lusec	Pirani[992]
Vacuum plant for isotope separation	Total leak rate	2×10^{-2} lusec	Mongodin[879a]
Plate valve	10 to 100 mm opening with shaft seals	$<1\times10^{-2}$ lusec	Leybold[762]
High vacuum system	Total leak rate as a general rule	1×10^{-3} lusec	Leybold[762]
Wilson shaft seal	3/8 in. shaft, in static position, after 8 hr pumping	1.2×10^{-3} lusec	Dawton[267]
Wilson shaft seal	3/8 in. shaft, in rotation 60 r.p.m.	4×10^{-3} lusec (see Sec. 51.74)	Dawton[267]
Shaft seal	Spring loaded lip seal (see Sec. 51.75)	5×10^{-4} lusec	Chaillot[201]
Flap valves	32–500 mm opening. Bellows sealed (see Sec. 61.32, 61.33)	$<1\times10^{-4}$ lusec	Leybold[762]
Welded seal	Max. leak at same hole	5×10^{-5} lusec	Mongodin[879a]
Ultra high vacuum valve	Bellows sealed, gold gasket closed (see Fig. 6.58 and Table 6.28)	10^{-5} lusec	Yates[1342]
Glass–metal seal	Leak rate max. 1 cm³ of He(STP) in 31 yr	8×10^{-7} lusec	Adam[4]
Ultra high vacuum valve	Knife edge on indium seal (see Table 6.28)	$<2\times10^{-8}$ lusec	Drawin[302]
Silicone rubber seal	110 mm dia. 6×6 mm (see Fig. 3.141)	0.2 lusec/cm (after 1 hr degassing at 150 °C)	Gale[420] Espe[354]
Copper gasket seal	Gasket 1 mm width, 78 mm dia. loaded with 10,000 kg (see Table 3.13)	3×10^{-5} lusec/cm	Armand[49]
Welded metal joints	Maximum permissible leak	2×10^{-5} lusec/cm	Garrod[434]
Plastic seals for low temp.	10^{-4} cm³ (STP) of He/hr per inch of seal at 1000 p.s.i. and 20 °K	8×10^{-6} lusec/cm	Weitzel[1293]
Rubber gaskets	72 mm dia. 4×4 mm (see Fig. 3.46)	5×10^{-8} lusec/cm	Kobayashi[703]
Elastomer gasket seals	Admissible range	6×10^{-5} to 3×10^{-8} lusec/cm	Jordan[646]
Ultra-high vacuum seal	Calculated requirement	1.5×10^{-10} lusec/cm	Boulloud[153]

ber to the atmosphere. Sometimes this transfer can be done by opening the vacuum chamber to atmosphere but in most of the applications it is required or preferable to do the transfer from the closed and evacuated chamber.

If the material to be transferred from or into the chamber is a gas, cut-offs, stopcocks, valves or leaks are used for the purpose (Section 6.1). Liquids are rarely transferred into the vacuum space, but if needed special techniques should be used (Section 6.2). For the transfer of solids, vacuum locks or similar devices are used (Section 6.3).

14.24 Radiant energy transmission. Radiation (or particles) are to be transmitted from the vacuum chamber to the atmosphere or from the surrounding space into the vacuum chamber, without opening the chamber. For this purpose sealed windows are used.

The range of windows on vacuum chambers and devices extends from the commonly used glass windows to transmit visible radiation, to the special windows for ultra-violet, infra-red or alpha, beta or gamma particles (Section 7.1). The techniques used to seal the windows extend from waxed seals to compression and elastomer or metal gasket seals (Section 7.2).

14.3 Special requirements

Very often the various vacuum seals should conform to special requirements. Among these requirements the most important ones are: resistance to high temperatures and/or low temperatures including temperature cycling, and resistance to chemical corrosion (or radiation damage).

14.31 Bakeability for ultra-high vacuum. The ultra-high vacuum (Table 1.2) work requires baking of the chambers to temperatures in the range 400–500 °C (Hoch[568], Vanderslice[1255], Adam[3], Munday[906], Roberts[1069a], Trendelenburg[1236, 1236a]). Consequently the seals used should not only be tight (with a very low leak rate, see Table 1.5) but they must maintain their tightness during baking and temperature cycling (Venema[1265], Caswell[197a]). This fact reduces the range of materials to be used to a very narrow one.

Small bakeable vacuum systems can be constructed of glass but any greased joint should be excluded. The metal bakeable systems are constructed using extensively the various stainless steels and similar alloys. Welded (argonarc, heliarc) seals are to be preferred in any place where this is possible.

Demountable seals used in bakeable ultra-high vacuum systems should conform to the severe requirements summarized as follows:

1. Leak rates lower than 10^{-6} lusec in the whole temperature range from room temperature to 500 °C (also Table 1.5).

2. The leak rate must not be influenced by repeated heating and subsequent cooling.

3. The seal should not contain materials having, even at 500 °C, vapour pressures higher than the ultimate pressure to be reached (e.g. 10^{-9} torr).

4. The joints should be simple to assemble and to dismantle.

5. The seal should be able to be re-used many times with the same gasket, or at least without the need to remake the finish of the flange faces.

6. The seal should be easily machined, and obviously at the lowest cost.

The seals developed for ultra-high vacuum systems conform to requirements 1–3 and more or less to the other requirements listed.

As in this book the ultra-high vacuum seals are described among the other seals according to the sealing technique, a cross reference list is felt to be useful. Table 1.6 lists the various ultra-high vacuum seals referring to the sections where they are discussed. The application of these seals in electrodes is described in Sections 4.2. Ultra-high vacuum motion seals are discussed in Sections 51.4, 51.5, 51.8, and 5.2. The various ultra-high vacuum valves are summarized in Tables 6.27 and 6.28 and the windows for this vacuum range are described in Sections 71.31, 71.32, 72.1, 72.2 and 72.9.

14.32 Low temperature (cryogenic) seals. Low temperature work requires vacuum to reduce heat transfer or to maintain the purity of the studied materials. Whether this vacuum has to be in a space around the refrigerant liquid or in an enclosure within the liquid, pressures of the order of 10^{-6} torr or less must often be maintained to ensure thermal insulation or thermal equilibrium.

For the techniques connected with the methods to reach and maintain low temperatures (from −50 °C to 4 °K) we refer to special literature, e.g. to White[1306], Scott[1114], Hoare[566], Vance[1254a], and to Brebner[165], Katan[656], Probyn[1012], Levantine[760], Baruch[93], Wexler[1299], Garrod[434], Butler[189], Ahn[11], Parker[967].

In seals for low temperatures some special problems arise since: liquid seals, waxed and cemented seals or elastomer gasket seals are no longer practicable because of the brittleness of these materials at low temperatures.

Although aluminium has excellent low temperature mechanical properties, its high thermal conductivity makes it undesirable for transfer lines which constitute thermal paths between low and high temperature regions. To exploit the advantages of aluminium for the isothermal surfaces, and at the same time allow the use of the insulating properties of stainless steels or Monel, these metals are joined with transition seals (Section 35.12), using *soldering* techniques. Low melting point solder seals are particularly useful also in making joints which can be broken and remade without raising the temperature of the adjacent metals to more than about 100 °C (see Table 3.4).

Epoxy (Araldite) resins provide good sealing agents for use over the low temperature range (White[1306]).

TABLE 1.6. BAKEABLE AND ULTRA-HIGH VACUUM SEALS

Seal	Reference	Remarks
Welded	see Sec. 22.32	Permanent seals except some constructions (Fig. 2.36) which can be easily opened and rewelded
Brazed	see Sec. 22.33	
Glass–metal	see Sec. 2.4	
Ceremic–metal	see Sec. 2.5	
Molten metal	Reynolds[1050], Blanaru[130], Haaland[494] see Sec. 37.6	In, Sn used in UHV valves; see Sec. 61.32b
Cooled elastomer "O" rings	Farkass[367, 369] see Sec. 38.41	Double "O" ring cooled with water under pressure in channel between the two grooves
Guard vacuum	Rivera[1067], Ehlers[331], Adam[3], Metcalfe[849], Kienel[672] see Sec. 38.23	Double chambers with intermediate space evacuated 0.1 to 0.01 torr
Metal "O" ring and flat flanges	van Heerden[527], Peters[986a] see Sec. 38.42	Gold or copper "O" rings
Corner	Caswell[197], Mark[803], Grove[489], Dreyer[304] see Fig. 3.94 and 3.96	Gold wire crushed in corner step
Cemented	Holden[578], Holland[582, 583], Comsa[227] see Sec. 38.46	Aluminium wire compressed and heated between flat flanges
Coined gasket	Milleron[865], Wheeler[1304] see Fig. 3.121	Special shape copper gasket seal
Diamond shape gasket	Hoch[568], Kienel[671] see Fig. 3.121c	Copper or gold gasket seal
X cross section gasket	Heathcote[526] see Fig. 3.136	Copper gasket seal with guard vacuum
Conical and cylindrical	Hall[503], Wishart[1327] see Fig. 3.126 and 3.134b	Seal between a curved and a conical surface
Surface friction	Brymner[179], Steckelmacher[1180], Marton[821] see Sec. 38.63	Conical seal based on surface friction
Shear	Lange[737], Henry[540], Wheeler[1304], see Sec. 38.55	Flat copper gasket sheared between stepped flanges
Knife edge	van Heerden[527], Hees[529], etc. see Table 3.39	Knife edges forced in flat gaskets
Groove and knife edge	Hintenberger[564], Paul[980], Papirov[966b] see Sec. 38.62	Ag, Cu, Al, Sn gaskets between flanges with mating knife edge and groove
Conflat	Wheeler[1304], Varian[1257a] see Fig. 3.149 and 3.150	Seals with conical surfaces on flat copper gasket
Foil	Ruthberg[1092] see Sec. 38.61	Thin Au, Ag, Cu, Al foil between plane flanges or ridges
Curvac	Ultek[1251a] see Table 3.39A	Half-toroidal ridges on flat copper gasket

Ground cone and socket joints can remain vacuum tight after cooling to liquid helium temperature if an Apiezon grease or glass forming liquid is used in the seal (White[1306]). A useful sealing compound can be made by dissolving 15 parts by weight of soap flakes in 85 parts of warm glycerine. This compound softens at 40–50 °C and forms a glassy solid at low temperatures (Hoare[566]).

In low temperature work it is desirable to use *hard glasses* because of their resistance to thermal shock (Table 2.11). *Glass–metal matched seals* (Section 24.32) or *Housekeeper seals* (Section 24.41) can be successfully used. Lane[733] found Kovar to Pyrex seals very satisfactory for use in the liquid helium region, but other authors (Corac[236]) point out the difficulties encountered, due to small leaks appearing in Kovar–glass seals cooled to low temperatures. Copper–glass Housekeeper type seals appear to be generally reliable at low temperatures (White[1306]) in sizes up to 25 mm diameter.

Metal gasket seals can be also used in low temperature work. For the liquid helium range Wexler[1298] recommends the gold gasket seals (Sections 38.44, 38.51). The various vacuum seals used at low temperatures are described in Sections corresponding to the sealing technique used. Table 1.7 lists the cross references to these various sections.

14.33 Corrosion resistance. During a long period after the beginning of vacuum engineering, mercury and sodium vapours were the single corroding

TABLE 1.7. VACUUM SEALS FOR LOW TEMPERATURES

Seal	Reference	Remarks
Welded	Section 22.32	
Brazed	Section 22.33	
Glass–metal	Section 2.4	
Soldered	Kaufmann[658], Scott[1114] (Section 35.1)	Pb–Sn alloys with high Pb content preferred (Fig. 3.12)
Epoxy resin	Lyons[790], Quarrington[1020], Gale[419], Balain[73], Lipson[770b] (Section 33.3)	Windows (Table 7.6) Electrodes (Section 43.1)
Molten metal	Hughes[606], Reynolds[1050], Martina[811] (Section 37.6)	
Teflon gasket	Nester[920], Scott[1114] (Fig. 3.54a)	
Metal gaskets	Spees[1167], Fraser[400], Hoare[566], White[1306] (Sections 38.33, 38.44)	Windows (Table 7.6)

agents in vacuum systems and devices. Recent applications in various vacuum technologies add many other materials to the list of chemically active ones treated in vacuum.

When mercury vapours are present in the vacuum system the main constructional metal remains iron, and metals like copper, aluminium, gold, silver, etc. are excluded. Alkaline metal vapours require the use of special glasses (e.g. Espe[354]). The strongest corrosion in vacuum technology plants is perhaps that of solutions containing fluorine as used in the isotope separation of uranium (Dore[288]), where the choice is restrained to a few materials.

14.34 Cost requirements. The ideal seal would be that made, connected and dismantled by semi-skilled personnel using a minimum of equipment. Hence the elements of the seal should be re-usable and have a low unit cost.

Experience shows that very few troubles arise more consistently in vacuum technique than those from the joints and seals (Stiff[1190], Berger[114]). These troubles lead to unsuccessful experiments, serious time losses, laborious leak hunting, cleaning or other unpleasant operations. Generally the cost of these operations exceeds many times the cost difference between a high quality and an improvised seal. Thus the total balance of cost shows the rentability of using from the beginning (design) more expensive but high quality seals.

CHAPTER 2

PERMANENT SEALS

2.1 PREFABRICATED VESSELS AND PIPES

21.1 Selection criteria for vessels and pipes

For the construction of vacuum systems or vacuum devices it is conventional to use metals, glasses, ceramics and some rubbers and plastics. The materials that become part of the vacuum system, forming the enclosure (vessels, pipes) must have sufficient *mechanical strength* to withstand atmospheric pressure (Section 21.11), must be *impermeable* to gases (Section 21.12), must have *low vapour pressures* (Section 21.13) and good resistance to the special working conditions.

21.11 Mechanical strength. Vacuum enclosures are made up from cylindrical, plane and hemispherical parts. All these parts tend to deform inwards as a result of the difference between external and internal pressures. This pressure difference is equal to one atmosphere as in most of the cases, since the outside pressure is generally atmospheric and the inside pressure is so low that it can be neglected compared with an atmosphere.

Cylindrical tubes and vessels tend to collapse at a theoretical pressure (Brownell[174a]) of

$$P_{\text{cyl}} = \frac{2E}{1-\mu^2}\left(\frac{t}{D}\right)^3,$$

where E is the modulus of elasticity, μ is Poisson's ratio, t is the wall thickness and D the mean diameter of the cylinder (Fig. 2.1). The modulus of elasticity and other mechanical properties of various materials used in vacuum systems are resumed in Appendix B.1.

This formula (corrected by the relevant safety factor) can be used for cylinders longer than the critical length L_c. The critical length is expressed as

$$L_c = 1.11D\sqrt{\frac{D}{t}}.$$

For the required D/t ratios (Table 2.1) it results that the critical length for metal and glass pipes is about 8 to 11 times the diameter, and for elastomers only about 2 to 4 times the diameter.

Using a common value of $\mu = 0.3$ (for glass $\mu = 0.20$–0.25, for metals $\mu = 0.25$–0.33) the maximum permissible pressure for a cylindrical vessel is given by*

$$P_a = \frac{K}{n} E \left(\frac{t}{D}\right)^3,$$

where n is the safety factor and K the correction factor (Strum[1290]) depending on the critical length L_c (Table 2.1). The values of K as a function of L/D for various D/t values are plotted on the graph from Fig. 2.2.

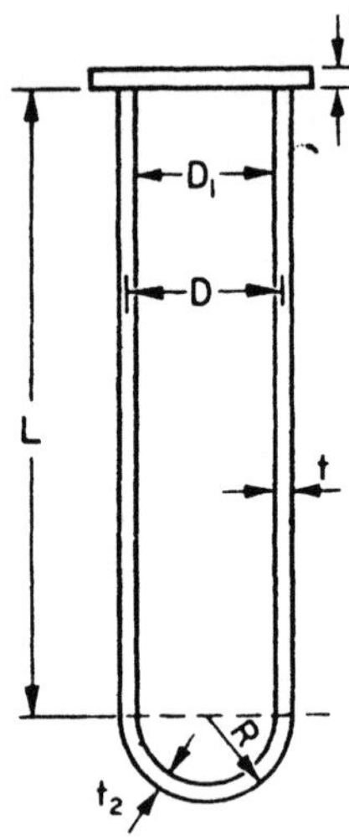

FIG. 2.1 Main dimensions for calculating the strength of cylindrical, plane and spherical shells

Table 2.1 lists the permissible values of D/t (Fig. 2.1) calculated for long cylinders ($K = 2.2$) of various materials, when P_{cyl} is 1 kg/cm² and a safety factor $n = 4$ is used.

Any eccentricity of the cylinder will reduce the strength of the vessel since under the action of external pressure the shape of elliptic section tends to change to an ellipse of greater eccentricity.

If w is the eccentricity, the circumferential compression stress σ for collapsing is given by Timoshenko[1224a] as

$$\sigma = \frac{P_a \cdot D}{2t}\left(\left|\frac{8w}{t}\right| + 1\right).$$

The influence of the eccentricity on the t/D ratio is illustrated by the graph from Fig. 2.3. The graph represents the minimum permissible wall thickness t of glass tubes with elliptic cross section (Morrison[893]) based on a design

* For other and more complex formulae see Strong[1207], Arnold[50] or Brownell[174a].

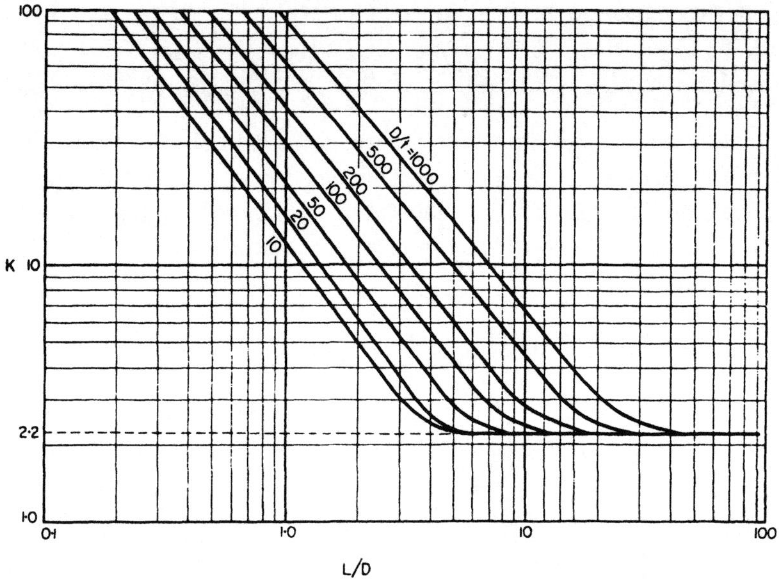

FIG. 2.2 Correction factors for the calculation of the collapsing pressure of cylindrica shells (after Strum[1209])

tensile stress of 1 kg/mm² and a safety factor 4. Here D and d are respectively the major and minor diameters of the ellipsis.

TABLE 2.1. PERMISSIBLE D/t VALUES AND CORRESPONDING L_c/D FOR CYLINDERS

Material	D/t	L_c/D	Material	D/t	L_c/D
Copper at 20 °C	84	10	Stainless steel at 20 °C	105	11.6
Copper at 500 °C	58	8.5	Stainless steel at 500 °C	89	10.5
Nickel at 20 °C	100	11	Glass (hard) at 20 °C	70	9
Nickel at 500 °C	90	10.5	Rubber (Neoprene) at 20 °C	2.5	1.7*
Aluminium at 20 °C	70	9	Teflon	12	3.8
Aluminium at 500 °C	62	8.7	PVC (Tygon)	3.7	2.1

* See Fig. 2.17.

For *circular end plates* the theoretical relationship between pressure P, stress σ or deflection at the centre δ is

$$P_p = \frac{16}{3}\sigma\left(\frac{t_1}{D_1}\right)^2 = \frac{256}{3(1-\mu^2)}\frac{E\cdot\delta\cdot t_1^3}{D_1^4},$$

if the plate is *clamped* at its edges, and

$$P_p = \frac{32}{3(3+\mu)}\sigma\left(\frac{t_1}{D_1}\right)^2 = \frac{256}{3(1-\mu)(5+\mu)}\frac{E\cdot\delta\cdot t_1^3}{D_1^4},$$

if the plate is *unclamped*. For the significance of t_1 and D_1 see Fig. 2.1. Table 2.2 lists the permissible D_1/t_1 values and the corresponding t_1/δ deflection ratios for clamped circular end plates. For unclamped circular end plates D_1/t_1 values are greater by a factor of approximately 1.2.

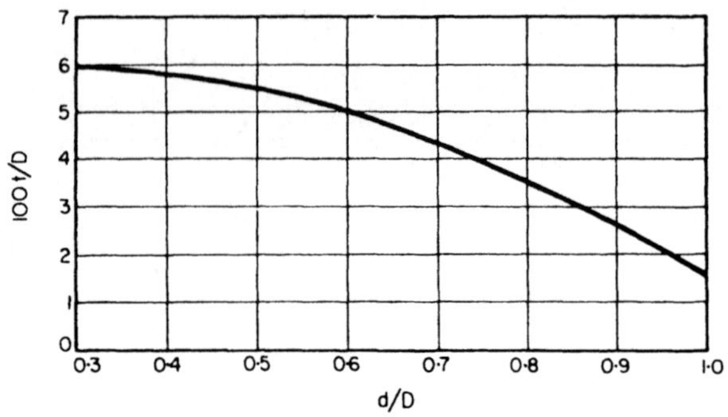

FIG. 2.3 Minimum permissible wall-thickness for glass tubes with elliptic cross section (after Morrison[893])

TABLE 2.2. PERMISSIBLE D_1/t_1 VALUES (FIG. 2.1) FOR CLAMPED CIRCULAR END PLATES AND THE CORRESPONDING t_1/δ DEFLECTION RATIOS AT CENTRE CALCULATED WITH SAFETY FACTOR 4

Material	D_1/t_1	t_1/δ	Material	D_1/t_1	t_1/δ
Copper	52	15	Glass (hard)	16	117
Nickel	73	8	Rubber (Neoprene)	10	0.2
Aluminium	37	57	Perspex (Plexiglas)	30	—
Stainless steel	89	3	Teflon	14	9
Beryllium	25	500	Mica	58	15

Minimum thicknesses of plates used as windows are listed in Table 7.5.

For *hemispherical ends* the theoretical collapsing pressure is given by

$$P_h = \frac{2E}{\sqrt{3(1-\mu^2)}}\left(\frac{t_2}{R}\right)^2.$$

For $P_h = 1$ kg/cm² and with a safety factor of 4, the permissible R/t_2 (see Fig. 2.1) ratios for hemispherical ends of various materials are those listed in Table 2.3.

If the end is ellipsoidal instead of hemispherical the required R/t_2 ratio is greater than for hemispherical by a factor x. This factor (Brownell[174a]) results from Fig. 2.4, where a/b is the ratio of the hemisphere and ellipsoid radii.

TABLE 2.3. PERMISSIBLE R/t_2 (FIG. 2.1) VALUES FOR HEMISPHERICAL ENDS

Material	R/t_2	Material	R/t_2
Copper	600	Stainless steel	830
Nickel	780	Glass (hard)	470
Aluminium	470	Rubber (Neoprene)	30

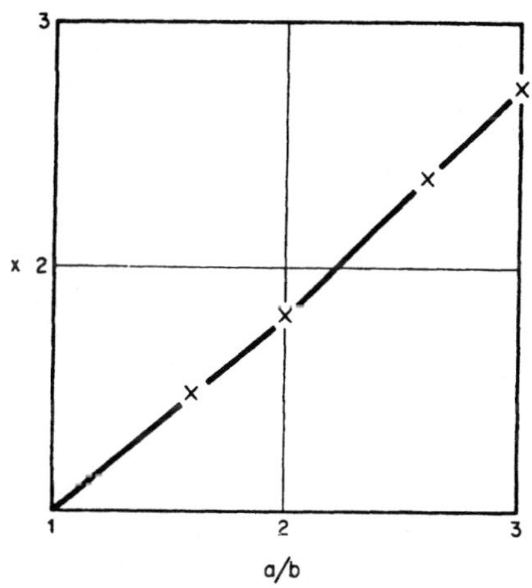

FIG. 2.4 Correction factors for ellipsoidal ends. (Plotted after data from Brownell[174a])

Conical ends can be interpreted as cylindrical ones or as plane end plates depending on the angle of the apex. If the apex is less than 45° the mechanical strength of the conical end can be calculated using the formula for cylindrical tubes. The equivalent cylinder diameter is in this case the large diameter of the cone, and the equivalent cylinder length is the axial length of the cone. If the apex is between 45° and 120° the conical end can be interpreted as a cylinder having both diameter and length equal to the large diameter of the cone. Finally if the apex is more than 120° the conical end can be seen as flat.

21.12 Permeability to gases. The metallic, glass or rubber walls of vacuum vessels or pipes are more or less permeable (see Section 12.2) to gases. The quantity of gas which permeates the walls can be really large as in the case of porous ceramics or castings or low as for the case of gas diffusion through "non porous" walls.

Excepting the cases when gas permeation is used to introduce controlled quantities or particular gases in the vacuum space (see Section 61.48), *the permeation is not wanted in vacuum systems.*

The permeation process (defined in Section 12.2) involves first the adsorbtion of the gas on the surface where the gas pressure is higher. After being dissolved in the outside surface layer the gas slides down the concentration gradient and diffuses to the vacuum side where it is desorbed and escapes (Norton[947]). The permeability (see Section 12.2) depends on the permeation mechanism, the material of the wall, its temperature and the kind of gas involved in the process. For detailed discussion of permeation mechanisms and processes refer to Jaeckel[623], Dushman[314, 315], Norton[944, 947], Waldschmidt[1283], Heinze[533], Barrer[84].

The permeation mechanism can be *atomic* or *molecular*. Hydrogen permeation through metals increases with the *square root* of the pressure (Nor-

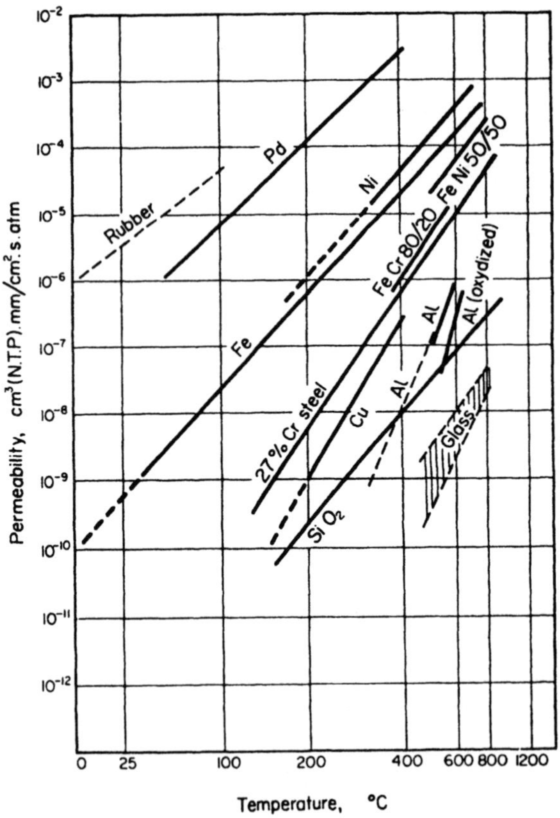

FIG. 2.5 Permeation of hydrogen through various materials. (Compilation of literature data from Jaeckel[623], Dushman[314], Norton[944, 947], Waldschmidt[1283], Heinze[533], Barrer[84])

ton[944]); this fact is explained by the dissociation of the hydrogen to *atoms* and their passage as such through the metal. Recombination occurs on desorption and on the low pressure side molecular hydrogen appears. In glasses and elastomers the gas permeation rate is *proportional* to the pressure. Here the permeation itself occurs in molecular form.

The permeation of atmospheric gases through metal walls does not include the rare gases (He, A, Ne, Kr, Xe), since *no rare gas diffuses through metals at any temperature* under purely thermal activation (Norton[947]). There can be penetration of rare gas ions under a potential gradient, or rare gases can be formed *in situ* in the metal interior by nuclear disintegration processes (Norton[944]).

In the construction of metal vacuum enclosures, rolled plates or forgings can be successfully used. Castings are generally porous (Avery[58]), excepting die castings which are reliable for small valve bodies (Millen[856]).

The permeability (see Section 12.2) of *aluminium* for hydrogen is very small

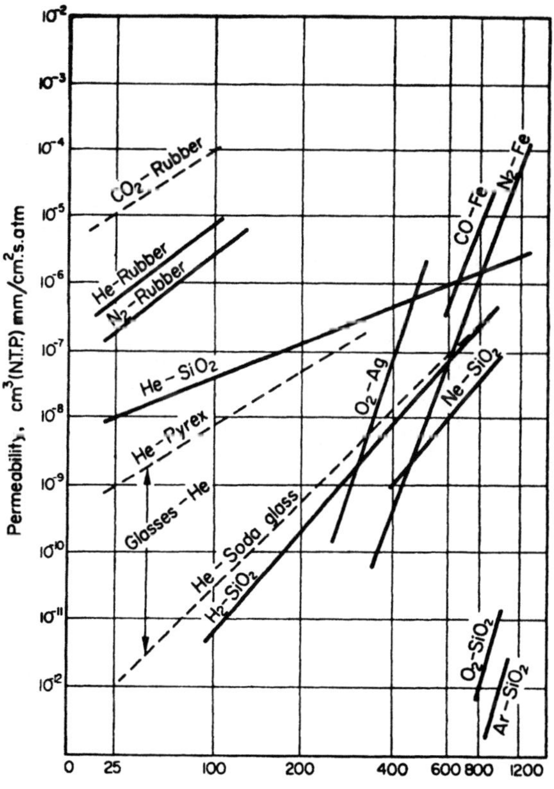

FIG. 2.6 Permeation of various gases through various materials. (Compilation of literature data from Norton[944, 947], Waldschmidt[1283], Jaeckel[623], Dushman[314], Mönch[878, 879], Heinze[533], Barrer[84])

(Fig. 2.5). It is negligible in all the cases except in ultra-high vacuum chambers or with chambers heated at high temperatures and having very thin walls.

Copper is a metal with low permeability for all the gases including hydrogen (see Fig. 2.5). *Nickel* has a higher permeability for hydrogen and therefore *for water-cooled chambers*, where the danger of hydrogen permeation is greater, *copper is to be preferred to nickel.* Due to its selective permeation of hydrogen against nitrogen, nickel is used in diffusion leaks (see Section 61.48). *Iron* vacuum containers have high permeability (Fig. 2.5) for hydrogen, especially if the hydrogen is in atomic form on the high pressure side due to chemical or electrolytic effects. Thus *the cooling of iron vacuum containers* should be made with *liquids which do not contain hydrogen ions* (e.g. oil) *or air cooling should be used.* The permeation of hydrogen through *steels* increases with increasing carbon content. Low carbon steels are thus preferred as vacuum containers.

The impermeability of technical glasses to gases *can be taken for granted* (Stanworth[1176]) when the vacuum system does not involve parts with pressures lower than 10^{-6} torr. For lower pressures the permeability of glasses to various gases and especially to helium is high enough to change the vacuum level inside the system (Alpert[18]). The helium accumulation in glass bulbs due to permeation from the atmosphere to the evacuated space is shown in Fig. 2.10.

The permeation is influenced by the kind of glass and the gas involved in the process. As a general rule the denser the structure of the glass and the greater the atom, ion or molecule of the gas, the less the permeation rate will be. This is the reason why gases permeate easier through fused silica than technical glasses. In technical glasses the open meshes of fused silica or other glass formers are occupied by network modifiers as sodium, potassium or barium ions. Figure 2.7 shows the increase of permeability of glasses (for helium) with increasing content of glass formers (Norton[944, 946]).

The permeability (see Section 12.2) for helium of various glasses is presented in Fig. 2.8 as a function of the temperature. The permeability for hydrogen and neon is lower than for helium. The *glasses* can be considered *practically impermeable* to argon, oxygen and nitrogen, the permeability for these gases being less than for helium by a factor of 10^5.

The permeation of helium through vitreous silica (or Vycor) is 10^7 times greater than through crystalline quartz. Vitreous silica has a considerable permeability also for other gases like hydrogen, nitrogen, oxygen and argon (Fig. 2.8).

In order to show the importance of the inflow of atmospheric gases through the walls of a vacuum vessel, Norton[947] gives some interesting examples. He considers a bulb of vitreous silica at 25 °C with walls 1 mm thick, surface area 100 cm^2 and volume 330 cm^3, and assumes that the walls have been completely degassed, that the initial pressure is 10^{-16} torr and that the steady-state flow is established at 25 °C. From the abundance (partial pressure) of the various gases in the atmosphere and from the permeation rates extrapolated to 25 °C (Table 2.4) the order of inflow of the gases (see Table 2.4) and the gas accumulation (Figs. 2.9 and 2.10) is established.

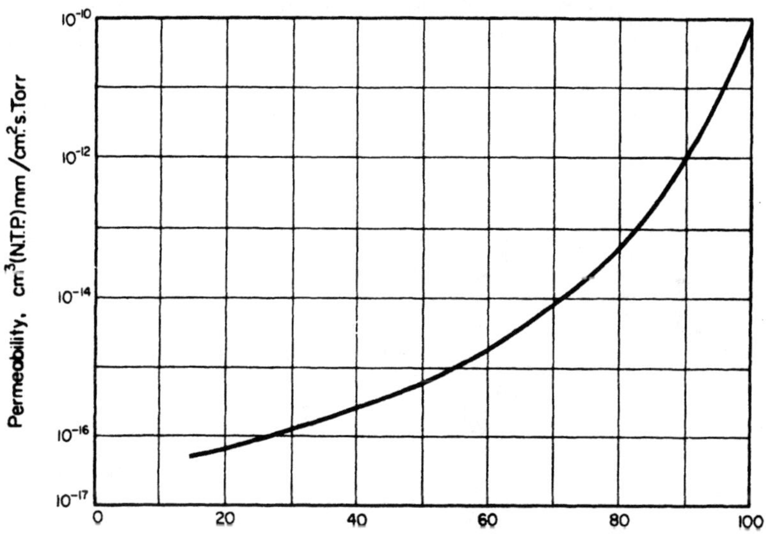

FIG. 2.7 Permeation of helium through glasses (at 100 °C) as a function of the glass-forming components (Diels[280], Kohl[706], Espe[354])

TABLE 2.4. ORDER OF FLOW OF ATMOSPHERIC GASES INTO SiO_2 BULB AT 25 °C (FOR 1 MM THICK, 1 CM² AREA)

Gas	Atmospheric abundance C (cm, partial pressure)	Permeation P (cm³ STP/sec for 1 cm Hg gas pressure difference)	Inflow $C \times P$ (cm³/sec)	Order of inflow	Atoms/sec
N_2	59.5	2×10^{-29}	1.2×10^{-27}	5	
O_2	15.9	1×10^{-28}	1.6×10^{-27}	4	
A	0.705	2×10^{-29}	1.4×10^{-29}	6	
Ne	1.8×10^{-2}	2×10^{-15}	3.6×10^{-17}	2	900
He	4.0×10^{-4}	5×10^{-11}	2.0×10^{-14}	1	500,000
H_2	3.8×10^{-5}	2.8×10^{-14}	1.0×10^{-18}	3	25

It can be seen that the gases of low abundance show the highest inflow and the order of accumulation in the silica bulb is (1) helium, (2) neon and (3) hydrogen. A big difference separates the succeeding gases oxygen, nitrogen and argon. The rate of gas accumulation with time is shown in Fig. 2.9. For the vitreous silica bulb the gases and pressures at the end of one year in air at 25 °C would be 10^{-4} torr helium, 10^{-7} torr neon, and 10^{-8} torr hydrogen. Only a few molecules of oxygen would have permeated even after a hundred years.

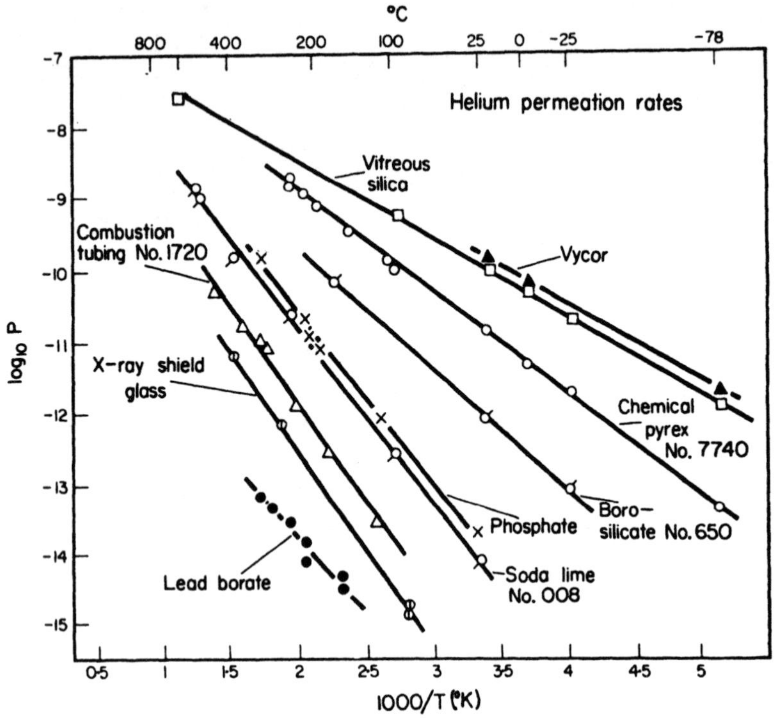

FIG. 2.8 Permeation rate (P) of helium through various glasses. P expressed in cm^3 of gas (STP) per sec. for 1 mm thickness of wall, per cm^2 area, per cm Hg of helium pressure difference. Reproduced from Norton[947] *(Courtesy of Pergamon Press)*

The increase in pressure in bulbs of silica and glasses is shown in Fig. 2.10 for helium permeating from the atmosphere. To reach a helium pressure of 10^{-6} torr (starting with 10^{-16} torr) requires, at 25 °C, 3 days for silica, a month for Pyrex and very long times for soda–lime glass or other glasses. From this, it is evident that if we are concerned with vacua in the pressure range of 10^{-9} torr, it is necessary to make the envelope of a glass of low permeability or surround it by a subsidiary evacuated chamber.

The literature is very poor in values of permeability values for *ceramic* walls. It is known that ceramics with vitreous coating are vacuum tight, and

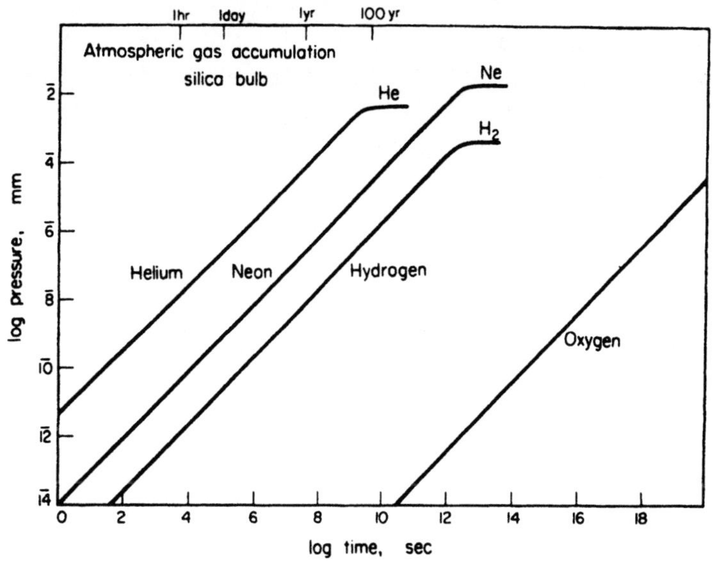

FIG. 2.9 Atmospheric gas accumulation at 25 °C, in a silica bulb 330 cm^3, 100 cm^2 wall area, 1 mm wall thickness. Reproduced from Norton[947] *(Courtesy of Pergamon Press)*

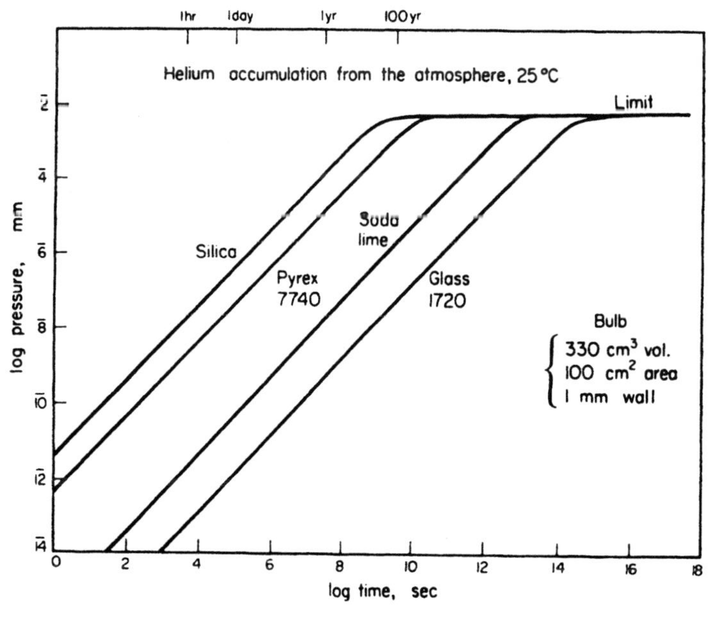

FIG. 2.10 Helium accumulation from the atmosphere in bulbs of various glasses, at 25 °C. Reproduced from Norton[947] *(Courtesy of Pergamon Press)*

that some ceramics are suitable for vacuum containers (e.g. Steatite 1.8 mm thick, Forsterite 0.5 mm thick).

The gas permeability of *mica* (perpendicular to its surface) is very low. The permeability for helium at 100 °C is about 10^{-16} cm^3 STP/cm^2.sec for 1 mm of wall thickness and 1 cm Hg pressure difference, and about 10^{-14} cm^3 STP/cm^2.sec at 400 °C, i.e. is lower than that of glasses.

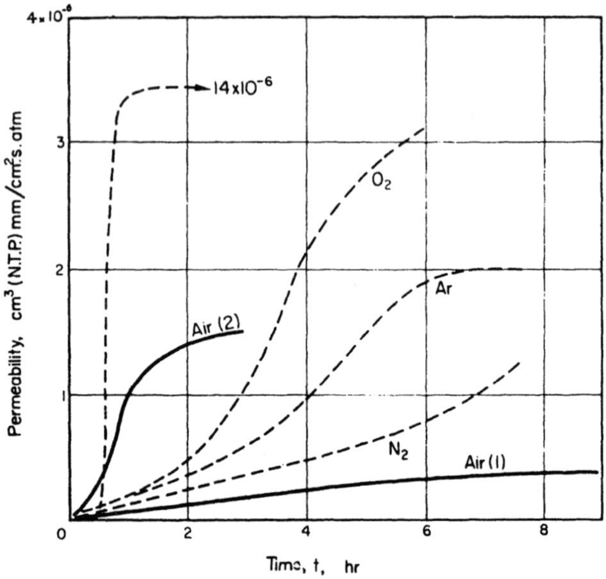

FIG. 2.11 Permeation of gases through elastomers; *air (1)* through Neoprene (1.6 mm thick); *air (2)* through natural rubber (1.6 mm thick); O_2, *A*, N_2 through natural rubber (3.2 mm thick). Compiled after Dawton[267] *(Courtesy of The Institute of Physics and The Physical Society, London)*

Organic polymers (rubbers, plastics) are permeated by all the gases, including the rare ones (Norton[944]). There are wide variations in permeability depending on the polymer and the gas involved in the process. The permeation rate of CO_2 through natural rubber is high (about 10^{-5} cm^3 STP. mm/cm^2.sec.atm), that of air is lower (10^{-6} cm^3 STP.mm/cm^2.sec.atm). In any case the range of permeation is the same as that of hydrogen through palladium (see Fig. 2.5). Natural rubber has higher permeability than Neoprene (see Fig. 2.11, and Norton[944, 946], Jordan[646], Kendall[661], Amerongen[27]).

Oxygen permeation through polymers is 3–4 times greater than that of nitrogen. Saran, polyethylene, and Kel-F have generally low permeabilities (about 3×10^{-7} cm^3 STP.mm/cm^2.sec.atm, at 25 °C, see Kendall[661]). Teflon presents in thin sheets abnormally high permeability values (Norton[944]).

21.13 Degassing. The walls of a vacuum system and the parts built in the system are always gas and vapour sources for the evacuated space. Even at highly evacuated spaces these gases and vapours remain as traces forming the *residual gas level* (Klapfer[648]).

The provenance of the gases and vapours contained in or laying on the walls or components, is connected with the history of the material, and can be *occlusion* during the molten state (metals, glasses) or *sorption* during the contact of the solid with the surrounding atmosphere.

Due to the low pressure created in the vacuum system the sorbed or occluded gases are slowly released (desorbed) giving in the first period of the evacuation the semblance of a leak (virtual leak, see Section 13.2). This virtual leak can be however differentiated from a real one by plotting the pressure rise vs. the time (Sections 13.2 and 13.3). The gas desorption increases considerably the time required to pump the system to a specific low pressure.

In order to reduce the time of evacuation or to reach lower pressures the sorbed or occluded gases must be removed from the material, by a process known as *degassing** using heat or other techniques (see also Section 13.2).

The degassing can only be achieved after careful selection of the materials and components since all of them must be subjected to a bakeout in vacuum at the required degassing temperature. The degassing without heating is also possible, but the efficiency is not satisfactory. For example, the bakeout at 400–450 °C and pumping during 10–20 hr results in an outgassing* rate lower by a factor of 10^5–10^6 than that produced after pumping for the same length of time without bakeout (Munday[906]).

Metals have always in their mass dissolved (absorbed or occluded) gases and adsorbed gases on their surfaces. The dissolved gas content depends on the nature of the metal, the metallurgical process used in its production and the pretreatment to which the metal was subjected. The adsorbed gas quantity depends on the surface area (apparent and true surface area, see Table 2.6), the kind of vapour present and the temperature of the material.

The *gas content of the metals* is generally expressed in terms of solubility i.e. cm^3 of gas (STP) per 100 g or per cm^3 of metal. Figure 2.12 shows the gas solubility in metals as a function of the temperature (Dushman[314, 315], Waldschmidt[1283] Jaeckel[624], Power[1003]).

The general rules concerning the gas solubility in metals can be summarized in the following way;

(a) Rare gases do not dissolve in any metal at any temperature;

(b) Hydrogen forms true solutions with aluminium, chromium, cobalt, copper, iron, molybdenum, nickel, platinum, silver and tungsten with the gas in atomic state;

* The *outgassing* is the spontaneous evolution of gas from a material in vacuum. The *degassing* is the deliberate removal of gas from the material. The outgassing rate per unit area is defined as the quantity of gas evolved per unit time at a given pressure and temperature; expressed in lusec/cm², torr l./sec and cm², etc. (see Appendix A.3).

(c) Nitrogen is insoluble in copper, gold and silver; it is soluble in aluminium, iron, molybdenum and tungsten;

(d) Oxygen is soluble in cobalt, copper and especially in silver.

Most of the available data (Dayton[268], Dushman[314], Geller[437], Santeler[1096]) on the *free outgassing rate* of ordinary (contaminated) metal surfaces at room temperature indicates that this rate varies approximately inversely as the first power of the time, for at least the first ten hours of pumping. The outgassing

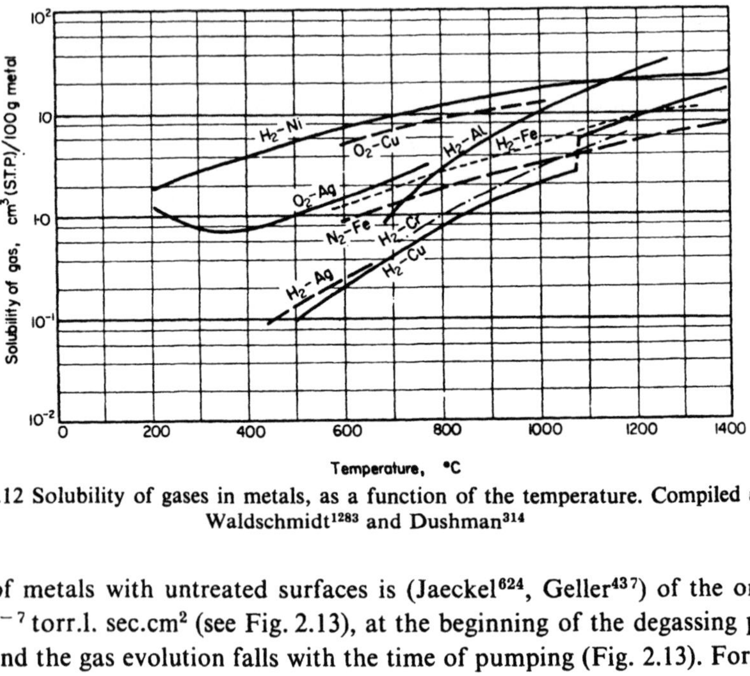

FIG. 2.12 Solubility of gases in metals, as a function of the temperature. Compiled after Waldschmidt[1283] and Dushman[314]

rate of metals with untreated surfaces is (Jaeckel[624], Geller[437]) of the order of 10^{-7} torr.l. sec.cm^2 (see Fig. 2.13), at the beginning of the degassing process and the gas evolution falls with the time of pumping (Fig. 2.13). For the outgassing rate K_n after the pumping time t_h, Dayton[268] gives the formula

$$K_n = \frac{K_1}{t_h^{\alpha}}$$

where K_1 is the outgassing rate at the beginning of the pumping (torr.litre/sec. cm^2), t_h the pumping time in hours, and a an exponent which varies from 0.7 to 2 but is frequently near to 1. The formula is rigorously valid for $t_h < 10$ hr. The pumping times can be shortened by the *pre-treatment* of the materials prior to their vacuum degassing.

The pretreatment includes degreasing, cleaning, firing, storing, handling and mounting. The influence of these treatments on the gas evolution from nickel samples (when heated to 850 °C) is summarized in Table 2.5 (Varadi[1256]). The metal has a high amount of gases evolved at the degassing process, if it was previously *degreased* in organic solvents, ultrasonically cleaned and

TABLE 2.5. GAS EVOLUTION FROM NICKEL SAMPLES (Arbitrary units)

Sample	Total	H_2	H_2O	$CO+N_2$	CO_2
Only degreased	3000	1050	340	2300	570
H_2 fired, stored dry	270	50	15	45	50
H_2 fired, touched by hand	6000	1800	348	*	1400
H_2 fired, touched by hand and degreased	590	220	15	110	70
H_2 fired, touched by rubber cots	800	300	250	300	110
H_2 fired, touched by freshly cleaned cotton gloves	1150	550	60	800	200
Air fired	1100	62	110	*	380
Acid cleaned	290	120	*	*	55

* Not measured.

washed in demineralized water (Table 2.5). The gas evolution from degreased nickel samples pretreated by *acid cleaning* (hydrochloric, nitric and acetic acid mixture) with a subsequent washing in demineralized water and drying in air was very low (Várady[1256]). For copper the ratio of gas evolution between non-treated, only degreased and degreased plus acid cleaned surfaces is 7:6:1 (Flecken[382]). By *firing* the nickel samples in wet hydrogen (1150 °C, 4 hr) the gas content was very much reduced (Table 2.5). Subsequent air firing (1050 °C) raises again the gas content, probably due to the oxide formed on the surface.

The *storage* in open air or container with a dessicant has shown no influence on the degreased and hydrogen fired samples, but the gas content of the acid-cleaned samples does increase by storage. The *touching* of the samples by the fingers increased the gas content (Table 2.5). Touching by rubber cots or cotton gloves increases the gas content by a smaller amount.

Occluded or adsorbed gases can be removed effectively by *heating* the metal *in vacuum* or in a gas which does not react with the metal. Hydrogen particularly is used because it simultaneously reduces the existing oxides and later can be driven out easily from the metal due to its higher permeation (see Figs. 2.5, 2.6). Actually the degassing can take place (a) during the melting process of the raw material (see Espe[354], Bunshah[181a]), (b) as a preliminary degassing of the individual components (in hydrogen or vacuum furnaces) or (c) by heating of the assembled system while being evacuated. As the permeation rate increases rapidly with the temperature, in order to reduce the time required by the degassing process it is desirable to do the degassing at the highest temperature possible. The upper temperature limit is determined by the mechanical strength at these temperatures or by the melting point of the metal and its vapour pressure (see Appendices B.2, B.4).

Dayton[268] shows the theoretical curves for the *desorption of water* at various temperatures (Fig. 2.14) between 25 and 150 °C. When the temperature is raised above 150 °C some of the physically adsorbed water becomes chemi-

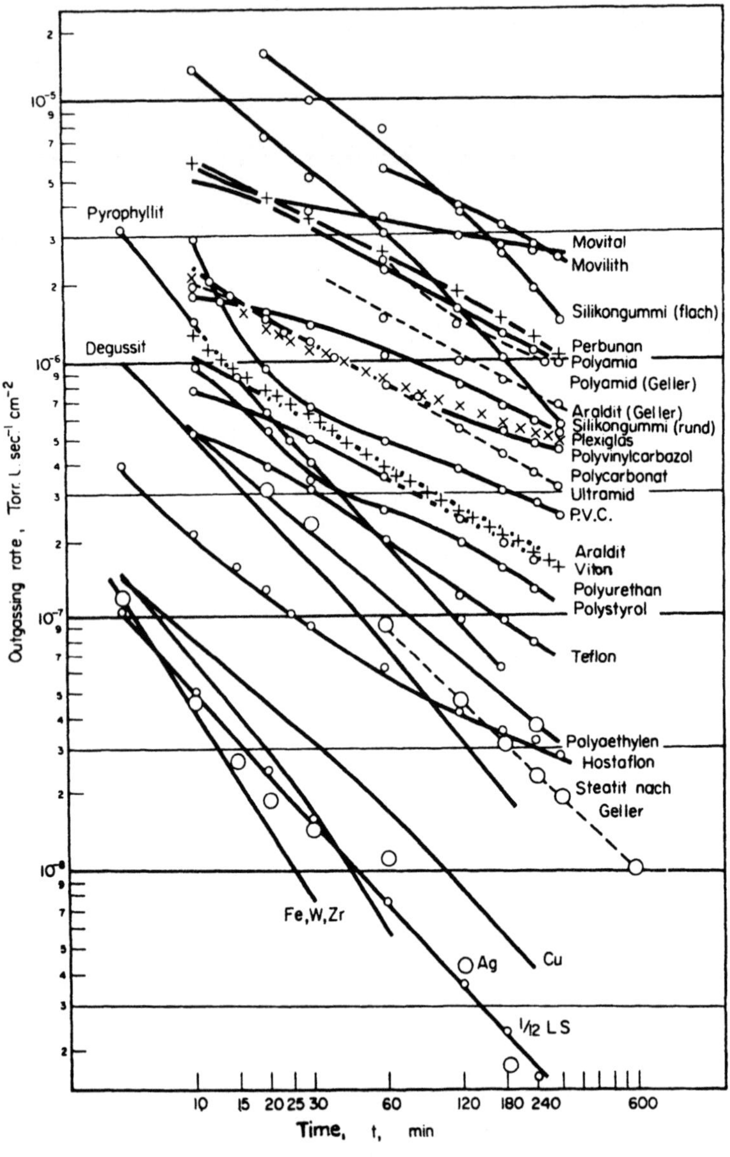

FIG. 2.13 Outgassing rates of various materials at room temperature as a function of time (min). Reproduced from Jaeckel[624] *(Courtesy of Pergamon Press)*

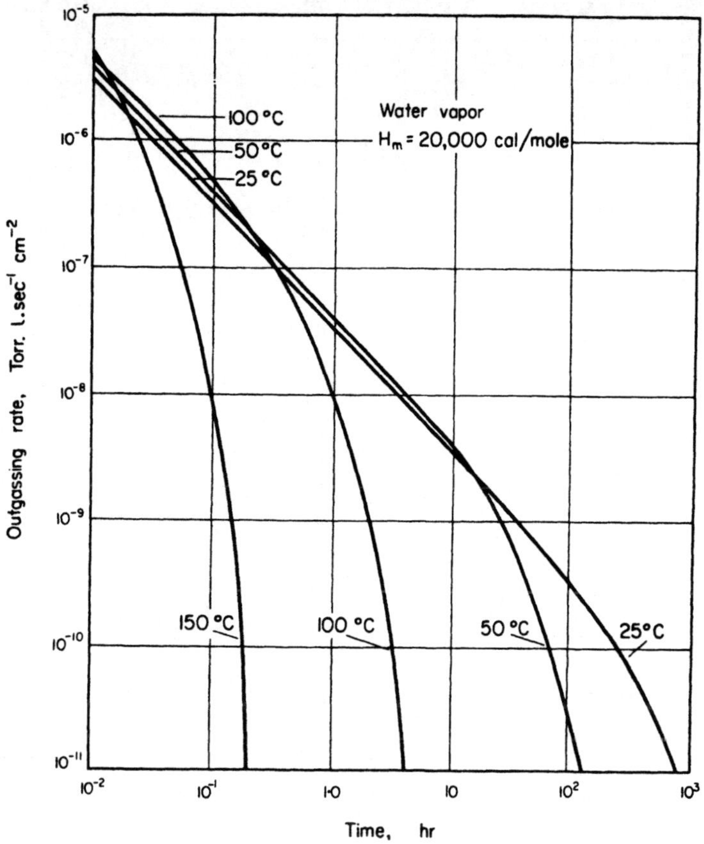

FIG. 2.14 Theoretical curves for the desorption of water at various temperatures. Reproduced from Dayton[268] *(Courtesy of Pergamon Press)*

cally sorbed. The latter begins to desorb at an appreciable rate only at temperatures in the vicinity of 500 °C. Using the values from Fig. 2.14 it must be kept in mind that the *true surface area* of metals is often very different from the apparent one. Table 2.6 lists some informative values in this respect (Morand[866], Schram[1107a]).

For a successful degassing the *temperature* of metal parts and enclosures is raised to: 400–500 °C for copper and aluminium and to 950 °C for iron, nickel and steel components. It is obvious that when steel containers sealed with copper or aluminium gaskets are to be degassed the latter materials determine the upper temperature limit.

The *time* needed for the degassing process using heating (baking) depends on the metallurgical history of the metal (Power[1003]) and decreases with increasing temperature. Figure 2.15 presents some general values for the required degassing time, in term of the time $t_{0.1}$ (in hours) required to degas

TABLE 2.6. RATIOS BETWEEN TRUE AND APPARENT SURFACE AREA OF VARIOUS METALS

Metal	Ratio	Metal	Ratio
Platinum foil	2	Copper (1 mm sheet)	14
Platinum foil (acid-cleaned)	3	Nickel, alternately oxidized and reduced	46
Nickel (rolled)	6	Nickel (acid cleaned)	50
Aluminium (in very thin foil)	6	Nickel (recently polished)	75
Stainless steel (1 mm sheet)	8	Aluminium foil (20 μ) anodized (Section 71.36c)	900
Nickel (polished for a long time)	10	Platinum foil, platinized	1800

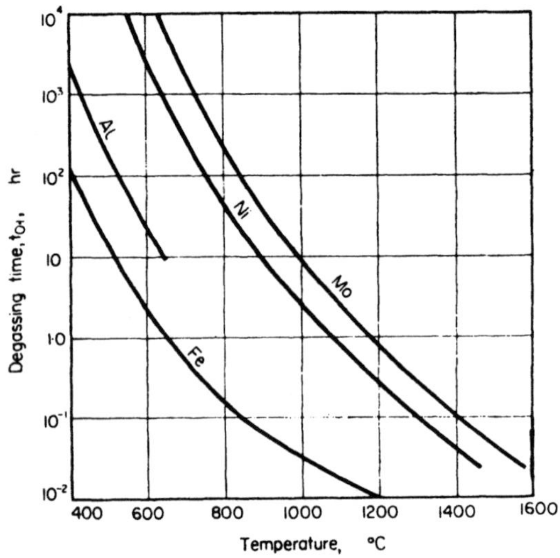

FIG. 2.15 Degassing time of 0.1 mm thick sheets, to reach 5 per cent of their initial gas content. After Jaeckel[623] *(Courtesy of Springer Verlag, Berlin)*

a 0.1 mm thick metal sheet reaching a gas content of 5 per cent of the initial one (Jaeckel[623]). For other thicknesses d (mm), the degassing time t_d (to reach the same per cent gas content) is given by the formula

$$t_d = \left(\frac{d}{0.1}\right)^2 \cdot t_{0.1}.$$

The heat required for the degassing is provided by a large furnace surrounding the vacuum system. For small diameter pipes also heating tape rolled over the pipe can be used. A method of heating by passing a controlled very high current (e.g. 2000 A) at low voltages through the system is mentioned by Ward[1286].

Glasses contain occluded and adsorbed gasses and vapours, especially water vapour. Besides water, also hydrogen, nitrogen, oxygen and carbon dioxide are contained in and on the glass.

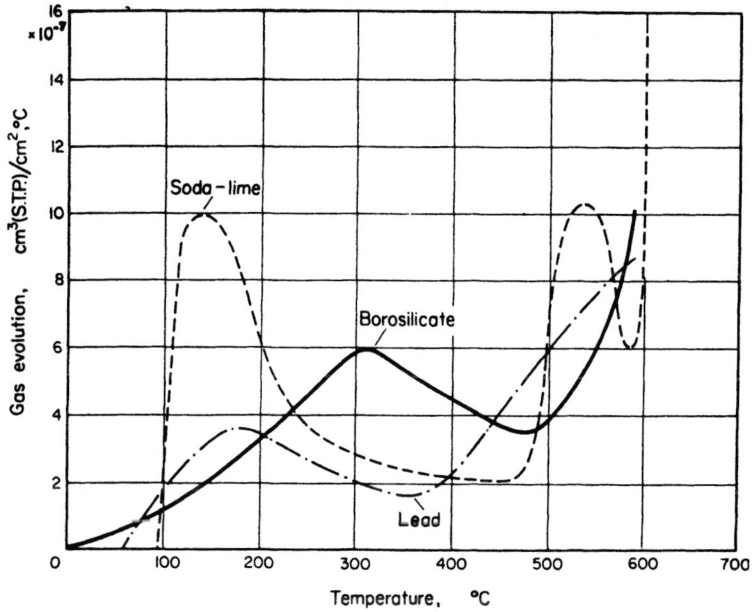

FIG. 2.16 The evolution of gas from various glasses as a function of the temperature (after Ardenne[47], Espe[354], Roth[1083])

The gas and vapour evolved from glasses is due only to the sorbed ones as the vapour pressure of the glass itself is extremely low (10^{-25}–10^{-15} torr).

By heating the glass to 150 °C in vacuum, the greatest part of the adsorbed gases and water vapour is given off. The curves representing the gas evolution (Fig. 2.16) have a maximum point at about 140 °C for soda–lime glasses (see Table 2.10), at 175 °C for lead glasses and at about 300 °C for borosilicate glasses. At still higher temperatures the gas evolution is reduced, but after the temperature range between 350 and 450 °C is exceeded additional gases are given off due to the decomposition of the glass.

Generally the gas evolution from glasses increases with increasing alkali content, thus lead glasses have a lower gas evolution (especially water) than the various others (Boettcher[144]). Obviously the gas content and evolution is influenced also by other factors as age, atmospheric influence during storage or method of cleaning. The usual procedure to make glass assemblies is to clean the glass pipes and to form and seal it by fusing. The procedure is not recommended (Donaldson[285]) since by fusing the glass, new contaminating particles are produced which are not destroyed by vacuum bakeout but only by cleaning. Thus the correct procedure should include a second cleaning after fusion and before bakeout.

It should be pointed out that the diffusion of water vapour from the glass into the evacuated space is a reversible process. For a given glass at an arbitrary temperature there is an equilibrium partial pressure of water vapour. If the partial pressure of water vapour in the surrounding space is greater than this equilibrium pressure, water diffuses back into the glass, if the partial pressure is less, water diffuses out. It results (Todd[1228]) that for water degassing, heating in *dry* atmosphere has the same effect as heating in vacuum.

As a general rule the degassing temperature of glasses is about 20–50 °C below the strain point (Section 23.11) of the glass, and maximum temperatures used for degassing are about 400 °C for lead glasses, about 500 °C for soda–lime and about 600 °C for borosilicates (Section 21.3).

Ceramics have generally very low outgassing rates, except when impurities are on the surface by lack of adequate cleaning. At high temperature (about 800 °C) and low pressures (less than 10^{-6} torr) a continuous source of oxygen evolution was observed in ceramics (Norton[946]) produced by the dissociation of Fe_2O_3 in these conditions.

Mica contains 18 per cent (by weight) of combined water and its surface can be regarded as saturated with adsorbed gas. The recommended procedures for the degassing of mica varies with temperature and time. Yarwood[1339] recommends heating in air or vacuum to 200 °C for 1–3 days, other authors want heating in vacuum to 450–500 °C for 8–15 hr or degassing at 675 °C for 16 minutes (Espe[354]).

Rubbers evolve a great amount of gases, especially when they are new and untreated, or have the form of pipes (containing higher quantities of plastifiers). The vapour pressure of natural rubber was found to be 10^{-3} torr at 20 °C (Cloud[218]), decreasing to 10^{-5}–10^{-6} torr at temperatures of −20 to −40 °C. Silicone rubbers have lower vapour pressures, and special vacuum rubbers (e.g. Hycar, Neoprene) even lower ones (1.10^{-4} torr at 20 °C).

The outgassing rates of some rubbers are plotted in Fig. 2.13 where they are compared with those of other materials, and vapour pressures are listed in Table 2.7 parallel with the outgassing rates.

In order to obtain lower outgassing rates the rubber must be *cleaned* with a solution of KOH (20 per cent) at 70 °C with subsequent washing with distilled water and drying with clean air, and/or degassing in vacuum at 70 °C for 4–5 hours. Using clean rubber the lowest pressure range attainable in the system is between 10^{-5} torr (Adams[6]) and 10^{-6} torr (Young[1347]).

Plastics have (like rubbers) high outgassing rates. Hence they are generally not included in vacuum systems for pressures lower than 10^{-4} torr. Excep-

TABLE 2.7. VAPOUR PRESSURES AND OUTGASSING RATES OF RUBBERS AND PLASTICS (Jensen[635], Saechtling[1092a], Normand[942], Hogg[576], Jaeckel[624], Markley[806], Diels[280], Geller[437], Young[1347], Holland[584], Thieme[1219a], Beckmann[102b])

Material*	Vapour pressure at 20 °C (torr)	Outgassing rate (after 3 hr pumping) (lusec/cm^2)
Neoprene rubber (Hycar, Perbunan, or Nitrile)	4×10^{-3}	1×10^{-3}–3×10^{-3}
As before but cleaned	1×10^{-4}	2×10^{-4}–3×10^{-4}
Silicone rubber	2×10^{-4}	2×10^{-4}–7×10^{-3}
Teflon, Fluon	3×10^{-5}	4×10^{-5}–3×10^{-4}
Hostaflon, Kel-F Viton	—	2×10^{-5}–6×10^{-5}
Plexiglass, Perspex, Lucite	1–2×10^{-4}	2×10^{-3}–4×10^{-4}
Polyethylene	5×10^{-5}	1×10^{-4}–3×10^{-4}
Polystyrene	4×10^{-5}	2×10^{-4}–9×10^{-4}
PVC, Astralon, Tygon	2–6×10^{-3}	4×10^{-4}–2×10^{-3}
Araldite	—	2×10^{-5}–6×10^{-5}

* See Table 2.14 which lists the trade names.

tions to this rule are the polytetrafluoroethylenes (Teflon, etc.) and the polytrifluorochloroethylenes (Hostaflon, etc.) as results from Table 2.7. Polythene (see Table 2.14) is also quoted (Martin[816]) as reliable in some particular applications to pressures in the range of 10^{-6} torr.

21.2 Metal vessels and pipes

Metals are extensively used as building materials for vacuum plants (pumps, connexion pipes, valves or vacuum chambers). The scope of this book is to point out the special requirements to be met by the metal components used in vacuum systems, and not to deal with the construction of pumps or to discuss the common technical problems connected with metal pipe construction.

"Vacuum vessels" include any evacuated container as: the envelope of electron tubes, mercury vapour rectifiers, spectrographs (Karlsson[655]), vacuum coating plants (Paul[979], Holland[581], Espe[354], Pirani[992], Methfessel[850]), vacuum furnaces (Bunshah[181a], Pirani[992]), vacuum distillation and drying plants (Holland-Merten[587]), particle accelerators (Willis[1320], Dreyer[304]), space simulation chambers (Spence[1168], Camack[191], Maker[796], Levantine[760]) etc.

The metals and alloys used for the construction of these vacuum vessels and that of the pipes connecting them to the pumps or gauges vary from brass and aluminium to stainless steels. The various groups of metals and other materials to be used in various vacuum ranges are listed in Table 2.8.

The advantages and difficulties in the use of metal vessels mainly arise from their mechanical strength, gas evolution, and corrosion resistance.

Obviously the first condition for any vacuum vessel should be to provide the necessary space for the parts to be built inside the evacuated volume (e.g. electrodes, evaporation systems, shutters, etc.). In this respect metal chambers are advantageous and extensively used, especially when large spaces are required. An example of such a vessel is described by Dreyer[304] (stainless steel vessel of 25,000 in^3, evacuated to less than 3×10^{-10} torr).

TABLE 2.8. USEFUL RANGE OF MATERIALS FOR VACUUM VESSELS, PIPES

Material	Pressure (torr)				
	760–1	1–10^{-3}	10^{-3}–10^{-5}	10^{-5}–10^{-7}	10^{-7}–10^{-10}
Iron, steels	good	good	good	only after degassing	only stainless steels
Cast iron, copper or aluminium	good	good	bad	bad	bad
Rolled copper or alloys	good	good	good	only after degassing	only OFHC copper
Nickel and alloys	good	good	good	good	good
Aluminium	good	good	only after degassing		not recommended
Glass, quartz	good	good	good with degassing	good	only thick-walled
Ceramics	good	good	only with vitreous coating		only special types
Mica	good	good	only after strong degassing		not recommended
Rubbers*	good	good	only degassed	bad	bad
Plastics*	good	only special types		only Teflon, Araldite	not recommended

* See Table 2.7.

With large vessels, the mechanical strength required to withstand collapsing leads to thick walls (Tables 2.1–2.3) or to the need for strong materials. Obviously metal vessels have the advantage of being robust and allowing machining to extreme accuracies.

As the chamber size increases, the weight of the chamber and the thickness of the wall increases rapidly. Above some critical size the baking of such a chamber becomes very difficult. A solution in such cases can be the *double chamber* (Simons[1140]). Here the outer chamber may remain cool and can be built from common metals (e.g. mild steel) with a wall thickness to withstand atmospheric pressure. The inner chamber (e.g. stainless steel) can be made quite thin since it supports practically no pressure difference, the space between the two chambers being evacuated to a "guard vacuum" level (Section 38.23). Thus the thermal mass of the inner chamber is much less and the heating process is simplified and quicker.

The material of the vessel should have a low vapour pressure at the maximum working temperature. Vapour pressure data are summarized in Appendix B.4 (see also Dushman[314, 315], Smithels[1157]). Some metals (Zn, Cd, Pb) have at 400–500 °C vapour pressures exceeding the pressures required in high vacuum systems, and therefore these metals cannot be used (Michaelson[854]). For ultra-high vacuum work the choice of metals are only: stainless steels, high nickel alloys and oxygen free high conductivity (OFHC) copper. Steel is in general undesirable for magnetic reasons but if admissible it should only be used with caution because of the hydrogen penetration (Norton[946], see Section 21.12).

The gas evolution from the metal surface should be low. To meet this requirement only previously degassed metals are to be used (Section 21.13), the true and apparent surface areas being (Table 2.6) kept to a minimum and cleaned thoroughly. Doré[288] recommends a solvent cleaning (to dissolve the mineral oils and greases) followed by cleaning in an alkaline solution (to remove the greases which can be saponified). Várady[1256] and Flecken[382] show the efficiency of acid cleaning (Table 2.5).

The vacuum vessel should not be permeable to gases. Thus only rolled or forged metal is recommended. Castings are generally not satisfactory for vacuum envelopes, excepting small components where die castings (Millen[858]), centrifugally cast metal or metal-sprayed or plated parts (Kronberger[721]) can be used.

The vacuum vessel should be corrosion resistant. This resistance extends from the light corrosion of the atmosphere of a laboratory or factory to the powerful corrosion encountered in metallurgy, biochemistry or nuclear research applications. The corrosion resistance of some materials used in vacuum systems is listed in Table 2.9. For detailed data refer e.g. to Molyneux[873a], Espe[354].

TABLE 2.9. CORROSION RESISTANCE OF SOME MATERIALS

Corrosion agent	Material											
	Cu	Al	Carbon steel	12 Cr steel	18/8 steel	Monel	Glass silica	Polyethylene	PVC	Saran	Teflon	Buna N
H_2SO_4 (10%)	F	F	P	P	P	G	E	E	E	E	E	E
NaCl (10%)	P	P	P	P	P	F	G	E	E	E	E	E
NO_3H (10%)	P	F	P	G	E	P	E	E	E	E	E	F
Acetic acid (10%)	F	E	P	F	G	G	E	E	E	E	E	E
NaOH (10%)	F	P	E	E	E	E	P	E	G	F	E	G
NH_4OH	P	G	E	E	E	P	F	E	E	P	E	E
Wet H_2S	P	E	F	F	G	G	G	E	E	E	E	E
Wet Cl_2	P	P	P	P	P	P	G	P	G	P	E	F
Wet SO_2	G	G	P	P	F	P	G	E	E	G	E	E
Hg vapour	P	P	G	E	E	P	E	G	G	G	G	G
Benzene	E	E	E	E	E	E	E	P	P	F	E	E
CCl_4	E	E	E	E	E	E	E	P	F	F	E	F
Acetone	E	E	E	E	E	E	E	P	P	F	E	G
Alcohol	E	E	E	E	E	E	E	P	E	E	E	E

N.B. E = excellent, G = good, F = fair, P = poor.

Metal pipes must fulfil more or less the same requirements as vacuum vessels. The pipe diameter should correspond to the gas throughput requirements (Section 12.1).

21.3 Glass (quartz) vessels and pipes

Glass is used as the envelope of many vacuum devices as incandescent lamps (Roth[1083], Ulmichek[1251]), electron tubes (Kohl[705, 706], Yarwood[1339], Hees[531], Milner[867]), X-ray tubes (Espe[354], Knoll[697]), photoelectric cells (Espe[354, 356]), discharge tubes and vessels (Neumann[923], Oranje[955], Ivanov[618], Roth[1083], Hogberg[575]), etc. Glass is also used as bell jars in small vacuum evaporation plants (Holland[581], Pirani[99] Laporte[741]) as reaction vessels and connexion pipes in laboratory and pilot vacuum plants (Mönch[879], Holland-Merten[587], Bachman[65]), for the construction of diffusion pumps (Strong[1207], Jaeckel[623], Reiman[1046], Mönch[878, 879]) and vacuum gauges (Dushman[314, 315], Pirani[992], Leck[752]).

Glass is a noncrystalline material that has no regular internal structure. It is rigid at ordinary temperatures and almost fluid at higher temperatures (Section 23.11). It has no definite freezing point but becomes solid because its viscosity increases progressively to values which, for all practical purposes are infinitely great.

Although silica (SiO_2) is the principal ingredient of most glasses, the addition of other melting agents and modifiers gives to glasses a wide range of properties. Depending on the choice of these additional constituents glasses can be classified into several groups having the same range of properies. A general classification divides glasses into *soft and hard* ones, corresponding to the temperature range in which the glass is soft enough to be worked (Fig.2.45). Table 2.10 presents such a classification with examples of some of the usual glasses. For details on glass fabrication and properties refer to e.g. Shand[1123], Jones[645], Kitaigorodsky[681], Marx[823], Colnot[223], Corning[239], Osram[958].

Glass is a fragile material. For this reason it is preferable to list its main characteristics according to the factors generally leading to its breaking. These are: mechanical stresses (due to tension, bending, impact), thermal stresses (due to expansion difference, heating, thermal shock) and composition changes (due to weathering, devitrification — see Section 23.21 — or chemical reactions).

Glass does not plastically deform before failure, it fractures only as a result of tensile stresses and not due to shear or compression.

The *useful strength* of glass is but a small fraction of its intrinsic strength because of stress concentrations due to surface imperfections, which accounts for the fact that glass is stronger under momentary loading than under prolonged stresses. In this respect the values listed as tensile stresses are statistical mean values (see Appendix B.l). The tensile stress which should be taken, providing an adequate safety factor, is 0.7 kg/mm^2 for annealed glass and 1.5–3.0 kg/mm^2 for tempered or thermally strengthened glass (Shand[1124], Morey[888], Korolev[712], Preston[1007], Springer[1171], Stanworth[1176]).

When glass is *suddenly cooled*, tensile stresses are introduced in the cooled surfaces and a compensating compression stress in the mass of the glass. *Sudden heating* leads to surface compression and internal tension. Since glass fails only in tension at the surface, *the temporary stresses from sudden cooling are much more dangerous than those resulting from sudden heating if all the surfaces are cooled or heated simultaneously* (Roth[1083]).

The transient thermal stresses σ increase directly with the expansion coefficient α and in a complex way with the glass thickness. The *thermal shock resis-*

TABLE 2.10. GLASSES USED IN VACUUM TECHNIQUE

Group	Type	Constituents % weight	Examples				
			Glass	Expansion coeff. (10^{-7}/°C)	Strain point (°C)	Annealing point (°C)	Manufacturer**
Soft glasses	Lead glasses	SiO_2; PbO= 40–50; alkali <10	Minos 1650[III]	88	–	415*	Jena
			Lead W2	82	–	415	Moosbr.
	Lead alkali silicate glasses	SiO_2; PbO= 20–35; alkali <10	Iron 153	123	380	396	Sovirel
			Lead N (915a)	100	–	425*	Moosbr.
			FeCr L 14	98	360	430	GEC
			Lead K 1 A	95	–	425*	Philips
			Lead 111	92	–	425*	Philips
			Soft lead L 1	91	340	430	GEC
			Copperclad C 12	91	380	435	BTH
			Soft glass 0010	91	397	428	Corning
			Soft glass 0120	89	400	433	Corning
			123a M	88	–	425*	Osram
			Lead GWB (GW2)	86	–	410	Chance
			Lead 3079[III]	81	–	480*	Jena
	Soda-lime silicate glasses	SiO_2; CaO= 5–12; alkali= 13–20	B8	96	460	530	GEC
			X 4	96	465	500	GEC
			FeCr C 19	95	–	530	BTH
			X 8	95	465	500	GEC
			Bulb 0080	92	478	510	Corning
			Magnezia 105	89	–	508*	Osram
			GWA (GW1)	87	–	530	Chance
Hard glasses	Alumino lime silicate glasses	SiO_2; Al_2O_3= 3–10; CaO= 6–12; alkali= 8–23	Iron R L 114	114	–	500	GEC
			Lime C 22	104	–	505	BTH
			Apparate Glas 584d	88	–	530*	Osram
			Thermometer 16[III]	80	495	537	Jena
	Alumino boro lime silicate glasses	SiO_2; B_2O_3= 3–8; Al_2O_3= 3–10; CaO= 6–12; alkali= 8–23	Amber Ma 1	75	400	580	GEC
			Mo. B.B.	47	563	594	Russian

(Table 2.10 Continued)

Group	Type	Constituents % weight	Examples				
			Glass	Expansion coeff. (10^{-7}/°C)	Strain point (°C)	Annealing point (°C)	Manufacturer**
Hard glasses	Alumino boro lime zinc glass	As before but CaO = 3–12; ZnO = 3–7 alkali = 8–14	Mo. 1447[III]	50	483	529	Jena
	Borosilicate	SiO_2; $B_2O_3 > 10$ $Al_2O_3 < 3$	Kovar C.40	48	455	505	BTH
			W seal, W 1	38	540	580	GEC
			Duran 3891 III	37	516	567	Jena
	Alkali-borosilicate glasses	SiO_2; B_2O_3 >10; Al_2O_3 <6; alkali $= 6–8$	Thermometer 7520	61	530	566	Corning
			Mo 637h	48	–	550*	Osram
			Neutrohm E(Mo)	48	–	505*	Baccarat
			FeNiCo 756	48	–	500*	Osram
			Mo. H.H.	47	500	590	GEC
			3072 (Gerätte 20)	46	–	558*	Jena
			Clear seal 7050	46	461	496	Corning
			Mo. C 11	45	500	575	BTH
			Uran 3320	41	497	535	Corning
			W glass C 9	36	480	525	BTH
			Hysil GH 1	33	513	556	Chance
			Pyrex 7740	33	515	555	Corning
	Alumino borosilicate glasses	SiO_2; $B_2O_3 =$ 5–20; $Al_2O_3 =$ 3–20; alkali < 6	Kovar GS 3	50	400	450	Chance
			Mo. H 26 X	46	600	725	GEC
			Supremax 3058[III]	33	–	738*	Jena
	Lead borosilicate glasses	SiO_2; $B_2O_3 =$ 15–18; PbO = 4–7	W seal 362a	39	–	522*	Osram
			Nonex 7720	36	484	518	Corning

* Transformation point, with viscosity $10^{13.3}$ poise (Section 23.11).
** BTH – The British Thomson-Houston Co. Ltd., Rugby, England.
Chance – Chance Brothers Ltd. Glass Works, Birmingham, England.
Corning – Corning Glass Works, Corning, N.Y., U.S.A.
GEC – Osram-G.E.C. Glass Works, East Lane, Wembley, Middlesex, England.
Sovirel – Sovirel Co., Bagneaux-sur-Loing, France.
Osram – Osram, Berlin, West Germany.
Moosbr. – Moosbrunner Glasfabrik, Vienna 4, Austria.
Jena – Jenauer Glaswerk Schott u. Gen., Mainz, West Germany.
Philips – Philips, Eindhoven, Holland.
Baccarat – Cristallerie de Baccarat, France.

tance is expressed as

$$\Delta T_s = A \frac{\sigma}{E \cdot \alpha} \sqrt{\frac{\lambda}{s \cdot \gamma}},$$

where A is a shape factor (for complicated shape A is less than for simple ones), E is the elasticity modulus, λ the thermal conductivity, γ the specific gravity and s the specific heat. Generally the thermal shock resistance is low in the case of soft glasses, higher for the hard glasses and very high for silica (Table 2.11).

If the glass is exposed to *steady temperature differences between the two faces*, thermal gradients are developed through the glass. These are even more dangerous to the integrity of the glass than sudden cooling. They are especially dangerous if added to stresses imposed by some parts of the glass upon others or by external mountings (Riley[1062]).

The temperature difference ΔT between the two surfaces of the glass produces a maximum stress σ (tension on cooler surface, compression on hotter surface) resulting from the formula:

$$\Delta T = \frac{2(1-\mu)\sigma}{E \cdot \alpha},$$

where μ is Poisson's ratio (for glass μ = 0.20–0.25), E modulus of elasticity, and α coefficient of thermal expansion. Table 2.11 list the values of ΔT that will produce a tensile stress of 0.7 kg/mm^2 (1000 p.s.i) on the cooler face, values known as *thermal stress resistance* of the glass. From these values it can be seen that a temperature difference of 17–19 °C (for soft glasses) or 40–48 °C (for hard glasses) is enough to produce the maximum permissible tensile stress.

Glass is generally inert to chemical action (see Table 2.9). It is attacked strongly only by hydrofluoric acid, hot concentrated phosphoric acid, and slowly by alkaline solutions (NaOH, KOH).

Glass has a low vapour pressure (Section 21.13) and is generally impermeable to gases (Section 21.12). These characteristics added to its transparency for light (Section 71.11) determine the extensive use of glass as envelope, vessel or chamber material in various vacuum techniques.

The good electrical insulation properties of glasses and their high dielectric loss factor (Section 41.1) make them useful for vacuum tight electrode seals.

Glass is available as prefabricated bell jars, bulbs, pipes or rods. Excepting the special shaped vessels and bulbs available, glass is extensively used to form vacuum systems or their components beginning with tubing or rod (Corning[239], Osram[958], Material Selector[824]). Glass tubing and rods are available in a large variety of dimensions, but the accuracy of these items cannot be as great as for metal tubing. Table 2.12 lists the normal tolerances of glass

TABLE 2.11. THERMAL SHOCK AND STRESS RESISTANCE OF GLASSES

Group	Glass	Thermal shock resistance ΔT, Sample thickness	Thermal shock resistance ΔT, °C	Thermal stress resistance ΔT (°C)	Manufacturer**
Soft glasses	Lead 123a M	DIN*	92	–	Osram
	Lead 0010	1/2 inch# 1/8 inch#	35 65	19	Corning
	Soda–lime 0080 and Lead 0120	1/2 inch# 1/8 inch#	35 65	17	Corning
	Magnesia 105	DIN*	99	–	Osram
Hard glasses	Apparateglas 584d	DIN*	89	–	Osram
	Normal 16^{III}	DIN*	110	–	Jena
	Mo 1447	DIN*	185	–	Jena
	Borosilicate 7050	1/2 inch# 1/8 inch#	70 125	34	Corning
	Uran 3320	1/2 inch# 1/8 inch#	80 145	40	Corning
	Pyrex 7740	1/2 inch# 1/8 inch#	100 180	48	Corning
	Mo. 637h	DIN*	171	–	Osram
	Duran 3891 W seal 362a	DIN* DIN*	215 220	– –	Jena Osram
	Nonex 7720	1/2 inch# 1/8 inch#	90 160	45	Corning
Silica	Silica 96% 7900	1/2 inch# 1/8 inch#	750 1250	200	Corning
	Silica (fused) 7940	1/2 inch# 1/8 inch#	750 1250	290	Corning

* Measured corresponding to German Standard DIN 52325/1953 by heating annealed glass rods of 30 mm length, 6 mm diameter, and plunging in cold water (20 °C).

\# Measured by heating annealed glass plates 150×150 mm of thickness mentioned and direct plunging into cold water.

** See footnote Table 2.10.

tubes and rods. The wall thickness of glass tubes can be determined mechanically at the ends of the tube or optically over its length (Wittwer[1328]).

TABLE 2.12. NORMAL TOLERANCES FOR GLASS PIPES AND RODS (in mm) (Shand[1124], Espe[354], Roth[1083], Thermal Syndicate[1218a])

Glass sort	Pipe				Rod	
	Diameter		Wall thickness		Diameter	
	Range	Tolerance ±	Range	Tolerance ±	Range	Tolerance ±
Soft glasses	3–12	0.4	0.7–1.5	0.15	< 2	0.15
	12–25	0.6	1.0–2.0	0.15	2– 6	0.25
					6–10	0.5
Hard glasses	3–10	0.5	0.5–1.5	0.3	< 2	0.3
	10–20	0.8	1.0–2.5	0.6	2– 6	0.6
	20–30	1.2	1.0–4.0	1.0	6–10	0.8
	30–50	1.5	2.0–6.0	1.5		
Silica (Quartz)	8–13	0.5	0.7–1.2	0.2	< 2	0.2
	13–18	0.8	1.0–1.4	0.3	2– 5	0.5
	18–30	1.2	1.1–1.7	0.3	6–10	0.8
	30–60	1.7	1.7–2.0	0.4		

21.4 Elastomer and plastic pipes

Elastomers and plastics are used in vacuum technique mainly as pipes and only rarely as specially shaped parts of the vacuum enclosures. Their use as sealing gaskets is discussed in Section 3.8. The elastomers utilized are classified in Table 2.13.

The field where elastomer and plastic pipes can be used is limited to backing lines (Fig. 1.1) or to dynamic vacuum systems (Section 11.3) because of the gas evolution (Table 2.7) of these materials (Jensen[635], Cook[232]) especially in compressed state (Biram[128]). The use of rubbers is also limited by their narrow temperature range, which extends towards high temperatures just to about 80 °C and towards the low temperatures to about −40 °C (see

Appendix B.2). At higher temperatures or after longer time rubbers present "ageing" effects in the form of their hardening. At low temperatures the rubbers become brittle. Silicone rubbers have a wider temperature range extending up to 180 °C and for shorter heating periods even up to 250 °C.

TABLE 2.13. CLASSIFICATION OF ELASTOMERS

Group	Name	Chemical comp.	Properties*				
			Electr. resistance	Flame resistance	Impermeability	Heat resistance	Cold resistance
Non-oil resistant	Natural rubber	Isoprene	G	P	F	F	G
	S.B.R. Buna S	Styrene/butadiene	G	P	F	F	G
	Butyl I.I.R.	Isoprene/isobutylene	G	P	E	G	F
	Polybutadyene	Butadiene	G	P	F	F	G
Oil and petroleum resistant	Thiokol	Organic polysulfide	F	P	E	G	F
	Nitrile, Philprene, Hycar, Buna N, Perbunan	Acrylonitrile/butadiene	P	P	E	G	F
	Polyurethane	Diisocyanate/polyester of polyether	F	F	G	G	G
	Neoprene	Chloroprene	F	G	G	G	F
	Hypalon	Chlorosulfonated polyethylene	G	G	–	G	P
Heat resistant	Silicone, Silastic	Polysiloxane	E	F	F	E	E
	Fluocarbon Viton	Vinylidene fluoride/hexafluoropropylene	E	G	–	E	F

* In comparison with the other elastomers: E = excellent, G = good, F = fair, P = poor.

A low sulphur content is essential when rubbers are to be used in vacuum technique. Before use the rubber must be thoroughly cleaned. Cleaning with hot (70 °C) 20 per cent NaOH or KOH solution followed by rinsing in distilled water and drying with hot air, is a satisfactory treatment (Angerer[34]).

Rubbers that are to be used in contact with mercury must be treated for at least an hour with the NaOH solution. After this treatment, the mercury will remain uncontaminated in contact with this rubber for years of active use (Pike[990]).

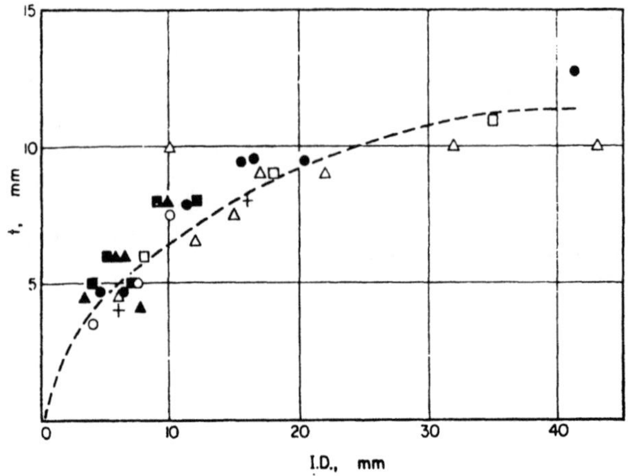

FIG. 2.17 The wall thickness (t) of rubber vacuum tubing as a function of the inside diameter. △ Pfeiffer[987]; ■ Edwards[329]; ● Welch[1294]; + Heraeus[542]; ○ Leybold[762]; ▲ Eisler[335]; □ Siemens-Schuckert

Elastomer pipes used as connexions have the advantage of being highly flexible and of permitting an easy connexion (Section 38.53) to other pipes (metal, glass). Thus these pipes are extensively used as anti-vibration connexions.

Because of their mechanical properties (see Appendix B.1) the wall thickness of elastomer pipes must be relatively large, to prevent collapsing (Section 21.11). Generally the wall thickness of vacuum elastomer pipes must be of the same order as their inside diameter. Figure 2.17 represents the wall thickness of various vacuum rubber tubes plotted vs. the inside diameter. For a discussion of the elastic deformation of rubbers refer to Section 38.31; other properties are summarized in Appendices B.1–B.6 and details can be found, e.g. in Dawton[267], Catton[198], Ennos[344], Auwarter[56], and for silicone rubbers in Nowak[948], Nitzsche[933], Wick[1311], Wacker[1276], Midland[856].

Thinner walled elastomer pipes can be prevented from collapsing by a helical wire spring inserted in the pipe (Fremlin[402]), which reduces the unsupported length of the rubber pipe to the spacing between turns, i.e. decreases the wall thickness required to resist collapsing (Fig. 2.2).

Plastics are also used to some extent in vacuum sealing techniques, especially pipes, castings or laminates (e.g. Epoxy resins, see Section 3.3), or thin films (e.g. Mylar, etc., see Section 71.36). The plastics used are listed in Table

2.14 with some examples of trade names for each group. For other details refer to Simonds[1139], Salsig[1094], Brady[159], Blackmon[129a].

Acrylics (as Perspex, Lucite, Plexiglas) have a relatively high outgassing rate (Table 2.7) and are not recommended for high vacuum use, except for low temperatures, dynamic systems or windows of small surface areas. These plastics are extensively used as glove boxes, windows (see also Section 7.1) and seals for electric wires (Section 43.1).

TABLE 2.14. PLASTICS USED IN VACUUM TECHNIQUE

Group	Chemical composition	Common trade names	Remarks*
Acrylics	Polymethyl metacrilate	Lucite, Perspex Plexiglas	Transparent, water resistant
Fluocarbons	Polytetrafluorethylene	P.T.F.E., Teflon, Fluon	Chemically inert heat, cold resistant
	Polytrifluorchlorethylene	P.T.F.C.E., Kel-F, Hostaflon	As P.T.F.E.
Polyethylene		Polythene, Alkathene, Hostalen, Alathon, Plaxpak	Flexible, water resistant, chemical resistant
Polystyrene		Styron, Lustrex, Polystyrol, Styrofoam	Radiation resistant
P.V.C.	Polyvinyl chloride, and acetate-chloride vinyl copolymer	Tygon, Koroseal, Vinylite, Astralon	
Vinylidene chloride		Saran, Velon	

* See Table 2.7 for outgassing rates.

Fluocarbons have a more extended temperature range than any other plastic. They are suitable for use at operating temperatures ranging from $-100\,°C$ to $+300\,°C$. Since their outgassing rate is very low (Table 2.7) they can thus be used for high vacuum system components. Teflon and similar plastics (Table 2.14) are available as pipes with inside diameters from 0.5 to 90 mm and wall thicknesses from 0.25 to 17 mm, in lengths up to 1–2 m or cylinders with i.d. 7–320 mm and o.d. 17–360 mm in lengths of 20–50 mm. Rods of 1–100 mm diameter as well as sheets are also available (Du Pont[312], Angus[34b]). The required shapes can be machined easily. It must be pointed out that the *powder of fluocarbons in contact with hot surfaces or flames produces very toxic gases with fluor content.*

For details on fluocarbons refer to, e.g. Kirby[677], Mehnert[841], Cornell[238], Horn[595], Schulz[1110]. The methods of bonding Teflon to itself and other materials are described in Section 32.2 (Croze[251], Riley[1063]) and their use as gaskets in Section 38.32.

Polyethylene (see Table 2.14) has an outgassing rate near to those of fluocarbons but a smaller useful temperature range. These plastics can be used up to maximum 80–100 °C. According to Hogg[576] the pressure in a chamber evacuated with a diffusion pump and cold trap with dry ice (solid CO_2) was only 20 per cent greater when Polythene was placed inside. Pressures of 10^{-4} torr can be held overnight in Polythene containers, when the plastic is clean (Duncan[308]). To assist the removal of occluded gases a low potential high frequency (Tesla) discharge coil can be used. The contact with naked flames is to be avoided as the melting point of polyethylenes is only 110 °C.

Polythene is available as tubing from 10 to 150 mm diameter. Tubing of 12 mm diameter and 3 mm wall thickness will withstand (Table 2.1) complete evacuation without collapsing (Duncan[308]).

Polythene tubing may be straightened or bent by inserting a strong coil spring as a support, immersing the tubing in a bath of hot water for 15 min, and fixing the tubing in the desired shape. After cooling the tubing retains its shape. The machining of Polythene is possible, but the accuracy is limited by the tendency to flow. Vessels up to 150 mm diameter may be made by sealing the ends of polyethylene tubing with discs. Smaller vessels are easier made by drilling from a rod. The sealing of Polythene to itself and other materials is described in Section 32.2.

Polystyrene has a low outgassing rate (Table 2.7) but is tough and brittle at room temperature and becomes rubbery and soft above 70 °C. Turner[1241] describes a technique for using rigid polystyrene foam (Styrofoam) in the construction of vacuum tanks (for particle accelerators). Styrofoam was used as the core on which after making its surface solid and smooth by filling it with plaster of Paris, a coating of fibre-glass cloth, saturated with epoxy resin (Section 3.3) was applied. After curing the core was dissolved out with acetone (Hanson[509a]).

Polyvinyl chloride (Table 2.14) is used in vacuum technique as transparent tubing in the backing line of high vacuum systems (Fig. 1.1) or in low vacuum systems where its outgassing rate (Table 2.7) is tolerable. PVC connexion pipes are available with thick walls (for vacuum) or with thinner walls and metal spring support inserted inside the tube (Balzers[444], Leybold[762]).

Vinylidene chloride (Saran, etc., Table 2.14) tubing can be used up to 80 °C. According to Sancier[1095] Saran is satisfactory even for vacuum in the range of 10^{-5} torr. For the method of sealing Saran see Section 32.2.

2.2 METAL TO METAL WELDING AND BRAZING

22.1 *Welding*

Welding is the generic term used to describe metal joining processes based on localized coalescence of metal, produced as a result of pressure, temperature or a combination of them.

The block diagram on Fig. 2.18 shows a classification of the welding methods which can be used for vacuum sealing. For detailed information on welding techniques refer to Davies[261], Soled[1164], Laughner[745], Espe[352], Jefferson[629], Udin[1248].

The *non-pressure welding processes* include the techniques of metal joining by the application of heat without the use of pressure. In these processes a mixture of molten metal is formed as a result of the local melting of the surfaces or edges. This liquid metal mixture (to which eventually a filler metal is also added) bridges the gap between the components to be welded to each other. After the source of welding heat has been removed, this liquid solidifies, thus joining the parts together (welding them).

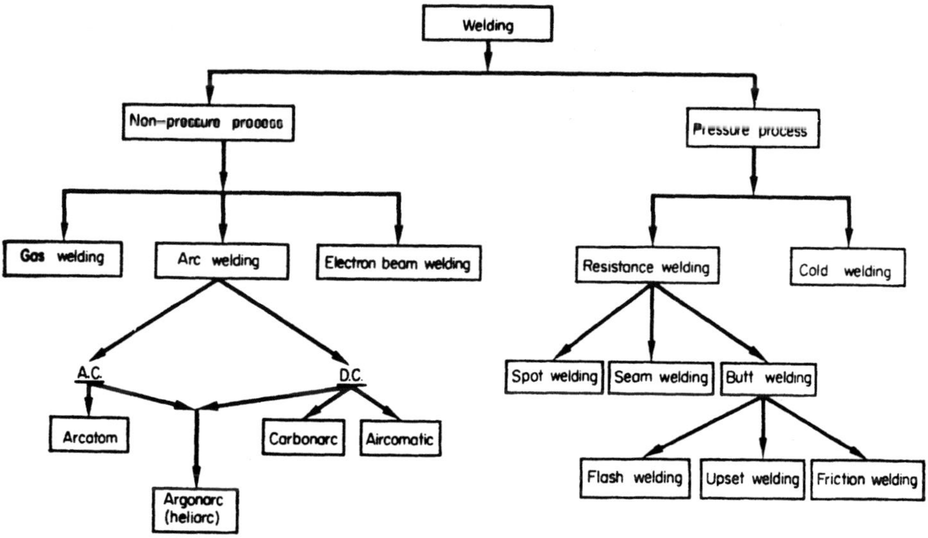

FIG. 2.18 Welding methods for vacuum sealing

The sources of heat used in non-pressure welding processes are: the flame (gas welding), the electric arc (arc welding) or the electron beam.

The *pressure welding processes* are techniques for joining metals using pressure with or without heating the components. This group of welding processes includes resistance welding and cold welding with their various techniques (Section 22.11).

22.11 Welding methods. (a) By *gas (torch) welding* the heat is furnished by the flame of a torch as a result of the combustion of a fuel gas (acetylene, hydrogen). The torch can be adjusted to produce a reducing, neutral or oxidizing flame. Neutral or slightly reducing flames are preferred in most operations.

The use of a reducing flame can be the cause of *porous welds* due to the occlusion of hydrogen in the weld, a fact especially evident when copper is the material to be welded (see e.g. Barrett[88]).

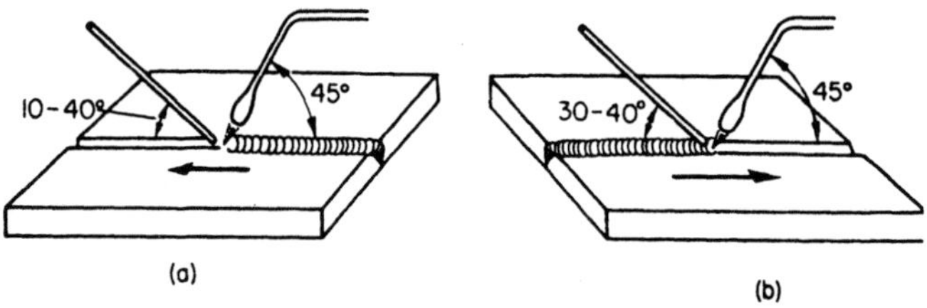

FIG. 2.19 Welding techniques: (a) forehand (leftward); (b) backhand (rightward)

The use of an oxidizing flame affects the strength of the weld introducing an oxide layer between the welded metal parts.

When required, filler metal is added by means of a filler rod. Torch welding requires the use of *fluxes* which are applied either to the work or to the filler metal. The necessity of a flux reduces the utilization of the gas (torch) welding methods in vacuum sealing only to some joints in the construction of very heavy copper or iron vacuum containers. In some recent welding systems this disadvantage was eliminated by introducing the flux in the gas itself, by conducting the gas through a vessel which contains boric acid and alcohol mixture ("Linde-Flux"). For details on torches, welding systems, etc. we refer to Kerwin[669], Davies[261], *Metals Handbook*[848].

The gas welding is applied using either forehand or backhand techniques (Fig. 2.19). Since the *forehand* technique provides a small welding metal puddle, which results in a smoother weld, it is generally preferred for the welding of thin metals (up to 3 mm). With *backhand* welding increased welding speeds, better control of the puddle and a better weld quality are possible, thus this technique is usually recommended for materials thicker than 3 mm.

(b) *Arc welding* is based on the heat obtained from an electric arc formed between the work and an electrode or between two electrodes. By the highly localized heat produced by the arc a molten metal pool is formed on the work. This pool is made to progress along the joint by manipulation of the electrode, resulting in the desired weld.

When consumable electrodes are used additional metal is obtained by melting the tip of the electrode, having usually a composition similar to the work to be welded. If non-consumable electrodes are used the additional weld metal may be supplied from a welding rod which is fused in the arc. The arc welding may be done using one or two consumable or non-consumable electrodes (Fig. 2.20). In the arc welding with one consumable electrode (Fig.

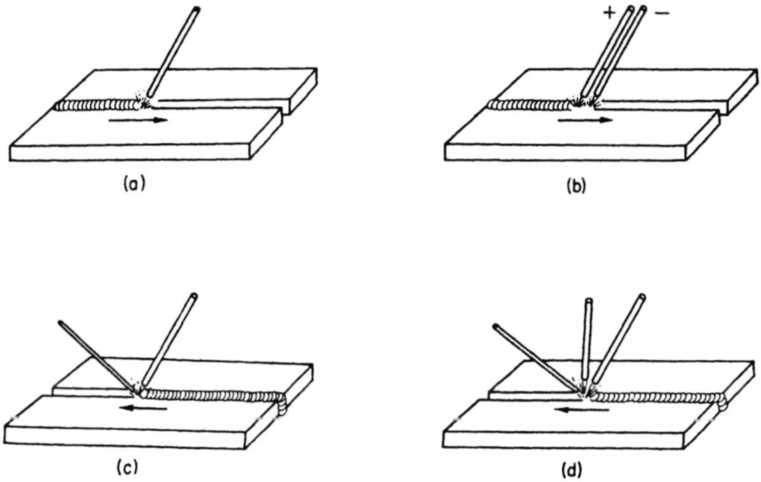

FIG. 2.20 Arc welding procedures: with (a) single consumable electrode; (b) a pair of consumable electrodes; (c) single non-consumable electrode; (d) a pair of non-consumable electrodes

2.20a) the droplets from the consumable electrode are added to the molten metal pool on the work. This procedure is used in the aircomatic welding technique. A rarely used procedure is that with two consumable electrodes (Fig. 2.20b). When a single non-consumable electrode is used (Fig. 2.20c) this electrode is made of a material with a high melting point, and high electron emissivity (carbon, tungsten). The procedure with a single non-consumable electrode is used in carbonarc and argonarc or heliarc welding. Finally the electric arc may be formed between two non-consumable electrodes (Fig. 2.20d), a procedure used in arcatom welding.

The power supply for the arc welding may be either alternating or direct current. In d.c. welding the electrode may be connected to the negative terminal of the power supply and the work to the positive, a technique known as *straight* polarity welding (Fig. 2.21a). When the electrode is connected to the positive terminal and the work to the negative the polarity is referred to as *reverse* polarity (Fig. 2.21b).

Since most of the heat is liberated at the positive terminal, the d.c. straight polarity welding is characterized by deep penetration and a narrow weld (Fig. 2.21a), and is suitable for heavy section welding. The reverse polarity welding results in a shallow penetration and a wide weld (Fig. 2.21b) and is suitable for thin materials.

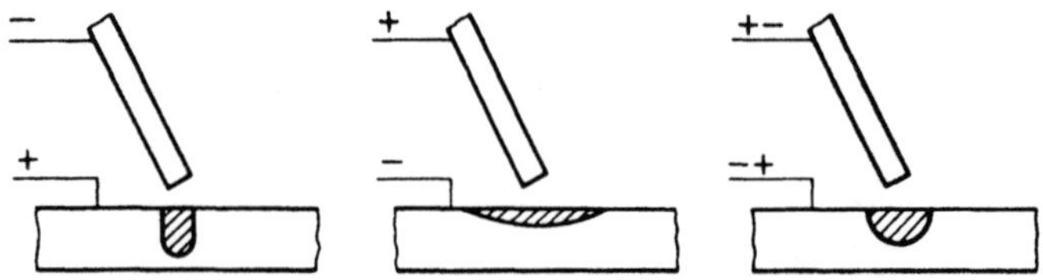

FIG. 2.21 Penetration of arc welds: (a) direct current, straight polarity (deep and narrow) weld; (b) direct current, reverse polarity (shallow and wide) weld; (c) alternating current (medium deep) weld

The d.c. welding is preferred for non-ferrous metals and alloys. The great disadvantage of this method is the *arc blow* i.e. the distortion of the arc from the intended path owing to magnetic forces of a non-uniform magnetic field. This effect appears especially when the welding is done in a corner or at the end of a groove.

With a.c. the resulting weld is of medium depth and narrow (Fig. 2.21c). The arc blow is greatly reduced and higher currents with larger electrodes may be used. The a.c. welding is adopted generally for iron and steel.

From the many commercial arc welding systems those used for vacuum sealing are: the atomic hydrogen (arcatom) process, the carbon arc, the air-comatic and the argon arc (heliarc) process.

For other information regarding arc welding refer, e.g. to Meller[843], Jefferson[629], Davies[261], Udin[1248], Young[1345].

The *atomic hydrogen arc welding process (arcatom)* is based on an alternating current arc between two non-consumable electrodes (Fig. 2.20d). Generally the current is 20–60 A, at 60–100 V with about 400 V by ignition (Dorrat[293], Cobine[219], Boyd[157]). The heat is obtained from the a.c. arc between two tungsten electrodes surrounded by hydrogen, and is transferred to the work by means of dissociation and recombination of the molecular hydrogen. The molecular hydrogen supplied through the electrode holder is dissociated in the arc to atomic hydrogen and recombines on contact with the cooler metal of the work. In this way heat is given up and temperatures up to 4000 °C can be reached (Langmuir[740], Weinmann[1291]).

The hydrogen also supplies a reducing atmosphere and shields the molten metal from oxygen and nitrogen. Welds are homogenous and smooth in appearance, because the hydrogen keeps the molten metal clean and the arc causes no turbulence in the molten pool (Dorrat[293]).

The arcatom process is used for the welding of difficult materials including aluminium and chromium (Table 2.16). Hydrogen, which is fairly suitable for the welding of iron and mild steels, is quite unsuitable for alloys containing nickel (as stainless steels) because the hydrogen is soluble in molten nickel and it escapes when the metal is setting, producing cracks and pores (Warmoltz[1287], Weinmann[1291]). The process is also not suitable for copper or copper alloys.

The *carbon arc welding process* uses a d.c. straight polarity arc (Fig. 2.21a) between the carbon electrode and the work, with 1–10 A and more than 100 V. An exception is the "twin" carbon arc process which uses the arc between two carbon electrodes positioned at an angle between each other.

The carbon arc can develop temperatures in the range of 3800–4800 °C on very concentrated surface areas.

Without using a shielding gas, this process is generally limited to the materials resistant to the contamination of oxygen or nitrogen. The process is used sometimes within a hydrogen atmosphere or liquid protection (Fansteel[366]). It may be used to weld iron, nickel, molybdenum, tungsten and also aluminium and copper (Table 2.16).

The *aircomatic welding* (Sohn[1163], Müller[903]) is a direct current straight polarity (Fig. 2.21a) process using a consumable electrode. The arc is shielded by a hydrogen stream and the heat developed by the burning and recombination of the hydrogen (see arcatom) is added to the heat of the arc.

The process is useful only for the welding of materials with not too high a melting point. It is used for aluminium and stainless steel (Benz[112]) but in these cases with an inert gas (argon, helium) protection instead of hydrogen (Table 2.16).

In *inert gas arc welding (argonarc, heliarc)* the heat is obtained from a direct or alternating current arc between the work and a tungsten (non-consumable) electrode (Gibson[449], Pumphrey[1019], Tuthill[1244]). The arc works in a shielded atmosphere of either argon or helium. Small d.c. plants operate at 45–75 V and 15–175 A, and large plants up to 300 A. In a.c. plants about 100 V and 250–300 A is used, but sometimes (as for aluminium welding) a high frequency, low voltage current is added to the welding current in order to permit the use of a longer arc.

The inert gas flows from an inverted cup around the electrode (Fig. 2.22) over the weld protecting the electrode as well as the molten metal from contamination by the oxygen and nitrogen of the atmosphere. The process is used generally with argon, but helium is regarded as superior for the d.c. welding of copper and stainless steel. The heliarc process must be restricted only to these cases since the helium consumption for the same shielding effect is about two and a half times (Schoonmaker[1105a]) more (due to the specific gravity of helium) and helium is much more expensive than argon. The argon or helium used must be pure and free of hydrogen (A more than 99.8 per cent, bal. N_2, He more than 99.5 per cent). Usually flow rates of 5–25 l./min are required.

The rare gases used are particularly suitable not only by virtue of their chemical inertness, but also because they are not soluble in the metals. This avoids formation of cracks or pores and allows clean leak-tight seals to be made.

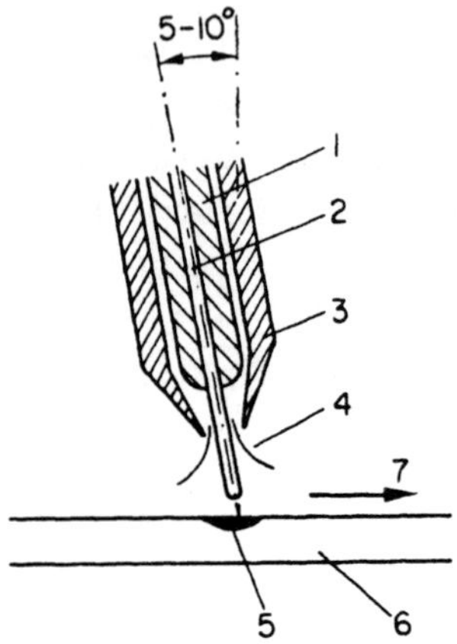

FIG. 2.22 Electrode for inert arc welding; (1) collet; (2) tungsten electrode; (3) nozzle; (4) inert gas; (5) molten pool; (6) work; (7) direction of travel of the electrode

The argonarc process is used with both a.c. and d.c., the choice of the current being determined by the metal to be welded. Aluminium, magnesium and their alloys are usually welded with a.c., while d.c. is used for steels, stainless steels (Thielsch[1219]), nickel, copper (Mortimer[895], Williams[1318a]), silver and titanium. As far as heat transfer is concerned, in d.c. welding the straight polarity (Fig. 2.21a) is by far the more efficient, but it cannot be used for welding aluminium and magnesium because of the absence of an oxide removing action (Huff[606]). In reversed polarity welding the electron emission of the metal in the area of the weld and/or the bombardment of the surface with positive ions removes the oxide film always existing on aluminium. Thus if aluminium is welded with d.c., only reversed polarity (Fig. 2.21b) should be used.

Inert gas arc welding is the most commonly used welding procedure for vacuum sealing for high and ultra-high vacuum (Kronberger[721], Hogg[577]). This is justified by the advantages of this procedure, which can be summarized (Chopinet[214]) as: (1) useful with most of the metals and alloys; (2) very concentrated source; (3) the electrode practically does not wear; (4) the joint is strong, smooth and tight.

Inert gas welding is not recommended for materials thicker than about 4 mm. When metals thinner than 0.5 mm are to be welded it is necessary to use copper coolers close to the edges to be welded. Nevertheless it is possible to weld thicknesses down to 0.15 mm without difficulty (Chopinet[214], Anderson[31], Mortimer[895]).

(c) *Electron beam welding*. Basically the electron beam welding process is carried out in a vacuum chamber within which a stream of electrons is accelerated through a high potential and focused on the work piece (Stohr[1195, 1196], Goebel[454a], Bas[93a], Schwarz[1112a], Wyman[1338]). The result is an extremely high heat input source. In the welding of very thin pieces it is possible to produce a point of impact of 0.1 mm^2, hence a very high welding

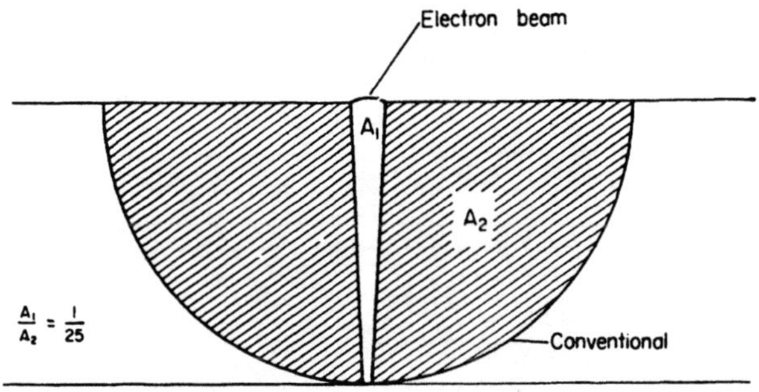

FIG. 2.23 Comparison of fusion zones; (A_1) produced by electron beam welding; (A_2) produced by conventional welding. Reproduced from Meier[856] *(Courtesy of Pergamon Press)*

accuracy. In welding thick pieces of metal, it is possible to produce a considerable energy dissipation on an impact area of 6–8 cm^2 e.g. 5×10^4 W (Stohr[1194]).

Since the welding is done under vacuum (about 5.10^{-5} torr) any oxidation is prevented. The electron beam as a tool has practically no mass and, therefore induces no impurities into the work being welded. In this respect the electron beam is even purer than any other type of electric arc process in a noble gas, since the purest gas still contains more impurities than would be found in a vacuum of 10^{-5} torr (impurity 1/100 million).

Because of the deep electron penetration, the shape of the fusion zone produced during electron beam welding is quite different from that produced by usual welding techniques. Figure 2.23 compares the typical fusion zone produced by conventional heating (A_2) with that produced during electron beam welding (A_1), and shows (Meier[856]) that to penetrate a given thickness of material only about 1/25 as much material must be melted during electron beam welding. Thus, only approximately 4 per cent energy is absorbed by the workpiece, which results in significantly less distortion and changes in the properties of the material.

The physics of electron beam penetration in the materials to be welded was described by Schwarz[1112a] and by Meier[856]. Some results of their measurements on the absolute and relative penetration depth are those presented in

Figs. 2.24 and 2.25. Figure 2.24 shows the effect of both voltage and current on penetration in 302 (18/8) austenitic stainless steel. It can be seen that penetration increases with both current and voltage (i.e. with power). Figure 2.25 presents the penetration depth for iron $X_{Fe}(M)$ as a function of the voltage. From the scale factor for other metals (listed on Fig. 2.25) it results that the penetration depth for aluminium is 2.8 times bigger, or in copper, only 0.88 of the penetration depth in iron.

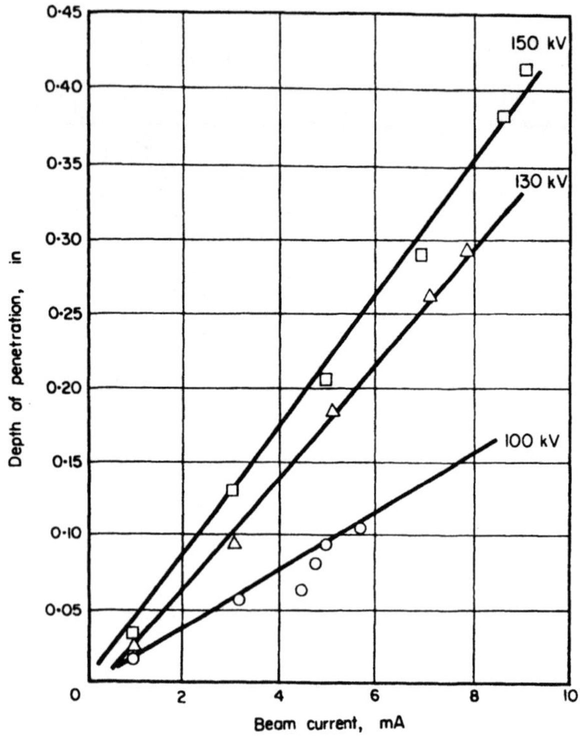

FIG. 2.24 Depth of penetration vs. beam current. Type 302 stainless steel; welding speed 27 in./min. Reproduced from Meier[856] *(Courtesy of Pergamon Press)*

Electron beam welding can be used for stainless steels, aluminium alloys, tungsten, molybdenum, titanium (Meier[842], Lander[732]), beryllium (Hess[550]) and tantallum (Ito[617]).

(d) *Resistance welding*. In resistance welding the joining of the pieces is produced by the heat obtained from the resistance set up in the metal parts by the passage of heavy amperage current, the parts being pressed one against the other.

Resistance welding is used as spot welding, seam welding and butt welding. The weldability of various metals is listed in Table 2.16.

In *spot welding*, overlapping metal parts are held between the electrodes which apply pressure while an electric current is passed through them. The spot weld is approximately the same size as the electrode tip (2–6 mm in diameter), thus this method is not used for vacuum-tight seals. *Seam welding* is generally performed on a resistance welding machine equipped with wheels instead of tips. These wheels are applied with pressure to the work pieces at

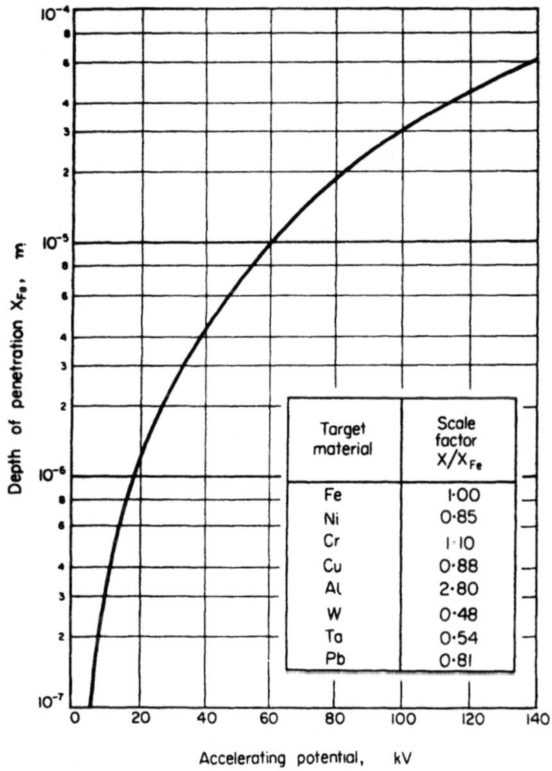

FIG. 2.25 Electron penetration in a constant density medium. Reproduced from Schwarz[1112a]
(Courtesy of Pergamon Press)

controlled speeds. While the application of pressure is steady and constant, the welding current is continuous or interrupted at rapid intervals to produce various types of seam welds (Fig. 2.26).

If the current flows continuously (Fig. 2.26d) or is interrupted at close intervals so that the fusion area caused by each pulse of current physically overlap (Fig. 2.26c) that of the previous one, the resulting weld can be vacuum-tight.

Butt welding is the joining of two parts placed with their ends abutting each other. The fusion of the ends is accomplished by *flash-butt* welding or by *upset-butt* welding. In flash welding, the metal pieces (tubes, rings) are placed in an electric circuit adjacent to each other, remaining slightly separated. An electric voltage is then applied which causes a current of sufficient magnitude to flow through the circuit and an arcing or flashing action between the adjacent pieces of metal. After the ends are heated by this flashing, pressure is applied. This results in expulsion of hot metal at the joint and by cooling the pieces are joined by the weld formed between them.

Upset welding is similar to flash welding except that the areas to be joined are pressed into contact and then heated by an electric current which is passed across the abutting surfaces.

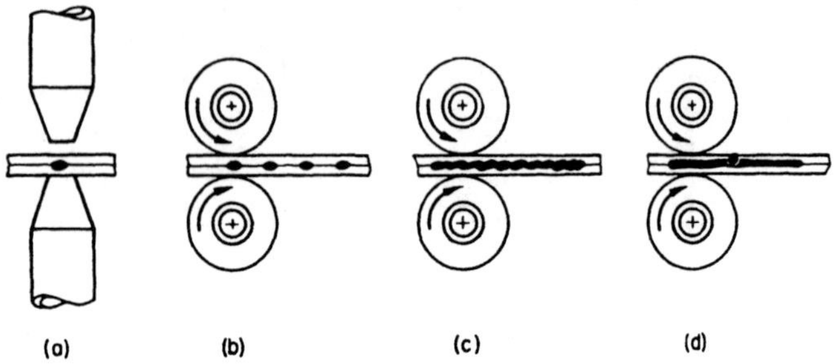

FIG. 2.26 Resistance welding: (a) spot weld; (b) roll spot weld; (c) overlapping seam weld; (d) continuous seam weld

The butt welding methods can be used to form vacuum-tight seals if the process is carefully controlled.

A butt welding method using the heat obtained from friction was described by Hollander[586]. The process seems to be limited to welding the ends of objects one of which must be rotated about an axis of symmetry. With this method nearly any metal can be welded to itself with good reliability, if the shape of the parts is that of tubes, studs, caps. For reliable welds the pressure sliding speed and time are critical factors. Table 2.15 lists some recommended values.

A method of ultrasonic welding of aluminium is described by Jones[644].

TABLE 2.15. RECOMMENDED VALUES FOR FRICTION BUTT WELDING

Material to be welded to itself	Diameter (mm)	Speed (rpm)	Pressure (kg/cm^2) during		Total time (sec)
			heating	joining	
Aluminium	18	3800	280	460	6
Carbon steel	12	3000	350	350	7
	24	1500	530	530	15
Copper	24	6000	350	700	18

(e) Cold welding. Certain metals such as copper, silver, aluminium, platinum, and some stainless steels can be welded together to give a vacuum-tight seal, by applying sufficient pressure (Loh[780], Sowter[1165], Hofmann[574a], Aitken[12], Steyskal[1189]).

After degreasing the surfaces to be joined, and removing the oxide layer (mechanically or chemically), the two parts are squeezed together with a suitable tool (Section 26.31 and Aitken[12], Neher[914]). The pressures required for cold welding are about 17–25 kg/mm^2 for aluminium, 50–75 kg/mm^2 for copper (Sowter[1165]) and about 200 kg/mm^2 for stainless steel (Hofmann[574a]).

The advantages of this process are the instantaneous sealing and the lack of heat, except that developed from the working of metal which is quickly conducted away (Loh[780]). The fact that the whole procedure is cold, and practically no gases are released during the sealing, makes this procedure adequate for the sealing-off of evacuated metal tubes (Section 26.31).

22.12 Weldability of metals and alloys. In common language, weldability means the possibility or impossibility to weld two metals or alloys to each other. In welding practice, *weldability* indicates the amount of precautions necessary for successful welding i.e. to make sound joints that are free of faults (such as porosity, non-metallic inclusions, cracks, hard zones) and with properties essentially the same as those of the materials being welded.

The main properties determining the weldability are: the metallurgic behaviour of the two metals or alloys, their thermal conductivity, thermal expansion and chemical behaviour at welding temperatures.

In the heat affected zone the metal must be heated and cooled through a range of temperatures large enough to cause changes in its structure with consequent changes in mechanical and other properties. When these changes in properties are intolerable, the original properties may be restored by heat treatment after welding (see Metals Handbook[848], Davies[261], Young[1345], Jefferson[629]).

It is difficult to weld metals and alloys of high thermal conductivity (as copper, aluminium, etc.) because the heat is conducted away from the weld so rapidly that it is not easy to raise the base metal to the fusion temperature. It is even more difficult to weld these metals to others, because of the unequal heating obtained in such cases (Table 2.16).

The thermal expansion of the metals to be welded determines the amount of distortion which will result. The base metal heated locally will expand and upset owing to the restraint imposed by the surrounding colder metal. This upset portion of the base metal will, upon cooling, contract beyound its original dimensions, and thereby set up internal stresses causing distortion. If the parts are not free to move cracks are produced. Thus the design of the parts must provide space for these contractions (Section 22.3).

The chemical behaviour of the metals heated to the welding temperatures,

TABLE 2.16. WELDABILITY

	Zr	W	Ti	Ta	Stainless steel	Pt	Ni	Mo	Kovar	FeNi
Ag	–	C	–	R_3	–	C R_3	C R_3	–	–	C R_3
Al	–	–	–	R_3	–	R_3	–	–	–	R_2
Au	–	–	–	–	–	–	–	–	–	–
Be	–	–	–	–	–	–	–	–	–	–
Brass	–	–	–	–	–	–	R_2	–	–	R_2
Co	–	–	–	–	–	–	–	–	–	–
Cr	–	–	–	–	–	–	–	–	–	–
CrFe	–	–	–	–	–	R_1	R_2	–	–	–
CrNi	–	C	–	C	R_2	C R_1	C R_2	R_2	–	C
Cu	R_3	C	R_3	–	(7)	C R_2	(5)	C (6)	(6)	C R_2
Fe	R_3	R_3	–	R_3	–	R_1	R_1	C R_3	R_3	R_3
FeNi	–	C	–	C	R_3	C	C	C	–	C, H P
Kovar	–	–	–	–	(7)	–	–	(6)	A,H(1) (4,5)	
Mo	R_2	C,A	–		R_2	C R_2	C R_2	A,C E(3)		
Ni	H R_2	C R_2	R_3	R_2	R_1	C R_1	C,H(5) M,R_1			
Pt	–	C R_2	–	R_2	R_1	C R_1				
Stainless steel	–	R_2	–	H R_2	H,M(5) E(3)					
Ta	–	–	–	–						
Ti	–	–	H,E(3) R_2							
W	R_2	C,A E								
Zr	H R_2									

OF METALS AND ALLOYS*

Fe	Cu	CrNi	CrFe	Cr	Co	Brass	Be	Au	Al	Ag
R_3	C	C R_3	–	–	–	–	–	–	–	C,H
R_2	–	–	–	–	–	–	–	–	A,C,H M, R_3	
–	–	–	–	–	–	–	–	T,P R_2		
–	–	–	–	–	–	–	E(3)			
R_2	R_2	R_2	–	–	–	C R_2				
–	–	–	–	–	A					
–	–	–	–	A, H (1)						
R_2	–	–	R_2							
R_2	C R_2	C R_3								
R_3	H(2),T P(5),R_3									
C,H T, R_1										

T torch weld. A arcatom. C carbonarc. H heli (argon) arc. M aircomatic. E electron beam. P—cold weld. R—resistance weld. 1 very easy. 2 good. 3 difficult.

* See Espe[352], Ardenne[47], Mönch[879], Roth[1083], Young[1345], Lander[732], Ito[617], Meier[856] Huber[605].

(1) After cleaning with phosphoric acid
(2) Only OFHC copper
(3) Limitation for large parts
(4) Use electrodes of CrNi-steel (18 Cr, 8 Ni)
(5) Vacuum-tight for many cycles up to 480 °C (Huber[605])
(6) Not recommended for vacuum sealing (Huber[605])
(7) Danger of leakage (cracks) (Huber[605])

determines the sort of gases which can be allowed on the weld during the process. Easily oxidizable metals cannot be welded in air, i.e. the weld must be made in a protective atmosphere (hydrogen, argon, helium) but metals sensitive to hydrogen (as copper) must be protected only by inert atmospheres.

A detailed discussion of the influences listed exceeds the scope of the present book. As a guide to the first choice of the welding method which can be used, Table 2.16 summarizes the recommended (or dangerous) solutions.

22.2 Brazing

The joining process of two metal parts with a third one having a lower melting point is known generally as *soldering*. When the solder has a melting point lower than 400 °C the process is known as *soft soldering;* if the solder melts above 500 °C the process is known as a *hard soldering*.

Brazing is defined as the metal joining process in which molten filler metal is drawn by *capillary attraction* into the space between the closely adjacent surfaces of the parts to be joined (British Standard 499/1952). The temperatures required for brazing are above 500 °C (American Welding Society) and they should be 50–200 °C lower than the melting point of the brazed parts (Espe[354]).

Generally soldered joints are considered to be permanent connexions. However, since soft solders melt at relatively low temperatures such joints are used also on regular access parts of some vacuum plants. For this reason, following the general classification of this book only brazing (hard soldering) is discussed here, the soft soldering techniques (semi-permanent seals) are described in Section 3.5.

For general or more detailed information on brazing techniques refer to e.g. Strong[1207], Angerer[34], Kohl[706], Knoll[697], Espe[350], Young[1345], Mönch[878, 879], Curtis[252], Garbe[424], Evans[363], Newitt[925].

The heat sources used to raise the temperature of the parts are the torch, the bath, the furnace, the arc, resistance or induction. The heat source to be used in a particular application is determined by the kind of material to be brazed, the brazing material and the shape of the joint. In any case the method should be such as to fulfil the following requirements:

(1) To heat both members to be joined to a temperature higher than the melting point of the brazing metal or alloy, heating the inner surfaces as quickly as possible without overheating the outer ones.

(2) To protect the surfaces from oxidation or other chemical effects that can occur at the brazing temperature.

(3) To hold the two parts being brazed in proper relation to each other, during both heating and cooling, ensuring that the two parts will not be moved until the braze is solid (otherwise the filler metal can crack).

22.21 Brazing methods. (a) Torch brazing uses single or multiple tip torches with various fuel as oxyacetylene, oxyhydrogen, oxygen–natural gas, etc. The flame should be directed to the parts away from the joint in such a way as to bring both members up to the brazing temperature simultaneously by indirect heating. A neutral or reducing flame (Section 23.23) can be generally used, except for brazing of copper where an oxidizing flame is used to avoid embrittlement (Barrett[88]).

Torch brazing needs fluxes which must be carefully cleaned from the joint after the completion of the brazing. The flux remaining on the vacuum side of the seal has a high outgassing rate (virtual leaks). Torch brazing can be used to apply local heat but the heat control usually requires skill and overheating is not uncommon.

(b) Dip or bath brazing is a method of joining parts by immersion in a bath of molten filler metal (metal bath brazing) or immersion of the assembled parts with the brazing alloy in position, in a molten flux bath (chemical bath brazing). This method has the advantage of quick heating of the parts, easy temperature control, and good protection from oxidation. It can be generally used only for small parts.

(c) Furnace brazing consists in heating an assembly of metal parts to be brazed in a furnace with a protective atmosphere. Vacuum or neutral gas free of oxygen and water vapours may be used for brazing copper without oxidizing or embrittling it. For other metals reducing atmospheres are preferred.

In the heating operation the brazing material melts wetting the clean surface (for cleaning see Section 22.31) and creeping on the surface until drawn into and through each joint by capillary attraction (Fig. 2.37). Any excess of brazing material usually forms a neat fillet at the outer extremity of each joint, which adds to the strength of the joint.

The advantages of furnace brazing are: uniform distribution of the brazing alloy, little or no distortion, clean appearence, flexibility to braze light parts to heavy ones and to join different metals. A disadvantage of this method consists in the fact that the entire assembly must be heated even though the surfaces being joined might represent a small fraction of the entire part (see also Bernhard[115]). The maximum temperature can be controlled using a fuse element on the electric circuit, made of the brazing alloy. When the furnace reaches the melting point of this fuse, it melts and breaks the heating current.

(d) Arc brazing is rarely used; the heat is obtained from an electric arc formed between the base metal and an electrode or between two electrodes (Fig. 2.20).

(e) Resistance brazing consists in holding the areas to be brazed between two electrodes. A low voltage and high current is passed through the circuit, heating the surfaces to be joined to the temperature at which the preplaced

brazing material (and flux) begin to melt. Pressure is maintained on the joint until the alloy is solid.

(f) Induction brazing utilizes a high frequency current (400–2000 kc) to heat the parts being brazed. The parts are placed in a specially fitted coil. For best results the coil shape and size should match the assembly to be heated (Section 22.33 and Babat[61], Reinartz[1047], Curtis[252], Kohl[706], Lupinek[789]).

Very short heating times can be achieved by a pulse technique, where the power is concentrated in the area of the joint for fractions of a second, so that critical components in close proximity of the joint, such as glass–metal seals, will not have time to heat up to a level to be damaged (Reinartz[1047]).

A rough calculation of the power P(watt) required to complete a given brazing process is (Reinartz[1047]):

$$P = 23.5 \frac{M \cdot c \cdot \Delta T}{t},$$

where M is the mass of the parts to be heated (in grammes), c is the specific heat of the material (cal/g), ΔT is the temperature raise (°C) required in the time t (sec). The factor 23.5 includes the usual losses.

The induction brazing can be used with a protective atmosphere or vacuum, since the coil can be outside the vessel containing the parts to be brazed.

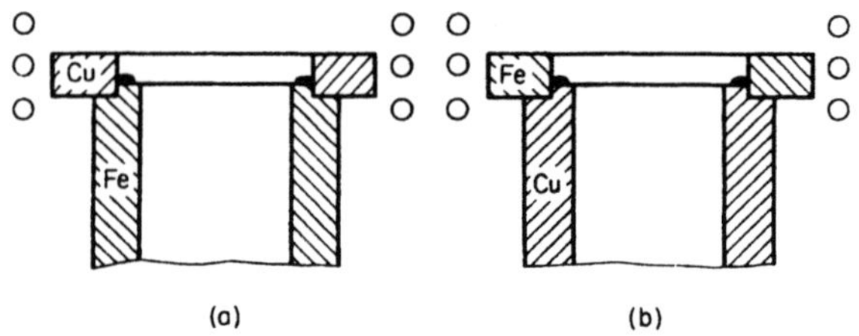

FIG. 2.27 Induction brazing of magnetic parts to non-magnetic parts. Arrangements for which (a) the induction brazing is possible; (b) the induction brazing is not possible

Induction heating usually offers good heat control and a high degree of accuracy applying the heat only to the surface which requires the brazing.

This brazing method has also some limitations, when parts of magnetic (iron) and non-magnetic metals (e.g. copper) are to be brazed to each other. The magnetic metal will be heated quicker than the non-magnetic one. In practice, joints as those represented in Fig. 2 27a can be brazed using induction heating, while arrangements like those in Fig. 2.27b (with the copper inside the iron) will not result in good brazed joints. Since the iron part will have a greater thermal expansion the clearance between the copper pipe and iron flange will increase and the brazing alloy will creep on the surface instead of being sucked between the parts by capillary action (see Fig. 2.37).

(g) Diffusion brazing. This method uses the ability of a thin layer of a suitable metal (gold, silver etc.) placed between the parts, to cause a union by diffusion in solid state (below the melting point, Law[747]). Such seals are vacuum-tight.

The metal used for the intermediate layer depends on the metals to be joined, but gold, silver and copper are suitable. The layer may be formed by electro-deposition on one or both of the members, or may be interposed as a foil of about 0.01 mm thickness (Metro–Vickers[851]). The surfaces to be joined should have two or preferably more circular ridges, in order to allow excessive localized pressure to be exerted on the surface. Wright[1331] describes this method used to join two copper flanges clamped together with a washer of gold wire (0.3–0.4 mm diameter) inserted between them. During the heating to 400–500 °C (about 15 min) the gold diffuses into the copper and constitutes a reliable vacuum seal.

The method can be applied for the sealing of flanges of other metals if the sealing surface is first electroplated with a copper layer (about 30 μ thick). It is recommended that such an electro-deposited copper layer be heated in hydrogen at 1030 °C for 20 min to ensure the adherence of the layer (Espe[354]). The method can also be used with a gold–copper (75/25) alloy, or by first electroplating one member of the seal with copper and the other with gold. It should be mentioned that the latter method requires higher sealing pressures. The results are bad if both members are gold-plated, since gold does not diffuse in gold at temperatures of a few hundred degrees C, used for this kind of seal.

To seal two copper parts by diffusion it may be enough to plate them with a layer of silver (some microns thick). The parts pressed together and heated to 780 °C are joined by the copper–silver eutectic formed at 778 °C (Fig. 2.29).

22.22 Brazing materials. The necessity to produce vacuum-tight seals (Section 14.1) by brazing involves the consideration of factors of either less or no importance to joints required only for mechanical strength. Thus for vacuum-tight brazed seals the factors determining the selection of brazing alloys will be: the vapour pressure, the purity, the ability to wet and flow at the brazing temperature, the alloying possibilities with the joined metals, chemical resistance and obviously mechanical strength (Goodman[459], Smith[1151], Hack[498]).

From the elements having *low vapour pressure* some also have high melting points. These elements and their alloys can be used as brazing filler material, but their use is obviously restricted to the brazed joint of high melting point metals (Table 2.18, Pos. 1–9) not often used in vacuum seals. Elements having vapour pressures higher than 10^{-3} torr at the outbaking temperatures (about 400 °C) as zinc, lead, cadmium or bismuth, cannot be used as components of brazing alloys for high vacuum seals. Thus the list of suitable elements to be used in brazing alloys for high vacuum seals is reduced in fact to copper, silver, gold and nickel. Indium and tin have low enough vapour pressures, but their melting point is too low to be used in bakeable systems.

The brazing alloy must be free of impurities. This means not only elements of high vapour pressure, but also such common impurities as dust, carbon, etc., which are often picked up during the casting of the original ingot and subsequent wire drawing or sheet rolling of the brazing alloy. These impurities are always the cause of voids built up in the seal leading to real or virtual leaks (Section 1.3). Purity is also important from the standpoint of the oxides, which may be present in the original alloy or may be built up during the brazing process. The complete removal of fluxes after the completion of the brazing is questionable (Hack[498]), and flux trapped in the final joint may form voids, detrimental from both strength and tightness standpoints. Thus the use of fluxes is not recommended in brazing vacuum seals; a protective atmosphere (hydrogen, argon, vacuum, etc.) is generally to be preferred to prevent oxidation or even reduce (hydrogen) minor amounts of oxides present.

Due to the fact that fluxes are not recommended in vacuum seals, the wetting and flowing qualities of the brazing alloys are of greater importance in these applications.

Wetting is the ability of the molten brazing alloy to adhere to the surface of the part being brazed, and to build up a strong bond to this part when

TABLE 2.17. WETTING ABILITY OF BRAZING COMPONENTS*

Brazing component	Brazed metal						
	Copper	Nickel	Iron	Molybdenum	Tungsten	Platinum	Tantalum
Copper	—	G	P	P	B	G	B
Nickel	G	—	G	G	G	G	G
Silver	G	B	B	B	B	—	B
Gold	G	G	G	B	B	G	P
Platinum	—	—	—	G	G	—	G
Tin	E	G	G	—	—	G	B
Zinc	—	G	E	G	G	—	G
Lead	—	G	G	B	B	—	B

* E = excellent, G = good, P = poor, B = bad.

cooled below its solidus temperature (Fig. 2.28). There is considerable evidence that in order to wet well, a molten metal must be able to dissolve, alloy with or diffuse into the metal on which it flows. The definition of wetting ability through the contact angle is given in connection with Fig. 2.103 and Table 2.44.

The main components for brazing alloys with their wetting ability towards some metals are listed in Table 2.17 (Steyskal[1189], Caldwell[190a], Countryman[240]). It can be seen that e.g. silver does not wet iron or nickel; copper wets nickel and gold wets both nickel and iron. Thus for brazing nickel to iron it is recommended to use copper–silver, copper–gold or gold–silver alloys instead of pure silver (Table 2.19).

Flow is the property of brazing alloys to travel away from their original position due to the action of capillary forces. Well flowing alloys may not have an appreciable increase in their liquidus temperature even if their composition is altered by the addition of the metal they dissolved. This is important because the brazing operation is carried out at temperatures just above the

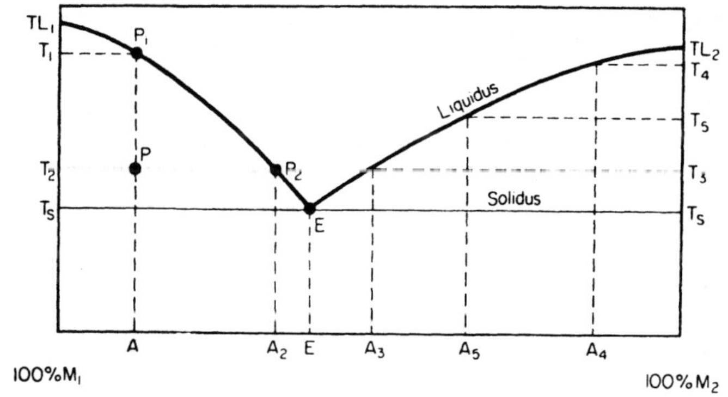

FIG. 2.28 Constitution diagram of a binary alloy system with eutectic

liquidus.* Figure 2.28 represents the constitutional diagram (principle) of a binary alloy system. The diagram is a plot of the liquid and solid states of all possible compositions of metals M_1 and M_2. The two curves TL_1-E and E-TL_2 represent the *liquidus*. Above this curves any composition is liquid. Point P_1 represents the liquidus temperature T_1 of an alloy with composition A_1. The horizontal line T_s–E–T_s is the *solidus*. Below this line the compositions are entirely solid. T_s is the solidus temperature of all the compositions. Composition E is the *eutectic*. When an alloy of this composition is heated it remains entirely in the solid state until T_s is reached. At T_s it becomes entirely

* The liquidus temperature is the temperature at which an alloy finishes melting during heating or starts freezing during cooling. Effective liquidus temperature is raised by fast heating and reduced by fast cooling (Campbell[193]).

liquid. T_8 is both the solidus and liquidus temperature of composition E. The eutectic is entirely liquid at a lower temperature than any other composition in the diagram. Unlike composition E, any other composition will not convert directly from solid to liquid at one temperature. Composition A_1 for example will be partly solid and partly liquid at any temperature between T_8 and T_1. At temperature T_2 (point P) composition A_1 is separated into a solid of composition M_1 (the pure metal) and a liquid of composition A_2 (point P_2). The liquid must be of composition A_2 because this is the only composition that can be liquid at temperature T_2. From the constitutional diagram of the alloy the following effects on the brazing process should be pointed out:

(1) Brazing alloys that are not of a eutectic composition present a danger if the parts being brazed are dislocated during the time the alloy passes through the part liquid–part solid range. Any motion during this period may produce cracks in the brazing alloy.

(2) The eutectic alloy flows quickly when it reaches the T_8 temperature (Fig. 2.28). Thus, for brazed joints with bigger clearances, where such a sudden flow is not desired, it is preferable to use a non-eutectic composition.

(3) Using a eutectic brazing alloy with a composition E at T_8 (Fig. 2.28) to braze parts of metal M_2 at temperature T_3, the alloy will have a composition A_3. If the rise in temperature is not rapid enough between T_8 and T_3 the alloy may dissolve enough M_2 to change its composition to A_4 with a resulting higher liquidus T_4. If the temperature has been set for T_5 the brazing metal will never become completely liquid and will not flow adequately.

(4) If the brazing alloy is allowed to remain at a temperature above the liquidus for a longer time in presence of metal M_2, then enough M_2 is dissolved to alter the composition. In this case a solid phase will slowly crystallize and weaken the brazed joint.

(5) If a non-eutectic alloy with composition A_1 (Fig. 2.28) is used to braze metal parts M_2, it will dissolve some of the metal and an alloy with lower melting point (e.g. A_2) will be formed. At subsequent heating (as for alloy A_1) the new alloy (A_2) will flow from the seal leaving voids. If copper is brazed with silver (at 980 °C) a copper–silver alloy is formed, with a lower melting point (Fig. 2.29). Consequently it can be recommended to use in this case the eutectic copper–silver alloy (Fig. 2.29), where by increasing or reducing the copper content the melting point is always increased. With alloys having no eutectic (as copper–gold, or gold–nickel, see Figs. 2.29, 2.30) the alloy with the lowest melting point can be used in the same way.

(6) The brazing alloy used should be selected to fit the maximum operation temperatures of the parts during the use of the system. The solidus temperature of the brazing alloy (Table 2.18) must be reasonably higher than the maximum operation temperature.

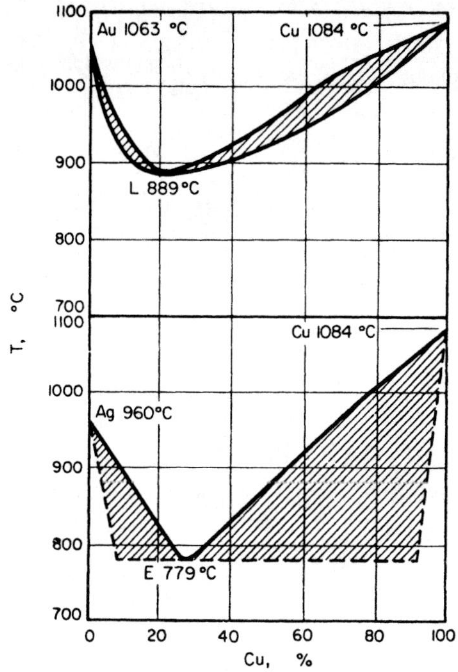

Fig. 2.29 Constitution diagrams of copper–silver and copper–gold alloys

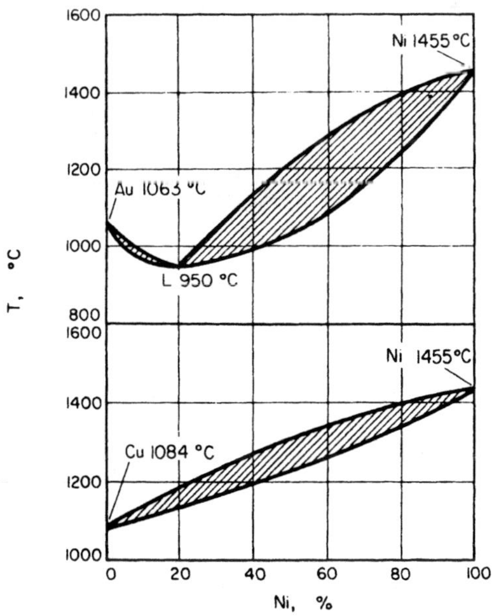

Fig. 2.30 Constitution diagrams of nickel–copper and nickel–gold alloys

TABLE 2.18. BRAZING METALS AND ALLOYS*

No.	Liquidus (°C)	Solidus	Type of alloy**	Composition (parts by weight)
1	3180	3180	P	Rhenium
2	2996	2996	P	Tantalum
3	2497	2497	P	Niobium
4	2427	2427	P	Ruthenium
5	2444	2444	P	Iridium
6	1966	1966	P	Rhodium
7	1950	1935	–	Rhodium (40), Platinum (60)
8	1852	1852	P	Zirconium
9	1770	1770	P	Platinum
10	1695	1645	–	Au (5)–Pd(20)–Pt (75)
11	1550	1550	P	Palladium
12	1452	1452	P	Nickel
13	1423	1423	L	Ni (36)–Fe (64)
14	1320	1320	E	Mo (46.5)–Ni (53.5)
15	1320	1290	–	Pd (30)–Ni (70)
16	1305	1260	–	Pd (13)–Au (87)
17	1300	–	–	Ni (51)–Mo (49)
18	1300	1230	H	Ni (45)–Cu (55)
19	1240	1190	–	Pd (8)–Au (92)
20	1238	1238	E	Ni (40)–Pd (60)
21	1232	1149	–	Mn (3)–Pd (33)–Ag (64)
22	1205	1150	H	Ni (25)–Cu (75)
23	1160	995	–	Pt (27)–Ag (73)
24	1135	1080	–	Fe (3)–Si (10)–Cr (19)–Ni (68)
25	1084	1084	P	Copper (OFHC)
26	1083	1083	–	Ni (3)–Cu(35)–W (62)
27	1065	1000	–	Pd (10)–Ag (90)
28	1063	1063	P	Gold
29	1060	1000	–	Ag (5)–Cu (95)
30	1050	1030	L	Ni(30)–Mn (70)
31	1035	1015	H	Au (30)–Cu (70)
32	1030	975	–	Cu (62)–Au (35)–Ni (3)
33	1025	970	–	Cu (97)–Si (3)
34	1025	960	H	Cu (95)–Ag (5)
35	1020	–	–	Fe (33)–Ni (56)–P (11)
36	1018	1018	E	Ni (40)–Mn (60)
37	1015	990	H	Cu (63)–Au (37)

(Table 2.18 Continued)

No.	Liquidus (°C)	Solidus	Type of alloy**	Composition (parts by weight)
38	1015	970	–	Cu (77)–Au (20)–In (3)
39	1010	985	H	Cu (60)–Au (40)
40	1005 993	996 976	– –	B (3.5)–Si (5)–Cr (16)–Ni (72.5)–Fe (3) B (2.9)–Si (4.5)–Ni (91)–Fe (1.6)
41	975	950	–	Cu (50)–Au (50)
42	971	960	–	Ag (85)–Mn (15)
43	962	–	–	Ni (3)–Ag (35)–W (62)
44	960	960	P	Silver
45	950	950	L	Au (82)–Ni (18)
46	–	946	–	Cu (85)–Sn (8)–Ag (7)
47	920	904	–	Au (58)–Cu (40)–Ag (2)
48	910	779	–	Cu (60)–Ag (40)
49	900	860	–	Au (60)–Cu (37)–In (3)
50	900	900	E	Cu (76)–Ti (24)
51	900	–	–	Cu (50)–Ni (10)–Mn (40)
52	900	714	–	Cu (95)–P (5)
53	896	885	–	Au (75)–Cu (20)–Ag (5)
54	889	889	L	Au (80)–Cu (20)
55	885	779	–	Ag (62)–Cu(32)–Ni (6)
56	–	800	–	Cu (50)–Ag (40)–Mn (10)
57	880	–	–	Ni (89)–P (11)
58	870	779	H	Ag (90)–Cu (10)
59	845	835	–	Au (60)–Cu (20)–Ag (20)
60	830	779	–	Ag (77)–Cu (21)–Ni (2)
61	821	794	–	Au (50)–Ag(30)–Cu (20)
62	779	779	E	Ag (72)–Cu (28)
63	770	714	–	Cu (93)–P(7)
64	721	640	–	Cu (87)–P (7.5)–Ag (5.5)
65	630	–	–	Au (80)–In (20)
66	636	–	–	Ag (50)–Cu (15.5)–Cd (18)–Zn (16.5)
67	625	(565)	–	Al (95)–Si (5)
68	570	(550)	–	Al (85)–Cu (3)–Si (12)
69	550	–	–	Ag (60)–Cu (23)–Sn (17)

* The alloys for soldering are listed in Table 3.4.

** P = pure metal; E = eutectic; L = lowest melting point alloy; H = alloy having higher liquidus temperatures if more of the main metal component is added.

TABLE 2.19. BRAZING

	W	Ti	Ta	Stainless steel	Ni	Monel	Mo	Kovar
Ag	–	–	–	–	62(43)	–	–	–
Au	–	–	–	–	–	–	–	–
Be	–	–	–	–	62	–	–	–
CrFe	–	–	–	–	–	–	–	–
CrNi	–	–	–	–	–	–	–	44,28
Cu	54,45 44[*8], 62[*3]	50	–	62,40	62,45 32,39 49	32	45,28 64,32	54,31,32 39,62[*4]
Fe	22	–	–	–	37,39	26	22	28,31,34 37,39
FeNi	–	–	–	–	–	–	–	–
Inconel	45	–	–	20,21 24,42	45	–	45	45
Kovar	45,32 54[*10]	–	–	28	45,28, 38,32,39	26,28,32 62[*6]	45(44)	28,39,54 62[*5]
Mo	22,23 32,45	–	–	19,27	27,32 45	32	3–14,23 32,45	
Monel	28,32	–	–	–	32	32		
Ni	45,27 32	50	–	27	17,31 32 28,40,41 62			
Stainless steel	19,20 21,24	–	–	15,19 24,30,36 40,45				
Ta	–	–	28[*6]					
Ti	–	44[*7]						
W	1–16 23,45							

POSSIBILITIES OF METALS*

Inconel	FeNi	Fe	Cu	CrNi	CrFe	Be	Au	Ag
—	62	—	62,64	62	—	—	—	54,62 64
—	—	—	—	—	—	—	62	
—	—	25*5	62*1	—	—	—		
—	—	25*6	—	—	25			
—	—	—	66*9	66*9				
62,45	62,54	31,39,48 54,62*2	29,33,39 34,46,47 54,62					
—	—	39,28,47 44,54						
—	28 62*4							
21,24 20,45								

* Numbers refer to alloy number from Table 2.18
*1 Short h.f. heating; alternative technique: braze Be to Fe using Cu, and Fe to Cu using Ag alloy (830 °C)
*2 Cu plated Fe, sintered in H_2
*3 W first etched (HNO_3/HF 1:1)
*4 Kovar first Cu plated
*5 H.f. heating in H_2 (5 min)
*6 H.f. heating in vacuum
*7 In vacuum, or with acetylene flame and AgCe flux (De Cecco[269])
*8 W first Ni plated
*9 Cd and Zn evaporates on heating in vacuo
*10 Heating in H_2, for 5 min at 920 °C

(7) On parts where successive brazings are required to complete the assembly, it is recommended that a eutectic or minimum melting point alloy be used (Figs. 2.29, 2.30). After the brazing process, the alloy has always a higher melting point than that of the original (eutectic) alloy. Consequently the next joint can be brazed with the same eutectic alloy, and this temperature will not destroy the previous brazing.

(8) If alloys with continuously increasing liquidus and solidus (e.g. nickel–copper, see Fig. 2.30) are used for successive brazing, the brazing may be done with compositions corresponding to lower and lower melting points.

In some cases of *physical incompatibility*, some brazing alloys are excluded from use with particular metals. So, Kovar (FeNiCo) cannot be brazed with silver, because the silver penetrates in the Kovar producing the splintering of the Kovar. Silver and gold brazing materials cannot be used in plants where mercury vapour will be present. Copper brazed joints become fragile in contact with mercury. The brazed joints with copper, silver or gold can be protected against mercury by nickel electroplating.

The brazing alloys are listed in Table 2.18 and the possibilities for using them for the sealing of various metals are summarized in Table 2.19.

Steel-to-steel brazing. Copper brazing alloys are highly satisfactory from the standpoint of both wetting and flowing on steel. The excellent flow characteristics are due to the fact that as the copper melts it takes into solution a small amount (2.8 per cent) of iron, thus raising the liquidus temperature by only 11 °C. Stainless steel joints can be brazed successfully using gold–nickel (82/18) or NiCrB respectively NiBSi alloys (Tables 2.19 and 2.18). See also Kohl[706] and Sistare[1141].

Nickel-to-nickel joints can be brazed using silver–copper eutectic alloy (Fig. 2.29) or gold–copper and gold–copper–nickel alloys (Tables 2.19, 2.18). A design of nickel joints brazed with silver alloys should avoid any stress in the joint, since nickel and nickel alloys show (like Kovar) brittleness after being in contact with molten silver. Silver-brazed nickel joints can be considered to be reliable only up to 250 °C. Nickel joints brazed with copper alloys can be used in air up to about 500 °C. Above this temperature they lose their strength due to oxidation. If oxidation resistance is required up to about 1100 °C some NiCrBSi alloys (see Nos. 24, 40 in Table 2.18) can be used.

Copper-to-copper brazed joints can be achieved using silver–copper eutectic (Fig. 2.29), gold–copper alloys or gold–copper–nickel alloys (Table 2.18). Pure silver is unsatisfactory, since the copper is highly soluble in silver and the melting point of the alloy is lower than that of the silver (see comments to Fig. 2.28). The same effect would occur with pure gold.

Copper can be joined to copper using gold diffusion seals (Section 22.21g).

Copper-to-nickel joints may be brazed with silver–copper eutectic (Fig.

2.29) or gold–copper alloys. For deep joints, the gold–copper–nickel alloys are recommended because of their better flow characteristic.

Copper–to–Kovar (FeNiCo) joints are to be brazed using gold–copper, gold–nickel and gold–copper–nickel alloys (Table 2.19). Gold–copper–nickel, and gold–nickel alloys flow better on Kovar than pure copper or gold–copper alloys due to the solubility of oxygen and nickel oxide in the molten nickel (Hack[498]). Silver alloys cause embrittlement of Kovar, or other alloys containing nickel.

For details on brazing of refractory metals (tungsten, molybdenum, etc.) refer to Kohl[706], Chatfield[208]. For the brazing of titanium and its alloys see e.g. De Cecco[269], Long[781], Clark[216].

Brazing fluxes are available generally together with the brazing alloys. Some flux compositions are given by Moss[897].

22.3 *Vacuum sealing by welding and brazing*

22.31 Cleaning of the metal surfaces. A vacuum seal cannot be successful if the surfaces to be joined are not clean. As cleaning is required for all surfaces exposed to vacuum (Dreyer[304], Hoch[568]) it is intended to describe in this section the metal cleaning methods used for surfaces to be welded or brazed as well as for other kind of seals (see Chapter 3).

Cleaning generally means the removal of undesirable materials laying on the surface of a solid. In vacuum technique, the cleaning must be regarded not only as the removal of the visible dirt from the surfaces, but including the subsequent removal of all the contaminants physically stuck on the surface (oil, grease, dust) or resulting from a chemical reaction (oxides, sulphides). The degree of cleanliness must be higher for higher vacuum.

The *oxides* and other similar surface layers can be removed by mechanical and/or chemical methods, as abrasive blasting, wire brushing or pickling and etching.

The cleaning of *oils and greases* depends on their nature, i.e. if they are soap-forming or not. The soap-forming oils and greases are those of animal or vegetable origin; the mineral oils do not form soaps. The soap-forming oils and greases can be removed by transforming them by hydrolysis in fatty acids and by reacting these acids with alkaline solutions to obtain water soluble soaps. The mineral oils can be removed by dissolving them in organic solvents (Scheflan[1103]), and in particular cases only they can be washed with alkaline solutions containing detergents. Since the nature of the contaminants is usually unknown a reliable cleaning must consist of two succesive steps (Doré[288]): (1) the degreasing with organic solvents, followed by (2) an alkaline degreasing.

The sequence of the cleaning operation begins generally with mechanical cleaning, followed by pickling, detergent cleaning and degreasing (Wallace[1284], Janecke[627a], Espe[354a]).

The *mechanical cleaning* methods are not specific for vacuum sealing, but are used previously to clean forgings, or stampings from scale, rust, etc. For this purpose abrasive blasting or wire brushing is usually utilized.

Pickling is the chemical removal of oxides and other surface layers, leaving the cleaned part with a metallic appearance with a smooth or rough finish, depending on the concentration of the solution and the pickling time. To obtain a good, uniform pickling the metal should be free of other impurities or protective layers (plastics, paints etc.). It is recommended to remove such layers prior to pickling. After pickling the part should be always thoroughly rinsed and subsequently neutralized by immersion in an alkaline bath, and drying with hot air (80 – 90 °C). The air drying is sometimes done after an alcohol rinsing. The acid pickling has a great influence on the gas evolution from the metal surfaces, reducing considerably the subsequent outgassing rates (Table 2.5). The pickling solutions to be used for various metals are listed in Table 2.20.

Electrolytic etching and polishing is the anodic (or cathodic) treatment of metal surfaces in various baths in order to produce clean (etched) or smooth (polished) areas. Electrolytic etching solutions are listed in Table 2.20. For electrolytic polishing not only the bath composition but also the electrical conditions are critical. Basically below a certain voltage the bath acts more like a pickling bath. A smooth electrolytic polishing is obtained in the optimum voltage range. At higher voltages the strong gas evolution leads to a non-uniform corrosion of the anodic surfaces. For details on electrolytic polishing see Jacquet[622], Bush[187], Dettner[276], Heyes[552], Kutzelnigg[724], Hancher[508].

Some recommended electrolytic polishing baths with the critical electric, temperature and time data are listed in Table 2.21 (Angerer[24], Espe[354], Steyskal[1189], Burkardt[182b], Bush[187]).

Alkaline detergent cleaning is performed either by immersion or as an electrocleaning process. The *immersion* cleaning is used usually with hot (60–85 °C) solutions. This cleaning method can be used to remove oil, grease, dirt or paint. The kind of the solution to be used is determined by the metal to be cleaned and/or the impurity to be removed. For ferrous metals and difficult cleaning operations stronger cleaners, containing sodium hydroxide (silicates, or phosphates), soaps and wetting agents are used, in concentrations of 10–40 g of each component per litre of solution. Non-ferrous metals and especially aluminium, are cleaned with inhibited alkaline cleaning solutions, i.e. sodium silicates, phosphates, carbonates with soap or synthetic organic detergents, at a concentration of 15–45 g/l. at 60–85 °C. *Electrocleaning* in alkaline solutions, can be used with the metal (to be cleaned) serving as the

TABLE 2.20. PICKLING SOLUTIONS

Metal (Alloy)	Pickling solution	Remarks
Aluminium	$NaOH$ (10 per cent sol.) saturated with $NaCl$ (If blackening appears, Al has Cu; in this case subsequent pickling in HNO_3 (20–30%) required; good washing)	At 80 °C, 15–50 sec. Subsequent immersion in HCl (10%) for shining surface.
	$NiCl_2$ (25% sol.) diluted 5:1 with HCl (1.16)	Burkard[182b]
	Electrolytic etching in sol. of 100 g H_3BO_3 in 1000 ml dist. water plus 0.5 g $Na_2B_4O_7$	50–100 V, up to 600 V Espe[354]
	Electrolytic etching in 5–10% sol. of chromic acid	3–15 mA/cm², 20–40 V about 30 min.
Beryllium	NaOH (or KOH) 50–100 g/1000 ml. dist. water at 20 °C, electrolytic etching	2.5–7 A/dm² Beach[101]
Constantan (CuNi55/45)	H_2SO_4 (10% sol) 50–60 °C	
Copper	250 ml. HNO_3 (1.40) with 600 ml. H_2SO_4 (1.83) and 20 ml. HCl (1.16) in 130 ml. dist. water	Bright dip
	500 ml. HNO_3 (65%) with 500 ml. H_2SO_4 (conc.) and 10 ml. HCl (37%) and 5 g carbon black	Immersion (2–3 sec) immediate rinsing
	1000 ml. HNO_3 with 1000 ml. H_2SO_4 and 15 g NaCl and 20 g carbon black. To be diluted 1:1 with dist. water 24 hr before use	Immersion (1–5 sec)
	10% $Fe_2(SO_4)_3$ in citric acid (0.1–1.0%) or in acetic acid (0.3–0.5%)	Bright dip Burkard[182b]
	HCl dil. immersion for 5 min followed by immersion in sol. of 100 g Cr_2O_3, 7 ml. H_2SO_4 (conc.) in 1000 ml dist. water	Espe[354]
	Sol. (1): 40 ml. H_3PO_4, 15 ml. HNO_3, 1.5 ml. HCl, 20 g NH_4NO_3, 45 ml. dist. water. Immersion for 3–4 min. After rinsing immersion in Sol (2): 65 ml. glacial acetic acid, 30 ml. H_3PO_4 (1.75), 5 ml. HNO_3 (1.42). Immersion about 1 min until gas evolves evenly all over	Sol. 1 at 35 °C Sol. 2 at room temp. Bright dip Turnbull[1240a]
Invar (FeNi 64/36	Cathodic etching in 1 vol. HCl (37%) with 1 vol. H_2SO_4 (96%) and 1 vol. HNO_3 (70%) in 1 vol. dist. water. 26 mA/cm². Carbon anode	Foley[388]

(Table 2.20 Continued)

Metal (Alloy)	Pickling solution	Remarks
Iron	50% sol. of HCl or 5–15% sol. of H_2SO_4. Recommended to add hydrogen evolution inhibitor (e.g. Ferrocleanol, see Espe[354])	
Iron-Chromium	500 g Cr_2O_3 with 5 ml. H_2SO_4 filled up with dist. water to 1000 ml.	Section 24.21
	Anodic etching in 335 ml. acetic acid with 240 ml. perchloric acid ($HClO_4$) in 100 ml. dist. water. 6 V, Cathode of graphite	8–10 min
Stainless steel (FeNiCr)	1–3% HNO_3 (conc) with 25% HCl (1.16) in dist. water. Temp. 65 °C	Bright dip Burkard[182b]
	7% H_2SO_2 (1.83) with 3% HCl (1.16) in dist. water. Temp. 65 °C	
	30 pbw Fe_2 $(SO_4)_3$ with 16 pbw HF (48–52%) in 380 pbw dist. water. Temp. 70 °C	Burkard[182b]
	27 pbw HCl (1.16) with 23 pbw H_2SO_4 (1.83) in 50 pbw dist. water. Immersion 60 min at 45 °C, followed by immersion for 20 min in a sol. of 11 pbw H_2SO_4 (1.83), 13 pbw HCl (1.16), 1 pbw NO_3H (1.40) in 75 pbw dist. water, at 60 °C.	Roth[1083]
Kovar (FeNiCo)	1 pbw HNO_3 (65%) with 1 pbw acetic acid (50%)	About 50 sec
	75 g $(NH_4)_2SO_4$ in 100 ml. H_2SO_4 (20%), at 50–100 °C	3–5 min
	1 vol. HCl (10%) with 1 vol. HNO_3 (10%) at 70 °C. Stirring required	2–5 min
	Electrolytic etching: 1% NaCl in HCl (10–15%) a.c. 10–12 V, 1.6 A/cm². Electrode graphite or Kovar	
Molybdenum	HF (40%) or HCl (5–8%)	
	2000 ml. H_2SO_4 with 37.5 g CrO_3, 100 ml. HF, 10 ml. HNO_3 (conc) at 90 °C	10 sec Chelius[210]
	10 g NaOH in 750 ml. dist. water with 250 ml. H_2O_2 (30–35%) at 40 °C	2–5 min
	Electrolytic etching in 20% KOH sol. d.c. 7.5 V. Carbon electrode	
Nickel	HNO_3 sol. 10%, at 70 °C	1–2 min
	150 ml. H_2SO_4 (1 83), 225 ml. HNO_3 (1.3), 3 g NaCl in in 100 ml. dist. water, at 20–40 °C	5–20 sec *Metals Handbook*[848]
	Electrolytic etching in sol. of 130 ml. H_2SO_4, 25 g $NiSO_4$ in 200 ml dist. water 6–12 V, nickel electrode	
	Acetic acid sol. 10%	
Tantalum	Hot HF	1–2 sec
	Anodic etching in 75–98% H_2SO_4 (or HCl) with 2–7% HF, in dist. water. 40–160 mA/cm²	Gall[421]
Tungsten	50 ml. HNO_3 with 30 ml. H_2SO_4 in 20 ml. dist. water	
	Boiling H_2O_2 (3% sol.)	
	Anodic etching in 250 g KOH and 0.25 g $CuSO_4$ in 1000 ml. dist. water	

TABLE 2.21 ELECTROLYTIC POLISHING

Metal (Alloy)	Bath composition (ml.)	Voltage (V)	Current density (mA/cm^2)	Temp. (°C)	Time (min)
Aluminium	40 ml. H_2SO_4, 40 ml. phosphoric acid 20 ml. dist. water	10–18	720	95	5
	165 ml. perchloric acid, 785 ml. acetic acid. 50 ml. dist. water	50–100	30–50	<50	15
	45 ml. perchloric acid, 800 ml. ethanol, 155 ml. dist. water	100–200	2000–4000	<35	30
Beryllium	100 ml. orthophosphoric acid, 30 ml. H_2SO_4, 30 ml. glyceryne, 30 ml. ethanol	–	2000 4000	–	–
Copper	670 orthophosphoric acid, 100 ml. H_2SO_4, 270 ml. dist. water	2–2.2	100	22	–
	7 g CrO_3, 22 g sodium dichromate, 7 ml. acetic acid, 6 ml. H_2SO_4, 58 ml. dist. water	20–60	–	–	–
Iron	530 ml. orthophosphoric acid, 470 ml. dist. water	0.5–0.2	6	20	10
Molybdenum	35 ml. H_2SO_4, 140 ml. dist. water	12	–	50	
Monel	200 ml. HNO_3, 400 ml. methanol	2.4–2.6	125–150	20–30	10
Nickel	60 ml. orthophosphoric acid, 20 H_2SO_4 2 dist. water	10–18	900	60	5
	210 ml. perchloric acid, 790 ml. acetic acid	22	180	20	–
Carbon steel	50 ml. H_2SO_4, 40 ml. glycerol, 2 ml. HCl. 8 ml. dist. water	10–18	50	10	60
	185 ml. perchloric acid, 765 ml. acetic acid, 50 ml. dist. water	50	40–70	<30	5–10
Stainless steel	50 ml. H_2SO_4, 40 ml. glycerol, 10 ml. dist. water	10–18	300–1000	30–90	3–9
	133 ml. glacial acetic acid, 25 g CrO_3, 7 ml. dist. water	20	900–2500	18	4–6
Tantalum	90 ml. H_2SO_4 (conc), 10 ml. HF	–	100	35–45	9
Tungsten	100 g NaOH, 900 ml. dist. water	–	30–60	20	20–30

cathode (cathodic cleaning) or as the anode (anodic cleaning). The tank is usually the second electrode. With anodic cleaning, oxygen is liberated on the surface of the metal being cleaned. This process requires 6–12 V d.c. at 50–100 mA/cm^2. With cathodic cleaning hydrogen is liberated on the cleaned surface. The cathodic cleaning processes requires 6–12 V d.c. for a maximum of 50 mA/cm^2. For steels the anodic cleaning is recommended, but for alloys containing Pb, Zn, Sn the cathodic cleaning should be used. A recommended solution for electrocleaning consists of: 1 pbw caustic soda, 1 pbw trisodium phosphate and 1 pbw soda ash, in water.

*Solvent cleaning** is done by using the solvent in a liquid or in a vapour state. Liquid cleaning can use inflammable products (benzene, xylene, etc.) if the necessary safety precautions are provided. Vapour degreasing is much more effective than liquid solvent cleaning. Here the solvent is heated to boiling. The parts to be cleaned are hung in the chamber in the hot vapour, which condenses on the metal surfaces, dissolves the oil and grease, and flows back to the solvent container. The solvents used are not flammable but they are toxic. (Maximum admissible concentration in air about 100 ppm.) In the increasing order of their boiling points the solvents are: dichlorethylene ($C_2Cl_2H_2$, BP = 55 °C), carbon tetrachloride (CCl_4, BP = 77 °C), trichlorethylene (C_2Cl_3H, BP = 87 °C), perchlorethylene (C_2Cl_4, BP = 121 °C). The most frequently used solvents are carbon tetrachloride and trichlorethylene. It should be pointed out that trichlorethylene vapours, in contact with open flames or hot surfaces, form hydrochloric acid and highly toxic gases. The trichlorethylene vapours are narcotic and toxic. Carbon tetrachloride is also toxic and produces irritation of the skin. Water, fatty acids, alcohol and aluminium must be kept out of these solvents, as they catalyse hydrolysis, which liberates hydrochloric acid. The same effect can be due also to the photochemical decomposition of these solvents. Consequently they must be kept in dark bottles.

Vapour degreasing does not always give clean surfaces; it does not remove mineral salts, the inert soil, or the metal soaps (see earlier in this Section).

22.32 Welded seals. The following remarks are intended just as an indication of the special precautions required in order to construct welded joints which are mechanically strong and at the same time leak-tight (Table 1.5). For details on general welding techniques see Davies[261], Soled[1164], Laughner[545], Espe[352], Young[1345].

In the design and construction of the welded seals for vacuum vessels and pipes the following points should be observed:

(1) The joints must be designed and welded with full penetration avoiding trapped volumes (crevices) in which contaminants may collect (Kronberger[721],

* For a review of the methods for cleaning see Janecke[627a], Espe[354a].

Brown[172], Dreyer[304], Mooz[885], Mark[804]). Some recommended arrangements of welded joints for vacuum seals are shown in Fig. 2.31 together with the corresponding poor constructions which are to be avoided.

Cornwall[239a] gives some design indications for butt, corner, *T* and edge welded joints for vacuum use.

(2) Whenever possible *single-pass* welds should be used (Horning[596]). The double-pass weld can be accepted only in very large vessels, especially in those subjected to thermal cycling (Farkas[369]).

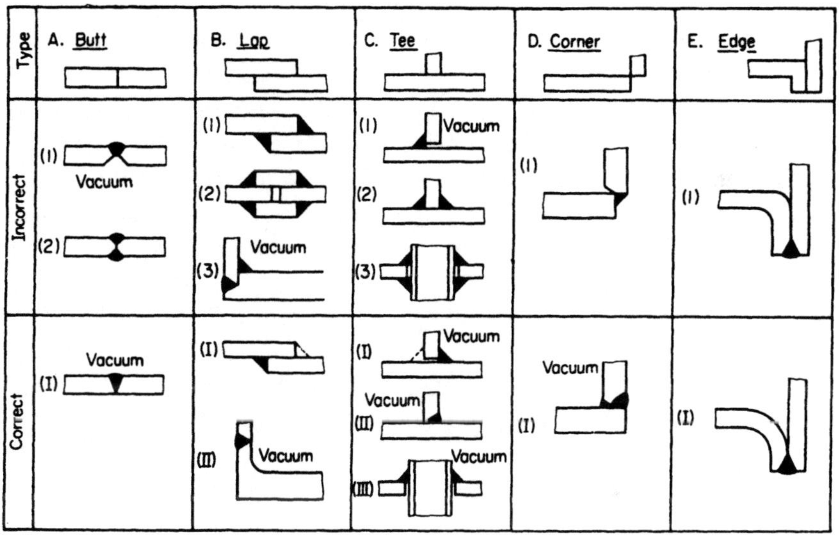

FIG. 2.31 Correct and incorrect arrangements of welded joints

(3) Whenever possible, welds should be made from the vacuum side of the vessel.

(4) When smooth inside surface is very important, an extra inside weld is made in addition to the outside weld (Fig. 2.31 D I) but in these cases it is recommended to make this inside weld leak-tight (Guthrie[491]) and to ensure the location of any leak by providing drilled and plugged holes on the outside weld.

(5) The use of backing strips is not recommended (Kronberger[721]). If structural welds are necessary inside the vessel, they should be made discontinuous to allow easy flow of the gases from any pocket (Horning[596]). These structural welds should not cross the sealing ones.

(6) In order to achieve long uninterrupted welds, it is recommended that the longest practical welding rod be used (Horning[596], Stoddard[1193a]). To avoid the use of too much welding metal (especially in the vessels to be heated)

it is advisable to use welding rods with small diameters, thus minimizing also the amount of heat required for the welding.

(7) The welded assembly should be designed so that a maximum number of welds could be tested separately in the construction stage and corrected prior to making the final assembly (Millen[858], Guthrie[491], Fleischmann[383], Miller[862], Kittle[682]). Sometimes, after a first leak detection, the welds are fired in dry hydrogen (at about 1100 °C) and then checked again to detect any leak that may have developed as a result of the burnout of impurities (Dreyer[304]).

For the leak detection of the welded joints prior to assembly (see Table 1.4 for methods) the welded part is attached to a vacuum box (Garrod[434]) using a double gasket seal with guard vacuum (Section 38.23). The maximum leak rate (air) permitted in welded seals is about 2×10^{-5} lusec/cm length of weld (Garrod[434]). If higher leak rates are found, the weld should be ground off to the base metal and a new weld must be made (Guthrie[491], Stoddard[1193a]). This also applies to welds which have cracked during service. The supposition that applying another layer of weld over the original (leaking) one, will correct the situation, is wrong. Leaks are seldom corrected this way, since stresses are likely to be set up, which will cause a new crack upon heating.

Butt welds, i.e. welds in which the edge of a plate (or tube) is brought in line with the edge of a second plate (or tube) and the joint is filled with welding metal (Fig. 2.31 A), are used for vacuum sealing, with *V* as well as with square or *U* grooves (Fig. 2.32). One side welding with the weld depth equal to the plate thickness (Fig. 2.31 A I) is preferred for these purposes to welds on both side (Fig. 2.31 A2), or to one side welds having the depth less than the thickness of the plate (Fig. 2.31 A 1). Butt welds are preferred to lap welds or corner welds. For example, the construction B 3 (Fig. 2.31) should be replaced by the arrangement B II (Fig. 2.31). Some recommended designs of butt welds, which can be used for vacuum sealing, are presented in Fig. 2.32.

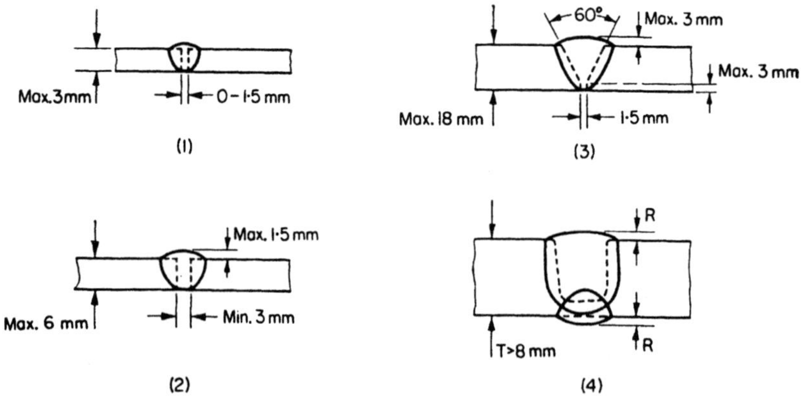

FIG. 2.32 Recommended designs for butt welds. $R = T/4$ for T up to 12 mm; $R = 3$ mm for T over 12 mm

FIG. 2.33 Butt welds used on large vacuum vessels of special shapes. After Mark[804] *(Courtesy of Pergamon Press)*

Figure 2.33 shows some butt welds as they are used in the construction of very big vacuum vessels of special shapes (Mark[804]). These are welds with full penetration, made with *V* or *U* grooves. When such welds are made by argonarc welding (Section 22.11b) the inside surface of the weld is of a very rough appearance if a separate argon backing stream is not used on the other side of the weld. When pipes are butt welded the backing argon can flow through the pipe itself. Hogg[577] describes a method to have argon just in the vicinity of the weld by using rubber balloons puffed inside the pipe to fill up the bore of the pipe in both sides of the weld.

Lap welds, i.e. welds in which the edges of a plate (or pipe) are lapped one over the other and the edge of one is welded to the surface of the other (Fig. 2.31 B) are not recommended for vacuum sealing. If necessary for the strength of the construction, lap welds are to be used only with a single pass and having the fillet inside the vacuum vessel (Fig. 2.31 B I). Double fillet lap welds (Fig. 2.31 B 1) or lap welds using backing strips (Fig. 2.31 B 2) are dangerous as vacuum seals because of the trapped volumes along the lapped parts. These parts can hide leaks and make them very difficult to be detected. Even the lapped corner design (Fig. 2.31 B 3) is to be replaced by a butt joint (Fig. 2.31 B II).

In *tee welded joints* for vacuum sealing, the trapped spaces (Fig. 2.31 C 1) are to be avoided by making the welding fillet only in the inside of the vessel (Fig. 2.31 C I and III) or by providing full penetration of the weld from the outside of the vacuum vessel (Fig. 2.31 C II). Figure 2.34 shows some recommended designs of the welded tee joints.

In *corner joints* welded according to the usual practice (Fig. 2.31 D 1) it is known that cracks are formed. During subsequent machining operations (such as facing flanges) machining oil can easily run down into these cracks. It has been found (Smith[1151]) that it is very difficult, if not impossible, to remove this oil by cleaning. To avoid this it is necessary to design these joints as full penetration welds (Fig. 2.31 D I).

The *edge weld* is especially convenient for sealing heavy sections. The edge seal can be made by creating a *lip* in the thick material which is then welded to a thinner part or by using thin metal sheats as the proper seal. Examples for the first method are shown by Fig. 2.35 (Grove[489], Farkass[369]). The lips are

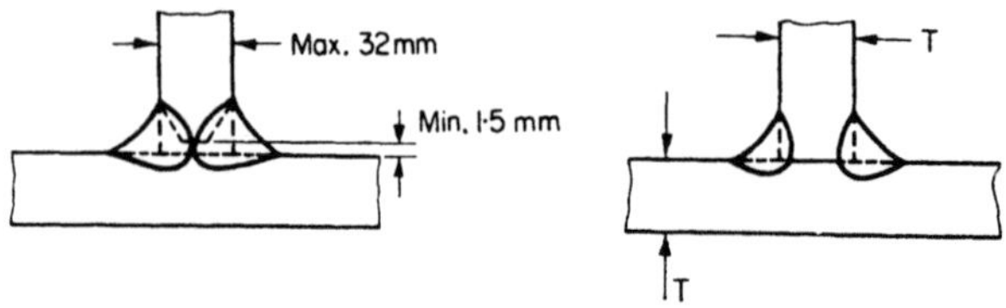

FIG. 2.34 Recommended designs for tee-welded joints

constructed by flaring or bending the edges of the material (Fig. 2.35a), by providing a groove near the edge to be welded thus leaving a lip with the required thickness (Fig. 2.35b), or by using the two methods together (Fig. 2.35c).

The edges to be welded can be straight or chamfered, forming between them a *V* (Stohr[1194]), but to avoid spaces trapping gases and dirt it is recommended that the weld penetrates to the inside surface (Fig. 2.31 E I).

If the edge weld is to be opened, the design from Fig. 2.35d can be adopted (Farkass[369]). The welded seal can be opened by grinding the surface of the weld. It can be rewelded many times.

Michelson[855] describes an edge weld used as a bakeable vacuum seal or to seal baked and non-baked parts. This consists of first welding thin metal rings to the heavy metal flanges, and then edge welding the rings to each other (Fig. 2.36). The thin metal rings can be edge sealed at the outside of the vessel (Fig. 2.36a) or towards the inside (Fig. 2.36b). For axial alignment of the flanges — and as radial support members — lugs are machined into the male flange (Fig. 2.36a), allowing the proper spacing between the flanges as required by the sealing rings. The inside seal (Fig. 2.36b) can be used only for large vessels where the sealing ring is accessible for the edge welding. These seals can be opened by using sharp files. Only 3–4 strokes of the file are usually required to cut through the weld. The file stroke should be along the seal weld, not across it.

The arrangement shown in Fig. 2.36c can be used for large diameters (up to 75 cm). This system uses a supporting ring placed between the two thin sealing members.

22.33 Brazed seals. In order to construct a leak-tight brazed joint the parts to be joined are assembled with a tight fit, allowing a clearance according to the brazing alloy and the configuration of the parts to be sealed. The brazing metal used is in the form of wire, rings, foils, strips or as an electrolytic

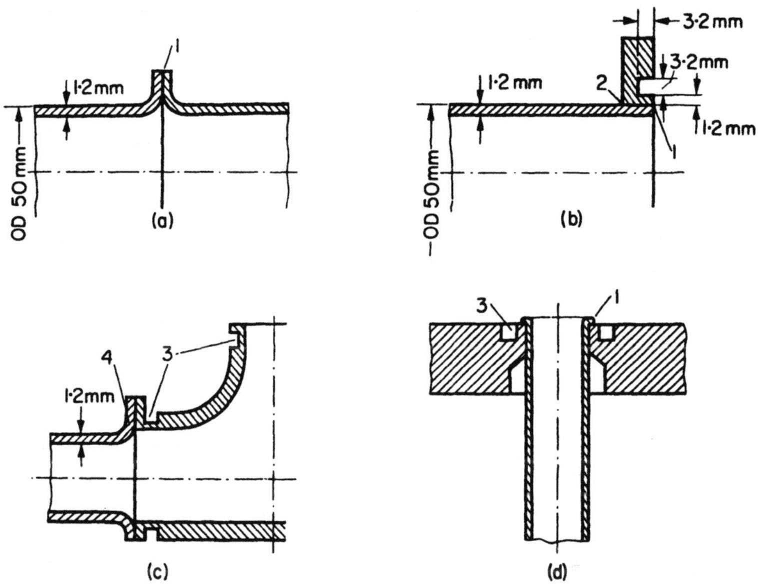

FIG. 2.35 Designs for edge welding: (1) fusion weld; (2) welded points; (3) grooves; (4) flared tubing; (a, b, c) after Grove[489]; (d) after Farkass[369] *(Courtesy of Pergamon Press)*

coating. The cleaned (Section 22.31) and assembled parts are heated (Section 22.21) at least up to the melting point of the brazing alloy. The brazing alloy flows in the interstices between the parts (capillary action), where it solidifies and joins the parts to each other.

Since the filler metal is usually weaker than the parts joined by brazing, the strength of a brazed joint depends on the amount of unchanged brazing alloy left in the joint* in the form of a thin coating. This depends, in turn, on the composition of the filler, on the material of the parts, on the time and temperature of the brazing process (Section 22.22) as well as on the clearance in the joint. Table 2.22 summarizes the recommended clearances to be used in brazed joints. To obtain reliable leak-tight brazed joints the points listed on pages 94–95 should be observed:

* The ultimate tensile strength of the brazed joints is really a linear function of the thickness to diameter ratio, rather than only of the thickness of the filler metal (Gross[486]).

(1) The smallest quantities of brazing alloy should be used. Experience shows that with small clearances and clean surfaces, better joints are obtained than when applying big quantities of brazing alloy.

(2) Wide or irregular spaces between parts are to be avoided.

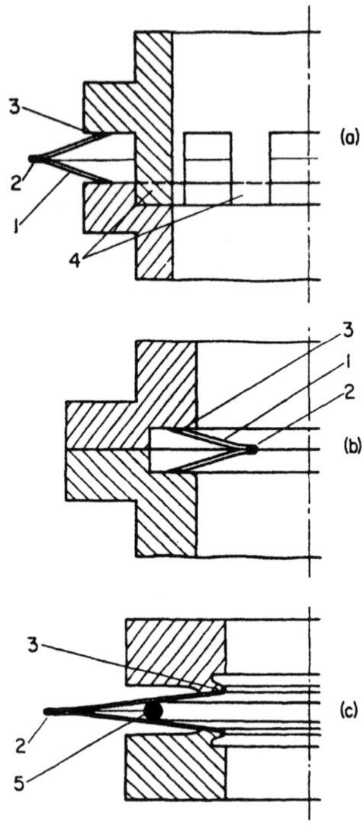

FIG. 2.36 Vacuum sealing by edge welding of thin metal sheets: (1) seal ring; (2) seal weld; (3) base weld; (4) lug; (5) ring (after Michelson[855] and Farkass[369])

(3) The overlap between the two parts brazed to each other must be a minimum 2–3 mm, in order to allow the capillary forces to suck in the brazing alloy.

(4) If metals with different thermal expansions are to be brazed, the assembly must be arranged to compress the brazing alloy during the cooling, i.e. the outer part must have the bigger thermal expansion (see Fig. 2.27).

(5) The flow of the brazing alloy can be controlled by the construction of the joint. The clearance at the corners determines how the brazing alloy will flow around these corners. Gaps in the joint may block (if so desired or not) the flow of the brazing alloy through the joint. Thus *square corners* (Fig. 2.37a) will give a good flow of the brazing alloy throughout all the joint Fig. 2.37b), the resulting joint being strong and leak-tight. *Round corners*

TABLE 2.22. CLEARANCES FOR BRAZED JOINTS

Brazing alloy	Brazed metal	Recommended clearance (microns)
Pure metals (Cu, Ni, Ag)	any	13–25
Copper (tubular, circular seals)	ferrous	10 μ per cm of diameter
Pure metals (Cu, Ni, Ag) electrolytically coated	any	10–15
Copper, nickel, silver, etc., eutectics (Fig. 2.29, Table 2.18) or with lowest melting point (Fig. 2.30)	as in Table 2.19	40–75
Alloys with raising melting point by adding more of the component (H in Table 2.18)	as in Table 2.19	50–100

stop the flow. If the first corner, from the side where the brazing alloy is applied, is round (Fig. 2.37c) the brazing alloy will not pass this corner (Fig. 2.37d). When only the second corner is round (Fig. 2.37e), the joint will be stronger and leak-tight (Fig. 2.37f). A *square edge* pressed *against a round corner* (Fig. 2.37g) would similarly stop the flow of the brazing alloy even though the clearance is right on each side of the corner.

(6) If the flow of the brazing alloy is to be avoided on a surface, the area must be coated with carbon or chromium.

(7) Butt brazed joints can be used but lap joints are preferred for vacuum seals. Butt joints (Fig. 2.38 A) can be satisfactory when carefully made, but it is recommended to use them only when no other alternative exists. The butt brazed edges must be square and straight in order to maintain a uniform space between the parts. The parts to be brazed must be held close together during brazing, to ensure even flow of the brazing alloy by capillary action. Brazed joint having a construction similar to those butt welded (Fig. 2.38 A 1,2) are not recommended. Greater strength can be obtained using *scarf joints* (Fig. 2.38 B) or *step joints* (Fig. 2.38 A I), which are in fact combinations of butt and lap joints. Eventually butt joints using straps (Fig. 2.38 A II) can be utilized in vacuum sealing.

Lap joints are to be preferred in vacuum sealing applications. In such joints, it is recommended to construct the overlapping length at least 1.5 times the thickness of the joined parts. The lap joints from Fig. 2.38 C1, C2, C3 do not correspond to this requirement. Instead of these joints the designs from Fig. 2.38 C I, II are recommended. Figure 2.39 shows some (good and poor)

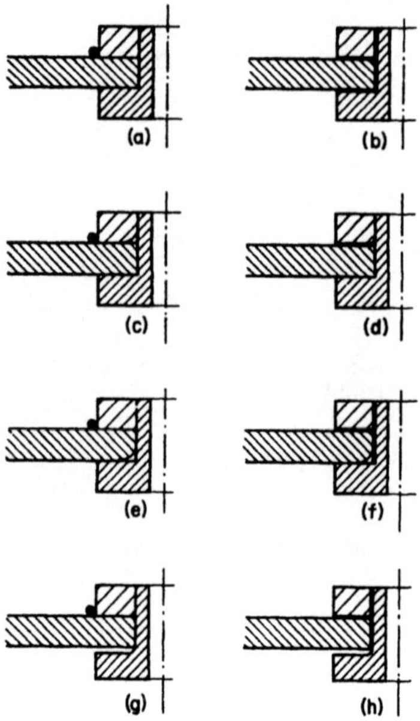

FIG. 2.37 Control of brazing by means of the gaps between parts

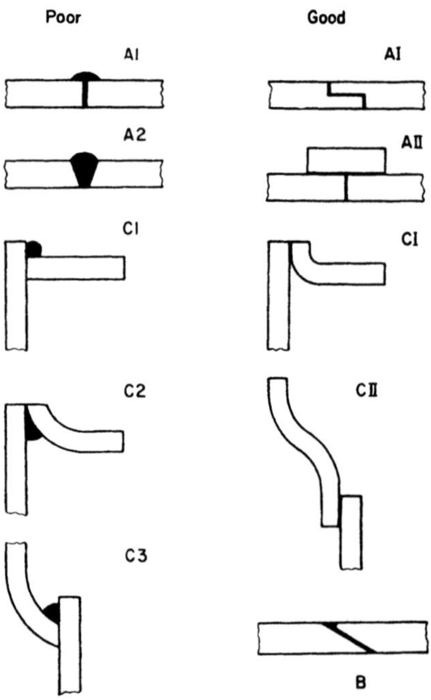

FIG. 2.38 Poor and reliable designs of brazed joints

designs of brazed joints between two pipes or between a pipe and a flange. The design on Fig. 2.39 A 1 is poor because the lapped surface is equal to the thickness of the flange, that on Fig. 2.39 A 2 lacks strength, and the alignment of the parts presents difficulties during heating. The joint design with a slight

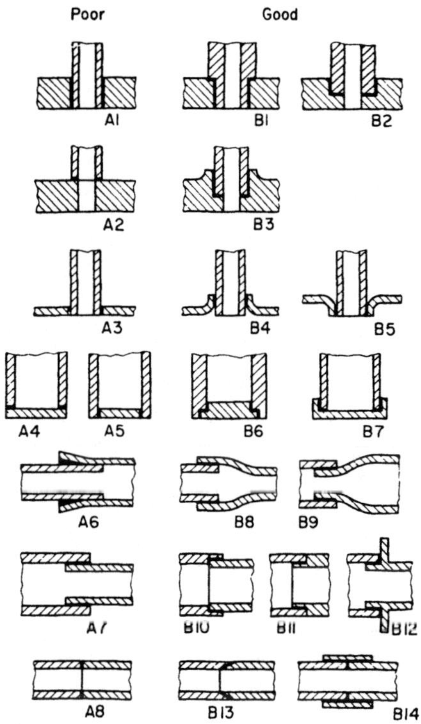

FIG. 2.39 Arrangements for brazing pipes and flanges

counterbore (Fig. 2.39 B 2) will give a much stronger assembly. The design B 1 will ensure good strength. The hub on the flange (Fig. 2.39 B 3) is recommended as an aid in heating, and is very important on heavy flanges. Without this hub the whole flange must be brought to the brazing temperature. This is satisfactory only in a furnace brazing (Section 22.21c) but not if other heating methods are used. Since most of the pipe joints are brazed using portable equipment the existence of the hub will help in many such brazed joints. When thin flanges are to be brazed to pipes the design from Fig. 2.39 A 3 should be replaced by one of those on Fig. 2.39 B 4, B 5. The design on Fig. 2.39 A 4 is of a poor construction; it does not provide any shear area in the seal. By inserting the disc in the pipe (Fig. 2.39 A 5) a stronger joint can be achieved. If possible, this design should be replaced by a construction like that on Fig. 2.39 B 6 or B 7. These designs assure a better alignment. The design B 7 is useful especially for pipes with thin walls. The constructions A 6 or a A 7 are to be avoided in the joints of pipes with different diameters. The

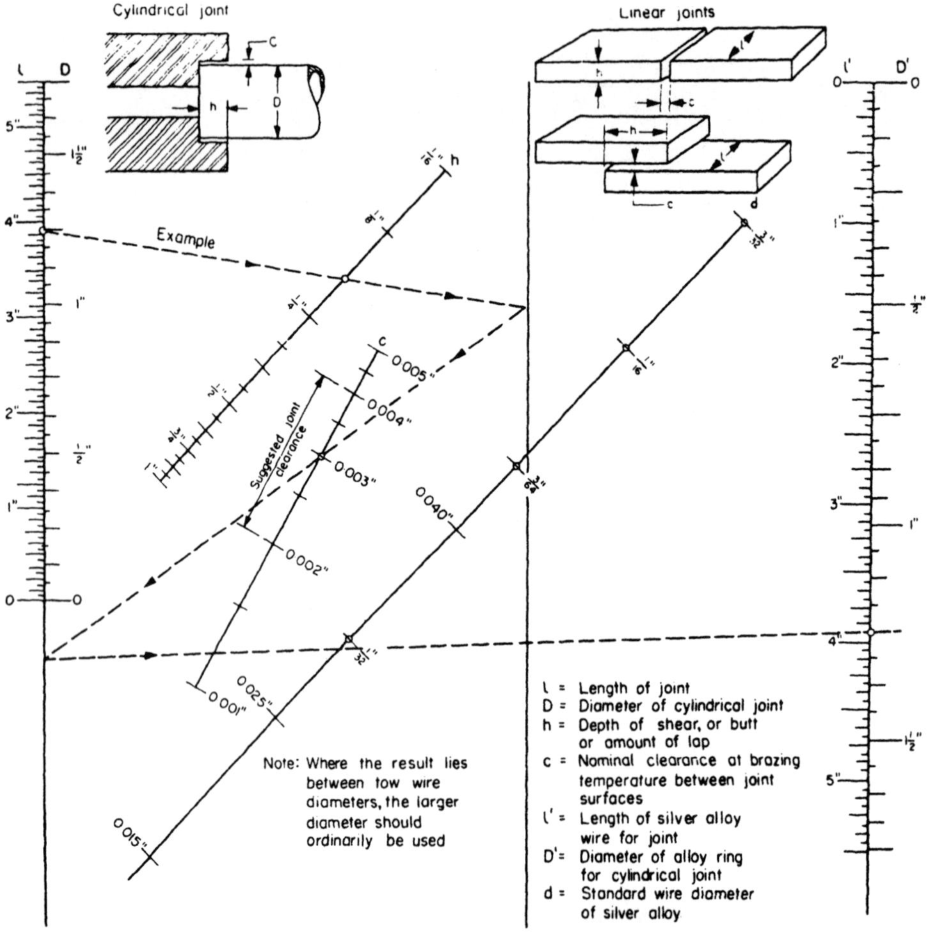

FIG. 2.40 Nomograph to select the size of wire for brazing. Reproduced from Handy-Harman[509] *(Courtesy of Materials in Design Engineering, New York)*

designs B 8–B 12 are recommended for such joints. For pipes with the same diameter the designs B 13 or B 14 are to be used instead of A 8.

(8) The brazing alloy is applied to the joint, usually in the form of wire, but brazing alloys applied in the form of thin sheets are also used. The brazing

alloy is added successively (as in welding) or the joint is preloaded with a wire ring of brazing alloy. The brazing with wire ring assures a perfectly uniform distribution of the alloy, especially when the ring is shaped as required to be inserted between the parts to be brazed. For this purpose the wires usually have diameters from 0.7 to 1.5 mm, and the shims have thicknesses between 0.05 and 0.1 mm. From the nomograph on Fig. 2.40 (Handy and Harman[509])

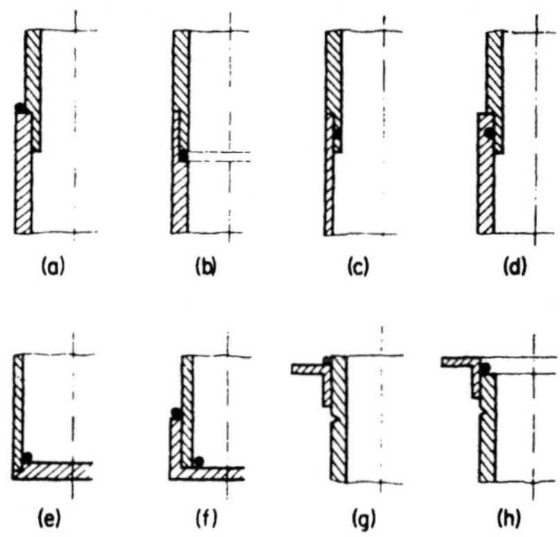

FIG. 2.41 Methods of applying the brazing alloy rings (wires) onto the parts

it is possible to select the wire size needed to silver braze a joint whose diameter (or length), shear depth (or lap), and clearance (Table 2.22) are known. The example presented on the monograph starts on the left, extending a line from the joint diameter on the D scale ($1\frac{1}{4}$ in.) through shear depth (3/16 in.) on the h-scale to intersect the centre axis. From this point the line runs back through the clearance (0.003 in.) on the c-scale to intersect the left axis. Lining up this point with the required ring diameter ($1\frac{1}{4}$ in.) on D'-scale, the intersection of this line with the d-scale shows the wire diameter (1/32 in.) to be used.

Wire rings can be placed on the parts in various ways (Fig. 2.41). When the ring is positioned as in Fig. 2.41a, the brazing alloy will flow throughout the lap joint when it is heated. The brazing alloy can be placed inside the joint (Fig. 2.41b) in which case the parts should be pressed together when the alloy begins to flow. A better design inserts the brazing alloy wire ring into a groove machined in the insert (Fig. 2.41c) or in the sleeve (Fig. 2.41d). The fluid brazing alloys (as silver) tend to flow more in the direction from which the heat is applied. This is an important fact especially when local heating methods are used (e.g. induction heating), where the direction of the flow of the brazing alloy can be influenced by heating the parts below or above the groove containing the brazing wire ring. The arrangements from Fig. 2.41e, f are satisfactory to join a cover to a pipe or a container. When flanges are to be joined to pipes, it is recommended to provide shoulders to hold the brazing alloy ring (especially in furnace brazing) as shown on Fig. 2.41 g, h and stakes to prevent slipping of the brazing alloy.

When small diameter pipes are to be brazed to flanges, swagging and brazing are used together. Some such joints are shown in Fig. 2.42.

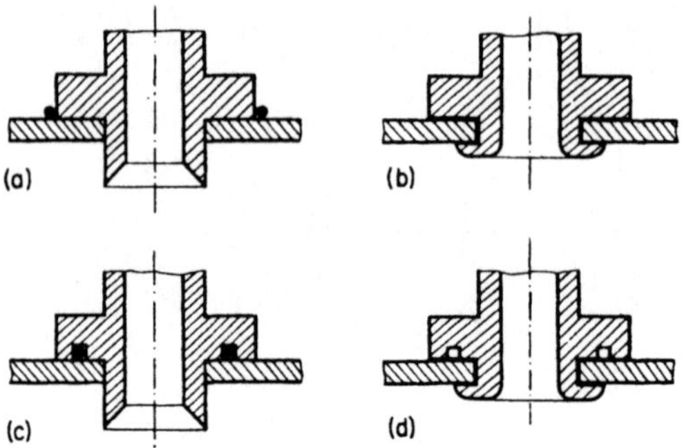

FIG. 2.42 Joining methods of flanges by swagging and brazing: (a, c) before; (b, d) after completing the joint

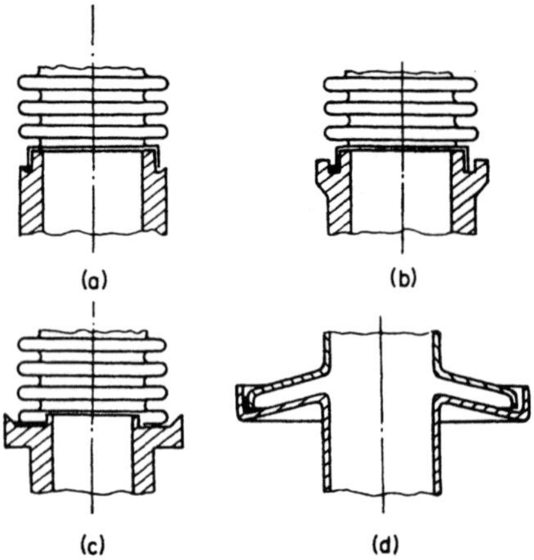

FIG. 2.43 Methods of brazing (a, b, c) bellows to pipes, (d) thin walled pipes to each other

For the joining of these walled tubes (e.g. bellows) to thick parts, it is recommended to provide some sort of groove on the thick part. In these grooves the edge of the thin walled tube (bellows) is inserted and brazed. On Fig. 2.43a, b, c one side of the joint is represented without the brazing alloy to show the details of the construction before brazing. Two cylindrical parts (with thin walls) can be brazed to each other using the construction on Fig. 2.43d. Here the two parts should have conical ends. The lower cone has a brim which is turned upwards, into which the brim of the upper cone—turned downwards—fits and is brazed (Medicus[838]).

Figure 2.44 shows some of the possibilities of placing the induction heater coils around the brazed parts. The coil can be placed outside the parts to be joined (Fig. 2.44a) or inside the assembly (Fig. 2.44b). In order to concentrate

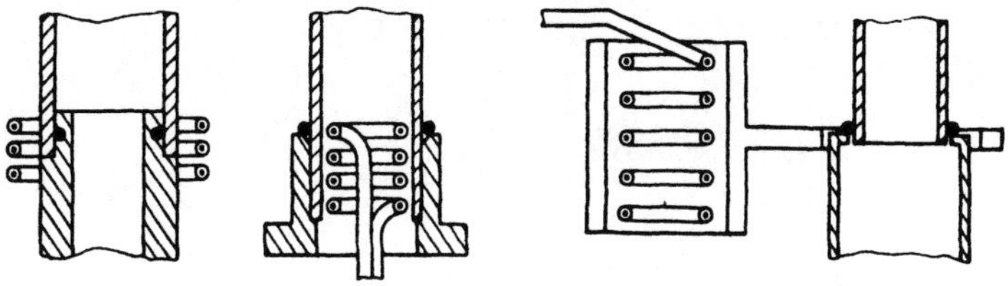

FIG. 2.44 Methods of placing the r.f. coil for brazing

the heat in the area to be brazed, without overheating the adjacent parts, concentrators are used (Fig. 2.44c). For details on concentrators see Babat[61], Kohl[706], Espe[354], Lupinek[789].

The brazing technique used to seal-off exhaust tubes is described in Section 26.33.

2.3 GLASS TO GLASS (AND QUARTZ) SEALS

23.1 Sealing principles

Glass to glass sealing is a part of "glass blowing technique" which includes the shaping of the glass parts to the required forms and the connexion of these parts by "welding" them to each other.

The permanent sealing of two glass parts is done by *heating* the ends to be connected, until they reach the needed viscosity, by *joining* them in this state and *cooling* them together. The quality of the seal is determined by the viscosity of the glass during the sealing process, the expansion matching of the two glasses sealed to each other, and by the cooling (annealing) process of the seal.

23.11 Viscosity and softening points. Glass is often described as an *undercooled* liquid* with a very high viscosity. Glass is unique concerning its viscosity which varies continuously with the temperature (Breadner[163], Shand[1124], Stanworth[1176], Stevels[1187], Fuchs[410a]).

At room temperature the viscosity of glass is so high that it can be considered to be infinite. As the temperature is increased the viscosity decreases† and the glass gradually assumes the character of a liquid.

* Glass is an inorganic melting product which has cooled off and solidified without crystallizing (A.S.T.M).

† A linear relation is found between the logarithm of the viscosity and the reciprocal of the absolute temperature.

The viscosity of glasses varies over an enormous range (10 to 10^{15} poises) which includes four distinct phases (see Fig. 2.45):

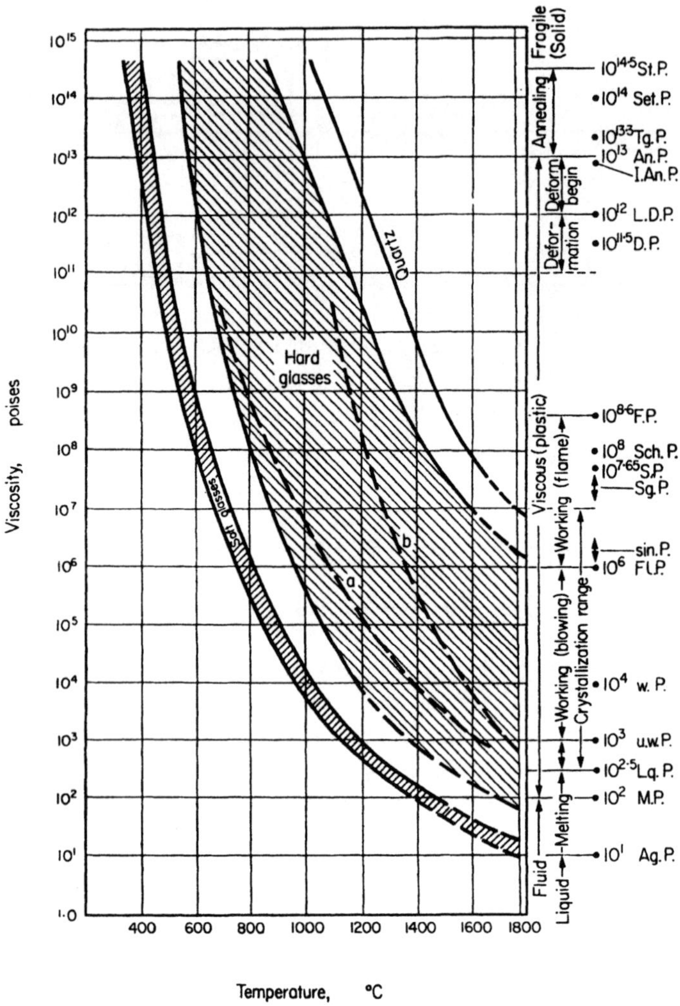

FIG. 2.45 Viscosity vs. temperature curves of glasses and their characteristic thermal points

(1) The *fragile or solid phase* which covers all the temperatures below the Strain Point. In this phase the glass cannot be subjected to excessive deforming stresses without fracture (Section 21.3).

(2) The *annealing phase* which covers the transition between the Strain Point and the Annealing Point. In this range of temperatures, the glass is virtually a *ductile* solid.

(3) The *plastic or viscous phase* which extends from the Annealing Point to the Melting Point.* This range includes the Deformation range and the various working ranges.

(4) The *fluid phase* extending from the Melting Point* towards higher temperatures.

The viscosity–temperature curves of the various glass groups are shown in Fig. 2.45, with the various characteristic points and ranges. Since most of these characteristic points are arbitrarily chosen and depend on the measuring methods adopted for them, they can be only explained with their original definition.

For the selection of the proper glass, as well as for the techniques of glass shaping, glass to glass sealing, glass to metal sealing, glass to ceramic sealing etc., the characteristic points of the glass determine the procedure to be used. It is recommended to make always clear, when consulting tabulated datas, the exact meaning of the given points. As a guide a complete list of these points is given below. The actual values of these characteristic points for various glasses can be found in Tables 2.10, 2.29 and Appendix B.3.

Strain Point (St.P). This is defined (A.S.T.M.[53], Lillie[767], Shand[1124]) as the temperature at which the internal stress in the glass is substantially relieved in 4 hours, and "absolutely" relieved in 15 hours. In practice the point is obtained by the extrapolation on the viscosity–temperature curve to an apparent viscosity of $10^{14.5}$ poises. The viscosity of 10^{15} poises is believed to be the limit for the solid (fragile) body (Winter[1326]), approximately equal to the viscosity of aluminium at room temperature.

The Strain Point represents the upper temperature limit at which annealed glass is serviceable. The tempered glasses begin to lose their temper even at temperatures below the Strain Point. At the same time the Strain Point is the lower end of the *Annealing Range* (Fig. 2.45), i.e. of the range of temperatures in which stresses in glass parts can be relieved at a commercially desirable rate (Kohl[706]). The upper end of this range corresponds to the Annealing Point.

Setting Point (Set.P.). The temperature at which the glass of a glass–metal seal can be considered as rigid (viscosity about 10^{14} poises). At temperatures below this point, the contraction rates of the metal and glass, are producing stresses in each other (Shand[1124]). The Setting Point is about 20 °C below the Annealing Point, and it has been arbitrarily defined as the temperature midway between the Annealing Point and the Strain Point (Hagy[501], Kohl[706]).

Transformation Point (Tg.P). This is the inflection point of the thermal expansion–temperature curve (Fig. 2.46) or of the electrical resistance–temperature curve (Berger[113]). The Transformation Point is defined as the inter-

* The "Melting Point" of glasses is defined very differently from that of other materials.

section of the mentioned curve with the bysector of the angle formed by the tangents above and below the inflection point (Fig. 2.46). It corresponds to a viscosity of $10^{13.3}$ poises. The dynamic Transformation Point established at a rate of temperature raise of 4 °C/min is about 10–30 °C higher than the above defined "static" transformation point (Mönch[879]).

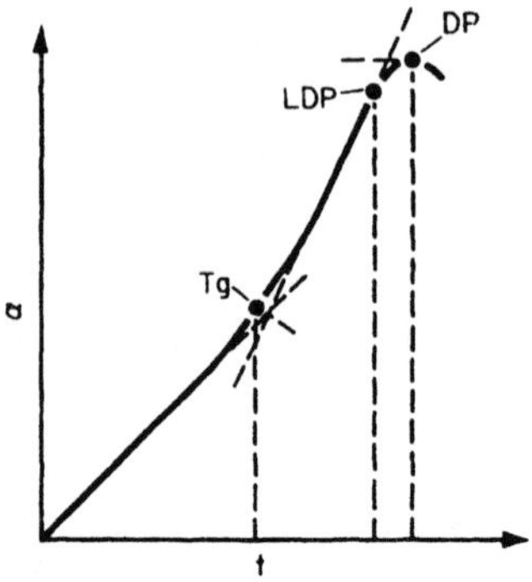

FIG. 2.46 Thermal expansion of glasses vs. temperature, and the method to determine the T_g point

Annealing Point (An.P.). The temperature at which the internal stress in a glass is substantially relieved in 15 min (Lillie[767]), determined at a rate of cooling of 4 °C/min on a glass fibre approximately 65 μ in diameter (Littleton[774]). The corresponding viscosity is 10^{13} poises.

In an annealing process the glass is heated somewhat above the Annealing Point (see Fig. 2.49) and slowly cooled to somewhat below the Strain Point. The glass can be heated to a maximum of 50—60 °C above the Annealing Point, since at higher temperatures the distortion of the glass becomes noticeable.

The Annealing Point (or the Transformation Point) indicates whether the glass belongs to "soft" or "hard" glasses (Table 2.10). This classification has nothing to do with the actual hardness of the glass at room temperature, despite the names used. It is more an indication of the fact that at the temperatures where soft glasses are quite workables, the hard ones are still in the fragile state. It is usual to call soft glasses those having their Annealing Point between 350 and 450 °C, and hard glasses those with the Annealing Point higher than 500 °C.

The *Instantaneous Annealing Point (I.An.P.)* is the temperature at which the initial stresses are practically instantaneously relieved (in 2 min). This point corresponds to a viscosity slightly lower than 10^{13} poises, being approximately 15 °C higher than the Annealing Point.

The *Lower Deformation Point (L.D.P.)* is the temperature — observed during the plotting of the thermal expansion curve (Fig. 2.46) — where (above the Tg.P.) the curve begins to change again its slope. This point corresponds to a viscosity of about 10^{12} poises (Fig. 2.45). The actual value of this point is obviously influenced by the kind of apparatus used for the measurement of the thermal expansion.

Deformation Point (D.P.). This is defined as the temperature observed during the measurement of the thermal expansion, at which the viscous flow of the glass exactly counteracts the thermal expansion (maximum point of the curve on Fig. 2.46). At this point the internal stresses are relieved in about 15 sec. The corresponding viscosity is $10^{11.5}$ poises.

The *Flame Working Point (F.P.)* is the temperature at which the viscosity is adequate for the working of the glass in the flame ($10^{8.6}$ poises). The *Flame Working Range* extends (Fig. 2.45) from this viscosity to about 10^6 poises. This range together with the second working range (10^6–10^3 poises, see later) form the total working range ($10^{8.6}$–10^3 poises). When this latest range extends to a larger temperature difference, the glass is known as a "long glass", (e.g. curve *a*, Fig. 2.45) in comparison with another glass (curve *b*) having this range on a shorter temperature difference, and known as a "short glass". The long glasses are formed more easily, thus they are preferred in glass blowing. In mechanical (automatic) production the short glasses are preferred.

Schott Point (Sch.P.). The temperature at which a glass fibre (0.3 mm diameter) heated on a length of 40–50 mm (temperature raise 10 °C/min) and loaded vertically with 1 g/mm^2, has an increase in length of 1 mm/min, is known as the Schott Point of the given glass. The corresponding viscosity is about 10^8 poises.

Softening Point (S.P.). This is the temperature at which a uniform glass fibre 0.55–0.75 mm in diameter, and 229 mm (9 in.) long, elongates 1 mm/min under its own weight when the upper 100 mm of its length are heated at a rate of 5 °C/min (Littleton[773]). The length is read each half minute and the temperature each half minute between the readings of the length. The difference in length for each half minute is plotted on a log scale vs. the temperature, a straight line drawn through the points, and the temperature at which the line intersects the 0.5 mm elongation is noted (Campbell[193]). This point corresponds to a viscosity of $10^{7.65}$ poises (for a glass of 2.5 g/cm^3 density).

At the Softening Point the glass deforms very rapidly and adheres to other bodies.

Sag Point (Sg.P.). This is the temperature at which a glass fibre of 0.2 mm ($\pm$0.05 mm) diameter, and 5 cm long (2.5 cm extending from the holder) starts to deflect on heating with a rate of 4 °C/min (Hirayama[565]).

Sintering Point (Sin.P.). The temperature at which the glass powder can be sintered (Section 42.3) to a compact piece by using a slight pressure. The viscosity corresponds to about 10^6 poises.

The *Flow Point (Fl.P.)* is the temperature at which the glass reaches a viscosity of 10^6 poises. It is measured by the heating time needed to draw down to a given index a glass fibre of 0.65 mm diameter having a controlled loading (Lillie[766]).

The *Working Point (W.P.)* is the temperature at which the glass is soft enough for hot working by most of the common methods (blowing, pressing). The viscosity is 10^4 poises.

The *Upper Working Point (U.W.P.)* is the highest temperature at which the glass is ready for working (viscosity 10^3 poises). This is the upper end of the *Working Range* in which glass is formed into ware. The lower end of this range corresponds to the temperature at which the glass is just sufficiently viscous to hold its formed shape (viscosity about 10^6 poises).

Liquidus Point (Lq.P.). This is the lowest temperature limit at which crystallization is not more possible. The *Crystallization Range* extends from this viscosity ($10^{2.5}$ poises) up to about 10^7 poises (Fig. 2.45).

The *Melting Point (M.P.)* of glasses is defined (differently to the usual definition) as the temperature at which the glass melt is degassed. This point corresponds to a viscosity of 10^2 poises. The *Melting Range* extends between viscosities of 10^1 to $10^{2.5}$ poises.

The *Aggregation Point (Ag.P.)* is the limit of the liquid behaviour of glasses. The viscosity is 10^1 poises.

Viscosity stages during glass to glass sealing. In order to seal two glasses to each other, they are heated to a temperature corresponding to their Flame Working Range (see Fig. 2.45). In this state they are joined to each other, and the joint is allowed to cool. During the cooling the joint passes through the following successive periods:

(1) The period in which both glasses are at temperatures above their Strain Points (see Fig. 2.45); during this period no stresses will develop in the joint even if the thermal contractions (expansions) of the two glasses are very different.

(2) The period in which the temperature is between the Strain Points of the two glasses. In this range the glass with the higher Strain Point is rigid, but the other can follow the deformations due to the thermal contraction, so that no remarkable stresses are developed in the joint.

(3) The period in which the temperature is lower than the lower Strain Point of the two glasses. Here the movement required by the thermal contraction cannot any more be equilibrated by the mobility of one of the glasses and stresses aie developed. As long as these stresses are smaller than the corresponding mechanical strength (tensile or compressive strength) of the glass, the joint will not break. If the stress exceeds the mechanical strength the joint will crack.

The intensity of the stresses is determined by the degree of expansion matching of the two glasses, and the annealing procedure applied (see following sections).

23.12 Expansion matching. When a temperature gradient is built up in the body of a glass part, a stress gradient arises from the unequal expansion or contraction. When a glass part, made up from a single type of glass, is cooled, the surface will cool more rapidly than the inner mass, because of the bad thermal conduction of the glass. Thus a temperature gradient is formed producing strains in the glass. To relieve this strain, i.e. to balance the stress ranging from a compression in the body to a tension on the surface, the glass part can be annealed (Section 23.13). When two different glasses are sealed to each other, the difference in contraction builds up stresses which act in the direction determined by the relative position of the two glasses in the seal. The strain produced in this case cannot be removed by annealing. The seals made up from two different glasses are also subjected to annealing but the purpose is just to remove the strains of the first kind (due to a too fast cooling) and not to add them to those produced already by the expansion difference between the two glasses.

The tangential stress P_θ in annealed glass to glass cylindrical joints is given (Lillie[766a], Rawson[1030]) by the simplified formula

$$P_\theta = \frac{\bar{E}\cdot\Delta}{2},$$

where $\bar{E}$ is the average of the elasticity modulus of the two glasses, and Δ is the difference in contraction between the two glasses from the Setting Point (Fig. 2.45) of the softer glass to room temperature. While a minimum stress will result when the two glasses have the same elasticity modulus, the difference in E does not introduce an important change in the value of the stress. For all practical purposes $\bar{E} = 7000$ kg/mm^2 can be taken as a mean value for all the glasses (including quartz).

The Setting Point can be expressed as 20 centigrade degrees less than the Annealing Point. The difference in the thermal contraction of the two glasses is

$$\Delta = (\alpha_1-\alpha_2)\cdot \text{Set.P.} = \left(1-\frac{\alpha_2}{\alpha_1}\right)(\text{An.P.}-20)\cdot\alpha_1 =$$

$$= \delta_\alpha\cdot(\text{An.P.}-20)\alpha_1,$$

thus

$$P_\theta = \frac{7000}{2}\cdot\delta_\alpha\cdot(\text{An.P.}-20)\cdot\alpha_1,$$

and the relative difference in expansion will be

$$\delta_\alpha = 1 - \frac{\alpha_2}{\alpha_1} = \frac{P_\theta}{3500 \cdot \alpha \cdot (\text{An.P} - 20)}.$$

Expressing the Annealing Point as a function of the thermal expansion coefficient the relative difference in expansion can be determined as a function of

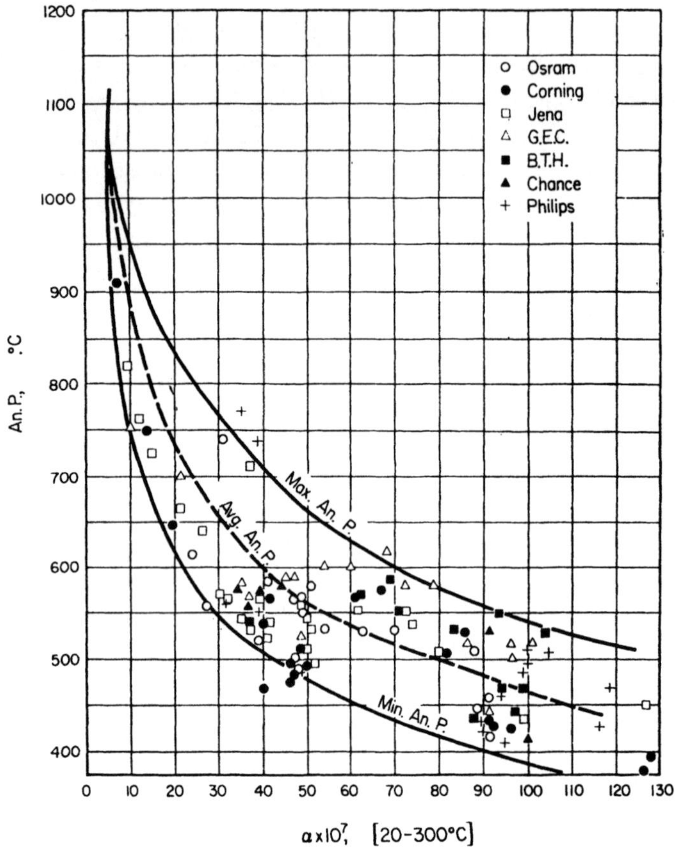

FIG. 2.46A Annealing Point of glasses as a function of the reference thermal expansion coefficient (20–300 °C)

the expansion coefficient and the stress allowed in the seal. For this purpose the Annealing Points of the commonly used glasses were plotted on Fig. 2.46a vs. the thermal expansion coefficient, and the extreme (minimum, maximum) and average curves were drawn. Since in glass–metal seals stresses of 0.5 – 1.5 kg/mm^2 can be tolerated, the values of the relative expansion difference were calculated, based on the three curves from Fig. 2.46A and on stress values of 0.5, 1.0 and 1.5 kg/mm². These values are plotted vs. the expansion coefficient in Fig. 2.46B. The curve avg. An.P., P_θ=0.5 kg/mm², should be used when seals of unfavourable shapes are considered. For simple and unloaded seals the avg. An.P., $P_\theta = 1.5$ kg/mm², can be used.

From the curves of Fig. 2.46B it results that when two soft glasses are sealed to each other ($\alpha = 90 \times 10^{-7}$) the expansion coefficients can differ by a

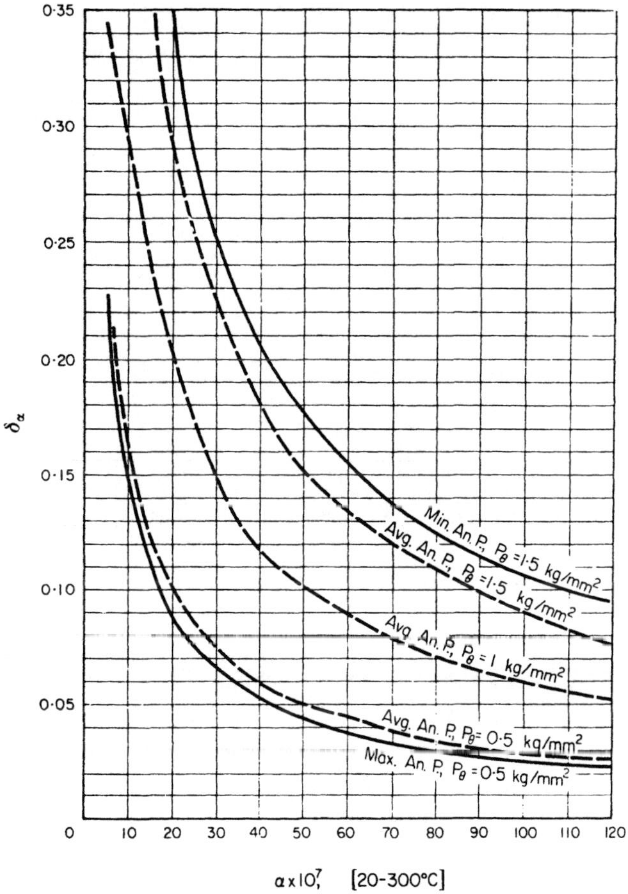

FIG. 2.46B The permissible difference (δ_α) between the thermal expansion coefficients of two glasses to be sealed to each other

maximum of 10 per cent ($\delta_\alpha \approx 0.1$); for hard glasses ($\alpha = 50.10^{-7}$) expansion differences of up to 15 per cent can be tolerated ($\delta_\alpha \approx 0.15$). If the difference in expansion is greater, graded seals must be used (Section 23.35).

The expansion matching of two glasses (intended to be sealed to each other) can be calculated from their tabulated expansion values (e. g. see Table 2.10, or Appendix B.3). Expansion coefficients can be determined accurately using the proper dilatometers (Bollenrath[148], Chevenard[212], Turnbull[1239], Kingston[675]) or by measuring the stresses produced when sealed to a known glass (Padmos[959]).

For a qualitative determination of the expansion matching the *ring method* can be used. A cylindrical ring of one glass is covered by a layer of the other glass, and the assembly is annealed (Section 23.13). After cooling the ring is scratched and cut axially. If the two glasses have matching expansions the slit will show no tendency to open cr to close. When the slit opens the glass from the outside has a greater expansion, when it closes the inside glass expands more.

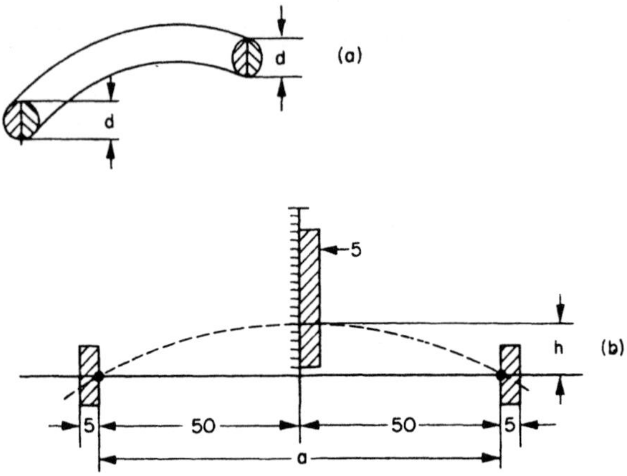

FIG. 2.47 Fibre test for expansion matching: (a) bent glass fibre; (b) graph arrangement for the test

For a quantitative determinations of the expansion matching the *fibre method* can be used. This method consists in sealing the two glasses to each other along a surface in such a way that the glasses do not mix. The seal is drawn (parallel to the surface) to a long thin fibre, which is held straight until it cools. Placed on a plane surface, the fibre bends due to the difference in expansion of the two glasses. In principle the fibre is built up of two halves (Fig. 2.47a) each consisting of one of the two glasses, sealed together along the cylindrical surface in the middle of the toroid.

For the measurement, a part (about 15 cm long) of the fibre is cut. It is recommended to choose a portion having a diameter of 0.1–0.8 mm, with a difference in diameter between the middle and the ends of maximum 10 per cent. This fibre (arc) is placed on a graph paper (Fig. 2.47b) provided with the proper scales to determine the sag. The graph paper is placed on a mirror. The hatched parts on Fig. 2.47b are areas where the graph paper is cut and exposes, the mirror from underneath. This is useful in order to avoid reading errors due to parallax. By measuring the sag *h*, the expansion difference can be calculated using the formula

$$\alpha_1 - \alpha_2 = \frac{5d \cdot h}{(a^2 + 4h^2) \cdot \Delta t},$$

where d is the average diameter of the glass fibre, a the length of the cord (Fig. 2.47b) and Δt the temperature difference between the Setting Point and room temperature. If $a = 100$ mm (as in Fig. 2.47b), the two glasses have matching expansions if the sag h is less than $0.7/d$. Knowing the expansion coefficient of one of the glasses, that of the other can be determined by this method with an accuracy of about $1 \times 10^{-7} \cdot 1/°C$.

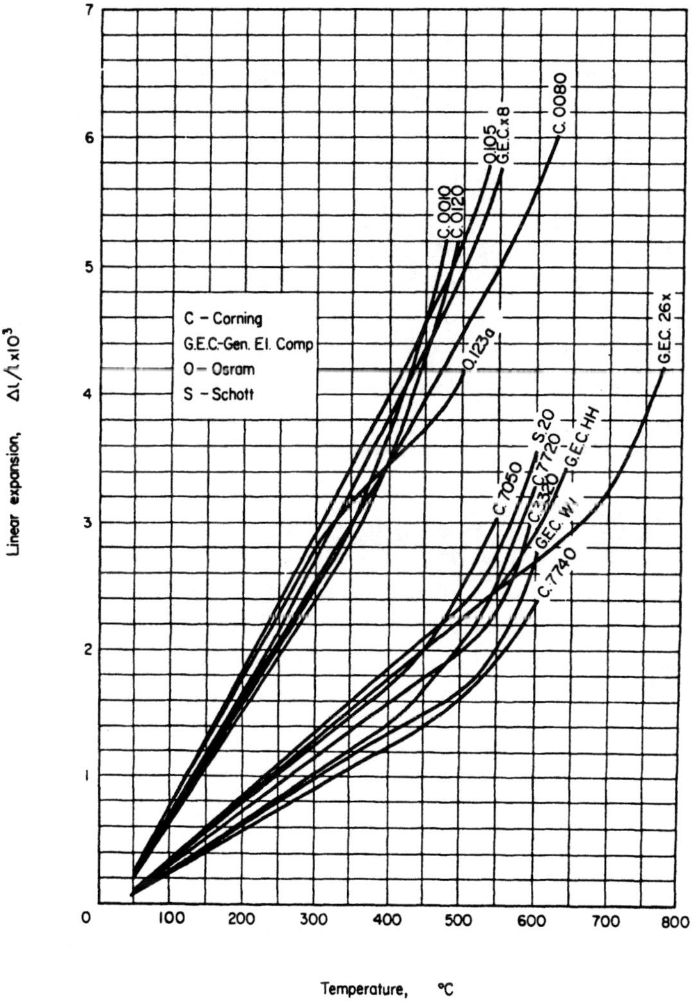

FIG. 2.48 Thermal expansion vs. temperature of various glasses. (Compilation of data from Pask[974], Hull[605], Douglas[297], Espe[350])

For a first estimate of the degree of the expansion matching, it is not enough to compare the tabulated values given for the mean expansion coefficient. If the expansion curves of the two glasses are plotted (as in Fig. 2.48) a comparison gives full information on the matching (Pask[974], Hull[609], Douglas[297]).

23.13 Annealing. Any temperature difference between the various parts of a glass piece, produces stresses, which if strong enough will cause subsequent cracking of the glass (Sections 23.11 and 23.12). The free cooling of a glass assembly is generally too fast to form a stress-free joint. If the glass has been allowed to develop stresses due to too fast a cooling, these may be released by a suitable annealing procedure. The annealing consists in warming up (see also Table 2.11) the glass carefully in a soft flame or in a furnace, to a temperature high enough to allow the release of the stresses by plastic flow. This temperature must be above the Annealing Point of the glass (Fig. 2.45). The heating period is followed by a progressive cooling. The part of the cooling period which is in the Annealing Range (Section 23.11) is carried out slowly enough to ensure that no fresh stress will develop. Below the Strain Point (Fig. 2.45) the cooling can be faster.

The annealing temperature should be chosen so as to require an annealing time, neither too short for proper control, nor too long (Adams[7,8], Reimann[1046], Tool[1232], Preston[1008], Murgatroyd[909], Shand[1124]). In the case of soft glasses and/or large (complex) forms it is advisable to extend the oven annealing time. When the sizes are small and/or the shapes are simple—especially with hard glasses—annealing in a flame is generally enough.

In flame annealing the work should be carefully heated to a temperature high enough to release the stresses (yellow flame for soft glasses, blue flame for hard glasses). The work is then cooled as uniformly and as slowly as possible, using a large flame. The careful cooling continues until the temperature decreases below the Strain Point (Section 23.11); after this it is allowed to cool more rapidly. It is common practice not to take the work out of the flame before it has been covered with a layer of soot. This layer is an indication of the temperature of the glass wherever it is formed, and serves also as a kind of thermal insulation during the cooling.

For furnace annealing, many schedules have been, proposed, varying from very simple rules to complex graphs.

A simple annealing schedule consists in (1) heating the work at the "high temperature" (some degrees above the Annealing Point), and (2) holding it at this temperature for 5–10 min; (3) then cooling the glass from the "high" to the "intermediate" temperature (about midway between Annealing Point and Strain Point) at a rate of 3 °C/min for glass tubing of 0.5 mm wall thickness, at 2 °C/min for 1 mm or at 1 °C/min for 3 mm wall thickness (Osram[958]); (4) cooling down from "intermediate" to "low" temperature (about the

Strain Point) at double the above rates; (5) cooling down from "low" to room temperature at any reasonable rate which can be accomplished without cracking the glass by excessive thermal shock (Table 2.11). The release of the stress can be checked examining the glass in a strain viewer (Section 24.5).

Another simple rule for the cooling rate of glass parts (heated to the Annealing Point) is to cool it down the first 100 °C from this temperature at a rate (°C/min) of less than $20/d^2$ for soft glasses and $100/d^2$ for hard glasses, where d is the wall thickness in mm (Lillie[765a], Morand[886]). For example; for a hard glass, 3 mm thick, the cooling rate will be $100/9 = 11$ °C/min.

The two schedules used in practice to determine more accurately the steps of the annealing are known as the *4-step* and the *5-step* schedules (Redston[1040], Adams[8], Lillie[765a], Kohl[706]). In these schedules the steps are determined by: the thermal expansion coefficient α of the glass, the Annealing Point (An.P.), the Strain Point (St.P.), the thickness of the glass d, and the heating (cooling) geometry (one side or both sides). Other parameters of the annealing, e.g. real shape, initial stress, specific temperature distribution in the furnace, etc., are not taken into account. For details on these see Lillie[765a].

The 4-step and the 5-step schedules are presented schematically in Fig. 2.49.

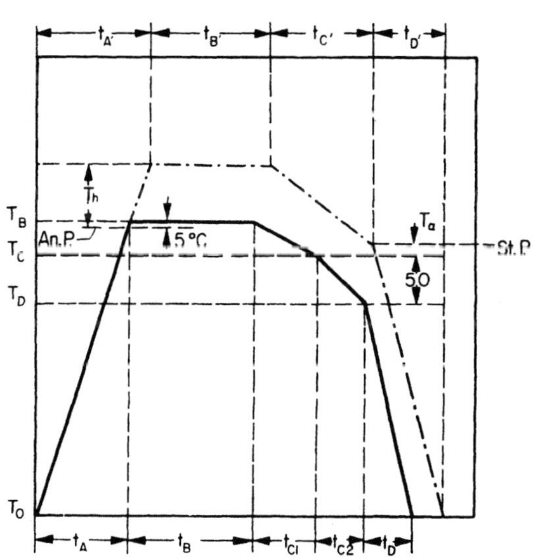

FIG. 2.49 Annealing schedules for glass, in 4 steps (dotted line) and 5 steps (continuous line)

The 4-step schedule (Redston[1040]) consists of:

(1) Heating of the glass at a rate $v_{A'}$ (Table 2.23 and Fig. 2.50) from room temperature T_0 to a temperature $T_{B'}$ higher (or lower) than the Annealing Point with a temperature difference T_h which depends on the cooling rate $v_{C'}$ (Fig. 2.51).

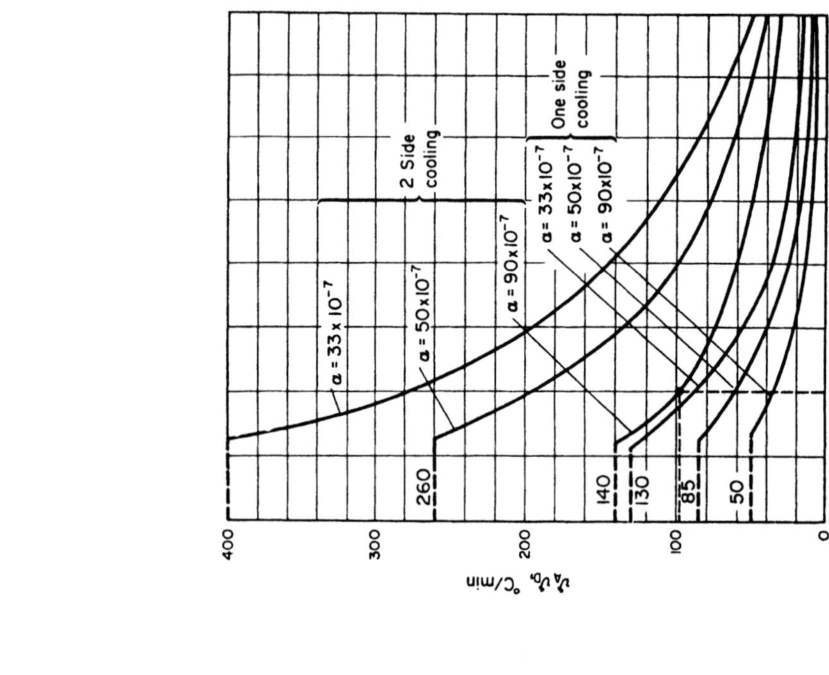

FIG. 2.50 Recommended heating rates for the annealing of glass. (Plotted after data from Redston[1040])

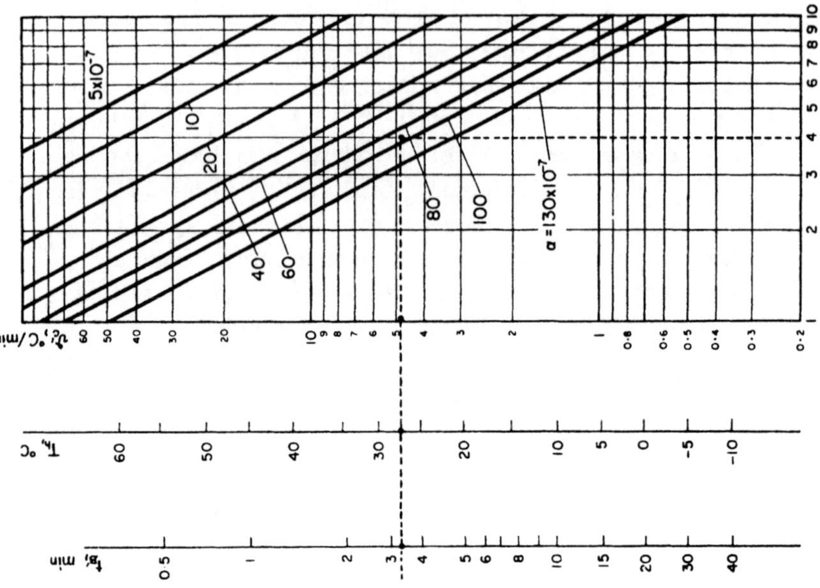

FIG. 2.51 Cooling rate, holding temperature and time in 4-step annealing schedule. (Plotted after data from Redston[1040])

TABLE 2.23. CALCULATION OF THE ANNEALING SCHEDULES FROM FIG. 2.49*

	A'	B'	C'		D'
v	$v_{A'}$ (Fig. 2.50) 95 °C/min	—	$v_{C'}$ (Fig. 2.51) 4.8 °C/min		$v_{D'} = v_{A'}$ (Fig. 2.50) 95 °C/min
T	T_h (Fig. 2.51) 27 °C $T_{B'} = \text{An.P}+T_h$ 428+27 = 455 °C $\Delta T_{B'} = T_{B'}-T_0$ 455−20=435 °C		$\Delta T_{C'} = T_{B'}$–St.P 435–397 = 39 °C		$\Delta T_{D'} =$ St.P–T_0 397–20 = 377 °C
t	$t_{A'} = \Delta T_{B'}/v_{A'}$ 435/95 = 4.6 min	$t_{B'}$ (Fig. 2.51) 4.6 min	$t_{C'} = T_{C'}/v_{C'}$ 39/4.8=8.1 min		$t_{D'} = \Delta T_{D'}/v_{D'}$ 377/95 = 4 min
	A	B	C_1	C_2	D
v	v_A (Fig. 2.50) 95 °C/min	—	v_{C1} (Fig. 2.52) 11 °C/min	v_{C2} (Fig. 2.52) 22 °C/min	$v_D = v_A$ (Fig. 2.50) 95 °C/min
T	$T_B = \text{An.P}+5$ 425+5=430 $\Delta T_B = T_B-T_0$ 430−20 = 410 °C	—	T_α (Fig. 2.52) 6.5 °C $T_{C1} = \text{St.P}-T_\alpha$ 397−6.5=390.5 °C $\Delta T_{C1} = T_B-T_{C1}$ 430 390.5= =39.5 °C	$T_D = T_{C1}-50$ 390.5–50= =340.5 °C	$\Delta T_D = T_D-T_0$ 340.5–20 = = 320.5 °C
t	$t_A = \Delta T_B/v_A$ 410/95 = 4.3 min	t_B (Fig. 2.52) 6.5 min	$t_{C1} = \Delta T_{C1}/v_{C1}$ 39.5/11 = 3.5 min	$t_{C2} = 50/v_{C2}$ 50/22 = 2.3 min	$t_D = \Delta T_D/v_D$ 320.5/95 = 3.8 min

* The numerical values are an example for a glass with $\alpha = 90.10^{-7}$ heater and cooled on both sides; d = 4 mm, St.P = 397 °C, An.P = 415 °C, T_0 = 20 °C

(2) Holding the glass at $T_{B'}$ for a time $t_{B'}$ (Table 2.23, Fig. 2.51).

(3) Cooling it at the rate $v_{C'}$ (Table 2.23, Fig. 2.51) to the Strain Point of the glass.

(4) Cooling the glass at the rate $v_{D'}$ (Table 2.23, Fig. 2.50) to room temperature.

The 5-step schedule (Corning) consists of:

(1) Heating the glass at the rate v_A from room temperature T_0 to a temperature 5 °C above the Annealing Point (Table 2.23).

(2) Holding the glass at T_B for a time t_B (Table 2.23, Fig 2.52)

(3) Cooling the glass at the rate v_{C1} to a temperature T_C within T_α deg. C below the Strain Point (Fig. 2.52).

(4) Cooling the glass at the rate v_{C2} to a temperature T_D, 50 °C lower.

(5) Cooling the glass at a rate v_D (Fig. 2.50) to room temperature.

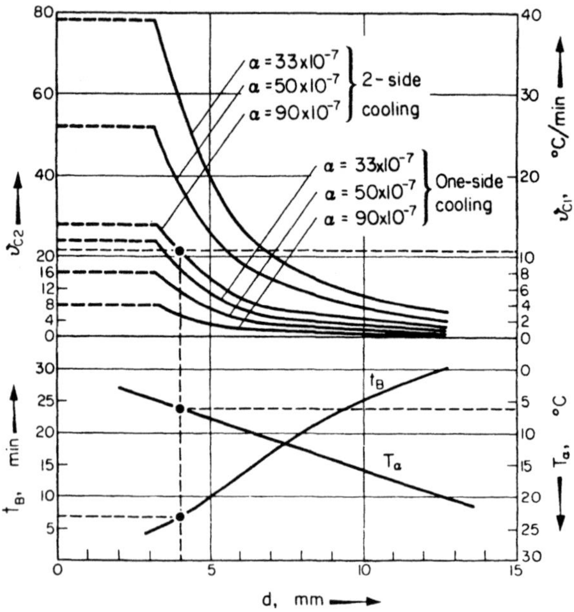

FIG. 2.52 Cooling rates, cooling temperature and holding time in 5-step annealing schedule. (Plotted after data from Corning[239])

The examples presented in Table 2.23 and Figs. 2.50, 2.51, 2.52 show that the total time required for the annealing is:

with the 4-step schedule $t_{A'}+t_{B'}+t_{C'}+t_{D'} = 21.3$ min

and with the 5-step schedule $t_A+t_B+t_{C_1}+t_{C_2}+t_D = 20.4$ min

23.2 Working procedures for glass and quartz

23.21 Cleaning of glass. In order to avoid inclusions in the glass to glass seals and to obtain inside surfaces sufficiently clean for vacuum purposes, the glass (or quartz) should be thoroughly cleaned before sealing.

If any inorganic material is left on the glass it will not burn off in the flame and will become embedded in the glass. This can lead to local strain in the joint or can prevent the two glasses from fusing together properly, leaving pinholes in the joint, or at least producing a matt and rough surface in the joint and near to it.

It must be noted that the rough appearance of the glass is often produced by *weathering* or sometimes by *devitrification.*

A glass, whose surface has undergone changes in composition due to the influence of the atmospheric vapors is said to be *weathered.* The process consists of the hydrolysis of the alkali silicates, forming alkali hydroxides and colloidal silicic acid. The alkali hydroxides react with the carbon dioxide from the air, forming a film of alkali carbonates, with separation of silica. To avoid undue weathering of the glass it is advisable not to store it for too long before use. The glass may be stored in a dry place, and particular care should be taken that the temperature does not fall below the dew point (of water). Wrapping paper sometimes contains materials which can attack glass, but in any case the paper acts as a conveyer of water to the glass surfaces. Thus *glass is best stored unwrapped,* or packed in plastics. When the weathering is not too advanced the matt surface can be cleaned by acid washing.

In the temperature range where the viscosity of the glass is between $10^{2.5}$ and 10^7 poises (Fig. 2.45) crystallization can occur. The process is known as the "*devitrification*" of the glass. The glass loses its transparency and becomes brittle. To avoid the devitrification, glasses must not be cooled too slowly through the crystallization range (Fig. 2.45), and the glass working operations must not be prolonged unduly. The appearance of the devitrified glass cannot be changed by cleaning its surface.

Any material left on the surface of *quartz*, causes, at higher temperatures, its reversion to the crystalline state. Even traces of such contaminations (particularly those containing alkaline metals, as tap water, sweat, etc.) will show up as permanent and unsightly markings or as a foggy haze on the otherwise bright, transparent surface of the quartz. Finger marks always cause such devitrification; the points extend considerably once the quartz is heated. Thus before heating any quartz, a cleaning with methylated spirit or alcohol is required.

As a general rule for glass blowing any glass, even if new and not weathered, should at least be given a superficial cleaning inside and outside. This cleaning is generally done before the heating of the glass. When seals are made for high vacuum purposes, it is recommended to clean the whole assembly again after it has been made (Donaldson[285]).

Normal chemical cleaning procedures are generally satisfactory for cleaning glass tubing before the glass blowing operations. Much of the dirt can be removed by the use of water, a detergent and a long (hair or plastic) brush (burette brush). It is desirable not to scratch the inside of the glass tube, thus it is advisable not to use wire brushes. The use of a brush can be avoided by pushing or pulling a piece of soft cloth or a cotton wool plug through the tube (Strong[1207]).

Glass vessels and pipes can be washed by immersing them in hydrochloric acid solution (1—5 per cent) for 3–10 sec (if they are not too dirty), followed by a subsequent rinsing in water (40–50 °C) and drying (Laug[744], Roth[1083], Angerer[34]). Tap water if quickly dried leaves salts on the surfaces. To remove these salts a second rinsing in distilled (or demineralized) water is recommended before drying (Hühn[608]).

For the washing of soda-lime glasses (Table 2.10) an acetic acid solution (3–5 per cent) is recommended instead of hydrochloric acid.

Chromic acid is satisfactory for cleaning glass (subsequent to a washing in water) provided that care is taken to make sure that the glass is free of mercury* (Weyl[1301]). The usual laboratory cleaning solution, known as chromic acid contains about 50 ml of saturated aqueous sodium dichromate in a litre of concentrated sulphuric acid. Another prescription for making up chromic acid is to use 100 g chromium trioxide in 30 ml. sulphuric acid (1.84) and to dilute it with water up to 1000 ml. (Rain[1024]). The chromic acid solution should be used only if it has its brown colour. If the colour is changed the solution is decomposed. After cleaning with chromic acid a thorough washing in water is necessary.

A solution which is much more effective than the chromic acid solution, consists of 5 per cent HF with 33 per cent HNO_3 and 2 per cent Teepol in 60 per cent water. The solution should be used cold (Crawley[245]).

Very dirty glass parts may be washed with a solution of NaOH or KOH, followed by washing with water and chromic acid. This procedure is not to be used for quartz.

Old glass (weathered) or very dirty quartz can be washed with hydrofluoric acid solution (40 per cent in volumes) by an immersion of 1–5 min. The glass is superficially attacked by this solution, but it remains smooth. A subsequent washing with (distilled) water, and neutralizing in NaOH solution is absolutely necessary. The final washing is made in distilled water (40 °C) and alcohol.

To remove very adherent dirt, glass balls (1–2 mm diameter) can be used with distilled water.

Organic solvents are adequate to remove *grease* from the inner surfaces of the glass tubes or vessels, where no other dirt is present. Silicone grease can be washed from glass surfaces by using dichlorethylene or kerosene with subsequent washing with a solution of 10g NaOH and 5g borax in 100 ml.

* If mercury is present a residue is precipitated with the chromic acid; this residue is difficult to remove.

distilled water or a solution of 10–15 ml. KOH (50 per cent) in 100 ml. ethyl alcohol (maximum immersion 10 min; see Roeben[1074]).

The *drying* of washed glass parts can be done by hot air. Care must be taken to see that the air (when obtained from a compressor or blower) is free of oil. While acetone or alcohol may be useful for drying glass parts, care should be taken that liquid or vapour from these solvents does not remain inside the glass when glass-blowing is commenced, because of the risk of an explosion (Scheflan[1103]).

In order to obtain extremely clean glass surfaces for very high vacuum, vacuum evaporation, adsorption studies, etc. a glow discharge is used (Varadi[1256], Dillon[280a], Holland[579, 584a]).

23.22 Cutting methods for glass. Glass tubing up to about 15 mm in diameter can be broken by scratching the surface (perpendicular to the axis) with a knife* and then breaking the tube by pressing forwards the thumbs and pulling the hands slightly apart (Fig. 2.53).

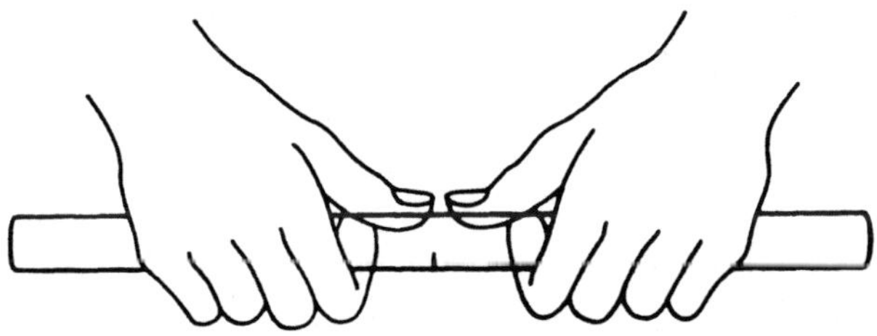

FIG. 2.53 Method of breaking thin glass tubes

It is recommended to make the scratch on the tubing with only one stroke. It is not necessary to try and cut through the wall of the tubing, since the knife is soon blunted when it is used like a saw (Strong[1207]). When scratching, it is better to hold the tube in the hands than to lay it on a plane (table).

Glass tubing having diameters in the range 15–30 mm can be cut by scratching strongly and placing it with the scratch uppermost, so that it rests on a copper wire (about 3 mm diameter) opposite and parallel to the scratch. The scratch is moistened and the ends of the glass tube are forced down strongly while the tube is balanced on the copper wire. This way a square break is obtained, especially when the tube has moderately thin walls (Robertson[1072]).

Glass tubes 5–30 mm in diameter can be cut by the *hot spot* technique. For this, a fairly deep and about 10 mm long scratch is made in the glass. The scratch is wet and the hot end of a glass rod (a few millimetres in diameter) is placed on the end of the scratch. The glass may crack right away, or it may

* It is recommended that only sharp knives be used; the knife should not be used for the forming of glass in the flame since this encourages loss of hardness.

crack when the hot spot is removed and damp cotton wool is applied to the glass. Usually with hard glasses (Table 2.10) the crack goes only part of the way round the tube; soft glasses often crack completely. When the crack stops, it can be continued by placing another hot spot against its end. This is repeated until the ends of the cracks meet. With borosilicate glasses the crack usually wanders, and the cut obtained is rarely square. A pin-point burner (Section 23.23) can be used instead of a hot spot.

A square cut is more easily obtained by the *hot wire* method. The glass tubing is first scratched right round. This may be done by hand, a line being first drawn around the glass with a wax pencil, and the scratch made along the pencil mark (checked if perpendicular to the axis). A scratch around the outside of the tube may also be made by rotating the tube against a rotating steel or carborundum wheel. An internal scratch may be made with a diamond, mounted on a long handle which can be inserted in the tube. In order to cut large number of glass pieces to accurate dimensions, a diamond cutter can be used. This consists of two parallel rollers on which the glass tube can be rotated about its axis, and a diamond-tipped peg carried on an arm which projects inside the glass tube to be cut. The setting of the diamond relative to the glass wall is critical: it must be perpendicular to the glass. The pressure on the diamond during cutting should be applied by hand and not mechanically (Breadner[163]).

After scratching (0.5–1.0 mm) nichrome wire is wrapped once around the scratch. When external it is recommended to wet the scratch. The wire is heated electrically to red hot (about 750 °C) and kept taut round the glass. The cutting device should be as shown on Fig. 2.54a so that when the tube is lowered the wire becomes slack (as in Fig. 2.54b) allowing the tube to be

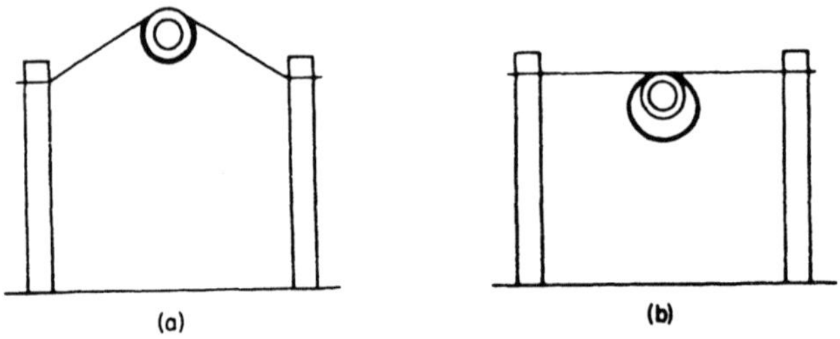

FIG. 2.54 Method of cutting glass tubes, with a hot wire

removed rapidly from the wire loop (Reimann[1046]). The tubing may crack immediately; if not it should be induced to crack by touching the scratch with wet asbestos after the hot wire has been in contact with the glass for about half a minute.

Glass bottles must be first heavily scratched all the way around, and then a very small flame is directed onto the scratch. A crack should start, and it can then be led around by the flame. Another method of cutting glass bottles consists of wrapping strips of wet blotting-paper or asbestos paper round both sides of the place where the crack is desired, leaving a gap of about 5 mm wide. Directing the flame onto this gap a square cut can be obtained. An alternative method of cutting large bottles is to spin them on a turn-table and to have a pin-point flame (Section 23.23) playing at the level of a small initial scratch. After a few minutes the flame is removed and a piece of wet cotton wool is pressed on the scratch.

Holes can be *drilled* in glass using hardened tools. Small holes can be drilled using triangular borers with turpentine oil lubrication. Larger holes are better drilled using copper tubes with sharp edges (preferably slotted longitudinally to pick up the abrasive) and silicon carbide suspension in water (Steyskal[1189], Ludwig[788], Breadner[163]). For accurate drilling of holes in glass ultrasonic techniques with boron carbide are used (Nepiras[918]). Threads can be cut on glass using thin diamond blades (Klein[686]).

Quartz tubes are cut by first grooving all round with silicon carbide or copper bonded diamond wheels (Eisler[335]). Then cuts are made right through the wall at three or four points around the periphery, leaving a series of connexions which hold the parts together. Finally these connexions are cut carefully to separate the two parts. The cut ends are then dressed on the side of a suitable wheel.

For drilling quartz, either copper tube drills are used with silicon carbide abrasive, or impregnated diamond drills are utilized. In both cases the cutting point should be cooled with water. To avoid breakage at the back surface either the drilling is performed from both sides or a glass plate is glued behind the plate to be drilled and the hole is continued in this plate too.

23.23 Heating methods for glass to glass sealing. The most important "tool" in glass to glass sealing and generally for glass blowing work is *the flame.*

The flame is produced by the burning of the combustible gas (or gas mixture) with the oxygen supplied in the burner and with that from the atmosphere. A flame consists of three conical zones, which are more or less distinct depending on the shape of the flame.

The inside zone *1* (Fig. 2.55) contains heated gas and air. In this zone no burning takes place. On the outer shell of this zone, a distinct blue zone can generally be seen. This layer *2* is the place where the reaction (burning) occurs between the combustible gas and the "primary" oxygen (i.e. the oxygen which flows from the burner, ready mixed with the gas). The third and external part of the flame *3* (Fig. 2.55) contains the products of the burning and the remaining unburnt gas, both at a high temperature. Here the remaining gases are burnt consuming oxygen from the surrounding atmosphere as well (secondary oxygen).

The temperature in the flame increases from the inside zone towards the outside zone. The coldest point of the flame is at the tip of the inner zone. The hottest point is at about 3–6 mm from the tip of the second zone (Fig. 2.55).

The various parts of the flame have not only different temperatures but also different chemical behaviour. The inside zone has reducing properties since it contains hydrogen and carbon monoxide at high temperatures. Due to the reducing character of this zone, lead glasses (Table 2.10) introduced in this part of the flame become black (the lead oxides are reduced on the surface

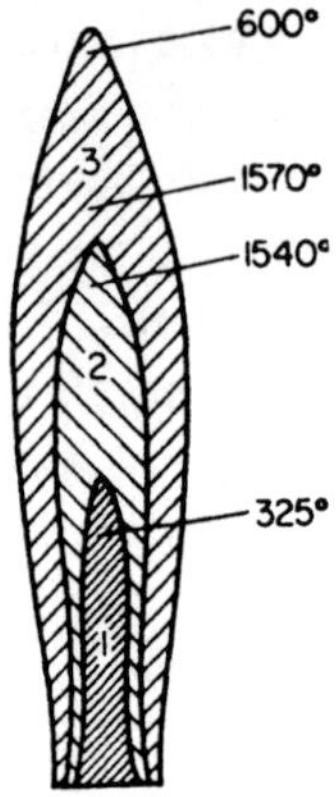

FIG. 2.55 The flame

to lead). The outer part (zone *3*, Fig. 2.55) of the flame, contains hot oxygen thus it has an oxidizing action. Lead glass blackened in the inside zone of the flame, will discolour if introduced in the outer part of the flame.

The reducing or oxidizing character of a flame depends also on the ratio of gas to air (oxygen). In the case of excessive "primary" air (oxygen) the flame is more oxidizing (normal air required for burning see Table 2.24). If the flame has less air (oxygen) than normal, it will be of a reducing character. It is obvious that oxidizing flames are always hotter than the reducing ones.

For lead glasses oxidizing flames are to be used, and for hard glasses reducing flames are recommended.

The length of the flame (L) depends on the diameter of the orifice in the burner (d), the rate of flow* (v) of the gas–air mixture and on the flame velocity (w). When the combustion is complete and the mixture of burning gas to air is the theoretically correct one, the length of the resulting flame is given by the formula

$$L = \frac{d}{2}\left[\frac{v^2 - w^2}{w^2}\right]^{1/2},$$

* v is the ratio between the volume of gas entering the flame per unit time and the area on which the burning occurs.

where L, and d are expressed in mm and v and w are in m/sec. The orifice should have a larger diameter than a critical value, since the flame propagation is impossible in a coal gas–air mixture when it is less than 2 mm, in methane–air mixture less than 3.6 mm and for hydrogen less than 0.9 mm.

The combustion characteristics of the commonly used gases are given in Table 2.24*.

TABLE 2.24. COMBUSTIBLE GASES

Property	Carbon monoxide	Hydrogen	Methane	Butane	Water gas	Coal gas	Natural gas
Composition (%):							
Hydrogen	—	100	—	—	50	46	7
Carbon monoxide	100	—	—	—	40	10	2
Methane	—	—	100	—	1	33	40
Hydrocarbons (various)	—	—	—	—	—	5	48
Carbon dioxide	—	—	—	—	5	3	—
Nitrogen	—	—	—	—	4	3	3
Air required for complete combustion of m^3 gas (m^3)	2.4	2.4	9.4	31.1	2.2	5.1	10.4
Burning products of m^3 gas (m^3)	2.9	2.9	10.4	—	2.9	6.2	11.4
Specific gravity (comp. to air)	0.97	0.07	0.55	0.58	0.52	0.42	0.85
Heat of combustion (kcal/kg)	3000	2550	8500	11 000	2500	4900	10 000
Theoretical temp. of flame (°C)	2030	1970	1830	1 973	2100	1930	2140
Flame velocity (m/sec)	1.2	4.9	0.7	—	3.0	0.9	0.8
Ignition temp. (in air) (°C)	650	585	715	550	680	630	640
Lower limit of flammability (% vol).	16	8	5	1.5	12	6	4
Upper limit of flammability (% vol.)	71	71	14	8.5	67	27	16

Theoretically the maximum temperature in a given flame is obtained when the air (oxygen) added to the flame is exactly that required for complete

* For flaring and stem sealing (Sections 23.3, 26.1, and 42.2) it is recommended to add sulphur dioxide (Richardson[1055]).

combustion (Table 2.24). In practice to obtain complete combustion of the gas, some excess air is needed.

Temperatures of 700 to 1150 °C can be obtained with the usual combustible gases and air; for temperatures up to 1700 °C oxygen must be added in the flame and if temperatures up to 2000 °C are required hydrogen plus oxygen should be used.

The shape of the flame is determined by the construction of the burner, the type of the gas used and the pressure (rate of flow). With the appropriate design, strong or weak, sharp or rounded, hard or soft, long or short flames can be produced. Their shape is determined by the work to be done. For manual work, hand torches, or blowing burners are used. For mechanical work the burners are mounted on glass blowing lathes, sealing machines, stem making machines etc., (Breadner[164], Richardson[1055], Espe[353, 354, 356], Dartnell[260], Eisler[335], Banki[79], Davies[261a], Kalpers[651], Roth[1083]).

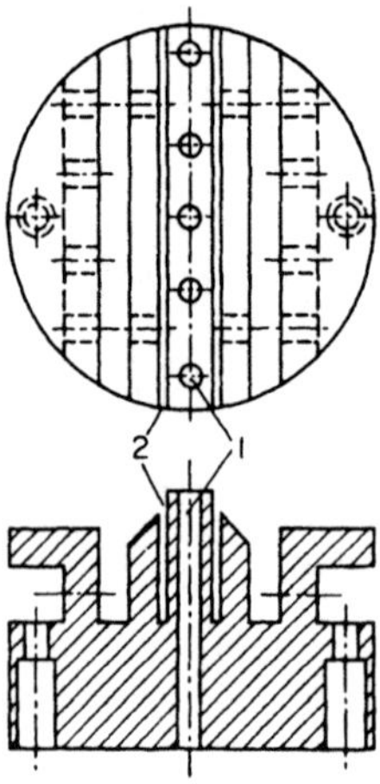

FIG. 2.56 Marshall burner

The *burner* usually used in glass blowing and glass sealing work is the Marshall burner and the pin-point burner. The Marshall burner is made in two halves, the burner head and the base containing the gas connection. The head consists of a tube which acts as a housing for the piece of rod in which the flame ports are machined. There are two sets of flame ports, the drilled holes (1, Fig. 2.56) through which the bulk of the gas/air mixture passes directly, and the slots (2, Fig. 2.56) on each side of this row of holes. The gas/air mixture reaches these slots at a low velocity after passing through a series of small orifices which are designed to restrict the flow of gas. The purpose of these slots is to stabilize the flames from the main orifices by providing a low pressure pilot flame on each side of the high pressure flames. In the housing of the burner two or more sheets of brass (or nickel) wire gauze (100 mesh, holes of about 1 mm) are secured by locking rings. The flames of the standard Marshall burners consist of long, thin jets with inner cones about 2–3 cm long. These jets are so close together as to form a single flat flame. To obtain much shorter flames the shrouded Marshall burners are used. In these burners the low pressure gas–air mixture which passes to the slots, is fed through a coiled strip of 100 mesh wire (Monel) gauze; the gas should pass here through twelve layers of tightly packed gauze forming the coiled strip.

The *pin-point* burner head has a central flame port with a diameter of about 1.2 mm surrounded by a number (e.g. twelve) of secondary flame ports (about 1 mm bore) through which the air–gas mixture flows at a low velocity, after passing a smaller hole (about 0.5 mm). This burner gives long thin flames.

For special applications, glass to glass seals can be made using electrical heating methods: resistance welding, high frequency induction or dielectric heating. For glass butt seals where the heating of a wide zone is to be avoided, a sort of *resistance welding* can be used (Guyer[493], Shaw[1127]). This consists in using two torches with sharp flames, placed opposite to each other on the two sides of the glass pipe. After the edges of the pipes which are to be joined have been first heated with the flame, a high frequency current (1000 V, 10^5–10^7 cycles) is passed through the torches. The gases from the flames are now used as current lead-throughs, and the preheated glass edges have enough conductivity (Fig. 4.1) to act as a resistor. The edges are heated quickly by the current, to a high temperature (1000 °C) without heating the rest of the glass tube. According to Courant[240a] the time required for the sealing by this method is only a tenth to a hundredth from that required in flame sealing. It must be noted that if the two glasses to be sealed have different electrical conductivities, the part having the higher resistance must be preheated at a higher temperature to obtain the conductivity required for the further electrical heating.

Instead of using flames, a *spark discharge* can be used, directed along the surface of the glass, in the region where the heating is required. When the resistance of the glass becomes smaller than that of the air gap along the surface, the discharge path will move into the body of the glass producing a uniform heating (Guyer[492], Whitney[1309]).

Descarsin[275a] describes a method of glass to glass butt sealing which can be completed even in vacuum. The glass edges are first preheated to about 700 °C and then for further heating the *dielectric losses* (Fig. 4.3) in the glass are used. For this heating method the seal is placed in a condenser of a high frequency generator (about 2 kW, 10^7–10^8 cycles).

Butt seals can be made by heating indirectly the ends of the glass tubes using metallic or carbon rings placed around the parts and heated by high frequency induction heating (Violet[1269]) or by inserting between the ends a very thin copper, silver, platinum or nickel foil (about 0.025 mm thick) and heating it by using its electric resistance (Zincke[1354]).

23.3 Glass to glass (and quartz) sealing techniques

Glass to glass sealing techniques are performed by the glass blower, working manually or using mechanized tools (glass blowing lathes, sealing machines, etc). The detailed description of all the operations carried out by the glass blower in his work exceeds the scope of this Section. For a detailed description of these techniques the reader is referred to Barr[81], Bachman[65], Frost[408], Friedrichs[406], Heldman[534], Robertson[1072], Strong[1207], Wheeler[1303], Reimann[1046], Parr[971], Jebsen[628a].

Together with the basic principles and methods of glass blowing techniques used for glass to glass sealing, only the common pitfalls in perfecting these seals and the necessary safety precautions will be given here.

(1) It is recommended that glass forming tools only be used for their specific purpose. For the flaring of glass or for forming it, use brass forceps, triangular tools or carbon plates, but never the glass cutting knives; the forming is inadequate and the knife is destroyed by the heating.

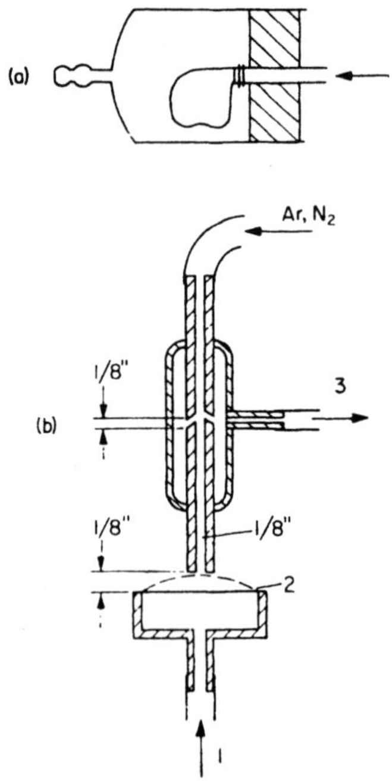

FIG. 2.57 Blowing devices; (a) for preventing contamination of the vessel; (b) for preventing contamination by radioactive or toxic materials from the glass parts. After Bleecher[135] *(Courtesy of The American Institute of Physics)*

For blowing the glass appropriate tubes should be used. A blow tube consists of 0.5–1.0 metre of rubber tubing (about 6–8 mm o.d.), having a mouth piece fitted on one end and the other end having an adapter to facilitate the connexion to the work. To enable the rotation of the blow tube without twisting it, a swivel joint is provided on the tube. For blowing into glass apparatus and to avoid that vapours from the glass blower reach the apparatus, or to prevent vapours from the work flowing towards the mouth of the glass blower, an intermediate vessel with a rubber or plastic bag (Fig. 2.57a) is to be used. If contaminated glass parts are to be blown (for repair) the device shown in Fig. 2.57b can be used (Bleecher[135]). When blowing through the mouth piece 1, pressure is exerted on the Mylar diaphragm 2 (about 25 μ thick), which alters the position of the diaphragm relative to the gas exit port directly opposite. Inert gas is admitted into the upper device, and the diaphragm acts as a valve to control the flow (pressure) of the inert gas through the exit 3, connected to the work.

For hard glass and especially for silica fusing works protective spectacles should be worn by the operator (a green glass is recommended).

(2) The glass parts should be handled with care. Place the glass parts only on a clean table covered with non-conducting material (wood, carbon, etc.), never in the vicinity of tools (hammer, screw driver, etc.). The contact of the hot glass with conducting materials (metals) cools it locally very rapidly, producing stresses. The hand tools can scratch the glass, which afterwards leads to breakages.

(3) The material used should correspond to the shape of the part to be made. When cutting the glass, leave plenty of glass for handles. For glass to glass seals use elements having about the same diameter and wall thickness.

(4) The shaping of the fused glass should be done by using successively the flame and the influence of both gravity and surface tension. Do not work the glass just at the softening point (Section 23.11); better work can be done when the glass is really hot. Make certain that the glass is heated uniformly in the flame of the burner. In making seals with tubing it is necessary to rotate the glass in the hottest zone (Fig. 2.55) of the flame. The rotating operation should be executed uniformly and with a good co-ordination of the two hands. The rotation of the glass pieces must continue even after having removed them from the flame, since the lower surfaces cool faster than the upper ones, as the latter ones are heated by the hot air rising from the parts below.

It is desirable that a minimum of glass be heated and fused in completing seals, because it is difficult to make perfect seals with wall thicknesses identical to that of the original glass (Barr[81]). If a large portion of the glass is deformed by fusion it will require more effort to build up a smooth seal.

When sealing lead glasses (Table 2.10) care should be taken to keep the work in an oxidizing flame (Section 23.23) to avoid blackening.

A controlled amount of sulphur dioxide added to the gas used, increases the resistance of the glass surface against the development of small cracks. This is especially useful where the glass is flared (Richardson[1055]).

23.31 Butt seals. In order to butt seal two glass tubes with *identical diameters* using a torch (burner) fixed to the working table, the two parts to be joined (having the ends cut perpendicularly to the axis) are brought simultaneously into the flame. They are rotated in the flame, initially at a distance of about one centimetre (Fig. 2.58a.1), but as the temperature of the ends rises, the

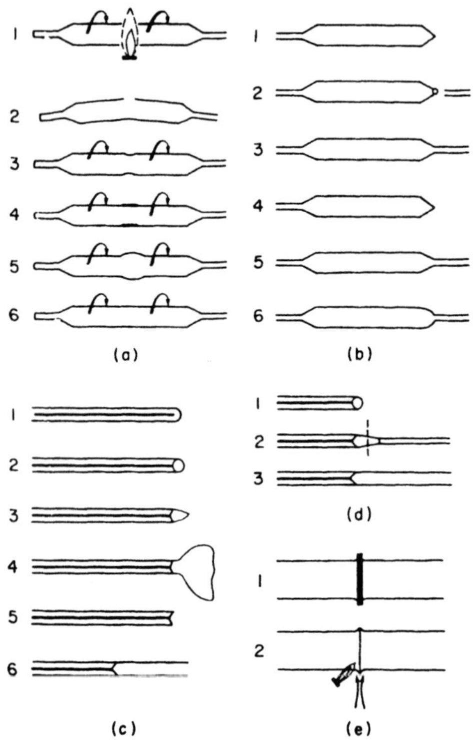

FIG. 2.58 Butt sealing of glass tubes

parts are brought closer together. After the circumferences of the ends are fused (bright yellow), the two tubes are brought in contact at an angle (Fig. 2.58a.2) and immediately the rotation is continued in a synchronized manner (Fig. 2.58a.3). The glass shrinks inwards and is allowed to thicken up to twice the normal wall thickness (Fig. 2.58a.4). In this phase some glassblowers take the work out of the flame and hold it in a vertical position to prevent sagging. After a slight pulling, the sealed place is enlarged by blowing (Fig. 2.58a.5), then drawn to a uniform diameter (Fig. 2.58a.6).

To butt seal two glass tubes of *different diameters*, the ends are initially worked to equal diameters. If one of the tubes has a very small diameter, the large diameter tube is first closed with a heavy wall cone (Fig. 2.58b.1), the apex of the cone is then heated, fused, and blown to a small bulb (Fig. 2.58b.2). This bulb is opened afterwards to the diameter of the small bore tube. The small tube is eventually flared and sealed to the other tube, with a tapered shoulder. If the difference in diameters is not too big, the flaring of the small tube is enough to make the seal. An alternative technique for the sealing of two glass tubes with very different diameters, consists in making first a light wall cone (Fig. 2.58b.4) on the end of the larger tube; This cone is opened, the small diameter tube is sealed to it Fig. 2.58b.5); and the walls of the cone are then thickened (by working it in the flame), bringing them to a round shoulder shape (Fig. 2.58b.6).

If a *capillary glass tube* is to be butt sealed to an other glass tube, the end of the capillary is first heated until the bore is closed (Fig. 2.58c.1), then the fused end is blown to form a small bulb (Fig. 2.58c.2). This bulb is heated (rotated) in the flame, fused at its outer end (Fig. 2.58c.3) and blown out (Fig. 2.58c.4). The excess glass is scraped away, and the end is fire polished (Fig. 2.58c.5). To this end the other tube can be butt sealed (Fig. 2.58c.6). An alternative technique is to fuse the end of the capillary tube, and to blow a little bulb (Fig. 2.58d.1). A glass rod is afterwards attached to the bulb in order to draw it out (Fig. 2.58d.2). It is cut at the required distance and then can be sealed to the other tube (Fig. 2.58d.3).

When the butt seal should be made on a *fixed system*, the same procedure is applied (Fig. 2.58) but instead of rotating the two glass parts simultaneously, the torch is moved around the glass so as to give uniform heating. It can be recommended to arrange the glass tube vertically since, in a horizontal butt seal, without rotation of the glass, the bottom wall will be always heavier due to the sagging of the tube and the running of the glass due to the gravity. A solution to compensate for the weight of the parts attached to glass systems, by means of a spring suspension is described by Richard[1051].

Large tubes must be flared first, if they are to be butt sealed without being rotated. After fitting together the flared ends (Fig. 2.58e.1) a section of the circumference is heated at a time, and the sealing is achieved by applying pressure with forceps (Fig. 2.58e.2). This is done around the circumference and subsequently the joint is locally softened — on a small section at a time of the circumference — and is worked by alternate shrinking and blowing until the wall is smooth. Finally the whole circumference is uniformly heated, aligned and annealed.

23.32 Side seals. For side seals (Fig. 2.59) the end of the side tube is prepared as for butt seals (Section 23.31). If a medium size side tube is to be sealed to the main tube, the procedure shown in Fig. 2.59a can be used for the side seal. On the main tube, at the point where the side seal is to be made, the surface is heated (Fig. 2.59a.1) and with a pin-point flame a circular area of the same diameter with the side tube is softened. A bulge is blown at this place (Fig. 2.59a.2), which is heated at the tip and blown out to a large bubble (Fig.

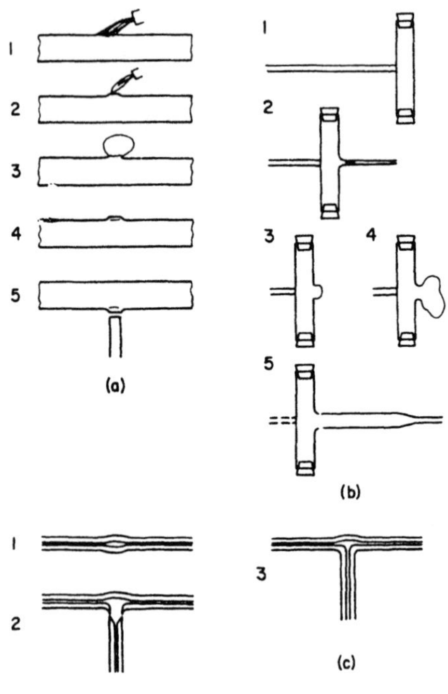

FIG. 2.59 Side sealing of glass tubes

2.59a.3). When the thin wall of the bubble is broken, a proper end to the side seal remains (Fig. 2.59a.4), which, after fire polishing, is sealed to the side tube (Fig. 2.59a.5).

If the side tube has about the same diameter as the main tube, to make the seal, a glass rod is first attached (Fig. 2.59b.1) at the point opposite to the place where the side tube will be attached, or a holding tool is used (Fig. 2.59b.2). The main tube is heated exactly opposite to the attached rod (or the handle of the holding tool) and the fused glass is drawn out with another glass rod (Fig. 2.59b.2). After this glass rod is cut, the drawn place is blown out first as in Fig 2.59b.3, afterwards as in b.4. The thin glass bubble is cut, and the sides of the remaining opening are fire polished. The side seal is completed (Fig. 2.59b.5) and the first rod is removed.

When the side seal is to be made on a *capillary* tube, this is rotated in the flame and the place where the seal should be is blown to a larger diameter (Fig. 2.59c.1). The place for the side seal is heated, drawn out to a cone (with a glass rod) and opened by blowing out (Fig. 2.59c.2). To this opening the side arm is attached (Fig. 2.59c.3).

23.33 End seals. To close the end of a glass tube, first a point is drawn at the end of the tube (Fig. 2.60–1) then the walls are made uniform (Fig. 2.60–2), the drawn point is cut with a flame, blown out to a small bulge (Fig. 2.60–4) then to a round end (Fig. 2.60–5).

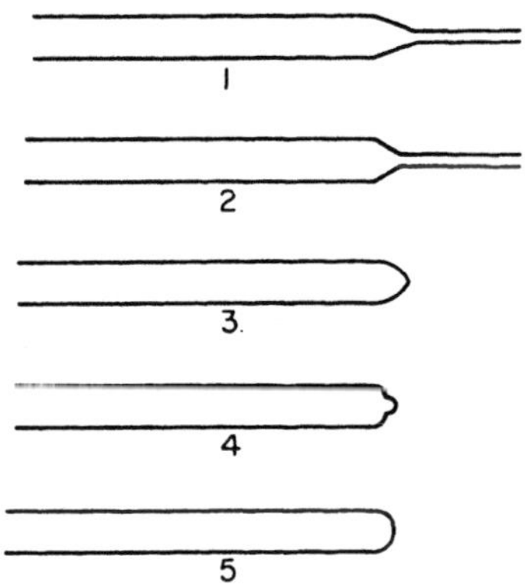

FIG. 2.60 End sealing of glass tubes

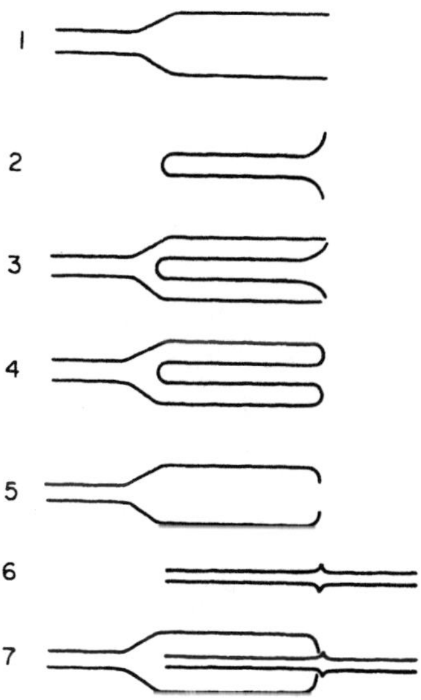

FIG. 2.61 Ring sealing of glass tubes

23.34 Ring seals. These are joints consisting of a smaller diameter tube sealed inside a larger one. The small diameter tube is flared to fit closely in the larger one (Fig. 2.61–1, 2) and then this tube is sealed inside the other (Fig. 2.61–3, 4). If the diameter of the inner tube is small compared with the outer one, the edges of the larger tube are bent inwards (Fig. 2.61–5). On the inner tube an annular protrusion is completed (Fig. 2.61–6) and the two tubes are sealed to each other.

23.35 Graded seals. Glasses of widely different expansion coefficients can be joined by graded seals. These seals consist of a number of segments of glass, having progressively slightly different expansion coefficients (see Fig. 2.46 B). Together the segments form a zone of gradual transition between high and low expansion (Riley[1062], Pequignot[982], Fraenkel[395]).

Lewin[761] calculated the stresses appearing in a three component graded seal (Fig. 2.62) and Rawson[1030] measured the stresses in such seals. From these studies it results that if the central tube is short the stresses in tube (3) are increased. Hence attention has to be paid to the length of the intermediate tube (2), especially if the diameter is large. It is believed that the ignorance of this fact explains the assumption, common among glass blowers, that large diameter graded seals are more prone to cracking.

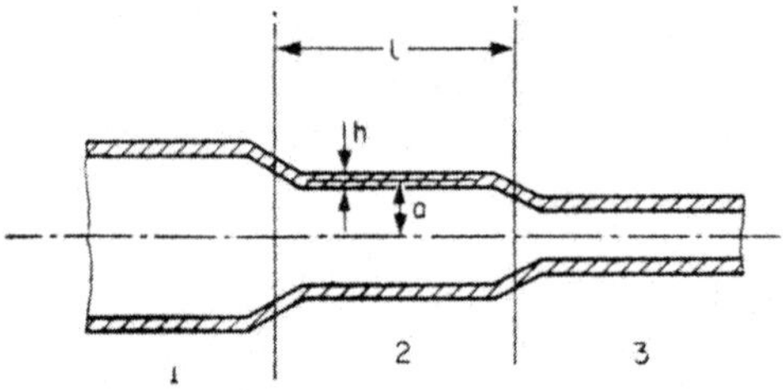

FIG. 2.62 Three-component glass seal

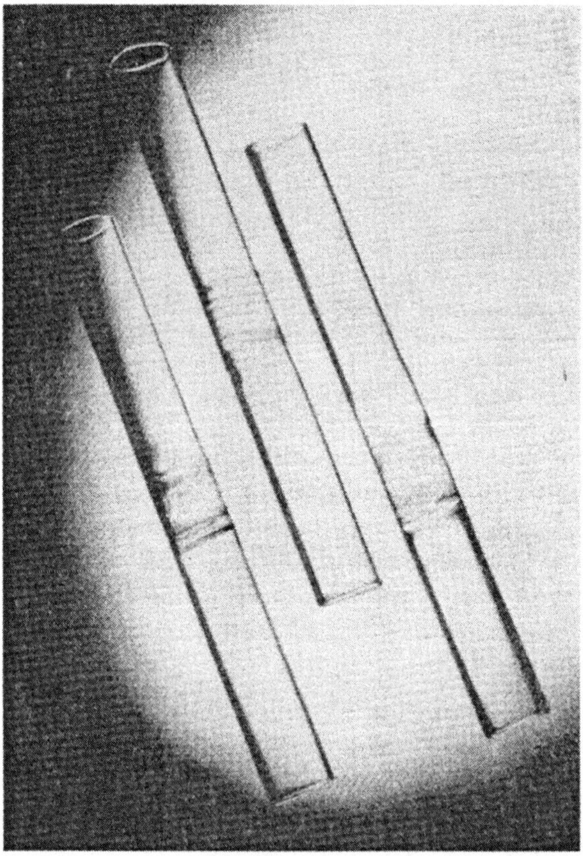

PLATE 1. Graded seals for joining Kovar-type borosilicate glasses to Pyrex *(Courtesy of Leybold*[762]*)*

TABLE 2.25. GLASS SERIES FOR GRADED SEALS

Schott**			G.E.C.**			B.T.H.**			Sovirel**			Corning**		
Glass	$\alpha \cdot 10^7$ (20–100 °C)	T_g^* (°C)	Glass	$\alpha \cdot 10^7$ (20–350 °C)	An.P.* (°C)	Glass	$\alpha \cdot 10^7$ (20–350 °C)	An.P.* (°C)	Glass	$\alpha \cdot 10^7$ (50–300 °C)	T_g^* (°C)	Glass	$\alpha \cdot 10^7$ (0–300 °C)	An.P.* (°C)
C 1	8	900	SiO_2	5	—	—	—	—	—	—	—	Silica	6	910
C 2	9	820	WQ 31	10	750	—	—	—	—	—	—	7900	8	910
C 3	12	760	—	—	—	R 48	13	—	—	—	—	GS 1	12	—
C 4	15	725	—	—	—	R 49	18	—	—	—	—	7230	14	750
C 5	21	663	WQ 34	21	700	R 50	23	—	—	—	—	7200	19	645
C 6	26	640	—	—	—	—	—	—	—	—	—	GS 4	28	—
C 7	30	562	H 428	32	800	—	—	—	—	—	—	7740	32	555
C 8	35	542	Pyrex	35	580	C 9	36	530	—	—	—	7720	36	518
C 9	37	534	W 1	37	560	C 11	45	585	CD 47	47	520	3320	40	535
C 10	48	558	HH	45	590	C 40	48	520	CD 50	50	540	7052	46	475
C 11	50	514	FCN	47	520	—	—	—	CD 53	53	540	7510	50	—
C 12	61	551	GS 1	54	600	—	—	—	CD 59	59	527	7520	61	566
C 13	62	562	GS 34	60	600	C 42	62	570	CD 63	63	527	—	—	—
C 14	73	555	M 6	68	620	C 43	72	550	CD 70	70	560	7530	71	—
C 15	73	548	GS 56	79	580	—	—	—	CD 72	72	520	7550	79	—
C 16	80	505	GS 67/L	87	520	C 44	84	530	CD 80	80	570	7560	86	536
—	—	—	L 1	91	435	C 12	87	430	CD 84	84	560	0010	91	428
—	—	—	X 4	96	520	C 19	93	550	CD 88	88	560	—	—	—
—	—	—	X 7	101	520	C 22	104	530	CD 92	92	510	—	—	—
—	—	—	—	—	—	—	—	—	CD 97	97	557	—	—	—

* T_g, An.P. see Fig. 2.45
** See Table 2.10

The minimum length of each tube section in a graded seal should be

$$L \approx 0.85(a \cdot h)^{0.5},$$

where a is the (mean) radius of the section (see Fig. 2.62) and h the wall thickness. In practice the length is 1/5 of the diameter (Zincke[1354]) and is very often even equal to the diameter.

Larger diameter graded seals often fail, because of inadequate annealing. Any graded seal should be oven annealed (Section 23.13) preferably through the annealing range (Figs. 2.45, and 2.49) of all the glasses used in the seal. The maximum temperature during the annealing should be 20 °C above the annealing point of the most refractory glass used in the graded seal.

Graded seals are readily available in various dimensions. A list of glasses used successively in graded seals by various manufacturers is given in Table 2.25. Plate 1 presents some graded seals joining Kovar type borosilicate glasses to Pyrex.

Using graded seals quartz can be sealed to hard or soft glasses (Table 2.25). This technique gives one of the methods to seal lead-in wires into quartz vessels (Section 24.42) or quartz windows to glass vessels (Section 71.31).

It seems that with the development of the sintered glass it is possible to build glass pieces having gradually increasing expansion cofficients (Knecht[696]). A graded seal can be made of powdered glasses pressed in layers, each of which has a different expansion coefficient, and fired to form a composite glass tube (Gleason[452]).

2.4 GLASS TO METAL SEALS

24.1 Sealing principles

To obtain a reliable, leak-tight* glass–metal seal, the following requirements should be fulfilled (Roth[1083]):

(1) To achieve a good bond (adhesion) between the metal surface and the adjacent glass (Section 24.11).

(2) To base the seal either on the matching of the expansion characteristics of the metal and of the glass, or on the plasticity of the metal (Section 24.12).

(3) To control the cooling process in order to minimize the stresses in the seal (Section 24.13).

(4) To choose the geometry (shape) of the seal so as to obtain the minimum and not dangerously oriented stresses (Section 24.14).

* According to Adam[4] (based on I. Comer, Electr. Mftg. p. 110, March, 1958) a glass–metal seal can be defined as vacuum-tight if the leak rate for helium is less than 1 cm^3 He, NTP in 31 years, i.e. 8×10^{-7} lusec (see Table 1.5).

Very often the various requirements (bond, expansion, shape) lead to contradictory solutions. In such case the priority is given to the requirement which if not fulfilled will result in the most dangerous stresses in the finished seal or during its use.

24.11 Glass–metal bond. A good bond between glass and metal can be obtained only if some interaction takes place on their contact surfaces. The nature of the interaction (its mechanism) is not theoretically explained for all the glass–metal seals. Experimental results show that in the usual glass–metal seals the adhesion is based either on a direct glass to metal bond or on a glass–(metal) oxide–metal bond.

In the *direct glass to metal bond* the metal surface adheres to the glass without any intermediate layer. This kind of seal can be vacuum-tight but the bond is usually not too strong. The glass–metal seals are mechanically stronger if between the metal and the glass an oxide layer is formed. In fact the oxide layer contains a graded series of oxide mixtures from the oxide of the metal to those forming the glass. The fact is best illustrated by the seal of iron–nickel alloys in lead glasses, where the iron takes the place of the lead in the glass layers near to the joint (Scott[1115]).

The reliability of the glass–metal bond is determined by the thickness of the oxide film, the uniformity of the layer and the kind of oxides present.

The oxide layer must be so thin as to be able to follow elastically the displacements of the glass, without building up stresses. Thick oxide layers have usually a porous outer shell. Even if the glass adheres to this shell, the joint will not be tight and the seal not strong enough (Huber[605], Kohl[706]).

The oxide layer should be continuous and should have a uniform thickness and the same composition over the whole seal.

The glass–metal bond is determined by the kind of the oxide formed in the seal, i.e. by the ratio of volumes of the oxide to that of the metal from which it is formed (Scott[1115]). If this ratio is not constant over the whole seal, the mechanical strength is not the same, and stresses will develop. Oxides with higher oxygen content are generally porous and less reliable for tight seals. The growth of the oxide layer can change the stress distribution not only during the processing of the seal but also after this. This happened, for example, in some tungsten seals heated somewhat below the glass annealing range (Section 23.11) where the radial tension was decreased and the tangential tension increased due to the growth of the oxide on the metal (Redston[1039]). As most of the metals can form various oxides with various volume growth ratios (Table 2.26) the conditions during the sealing process should allow the build-up of just the desired kind of oxide. The kind of oxide formed is determined by the composition of the metal and of the glass, the impurities present in the seal (kind of atmosphere, flame, etc.) and the temperature during the sealing. As is shown in Table 2.26 the difference in volume growth between the various oxides of the same metal is small for some metals (e.g. copper) but very important for other metals (e.g. iron). The latter category of metals can be sealed to glass only if the very critical conditions required by the seal are fulfilled.

TABLE 2.26. VOLUME GROWTH RATIO OF OXIDES
(Scott[1115], Roth[1084], Kubaschewski[721a], Holland[581])

Oxide	a*	Oxide	a*
Cu_2O	1.64–1.67	MoO_2	2.1 –3.0
CuO	1.7	MoO_3	3.2 –3.3
CoO	1.74–1.86	WO_2	1.87–2.08
Co_3O_4	2.01	WO_3	3.4
		W_4O_{11}	3.03
NiO	1.51–1.65	Ta_2O_5	2.54
FeO	1.77	TiO	1.2
Fe_2O_3	2.14	Ti_2O_3	1.48
Fe_3O_4	2.1–2.3	Ti_3O_5	1.64
		TiO_2	1.73
Cr_2O_3	1.94–2.07	ZrO_2	1.56

$$*a = \frac{M_0 \cdot \varrho_m}{n \cdot M_m \cdot \varrho_0}$$

where M_m and M_0 are the atomic weights of the metal and the molecular weight of the oxide respectively, and ϱ_m and ϱ_0 the densities of the metal and oxide respectively; n is the number of the metal atoms in the molecule of oxide.

The impurities from the flame or the surrounding materials can significantly change the regime of the oxidation process; especially carbon dioxide, water vapour and sulphur compounds have a strong effect. For example, if during the oxidation process of copper the surrounding atmosphere contains only 0.1 per cent of sulphur dioxide, the speed of oxidation increases by a factor of twenty (at 400 °C) (Scott[1115]). The influence of the sulphur dioxide is specially strong in flames containing less than 4 per cent oxygen (Zincke[1354]).

Obviously the oxidizing period has important consequences as well, on the oxidation of the metals to be sealed to glass. Too long an oxidation process will result in a thick oxide layer, which as mentioned before, is not adequate for a good seal. Some metals, like molybdenum, are very heavily oxidized at the temperature needed to make the seal (Section 24.32). To overcome this effect a protective coating, a special moulding or a prebeading is used (Liebson[765]).

The colour of the glass–metal seal is generally a good indication of the bond (the quality of the seal), as most of the metal oxides give strong colours when dissolved in glass. For most of the seals the colour indication can show if the seal is reliable, but in some cases this is a not definite indication whether the seal contains or does not contain the desired oxide (e.g. tungsten).

Generally over-oxidized seals have a dark colour, and under-oxidized seals a very light appearance. In copper to glass seals the colour of the seal shows without doubt if the bond is made by means of the metal (gold–red), the cuprous oxide Cu_2O (purple) or the cupric oxide CuO (black), but the colour does not show if the layer is thin or thick, i.e. if the joint is tight and strong.

The composition of the glass can also influence the colour of the seal. Thus, tungsten sealed in glass which contains lithium does not shows the characteristic brown–yellow of the tungsten–glass seals, but a bluish colour due to the lithium tungstate formed.

The colour indication is very useful in those cases where the colour differs visibly according to the degree of oxidation. Thus the FeNiCo (Table 2.33) seals show a silverish metallic colour if the oxidation is not enough strong for a good seal, a grey to brown colour for the correctly oxidized seals, and a black appearance if the metal was over-oxidized.

A short review (see also Partridge[972], Dartnell[260], Zackay[1353], Fulrath[413], Roth[1084], Kohl[706], Katz[657], Volpe[1272], Gordon[454], Pask[973]) of the adhesion* of the metals and alloys to glass occuring in glass–metal seals is considered to be useful. *Platinum* can be sealed in glass only without oxides, since it does not form them; thus the platinum–glass seal has a metallic appearance and a limited strength. *Copper* can give very adherent seals, if the oxide is Cu_2O and its thickness is the proper one. In an oxygen-free atmosphere, copper can be also sealed to glass by "metallic bond" but these seals are never as strong as those with oxide layers. The correct colour of copper to glass seals varies thus from gold–yellow to purple. *Nickel* can adhere to glass, with a metal or oxide bond. The proper colour of an oxide bond is for nickel–glass seals, green–grey. *Iron* is a very promising metal for glass–metal seals, but its use is restricted due to the difficulties presented by its various oxides. The difference in the volume growth ratio of the various iron oxides is important (Table 2.26), and if the sealing conditions cannot be kept absolutely constant and are not reproducible, the change from one oxide to another gives dangerous variations in volume, leading to cracks in the oxide layer. *Tungsten* and *molybdenum* can give metallic bonds, but the seals having an oxide layer

* A good (but destructive) test for the adhesion is to break the seal; if a thin layer of glass remains on the metal, the adhesion is strong enough for a reliable glass-metal seal (i.e. seals of type M_2 or G_2, from Table 2.31, can also be used).

are preferred despite the difficulties in constructing them. The colour of an adequate tungsten–glass seal is from golden yellow to brown if sealed in glasses containing sodium or potassium, blue if the glass contains lithium and grey-brown in lead glasses. The molybdenum seals are generally brown. *Chromium* forms an oxide bond and produces very strong seals, which are dark green-in colour. The FeCr glass seals are brown–green, those with FeNiCo grey or blue and FeNiCr seals are brown or brownish–green. In order to make successful seals with iron containing alloys, the sealing technique tends to form on the surface of the alloy *only* chromium or nickel oxides since the iron oxides present the difficulties discussed before. These oxides are easily formed* in the seal when FeCr of FeNiCo is used, but with FeNiCr alloys the desired oxides must be formed by a previous chemical treatment (preheating in wet hydrogen).

The glass–metal bond is also influenced by the gases occluded in the metal. At the temperature of the sealing process, these gases evolve from the metal, but they cannot leave the glass, and are trapped on the glass–metal border, forming bubbles (Section 24.5).

24.12 Expansion matching. The matching between the expansion of the metal and the glass can avoid stresses in the seal or can just limit the stresses to values which are not dangerous for the integrity of the seal. In seals, even where some stresses are tolerated, the tendency must always be to form compression and to avoid tensile stresses (Section 24.12 and Table 2.28). In order to satisfy this requirement either "matched" or "unmatched" seals are to be used.

The name "*matched seals*" refers to those seals where an attempt is made to have partners of equal thermal expansion coefficients (Partridge[972], Went[1295], Stanworth[1177], Tolliver[1230]).

The absence of such an agreement in the expansion characteristics of the partners is compensated for in the "*unmatched* seals" by the plastic deformation of the metal or by a great compression exerted on the glass (Section 24.4). In some applications the metal to be sealed into the glass is plated with a layer of another metal in order to bring the seal closer to the matching conditions. This layer is electro-deposited on the metal, mechanically fitted on to it (clad wires) or it is obtained by the evaporation of an oxide and its subsequent reduction (Katz[657]). The seals using cladded wires are generally matching seals in radial direction and unmatched seals in the axial direction. If the metal should be sealed to a glass with a very different expansion, graded seals can be used (Section 23.35).

* It has been established that by the oxidation of FeCr alloy with 28 per cent Cr the obtained oxide is $(FeCr)_2O_3$ plus $(FeCr)_3O_4$ containing more than 55 per cent Cr and only 6 per cent Fe (Fejna[372a]).

The degree to which the expansion of the metal should match that of the glass depends on the adhesion of the two components, on the elasticity of the glass and the metal and also on the plasticity of the metal. For glass–metal pairs forming a strong bond and for more plastic metals, a greater difference of expansion can be tolerated. Experiments have shown that the real stress measured in glass to metal seals is generally smaller than that calculated from the expansion difference (Table 2.29). This is due to the plasticity of the metal part, and to the fact that a certain viscous flow takes place also below the setting point (Littleton[772], Redston[1038], Hagy[501], Martin[813]). On the other hand, the nominal tensile strength of the glass cannot be reached by the stresses, due to the imperfections existing on the surface of the glass, which permit the cracking of the glass before the nominal values are reached. In glass–metal seals tensile stresses of up to 0.5 kg/mm^2 are not considered

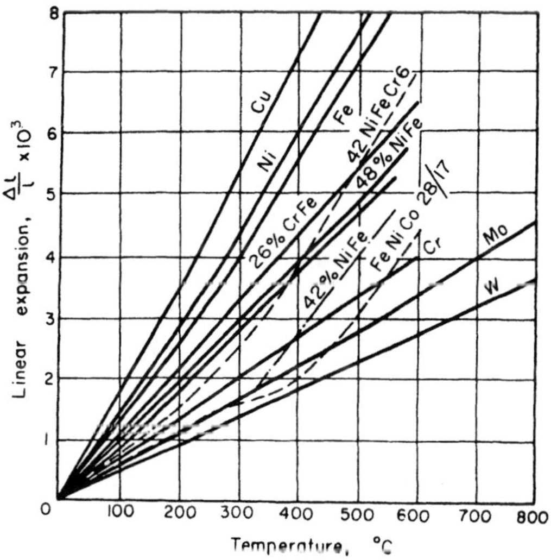

FIG. 2.63 Thermal expansion of metals and alloys, used in glass–metal seals

dangerous, and stresses up to 1.5 kg/mm^2 are considered as permissible values in well-annealed seals.

When considering the expansion matching of a metal and a glass, it is not enough to know the mean value of the expansion between given temperature limits, but the whole expansion characteristics of the metal and the glass should be compared (Engel[342], Hagy[501], Bruckner[175], Porubsky[1001]).

The expansion of glasses is linear with temperature up to the inflection point (Fig. 2.46 and Fig. 2.64). Higher up the expansion increases more rapidly with the temperature. The expansion of pure metals is constant with temperature (Fig. 2.63) and that of some alloys used in glass–metal seals is

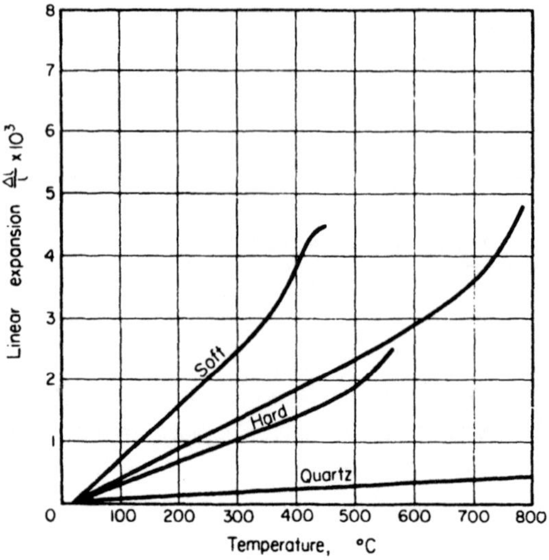

FIG. 2.64 Thermal expansion of glasses

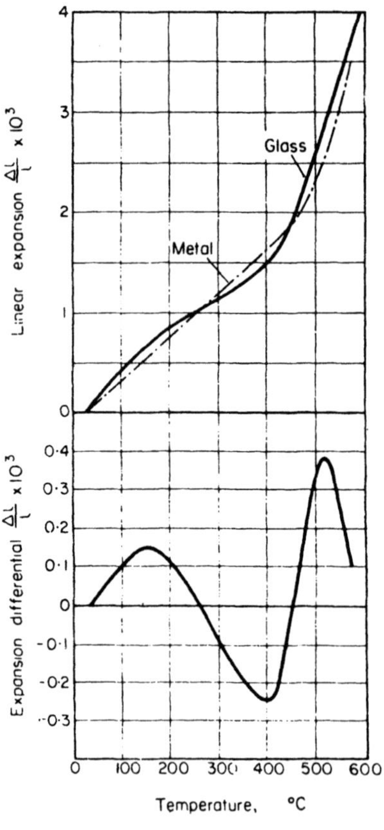

FIG. 2.65 Expansion differential in glass–metal seals

linear up to the inflection point (Table 2.33 and Fig. 2.63) after which their expansion increases more rapidly.

The expansion differences between glass and metal are best presented by differential curves plotted as in Fig. 2.65 derived from the expansion curves of the metal and the glass (Fig. 2.65). Such differential curves can also be computed* from the strains, determined by optical methods (Section 24.5). The differential expansion shows stresses of the same kind (e.g. tensile stresses) during the cooling from the sealing temperature to room temperature (as in Fig. 2.66) or the change in stress (from tensile to compressive) during the cooling from the sealing temperature. Hull[609] defines the *sealing point* as the temperature given by the intersection of a straight line with a slope 15 per cent greater than the average expansion coefficient of the glass between 0–300 °C with the actual expansion curve of the glass. This point coincides closely with the temperature used for the sealing of borosilicate glasses (Section 21.3).

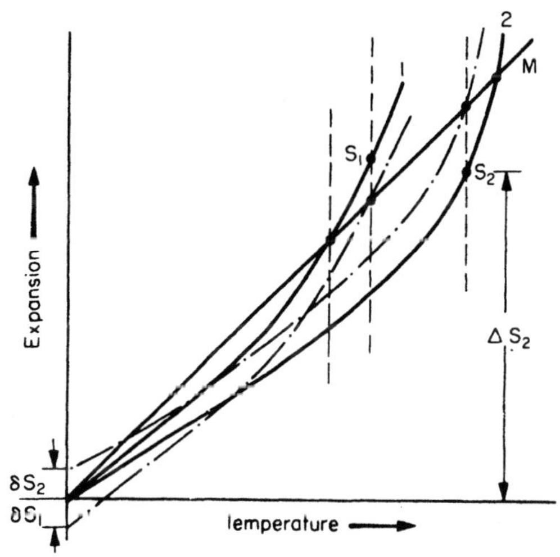

FIG. 2.66 Graphical determination of the expansion differential in a glass–metal seal

Consider glasses 1 and 2 (Fig. 2.66) having the setting points S_1 and S_2 respectively during their sealing to the metal *M*, the stresses resulting in the seal will correspond to an expansion differential δS_1 and δS_2 respectively. This can be obtained graphically (Fig. 2.66) by shifting the expansion curves of the respective glasses parallel to the vertical, in such a way that the Setting Point of the glass moves on to the expansion curve of the metal. If the room temperature side of the shifted curve is above that of the metal (like δS_2

* Comparing the two differential curves (those obtained from expansion to those from strains) some differences were observed at temperatures above the Setting Point, due to the hysteresis of the stress curve (Herrmann[546]).

from Fig. 2.66) the seal is of type M (Table 2.28); in the opposite case the seal is of type G. Values of δS for various glasses and metals are given in Table 2.29.

The example on Fig. 2.66 shows the case of a seal with a pure metal, having a linear expansion characteristic curve. The expansion curve of the alloys used for glass–metal seals has an inflection point (Fig. 2.65) in the range of 350–500 °C (Table 2.33), occurring very sharply at a temperature interval of some degrees*. These alloys can be sealed only with glasses having the Setting Point below the inflection temperature of the alloy. If not, the stresses formed are usually so big that the glass cracks or the bond to the metal is destroyed.

The expansion coefficients of the oxide layers differ markedly in some cases (e.g. copper) from those of the metal. It is thus recommended that the metal should not be allowed to cool after the oxide has been formed, until the glass has been laid down on the oxide and the assembly of the seal completed.

24.13 Control of the cooling; annealing. The cooling rate has a considerable influence on the quality of glass-metal seals. At high temperatures the viscosity of the glass (Fig. 2.45) does not permit the build up of stresses. When during cooling the Setting Point** is reached any difference in contraction between adjacent parts produce stresses. These stresses change their kind and magnitude during the cooling (Fig. 2.65) and the seal finally reaches room temperature with a given stress.

The transient stresses which appear during the cooling can be incomparably greater than the final stress at room temperature. These transient stresses may be allowed to be high as long as they do not exceed the limit where the glass breaks or the bond between glass and metal (Section 24.11) is destroyed, but the final stresses in the seal must be kept as small as possible.

Glass is a bad conductor of heat, and the temperature gradients developed in a glass part during cooling can be important, if the rate of cooling does not correspond to that established for the annealing of the glass (Section 23.13). If the stresses which appear because of the temperature gradient in the glass itself can be avoided by a proper cooling, the stresses remaining in the glass-metal seal are only those due to the contraction difference between glass and metal. These stresses cannot be influenced by the cooling rate but only by the metal–glass pair used and the geometry given to the seal. Due to their

* The expansion curve is not reversible; it follows a different course on heating than on cooling. The expansion characteristic of some metals (Ti, Zr) change appreciably on prolonged heating in air; with these metals it is necessary to keep the sealing time as brief as possible (Rawson[1033]).

** The Setting Point (Section 23.11) is influenced by the cooling rate. With rapid cooling the Setting Point is higher. Usually the Setting Point is established by optical strain measurements at a cooling rate of 2 °C/min.

geometry the glass–metal seals should be generally cooled at even slower rates than the corresponding glass to glass seals. Table 2.27 lists some generally recommended values for the cooling and annealing stages of glass–metal seals.

If uniform cooling was assured to all the parts of the seal, the stress distribution in the seal would depend only on the geometry of the seal (Section 24.14).

TABLE 2.27. THE ANNEALING OF GLASS-METAL SEALS*

Glass		Holding time at the annealing temperature (min)	Cooling rates
Type	Thickness (mm)		
Hard glasses	<3	10	From 600 °C max 15°/min (for perfect annealing 1 °C/min Hull[610])
	3–12	10–20	3 °C/min down to 300 °C, then natural cooling
Soft glasses	<2	10	From 500 °C, 10–15 °C/min
	2–10	10–20	3 °C/min down to 250 °C, then natural cooling

* See also Table 2.23

24.14 The shape of the seal. The influence of the shape of a glass–metal seal can critically influence its reliability. Since the shape of a glass–metal seal varies considerably, it is only possible to discuss its influence by reducing it to some basic geometrical forms. The shape of a real seal will then consist of one or more of such elements. The basic elements are: the flat seal, the circular internal (rod) seal, and the circular external (window) seal (Fig. 2.67).

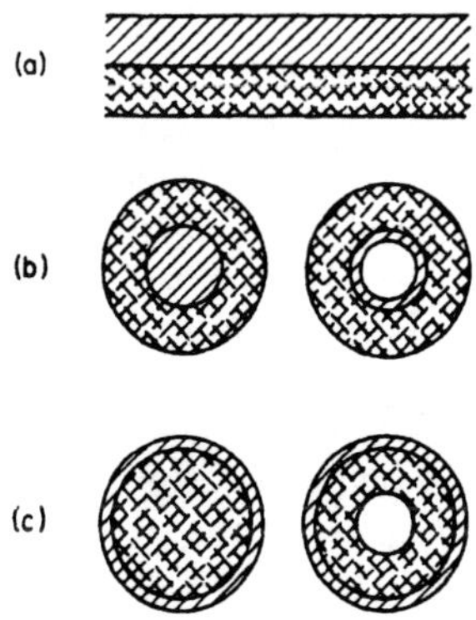

FIG. 2.67 Elementary shapes of glass-metal seals

In the *flat seal* the contact surface between metal and glass is plane. The circular *inside (rod) seal* has the metal inside the glass (Fig. 2.67b) and the circular *outside (window) seal* has the metal outside the glass (Fig. 2.67c) (Monack[877], Roth[1084]).

Considering the relative expansions of the metal and the glass, the seals are either of type *M* (expansion of the metal greater than that of the glass) or of type *G* (expansion of the glass greater than that of the metal).

In flat seals the stresses can be normal or tangential to the contact surface. In cylindrical glass–metal seals the stresses can be reduced to: axial (longitudinal), radial (normal to the contact surface) and tangential (on the circumference) stress components. Each of these can be either tensile or compression. The kind of stress which tend to develop in the various basic geometries and *M* or *G* type glass–metal seals, is indicated in Table 2.28.

TABLE 2.28. STRESSES IN GLASS–METAL SEALS

No.	Seal geometry	Type *M*			Type *G*		
		Normal (Radial)	Tangential	Axial	Normal (Radial)	Tangential	Axial
1	Flat	nil	compr.	—	nil	tens.	—
2	Rod	*tens.*	*compr.*	*compr.*	*compr.*	*tens*	*tens.*
3	Window	compr.	compr.	compr.	tens.	tens.	tens.

As the compression strength of glass is about ten times greater than the tensile strength, compression should be always preferred to tension. From the possibilities presented in Table 2.28 the best combinations are obviously M_1 and M_3 where the stresses produce compression in all the directions. The seals of type M_2 and G_2 are also tolerable if the expansion difference between the metal and the glass is not too big and/or a very strong metal–glass bond exists. Seals of type G_1 and G_3 are to be avoided.

Generally the value of the stress in a glass–metal seal is given (Lillie[765a], Rawson[1029, 1030]) by the relation:

$$P = F \cdot \delta,$$

where δ is the expansion difference $\Delta l/l$ between the glass and the metal at the Setting Point of the glass, and F is a function of the dimensions and the mechanical characteristics of the metal and the glass (see Fig. 2.68).

The expansion differential δ can be determined from the difference in elongation of the metal Δ_m and the glass Δ_g in the temperature range between the setting point (T_s) of the glass and room temperature (T_0).

$$\Delta_g = \alpha_g(T_s - T_0),$$
$$\delta = \Delta_g - \Delta_m = \Delta_g - \alpha_m(T_s - T_0).$$

Δ_g is a constant for each glass (at a given cooling rate, Section 24.13). This value is represented graphically in Fig. 2.66 (and marked ΔS_2), and their numerical values for various glasses can be found in Table 2.29. α_m is the mean value of the thermal expansion coefficient of the metal for the temperature range T_s to T_0. Table 2.29 shows also the values of δ for various glass–metal pairs. The letters M and G refer to the type of the seal as shown in Table 2.28, and the following figures show the values of the differential expansion δ multiplied by 10^4. Although a zero expansion differential provides a completely unstressed seal, some expansion differential is always permissible. Generally an expansion differential less than 1×10^{-4} (marked in Table 2.29 as 1) results in excellent stress conditions in the seal. Differentials of $1–5 \times 10^{-4}$ result in satisfactory stress conditions for most medium size seals. Differentials of $5–10 \times 10^{-4}$ result in critical stress conditions (Gleason[452], Kohl[706]) when the stress is a tensile one (Table 2.28).

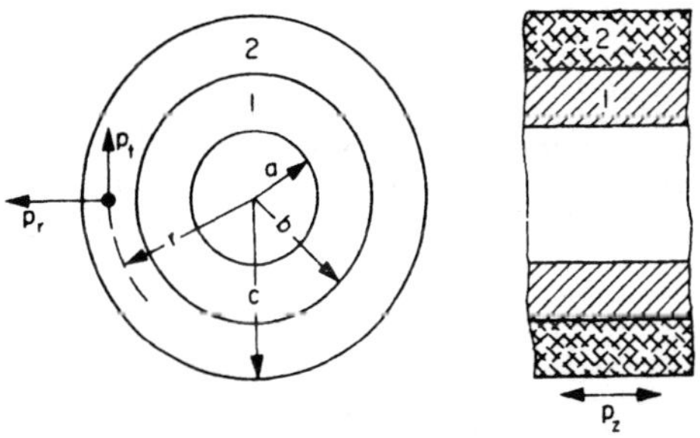

FIG. 2.68 Dimensions for calculating stresses in glass–metal seals

The value of F depends on the shape of the seal (Table 2.30) and the direction of the considered stress (radial, tangential, axial). From the possibilities presented in Table 2.28 it is worthwhile to consider only those with a tensile stress, since the compression is generally not dangerous. Table 2.30 gives the formulae to be used in these cases to calculate the value of F.

Example. A Kovar ring is sealed inside a Corning glass 7040 tube. The dimensions of the seal (Fig. 2.68) are $a = 17$ mm; $b = 18$ mm; $c = 20$ mm.

From Table 2.29 it results that the differential expansion δ in this case is G 0.8×10^{-4} i.e. the seal is of type G_2 (Table 2.28). In such a seal tensile stress appears in the tangential and the axial directions (Table 2.30). Here $E_1 = E_{\text{kovar}} = 1.4 \times 10^4$ kg/mm², $E_2 = E_{\text{glass}} = 0.6 \times 10^4$ kg/mm², and $\sigma = 0.3$. To calculate the stresses on the contact surface between the glass and the

TABLE 2.29. DIFFERENTIAL EXPANSION

Glass	7740 Pyrex	C 38	8330 Duran	C 9	Bluesil	7720 Nonex	712/b	W 1	GSD
Manufacturer*	C	BTH	J	BTH	P	C	O	GEC	Ch
Setting Point (°C)	530	485	500	505	520	500	540	550	550
$\Delta g \cdot 10^4$	19	20	20	20	21	22	22	22	22
With *Metal***									
Tungsten	M5	M2	M2.5	M2.5	M3	M0.4	M3	M3.5	M3.5
Vacon 12	M15	M9	M9	M9	M11	M7.5	M13	M13	M13
Nilo 42, Driver 42, Carpenter 42, AL 42	M23	M16	M18	M18	M18	M16	M19	M19	M19
Nilo K, Sealvac A, Rodar, Therlo	M14	M7	M9	M9	M10	M8	M11	M15	M15
Fernico, Vacon 10, Sivar 48, Superior 42, Kovar	M15	M9	M11	M11	M12	M10	M12	M16	M16
Molybdenum	M10	M7	M8	M8	M9	M5	M8	M9	M9
Vacon 20, Tantalum	M15	M11	M12	M12	M12	M11	M12	M12	M12

Glass	0120	C 12	L 1	0010	GWB
Manufacturer*	C	BTH	GEC	C	Ch
Setting Point (°C)	413	415	410	408	390
$\Delta g \cdot 10^4$	37	37	37	38	38
with *Metal***					
Vacodil 42, Vacodil 43	G10	G10	G10	G11	G13
Vacovil 46, Driver 46	G5	G5	G5	G7	G9
Vacon 70	G6	G6	G6	G7	G9
Nilo 48	G2.3	G2.2	G2.4	G3.5	G5
Vacovit 426, Driver 14, Sylvania HC-4, Sealmet 4, Carpenter 426	M2	M2	M4	M2	M1
Platinite, Driver FeNi	G0.6	G0.5	G0.9	G2	G4
AL 4750, Vacovit 501, Carpenter 49	M0.6	M0.7	M0.3	G1.3	G3.3
Platinum	M1	M1	M0.7	G0.5	G1.7
Nilo 50	M1.4	M1.6	M1.8	G0.1	G1.7
Sealmet 1, Carpenter 27, Chrom-iron, Dilver 0, FeNiCr, Vacovit 511, Driver 79/26	M4	M4	M4	M3	M2
Telemet, Dilver T, Novar B, Vacovit 540, Vacovit 025	M8	M8	M8	M7	M4

* C — Corning Glass Works, Corning, N.Y.; Ch — Chance Brothers Ltd. Glass England; J — Jenauer Glaswerk Schott & Gan. Mainz, W. Germany; P — Plowden-Works, East Lane, Wembley, Middlesex; K — Kimble, Owens Illinois Glass Co.

** For the composition, characteristics and manufacturers of the metals (alloys) see

OF GLASS–METAL PAIRS

7050	GSB	7040	Dial 43	C 11	637/h	3072 G 20	Ko-dial	C 40	K650	7060	HH	362a	1447	GS4
C	Ch	C	P	BTH	O	J	P	BTH	K	C	GEC	O	J	Ch
476	430	464	510	550	525	530	480	485	485	475	570	500	500	600
23	24	24	25	26	27	27	27	27	27	27	28	29	30	36
							$\delta \times 10^4$							
G2.5	G5	G4	G2	G1	G3	G3	G5.5	G5	G5	G5.6	G1.8	G6	G7.5	G8.4
M3.5	G0.8	M1	M5.5	M11	M6	M6	G0.2	M0.2	M0.2	G0.5	M12	M0.5	G0.5	—
M7	M2	M7	M14	M18	M16	M16	M7	M7	M7	M7	M18	M9	M8	—
M4	G2.5	M1	M5	M8	M5	M5	M0.7	M0.4	M0.4	G0.4	M11	M0.8	G0.2	M9
M4.5	G1.2	G0.8	M7	M12	M9	M8	M0.8	M2	M1.6	M0.1	M12	M1.5	M2	M11
M2	G0.3	M1	M3	M5	M3	M3	G0.2	M0.2	M0.2	G0.4	M4	G1	G2	G1
M7	M3	M5	M8	M9	M7	M7	M4	M4	M4	M3	M7	M4	M2	M2

526	C 31	0240	M 6	16 III	0080	GWA	584d	X 8	C 19	C22	105
O	BTH	C	GEC	J	C	Ch	O	GEC	BTH	BTH	O
540	422	405	560	530	490	490	500	500	510	485	480
40	41	42	42	43	47	48	50	50	51	52	54
						$\delta \times 10^4$					
M6	G13	G16	M8	M1.5	G8	G9	G10	G10	G9	G13	G14
M8	G9	G11	M9	M5	G4	G5	G6	G6	G5	G10	G12
G0.1	G9	G12	M0.6	G4	G12	G13	G14	G14	G14	G17	G18
M8	G5	G8	M9	M4	G4	G5	G6	G6	G11	G10	G11
M22	M2	G1	M25	M19	M4	M6	M6	M6	M7	M1.3	G0.7
M11	G3	G6	M12	M6	G2.5	G3.5	G4	G4	G3.5	G8	G9
M14	G1.7	G5	M15	M10	M0.5	M0.1	G1	G1	G0.5	G5	G8
M11	G1.7	G4.7	M11	M8	G2	G2	G2.5	G2.5	G1.5	G6	G8
M11	G1.7	G4.3	M10	M6	G1.4	G2.4	G3	G3	G3.5	G7	G9
M13	M2	G0.8	M13	M9	M3	M2	M1.5	M1.5	M1.5	G1	G3
M19	M5	M2	M20	M14	M7	M6	M5	M5	M5	M1	G1

Works, Birmingham, England; BTH — The British Thomson–Houston Co. Ltd. Rugby, Thompson, Stourbridge, England; O — Osram, West Berlin; GEC — Osram G.E.C. Glass

Table 2.33.

TABLE 2.30. EXPRESSIONS TO BE USED FOR F.
(See Table 2.28 and the formulae listed below this Table)

Seal geometry	Stress direction		
	Radial	Tangential	Axial
Flat	—	Type G_1 $F = \frac{\bar{E}}{2}$	—
Rod	Type M_2 $F = F_r$	Type G_2 $F = F_t$	Type G_2 $F = F_z$
Window	Type G_3 $F = F'_r$	Type G_3 $F = F'_t$	Type G_3 $F = F'_z$

$$F_r = \frac{\alpha(1-a^2/r^2)}{(2\sigma-\beta)/E_2+[\alpha(\gamma-2\sigma)]/E_1}, \quad F'_r = \frac{1-c^2/r^2}{(2\sigma-\beta)/E_2+[\alpha(\gamma-2\sigma)]/E_1},$$

$$F_t = \frac{\alpha(1+a^2/r^2)}{(2\sigma-\beta)/E_2+[\alpha(\gamma-2\sigma)]/E_1}, \quad F'_t = \frac{1+c^2/r^2}{(2\sigma-\beta)/E_2+[\alpha(\gamma-2\sigma)]E_1},$$

$$F_z = F'_z = \frac{\alpha E_1 E_2}{E_1-\alpha E_2} \cdot \frac{\beta E_1-\alpha\gamma E_2}{(2\sigma-\beta)E_1+\alpha(\gamma-2\sigma)E_2}$$

where $$\alpha = \frac{b^2-c^2}{b^2-a^2}, \quad \beta = 1+\frac{c^2}{b^2}, \quad \gamma = 1+\frac{a^2}{b^2}.$$

For a, b, c, r see Fig. 2.68; σ = Poisson's ratio E = modulus of elasticity.

metal, $r = b$. From the formulae from Table 2.30 the following values are obtained:

$\alpha = -2.18$; $\beta = 2.24$; $\gamma = 1.88$; and $F_t = 0.88 \times 10^4$; $F_z = 0.94 \times 10^4$.

Hence the tangential stress (tensile) will be

$$P_t = F_t \cdot \delta = 0.88 \times 10^4 \cdot 0.8 \times 10^{-4} = 0.7 \text{ kg/mm}^2,$$

and the axial stress (tensile):

$$P_z = F_z \cdot \delta = 0.94 \times 10^4 \cdot 0.8 \times 10^{-4} = 0.75 \text{ kg/mm}^2.$$

The values of $F_r = F'_r$, $F_t = F'_t$ and F_z for $r = b$ (Fig. 2.68) and $\sigma = 0.3$; E_1 (metal) $= 1.8 \times 10^4$ kg/mm² and E_2 (glass) $= 0.65 \times 10^4$ kg/mm² are plotted on Fig. 2.69 as a function of c/b for various a/b ratios. It can be seen from this diagram that when a/b and c/b are large, the value of F is low, i.e. the stresses in the seal will be low for a given value of δ. This explains why it is possible to seal thin walled metal tubes (a/b near to one) through glass even when the two materials have widely different thermal expansions (δ is big). The curves marked $a/b = 0$ show the values of F when massive rods are sealed into glass.

The direction and the distribution of the stresses in glass–metal seals is determined by the shape of the seal. Figure 2.70 shows some glass–metal seals, and illustrates the fact that the "stress lines" are concentrated in the vicinity of sharp edges or thin parts of the glass. The most sensitive part of a

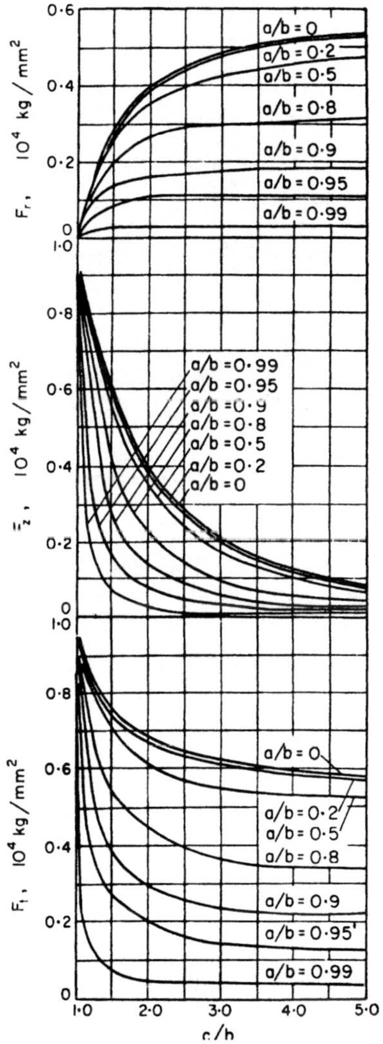

FIG. 2.69. Values of F_r, F_t, F_z as a function of c/b and a/b. After Rawson[1029] *(Courtesy of The Institute of Physics and The Physical Society, London)*

glass–metal seal is always the metal–glass–air border line, since the surface of the glass always contains small cracks which reduce the tensile strength.

The knowledge of the stresses appearing on this border line is very important, since it has been established that in most of the unsuccessful glass–metal seals the cracking starts from this line. The type of stress appearing on this line is determined by the basic geometry of the seal, the type of the seal

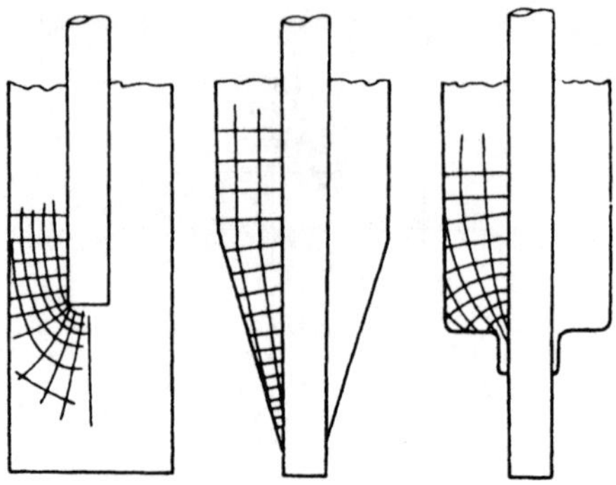

FIG. 2.70 Stress distribution in glass–metal seals

and the angle formed between the glass–air surface and the surface of the metal. Table 2.31 summarizes the type of stresses occuring in various combinations of these factors.

TABLE 2.31. STRESSES NEAR THE GLASS–METAL–AIR LINE

No.	Seal geometry	Type *M*			Type *G*		
		Contact angle					
		acute	right	obtuse	acute	right	obtuse
1	Flat	compr.	(tens.)	tens.	tens.	(tens.)	compr.
2	Rod	compr.	tens.	tens.	tens.	compr.	compr.
3	Window	*compr.*	compr.	tens.	tens.	tens.	*compr.*

The combinations recommended (from Table 2.31) are obviously those which result in compression stresses, i.e. for seals of type *M* those having acute contact angle and for seals of type *G* those having obtuse angles.

The metal parts protruding from glass–metal seals are often subjected to mechanical loading; usually they are bent laterally. Due to such loading the metal peels off from the glass if the adhesion is weak, or the glass will crack if the bond is strong. The supplementary tensile stress produced by these mechanical loads is well balanced if initial compressions stresses exist in the seal; thus the seals of type M_3 or G_2 can be recommended (Table 2.31). The same technique of compensation should be used for the stresses which will be set up in lead-in seals when an electric current is transmitted through the metal and it heats up (Section 41.2).

24.2 Heating and working techniques

24.21 Steps taken in completing glass–metal seals. The procedures used to construct glass–metal seals consist in:

(1) The preparation of the metal parts (shaping, cleaning, degassing, oxidizing, beading).
(2) The preparation of the glass parts (cutting, forming, cleaning).
(3) The sealing of the metal to the glass.
(4) The treatments of the completed seal (cleaning, annealing).

This Section gives the general features of these steps; the particular techniques for specific metal–glass pairs are described in Sections 24.3 and 24.4.

a) Preparation of the metal parts. The metal parts are usually purchased already in the shape needed for the seal (wire, rod, tubing), and the mechanical preparation of these parts consists in cutting them to the required dimensions and in mechanically cleaning their surfaces. This is done by turning, grinding or sand blasting according to the specified shape, the mechanical properties of the metal and the quantities involved. Some special techniques used or specific shapes needed in various glass–metal seals are given in Sections 24.3 and 24.4.

In any reliable joint, the surfaces of the joined materials must be free of impurities. To clean the metal parts for glass–metal sealing, chemical cleaning methods are used (Section 22.31). In the cases where special cleaning methods should be used, they are mentioned in Sections 24.3 and 24.4. Independent of the cleaning method used, it is recommended not to touch the metal with the fingers after the cleaning.

The glass–metal seal should be *free of bubbles* (Section 24.5), since a seal having gas bubbles included on the junction between metal and glass is weak and can be leaky. The bubbles in the seal are produced by the gas dissolved in the metal (Section 21.13) which evolves during the heating of the metal in the seal. Especially gassy are nickel, iron and their alloys, if they were not vacuum melted. A *degassing* of the metal parts in vacuum or hydrogen (not to heat copper in hydrogen!) at a temperature above that used in the sealing process, generally results in a metal part which can give bubble-free seals. Sometimes degassing is achieved by simple heating in air, in order to oxidize the surface of the metal part. In any case the oxidized metal gives fewer bubbles in the seal than the clean but not degassed metal.

A source of gas is very often produced by the carbon content of the metal or alloy, which forms carbon oxides during the sealing. The decarbonizing of the metals can be done by heating in wet hydrogen for 1–2 hours at 900–1100 °C (Kleinteich[678a]).

Most of the metals should have their surface oxidized, in order to produce a strong bond (Section 24.11) and a tight seal. To obtain the required oxide layer the metal part is heated in a burner flame, in an appropriate furnace or by electrical heating methods (also Section 22.21). The temperatures, the heating times, and the atmosphere used are critical in some cases; the recommended techniques are described in Sections 24.3 and 24.4 together with the chemical oxidizing treatments.

It is often useful to make the seal in two stages: first to complete a *bead* and then to *seal* the beaded metal into the glass apparatus. This is done especially with metals where the oxidation rate is critical and difficult to control (tungsten, molybdenum) or with seals where the shape requires such a two-stage sealing. The procedure used for the beading of metal in glass depends on the shape and type of the metal. Metal rods are beaded by coiling a glass rod on the metal (Figs. 2.72, 2.74) or by slipping a thin glass tubing over the rod (Fig. 2.74). In beading with a glass coil, the metal rod is rotated slowly. The softened end of the glass rod touching the metal and by successive heating parts of the glass rod are joined to the metal forming a coil. If a glass sleeve is used, this must closely fit the diameter of the metal rod. It is recommended that the heating of the sleeve is begun at one end and the sealing is continued towards the other end, in order to permit the escape of the air and gases formed. The sharp edges of pipes are beaded (Fig. 2.79) by sealing, outside or inside the pipe, a piece of close-fitting glass tube, and then flaring it around the edge of the metal pipe.

In some cases (Section 24.3) the beading is done by spraying powdered glass suspension on the metal surface and then forming a thin glass layer by baking.

b) The *preparation of the glass* parts for glass–metal seals is similar to the procedures used for glass to glass seals (Section 23.2).

c) The techniques connected with the *sealing process* are described in Sections 24.3 and 24.4 for the various kinds of glass–metal seals.

24.22 Heating methods used in glass–metal sealing processes. The methods used to heat the glass and metal parts during their sealing, pre-heating or annealing are either the flame or the electrical heating techniques.

Flame heating is extensively used in hand work or glass–metal sealing done by machines some of which are automatized to a high degree (those used in the production of electric lamps, electron tubes, etc). Regarding the kind of flames used, the reader is referred to Section 23.23.

Electric *resistance heating* can be used for glass-metal sealing only in the cases where the shape of the metal allows a uniform heating (e.g. disc seals). To produce a seal by this method, the metal part (disc) must have conducting connexions in two diagonal points. These connexions should have less electrical resistance than the ring itself in order to ensure only the heating of the ring.

Induction heating is also used to complete glass-metal seals (Haase[497], Handen[508a], Housekeeper[599], Kohl[706], Pentland[981a], Bleuze[138], Machlett[792],

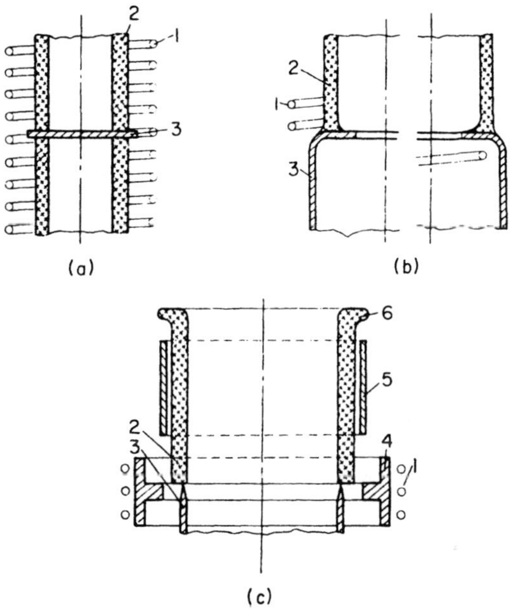

FIG. 2.71 Sealing glass to metal, by induction heating

Adam[4], Malic[797], Snyder[1162a]). This kind of heating is based on the creation of eddy currents in the metal part to be sealed to the glass. The metal part heats up the glass adjacent to it; this become soft and flows under the action of the gravity or by external pressure on the metal. Induction heating is particularly useful in completing disc seals or seals where the metal part has a more complicated form.

The induction coil is usually constructed of circular or square cross section copper tubing, bent in order to conform as closely as possible to the shape of the metal part. The coil is water cooled. Any asymmetrical position of the coil relative to the metal part will produce unequal heating in the seal, which leads to asymmetrical strains.

Figure 2.71 shows some arrangements which can be used to seal discs and cylindrical metal parts to glass by induction heating. The induction coil must be well centred and placed at the appropriate height to heat up the metal parts exactly in the region where it is to be sealed to the glass (Fig. 2.71a, b). Figure 2.71c shows schematically an arrangement for sealing a glass tube (2) to a metal tube (3). The metal tube (3) is heated by the high frequency coil (1) using an iron ring (4) as a concentrator (see also Fig. 2.44). The glass tube (2) descends under its own weight while the edge of the metal tube (3) penetrates in to the softened glass. The glass tube should be guided (5) and on its end a stop (6) should be provided to limit the depth of the seal (Lupinek[789]). Snyder[1162a] used this technique to seal glass to molybdenum in an argon atmosphere. A method for using induction heating to complete glass–metal compression seals is shown in Fig. 2.89.

24.3 Matched seals

24.31 Matched glass–metal seals with soft glasses (Section 21.3) are used especially with platinum, FeNi, FeCr and FeNiCr alloys. Table 2.32 summarizes the various glasses corresponding to these and other metals. For more details on the expansion matching of some of these glass–metal pairs Table 2.29 can be consulted. The composition and characteristics of the metals and alloys available for glass–metal seals are listed in Table 2.33.

a) Sealing of platinum. The platinum matched* seals should be made with glasses having a corresponding expansion (Table 2.32).

The platinum should be first cleaned by dipping it into hot aqua regia (HNO_3 plus HCl, 1/3) followed by a thorough washing in water. The glass is then collapsed onto the platinum and the seal heated to 1000–1100 °C (to a yellow–orange colour). To ensure proper wetting of the metal by the glass, a high uniform heating is necessary. It is obvious that platinum does not form an oxide bond to the glass, but the wetting of the glass is influenced by the absorption of oxygen and hydrogen (Volpe[1272]).

The steps taken to complete a platinum–glass seal can be summarized in the following. A small glass sleeve (Fig. 2.72) is first slipped over the platinum wire (1, Fig. 2.72). This sleeve is fused progressively from one end to the other (2, Fig. 2.72) avoiding any trapped air pocket in the seal. By heating the cylindrical bead it is transformed into a ball (3, Fig. 2.72) of about 4 mm in diameter. An alternative technique consists in fusing a glass rod over the wire while winding it in a coil (4, Fig. 2.72). The coil is then fused to a cylindrical beading (5) which is transformed subsequently into a glass ball (6). After the end of the glass tube has been properly shaped (7, Fig. 2.72) the beaded wire is sealed to the tube at the end (8) or at the side (9). The opening in the tube prepared for receiving the bead, must have a diameter smaller than that of the ball.

* For unmatched seals using platinum see Table 2.35.

TABLE 2.32. SOFT GLASSES FOR GLASS–METAL SEALS*

Metal**	Glass***	Remarks
Platinum	C : 0280, 0041, (7550), 7570, 7560, 0050, 0080, 0010 Ch : GW2 (GW1), PWD, PWL BTH : C 12, C 19, C 94 J : 16, 2962 O : 301b GEC : X 4, L 1 (L 15) P : DIAL 444 K : RS, R6	Wichers[1310]
FeNi (50/50 or 46/54)	C : 0010, 0120, 0080 BTH : C 12 J : 16 III Ch : PWD, PWL GEC : L 1	wires with 5 mm max. diameter
Dumet	C : 0050 J : 2962 III Ch : GW2, PWD, PWL O : 352, 743g, 123a BTH : C 12, (C 19, C 94) GEC : L 1, K : R 5, R 6	Fink[377] Eldred[336] see Sec. 42.2
Fe Cr (74/26 or 80/20)	C : 0050, 0060, 0080 (9019, 9010) Ch : GW2 (GW1), PWD BTH : C 31, (C 12) GEC : L 14, X 8, (L 1) O : 123a	Breadner[163] Hull[610] Espe[351]
FeNiCr	C : 8870, 0080, 0014, 0120, 0010, 0050 Ch : PWD, PWL BTH : C 12 GEC : L 1 K : KG 12	Stanworth[1177]
FeNiCoCr (37/30/25/8)	C : 0050, 0080 Ch : GW2, PWD K : R 6	
Nickel	GEC : NSG 2	Oldfield[954] Kohl[706]
Iron	C : 7290, 1990 (1991) GEC : R 16, ISG 20, NSG 2 BTH : C 76, (C 41) J : 4210	Kohl[706] Monack[874]
Copper	C : 7295 GEC : CSG 3	Oldfield[954]
Titanium	BTH : C 77, C 78	Rawson[1033]

* See also Kalsing[652], Kohl[706], Mönch[879], Ardenne[47], Espe[350, 356], Roth[1081, 1083], Colnot[223], Zincke[1354].

** See Table 2.33.

*** For the signifiance of the abbreviations see footnotes of Table 2.10 and Table 2.29.

b) Sealing of iron–nickel alloys. The FeNi alloys should preferably be heated to about 950 °C in hydrogen. This treatment cleans the surface and degasses the metal. It also removes any machining marks without causing excessive grain growth.

During the oxidizing of FeNi alloys great care must be taken not to oxidize them too heavily, since the oxide layer has a tendency to flake off from the metal.

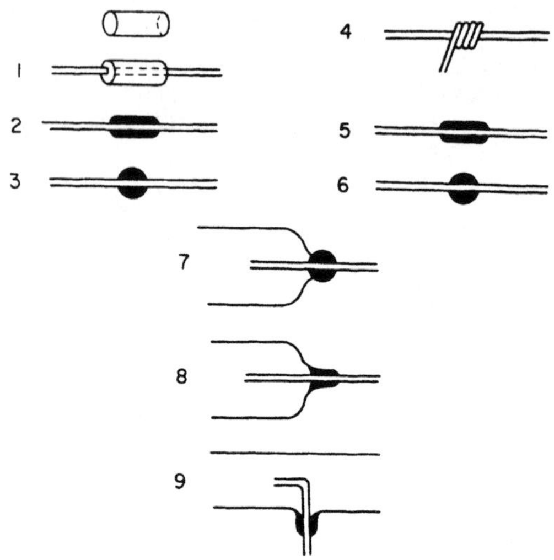

FIG. 2.72 Stages in completing a platinum-to-glass seal

Ferrous alloys have a tendency to reduce the lead oxide from the glass. Thus when FeNi alloys are sealed into lead glasses it is recommended to prevent the direct contact between FeNi and lead glass by plating the FeNi with platinum or copper or using sleeves from a glass without lead. Thin electro-plating (about 0.05 mm) with copper will also avoid the danger of over-oxidizing the FeNi. If the copper clad is thick the sealing wire known as *Dumet* is obtained (Section 42.2). Instead of copper Foley[388] used silver-plated FeNi wires. For glasses having the linear expansion coefficient over 80×10^{-7}, Went[1295] proposes the use of FeNiCu alloys, containing at most 54 per cent Fe, at least 1 per cent Cu and at most $(56+p/3)$ per cent Ni; p being the Cu content in per centage. He recommends alloys containing 1–7 per cent Cu, 48–56 per cent Ni plus Cu, and the remainder Fe.

c) Sealing of iron–chromium alloys. FeCr alloys can be readily machined, and can be sealed to glass as rods, tubes or discs. The mechanical cleaning of FeCr is done by sand blasting and/or by tumbling. Silicium carbide should be avoided in these operations since small particles penetrate into the FeCr surface and later during the sealing produce carbon dioxide bubbles in the seal.

The oxidation of FeCr is very reliable; it does not present the danger of being over-oxidized since the first thin oxide layer has a strong adhesion to the alloy and protects it from further oxidation. The oxidation can be done by heating in air, in which case an oxide rich in iron is formed (Section 24.11). To obtain the desired chromium oxide it is recommended to oxidize the FeCr alloys by heating them (at 950–1100 °C, for 15–30 min) in wet hydrogen (bubbled through distilled water) or by immersion in hydrochloric acid (20 per cent solution) for about two minutes, followed by a washing. In the seal the formed oxide can be identified since the iron oxides give a grey colour to the seal while the desired chromium oxides have (in the seal) a greenish tint.

The cleaning of the FeCr parts protruding from the glass–metal seal can be done (after the seal has been completed) by electrolytic pickling (Table 2.20).

If the Fe Cr has not been pre-oxidized (as described before) it should be thoroughly heated in the flame before the glass is engaged with it.

Hrzek[604] gives a method for the sealing of FeCr rods having small cracks at their surface, consisting in first coating the rod with a vitrifier capable of dissolving metal oxides at high temperatures and having a low surface tension (e.g. an aqueous solution of $Na_2B_4O_7$ containing the metal oxides of the rod as filler material.

Hull[610] considers that FeCr seals in glass are best when they are not cooled too slowly (see also Section 24.13).

d) Sealing of iron–nickel–chromium alloys. The pre-oxidation of the FeNiCr alloys (Table 2.33) must be done in such a way as to form on the surface a definite thickness of chromium oxide. The procedure to obtain this consists in heating the alloy in wet hydrogen at 1050–1250 °C. With this pre-oxidized surface a strong bond is obtained in the seal.

The sealing process is the usual one for glass–metal seals (e.g. Figs. 2.74, 2.76). The cooling rates are generally not critical, due to the good matching of the alloys and glasses used (see Table 2.29).

e) Nickel can be sealed to the specially developed soft glasses (Table 2.32). The nickel must be first chemically cleaned by an immersion in a solution containing 750 cm³ sulphuric acid, 1000 cm³ concentrated nitric acid, 50 g sodium chloride and 900 cm³ distilled water. The pickling in this solution is followed by a thorough washing and drying and by a heat treatment (degassing) in wet hydrogen (30 min at 1050 °C).

The pre-oxidation of nickel can be done by heating it in a hydrogen flame (1–5 sec at 1000°) followed by cooling in air (to about 650 °C).

The glass–nickel seal is preferably made in an oxidizing flame (Section 23.23) at about 650 °C.

TABLE 2.33. METALS AND ALLOYS

Metal or Alloy	Composition (%)				Specific gravity	Expansion coefficient $\alpha \cdot 10^7$ (1/°C) from 20 °C to (°C)					
	Ni	Cr	Co	Fe		100	200	300	400	500	600
Tungsten	–	–	–	–	19.2	–	–	45	–	46	46
Nilo 40	40	–	–	60	8.1	–	41	–	–	–	–
Vacon 12	28	–	18	54	8.3	58	53	48	47	59	76
								(from 0 °C)			
Nilo 42 Driver 42 Carpenter 42	42	–	–	58	8.2	52	48	49	56	76	–
A L 42	42	–	–	58	8.2	47	47	47	55	78	93
								(from 0 °C)			
Dilver P	29	–	17	54	8.5	56	52	48	47	60	–
								(from 0 °C)			
Nilo K Sealvac A Rodar Therlo	29	–	17	54	8.3	58	51	47	46	60	
Vacon 10 Fernico I	28	–	18	54	8.3	60	56	51	50	61	78
Sivar 48	29	–	17.8	53.2	8.3	59	57	52	50	63	77
							(from 30 °C)				
Kovar A	28.7 29.2	–	17.3 17.8	52.9 53.4		43–53	44–52	45–51		57– 62	79
Fernico II	31	–	15	54	8.2	–	57	54	53	65	–
Molybdenum	–	–	–	–	10.3	49				55	
Vacodil 42	42	–	–	58	8.2	52	52	53	63	81	95
Vacodil 43	43	–	–	57	8.3	64	61	60	65	81	95
Vacon 20	28	–	21	51	8.3	71	68	65	63	63	76
Vacodil 46 Driver 46	46	–	–	54	8.2	74	74	73	75	88	99
Vacon 70	28	–	23	49	8.3	85	80	77	74	71	81
Nilo 48	48	–	–	52		87	83	83	83	88	–
Vacovit 426 Driver 14 Sylvania HC-4 Carpenter 426 Sealmet 4	42	6	–	52	8.2	69	72	83	101	114	124

FOR GLASS–METAL SEALS

Inflexion Point (°C)	Electrical resistivity (Ohm-cm. 10^6)	Thermal conductivity (cal/cm.sec. °C)	Manufac-turer*	Remarks
–	5.5	0.5	–	–
330–250	62–70	0.025	W	Curie Point 330 °C M.P. 1450 °C
430	45		V	
340	60	0.025	W D CS	Curie Point 375 °C M.P. 1450 °C
340	65	0.026	A	
	45	0.042	I	Curie Point 425 °C
430	44–50		W VM WD D	Curie Point 453 °C M.P. 1450 °C
425 430	45 50		V G	
423	–		M	
435	49	0.044	St	Curie Point 453 °C M.P. 1450 °C
420	44		–	
–	58	0,3		
355	60–100		V	
370	60		V	
480	45		V	
400	60–100		V D	
515	45		V	
435	50	0.030	W	Curie Point 450 °C
265	95	0.033	V D S CS A	

(Table 2.33 Continued)

Metal or Alloy	Composition (%)				Specific gravity	Expansion coefficient $\alpha \cdot 10^7$ (1/°C) from 20 °C to (°C)					
	Ni	Cr	Co	Fe		100	200	300	400	500	600
Platinite	49	–	–	51	8.2	87	(from 0 °C) 88	88	87	92	–
Driver FeNi	48	–	–	52		79	86	88	88	93	–
A L 4750	47	–	–	53		84	88	89	91	98	108
Vacovit 501 Carpenter 49	49	1	–	50	8.2	91	91	91	89	97	107
Platinum	–	–	–	–	21.4	89		(from 0 °C) 92		96	
Nilo 50	50	–	–	50	8.2	93					
Driver 52	52	–	–	48						100	
Sealmet 1	–	28	–	72	7.6	84	93	98	102	105	108
Chrom Iron	<0.4	23.5 27	–	76– 73				99	104	108	
Dilver O	–	28– 25	–	72– 75	7.5	84	93	98	102	105	108
FeNiCr	47	5	–	48				88			
Vacovit 511	51	1	–	48	8.2	101	101	101	101	102	109
Driver 74/26	–	26	–	74		86	97	102	106	112	–
Telemet	–	16– 23	–	84– 77		–	–	104	–	110	
Dilver T	–	×	–	×	7.6	93	(from 0 °C) 99	104	108	110	–
Novar B	0.3	19	–	80 (Nb 0.5–1		96	99	106	108	110	112
Vacovit 540	54	–	–	46	8.2	106	106	107	107	108	113
Vacovit 025	–	25	–	75		103	105	107	109	111	112
Fernicochrom	30	8	28	37		88	91	94	101	115	–

* A – Allegheny Ludlum Steel Corp. Brackenridge, Pa., U.S.A.; Ph – Philips Eind-
AG., Hanau, Germany; CS – Carpenter Steel Comp. Reading, Pa., U.S.A.; I – Acièries
Sylvania Electric Prod. Inc., Bayside, New York, U.S.A.; St – Stupakoff Ceramic %
see, Reute-Tirol, Austria; H – Stahlwerk Hagen, Germany; VM – Vacuum Metals Corp;

Inflexion Point (°C)	Electrical resistivity (Ohm-cm. 10^6)	Thermal conductivity (cal/cm. sec. °C)	Manufacturer*	Remarks
–	50	0.038	I	Curie Point 480 °C
			D	
425	50	0.037	A	
			V	
445	58		CS	
–	9.8	0.17		
470	41–47	0.032	W	Curie Point 480 °C M.P. 1450 °C
	43		D	
		0.059	A	M.P. 1480 °C
		0.04	Ph	M.P. 1490 °C
	65	0.029	I	Curie Point 570 °C
340			W	
480	51		V	
			D	
	60	0.057	A	
	65	0.029	I	Curie Point 640 °C
			H	
550	35		V	
	70		V	
380	–	–	–	Hull[610]

hoven, Holland; D – Driver Harris Co. Harrison, N.J., U.S.A.; V – Vakuumschmelze d'Imphy, Nièvre, France; W – Henry Wiggin & Co. Ltd., Birmingham, England; S – Mfg. Co. Latrobe, Pa., U.S.A.; G – General Electric Co., U.S.A.; M – Metallwerk, Plan- WD – Wilbur W. Driver Co.

f) Iron cannot be sealed directly to lead glasses, since it reduces the lead oxide contained in them, to metallic lead. For such seals a prior copper plating or the use of an intermediate glass (without lead) is necessary.

If pure iron or easily oxidizing iron alloys are to be sealed to glass it is recommended that their surface be decarburized in order to avoid gas bubbles being formed in the seal. The decarburizing is done by heating the iron in wet hydrogen (30 min at about 1050 °C). It is also advisable to plate the iron with copper, nickel or chromium.

g) Copper regularly used to be sealed to glasses utilizing the Housekeeper technique which does not require the matching of the expansions (Section 24.41), but there are also a few glasses (Table 2.32) which can be sealed to copper using a matched seal. According to Oldfield[954] the seal can be made using "any convenient size and shape" and "normal glass engineering techniques".

A very light cuprous oxide layer can be formed by the usual borating technique (Section 24.41) or by using controlled oxidation to produce "a pale red" glass-to-metal interlayer. To keep the cuprous oxide layer thin enough it is necessary to prevent excessive oxidation of the copper prior to engagement with the glass. It is important that the oxide be the red (cuprous) and not the black (cupric) one, since if there is any appreciable proportion of cupric oxide the seal may leak. An ideal seal is characterized by a brilliant red, almost scarlet colour. However seals of other colours (pink, sherry, honey) are generally also sound. These colours correspond to greater degrees of incorporation of the cuprous oxide within the glass.

According to Wroughton[1334] conducting parts for sealing to soft glasses can be made by pressing a mixture of 50 per cent copper powder and 50 per cent (by volume) molybdenum powder to a rod, and firing this in dry hydrogen for two hours at 1050 °C. If the content of molybdenum is varied from 40 to 80 per cent alloys having expansion coefficients of 110 to 88. 10^{-7} can be obtained.

h) Titanium (and zirconium) may be sealed to some glasses (Table 2.32) without the need for any special sealing techniques. If a flame sealing method is used it is necessary to keep the sealing time as short as possible since the expansion characteristics of titanium change on prolonged heating in air (Rawson[1033]).

24.32 Matched glass–metal seals with hard glasses. Hard glasses can be sealed to tungsten, molybdenum, FeNiCo alloys and some other metals (Table 2.34).

a) Tungsten has a fibrous structure and develops longitudinal splits if it is not handled properly. These splits are produced especially by bending or cutting operations. Thus bending of tungsten should be avoided, and instead of cutting, the required length should be ground off.

TABLE 2.34. HARD GLASSES FOR GLASS–METAL SEALS*

Metal**	Glass***	Remarks
Tungsten	C : 3320, 7720, 7780 (7070, 5420, 7741, 7252, 7750, 7331, 7050) Ch : GS 1 (Intasil), (GH 1 Hysil) BTH : C 14, (C 9) J : 1646, 8212 (3891, 8330, 2955, 8409) O : 712b, 712h, 742c (362a) GEC : W 1 K : K-772 P : Bluesil, Dial 36	
Molybdenum	C : 7040, 7052, 7050, 7042, 7510, 8830, (7750, 1720, 7331, 7055, 7720) Ch : GS 4, GSB BTH : C 14 (C 11, C 37, C 46) J : 1639, 2877, 2954 (8401, 1447) O : 637h, 637n, 637x, 906c, 632a GEC : HH, (H 26) K : 51-26 P : (Kodial)	Hull[610] Bertele[116] Adam[2a]
FeNiCo	C : 7052, 7040, 8800, 7520, 7055, 7050, 7750, 7340, 7060, 1720 Ch : GS 3 BTH C 40 J : 1447, 8243, 8401, 8482 O : 756b, 911b GEC : FCN, SBN 124 K : K 650, K 705, EN 1 P : Kodial	see Table 2.33 Herrmann[547] Redston[1037] Graham[468] Heylen[553] Scott[1115] Mairs[795]
FeNiCoCr	C : 0080, 0050 Ch : PWD K : R 6	see Table 2.33
Rhenium	GEC : HH, H 26 X,	
Zirconium	C : 7052 BTH : C 40	Corak[236] Rawson[1033]
Tantalum	C : 7052, 7720 J : 1447 Ch : GS 4	
Silver	O : 424d	

* See also Kalsing[652], Kohl[706], Mönch[879], Ardenne[47], Espe[350, 356], Roth[1081, 1083], Colnot[223], Zincke[1354].

** See Table 2.33.

*** For the significance of the abbreviations see footnotes of Tables 2.10 and 2.29.

For sealing to glass the tungsten rods must first be surface ground in order to remove longitudinal channels which usually exist on the surface of this metal. The surface channels can be made more visible by electrolytic etching (at about 7 A/cm^2) in a solution of KOH using the tungsten as the anode against a cathode of copper (see also Table 2.20). Reimann[1046] recommends that during the etching the current be changed from direct to alternating for a moment and then back again to direct current. The change in current density is accompanied by changes in the noise produced by the bath, and the noise can assist in finding the proper conditions for the etching. The current density can be controlled by changing the depth of the tungsten rod in the bath.

For the surface cleaning of the tungsten prior to the sealing in glass, one of the pickling solutions from Table 2.20 can be used. If sodium nitrite or potassium nitrite is used the tungsten is heated to a dull redness and a small quantity of the salt is rubbed on the surface. The salt melts and flows over the whole surface heating up the tungsten to bright redness by chemical reaction. After washing off the salt the surface remains clean and bright. If a solution of nitrite is used, the tungsten must be heated prior to repeated immersion in the solution.

It is recommended that the tungsten be oxidized to a blue–green colour. The oxidation is limited to the desired degree, by slipping over the tungsten rod a closely fitting glass sleeve and (during heating) quickly collapsing it to the metal. The beaded wire should then be heated to bright red, until on cooling a gold–yellow colour is obtained (see also Section 24.11).

Reimann[1045] developed a method of sealing tungsten to glass by first providing the tungsten with an adherent layer of copper and then sealing it to glass in the same way as used by Dumet (Section 42.2). The copper (or copper plus nickel) is first electro-plated on to the tungsten* rod and then fused in hydrogen (10 min at 1130 °C) to flow into and fill the cracks. After this firing process, the copper layer is thickened by further electro-plating up to about 0.03 mm. Finally the coating is sintered in hydrogen, burnished, borated (Section 24.41), and embedded in the suitable glass (Fig. 2.73). The expansion coefficients resulting in radial direction by copper plating the tungsten are plotted in Fig. 2.73.

Tungsten seals usually develop leaks due to the humidity especially when this is combined with heat (tropical climates) (Chilcot[213], Kohl[706]). To avoid this effect it is recommended (Trebuchon[1235]) that the glass surfaces near the seal be flame polished, or the tungsten be plated with a thin layer of nickel covered by a second layer (chromium).

* The electroplating was done (after cleaning in a sodium nitrite solution) in a bath containing 29 g sulphuric acid, 160 g copper sulphate, and 1000 cm^3 distilled water, using 25–35 mA/cm^2.

Where the seal must be resistant to alkaline vapours (e.g. ceasium), chromium plating is recommended, since these vapours reduce the oxides of tungsten, (Fincke[375], Kesselring[670]).

When completing a tungsten–glass seal, the tungsten wire is first degassed by heating it in a reducing flame (Section 23.23) in order to avoid bubbles in the seal. After cleaning the wire in sodium or potassium nitrite (see before)

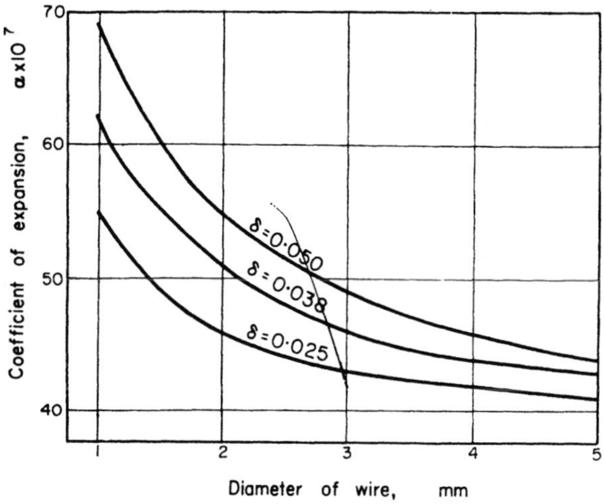

FIG. 2.73 Optimal thermal expansion of glasses for various copper plated tungsten wires (δ-thickness of the copper clad, mm)

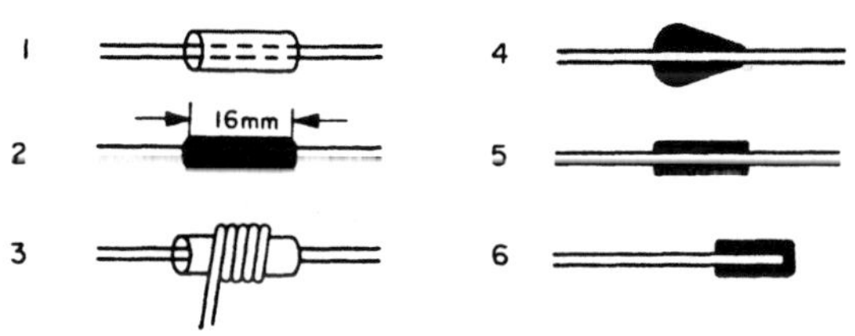

FIG. 2.74 Stages in completing a tungsten wire–glass seal

it should have a uniform bright surface. If dark portions appear on the tungsten, it must be cleaned again. The fitting glass (Table 2.34) tube is drawn to a diameter close to that of the tungsten rod, and a length of about 15 mm is cut from the drawn, thin walled tube (1, Fig. 2.74). After the tungsten has been oxidized (see before) the glass sleeve is slipped over the rod and it is fused to the rod first at one end and continued progressively to the other end (2, Fig. 2.74). Over the sleeve, more glass is then applied by winding it from a glass rod (3, Fig. 2.74), and fusing it to a pear shape (4, Fig. 2.74). If the tungsten rod (wire) is to be sealed into a pinched stem (Section 42.2) a thin beading is enough (5, Fig. 2.74), and if its designation is an end seal the thin bead should also cover the end of the rod (6, Fig. 2.74).

b) Molybdenum, unlike tungsten, has a surface which is usually free from cracks. Molybdenum oxidizes much faster than tungsten, and the oxide formed is volatile (above 700 °C it appears as a white smoke). The oxide layer quickly becomes thick and forms a non-adhering layer between the molybdenum and the glass. Thus the molybdenum should initially be oxidized only to a slight yellowish or at maximum to a blue–green colour.

To prevent "boiling" of the glass and excessive oxidation of molybdenum, Kohl[706] recommends that the combustible gas should be bubbled through a 5 per cent solution of boric acid in methyl alcohol or to pass the gas (or a part of it) through ethyl orthosilicate. Rawson[1032] and Stanworth[1178] prefer to place the molybdenum in hydrogen saturated with silicon tetrachloride (at 1100 °C) or to expose it packed in silicon powder to hydrogen containing hydrochloric acid vapour (30 min at 1000 °C). Hillier[563] prepares the molybdenum parts by first furnacing them in wet hydrogen at 900 °C and deplating the edges to be sealed in a weak chromic–sulphuric acid (Table 2.20). The parts are then oxidized for 5 min at 580 °C in air or oxygen, and the amount of oxide formed is evaporated by heating the molybdenum in dry argon (at about 1000 °C) so as to leave just the required layer of adherent oxide. Particular care should be taken not to contaminate the oxidized surface (e.g. by the fingers).

During the sealing process the molybdenum oxide condenses on the parts protruding into the vessel and on the walls of the vessel. To avoid this deposition one of the following techniques can be used:

(1) The molybdenum may be butt welded to another metal (e.g. nickel) which after sealing, projects alone into the vessel; a part of the nickel is also embedded in the seal.

(2) A long, close-fitting glass sleeve can be made for the beading, and oxygen should be excluded as much as possible during the sealing process (Barr[81]).

(3) The bead can be sealed to the vessel at one end, so that the whole length of the bead projects inwards (Reimann[1046]).

(4) The molybdenum projecting inwards may be cleaned after beading (Table 2.20) and protected from re-oxidation by making the sealing to the vessel in a neutral atmosphere (Reimann[1046]). Nitrogen cannot be used as the neutral atmosphere since it causes the embrittlement of the molybdenum over 1100 °C (Hillier[563]), thus dry argon should be used.

(5) The molybdenum parts not covered by the bead can be covered by close fitting nickel sleeves and these beads sealed to the vessel. After sealing the nickel sleeves are removed and the molybdenum cleaned (Liebson[765]).

(6) To plate the molybdenum with another metal, by using electro-plating techniques is reported (Hees[531]) as being unreliable. Hees[531] had "excellent results using the Chromallizing process. This is a vapour phase method of

depositing a dense layer of chromium on molybdenum". At the high temperatures employed in this process, diffusion occurs between the metals, forming a chromium case on the molybdenum. The bond between the case and the molybdenum core is strong. Oxidizing the chromium layer in wet hydrogen at 1100 °C for 10–15 min gives the rod (wire) adequate resistance to oxidation so that the surface can withstand direct flame at 1200–1300 °C. The chromallizing process should be applied after heating the molybdenum parts first in dry hydrogen and then in vacuum at 950 °C.

The sealing process of molybdenum is similar to that of tungsten (Fig. 2.74).

c) Iron–nickel–cobalt alloys (Table 2.33) should be oxidized either before or during the sealing. To ensure freedom of gas bubbles in the seal it is useful to decarbonize the surface of these alloys by heating the parts in wet hydrogen (about 4 hr at 900 °C or 1 hr at 1100 °C), or better still in vacuum since the remaining hydrogen can produce bubbles as well (Partridge[972]).

The cobalt renders the oxides more fusible and soluble in the glass, forming a strong bond both to the glass and to the metal. According to Scott[1115] a bright metallic appearance of the seal indicates a weak bond, a grey (to blue or brown) colour is a guarantee of a good seal and a black appearance shows that the metal is over-oxidized and the seal is not reliable. According to Pask[973] an excellent adherence of Kovar to glass is obtained with an oxide layer corresponding to a gain in weight of 0.3 to 0.7 mg/cm^2. This can be obtained by heating the Kovar in air for 17 min at 800 °C, or 3 min at 900 °C, or 1 min at 1000 °C.

For FeNiCo seals generally no beading is required. McCarthy[828] recommends that the Kovar–glass seal be made in the non-oxidizing part of a bushy gas flame (Section 23.23). A funnel of stainless steel is placed on top of the grating of the gas burner. The Kovar piece is first heated in the non-oxidizing central part, to orange red, then cooled in air for oxidation. A glass piece is placed on top of the Kovar and the assembly is reheated to the sealing temperature.

According to an other technique (Anon.[35]) the oxidized Kovar surface is covered with powdered glass suspended in a suitable solvent and sprayed on the Kovar. The suitable solvent for powdered glass to be sprayed is (Pask[973]) water or alcohol, having added a few drops of $LiNO_3$ dissolved in NH_4OH to keep the glass particles from settling out into a hard mass. After spraying, the assembly is placed in a furnace and heated in air (6 min at 1000 °C) until the powdered glass is fused to a smooth film on the Kovar. This glazed Kovar is then sealed into the glass vessel.

Turnbull[1240] first plated the Kovar parts with copper to a thickness of about 32μ and then fired them in hydrogen. Subsequently the parts were plated with chromium (very thin) and fired again in hydrogen. Finally the parts were oxidized in a hydrogen atmosphere. The thickness of the copper plating is limited by its effect on the stresses produced in the glass. The chromium layer should be thick enough to prevent oxidation throughout its whole thickness, and this way to endanger its adhesion to the copper. Figure 2.75 shows the expansion obtained in the radial direction by plating the Kovar with copper.

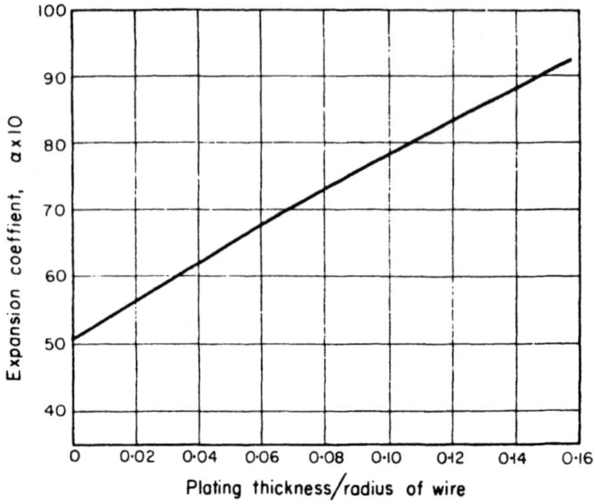

FIG. 2.75 Thermal expansion coefficients (in radial direction) of copper plated Kovar rods

Levin[761] considered the design of a short Kovar tube sealed at one end to a glass tube while its other end was welded to a flange made from a high expansion metal (e.g. stainless steel), and calculated the minimum length of the Kovar tube to reduce sufficiently the influence of the welded side onto the Kovar to glass seal. It was found that the design criterion for a safe seal is in this case

$$l \geqslant 3.5(a \cdot h)^{0.5},$$

where l is the length of the Kovar part, a its mean radius and h its wall thickness.

Figure 2.76 shows the typical steps in completing a Kovar–glass seal. A sleeve of glass is slipped over the oxidized Kovar tube (1, Fig. 2.76) and the glass is fused to build up a ring on the oustide of the tube (2, Fig. 2.76). The end of a glass tube is shaped as required (3, Fig. 2.76) and the glass tubing is sealed to the glass ring. An alternative technique consists in sealing the ring so as to extend to the end of the Kovar tube (4, Fig. 2.76) or even to cover the rounded edge of the Kovar tube (5, Fig. 2.76).

Barr[81] recommends the annealing of the Kovar–glass seals, by heating them to 484 °C for 20 min and decreasing the temperature at a rate of 1 °C/min down to 450 °C, followed by a cooling at 7–10 °C/min to room temperature.

During the glass–metal sealing process, the Kovar parts near the seal are always oxidized. They can be cleaned by immersion (for 10–60 min) in a hot solution (60–80 °C) of 50 g ferric ammonium sulphate, 125 cm^3 sulphuric acid (1.84), 150 cm^3 hydrochloric acid (1.16) made up with water to 1000 cm^3.

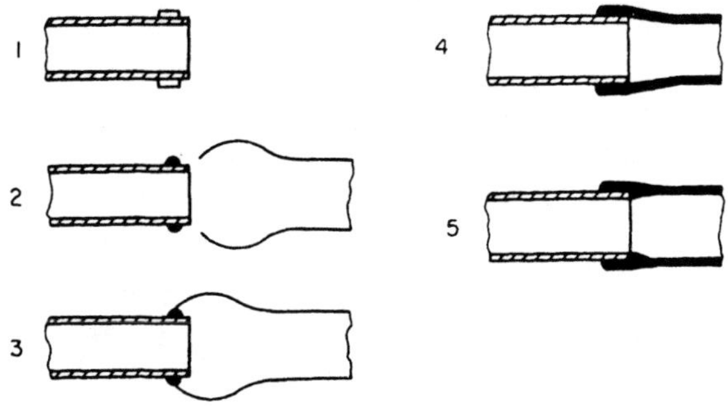

FIG. 2.76 Stages in completing a Kovar-to-glass seal

d) Zirconium was also sealed to glass (Corak[236]), using a zirconium tube of 25 mm diameter and 1.5 mm wall thickness, or a zirconium wire of 0.6 mm diameter. Good vacuum-tight seals were made with zirconium wire without no special treatment of the wire other than rubbing the wire with fine emery paper and heating for a short time in the flame. Zirconium seals are useful where chemical resistance or lack of megnetism is important.

24.4 Unmatched seals

The sealing of metals and glasses having expansion coefficients too different from each other to qualify for a matching seal (Section 24.12) can be accomplished by using some special techniques (Housekeeper[599], Partridge[972], Monack[875, 877], Reimann[1046], Kohl[706], Skinner[1146], László[742], Ronci[1075]).

The glass–metal seals known as "unmatched seals" (see also Section 24.1) are based either on the fact that the stresses developed in the glass are minimized by the deformation (elastic or plastic) of the metal or on the fact that the developed stresses are only compression.

The first kind of seals are represented by the seals known as Housekeeper[599] or Kruh seals, and the second kind by the compression seals. The Housekeeper seals are discussed in Section 24.41 (glass–metal) and Section 24.42 (quartz–metal), and the compression seals are described in Section 24.43.

*24.41 Housekeeper seals** can be made with any kind of glass if the materials used fulfil the following requirements:

(1) The expansion of the metal part must be larger than that of the glass (seal type *M*, see Table 2.28) in order to assure that the stresses normal to the glass–metal interface will always be compressive.

(2) The metal used must be soft enough and thin enough to be able to deform plastically, in order to allow dimensional changes which take place in the glass as it cools.

(3) The metal must form a strong bond with the glass (Section 24.11).

(4) The shape and dimensions of the seal must be designed so, as to provide a large contact area between metal and glass, avoiding at the same time any tensile stress perpendicular to the surface or too great stresses in any direction.

The metal fulfilling these requirements is copper, but in some such seals platinum, iron (steel) and molybdenum can also be used (Table 2.35). Using one or the other of these metals, the seal can be made in various shapes: wire seal, ribbon seal, feather-edge seal or disc seal. Table 2.35 and Fig. 2.77A summarize the dimensions used for the various shapes of Housekeeper seals for the various metals.

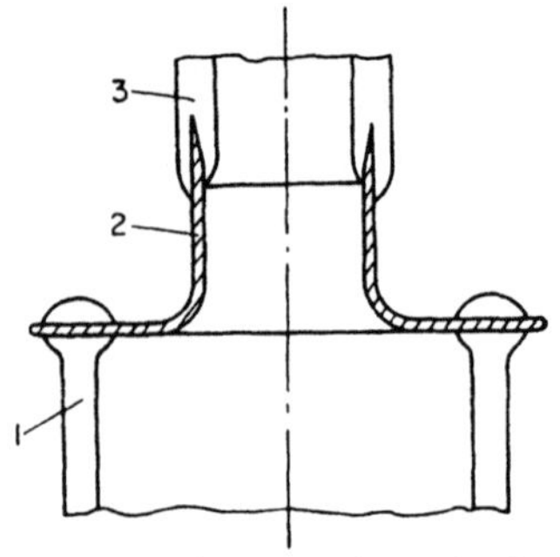

FIG. 2.77 Housekeeper seal, connecting soft glass to hard glass

a. Housekeeper seals using copper, are based on the plastic deformation of this metal, i.e. in the seal the yield strength of the copper is exceeded. Thus these seals are sensitive to sudden temperature cycling; e.g. the feather edge seals of copper can resist only about 300 cycles from 400 to 50 °C, while cup seals of FeNi can resist about 1000 cycles and FeNiCo cup seals over 5000 cycles of heating and cooling. From this stand-point, the copper Housekeeper

* Kohl[707] made the remark that the Housekeeper seal is an example of a technical solution delayed for many years, in spite of the fact that the principle was being used in this field. The remark refers to the Dumet seal which makes use of the low yield point of copper to release stresses caused by the expansion differences between metal and glass, when sealed to each other. "To extend this thin coating to a thin feather edge, required a bold leap in the imagination of the inventor."

TABLE 2.35. CRITICAL DIMENSIONS OF THE UNMATCHED SEALS

Shape	Copper	Platinum	Iron (steel)	Molybdenum
Wire (1) Fig. 2.77 A	d_{max} = 0.05 mm Barr[81]	wire: d = 0.2 mm flattened to 0.1 mm Barr[81] tubing: d = 2 mm wall 0.075 mm see Fig. 2.81 Wickers[1310]	stainless steel d = 0.5 mm Colgate[221a]	–
Ribbon (2) Fig. 2.77 A	a=25 mm b=0.4 mm Barr[81] α=8–10° a=4 mm b=0.1 mm Reimann[1046]	a=0.1 mm b=0.008 mm Davis[263]	–	a=1–3 mm b=0.01–0.05 mm see Section 24.42
Feather edge tubular seal (3) Fig. 2.77 A	d | t* | α | s 10 | 0.07–0.09 | 2–3 | 2.5–3.0 11– 50 | 0.07–0.09 | 2–3 | 3.0–3.5 51–100 | 0.11–0.13 | 2–3 | 3.5–4.0 101–125 | 0.13–0.15 | 3–4 | 4.5–5.0 Düsing[317] 27 | 0.08 | 1 | – Corak[236] – | 0.02–0.05 | 5 | – Kohl[706]	α=1–1.5° see Fig. 2.82 Graves[470]	$\alpha\approx$1° see Table 2.36 Benbenek[106]	–
Disc (4) Fig. 2.77 A	$t\approx$ 0.4 mm Barr[81] t = 0.3 mm Zincke[1354]	–	–	see Section 24.42

* t (mm) at a distance of 1 mm from the edge. d, s in mm; α in degrees.

seals are recommended for applications where the temperature is not high and temperature cycling occurs only exceptionally.

Housekeeper feather edge seals are not recommended for vibration service or in corrosive atmospheres, because of their weakness.

It is not recommended to use Housekeeper disc seals, by juxtaposing hard and soft glasses on the two sides of the copper disc. If the seal should be used as a connexion between hard and soft glasses it should be constructed as shown in Fig. 2.77. Here one of the glasses (1) is sealed to a flange (2) as a disc seal and the other glass (3) is sealed to the edge as a tubular seal.

As the Housekeeper seal is based on the thinness of the metal part sealed into the glass, on thicker ribbons, tubes or discs a tapered edge should be provided (Fig. 2.77A). It is recommended that the taper be made by rolling or machining against a steel backing, and not by turning. The surface of the copper at the tapered edge must be highly polished.

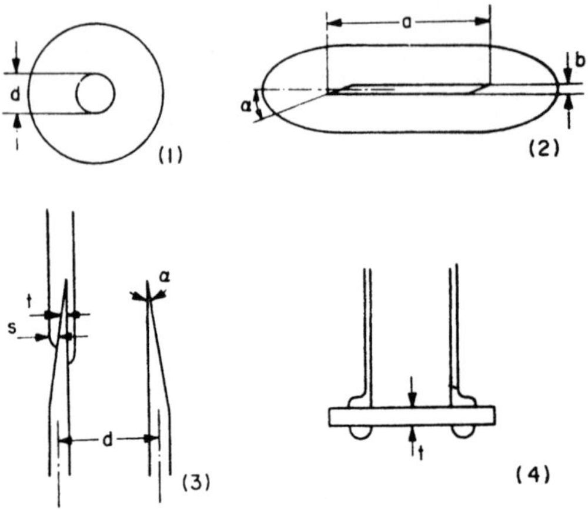

FIG. 2.77A Dimensions of Housekeeper seals (Table 2.35)

It is always advantageous to anneal the metal before using it for Housekeeper seals. This is done generally by a heating (in vacuum or hydrogen) which serves also to degas the metal and avoids bubbles in the seal. Corak[236] quotes heating times of about 3 hr at 950 °C.

For a good copper–glass bond a cuprous oxide layer is required (Section 24.11). The oxidation is done by heating the thick part of the copper, and preventing in this manner the thin portion from burning, since it is heated only by conduction from the thick part.

According to Kohl[706] the copper discs should be chemically cleaned (Table 2.20), hydrogen fired and oxidized by induction heating in carbon dioxide or nitrogen. It is advisable to accomplish the seal fairly soon after the cleaning (the same day).

A cleaning process for the copper part consists in washing twice in carbon tetrachloride* and rubbing thoroughly the sealing surface with cotton wool soaked in a chromic–sulphuric acid solution**. After this, the parts are washed in water, dried in hot air and borated.

* Excess solvent should be shaken or wiped off after each washing.

** The solution is made from 250 cm^3 chromic trioxide dissolved in a solution of 50 cm^3 concentrated sulphuric acid in 1000 cm^3 of distilled water.

The borax layer protects the copper from excessive oxidation, prevents the formation of the wrong (cupric) oxide and the incorporation of the borax in the glass results in a reduction of the softening temperature of the glass next to the copper. This reduces the temperature range through which the copper has to accomodate itself to the glass. See also McCarthy[827].

In order to borate the copper it is dipped into a saturated solution of borax ($Na_2B_4O_7 \cdot 10\ H_2O$) and subsequently drawn slowly through an oxidizing flame (Section 23.23) each part in turn being raised to about 900 °C (bright red). When cooled the borated surface should appear smoothly glazed with a red (sherry) colour. During the melting of the borax, the copper gradually becomes oxidized by the diffusion of oxygen through the molten borax. If after cooling the surface is too pale the heating was not sufficient; if the surface is matt the heating time was too long. After borating, any impurities left on the surface (from before) will show up as dark spots.

Malic[797] recommends that for a more uniform oxidation the parts should be heated after borating, to about 860 °C by induction heating.

Copper wire can be sealed into glass, based on the Housekeeper technique, if the wire is very thin or is flattened, so that the portion which is sealed into the glass is not over 0.05 mm thick. The clean flattened wire is pre-oxidized by heating and borating, and sealed into the glass by pinching it without any preliminary beading. The pinching consists in heating the glass and pressing it to the wire.

Copper ribbons up to 25 mm wide and 0.4 mm thick can be used for Housekeeper seals. The ribbon is sealed in a similar manner to the flattened wire, but it requires an additional operation. The edges of the ribbon should be rolled (or filed) until the two sides have angles (Fig. 2.77 A) of 8–10° (Barr[81]). To bead the copper ribbon Reimann[1046] recommends that it be supported vertically by an attached wire, the ribbon being inserted in the flattened end of a glass tube also held vertically. After softening the edge of the glass and pushing it into contact with the copper, the glass is sucked down progressively onto the copper, in such a way as to avoid the entrapment of air bubbles. When the assembly is allowed to cool, a crack develops in the region where the ribbon ends. The unengaged glass may easily be cracked off and removed. The beaded ribbon is then carefully re-heated and the region where the cracked glass was removed is heated locally to a sufficiently high temperature to allow surface tension forces to round off the glass remaining on the ribbon (Fig. 2.77B).

In its original form the Housekeeper[599] seal consisted simply of a copper ribbon sealed through the glass. The ribbon may be cut from a sheet or prepared from a previously annealed wire by hammering or rolling. The ribbon must be thin enough (Table 2.35) in order to avoid cracks developing in the glass. Housekeeper[599] found that such cracks always start at the edges of the ribbon, indicating that there the stresses in the glass are greatest. This suggested to Housekeeper the idea of making the ribbon thinner in this critical region.

Reimann[1046] made satisfactory lead–glass seals using ribbons 0.4 mm thick and as narrow as 3–4 mm. With hard glasses the maximum safe thickness is about 75 μ (if the angle θ is acute; see Fig. 2.77B). Such seals are difficult to make, due to the danger of burning the thin copper at the high temperatures required for hard glass seals.

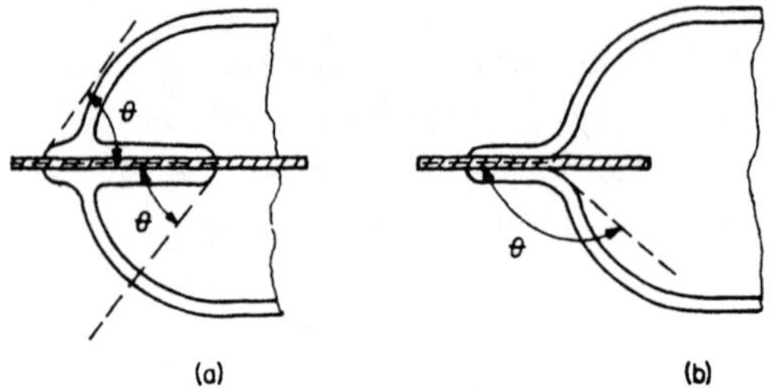

FIG. 2.77B Correct (a) and incorrect (b) ribbon Housekeeper seals

In order to take full advantage of the copper thickness it is essential (Reimann[1046]) that the angle at which the surface of the glass meets the copper (Fig. 2.77B) be acute or maximum a right angle (see also Section 24.14), the most favourable angle being about 40°. If the contact angle *6* is more than 90° the bulk of the glass will tend to tear away from the copper, leading finally to break-down of the whole seal. For this reason it is not permissible to make a seal of the form shown in Fig. 2.77B.b.

Tubular Housekeeper seals can be constructed:

(a) having the glass outside the metal (external seals, Fig. 2.78a),

(b) having the glass only inside the metal (internal seals, Fig. 2.78b),

(c) having the glass inside and outside the metal (double sided seal, Fig. 2.78c). In all these seals the copper tube is either cylindrical (Fig. 2.78a, c) or flared outwards (Fig. 2.78b).

With all three types of seals it is necessary to avoid sharp bends in the glass immediately beyond the edge of ıe copper (as on Fig. 2.78e); the seals should be constructed as shown in Fig. 2.78 d. It is recommended that copper not thicker than 0.075 mm at the edge be used for soft glasses and 0.050 mm for hard glasses. For more detailed dimensions see Table 2.35. The taper should be 2–5° (Table 2.35).

It is usual for the thin-walled tapered copper part to extend at least as far behind the seal as the length of the seal itself, but this condition is absolutely necessary for seals larger (in diameter) than 10 mm.

External tubular seals (Fig. 2.78a) are the least satisfactory, since in these seals the glass is in a strong radial tension. Nevertheless, such seals are usually successful when a soft glass is used, or if (with hard glass) careful attention is given to the profile of the glass and the seal is not submitted to heating and cooling cycling. The profile should be shaped as in Fig. 2.78f, rather than left with sharp edges as in Fig. 278g, since in these edges a dangerous

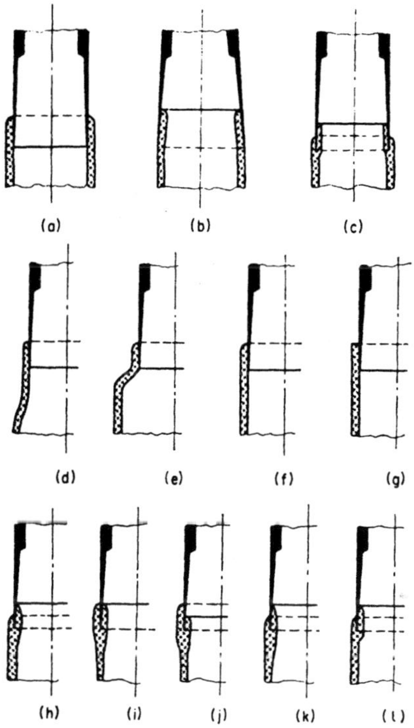

FIG. 2.78 Shapes of tubular Housekeeper seals. The use of arrangements e, g, j, l must be avoided

concentration of stresses appear (Fig. 2.70). In the case of soft glasses the profile usually acquires the required form on its own due to surface tension. Repeated heating and cooling cycling, lead to the hardening of the copper, which subjects the glass to progressively increasing stresses. These stresses finally become greater than the strength of the glass and lead to the cracking of the glass (regularly during a cooling cycle).

If made, external seals should have a glass wall of about 0.6 to 1 mm and the glass to copper overlapping by 2–3 mm.

Internal tubular seals (Fig. 2.78b) are much better than external seals, since on cooling the copper presses inwards, producing a compression in the glass. During the re-heating of such a seal the glass is subjected to tension but the copper being hot, can flow much more readily than it could if it were cold (in external seals). This results in a considerable reduction of the tensile stress developed in the glass during the heating (Reimann[1046]).

Internal seals should have wall thicknesses (glass) of 0.6 to 1 mm for smaller diameters, up to 1–1.8 mm for larger sizes. The glass–copper overlap should be 4–6 mm. The overlap increases inversely with the copper wall thickness, i.e. the larger overlap figures should be used for the thinner copper tubes.

Double-sided tubular seals (Fig. 2.78c) are particularly suitable for applications with temperature cycling. In the construction of these seals, it is important that the internal glass overlap should be longer than the external. Figure 2.78 h-j shows a good design (h) and two bad designs (i, j) of such a seal. If the external overlap of the glass was longer there would be a portion of the copper held unfavourably, as in the external seals. If the overlap is equal on the two sides (Fig. 2.78 i) the seal is stronger but not strong enough for temperature cycling.

In double-sided seals the glass should be pulled out so as to avoid any sharp transition in the wall thickness (as e.g. in Fig. 2.78 l); the correct design is that in Fig. 2.78 k.

In these seals the glass thickness should be 0.5–1.0 mm outside and 1–2 mm inside. The suitable glass–copper overlap is 2–3 mm outside and 4–6 mm inside, the overlap being larger for thinner copper tubes.

In order to construct a double-sided Housekeeper seal, the thin edge of the copper tube is heated red hot in a flame and a glass tube is pushed over the feather edge, (Fig. 2.79a). The glass is fused to the copper (without occluding bubbles) keeping the flame away from the thin edge, and the assembly is allowed to cool. The glass tube is cut resulting in an end as shown in 2 Fig. 2.79 a. The glass is heated and bent inwards (3, Fig. 2.79a) and then formed over the edge of the tube (4, Fig. 2.79a). At this stage the shape of the glass covering the feather edge should correspond to the condition required for the overlap (Fig. 2.78h). The glass ring covering the edge of the copper is butt sealed to a glass tube (5, Fig. 2.79a). This technique can be used to join two glass tubes of different expansions (e.g. a soft lead glass and Pyrex) with a portion of copper tube. Such an assembly is represented by 6, Fig. 2.79a.

Heavy copper conductors can be sealed into glass vessels using the same technique, giving bakeable seals for the transmission of an electric current into highly evacuated vessels (see also Section 4.2). The copper rod may be brazed or welded to the copper part which is sealed to the glass (1, Fig. 2.79b), assembled in a pressed fit (2, Fig. 2.79b) or joined by a screw thread (3, Fig. 2.79b). Such a seal can be made following steps 1–4 (Fig. 2.79a) or by the technique shown in 4–6 (Fig. 2.79b). Here, in the copper part (machined, cleaned, oxidized: 4, Fig. 2.79b) a glass tube flared (5, Fig. 2.79b) is inserted. The flared end of the glass tube is heated and bent over the edge of the copper (6, Fig. 2.79b).

Whitehead[1307a] describes a simple tool which can be used to facilitate the glassing of the feather edge. The tool (Fig. 2.80) is an expansion arbor machined from graphite rod. One end is sawed lengthwise up to a relief hole drilled through the rod, perpendicular to the cylinder axis. A set screw (Fig. 2.80)

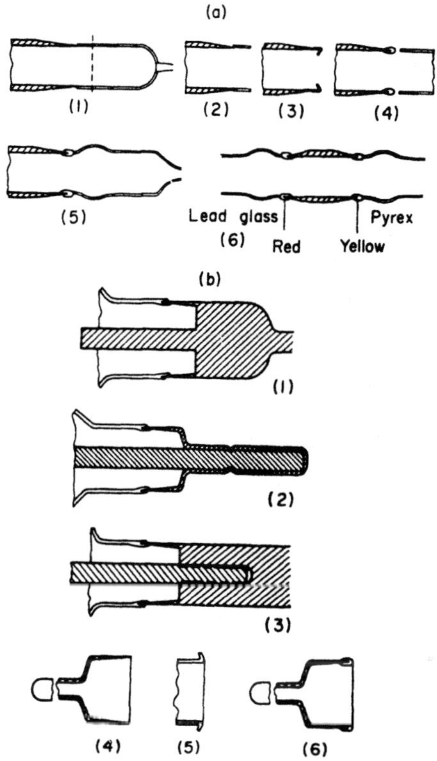

FIG. 2.79 Operations in completing a tubular Housekeeper seal

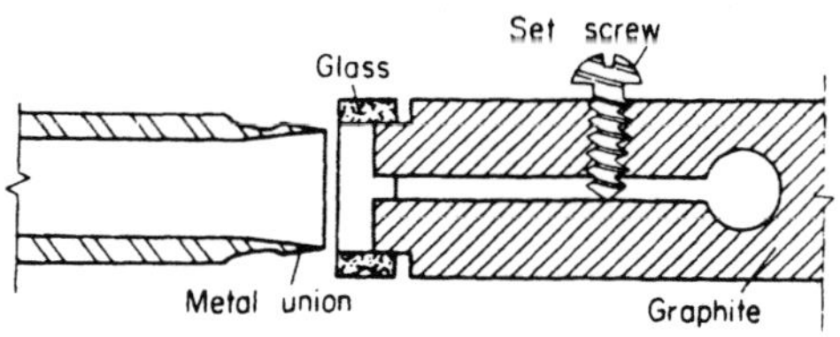

FIG. 2.80 Tool for making glass beads for Housekeeper seals. Reproduced from Whitehead[1307a] *(Courtesy of Pergamon Press)*

permits a variable force to be exerted between the jaws, while holding the glass ring on the end of the mandrel. The protruding end of the glass ring is positioned over the feather edge, is heated and sealed to the metal. The clamped end of the ring is then heated to softness and the graphite mandrel is withdrawn. The wrapping operation is next completed as usual (3, 4, Fig. 2.79a).

Malic[797] completed tubular seals using induction heating. The glass was placed in contact with the (borated, oxidized) copper part. The copper was heated by induction heating and kept at 800 °C for 4 min then at 900 °C for 5 sec. At this latter temperature the seal was made by pressing the glass onto the hot copper. By combining centrifugal action with induction heating, vacuum tight seals were made quickly without overheating the copper. In a particular case, the glass tube was inserted inside the tapered edge of the copper tube, and while applying induction heating the assembly was rotated at about 6000 r.p.m. At 920 °C the glass (Pyrex) was soft enough to flow along the copper surface by centrifugal action.

Disc seals were developed by Housekeeper[599] in order to construct sealed lead-throughs for heavy currents. For this purpose a copper wire passing through a hole in the centre of a copper disc is brazed to the disc resulting in a vacuum-tight joint. The copper disc is cleaned, borated, pre-oxidized, and the end of a glass tube is fused to the face of the disc. In order to reduce the stresses in the seal, a ring of the same glass should also be sealed on the opposite face of the copper disc (see also Fig. 2.77, the seal to glass 1). Cooled at room temperature the seal is robust but it would not withstand subsequent heating cycles except if the disc is thinner than 0.75 mm (for lead glasses).

b. Housekeeper seal with platinum. Thin platinum wires (0.2–0.3 mm diameter) can be sealed in any soft glass, resulting in a reliable seal despite the difference in the expansion characteristics, due to the ductility of the platinum. With lead glasses, thicker platinum wires can be sealed. Thus if the apparatus is to be constructed from soda lime or other soft glass, it is useful to initially embed the platinum wire in a sleeve of lead glass and then to seal this in the other glass.

In order to seal platinum wire into hard glasses it must be flattened. According to Barr[81], 0.2 mm platinum wire flattened to about 0.1 mm and having the edges formed in a V-shape can be successfully sealed into Pyrex or similar glasses. For this seal, the thin platinum ribbon is embedded first in soft glass (pressing strongly the soft glass to adhere to the platinum) and this embedded wire is sealed into the hard glass. Davis[263] describes the seal of a platinum tape ($12 \times 0.1 \times 0.008$ mm) embedded in a short length of glass and sealed through a small hole in the wall of a vacuum tube.

Wickers[1310] sealed platinum tubing to hard glass, using the principle of a Housekeeper seal, based on the fact that the metal has a sufficiently low yield point. A platinum tube (1. Fig. 2.81) 2 mm diameter and 0.075 mm wall thickness was sealed in the glass part (2).

Graves[470] modified the original Housekeeper seal in order to make tubular seals using platinum. This modification was necessary because platinum has a weaker adhesion (Section 24.11) to (Pyrex) glass than does copper, and by using the original design it was found that the platinum invariably separated from the Pyrex resulting in cracks and leaky seals (especially when platinum tubing 10–32 mm diameter and 0.075–0.10 mm wall thickness was used). The modification consists in making the angle of the tapered portion of the feather edge 1° to 1.5°. This small angle reduces the tension necessary to cause the stretching of the platinum to a level below the strength of the weak platinum–glass bond. The most satisfactory seals have been obtained with feather edges rolled out on a closely fitting rotating mandrel, using a steel roller or an agate burnishing tool.

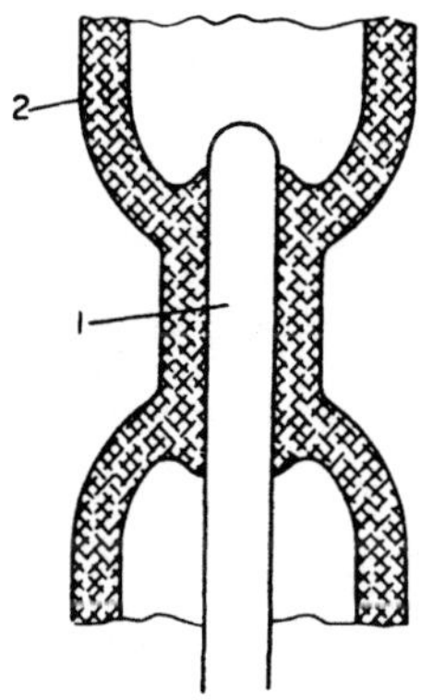

FIG. 2.81 Platinum tubing sealed to hard glass (after Wickers[1310])

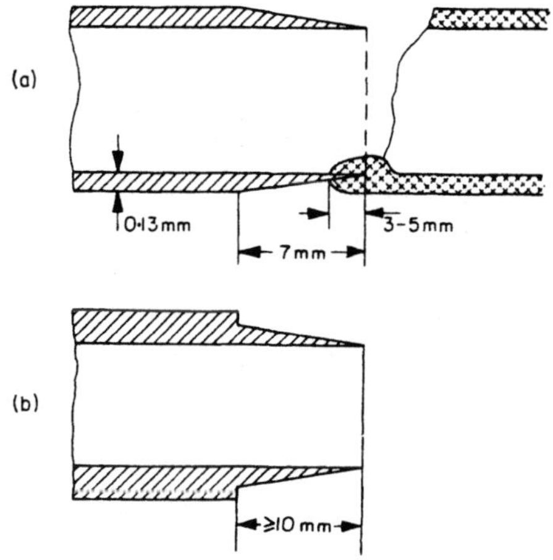

FIG. 2.82. Platinum feather-edge cylindrical Housekeeper seal. After Graves[470] *(Courtesy of The American Institute of Physics)*

In platinum tubular seals, the metal part may have at its edge a thickness of up to 0.025 mm, and the whole sealing surface should be smoothly polished. With tapers having ragged edges, the seals may appear satisfactory at first, but usually develop cracks or leaks after several hours of service.

Graves[470] recommends platinum–glass overlaps (Fig. 2.82a) of 3–5 mm both inside and outside, keeping the overlap inside as nearly equal to that outside as possible, but never shorter.

If the seal is to be made on the end of a thick walled tube, it is not necassary to produce a taper throughout the entire wall thickness; a tapered portion of 10 mm is definitely satisfactory (Fig. 2.82b).

c. Housekeeper seal with stainless steel. In order to make a tubular seal, the stainless steel is first fired in dry hydrogen at 1065 °C for 15 min to eliminate strains resulting from previous machining (Benbenek[106]). The strain-free stainless steel tubing is machined to the shape shown in Fig. 2.83. After a first cut from the inside of the tube has been made, to allow for centreing of the part, the taper is made on the outside. The modification introduced by Benbenek[106] as compared to the original Housekeeper[599] copper seal, consists in the second cut on the taper (Fig. 2.83) having a depth of about

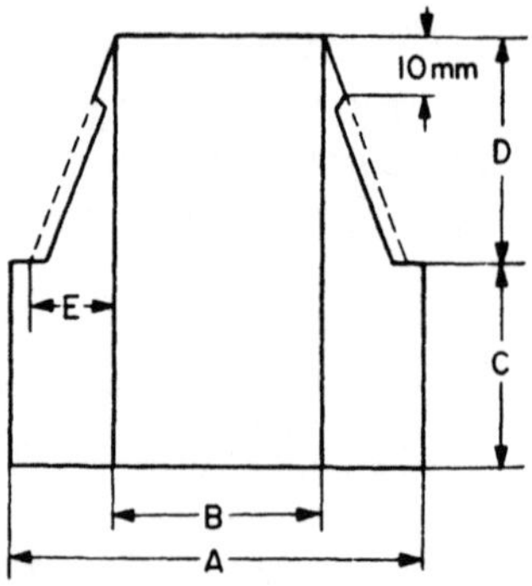

FIG. 2.83 Stainless steel feather-edge cylindrical Housekeeper seal (Table 2.36). After Benbenek[106] *(Courtesy of The American Institute of Physics)*

0.025 mm. This gives the required flexibility to the tapered end. The edge should be 0.025–0.035 mm thick. The other dimensions as recommended by Benbenek[106] are given in Table 2.36.

TABLE 2.36. CRITICAL DIMENSIONS FOR CYLINDRICAL STAINLESS STEEL HOUSEKEEPER SEALS (see Fig. 2.83)

Dimension (mm)	A	B	C	D	E
Size I	3.15	2.95	1.0	1.90	0.3
Size II	4.42	4.10	0.76	2.54	0.3

After machining, the stainless steel part is degreased and then fired* in line hydrogen at 1065 °C for 20 min.

Benbenek[106] used a borosilicate glass with an expansion coefficient of 46×10^{-7} which therefore matches the metal better than Pyrex. Colgate[221a] obtained good results by sealing Type 303, 304 stainless steel and Inconel directly to Pyrex. For the sealing operation the stainless steel tube was chucked in a lathe and wrapped (to a distance of 10 mm from the end to be sealed) with wet asbestos paper tape to keep it cool. The glass tube was shaped to overlap the outside of the feather edge for about 1 mm and fused to the metal, by pointing the flame on the glass rather than on the metal to prevent over-

* Colgate[221a] found that the seal can be made without firing if the metal is cleaned with reagent grade degreasing solvents.

heating the thin metal edge. The glass was flame-cut at a distance of about 3 mm from the edge of the metal and then rolled inside to complete the internal beading of the stainless steel (see Fig. 2.79a). To this beading the glass tube is then sealed.

d. Molybdenum Housekeeper seals consist usually of ribbons sealed in hard glasses and quartz. Instead of molybdenum also tungsten can be used, but molybdenum is preferred (Section 24.42) because of its easier mechanical workability.

24.42 Quartz to metal seals. The method used for sealing metals to quartz (fused silica) are based on one of the following principles:

(1) The sealing of thin metal foils or discs (Housekeeper seal).

(2) The use of ground joints sealed with plastic or liquid (metal or wax) layers.

(3) The use of graded seals.

Because the Housekeeper type of quartz–metal seal is the most widely used (Oranje[955]) and in order to keep the discussion of these seals together, the other two methods are also treated in this section, despite the fact that methodically they belong to techniques treated elsewhere in this book (Sections 3.1, 3.6, 23.35).

a. Thin metals sealed into quartz. The most commonly used quartz–metal seal is the molybdenum ribbon lead-through. The vacuum tightness of this seal is based on the fact that silica wets clean molybdenum and that the adhesion between the two is very strong, especially if the surface of the molybdenum is roughened by chemical etching (Partridge[972], Pequignot[982]).

The molybdenum ribbon required for the seal should be cut from foil using a hardened steel wheel, the foil being laid on a glass plate. The ribbon may also be rolled from molybdenum wire.

The etching of the molybdenum ribbon may be carried out either by pickling in a solution of sodium or potasium hydroxide or by an immersion in a mixture of nitric and sulphuric acid, followed by a thorough washing (see also Tables 2.20, 2.21).

The ribbon should be thin enough to avoid tensile stresses appearing in the quartz in a direction perpendicular to the plane of the ribbon. At the same time it should be thick enough to withstand the longitudinal stress set up in it, during the cooling of the seal. A molybdenum ribbon 0.01 mm thick withstands this stress if its width does not exceed 2 mm; with a thicker ribbon the width may be made greater. It is possible to make successful seals using ribbons as thick as 0.1 mm if the thickness is suitably reduced for a distance at least 0.5 mm from each edge (e.g. by electrolytic etching; see Table 2.20, 2.21), so that the cross section be elliptical with the thickness near the edge not exceeding 0.08 mm (Partridge[972]). Pequignot[982] recommends ribbons with a thickness of about 0.02 mm and a width of maximum 5 mm, or thicker ribbons having the edges thinned down to about 0.015 mm, (molybdenum ribbons of 15×0.0156–4×0.0156 mm are available from e.g. Thermal Syndicate[1218a]).

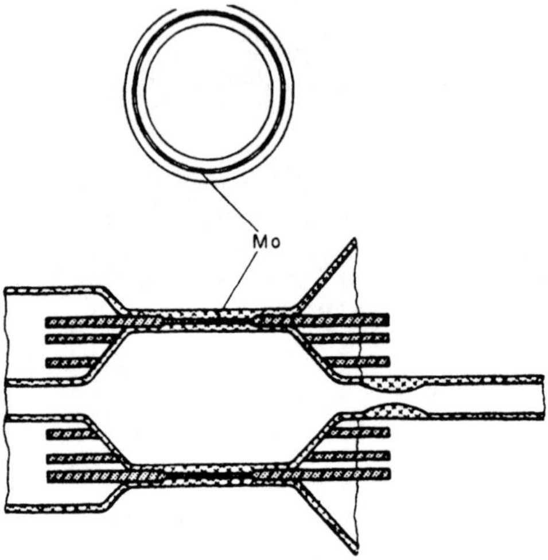

FIG. 2.84 Concentric arrangement of molybdenum ribbons sealed into quartz. After Pequignot[982] *(Courtesy of Société Française des Ingénieurs et Techniciens du Vide)*

The ribbon (with attached leads) is first inserted into a silica tube of internal diameter only slightly greater than the width of the ribbon, and with wall thickness not less than 1 mm. The silica tube is closed at one end, and then evacuated and heated for about one minute to bright redness. This procedure degasses both the quartz and the molybdenum ribbon. Finally the silica is collapsed on to the molybdenum and sealed to it.

Since the cross section of the ribbon is limited, for heavy currents a number of ribbons may be connected in parallel, disposed in a circle around a silica rod or tube contained within a larger tube. The outer tube is then collapsed so as to seal the ribbons between the inner and outer silica (Fig. 2.84a). Pequignot[982] reports on a seal having twelve ribbons sealed between the two concentric silica tubes (Fig. 2.84b), fitting closely in each other. After heating the assembly in vacuum first at 950 °C (for several hours) and then to about 1500 °C (to degas the quartz and the metal) the silica was fused slowly in order to allow for the elongation of the ribbons. The molybdenum should reach the maximum length at the moment of the pinching between the two walls. The pinching is made by the pressure difference between the space, where the ribbons are placed and the outside atmospheric pressure.

Besides the ribbon seal widely used in the construction of quartz lamps, a series of other quartz–metal seals based on thin sheets have been proposed or constructed. Nelson[917] describes a quartz-to-metal seal consisting of a pair of thin metal sheets juxtaposed around a metal rod and sealed to the rod (brazed) at one end. The other end of the sheets is sealed into the surrounding quartz tube. A similar technique using a thimble sealed to the rod and having a thin cylindrical wall tapering with the edge embedded in the quartz, is described by Greiner[482]. Power[1002] suggests making a seal using a metallic disc, and sealing the edge of this disc between two properly shaped silica tubes.

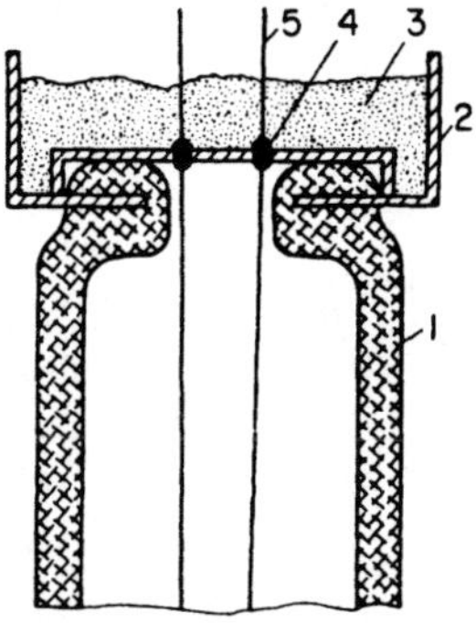

FIG. 2.85 Quartz-to-metal seal using a thin disc; (1) quartz tube; (2) molybdenum cupi (3) sealing powder; (4) insulating seal (Autostic); (5) thermocouple wires. After Priem[1010] *(Courtesy of The Institute of Physics and The Physical Society, London)*

Priem[1010] used a molybdenum disc (cup) of 1 mm (thickness) sandwiched between a quartz tube and a quartz rod (Fig. 2.85). After the sealing of this molybdenum cup (2) the end of the quartz tube (1) was pierced and a second molybdenum disc holding the conductor (thermocouple) wires was sealed into the cup by filling it with sealing powder* and baking it at 850 °C in a vacuum furnace.

b. Ground or tight fit quartz–metal seals. One of the first techniques used for making quartz–metal vacuum tight seals was based on Invar electrodes ground into silica (Fig. 2.86a). In these seals the tungsten rod (1) is inserted with the closest possible mechanical clearance in the quartz tube, but the vacuum tightness is assured by the mercury (2), closed with the Invar plug (3) grounded in the silica cone, and sealed eventually again with Piceine (4) (see also Section 31.2).

In another technique (Fig. 2.86b) the sealing is done using the ability of lead to wet silica and molybdenum.‡ The lead can form a vacuum-tight joint between the molybdenum wire and the silica wall of the tube. In order to complete the seal‡, a molybdenum wire is inserted through the capillary part (1 Fig. 2.86b1) of the silica tube and a clean piece of lead (2) is placed in the container situated above the capillary. After the silica tube has been

* Sealing powder 625/2011 from Ferro Enamels Ltd.

‡ See H. J. S. Shand, *Proc. Phys. Soc.*, **26**, 127 (1914).

closed at the top it is evacuated. The silica is then heated and collapsed on to the molybdenum wire and then the lead (2) is heated well above the melting point and allowed to flow into the space near the molybdenum (Fig. 2.86b2). Next the silica tube is cut and the upper end broken off while the lead is still molten, thus allowing the atmospheric pressure to force it against the silica and the molybdenum. Finally a tinned wire is introduced into the lead (Fig. 2.86b2) and the lead is allowed to solidify. This seal can be used only up to about 300 °C (melting point of lead: 327 °C).

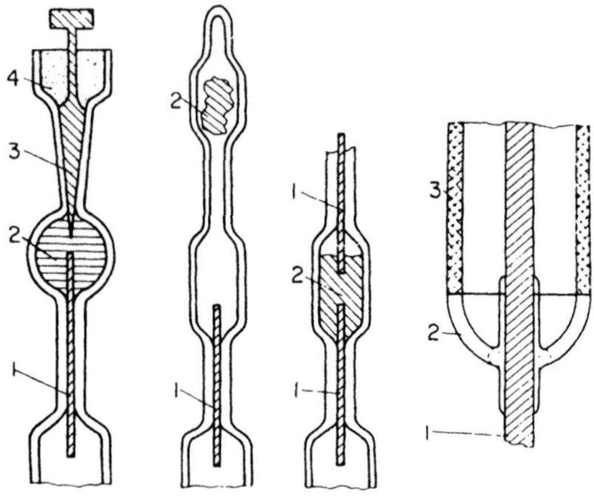

FIG 2.86 Quartz-to-metal seals; (a) with mercury; (b) with lead; (c) with glass sleeve

Instead of lead, Morris[889] used a lead–tin solder (Section 35.11). To make this seal an axial hole of about 8 mm diameter was drilled through the rod (of about 12 mm diameter). The rod was of a FeNi alloy (36 per cent Ni). The drilled end of the rod was turned to 1/10 taper which is then used as the female part of the seal. The inside of the tapered part was cleaned and tinned with antimony free lead–tin solder. The silica tube was made so as to fit as a male part into the metal cone. After heating up the two parts just above the melting point of the solder they are pushed gently together. On cooling a strong vacuum-tight seal was obtained (Morris[889]), which can withstand temperatures from −196° up to 200 °C.

In an attempt to make the seal stronger, Morris[889] used also a platinizing procedure. The silica surface, first roughened with carborundum was platinized twice by painting with a solution of platinum chloride in lavender oil and then heating gradually to a red heat. On the platinized surface a layer of copper was electro-plated, and the surface was then tinned and moistened with a zinc chloride solution (to avoid oxidation). The silica part was then soldered to the FeNi part.

Another possibility for sealing metal into quartz is that shown in Fig. 2.86c. It consists (Partridge[972]) in laying down on a tungsten rod (1) a thin (0.25 mm) layer of alkali-free glass (e.g. GEC 1404, Corning 7230) having a thicker portion in the centre formed to a button (Fig. 2.86c). This is sealed with a smooth rounded transition (2) to the silica tube (3) which has also been built up with sealing glass (e.g. GEC 2023). Despite the expansion differential which is in this seal many times greater than normal in common seals, (Section 24.12), this assembly can be successful, due to the strong glass-metal bond and the small thickness of the relatively long glass coating.

b. Graded quartz–metal seals. Metals can be sealed to quartz using graded seals consisting of a series of glasses with expansions varying from that of quartz to that matching the metal (Table 2.25). These graded seals have generally the disadvantage of being relatively long (Section 23.35).

Shorter graded seals than those formed by steps are the seals known as "multiform graded seals" and those known as "impregnated graded seals".

The *multiform* graded seals consist of rings built up of powdered glass mixtures, granulated with a binder, fired in a furnace and fused together to complete the grade. These graded seals are made (Corning Glass Works) to match tungsten or Kovar.

The *impregnated* graded seal is obtained by dipping a porous Vycor (Corning) glass into a solution containing alkali borates. After drying and firing higher expansion glasses are obtained. As the solution is sucked at various rates to various distances from the dipped end, the part produced has continuously varying expansion coefficients.

Lorenz[782] describes vacuum-tight conducting *cermets* melted into silica glass, produced from metallic powder (e.g. W, Mo) and silica powder, mixed, pressed and sintered.

24.43 Glass–metal compression seals. A poor match between glass and metal may be overcome in some cases by designing the seal to produce compressive stresses in the glass. In practice (Hull[609a], Adam[1, 4]) the compression seals consist of a metal ring (1, Fig. 2.87) surrounding a glass window (2). The window may have metal rods (3) or pipes (4) sealed through it.

In the window seals (Fig. 2.87a, c) the expansion coefficient of the outside metal ring should be always greater than that of the glass. In the rod (Fig. 2.87b, e) or pipe seals (Fig. 2.87d, f) the expansion of the metal ring (1), glass (2) and rod (or pipe) (3) must be in such a ratio so as to develop only compression stresses in the glass part. The most important cases are:

$$\alpha_1 = \alpha_2 > \alpha_3.$$

The expansion of the metal ring is equal to that of the glass (matched seal) but is greater than that of the rod or pipe sealed in it.

$$\alpha_1 > \alpha_2 = \alpha_3.$$

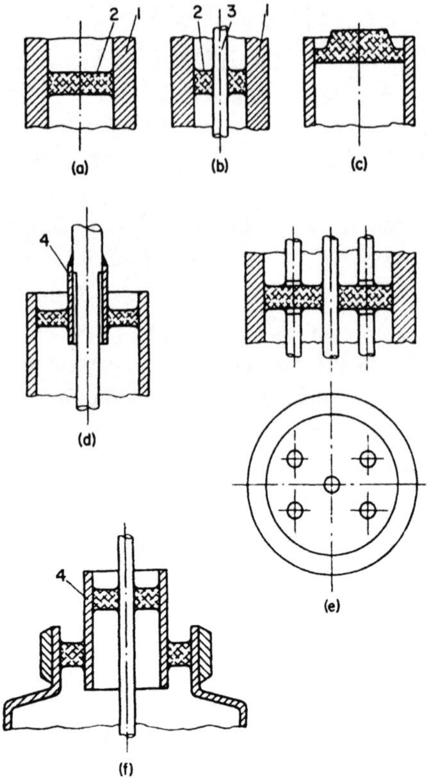

FIG. 2.87 Compression seals; (a) window seal; (b) rod seal; (c) specially shaped window seal; (d) pipe seal; (e) multiple rod seal; (f) rod and double ring seal. (1) metal ring; (2) glass window; (3) metal rod; (4) metal tube

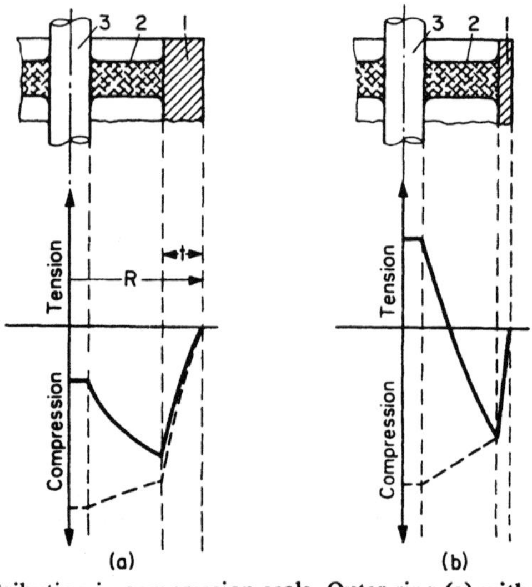

FIG. 2.88 Stress distribution in compression seals. Outer ring (a) with thick wall; (b) with thin wall. (1) metal ring; (2) glass; (3) metal rod

The expansion of the ring is greater than that of the glass, but the expansion of the rod or pipe sealed in is equal to that of the glass (matched seal) (see dotted lines Fig. 2.88).

$$\alpha_1 = \alpha_3 > \alpha_2$$

The expansions of the ring and rod (or pipe) are the same, but greater than that of the glass. Table 2.37 shows some metal–glass combinations which can be used in compression seals. Theoretically also cases like $\alpha_1 > \alpha_2 > \alpha_3$ or $\alpha_1 > \alpha_3 > \alpha_2$ are possible, but they have not yet been used in current practice.

TABLE 2.37. COMBINATIONS USED IN COMPRESSION SEALS

Ring	Glass	Rod or pipe
Mild steel	Lead glass (soft)	–
Mild steel (ratio of inner diameter to outer diameter = 0.75)	Lead glass	Single, mild steel rod, preferably 5 mm diameter, and max. 15 mm diameter
Mild steel	Glass matching rod or pipe (Table 2.32)	FeNiCr (Table 2.33)
Mild steel	Glass matching rod or pipe (Table 2.34)	FeNiCo (Table 2.33)
Mild steel	Glass matching Mo, W (Table 2.34)	Molybdenum Tungsten
Mild steel (diameter of the ring twice that of the rod; wall thickness max. 2 mm)	Glass matching FeNi or FeCr (Table 2.32)	FeNi or FeCr (Table 2.33)
Mild steel (or iron alloy) 40 mm OD; 7 mm wall thickness	Lead glass	Copper pipe (borated) 9 mm OD; 1 mm wall thickness

If the outer ring and the rod sealed in the glass are made of the same material (e.g. mild steel) the ratio between the outer radius R of the ring and its wall thickness t is given by the relation:

$$\frac{t}{R} \geqslant 1 - \left[\frac{E_1 - E_2}{(1-2\mu)E_2 + E_1}\right]^{0.5},$$

where E_1 and E_2 are the elasticity modulus of the ring material and of the glass respectively, and μ is Poisson's ratio (for the glass). Using $E_1 = 21000$ kg/mm^2, $E_2 = 7000$ kg/mm^2 and $\mu = 0.24$ it shows that the wall thickness of the ring must be equal or greater than one quarter of the outer radius of the ring ($d \geq 0.25R$). If this condition is not respected, tensile stresses will develop in the glass near to the rod (Fig. 2.88b). If the wall is thick enough, only a compression stress will appear (Fig. 2.88a). The dotted lines on Fig. 2.88 represent the distribution of the stresses if the seal between the rod and the glass is a matched seal.

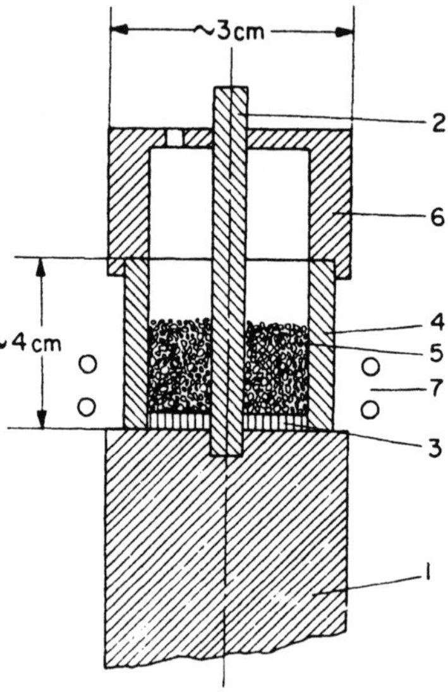

FIG. 2.89 Method of completing a compression seal of granulated glass. (1) cast iron support; (2) metal rod to be sealed to glass; (3) graphite washer; (4) mild steel ring; (5) granulated glass; (6) centring cup; (7) induction coil. After Adam[4] *(Courtesy of Rudolf A. Lang Verlag, Esch/Taunus, Germany)*

In multiple rod seals (Fig. 2.87e) the distance between rods or between the rods and the outer ring must be larger than the diameter of the rods.

Compression seals up to 20 mm diameter may be made using the powdered glass technique (see also Section 42.3). For larger seals granulated glass can be used. Figure 2.89 shows an arrangement to make a compression seal using induction heating (Adam[4]). It is recommended to heat the assembly in such a way so as to keep the internal rod at a high enough temperature without overheating the outer ring. For this purpose the cup (6) assures that the rod is kept hot and also centres it. The glass is used in a granulated form or as small rods packed between the rod (2) and the ring (4).

24.5 Faults in glass–glass and glass–metal seals and their detection

Unfortunately the kinds of faults occurring in glass-to-glass (quartz) and glass-to-metal seals are numerous. The drastic result of these faults is obviously the breakdown (cracking) of the seal, but for vacuum seals any cause leading immediately or after some time to a leak rate higher than that specified (Table 1.5), must be considered as a fault.

The faults can have their origins in the design of the seal, in one of the materials used (glass, metal), in the preparation of the materials for the seal, in the technique used during the sealing or in the treatment after the sealing. The main causes of faults in glass–glass and glass–metal seals are summarized in Table 2.38. For details of the various causes and the methods to avoid them the reader is referred to the Sections mentioned in Table 2.38.

TABLE 2.38. FAULTS IN GLASS–GLASS OR GLASS–METAL SEALS

Fault		Reference
Nature	Cause	
Mechanical	Excessive pressure difference on the glass walls (collapsing, bursting)	Tables 2.1, 2.2 and 2.3
	Excessive bending of the glass tube	Sec. 21.3
	Knocking with hard materials	Sec. 23.23
	Scratching with hard materials	Sec. 23.22, 23.23, 23.3
	Inclusions of heterogenous grains or bubbles in the glass	Sec. 23.3, 24.21
	Lack of adhesion (bond)	Sec. 24.11
Thermal	Excessive thermal gradients	Sec. 21.3
	Excessive thermal shock	Sec. 21.3 23.23
	Expansion differential (between two glasses, or between glass and metal)	Sec. 23.12 24.12
	Inadequate annealing	Sec. 23.13, 24.13
	Incorrect shape (of the seal)	Sec. 23.3, 24.14, 24.3, 24.4
Chemical	Weathering	Sec. 23.21
	Devitrification	Sec. 23.21
	Electrolytic effects	Sec. 41.14
	Chemical attack (flame, solutions, vapours)	Sec. 23.23, 23.21. 24.11
	Radiation damage	Sec. 71.3

The tests to be made in order to reduce the number of unsuccessful seals, or to avoid the use of unreliable seals on vacuum plants, devices, etc. depend on the kind of the production (single seals, mass production) and on the importance of the seal. A complete test schedule includes:

(a) The testing of the materials to be used in the seal.

(b) The destructive testing of some sample seals.

(c) The non-destructive testing of the seals.

The *testing of the materials* to be used in the seal refers especially to the expansion and thermal characteristics of the glasses and metals (Sections 23.1 and 24.1). These and other data can be found from the tabulated figures published by the manufactures, but if the materials are used in mass production, the tolerances must be always checked.

The *destructive tests* of the seals consist in heating or heat cycling the sample seals between the maximum and minimum temperatures possible during the utilization of the seals, and checking their behaviour. Simultaneously or separately, the seals can be exposed to humidity, corrosive atmosphere etc., according to the most difficult conditions envisaged for the seal.

Graves[470] described this test for Housekeeper seals (platinum). After the seal had cooled, the platinum is heated to a red heat at a distance of 1–1.5 times the diameter of the tube from the seal. The seal, if reliable, must not suffer any damage due to this treatment. If cracks appear, or any change can be seen with a magnifying glass after this treatment, the seal cannot be considered as reliable. Obviously it is advisable to anneal each seal after this test has been applied.

The *non-destructive testing* of the seals consists in visual inspection for shape imperfections, inclusions, colour, etc. using optical instruments (magnifying glasses, microscopes).

The stresses in glass can be detected before the appearance of the cracks. The method used for this detection is known as the "strain analysis in polarized light" or as the "stress birefringence method", and the literature on the subject is extensive; see e.g. Brewster[166], Hull[611], Partridge[972], Padmos[959, 960], DeVries[1274], Porubsky[1001], Espe[354], Kohl[706], Bartels[90], Roth[1082], Herrmann[545, 546], Hetenyi[551], Honigmann[592], Haase[496], Balmforth[78], Douglas[298], Fortey[391], Mylonas[910], Read[1036], Martin[813], Engel[343].

This method is based on the fact that the light which crosses strained glass (or birefringent crystals) is split into two components known as "the ordinary" and the "extraordinary" ray. Considering light as consisting of electromagnetic vibrations, the two rays have their electrical vibrations in planes at right angle to each other, and they travel at different speeds. If the ordinary ray travels faster, the crystal is called positive, in the opposite case the crystal is negative. Glass in tension behaves as a positive crystal, in compression as a negative one. The stresses in the glass can be determined by the optical path difference (retardation) Δ, produced between the two rays.

The practical arrangement for the determination of the stress is based on the measurement of the retardation and consists basically in a source of light, a polarizer and an analyser (Fig. 2.90). The light source *L* is an incandescent lamp (giving diffused light) or a monochromatic source of light depending on the kind of the measurement to be made. When the detection is to be based on colour comparison the incandescent lamp is used, when black and white extinction is wanted, monochromatic light should be used. The light crossing the polarizer *P* (Fig. 2.90) is polarized in a given plane* (say vertical). The

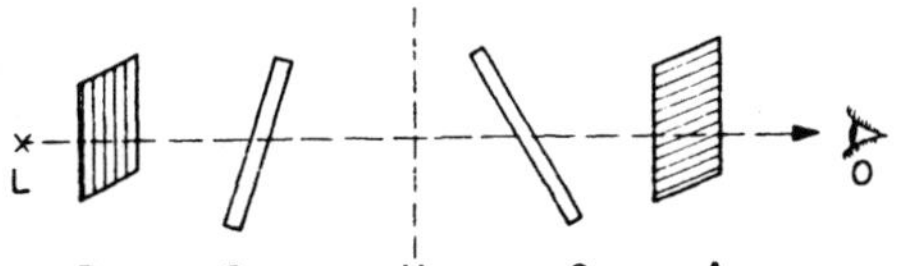

FIG. 2.90 Principle of strain analysis in polarized light. L = light source; P = polarizer; S = stressed glass; M = mica sheet; C = compensator; A = analyser; O = observer

plane in which the analyser *A* transmits the light is perpendicular to that of the polarizer (say horizontal). If the apparatus is composed only of *L*, *P* and *A* the field of the apparatus will appear dark for an observer *O*.

The polarizer *P* can be a black mirror receiving the incident light beam at an angle of 56–63° with the normal, or a polarizing filter (polaroid***). The analyser A is a Nicol or Glan-Thomson prism** or another polarizing filter (polaroid***).

When interposing between *P* and *A*, a mica sheet *M* (Fig. 2.90) the field appears coloured, the colour depending on the thickness of the mica. The most commonly used arrangement is that in which the mica gives a retardation of 560 mμ (the filter known as "red I").

Interposing the stressed glass *S* in the system (Fig. 2.90), the retardation produced in the glass is added or subtracted from the retardation produced by the mica. Due to this effect the differently stressed parts appear to the observer *O* in various colours. The value of the stress *P* (kg/mm^2) is given by

$$P = \frac{\Delta}{C \cdot d}$$

* The "plane of polarization" is normal to the plane in which the electrical vibration of a plane-polarized light beam takes place; i.e. the plane of polarization is that containing the magnetic vector.

** Calcite prisms, glued obliquely together; the ordinary ray is totally reflected on the contact surface, and absorbed in the blackened side of the prism, so that only the extraordinary ray can leave the prism.

*** The sheet polarizers are thin plastic foils containing or coated with small herapatite (quinine sulphate periodine) crystals, which have polarizing properties. These plastic foils are usually inserted between two, stress-free, glass plates (Haase[496]). Sheet polarizers are available e.g. from Sheet Polarizer Co. Union City, N.J., U.S.A.; Carl-Zeiss, Jena, Germany.

where Δ is the retardation in millimicrons (mμ), d is the thickness crossed by the light in the stressed glass, C is the photoelastic constant** of the glass (Table 2.39).

TABLE 2.39. PHOTOELASTIC CONSTANTS OF VARIOUS GLASSES (Brewster)

Glass	Manufacturer* or reference	C	Glass	Manufacturer* or reference	C
Lead alkali (80% PbO)	Shand[1124]	−1.05	7060	Corning	3.37
Lead alkali (73% PbO)	Shand[1124]	0.24	7740	Corning	3.43
Lead alkali (60% PbO)	Shand[1124]	2.05	3320	Corning	3.47
0080	Corning	2.40	Silica	Shand[1124]	3.47
0120	Corning	2.48	GSC	Chance	3.50
X 8	GEC	2.55	H H	GEC	3.54
X 4	GEC	2.57	GS 4	Chance	3.55
GWA	Chance	2.60	W 1	GEC	3.56
0060	Corning	2.66	GH 1	Chance	3.56
B 8	GEC	2.68	7050	Corning	3.60
C 19	BTH	2.70	7750	Corning	3.60
0010	Corning	2.73	C 40	BTH	3.60
K 1a	Philips	2.80	C 9	BTH	3.70
L 1	GEC	2.85	GSD	Chance	3.75
0050	Corning	2.85	GHA	Chance	3.80
GWB	Chance	2.90	GS 1	Chance	3.82
C 12	BTH	2.90	GSB	Chance	4.10
N 542	Philips	3.10	GS 3	Chance	4.26
7720	Corning	3.30	7070	Corning	4.36

* See footnotes of Tables 2.10 and 2.29.

** The photoelastic constant is expressed in Brewsters; 1 Brewster is equal to

$$1\,\frac{m\mu}{cm}\,\frac{kg}{cm^2} = 1\,\frac{Å}{mm}\Big/\frac{kg}{cm^2} = 100\,\frac{m\mu}{cm}\Big/\frac{kg}{mm^2} = 10\,\frac{m\mu}{mm}\Big/\frac{kg}{mm^2} = 0.175\,\frac{m\mu}{in}\Big/\frac{}{n^2}$$

For example, a 3 mm thick Corning glass 0050 ($C = 2.85$ Brewster), presenting a retardation of $\Delta = 150$ mμ has a stress of

$$P = \frac{150}{2.85 \times 0.3} = 175 \text{ kg/cm}^2 = 1.75 \text{ kg/mm}^2.$$

This example is represented also on Fig. 2.91, which can be used to evaluate the specific stress P_0. For the example plotted on the diagram it results $P_0 = 5.25$ kg/mm$^2 \times$mm, i.e. for the glass having a thickness of 3 mm

$$P = P_0/3 = 1.75 \text{ kg/mm}^2.$$

The graph can also be used, based on the colour appearing with an inserted Red I mica sheet. On the diagram in Fig. 2.91, Δ and P_0 are represented with

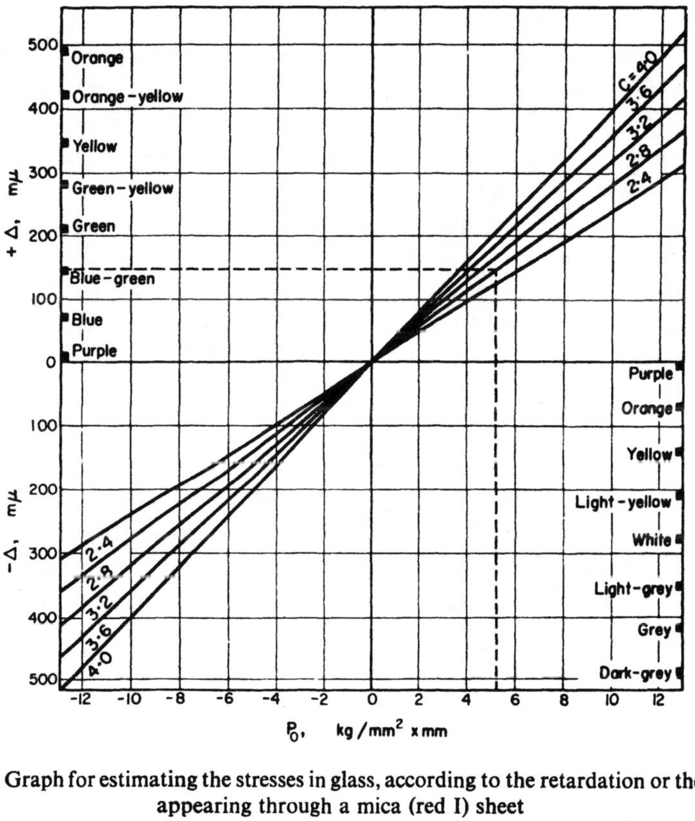

FIG. 2.91 Graph for estimating the stresses in glass, according to the retardation or the colour appearing through a mica (red I) sheet

plus and minus signs. If a plus sign represents tension, the minus represents compression and vice versa. In order to establish if in the given seal actually tension or compression occurs, a calibration is necessary. The calibration may be done by bending a glas rod placed at 45° with respect to the analyser and observing the succession of the colour appearing on the concave part compression) and on the convex side (tension). The calibration can also be done by sealing a metal (wire) into a glass with a known, very different expansion. If the expansion of the glass is greater (e.g. tungsten sealed in soft glass) the axial stress will be tensile, and if the expansion of the wire is greater (e.g. platinum sealed in hard glass) the axial stress will be always compressive. Observing the colour succession in such a seal, the kind of stress can be defined in any stressed glass inserted in the polarized light at the same angle as the calibrated seal.

The photoelastic constant of a glass can be determined by measuring the retardation produced by a given thickness of the glass (rod) subjected to a known loading. The retardation is measured in a direction perpendicular to the direction of the tensile force, and the photoelastic constant is calculated using the formula C (Brewster) $= \Delta(\mathrm{m}\mu)/\mathrm{d(cm)} \cdot P(\mathrm{kg/cm^2})$.

For the proper measurement of the stresses *calibration scales* or *compensators* (C, Fig. 2.90) are used. The simplest calibration scale consists of a series of gypsum sheets having gradually greater thicknesses in steps representing the same fraction (say 1/8) of the wave length of the incident monochromatic light (say 560 mμ). This calibration scale inserted at 45° with respect to the polarizing plane, can be used to choose the step which gives the same retardation as the stressed glass. For example if the stressed glass shows the same appearance as the seventh step of the calibration scale, then the retardation of the stressed glass will be $560 \times 7/8 = 500$ mμ. The value of the stress itself can be determined by using the diagram on Fig. 2.91.

The compensators are devices able to produce a variable retardation, in order to compensate for the retardation due to the stressed glass. The compensator may consist of two quartz wedges which can be displaced with regard to each other (compensator of Babinet), or it is a calcite plate (0.1 mm thick) which can be tilted to an angle controlled by a micrometer screw (compensator of Berek). With common quartz compensators the retardation can be determined to an accuracy of about 5 mμ; using Berek compensators an accuracy of about 2 mμ can be reached, and by using very refined methods (Padmos[960]) even an accuracy of 0.05 mμ has been obtained.

Hull[611] describes a simple method of measuring the retardation by placing the seal at 45° to the plane of polarization, between crossed Nicol prisms, and inserting a thin quartz wedge (with an apex angle of about 1/4°) between the seal and the analyser. The field of view is then crossed by parallel dark lines, each representing an integral number of wave-lengths of retardation in the quartz. These lines are undeviated by the seal if it is free from stresses; but if the glass of the seal is stressed the dark lines will be displaced by an amount proportional to the difference of path (retardation) through the glass. The deviation will be towards the thick edge of the quartz for tensile stress and towards the thin edge for compression, in the case of a quartz wedge cut with its edge parallel to the optic axis, and vice-versa if the side of the wedge is parallel to the optic axis. The calibration of the device should be done by substituting for the glass seal a glass rod loaded with a weight.

The proper measurement of stresses can be done only if the stressed part (glass) can be placed in such a way on the path of the polarized beam, so that the polarized light cross only points with stresses of about the same magnitude and direction. The effect of the stresses on the polarized light depends on the summation of the stresses through the thickness of the glass crossed by the polarized light. Thus, as a limiting case, it is possible that the retardation produced in the first half of the glass part be compensated by that produced in the second half crossed by the polarized light, so as the final retardation (for the observer) be zero. To avoid this kind of mistake the strain analysis should be done either as a proper measurement on samples especially shaped for the purpose, or just as a comparison between various samples of the same shape taken from mass production.

In the case of glass–glass seals, the samples are prepared by butt sealing (Section 23.31) the ends of two rods (about 5–7 mm diameter). The cut ends of the rods are ground perpendicular to the axis and the seal is made so as to obtain a minimum flow of one glass into the other (sharp flame). The seal is well annealed (Section 23.13) and the measurement of the retardation is made as shown by Fig. 2.92, keeping the axis of the rod at 45° with respect

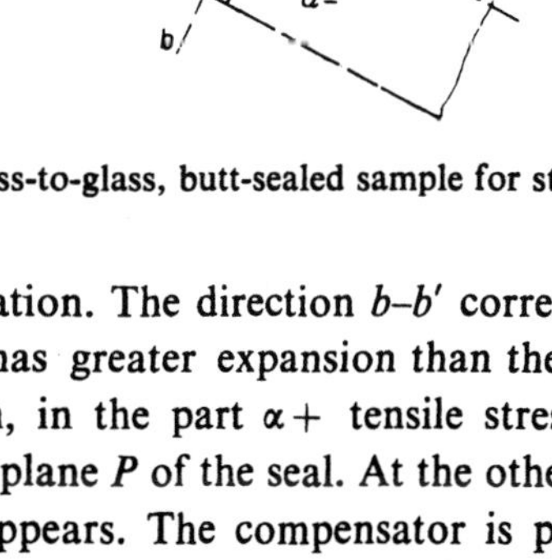

FIG. 2.92 Glass-to-glass, butt-sealed sample for strain analysis

to the plane of polarization. The direction *b–b′* corresponds to the plane of the seal. The part $\alpha+$ has greater expansion than the part $\alpha-$. Due to this difference in expansion, in the part $\alpha+$ tensile stress will appear parallel to *b–b′* and near to the plane *P* of the seal. At the other side of *P* in the same manner compression appears. The compensator is placed parallel to *b–b′*, the polarized beam being normal to the plane of the figure. The compensator is moved so as to compensate for the retardation appearing on one side of *P*. Then, from the value of the retardation determined in this manner, the value of the stress can be calculated (Fig. 2.91).

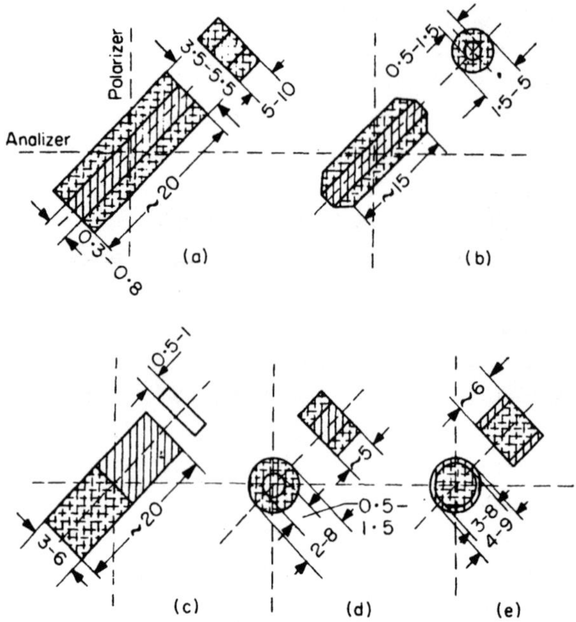

FIG. 2.93 Recommended shapes and dimensions of glass–metal samples for strain analysis; (a) metal plate sealed between two glass plates; (b) wire sealed into glass tube; (c) butt-sealed metal and glass plates; (d) metal wire sealed into glass ring; (e) glass window sealed into metal ring

TABLE 2.39A. IMMERSION LIQUIDS FOR STRAIN ANALYSIS

Immersion liquid	Index of refraction n	Glass (Corning) (n, for $\lambda = 589.3$ mμ)
Ethyl alcohol	1.36	
Carbon tetrachloride	1.46	7060 (1.48); 7740 (1.47); 3320 (1.48); 7050 (1.48); 7750 (1.47)
Iso-amyl-phthalate	1.48	7070 (1.47); Quartz (1.458)
Benzol	1.50	0080 (1.51); 7720 (1.49)
Xylol	1.50	
Chlor-benzene	1.52	0010 (1.54); 0120 (1.56) 0050 (1.55)
Tetraline	1.55	
Carbon disulphide	1.63	Heavy lead glasses (flint) (1.56–1.65)
Monochlor-naphtalene	1.63	
Monobrom-naphtalene	1.66	

In glass–metal seals the strain analysis gives even greater advantages as in glass–glass seals. The stresses resulting in a glass–metal seal depend on the expansion differential in the range between the setting point (Section 23.11) and room temperature. The nominal stresses in a glass–metal seal can be calculated (Sections 24.12, 24.14) but the real stresses appearing are somehow different from those calculated, due to various factors which cannot be included in the calculation. The real stresses can be determined only by strain analysis. For the determination of the setting point, polarized light is used, during the heating or cooling of the seal. The setting point is thus the temperature where the stresses disappear (on heating) or begin to appear (on cooling).

The recommended shapes of glass–metal sample seals for strain analysis are shown in Fig. 2.93. The best arrangement is that in Fig. 2.93d, since here the stress can be measured in direction *a* as well as in direction *b*.

Where the shape of the seal is such that the surfaces are not plane and parallel to each other, the refraction of the light must be avoided by examining the sample immersed in a parallel-sided cell containing a liquid with an index of refraction equal or near to that of the studied glass. Table 2.39A lists some liquids used for this purpose.

2.5 CERAMIC TO GLASS AND CERAMIC TO METAL SEALS

25.1 Sealing principles and materials

The ceramic–glass and the ceramic–metal seals are based on a single feature: the intermediate bonding layer between the two materials. The various aspects of the bond and the resulting bonding techniques are discussed in the extensive literature on the subject; see Vatter[1259], Pulfrich[1016a], Nolte[936a], Jenkins[630], Bondley[151], Mitchell[870], Palmour[965, 966], Williams[1317], van Houten [600, 601, 602], Kohl[705, 706], Espe[356], Tank[1213], Litton[775], MacDonald[791a], Coxe[243], Chick[212a], Brand[160], Hare[510], Marsden[808], Belser[105], Gibbons[448], Grove[489], Pryslak[1016], Goldstein[456], Godron[454], Beggs[103], Kühner[721b].

25.11 Sealing principles. The bond is a result of chemical and physical interaction in the very thin interface between ceramic and metal or ceramic and glass used as an intermediate material from ceramic to metal (Kingery[674], Pincus[991], Burnside[184], Pulfrich[1016a]). Various techniques were developed in order to obtain the best bond in these seals; Table 2.41 gives a list of these techniques, referring to the sections where they are described.

The tendency in ceramic–metal seals (as in glass–metal and ceramic–glass) is to choose metals having expansions near to that of the ceramic part. Even if the ceramic and the metal are chosen to have matching expansion coefficients on the whole range of temperatures occurring during the sealing opera-

TABLE 2.40. CERAMICS USED FOR

	Ceramic	Manu factu- rer**	α (10^{-7}/°C)	Ten- sile strength (kg/mm²)	Com- press. strength (kg/mm²)	Ben- ding strength (kg/mm²)
Forste- rites	Alsimag 243	AL	(25– 700 °C) 112	7	60	14
	Frequenta M	SM	(20– 800 °C) 106	4	85	11
	Rosalt 7	RI	(20–1000 °C) 90	5	90	15
	Forsterite 352	HC	from 125–85	7	60	13
	BN 3054	GC	(20– 400 °C) 105	7	60	14
	Frequentite S	SPP	(20– 700 °C) 111	—	70	14
Steatites	Alsimag 196	AL	(25– 700 °C) 86	7	63	14
	Almanox 13889	FP	(25– 700 °C) 73	5	52	12
	Steatit	SM	(20–1000 °C) 90	5	88	13
	Calit	H	(20–1000 °C) 85	5	95	15
	Steatite	SPP	(20– 700 °C) 85	5	85	13
Alumina	(Al_2O_3					
	Alsimag 576 (85%)	AL	(25– 700 °C) 75	14	98	28
	AD-85 (85%)	CP	(25–1000 °C) 79	12	140	—
	Stemag A 16 (90%)	SM	(20– 800 °C) 85	15	170	—
	Almanox 4462 (94%)	FP	(25– 700 °C) 73	11	131	25
	Alsimag 614 (96%)	AL	(25– 700 °C) 79	18	280	54
	Alsimag 652 (98%)	AL	(25– 700 °C) 80	18	294	44
	AD-99 (99%)	CP	(25–1000 °C) 92	—	220	—
	Degussit Al 23 (99.5%)	D	(0–1000 °C) 83	26	300	—
Zircon	Alsimag 475	AL	(25– 700 °C) 41	8	70	13
	Almanox 3569	FP	(25– 700 °C) 45	5	44	12
	ZI-4	CP	(25–1000 °C) 57	—	55	—
	Zr Porcellan	SM	(20–1000 °C) 52	5	80	11
Al-sili- cates	Hartporcellan	SP	(20– 700 °C) 38	3	42	8
	Porcelain	SPP	(20– 700 °C) 39	3	50	8

* See Jenkins[630], Kohl[705, 706], Williams[1317], Pryslak[1016], Nolte[936a], Gross[485b], Bondley[151],

** AL – American Lava Corp., Chattanooga 5, Tenn., U.S.A.; FP – Frenchtown D – Degussa, Frankfurt a.M., Germany; SM – Steatit-Magnesia A.G., Lauf/Pegnitz, Hermsdorf, Thuringen, Germany; SP – Staatlichen Porzellanmanufaktur, Berlin. Ger- N.J., U.S.A.; SPP – Steatite & Porcelain Prod. Ltd., Stourport on Severn, Worcs., England

\# For the significance of abbreviations see Tables 2.10, 2.29.

tion, the differences in heat capacity between metal and ceramic, makes it practically impossible to obtain a real matched seal. For this reason the

SEALS WITH GLASS AND METAL*

Safe temperature (°C)	Softening temperature (°C)	Specific gravity	Thermal conductivity (cal/cm.sec. °C	MATCHING Glass#	MATCHING Metal**
1000	1440	2.8	0.008	BTH C22, C 19, GEC RL 16, X 4, B 8	FeNi (46–51 % Ni); FeCr (16% Cr); Fe; FeNiCr
1100	–	2.8	0.006		
–	1460	2.7	0.006		
1000	–	2.9	–		
1000	–	3.0	0.008		
–	–	2.9	–		
1000	1440	2.6	0.006	GEC L 1; Corning 0010, 0080; Osram 562m, 850, 584d; Jena 2954, 16 III; BTH C 12;	FeNi (42–46% Ni) (FeNiCo)
–	1388	2.7	0 006		
–	–	2.7	0.006		
–	1470	2.7	0.006		
–	–	2.6	–		
1100	1440	3.4	0.040		Ni; FeNi, (FeNiCo)
1400	–	–	0.034		
1400	–	3.5	0.010		
1500	1920	3.5	0.018		
1550	1650	3.7	0.045		
1600	1700	3.8	0.045		
1725	–	–	0.070		
1900	2030	3.8	0.012		
1100	1440	3.7	0.012	Mo seal glass; Corning 3320, 7050; GEC H26X; HH; BTH C 11; Jena 3072	Molybdenum, Kovar
–	1550	3.1	0.007		
1300	–	–	0.010		
–	1500	3.0	–		
1100	–	2.5	0.004	Corning 7740, 7720; Jena 2950, 8330; GEC W 1; BTH C40, C9; Osram 394b, 3891;	
–	–	2.4	–		

Hynes[614].
Comp. Trenton, N.J., U.S.A.; CP – Coors Porcelain Comp., Golden, Colorado, U.S.A.; Germany; RI – Rosenthal-Isolatoren GmbH, Selb/Oberfranken, Germany; H – Hescho, many; HC – Hackney & Co. Ltd., England; GC – General Ceramic Corp., Keashey,

** See Table 2.33.

following principles are to be observed also when the expansion matching is good (see Table 2.40 and Fig. 2.94).

(1) The ceramic part should be designed having its wall thick enough to withstand the stresses and mechanical actions.

(2) It is recommended to use soft metals with high ductility, which cannot develop stresses exceeding the strength of the ceramic part.

(3) The expansion coefficient of the combined bonding and brazing layer should be kept near to that of the ceramic part, especially when the metal is sealed inside the ceramic part.

(4) The metal parts connecting ceramic to ceramic should be kept thin, in order to assure the elasticity of the assembly.

25.12 Materials used in ceramic–metal seals. The word *ceramic* refers generally to inorganic materials having their shape and hardness obtained by high temperature firing. The ceramic materials used in vacuum-tight seals (with glass or metal) represent just some small groups from this great family, namely the ceramics based on aluminium silicates, magnesium silicates or pure oxides. These ceramics are listed in Table 2.40, with their characteristics pertaining to the seals.

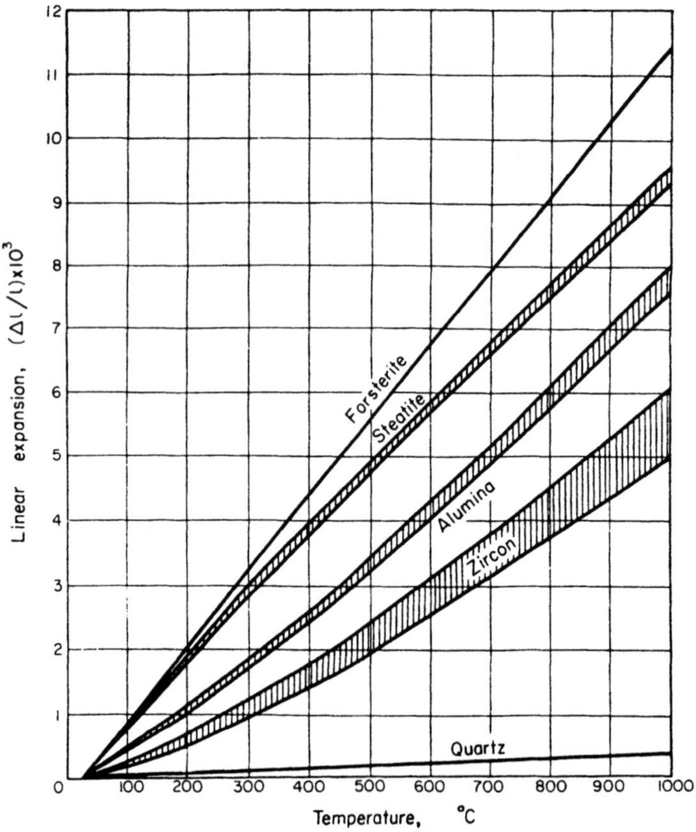

FIG. 2.94 Thermal expansion of ceramics

Ceramics are generally used in seals where their properties correspond better to the working conditions than those of glass. These advantageous properties of the ceramics are: high electrical resistance and low loss factor (Section 41.1), mechanical strength at high temperatures, low vapour pressure at high temperatures, bakeability, stability of geometric shapes and accuracy of their dimensions, no special annealing required (Velte[1263], Vlaardingerbroek[1270], Gibbons[448]). Although ceramics are more susceptible to radiation damage than metals, they can be heat treated to anneal the damage; in most cases nearly complete recovery is achieved (Koenig[704]).

For details on ceramics the reader is referred to: Hippel[564a], Marx[823], Kotowsky[713], Colnot[223], Johnson[636], Hynes[614], Economos[321], Mitchell[870], Kohl[706], Espe[354]. Porous ceramics used as calibrated leaks are treated in Section 61.44.

The ceramic parts are usually ordered in the required shapes and dimensions, but the ceramics can be ground (Rea[1035]) with tolerances of about 5 μ, and even threads can be cut in (or on) the ceramic parts, using diamond blades (Klein[686]).

To form an adherent bonding layer on ceramics, the surface must be free of any contamination. Since the contaminants are ordinarily unknown, a single specific cleaning procedure may be insufficient. Usually suitable cleaning is obtained by firing the ceramic parts in air at 800–1000 °C (Burnside[184]) Alternatively an alkaline cleaning solution (Section 22.31) can be used, followed by immersion in dilute nitric acid (2–5 min). Chromic acid or other glass cleaning solutions are also satisfactory (Section 23.21).

The metals used in ceramic–metal seals should not have a melting point higher than that of the ceramic (as in the case of the glass–metal seals), but just higher than the joining material (glass, glaze, etc). The metals and alloys used in ceramic-metal seals are Mo, W, Ta, Ti, FeNi, FeCr, FeNiCr, FeNiCo and also Cu and Ni (see also Table 2.40).

25.2 Sealing techniques

The various techniques used to make ceramic–metal seals are summarized in Table 2.41.

25.21 Ceramic–glass and ceramic–glass–metal seals. The glass to ceramic seals are used to join ceramic parts to glass equipment or ceramic parts to metal systems utilizing the glass as an intermediate part between the ceramic and metal.

(a) Glass–ceramic seals. The rules for glass to ceramic seals are similar to those of glass–glass seals, since both materials have a high compression strength and a much weaker resistance to tensile stresses. As with glass–glass seals (Section 23.12) the vacuum-tight sealing of a glass to a ceramic material is

TABLE 2.41. SEALING TECHNIQUES FOR CERAMIC–METAL SEALS

Sealing technique	Procedure	Reference
Ceramic–glass–metal	Section 25.21	Strong[1207], Roth[1083], Mönch[879], Anderson[32]
Glazed seals	Glaze in thin layer; Powdered glass Section 25.21; 35.2	Litton[775], Burnside[184], Spindler[1170]
Soldered glazed seals	Glaze with metal salt Section 25.21	Burnside[184], Bondley[151]
Sintered metal	Moly-manganese or W-Fe process (Section 25.22) Platinized porcelain (Section 25.25) Metallic oxide seals (Section 25.25)	Table 2.42 Strong[1207], Martin[819] Zincke[1354] Burnside[184]
Active metal processes (Section 25.23)	Hydride process Powdered Ti and Ni Carbide process Active metal (massive) Active metal "core" Active metal alloys	Bondley[151], Brand[160], Pearsall[981], Marsden[808], Stoeckert[1199] MacDonald[791a] Beggs[103] Bender[108], Velte[1263], Coxe[243] Kotowsky[715], Muller[904a]
Diffusion seals (Section 25.24)	Pressure seals (ram seal) Heated pressure seals	Tank[1213], Zollman[1357], Levin[761], Grove[489], Comsa[227]
Graded powder seal	Section 25.25	Knecht[696]
Electro-formed seals	Sealing by copper plating in electrolytic bath	Hare[510] (Section 25.25)
Various seals (for room temperature)	Graphite and metallized layer (Section 25.25) Lead seal (Fig. 7.12)	Jamieson[628] Marsden[808]

possible only if the expansion differential is not greater than 10 per cent; in exceptional cases this range can be raised to a maximum of 30 per cent if the compression stress is in the glass part and the tension in the ceramic part (Table 2.28). In this case, if glass with a lower expansion than the ceramic is used, it must be inside the ceramic part, but if the glass must be outside it must have a greater expansion than the ceramic part. Some glasses which can be sealed to ceramics are listed in Table 2.40.

Porcelain* may be sealed directly to Pyrex (Corning 7740) in diameters less than 12 mm, or it may be sealed to Pyrex in larger diameters with an intermediary glass ring of Nonex (Corning 7720) (Strong[1207]). Vacuum-tight seals between Mullite** tubes and Pyrex glass can be readily made using an intermediate ring of W 1 glass (GEC). Seals of this kind up to 50 mm diameter can be made with the bench blow-pipe; for larger diameters the glass-blowing lathe is to be used (Kent[662]). The seals are made by heating the end of the refractory tube to red heat, by rotating it in a large flame and then applying a ring of W 1 glass to the end, from a rod of about 3 mm diameter. After pressing this glass ring onto the ceramic tube to eliminate pinholes, the seal is completed by joining the Pyrex tube to the W 1 glass ring. The ceramic tube retains its heat longer than the glass, thus during the cooling the glass side of the seal must be heated with a large soft flame. In larger seals‡ made on a lathe it is convenient to use W 1 glass tubing and make a butt seal.

An alternative technique for glass–ceramic seals consists of pre-coating the ceramic surface with a glass layer, obtained from a powdered glass suspension in water or any other liquid. The suspension is painted on the surface of the ceramic part and after drying it is melted by heating the ceramic part in an adequate furnace. The glass part is then joined to this pre-coated ceramic surface. For this technique, torch heating can be used if the ceramic is not too thick (up to about 25 mm diameter), but furnace heating always gives more reproducible results. Ceramic plates cannot be sealed by using torch heating; for this purpose a furnace is always necessary. Figure 2.95 shows the typical shapes of ceramic–glass seals.

(b) Ceramic–glaze–metal seals. The ceramic–metal seals can be made by using solder glasses (Section 35.2).

For this process it is useful to have closely fitting ground or polished surfaces both on the glass and ceramic parts. The ceramic surface is coated with a solder glass or an enamel, applied usually as a suspension of powdered glass in an organic binder. On heating the ceramic component, the binder is eli-

* "Insulite" produced by Stupakoff Co., Pittsburg, Penn., U.S.A.

** Thermal 525 Mullite or H5 Mullite from Thermal Syndicate[1218a]

‡ In order to avoid the spreading of the cracks formed by cutting with an abrasive wheel, large Mullite tubes are best cut by sawing with a NiCr wire charged with carborundum-glycerine paste (Kent[662]).

minated and the glaze coating transformed into a uniform layer. This layer (about 0.2 mm thick) is ground to the required dimensions. The ceramic and the glass parts are then brought into mechanical contact and the so-formed envelope (see also Fig. 2.112) can be evacuated. The seal is effected in a furnace (Comp. Gén. Télégraphie Sans Fil[225]).

Sapphire–glass seals can be made using powdered glass (Spindler[1170]) or by direct sealing in a furnace (Anderson[32]).

Metal-to-ceramic seals can be made with glaze materials, if means are taken to control the oxidation of the metal so as to form a strong bond (Burnside[184]). Litton[775] describes such a seal made by applying a (25–50 μ thick) layer of

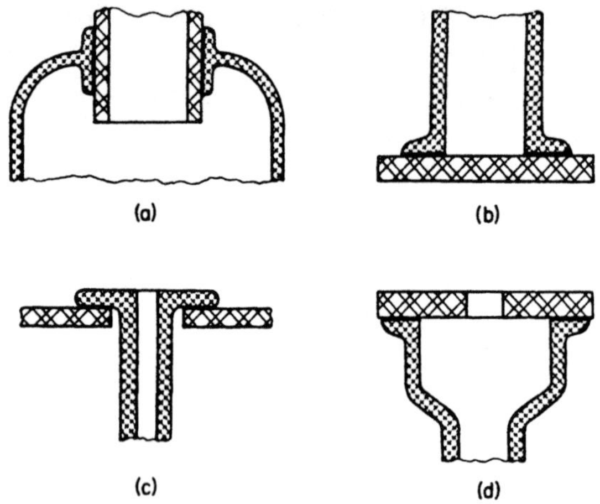

FIG. 2.95 Ceramic-to-glass seals; (a) ceramic tube sealed into glass tube; (b, c, d) ceramic plates sealed to glass tubes

ceramic enamel between the ceramic and the metal part, brought together with a ground fit. Due to its small thickness, the glaze layer is incorporated into the ceramic–metal structure, so that after the seal is first completed it can be heated to temperatures higher than the softening point of the glaze.

An intermediate solution between the glazing processes and the metallization processes is reported by Burnside[184]. This consists of firing on the surface of the ceramic, a mixture of a suitable glaze and metal powder mixture (e.g. 2 parts of glaze and 8 parts of Ni powder) made into a suspension for brushing or spraying with nitrocellulose dissolved in ethyl acetate. The firing is done in a protective atmosphere (e.g. 15 per cent H and 85 per cent N).

According to Bondley[151] another method to join metal to ceramics consists of coating the ceramic part with a paint or paste made from a low melting glaze and a salt of gold, silver or platinum. When this mixture is heated in air to 450–600 °C the glaze melts and wets the ceramic. The precious metal then floats to the surface of the glaze and forms a continuous conductive layer. After electro-plating on this conductive layer a thicker metal layer the metal part can be sealed to this by soldering.

(c) Ceramic–glass–metal seals. Metal rods (FeNiCo) are sealed through ceramic plates using the glass as an intermediary material. For this, the metal rod (wire) is embedded in the appropriate glass (Section 24.31) and the inside surface of the hole through the ceramic is coated with glass from a powdered glass suspension (Section 25.21b). The beaded rod is placed in the coated hole (Fig. 2.96a) and the assembly is heated in a furnace until the glass softens

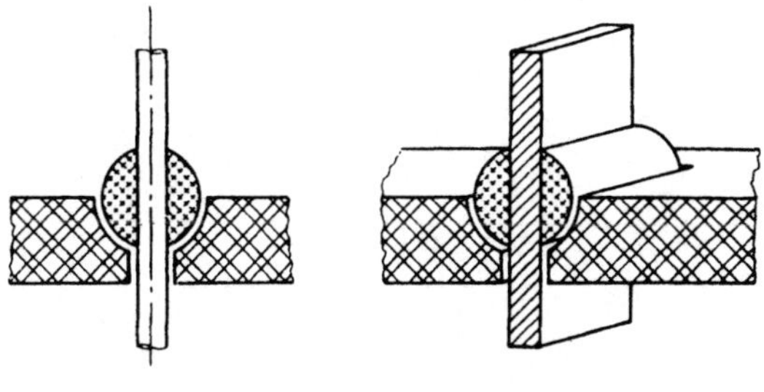

(a) (b)

FIG. 2.96 Ceramic–glass–metal seals; (a) rod seal; (b) plate seal

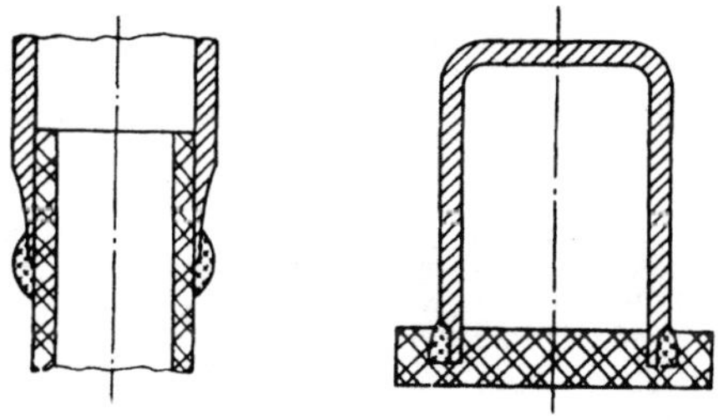

(a) (b)

FIG. 2.97 Cylindrical ceramic–glass–metal seals; (a) with ceramic cylinder; (b) with ceramic disc

and seals. To guide the rods and to limit their final position in the seal, the ceramic plate is placed during the sealing on a refractory block provided with holes receiving the protruding sides of the rods.

Thin metal plates can be sealed through ceramics in the same way (Fig. 2.96b).

Cylindrical metal–glass–ceramic seals can be made using a modification of the Housekeeper technique (Section 24.41). For this purpose a thin-walled (0.1 mm) copper tube is brought on the ceramic tube at a tight fit. The edge of the copper tube is then sealed to the ceramic using a glass seal (Fig. 2.97a).

TABLE 2.42. SINTERED

Technique	Powder mixture	Suspension liquid	Mixing	Coating
Moly-manganese	160 g Mo (200 mesh) 40 g Mn (150 mesh) particle size 3–10 μ (Marsden[808])	50 cm^3 amylacetate 50 cm^3 acetone 100 cm^3 pyroxyline binder (Du Pont 5511)	Ball milling 24 hr	Brushed, or sprayed 25–50 μ layer
Moly-iron	40 g Mo 0.8 g Fe particle size 5–8 μ	10 g nitrocellulose sol in amylacetate 150 cm^3 lacquer thinner from: 90 cm^3 aromatic naphta 9 cm^3 ethyl (buthyl) alcohol 27 cm^3 ethyl acetate (85–88%) 24 cm^3 normal buthyl acetate (83–92%)		Brushed 15–20 μ (on Steatite)
	40 g Mo 1.6 g Fe (carbonyl) particle size 3 μ	100 g nitrocellulose sol from 10 g nitrocellulose 90 g ethyl acetate		Brushed about 100 μ
Tungsten-iron	90 g W 10 g Fe particle size 1–4 μ	nitrocellulose in ethyl acetate	—	Thickness 25–50 μ
	90 g W 10 g Fe	not specified (shellac 2% sol in alcohol)	—	—
Moly-manganese-iron	200 g Mo (400 mesh) 40 g Mn (400 mesh) 10 g Fe (H_2 reduced) 2 g silicic acid powder 2 g calcium oxide (200 mesh)	55 cm^3 acetone 25 cm^3 methyl ethyl ketone 50 cm^3 ethyl ether 45 cm^3 nitrocellulose lacquer* (600–1000 sec)	Ball milling 100 hr	—
Activated Mo–Mn	176 g Mo (200 mesh) 44 g Mn (200 mesh) 9 g titanium hydride	same as above	as above	Coated and fired in two layers
Activated Mo–Mn–Fe	200 g Mo (400 mesh) 40 g Mn (400 mesh) 10 g Fe (H_2 reduced) 2 g silicic acid powder 8 g alumina powder (90 mesh) 8 g titanium hydride	160 cm^3 acetone 30 cm^3 methyl ethyl ketone 40 cm^3 ethyl ether 40 cm^3 nitrocellulose lacquer* (600–1000 sec) 20 cm toluene	as above	Painted or sprayed

* Nitrocellulose lacquer: 40 g nitrocellulose, 165 cm^3 toluene, 75 cm^3 ethyl alcohol.

CERAMIC–METAL SEALS

Sintering	Plating	Brazing	Remarks References
1300–1400 °C, in hydrogen or dissociated ammonia 15–30 min	Cu, or Ni or Cu-Ni fired at 1000 °C 10 min in reducing atmosphere	Cu–Ag eutectic	Bond strenth 2.7–6 kg/mm^2 (Kotowsky[715]) van Houten[601] Nolte[936a]
1250 °C, 20 min in hydrogen–nitrogen (28/72)	Sprayed with Ni suspension (4 μ particle) 15 μ thick layer; fired 1100 °C, 15 min in wet hydrogen	Butt sealed to Kovar, plated with 1.5 mg/cm^2 Cu, by interposing between Kovar and metallized ceramic an Ag washer (25 μ) and heating to 1000 °C for 10 min.	Chick[212a]
1400 °C, 30 min hydrogen–nitrogen (30/70)	Brushed with 40 g Ni powder in 10 g nitrocellulose in 90 g ethyl acetate; Fired H_2, 1000 °C, 15 min		Williams[1317]
1340–1360 °C (Zircon) 1350–1400 °C (Alumina) 15–30 min, in hydrogen–nitrogen (15/85)	–	–	Burnside[184]
1400 °C in hydrogen	After wire brushing plated Ni or Cu	Ag–Cu eutectic	Bennett[110], Mönch[879], Pryslak[1015]
For high alumina ceramics	–	–	LaForge[729b]
1525 °C	Ni plated to total thickness of 25 μ	Ag–Au eutectic, brazed to copper parts	Tensile strength about 6 kg/mm^2 Kohl[706]
1500 °C 30 min or 1250 °C 45 min in wet hydrogen	Ni plated 5–8 μ	Good seal to OFHC stacked Cu discs (0.25 mm thick) by brazing with Ni–Cu–Au alloy or Ag–Cu eutectic	Kohl[706]

60 cm^3 ethyl acetate.

In the same way a metal tube can be sealed to a ceramic disc, using the glass as the sealing material (Fig. 2.97b).

25.22 Ceramic–metal sintered seals. This process consists basically in covering the ceramic part with a layer of molybdenum (or tungsten) powder with a slight addition of manganese (iron or titanium) and in sintering the layer at a high temperature. After an eventual coating (plating) with a layer of nickel (or copper) the ceramic part is ready for brazing to the metal part (Sections 22.2 and 25.3). The sequence of the various layers in a ceramic–metal seal with sintered metal is shown in Fig. 2.98.

The "sintered metal" sealing technique evolved from a tungsten-to-glass seal. Vatter[1259] and Pulfrich[1016a] developed a molybdenum–iron seal and Nolte[936a] substituted manganese powder for the iron powder and developed the final technique for the current molybdenum–manganese bonding process (van Houten[601]). La Forge[729b] reports on a combined method called "activated molybdenum–manganese" technique (Table 2.42).

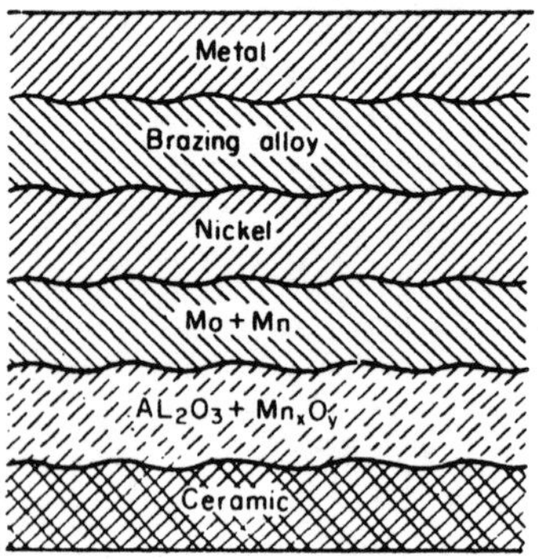

FIG. 2.98 Sequence of layers in a ceramic–metal sintered seal

In these techniques (Table 2.42) the coating is done with a metal powder or mixture of powders in a suspension liquid containing a binder.

The powders can consist only from those metals which are not dissolved or only slightly dissolved by the brazing alloys used in this technique (Section 25.3) and which can form a strong bond with the ceramic part. Thus the metals used are molybdenum, tungsten, manganese, iron, chromium, copper, nickel, rhenium. To the metal powders, sometimes a small amount of oxides is also added (e.g. manganese oxide) to facilitate the oxidation process, necessary in the bond. Pincus[991] quotes the use of molybdenum oxide instead of molybdenum powder. Tweeddale[1244a] used a molybdenum oxide–manganese oxide (20:1) mixture.

The metal powders are ball-milled (Table 2.42) to an average particle size of 3–10 μ (Marsden[808]). The particle size is not critical if kept within the required range. Bender[108] prepared the molybdenum powder (1–2 μ) from molybdenum oxide ball-milled for 10 hr and fired in very dry hydrogen at 550 °C for 56 hr and then at 700 °C for 24 hr.

The *suspension liquid* must meet the following requirements:

(1) The viscosity of the liquid must be high enough to avoid sedimentation of the powder particles;

(2) The solid residue of the liquid, if fired, should be very small;

(3) The liquid should dry at the temperature and in the time required by the coating process;

The suspension liquids used are mainly: acetone, amylacetate, xylol, buthylacetate (Table 2.42) and the binder may be nitrocellulose, metachrylates, glyptal. The vapour pressure of the suspension liquid used allows the control of the desired drying time, and the quantities (proportions) used should give the needed viscosity of the suspension for brushing or spraying. Nolte[963a] quotes a viscosity reading of 22 sec with a Zahn viscosimeter, and Bender[108] requires a viscosity of 13 poises.

The viscosity of the prepared suspension is quoted to be 13–80 poises (Zincke[1354]).

The suspension is applied on the surface of the ceramics by brushing or spraying, but also some different techniques have been used for the purpose. Doolittle[287] developed a technique which utilizes a metallizing self-supporting plastic tape which contains the metallizing Mo–Mn powder mixture in poly (n-buthylmethacrylate 6 per cent and sucrose-acetate iso-butyrate) carried on a polyethylene film (thickness of the deposit 50 μ). The metallizing tape can be applied to the ceramic part by first separating it from the carrier film and using it as a self-supporting film, or applied to the ceramic first and then separated from the carrier.

If suspensions are used, they must be mixed thoroughly (Table 2.42) to obtain a homogenous distribution of the solid particles in the solution.

The ceramic surfaces to be metallized must be very clean; any included particles of dust or dirt can stop the tight bond from forming. According to Williams[1317], a cleaning of the ceramic surface in nitric acid solution (1:1) for about 1 min, followed by a thorough rinsing, is necessary.

The thickness of the coating seams to be very important (Chick[212a]) and has to be measured to detect the parts, where it is above or below the required value (Tables 2.42, 2.43). According to Chick[212a] extreme accuracy in both the smoothness of the coating and its thickness is important, and for the particular granular sizes used (Table 2.42) these limits cannot be varied by any substantial amount. The influence of the thickness of the coating on the behaviour of the metallized layer towards the bonding strength, uniformity of the layer and the ability to be brazed is summarized in Table 2.43. The best thickness range is between 20 and 60 μ. In this range the layer is thick enough to make an excellent bond, without having excess material, which can weaken the bond and disturb the uniformity of the layer. Kotowsky[715] measured the bonding strength of "sintered metal" seals with a layer of 50 μ, and obtained values of 2.7 kg/mm^2 for Mo–Mn on Frequenta fired 30 min at 1275 °C and 4–6 kg/mm^2 for Mo–Mn on 85 Al_2O_3 fired 50 min at 1525 °C.

TABLE 2.43. BEHAVIOUR OF METALLIZED COATINGS OF VARIOUS THICKNESSES

Thickness (μ)	Bond	Uniformity	Brazing
8–20	good	good	good
20–60	excellent	excellent	excellent
60–75	good	good	excellent
75–130	poor	problematic	good

If the metallization is applied on a circular ceramic part, it is recommended that the part be rotated during the brushing or spraying. A rotation of about 200 r.p.m. is recommended (Williams[1317]). The coating should be

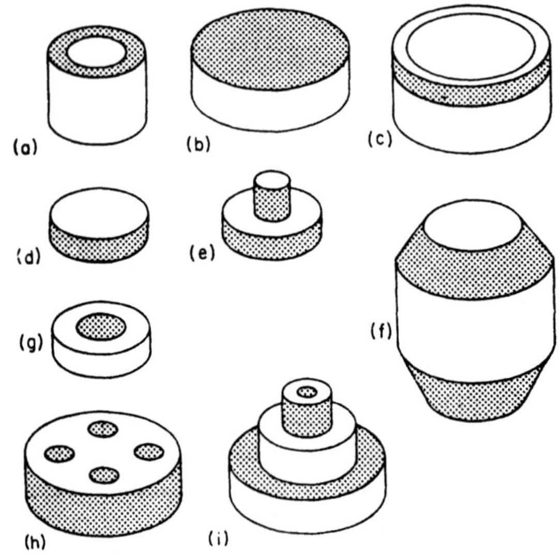

FIG. 2.99 Metallized ceramic parts; (a) cylinder, metallized on edge; (b) disc, metallized on face; (c) cylinder, metallized on sides; (d) disc, metallized on side; (e) composed ceramic part, metallized on sides; (f) cylinder, metallized on taper; (g) cylinder, metallized inside; (h. i) ceramic parts metallized both inside and outside

applied only on the surfaces where the seal is to be made. Some examples of metallized ceramic parts of various shapes are shown in Fig. 2.99.

The metallizing coating must be *sintered* in a protective atmosphere. Generally hydrogen or a hydrogen–nitrogen mixture is used (Table 2.42). In the techniques, where the coating must form an oxide, the gas used must be wet. In the molybdenum–manganese technique the manganese must be oxidized but the molybdenum not. For this purpose the protective gas must be saturated with water vapour at 40–60 °C (Marsden[808]).

If a strongly reducing atmosphere is used (e.g. dry hydrogen) some ceramics are coloured due to the reduction of the metallic oxides contained in the ceramic. In this way, blackening appears as a result of the reduction of the iron oxides or partial reduction of silica, or a brownish-red colour can appear due to the manganese. In such cases the use of wet hydrogen is recommended. Some authors recommend also the use of an atmosphere of dissociated ammonia (Table 2.42).

The sintering temperature must be high enough to produce a strong bond and a vacuum-tight seal between the metal and the ceramic part. This temperature is generally between 1100–1700 °C depending on the ceramic as well as on the metallizing composition. According to Velte[1263] the best temperature range is between 1400 °C and 1600 °C (see also Table 2.42). For the lowering of the sintering temperature van Houten[602] recommends the addition of alkali and earth–alkali carbonates or oxalates, sodium-fiuosilicates (Na_2SiF_6) or fluoberyllates (Na_2BeF_4). These materials are especially required for ceramics with high alumina content, due to the high sintering temperature needed for these ceramics. For the same purpose powdered glass (borosilicates) also was added by some workers. In any case, the added quantity should not be so great as to change the metallic character of the mixture, but should be just enough to bring the sintering temperature to the required lower level.

The sintering time must be long enough to allow corresponding diffusion of the bonding metal in the ceramic, but not too long to permit oxidation of the molybdenum. The sintering time varies between 10 and 60 min (Table 2.42).

After sintering the ceramic is generally plated with Ni or Ni–Cu (Table 2.42) and then the metal part is brazed (Sections 22.2 and 25.3) to this assembly.

25.23 Active metal processes. The "active metal or alloy" sealing process (known also as "reactive metal method") consists in using a brazing alloy with a reactive metal (such as titanium or zirconium), which when melted wets the ceramic, (Bondley[151], Brand[160]). The activity of titanium (or zirconium) is based on the fact that under favourable conditions they can reduce the oxides from the ceramic, forming in this way a strong ceramic–titanium (or ceramic–zirconium) bond. On the other side titanium (or zirconium) forms stable alloys with some brazing metals and alloys, and can be bonded to metal parts by brazing.

There are several methods of blending the active metal with the brazing alloy:

(a) by application of a metal hydride layer on the ceramic,
(b) by application of a titanium–nickel powder mixture,
(c) by using a mixture of carbide powders,
(d) by using a brazing rod with a reactive metal core,
(e) by using washers of active metals,
(f) by using washers of active metal alloys.

During the heating (in an inert atmosphere or vacuum) the brazing alloy melts, combines with the reactive metal and forms an alloy which wets the ceramic and builds up a strong joint between ceramic and metal. The main advantage of the active metal process consists in the need of only a single firing (sintering) operation.

(a) The hydride process consists in coating the ceramic* with the hydride of an active metal (suspension in liquid) and assembling it with the metal part to which it is to be sealed, along with the brazing metal. The assembly is then heated to the melting temperature of the brazing metal. Simultaneously with the brazing, a strong vacuum-tight bond is formed between ceramic and brazing alloy.

For this process Bondley[151] used titanium hydrides, Pearsall[981] and Brand[160] recommend zirconium hydrides. The hydrides are used in form of powders, with particle sizes of 7–10 μ. The purity of the hydrides is critical for the reliability of the bond. The powder is suspended in amylacetate with nitrocellulose (Bondley[151]) or other appropriate liquids (e.g. methyl metacrylate).

The hydride suspension can be applied on the ceramic parts either by brushing or by spraying. When two ceramic parts are to be joined to each other, the joining surfaces of both the ceramic parts should be painted with the suspension. The thickness of the coating must be of about 25 μ. The hydride process exhibits extremely high solder fluidity at the alloying temperature and therefore this process requires a careful coating in order to insure a layer of uniform thickness.

Over the coating of hydrides, another coating should be made, consisting of a suspension of silver powder, or a washer of silver, nickel, copper or silver–copper alloy should be placed over the hydride coating. Brand[160] sealed a ceramic cylinder to a nickel–iron (Kovar or molybdenum) using a ring of silver wire placed on the part coated with ZrH_4, and heating the assembly

* Grove[489] used the method to seal sapphire window on FeNi cylinder (Section 7.2). Knoll[698] used the process to seal quartz crystals soldered with Pb–In–Cu (97/97/6) alloy.

in vacuum. Hume[612] used a ring or washer of silver placed over the area painted with zirconium hydride, and made a tight bond by baking the assembly in vacuum at 485 °C. This was used to make a vacuum-tight bond to aluminium. For this (Hume[612]) the tinned ceramic was placed in a 1100 Al alloy bushing; upon heating in vacuum, the Al bushing melts and alloys with the tinned area. The bushing can then be joined by brazing or inert arc welding it to a massive metal part.

Van Houten[601] describes a technique where the hydride is mixed with copper powder in proportions to yield the copper–titanium eutectic (74 per cent Cu and 26 per cent Ti). This forms by firing at 925 °C in vacuum a metal layer which can be subsequently built up by electro-plating.

The parts coated with hydrides and brazing metal are pressed together, jigged into place and sintered. The sintering (firing) is done in vacuum furnaces (pressure less than 2×10^{-5} torr) at temperatures of 900–1000 °C, for about 30 min, followed by cooling in vacuum*. Hume[612] recommends heating from 500 °C up to 1080 °C in 10 min and cooling in vacuum in 30 min.

With this treatment, the hydride dissociates and the resulting pure metal (titanium) forms with the second metal (silver) an alloy, and reacts simultaneously with the ceramic part, resulting in a strong bond. In fact the bond is stronger than the ceramic itself (Bondley[151]); destructive tests always reveal a layer of ceramic clinging to the metal.

The decomposition of the titanium hydride is slow up to 350 °C, and begins to accelerate in the range 350–450 °C, and between 450–650 °C about all the hydride will be decomposed. To decompose the remaining hydride, heating up to 1000 °C is required. The hydrogen evolved during the heating is in atomic state and therefore is chemically extremely active. The presence of this active hydrogen is very beneficial in cleaning the surfaces of the parts, which helps to better soldering.

In conclusion the hydride process yields a strong bond, requires a sintering temperature which is well below the softening point of most ceramics and needs a single operation in which the joint is made. Since the brazing can be done in this procedure in vacuum, a better outgassing can be achieved, and metals or alloys sensitive to oxidation can be brazed to ceramic parts.

When silver or copper is melted in hydrogen, they absorb large amounts of gas which is later released and the resulting open structure is often a source of very small leaks. With vacuum brazing this danger is avoided.

(b) Active metal powders are less used to make up sealing layers, due to their sensitivity to humidity. These powders are more difficult to handle than the hydrides.

* Final stages of the cooling can be accelerated by introducing a non-oxidizing gas into the vacuum chamber.

Stoeckert[1199] describes a technique in which a blended powder of titanium and nickel (35/65) were pressed on the lead-through wires, using nickel and chromium plated molybdenum washers. The lead wires were then placed in the holes of the ceramic plate, the washers resting on the top surface. By heating (at 1250 °C in vacuum) the pressed powder melts and flows into the interspace, making a vacuum-tight seal. The blended powder (200–325 mesh) should be annealed and outgassed before use.

(c) Carbide seals use a mixture of two carbides; one of them is a powder of tungsten or molybdenum carbide, and the other, a powder of an active carbide (e.g. titanium carbide). The mixture contains the two powders in a ratio of about 9 parts inert (W, Mo) carbide to 1 part active carbide. A suspension of this powder mixture in an organic liquid is painted on the ceramic and sintered at 1650 °C in hydrogen or dissociated ammonia. This forms an adherent layer on the ceramic surface on which a metal part (Ni, Co, Fe) can be brazed (e.g. MacDonald[791a]).

(d) Active metal core seals. This method consists of using a brazing filler rod (Section 22.22) made up from an active metal core (Ti, Zr, Th, Hf) and a surrounding clad of a ductile metal or alloy. The clad metal should have a melting point lower then the core and should be capable when molten of dissolving some of the core material, forming an alloy which wets the ceramic surface (Coxe[243]).

A titanium core, clad with nickel, copper or a silver–copper eutectic, is used for this sealing technique (Marsden[808], Bender[108]). The ratio of the titanium core weight to the cladding metal is kept in a certain range so that after melting it reaches the required alloy, which wets the ceramic. The quantity of titanium is not very critical and ought to be 6–8 per cent of the total weight of the rod (Velte[1263]). If the cladding of the rod is of an Ag–Cu eutectic (78/28), the titanium content should be about 8–14 per cent*. If the titanium content is below the quoted values the joint might not be achieved; if it is above, the wetting would be excessive and this would possibly cause leaks in the seal. For ceramic–ceramic seals brazing rods with 3–5 per cent titanium core, are recommended.

The cored-wire is placed on the surface of the ceramic or in grooves made for this purpose. The wire can be flattened, but this raises the risk of changing the uniformity of the clad around the titanium core. In some cases, sandwiched titanium (50–100 μ) foils with silver–copper eutectic foils (250 μ) are used.

The sealing method using active metal cores has the same difficulty as that

* Such a Ti-cored Ag–Cu alloy is the BT–Ti of Handy and Harman[508b] Diameters of cored wire 0.25–0.40 mm; with Ti core 3, 5, 8, and 12 per cent.

Ni-clad Ti wire is available from Little Falls Alloys Inc., 189 Caldwell Ave, Paterson 1, N.J., U.S.A.

using active metal sheets, i.e. the control of the alloy formed is difficult because the eutectic is built up successively during the heating.

(e) Seals with massive active metal parts consist in joining an active metal part (titanium) by brazing it to a ceramic part. According to Beggs[103], titanium and ceramic parts are attached with inserted rings of brazing alloy (nickel, copper, silver), and the assembly is heated in vacuum. The brazing alloy melts and alloys with the titanium, and the new alloy wets the ceramic and forms a tight and reliable seal. To make a reliable seal with this technique, very accurately abutting surfaces are required.

The thin sheets of Ni, Cu, Ag inserted between the parts to be sealed should be about 10 μ thick; Beggs[103] used sheets of up to 120 μ thick. The sheets are placed with a tight fit (clearance less than 15 μ) between the ceramic and the titanium part. By heating the assembly in a vacuum furnace to a temperature of about 70 °C above the eutectic point* of the titanium–metal alloy, the needed eutectic is formed.

According to Beggs[103] it makes no difference whether the active metal is supplied by the sheet or by the part to be sealed or if the active metal is or is not in contact with the ceramic. The sealing occurs as soon as the active metal is present in a liquid phase, when it can readily come in contact and react with the ceramic. The amount of alloy formed depends on the metal pair used, the thickness of the shim and the sealing temperature. The sealing temperatures used (Beggs[103]) are 875–910 °C for copper parts and Ti shims (or Zr shims) and 940–1050 °C for Ti (or Zr) parts and Ni or Fe shims.

By placing# over the ceramic a sheet of zirconium, over this a sheet of nickel, and over this a layer of gold on which the metal part is seated, and heating the assembly to about 1000 °C in a protective atmosphere or vacuum, a Ni–Zr eutectic is formed at 961 °C and an Au–Ni eutectic at 950 °C. The mixture of these alloys makes a strong seal between the ceramic and metal.

(f) Seals with active metal alloys. The use of titanium–alloys already made (Ag–Ti alloy with 5 per cent Ti, M.P. 950 °C) has some advantages compared with other methods. The brazing can be done at lower temperatures, a fact especially useful when silver-titanium alloys are utilized, since it diminishes the danger of the evaporation of the silver during the brazing. At the same time, by using lower brazing temperatures the expansion differences between the ceramic and metal parts have less influence on the seal formed. The active metal–inert metal alloys (e.g. Ti–Ag 15/85 or Zr–Ag 15/85) wet the ceramics with a high alumina content (Table 2.40) especially well. By the use of these alloys with alumina ceramics, the quantity of alloy used must be kept to

* The eutectic points are: Ti–Ag, 850 °C; Ti–Cu, 875 °C; Ti–Ni, 955 °C; Ti–Fe, 1080 °C; Ti–Mn, 1185 °C; The Ag–Cu eutectic alloys with titanium, at a temperature very near to its eutectic point (779 °C).

The process was developed by Cerberus Co., Mönnedorf-Schweiz.

the minimum possible, because of the tendency of the alloy to spread over the ceramic surface. Therefore in these seals capillary considerations control the design. The process has the advantage of good reproducibility and simplicity.

Kotowsky[715] mentions a Cu–Ti (70/30) alloy, used to seal metal parts to high alumina ceramics. The sealing was done at 1050 °C for 10 min, and a bond strength of 5 kg/mm^2 was obtained.

The titanium content of the alloys used in these techniques varies between 5 per cent (quoted as a minimum by Marsden[808]) and 30 per cent (quoted as the highest useful titanium content by Kotowsky[714]). When using alloys with very high content of titanium (say 60 per cent), a very weak bond was obtained. The silver–copper–titanium alloy is hard and brittle, and therefore it is recommended only in the cases of ceramic–metal seals where the two parts have a good expansion matching or where the ceramic is subjected only to compression.

For applications where the seal will not be used at temperatures higher than 250 °C, titanium–lead alloys (1.3–5 per cent Ti) or silver–titanium–tin (88/2/10) alloys (Muller[904a]) can be used, the soldering being done usually in a hydrogen atmosphere.

25.24 Compression seals. The diffusion of the metal in the ceramic, under high pressure, with or without heating, can be used as a sealing method, when the parts to be sealed are strong enough to withstand the required pressures*.

These seals are known as "ram seals" or "crunch seals", and the techniques based on this process are described by Zollman[1357], Grove[489], Tank[1213], Belser[105], Levin[761].

Cylindrical ceramic parts (with a high content of alumina) having diameters up to 50 cm can be sealed to steel pipes with the inside diameter slightly smaller than the outside diameter of the ceramic cylinders. For this purpose the end of the ceramic cylinders is ground conically (Fig. 2.100) and the steel pipes are copper clad. The two pipes are pressed together with a high pressure. If the assembly is heated to about 1000 °C, a pressure of 1.5–2.0 kg/mm^2 applied for about 2 hr is necessary to complete the seal (Burnside[184], Tank[1213]).

These seals are usually completed in vacuum or in a neutral atmosphere.

The method can also be used to seal together metallized ceramic parts subjecting their assembly to pressure at moderate temperatures (Kohl[706]).

Zollman[1357] describes such seals, constructed with diameters up to 50 cm. Figure 2.100a shows a compression seal assembly in which the seal is made on the outer diameter of the ceramic only. This construction places the ceramic cylinder in hoop compression; however a longitudinal bending stress

* For ceramic-metal knife edge seals see Section 38.56 and Hees[530], Cohen[220].

which has a tensile component is also induced in the ceramic part. For the contact length l_c (Fig. 2.100a) between ceramic and metal Levin[761] gives the formula

$$l_c = 0.38(a \cdot h)^{0,5},$$

where a is the mean radius of the ceramic tube and h its wall thickness.

Figure 2.100b shows a coaxial or inside–outside seal. The forces on this seal are somewhat balanced so that the ceramic is always kept in compression. Figure 2.100c shows a sapphire or other ceramic window seal. This type is a variation of the outside seal; however the danger here is not the longitudinal bending stress but the possibility of buckling.

From Fig. 2.101a it can be seen that the outside diameter of the ceramic part is larger than the inside diameter of the metal sleeve. The seal is made by the application of an initial force, which may be as high as 20 tons, along

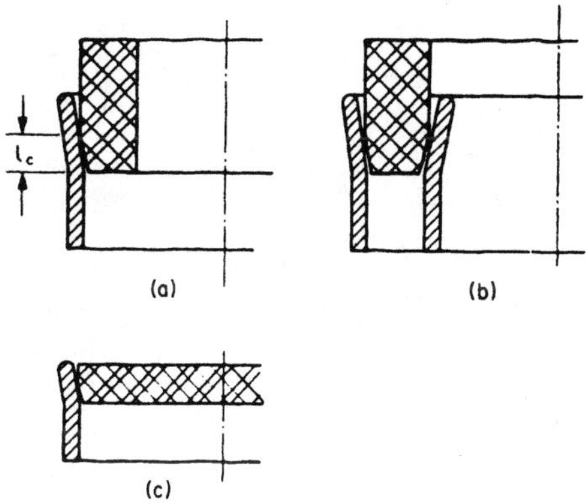

FIG. 2.100 Compression seals; (a) inside seal; (b) inside–outside seal; (c) window seal. After Zollman[1357] and Levin[761] *(Courtesy of Pergamon Press)*

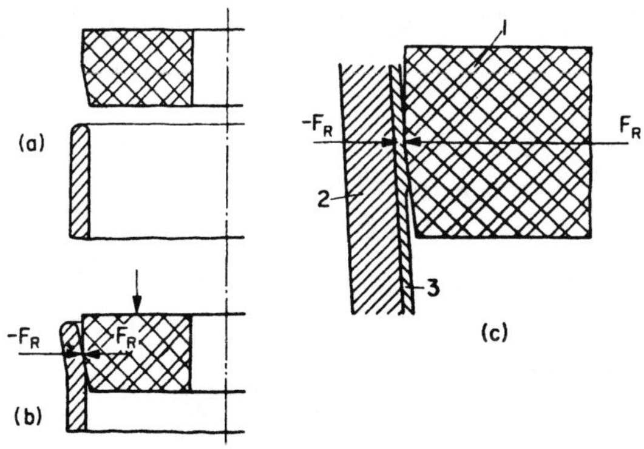

FIG. 2.101 Completing a compression (ram) seal; (a) parts before sealing; (b) the completed seal; (c) detail of the seal: (1) ceramic; (2) metal tube; (3) ductile plating. After Zollman[1357] *(Courtesy of Pergamon Press)*

the axis of the members to be sealed. The wedging action creates a radial force F_r, deforming both the metal and the ceramic cylinders (Zollman[1357]). In most cases the deformation of the ceramic part is kept below 0.1 mm on the diameter. The force F_r is confined to a relatively narrow area of contact near the top of the taper. This high concentration of stress caused by the fit during the sealing, creates plasticity in the ductile plating, so that it flows around the ceramic to form a vacuum–tight seal (Fig. 2.101c).

Although the force maintained by the metal member on the ceramic part is at a maximum at room temperature, it must exert sufficient force at the maximum temperature to ensure a vacuum-tight seal also during heat cycling.

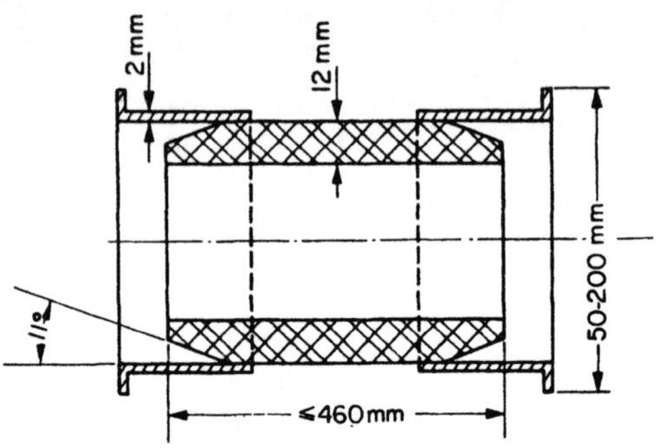

FIG. 2.102 Ram seal. After Grove[489] *(Courtesy of Pergamon Press)*

Grove[489] tested some ram seals and found them tight for ultra-high vacuum use after 15 heat cycles between room temperature and 450 °C. The seals tested were of the type shown on Fig. 2.102, consisting of Inconel caps forced over the conical ends of high alumina ceramic pipes (6 in. diameter and 18 in. long). The Inconel caps were heat treated for maximum strength and coated with 25–50 μ of copper on the inside. With an interference fit calculated to take the metal just beyound the elastic limit as it passes over the full diameter corner (Fig. 2.102), to make a seal 6 in. in diameter requires about 15 tons.

Comsa[227, 227a] describes a technique of sealing aluminium to ceramic parts, by first coating the ceramic part with a thin film of an aqueous solution of sodium silicate. After drying and baking at 1000–1100 °C, a compression seal with the metallic aluminium is made at 530 °C by applying high, local pressures along the edge of the aluminium pressing on the ceramic. Satisfactory results are obtained only if the silicate layer film is extremely thin; thick layers lead to seals with poor mechanical resistance.

25.25 Various ceramic–metal seals. Besides the seals described in the previous Sections, some other sealing techniques have also been developed, using oxides, noble metal foils, graphite coating or electro-forming.

The *oxide seal* consists of applying e.g. copper oxide to the ceramic part and by heating the coated part in a reducing atmosphere, a pure metal, thin and adherent layer is obtained (Burnside[184]). According to Zincke[1354], ceramic parts (steatite, forsterite, zircon) can be tightly sealed to metal parts, using a mixture of fine powders of silver oxide (or silver) and copper oxide (or copper). The mixture is made using a weight ratio of Ag_2O to Cu_2O between 20:1 and 10:1. The parts should be coated with the mixture, and then the assembly heated (in air, nitrogen or vacuum) to above 945 °C.

Knecht[696] suggests that layers of metal powder be gradually enriched with a ceramic powder until layers of pure ceramic are reached. By this procedure a graded seal could be produced from the metal to the ceramic.

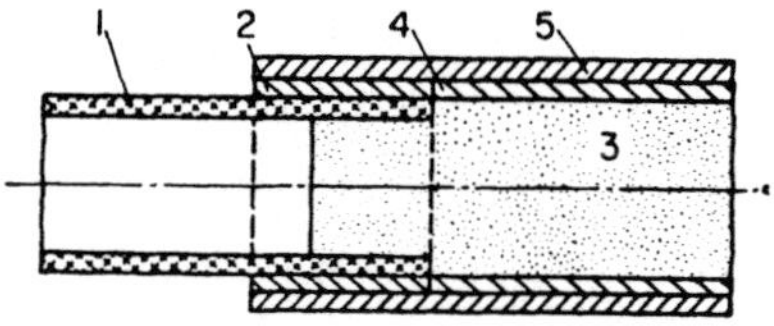

FIG. 2.102 A Electroformed seal. After Eckstein[320] *(Courtesy of The American Institute of Physics)*

Strong[1207] mentions that platinum film deposited on porcelain from a 10 per cent H_2PtCl_6 solution, and burned onto the porcelain, may be electroplated with copper and soldered to metal parts. Martin[819] recommends the use of silver plating, because the copper peels off too easily.

It has been found (Jamieson[628]) that copper and other metals can be electro-deposited on a ceramic or glass surface made electrically conductive by a film of graphite deposited from aquadag, painted on the clean surface and baked at about 200 °C for about 1 hr. On rough surfaces the adhesion is better.

The *electroformed seal* consists (Hare[510]) of pre-metallized, copper-plated ceramic cylinders stacked with disc electrodes inserted between the abutting surfaces. The assembly is copper-plated in a bath (for several days), where the copper required for the vacuum-tight seal is deposited (minimum of 0.2 mm thickness).

A similar technique is described by Eckstein[320]. The glass tube (1 Fig. 2.102A) is coated on one end with a ceramic type silver paint, 2, (e.g. Du Pont No. 4666). This is then fired on the tube, producing a strong bond, and making the coat conducting. A wax form, 3, is inserted in the coated end of the tube, in order to form (support) the shape of the future connecting copper tube. The wax form is then made electrically conductive by coating it with a layer, 4, of air-drying silver paint. Then a layer of copper, 5, is electro-plated over the whole surface. After the wax form has been removed by heating the assembly, a vacuum-tight joint is obtained between the glass (or ceramic) pipe and the electro-formed copper pipe. The same technique can be used to make up vacuum-tight joints between quartz and Pyrex tubings, or ceramic and metal or glass tubing, or to seal glass windows on metal cells.

25.3 Possibility of brazing ceramic–metal seals

As the general brazing technique for vacuum sealing is treated in Section 2.2, only the special features regarding the brazing of ceramic–metal seals will be pointed out here.

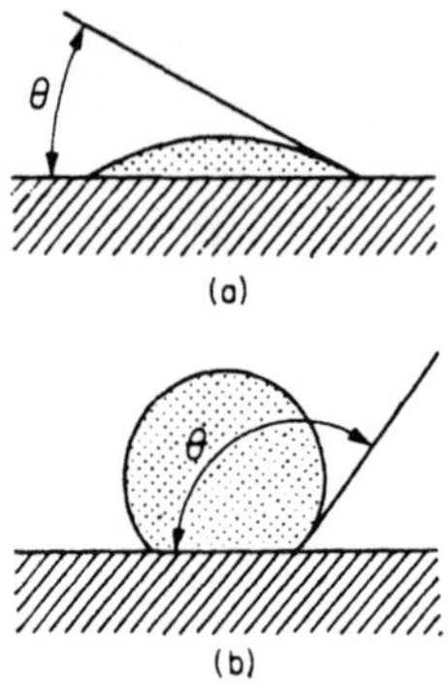

FIG. 2.103 Contact angle; (a) wetting; (b) non-wetting

The basic requirement of the brazing alloy for ceramic–metal seals is to *wet* the surfaces. If there is more than one brazing alloy, which wets the surfaces of the materials to be sealed together, the final choice is made after taking into consideration the specified melting temperature, the brazing configuration and costs.

The degree of wetting of a liquid on the surface of a solid is expressed by the contact angle θ (Fig. 2.103) of a drop of the liquid on the given surface. When the contact angle θ is less than 90° it is considered that the liquid wet the surface (Fig. 2.103a); when this angle is greater than 90 degrees the liquid is considered to be *non-wetting* (Fig. 2.103b) for that surface. An absolute wetting means an angle of contact of zero degrees, while a complete non-wetting occurs at 180°.

Table 2.44 lists the contact angles of various brazing metals and alloys used in ceramic–metal seals, on alumina ceramics directly or on alumina coated with various metallizing layers. It shows (Williams[1319]) that the original ceramic (alumina) surfaces are wetted only by titanium (active metals); the surfaces metallized with Mo, W, MoMn, MoFe or WFe and coated with Ni (Table 2.42) are wetted as well by Cu and Ag as by an Ag–Cu eutectic or an Au–Ni eutectic. Pure copper wets surfaces metallized with MoMn while an Au–Ni eutectic wets only surfaces metallized with MoMn.

TABLE 2.44. DEGREE OF WETTING OF ALUMINA BY VARIOUS BRAZING METALS AND ALLOYS (in helium at 100 torr)

Brazing metal or alloy	Temp. (°C)	Ceramic metallized with	Contact angle (deg)	Time to reach equilibrium (min)
Cu	1085	Mo–Ni, MoMn–Ni, MoFe–Ni, W–Ni, WFe–Ni	0	1
Cu	1120	MoMn	0	4
Ag–Cu eutectic	800	Mo–Ni, MoMn–Ni, MoFe–Ni, WNi. WFe–Ni	0	1
Ag	980	as above	0	1
Au–Ni eutectic*	950	as above	0	1
Au–Ni eutectic*	1000	Mo	0	18
Au–Ni eutectic*	1000	MoFe	23	30
Ti (8%) Ag–Cu	950	non-metallized	28	1
Ti (5%) Ag–Cu	1050	non-metallized	30	20
Au–Ni eutectic*	1025	W	32	20
Au–Ni eutectic*	–	WFe	36	35
Ti–Ni	1050	non-metallized	40	15
Ag–Cu eutectic	920	MoMn	48	15
Cu	1240	WFe	80	15
Ag–Cu eutectic	1000	Mo	86	20
Cu	1140	WFe	100	15
Cu	1125	Mo	118	10
Ag–Cu eutectic	950	WFe	118	32
Cu	1125	MoFe	128	1
Ag–Cu eutectic	935	MoFe	128	30
Ag	1050	MoMn	130	25
Ag or Cu	1000 1100	non-metallized	140	3
Cu	1100	W	144	2
Ag	1020	Mo	150	2
Ag–Cu eutectic	950	W	152	10
Au–Ni eutectic*	1050	MoMn	152	3
Ag–Cu eutectic	950	non-metallized	156	2
Au–Ni eutectic*	1000	non-metallized	156	2
Ag	1040	W, MoFe, WFe	156–160	2–10

* The Au–Ni alloy called here eutectic means the lowest melting Au–Ni alloy (Fig. 2.30).

25.4 Ceramic–metal seal shapes

Ceramic-to-metal seals are used in a limited variety of shapes, i.e. butt seals and cylindrical (inside or outside) seals.

25.41 Butt seals. Metal–ceramic butt seals are used mainly where the space is limited (e.g. stacked disc seals as in Fig. 2.104c). Figure 2.104 shows some seal arrangements. In order to increase the strength of such butt seals, it is

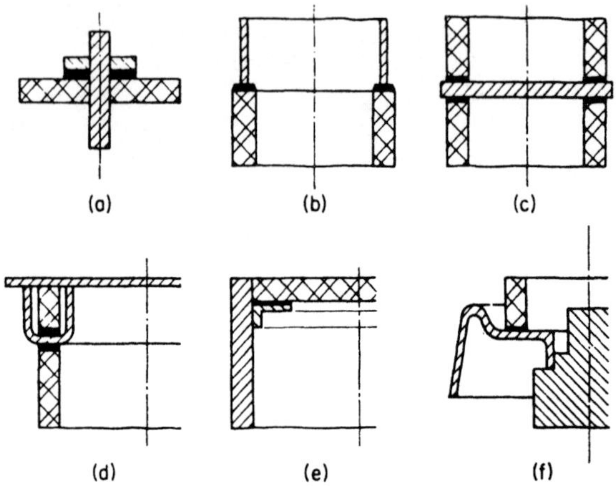

FIG. 2.104 Ceramic–metal butt seals: (a) rod seal; (b) cylindrical seal; (c, d) disc seals; (e) window seal; (f) elastic seal

useful to seal a ceramic part on both sides of the metal (Fig. 2.104c, d) even if this second ceramic part (Fig. 2.104d) has no other function in the device. To avoid excessive stresses, butt seals are made on thin elastic metal parts (Fig. 2.104c, d). When massive metal parts are to be sealed to the ceramic part, or big expansion differences exist between the metal and the ceramic, an intermediate thin metal part is to be used (Fig. 2.104e, f).

25.42 Cylindrical seals. Outside cylindrical seals, i.e. seals having the metal part outside the ceramic, are used mainly with metals having greater expansion than the ceramic. With this arrangement a compression stress is induced in the ceramic part.

Figure 2.105 shows some typical arrangements used for sealing together ceramic and metal cylinders, to seal ceramic windows to metal cylinders or metal windows onto ceramic cylinders.

For cylindrical seals, where the metal is inside the ceramic, it is recommended that only metals having lower expansion than the ceramic be used. When in such seals a metal with greater expansion (than the ceramic) is used, the metal tends to shrink away from the ceramic, which puts the interface in tension, leading (eventually) to the failure of the ceramic part.

Rod seals inside ceramic plates (Fig. 2.106b, d) are successfully made by using Mo, W, Ta, Ti rods. Butt seals are to be preferred if other metal rods must be sealed through ceramic plates sealing the rod indirectly by means of a disc joined to the rod (Fig. 2.104a). Small tubular ceramic–metal seals

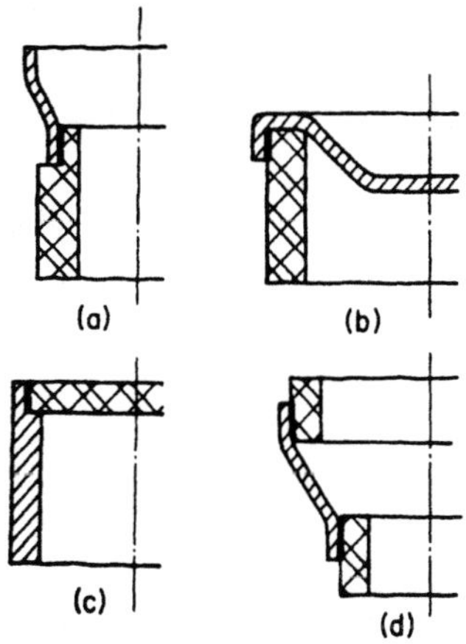

FIG. 2.105 Cylindrical ceramic–metal outside seal: (a) tubular seal; (b) metal window seal; (c) ceramic window seal; (d) elastic seal

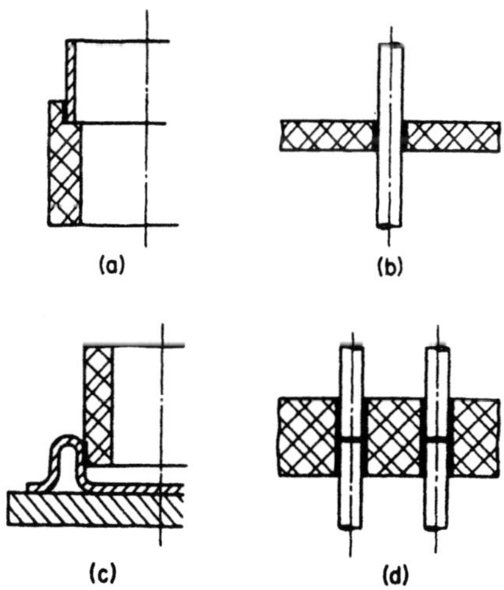

FIG. 2.106 Cylindrical ceramic–metal inside seal: (a) tubular seal; (b) rod seal; (c) elastic seal; (d) multiple rod seal

can be made as a simple assembly (Fig. 2.106a) but for larger diameters the elastic construction shown in Fig. 2.106c is recommended.

When combined inside–outside cylindrical ceramic–metal seals are to be made (Fig. 2.107), the metal placed inside the ceramic should have a smaller expansion than the ceramic, and the metal part placed outside the ceramic should have a higher expansion than the ceramic (Table 2.40). In this way the ceramic part can be kept under compression (Williams[1317]).

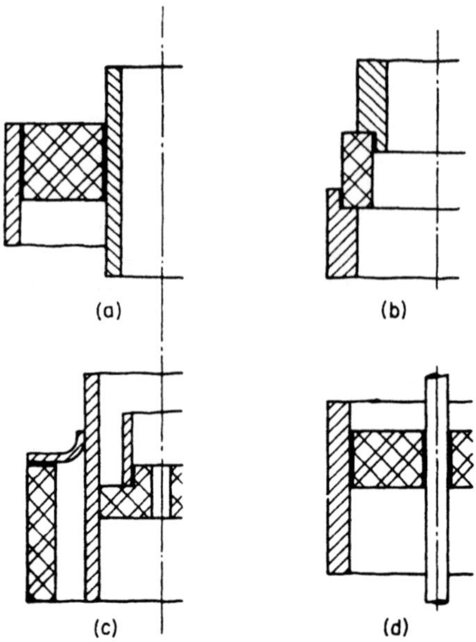

FIG. 2.107 Complex ceramic–metal seals: (a) concentric tubular seal; (b) tubular step seal; (c) elastic seal; (d) electrode seal

25.5 Testing of glass–ceramic–metal seals

Glass–ceramic seals can be tested in a manner similar to that for glass–metal seals, using polarized light (Section 24.5).

Glass–ceramic seals can also be tested by exposing them to heat cycling over a range of 80–100 °C, and then to check the surface of the glass in the vicinity of the ceramic. If the expansion of the glass is higher, the tensile

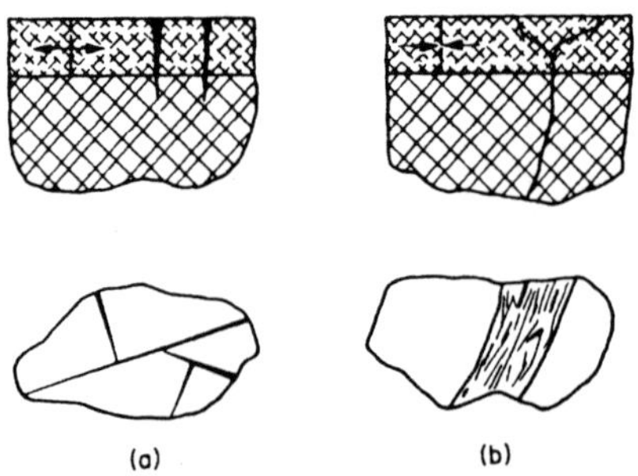

FIG. 2.108 Cracks developed in glass–ceramic seals: (a) expansion coefficient of glass greater than that of ceramic; (b) expansion of glass smaller than that of ceramic

stresses in the glass will produce small cracks normal to the plane of the seal; the cracks appear as a network (Fig. 2.108a). If the cracks are produced by compression stresses (the expansion of the ceramic is higher), the cracks are usually parallel to the plane of the glass–ceramic interface (Fig. 2.108b).

Unfortunately, for ceramic–metal seals the polarized light cannot be used to test the seal. These seals can be tested only by leak detection methods (Section 13.3) after or during heat cycling. Methods of testing ceramic–metal seals are described by Brook[170b], Falce[364a].

2.6 SEALING AND SEALING-OFF

26.1 Sealing vacuum devices

The *sealing* operation is known in the electric lamp, radio tube, TV tube, and similar technologies as the technique of closing the assembled "stem" (Section 42.2) or electrode system (Section 42.5) in the envelope (bulb, tube). The two techniques used regularly for this purpose are the *drop sealing* and the *butt sealing;* to these techniques the *diffusion sealing* was added recently.

26.11 Drop sealing. The assembled stems of electric lamps, electron tubes, or other vacuum devices of similar form are sealed into their bulbs (or tubes), using the bulb sealing machines. The stem and bulb are arranged (manually or automatically) in the proper position with regard to each other, then the bulb is lowered over the stem (Fig. 2.109a) or the stem is raised inside the bulb. The assembly passes through a series of flames* while rotating conti-

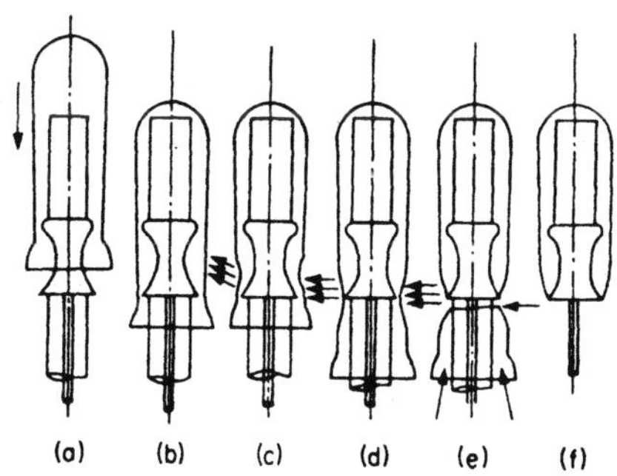

FIG. 2.109 Drop sealing process: (a) lowering the bulb over the stem; (b, c) heating with soft flames; (d) sealing the bulb to the flared stem; (e) cutting (dropping) the excess glass; (f) sealed device, ready for pumping

* Due to their relative position during the heating and sealing process, the bulb glass is heated more than the stem, thus the bulb (tube) must be harder than that of the flared part of the stem, the two glasses having in the same time matching expansions (Section 23.12). Since it is impossible to adjust the flames for each individual stem and bulb, constancy of the glass composition and dimensions is absolutely necessary (Table 2.12).

nuously. After the neck of the bulb has been softened enough by the heating, first in a soft flame (Fig. 2.109b) and then successively in stronger flames, the weight of the excess glass (or the weight attached to the neck of the bulb) will cause stretching of the neck (Fig. 2.109d). The diameter of the neck decreases until the bulb contacts the flare of the stem and is sealed to this. In the next sequence of the machine, the excess glass is flare cut (Fig. 2.109e) using sharp flames from outside and air pressure from inside the neck. In some cases the seal is formed in the next sequence by blowing it again (through the exhaust tube) in suitable moulding forms.

26.12 Butt sealing. This seal is used to join the lead-throughs holding discs (see Sec. 42.5) to the bulbs or glass tubes. For butt sealing the bulbs have no excess glass (as in glass sealing) and the disc is positioned to have its edge on the edge of the bulb.

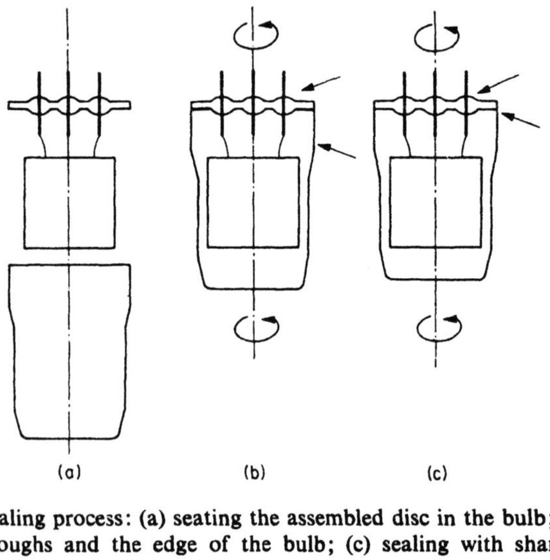

FIG. 2.110 Butt sealing process: (a) seating the assembled disc in the bulb; (b) preheating the lead-throughs and the edge of the bulb; (c) sealing with sharp flames

The disc together with the edge of the bulb is successively heated during one cycle of the sealing machine, (Fig. 2.110). Figure 2.110b, shows the arrangement in this step; in effect the pre-heating is done in 4–6 steps of the machine. The proper seal is made by sharp flames (Fig. 2.110c) followed by softer ones allowing it to cool at the required rate, to assure proper annealing (Section 23.13).

Butt seals can also be made using indirect heating from metal or carbon rings placed around the edge of the bulb (Violet[1269], Section 24.22).

Thin-walled glass tubes may be sealed together by heating them with a number of flames placed radially (Fig. 2.111a). The hot edges are pressed together, producing a joint (Fig. 2.111b). By immediately stretching the seal slightly a joint is obtained which does not differ from the rest of the envelope (Fig. 2.111c). Edens[322] mentions that the sealing of two thick-walled (5–6 mm) sections by the same method does not produce a good joint, since a notch is formed on the inside (Fig. 2.111d), due to the temperature difference on the two sides of the wall. Edens[322] proposes a method of sealing in which the edges to be sealed are heated with flames directed perpendicular to these edges (Fig. 2.111e).

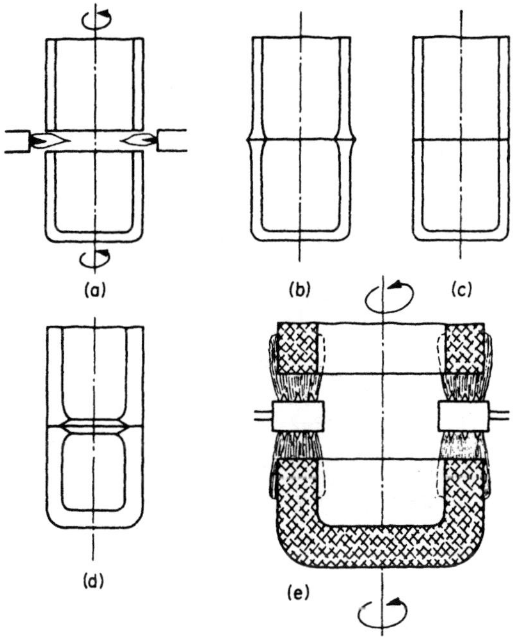

FIG. 2.111 Butt sealing of tubes: (a, b, c) with thin wall; (d) with thick wall (weak seal) (e) perpendicular heating for thick walled tubes (reliable seal). After Edens[322] *(by permission from Philips Techn. Rev. 19,318,1957/58)*

26.13 The diffusion seal between optically polished glass surfaces is described by Danzin[259], Bleuze[138]. In this technique the edge of the glass tube (bulb) and of the disc to be sealed to it are spherically polished (Fig. 2.112a, b). The polished surfaces are optically tested using the interference fringes, which appear when the surfaces are placed against standard polished surfaces of the same shape. The disc is placed on the tube (bulb) as shown in Fig. 2.112c and the bulb is evacuated. The optically polished edges can assure a tight enough seal to hold a vacuum of about 10^{-5} torr. The assembly is placed in a furnace and heated to a temperature *below* the softening of the glass (heating at 300–400 °C for about 15 min). The edge to be sealed is heated for a very short time (about 30 sec) to a higher temperature using a graphite ring heated by high frequency induction coils (Fig. 2.112c). By this process, the seal is made on the surface (Fig. 2.112d) due to the diffusion of molecules from one part to the other, without the necessity of heating the glass up to the softening point.

The diffusion seal is especially recommended in cases where the heating of the seal can damage (poison) some inner part of the tube.

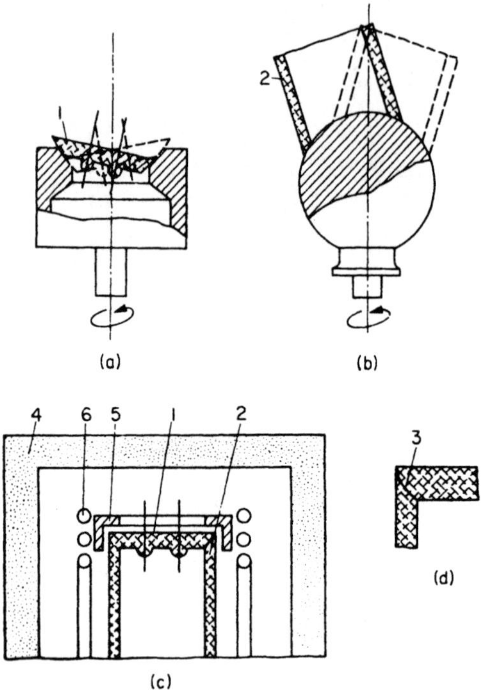

FIG. 2.112 Diffusion seal: (a) polishing the edge of the disc; (b) polishing the edge of the tube; (c) the sealing process; (d) the sealed edge. (1) disc; (2) tube; (3) optically polished contact edges; (4) electric furnace (200–400 °C); (5) graphite ring for local heating; (6) high frequency induction coil. After Bleuze[138] *(Courtesy of Société Française des Ingénieurs et Techniciens du Vide)*

Macembo[789a] describes a similar seal, between two optically polished glass flanges, but without using any heating (Section 36.1).

The sealing of the disc to the tube can also be done, by using solder glasses (Section 35.2).

26.2 Sealing-off glass exhaust tubes

26.21 Sealing-off small diameter tubes. The sealing-off, i.e. the separation of the evacuated devices from the pumping system, is done by sealing (closing) the exhaust tube connecting them to the pump.*

* The evolution of the seal-off technique is reviewed by Martin[813a].

For standard glass devices, generally small bore glass exhaust tubes are used (with inside diameter 5–7 mm). The sealing-off of such exhaust tubes is done with hand torches or automatically by the cross-fires of the exhaust machines. During the sealing-off, the evolution of gases produced by the decomposition of the glass (see Fig. 2.16) at the heated portions, cannot be avoided, but the evolved gas quantities can be minimized by completing correctly the sealing-off operations. The precautions to be observed in order to obtain a strong and vacuum-tight seal and to minimize the degassing rate, can be summarized as follows:

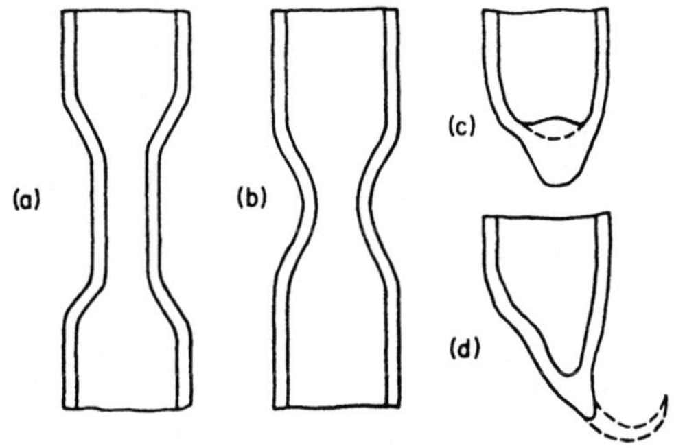

FIG. 2.113 Sealing-off glass exhaust tubes: (a) correct shape (thick-walled, long constriction); (b) incorrect shape (thin-walled, short constriction); (c) good seal; (b) bad seal

(1) The shape of the constriction made on the exhaust tube should be as shown by Fig. 2.113a, i.e. long and with thick walls. The thin-walled and/or sharp constrictions (as in Fig. 2.113b) are to be avoided. Very small bore constrictions drastically decrease the pumping speed.

(2) It is recommended that the exhaust tube be preheated along the whole portion where the seal will be made, and to allow the evolved gases to be exhausted before sealing-off the tube.

(3) For a correct sealing-off (closing) operation the heating should be started from the nearest point to the vacuum device and continued towards the pump.

(4) The last stages of the sealing-off operation (after the initial collapse of the exhaust tube) should be done quickly; this is made easier by pulling axially, particularly when previous constrictions were not made on the exhaust tube.

(5) The sealed end of the tube should be kept in the flame long enough to obtain a rounded seal (as in Fig. 2.113c). Seals ending in thin pointed shapes (as in Fig. 2.113d) are not safe, even if immediately after sealing-off, they are vacuum-tight.

26.22 Sealing-off large diameter tubes. The usual exhaust tubes have a conductance of 1–4 l./sec, which in comparison with the pumping speed of the diffusion pumps (more than 10 l./sec) leads to a very low pumping efficiency, if the device must be pumped to high vacuum. For better pumping efficiency a wide bore exhaust tube would be necessary; but the sealing-off of such an exhaust tube by the usual methods is impossible. Generally when attempts are made to seal off wide bore glass tubes with torches, cracks develop due to the great quantity of glass which must be sealed together.

A method to obtain higher efficiencies is obviously to use two or more exhaust tubes in parallel; this method usually presents constructional difficulties and leads to an increased possibility for leakage, due to the multiplicity of the sealed-off points.

Adam (in Auwärter[56]) describes some experiments to seal off glass exhaust tubes of about 20 mm diameter; this is carried out using a tubular heating furnace (40 mm inside diameter, 150 mm long) slipped over the exhaust tube. After the pumping process, the furnace is heated, and the exhaust tube softens and collapses. After cooling, the furnace is taken away and the collapsed (sealed-off) tube is cut from the pumping connexion. The disadvantage of this method is the fact that the sealed-off part of the exhaust tube must be long (about 150 mm), while the collapsed portion is about 40 mm. If the exhaust tube is not heated along a long portion (the collapsed part is not long), the seal develops leaks and cracks. For sealing-off wide bore quartz tubes (up to 65 mm bore) Hill[560a] recommends the use of an "expandable silica plug filled with air". This plug which is in fact a quartz tube of smaller diameter than the tube to be sealed off, is sealed at both ends, while containing air at atmospheric pressure. It is then placed inside the quartz tube to be sealed off and, by heating the latter, is collapsed over the plug and sealed to it. The advantage of this method is that it gives very reproducible results, especially when glass tubes with calibrated wall thicknesses are used.

26.23 Sealing off quartz exhaust tubes. To seal off quartz exhaust tubes strong flames are necessary (Section 23.23).

Schneidemesser[1105] describes a sealing-off procedure for quartz tubes using an electric arc. The arc is generated between two carbon rods. The power is supplied from an a.c. or d.c. source and adjusted with a suitable resistance to give a voltage drop of about 60 V across the carbon rods. The sealing-off tool has the shape of tongs, the ends of which are fitted with porcelain tubes (A, Fig. 2.114) to insulate the carbon rods *B* from the body of the tongs. Into these porcelain tubes, copper tubes *C* (of about 8 mm internal diameter) are inserted, having one end flanged and the other threaded and slit to form a simple chuck. A nut *D* tightens the chuck around the carbon rods, which have nearly the same diameter as the copper tubes, thereby ensuring good electrical contact and locating the copper tube firmly in the porcelain. The ends of the carbon rods protruding from the chuck are supported by the copper springs *E*. The handles of the tongs are forced apart by the steel spring *F*.

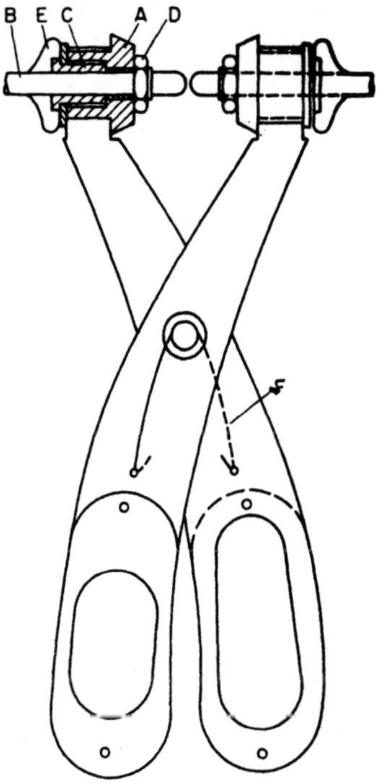

FIG. 2.114 Tool for sealing-off quartz exhaust tubes. Reproduced from Schneidmesser[1105] *(Courtesy of Pergamon Press)*

26.3 Sealing-off metal exhaust tubes (vessels)

Metal vessels (tubes) may be sealed off using either cold welding (Section 22.11) or regular brazing (Section 22.2) or soldering (Section 3.5).

26.31 Cold sealing-off. The cold sealing-off technique consists in pinching-off the metal exhaust tube (Loh[780], Neher[914], Varian[1257a]). For this purpose the tube is flattened between two jaws (Fig. 2.115) until the two opposite walls of the flattened tube are cold welded to each other (see also Section 22.11). The jaws are mounted on a hand tool* (Plate 2), actuated by a hydraulic system* or even actuated by the force of an explosion (Loh[780]).

This technique can be used to seal off copper (OFHC), annealed electrolytic nickel, and with some difficulties iron–nickel, exhaust tubes. For a good

* Cold sealing-off tools are available e.g. from Varian Associates, 611 Hansen Way, Palo Alto, Calif., or Kane Engineering Labs., 845 Commercial St., Palo Alto, Calif., U.S.A.

seal, the inside surface must be very clean, free of dirt and oxides. If the tubing is cleaned with steel wool or emery paper, it is essential that no particles remain to interfere with a vacuum-tight pinch-off.

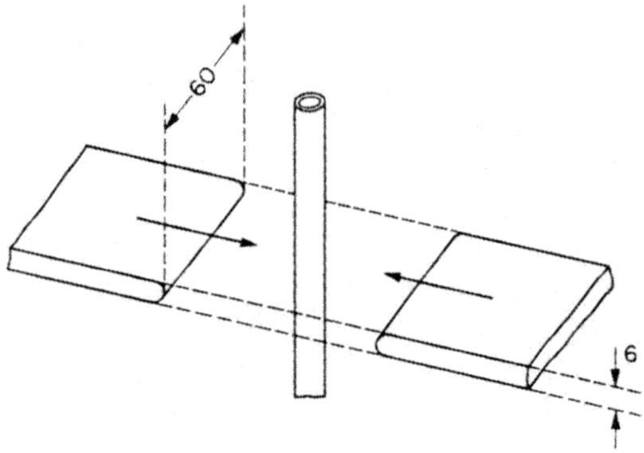

FIG. 2.115 Sealing-off by cold welding

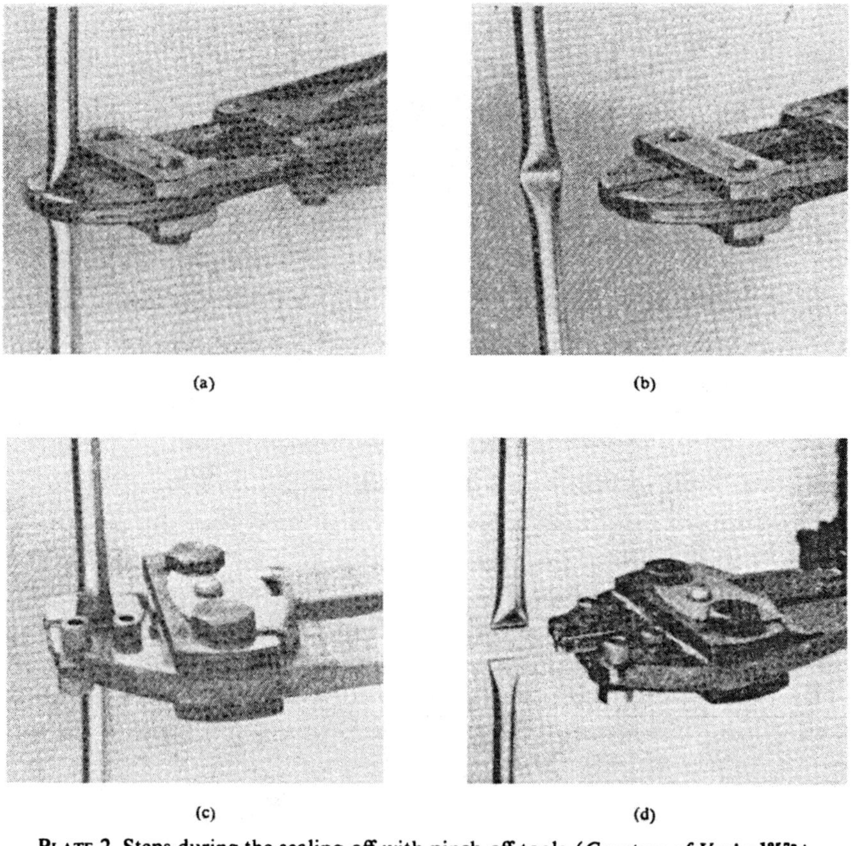

(a) (b) (c) (d)

PLATE 2. Steps during the sealing-off with pinch-off tools *(Courtesy of Varian*[1257a]*)*

The sealing-off by cold welding can be used on tubes having diameters in the range 8–25 mm and wall thicknesses between 0.5 and 1.5 mm.

When small diameter tubes (6–10 mm) are sealed, the sealing-off process can be completed in a single step, but for larger diameters it is recommended to complete the sealing-off using first a flattening step (Plate 2a, b) followed by a pinch-off step (Plate 2c, d). For the flattening step, rounded-edge jaws should be used having a thickness between half and one diameter of the exhaust tube. The width of the jaws should be at least twice the diameter of the exhaust tube. For the pinch-off, the jaws must be centred around the exhaust tube to ensure that the flattened tube will not extend beyond the width of the jaws.

During the sealing operation both ends of the exhaust tube should be held rigidly. If one end is free to move, the tube has a tendency to break prematurely at the site of the sealing and results in a leaky seal (Loh[780]).

The flattening or pinching-off should be done in one continuous stroke, without stopping part of the way and starting again.

If a brazed joint exists of the exhaust tube, the pinch-off should be made no closer than two diameters from the brazed area, allowing a smooth transition from the round tubing to the flat pinch-off.

To ensure protection against corrosion, which can produce leaks in the pinched-off edge, it is useful to dip the pinched-off end in molten tin. To protect the pinch-off from accidental damage, a PVC protector cap should be slipped over the sealed-off end of the exhaust tube.

The pinched-off place can be opened by squeezing it between the jaws of a plier or a vice. Neher[913a] describes a special tool constructed for this purpose. It is not recommended to make another cold seal on the same place where the previous one was opened.

The cold sealing-off procedure has the advantage of not releasing gases, and results in very reliable seals.

The cold sealing-off was used by Aitken[12] to seal irradiation containers under vacuum, by squeezing together the two parts of the container (Cu, or Al) placed in the vacuum vessel, the sealing motion being actuated through bellows (Section 51.4).

26.32 Sealing-off with valves. Vacuum vessels (pipes) can be sealed-off using a plug, which since it must be closed under vacuum, should be applied from a sealing-off valve (Fig. 2.116).

The seal-off valve (see also Section 61.35a) allows a disc (3, Fig. 2.116) to close the port, and then to remove the valve bonnet and stem. When the desired vacuum is obtained the forked part of the valve stem (1) is engaged in the two ports in the movable disc-holder (2); the valve handle is then turned until the the disc (3) closes the port. The bonnet screws are loosened, and all the parts except the movable closing assembly (2–3) are removed as a

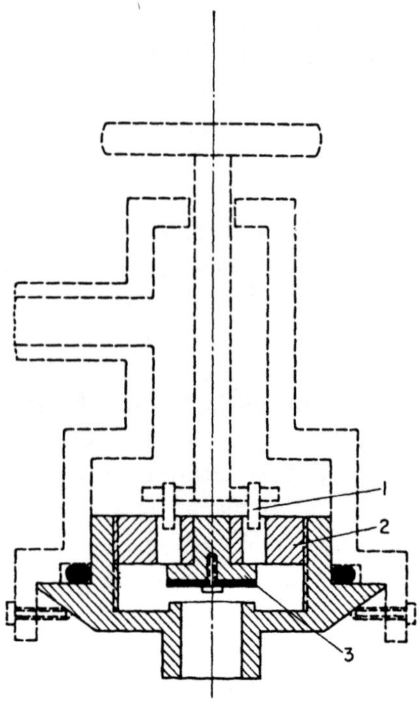

FIG. 2.116 Seal-off valve. After Richards[1053] *(Courtesy of The American Institute of Physics)*

unit. When the port is to be opened again, the unit is replaced again over the port, the bonnet screws tightened, and after pumping the space in the valve, the valve handle is actuated to open the valve. This type of valve is described by Richards[1053] and Klipping[689]; similar valves are manufactured by Leybold[762] (Fig. 6.90).

26.33 Brazed seal-off. Metal exhaust tubes may be sealed-off by pressing (flattening) the exhaust tube in the portion where a brazing (soldering) alloy insert was previously placed in the tube, and by simultaneously heating this section (Fig. 2.117).

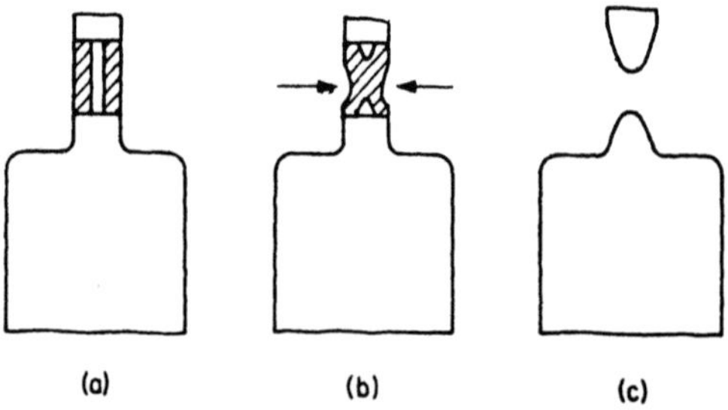

FIG. 2.117 Sealing-off with solder insert: (a) solder inserted in the exhaust tube; (b) sealing-off; (c) pinched tube

After the tube has been closed by flattening and heating it can be cut mechanically. The heating of the inserted soldering plug can be done by flame or by resistance heating (Vatter[1258]).

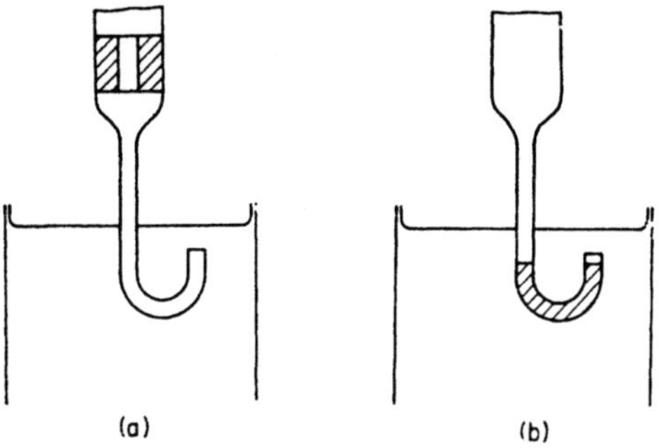

FIG. 2.118 Sealing-off with trapped solder: (a) before and (b) after sealing-off

Another sealing-off technique using soldering alloys is based on the flow of the molten alloy to a place where it closes the tube. An arrangement for such a technique is shown on Fig. 2.118. The sealing alloy is inserted in the exhaust tube (Fig. 2.118a) and is kept cold during the pumping. To seal off, the solder is melted by heating the tube, the alloy flows on the bottom (U-shaped) part of the tube and when cooled closes the tube (Fig. 2.118b).

CHAPTER 3

SEMI-PERMANENT AND DEMOUNTABLE SEALS

3.1 WAXED SEALS

In early literature (Stohr[1197], Strong[1207], Walden[1282]) the usefulness of waxed seals was stressed, but more recently wax has been replaced in many cases by other seals (epoxy resins, elastomers). Nevertheless waxed seals should be briefly described since they are still used in unique or temporary applications. These seals can be used in applications where heating is inconvenient or not possible. Waxed seals can be used to join temporarily, metal, glass, quartz, ceramic and sometimes elastomer parts, or to seal temporarily pin-holes or leaky joints. It is recommended to avoid the use of waxed seals in any permanent or for long term vacuum work.

31.1 Sealing waxes

Waxes are compounds, which when warmed are plastic but become rigid at room temperature and this can easily be used to make vacuum seals.

Table 3.1 summarizes the waxes which may be used for temporary vacuum sealing, giving the range where they can be used with respect to temperature and solvents.

31.2 Vacuum sealing technique with waxes

A good waxed seal is obtained only with clean surfaces, assembled with a minimum amount of wax, applied at temperatures high enough for the particular wax.

The surfaces to be joined by the wax must first be cleaned and especially degreased using apropriate solvents (Sections 22.31 and 23.21). The surface must be dried since wax applied on even a slightly wet surface does not adhere and the seal will leak.

The parts to be sealed should be heated with a flame or on a hot plate, until the part is hot enough to melt the wax so that it flows on application i.e. to form a meniscus as in Fig. 3.1b. Procedures consisting of applying hot wax on colder surfaces, do not give reliable seals. Such seals made at too low a temperature can be easily recognized (Fig. 3.1a).

After the surfaces, which are to be in contact, are covered with wax, the parts are pressed together, while still hot, with a slight rubbing action, sufficient to expel the air bubbles. Then the assembly should be allowed to cool.

It is recommended to handle the wax without touching it. The waxes are supplied in sticks (Table 3.1) which are usually too thick to use directly. Before use the stick of wax is softened and pulled gently to a diameter of

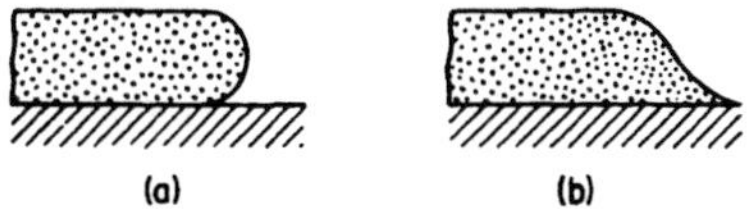

FIG. 3.1 Waxed seals: (a) incorrect technique (wax applied on cold surfaces); (b) correct shape

3–6 mm (about 20 cm in length); in this shape the wax can be used easily. An alternative method is to use a greaser from which, after slight heating, the wax can be extruded as a cord of 2–3 mm diameter. A tool for general waxing use consists of a triangular plate (side about 7 cm) fitted with a tang. The tool is heated and the wax, held in contact with the hot metal surface, will soften and run into any place at which the tool point is held.

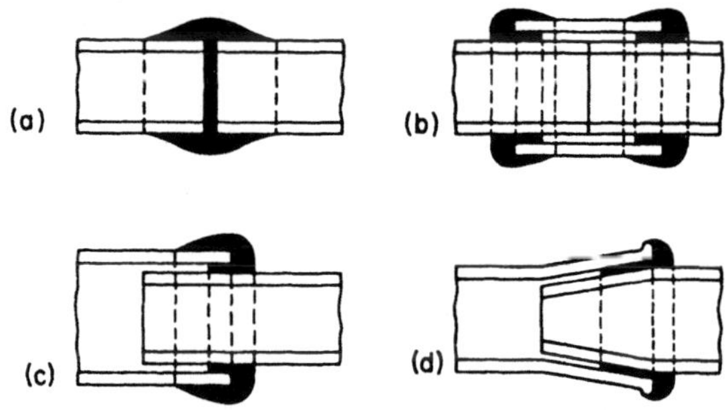

FIG. 3.2 Methods of joining pipes by waxed seals: (a) butt joint; (b) butt joint with sleeve; (c) telescopic joint; (d) conical joint

It is recommended to use the wax just as the element which makes the joint vacuum-tight, and not as the structural element of the joint. For this the parts to be joined should have as good a mechanical fit as possible.

Pipes can be joined temporarily by forming a waxed butt joint (Fig. 3.2a, b) a telescopic joint (Fig. 3.2c) or a conical joint (Fig. 3.2d). The procedure for making a butt joint is to wrap a soft strip of wax around the butted pipe ends, after heating them to a temperature high enough to melt the wax (Table 3.1). The wax is then moulded as in Fig. 3.2a.

If the joint is to be subjected to slight bending or other mechanical efforts, it is best made using a third pipe (sleeve) over the joint, and sealing this

TABLE 3.1.

Wax	Composition	Softening	Max. safe	Vapour pressure (torr)	Solvents (in up to 24 hr)
		temperature (°C)			
Soft red wax	bees-wax (5 pbw) Turpentine (1 pbw) dyestuff	55–60 (wetting)	25	10^{-5} (25 °C)	Acetone, alcohol, benzene, chloroform, ether, turpentine, xylene
Faraday wax	rosin (5 pbw) beeswax (1 pbw) Venetian red (1 pbw)	60–75 (75–95 wetting)	–	–	Acetone, alcohol, benzene, ether, xylene
Beeswax–rosin	Rosin (1 pbw) Beeswax (1 pbw)	47	40	5.10^{-6} (25 °C)	Mixture of carbon tetrachloride and alcohol (1:1)
Celvacene heavy	dark yellow-brown transparent, greasy	130	–	10^{-6} (25 °C)	Chloroform, acetone
Shellac	Insect and tree resin (India), mixture of polyhydroxy acids and esters. Used dissolved in warm spirit*	60–80 (100–125, wetting)	–	–	Acetone, alcohol, chloroform, ether, butyl-phtalate
Red sealing wax	Shellac, Venice turpentine, Vermillon or Chinese red	60–80 (100–125 wetting)	–	10^{-5} (25 °C)	Acetone, alcohol, benzene, chloroform, ether, xylene
De Khotinsky cement	Shellac and Caroline (wood) tar	85–100 (95–150 wetting)	40	10^{-3} (25 °C)	Acetone, alcohol, benzene, chloroform, ether, xylene
Sealstix	De Khotinsky type	–	–	–	ethyl alcohol, acetone
W. E. Wax 6	Shellac base, brown wax	80	–	–	alcohol

SEALING WAXES

Non-solvents (up to 60 hr)	Attacked (A) Resistant (R)		Remarks	References Suppliers+
	acids	alkalines		
water	as shellac	A by potassium sol. R to sodium sol.	Slightly harder than plasticine. Loses its plasticity by oxidation	Walden[1282], Strong[1207] Zabel[1352a]
water	as shellac	R (slight discolouration)	Characteristics may be changed by various mixtures	Walden[1282]
–	–	–	Good adhesion to cold metals	Strong[1207] Zabel[1352a]
–	–	–	Vacuum-tight bond for rubber-to-metal or-glass	CVC
Benzene, turpentine, xylene, water, most oils	A by HNO_3 and H_2SO_4 R to HCl	A	Moderately tough resin. Desintegrates and sol. no more suitable for sealing. By heat (30 hr at 90 °C, or 3 hr at 150 °C) becomes harder due to polymerization	Walden[1282] Young[1345]
oils, water	as shellac	A	Bends or prone to slow changes; brittle when dropped	Walden[1282]
Petroleum, turpentine oil, water,	R	A	Tough but very slightly plastic. Polymerizes at room temperature in 6 month#	Cenco, Zabel[1352a] Walden[1282]
water	A by chromic acid	A	Polymerizes (6 months) Cleaning off with chromic acid	Cenco
benzene, toluene	–	–	–	Edwards

(Table 3.1 Continued)

Wax	Composition	Softening	Max. safe	Vapour pressure (torr)	Solvents (in up to 24 hr)
		temperature (°C)			
Picein	Hydrocarbons from rubber, shellac, bitumen. Black wax	80 (90 wetting) 90 (105 wetting)	50 60	1.10^{-6} (−25°C) 4.10^{-4} (25 °C) 5.10^{-3} (50 °C)	Benzene, benzine, chloroform, ether, turpentine, xylene
Wax V	Solid high molecular weight hydrocarbons, fine inorganic powder, rubber	183 (drops)	30	10^{-4} (25 °C)	–
White sealing wax	Shellac with resins and heat resistant minerals	106 (drops)	50	10^{-3} (25 °C)	Petroleum, benzene, turpentine, alcohol (water)
Apiezon sealing compound Q	Graphite, grease or parrafin oil distillation products	45 60 (wetting)	30	10^{-4} (25 °C) 2.10^{-4} (70 °C)	–
Vacoplast	similar to Wax Q	–	–	–	–
Apiezon Wax W-40 (soft)	Black wax in sticks	45	30	10^{-6} (25 °C) 10^{-3} (180 °C)	xylene
Apiezon Wax W-100 (medium)	Black wax in sticks	55	50	as W-40	xylene
Apiezon Wax W (hard)	Black wax in sticks	85 (100 wetting)	80	10^{-7} (25 °C) 10^{-3} (180 °C)	xylene (benzene, chloroform)

* Shellac used alone is brittle and tends to form hair-line cracks. It is therefore rein-
A mixture which after melting can be poured into moulds (to form sticks) consists of: (Walden[1282]).

+ Edwards High Vacuum International Ltd. (Edwards[327–329]); Leybold's Nachfolger, solidated Vacuum Corp.[231]; Cenco[199]; CGR – Comp. Gen. de Radiologie[224]; JGB –

\# The actual procedure of sealing is that for any sealing wax, care being taken to avoid quickly and become quite difficult to be removed.

Non-solvents (up to 60 hr)	Attacked (A) Resistant (R) acids	alkalines	Remarks	References Suppliers+
acetone, alcohol water	A by H_2SO_4 R to HCl, HNO_3 (and chromic)	R	Suitable for metal, glass sealing. Vibration resistant; not brittle; Available in 2 grades; Separation of joints by heating.	Strong[1207] Edwards, Leybold, CGR (Mastic P)
–	–	–	For sealing ungrounded joints	Leybold
–	R	R	Adheres well to glass and metal	Leybold
–	–	–	Consistency of plasticine; Temporary sealing or blanking during leak detection	Edwards Shell JGB
–	–	–	–	CGR
–	–	–	Application where it is required to flow the wax in or round the joint. For joints subjected to vibration, but not heated	DCC Edwards Shell JGB
–	–	–	Safe for cracks in joints subjected to vibration	Edwards Shell JGB
acetone, alcohol (water)	A by H_2SO_4 R to HCl, HNO_3	R	High vacuum work where the parts tend to warm up in use. Brittle to shock.	Edwards Shell DCC JGB

forced with other materials making it more readily fusible, more adhesive and stronger. shellac (50 pbw), wood creosote (5 pbw), turpentineol (2 pbw) and ammonia 0.88 (1 pbw)

(Leybold[762]); Shell Comp., (Shell[1129]); DCC-Dow Corning Corp.[299]; CVC – Con-
James G. Biddle Co., Plymouth Meeting, Penn., U.S.A.

melting of deKhotinsky wax in the heating flame, since due to overheating it decomposes sleeve to the ends of the pipes (Fig. 3.2b). Using a connecting sleeve of rubber (Norbury[937] and Fig. 5.2b) the joint can be made flexible.

Pipes having slightly different diameters may be connected by a telescopic joint (Fig. 3.2c); if the difference in diameters is too big the joint should be made by interposing a third pipe or a plug (Champeix[202], Mönch[879]).

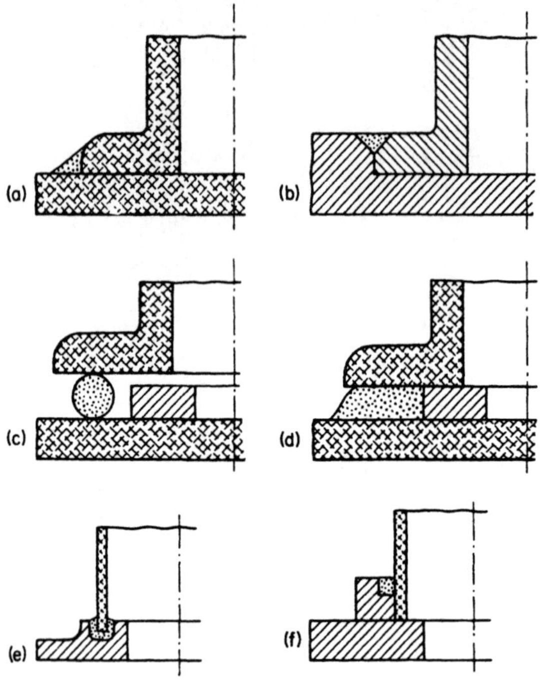

FIG. 3.3 Waxed seals between flanges, pipes and plates: (a) outside seal; (b) interval seal; (c) spacer seal before sealing; (d) spacer seal after sealing; (e, f) groove seals

Unground conical joints may be sealed with wax, allowing it to flow halfway down the inside of the joint (on the atmospheric side, see Fig. 3.2d). In this way the smallest possible surface of wax is exposed to the evacuated space. If ground joints (Section 36.2) are to be sealed with wax, it should be applied just around the outside of the joint, without allowing it to flow in the ground area.

Pipes ending in flanges or bell jars can be sealed onto plates or other flanges by the arrangements shown in Fig. 3.3. The wax is placed *outside* the periphery of the flange (Fig. 3.3a), in the space between the side of the flange and that of the plate (Fig. 3.3b) or between the face of the flange and the plate (Fig. 3.3d). In the last case it is recommended that a spacer between the flange and the plate be used, in order to minimize the area of wax exposed to the evacuated space and to facilitate the opening of the seal. The spacer is surrounded by a cord of wax (Fig. 3.3c) having a diameter greater than the height of the spacer; the flange is placed on the top, and by heating and applying a slight pressure the required seal is obtained (Fig. 3.3d). To open the seal the flange is raised by an inserted tool (Strong[1207]).

Pipes can be sealed to flanges or plates by filling with wax a groove provided in the plate (Fig. 3.3e) or the groove formed between the flange and the pipe (Fig. 3.3.f).

When electrodes are sealed for temporary use, the wax may simultaneously seal the insulating part (2, Fig. 3.4a) to the wall of the chamber (1) and the electrode (3) to the insulating part. If compression seals (Section 24.43) or

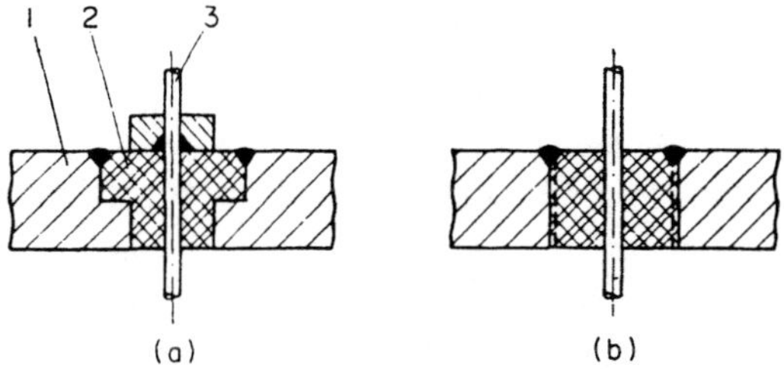

FIG. 3.4 Waxed electrode seals: (a) double seal; (b) single seal

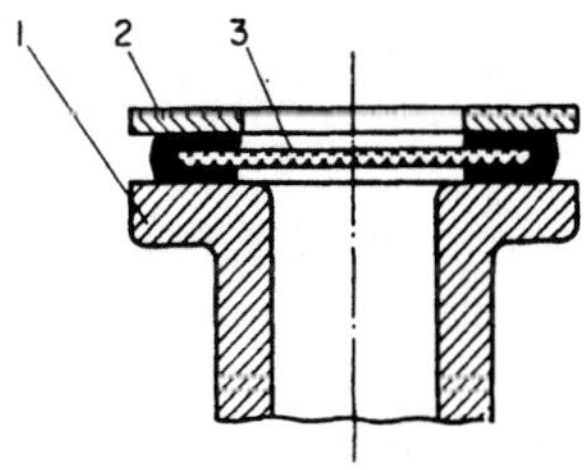

FIG. 3.4A Waxed window seal

ceramic–metal seals (Section 25.42) are used, they can be provided with a screw and can be screwed into the wall of the chamber (Fig. 3.4b), the wax being used only to seal the screw.

Scott[1114] describes a technique to seal thermocouple wires with wax (Fig. 4.27a).

Windows may be wax sealed (Section 72.3) between two flanges; one of these flanges (1, Fig. 3.4A) is usually the port, and the other (2) is a covering flange. In order to seal the window, both flanges are first covered with wax on the surfaces where the seal will be made. The window (3) is placed between the two flanges and the assembly is fixed in a jig or clamps. Then the assembly is heated to the wetting point of the wax (Table 3.1). The diameter of the flanges and of the ring of wax applied on their surfaces should be greater than that of the window, so that by melting the wax the edge of the window will be embedded in wax (Fig. 3.4A). An alternative technique (Strong[1207]) consists of clamping the window in the required position heating it to the required temperature, then applying the wax to the outer edge of the window, from where it will be drawn between the window and the port by capillary action.

3.2 SEALING WITH PAINTS AND PLASTICS

32.1 *Irreversible adhesives*

This category of sealing compounds consists of adhesives mainly used for sealing pinholes or porous walls (sealing lacquers), or of adhesives used for proper sealing purposes (sealants). Table 3.2 summarizes the characteristics of the adhesives of these two groups.

The sealing lacquers (e.g. Glyptal), painted on to the surfaces in order to seal pinholes, are drawn into the orifices, and as the volatile solvent is removed, the residue plugs the orifice. Since these plugs carbonize during the softening of the glass and change the composition of the glass, they can present a problem if more permanent repair is to be made later. With metal structures the difficulty will be in cleaning enough of the residue away to be able to do a good brazing or welding job (Bachman[65]).

The sealing lacquers may also be used to seal conical joints, or screwed fittings (Strother[1208]).

Recently the lacquer called *anaerobic permafil* was developed (Burnett[183]). Its mode of setting is in complete contrast to that of the usual paints or lacquers. With a large supply of air, the permafil remains fluid, but when oxygen is excluded it sets to a solid of low vapour pressure. It stays fluid while it penetrates deep into the pore and there it solidifies. Anaerobic permafil is prepared locally where used by combining two ingredients: about 4 per cent of a solid promoter with 96 per cent of a liquid monomer, and aerating this solution for several days to obtain the desired degree of chemical reactivity. Storage of the activated liquid polymer is either by continuous aeration at 20 °C or less, or by refrigeration at the dry ice temperature without aeration.

Anaerobic permafil is applied by brushing it over the surface of the porous object, which can be evacuated to aid the penetration of the liquid polymer if the pores are extremely fine. The pores must be first freed of oil, water, greases, etc. Rinsing the surface with a pure solvent is a poor way to clean out the pores; however, heating the object to 300–500 °C is a better way of cleaning by burning out the contaminants. After applying the liquid, the treated surface is allowed to remain at room temperature overnight or longer to permit the anaerobic permafil to solidify completely. Mild warming will speed up the process. The permafil remaining on the surface stays wet for

TABLE 3.2. IRREVERSIBLE ADHESIVES

Lacquer or sealant	Composition	Max. safe temperature (or range °C)	Characteristics	Remarks	Reference Manufacturer*
Sealing lacquer	Mixture of high polymers in solvents	−20 80	After drying insoluble in oils, petroleum, water		Leybold
Glyptal lacquer	Alkyd resin, formed by condensation of phtalic anhydride, glycol	100	Vapour pressure 10^{-6} torr (−25 °C), 2×10^{-4} (25 °C), 10^{-1} (70 °C). Soluble in acetone, xylol or benzine. Insoluble in mineral oils, alcohol, water	Good fluidity and wetting, also Al, and Plexiglass; Drying 8 hr at room temp. Polymerizes 1–2 hr at 140 °C	CGR Strong[1207] Ardenne[47] Espe[354]
Helmitin sealing compound	Polyester to be mixed with hardener (600 C) (Hardener irritant of mucose!)	100	Soluble in methylene chloride, methanol, water. Insoluble organic solvents. Tensile strength 5 kg/mm² (25 °C)	Hardens: 1 hr (180 °C), 15–20 hr (20 °C)	Leybold
Desmodur	Ester of diysocyan acid. Two solutions are to be mixed max. 3 hr before use	110		Seals glass, metals, ceramics, rubber. Needs pressure 3–4 atm for 2 hr at 90–130 °C or 8–10 hr at room temperature.	IGF Ardenne[47]
Anaerobic permafil	Polymer of polyglycol dimethacrylate	—	Sets by exclusion of oxygen (vacuum)	Applied by brushing on the outside of porous walls.	Burnett[183]
PVA solution sealant	Polyvinil acetate (1 pbw) solution in toluol (10 pbw) or PVA in acetone	65	Vapour pressure 1×10^{-5} torr (25 °C)	After drying the coated surface is heated for 30 min at 150 °C	Espe[354]

(Table 3.2 Continued)

Lacquer or sealant	Composition	Max. safe temperature (or range °C)	Characteristics	Remarks	Reference Manufacturer*
Loctite sealant	Thin liquid	−40 150	Liquid soluble in trichlorethylene. Sets by exclusion of air.	Bonds metal, glass, ceramics. 4–12 hr (25 °C), 10 min (100 °C) 5 min (180 °C)	ASC
Natural rubber sealent	Natural rubber (1 pbw) in benzol (2 pbw)	70	Applied on rough surfaces.	8 hr drying, then pressed	Medicus[837]
Silicone rubber sealant	RTV-102 silicone rubber adhesive	−60 150	Air cure at room temp.; first forms a surface skin; completely cured 24 hr (3 mm layer)		GE Gondet[457a]

* ASC – American Sealants Co., Hartford 11, Conn., U.S.A.; CGR – Compagnie Gen. de Radiologie[224]; Leybold – Leybold[762]; GE – General Electric Co., Waterford, New York, U.S.A.; IGF – I.G. Farben-Industrie, Germany.

weeks since it is exposed to the oxygen of the air. The excess liquid should not be wiped off for at least a day, because it acts as a reservoir to feed more liquid polymer into the pores and take care of the shrinkage (about 10 per cent) which occurs during polymerization in the absence of oxygen.

Young[1351a] reports the use of a silicone resin (G.E. SR-82) which has been found to be capable of sealing leaks in metal and glass systems. This resin withstands bakeout to 400 °C and has a very low vapour pressure, so that it can be used down to 2×10^{-10} torr.

Reeber[1041] reports on the use of Loctite (Table 3.2) for high vacuum seals in cryogenic work. After cleaning (degreasing) the flanges, the groove (3 mm wide, 0.05 mm deep) of one of them was filled with Loctite, and then placed in contact with a second (plane) flange, and allowed to set overnight. The seals obtained when the flanges were made from the same material were vacuum-tight down to liquid helium temperatures, but if the flanges were made from materials having different contractions (as stainless steel and copper) the seal was successful just down to liquid nitrogen temperatures. The seal can be opened by inserting a screwdriver between the lips of the flanges.

In order to seal electrical leads, loctite requires a hole about 0.1 mm larger in diameter than the wire.

32.2 Sealing of plastics and the use of plastics for sealing

Polythene (Table 2.14) can be joined to metal or glass by heating with a jet of hot nitrogen, i.e. nitrogen circulated through a pipe heated to about 200 °C (Duncan[308]). Polythene tubes can also be joined to metal, glass or ceramic tubes by heating the end of the second tube to above the softening point of the polythene and inserting it into the Polythene tube. The junction is then warmed slowly and the Polythene pressed with a cold flat tool to ensure a vacuum-tight seal.

Small holes in Polythene tubing can be sealed by melting fresh Polythene and pressing it with a cold flat tool onto the leak.

Saran (see Table 2.14) tubing heated to the softening point is applied over the hot metal or glass tubing with a tight fit. Further heating (in the hot gases of a flame) and pressure completes the bond between Saran and the rigid tube. With metal tubing it is better to heat the tube and use the thermal co nduction of the metal to heat up the plastic (Sancier[1095]).

Besides their use as tubing, recently some plastics were used as sealing tapes. George[443] joined tubes by a telescopic arrangement, using Polythene tape wrapped around the tube with the smaller diameter. After inserting the wrapped end into the larger tube, the assembly is heated to about 120 °C, when the Polythene just melts. On cooling the Polythene forms a vacuum, tight joint. It is recommended to keep the tubes vertical during the sealing, and to use a ring of asbestos cord placed below the Polythene to prevent it flowing out from the joint.

Vacuum sealing tapes with polyethylene basis (Cenco[199]) are recommended to seal connexions of glass, metal, rubber, polyethylene or vinyl tubes with about equal diameters, or to seal complex contours. If the joint is to be used for a long time a coating of Shellac (Table 3.1) or Glyptal (see Table 3.2) may be applied over the tape seal to protect it.

Hickman[557] reported a seal for high vacuum using a plastic electric tape. In this seal a turn of a 12 mm wide tape was applied under slight tension around a glass bell jar standing on a clean plate, so that about half the width of the tape lies flat on the plate and half upright against the wall of the bell jar. At the junction of the tape ends a 25 mm overlap was made, the overlap area and about 12 mm in each side was generously smeared with heavy grease (Celvacene Heavy, see Table 3.1). The grease was momentarily melted with a miniature hand blow torch, and allowed to re-harden in gel form (important). The vacuum inside the jar pulls the tape into the closest contact. A paint of polyether (diffusion pump fluid) was spread evenly around the tape, overlapping both plate and bell jar. The seal, if soundly made, is successful down to 10^{-10} torr (Hickman[557]) part of its success is due to the very low admittance of residual sealant vapours through the crack between jar and plate.

To dismount, the tape is pulled with a pair of sharp-nosed pliers and the jar lifted off.

Teflon (Table 2.14) cannot be bonded with usual gluing or adhesive techniques. Thin (less than 0.1 mm) Teflon sheets can be bonded only after treatment which chemically etches its surface, permitting adhesives to form a mechanical bond (Anon.[38]). The treatment consists in the immersion of Teflon in a bath of metallic sodium dissolved in liquid anhydrous ammonia, followed by quenching in cold water.

Fluocarbons (TFE) can be welded to themselves with contact heaters at temperatures* of about 370 °C and pressures of about 2–3 atm (Riley[1063]). Sections about 1.5 mm thick provide the optimum results since they have the strength necessary for fabrication, and are thin enough to be brought to the required temperature in a relatively short time. After joining the assembly should be stress-relieved by heating it for 2 hr at about 30 °C above the maximum service temperature. Scarf joints are the most satisfactory, but feather edges on the scarf should be avoided as they may warp when heated. During the sealing, the temperature and pressure should be uniform and carefully controlled; cold spots result in poor bonds which must be resealed. The joint can be improved by using a flux composed of 65 pbw of fluocarbon oil and 35 pbw powdered TFE. The flux should be mixed and applied at about 65 °C.

3.3 SEALING WITH EPOXY RESINS

Epoxy resins (known also as ethoxyline resins) are used widely in many industries to join almost all materials; recently they have had more and more applications in the various vacuum sealing procedures. For details on these resins and their application in various fields see, e.g. Skeist[1143], Schrade[1107], Lee[753], Moss[898, 899], Andre[33], Leutner[759], Preiswerk[1005, 1006], Miller[859], Anon[41] Balain[73], Schäfer[1102], Perry[984], Martin[814], Stivala[1192], Salsig[1094], Meriam[847], Bachman[69].

33.1 Epoxy adhesives

The epoxy adhesive resins are available under various trade names from various suppliers, e.g.

Araldite (Ciba Ltd., 141 Klybeckstrasse, Basle, Switzerland).
Epon (Shell Chem. Co., 380 Madison Ave., New York, N.Y., U.S.A.).
Epikote (Shell Chem. Co., Malborough House, London, England).
Gen Epoxy (General Mills Inc., Kankakee, Ill., U.S.A.).
DER (Dow Chemical Co., Midland, Mich., U.S.A.).

* Teflon gives off toxic fumes at these temperatures (Section 21.4).

Epi Rez (Jones-Dabney Co. 1481 South 11th Street, Louisville, Ky, U.S.A.).
ERL (Bakelite Co. 30 East 42nd Street, New York, N.Y., U.S.A.).
Epoxy-Stycast (Emerson & Cuming Inc., Canton, Mass., U.S.A.).
Epoxydharz (VEB Leuna-Werke, Leuna, Germany).
Torr Seal (Varian Ass., 611 Hansen Way, Palo Alto, Calif., U.S.A.).
UHU-plus (UHU-Werk, H.u.M. Fischer, Bühl, Baden, Germany).
Epotuf (Reichhold Chem. Comp., 525 North Brodway, White Plains, N.Y., U.S.A.).
EKR (Union Carbide Corp., 270 Park Avenue, New York, N.Y., U.S.A.).

The choice of the adhesive to be used in a particular application depends on the *shape* of the seal, the *thickness* of the joint, and if *heating* is desired or available for curing.

Heat curing resins (Table 3.3) are preferred for sealing similar materials or materials with small differences in their coefficients of thermal expansion.

For the sealing of materials with very different thermal expansion, or for heat sensitive materials it is better to use adhesives which can be cured at room temperature (Table 3.3).

Epoxy adhesives were found to have many physical and chemical properties useful in applications for high vacuum seals. These adhesives are generally very stable. By heating, no breakdown, separation or sensible gas emission occurs (Stivala [1192], Balain [73]).

The epoxy resins have after curing, a high enough shear strength (Fig. 3.8). For example, when cured for optimum strength Araldite CN 502 can withstand a force equivalent to 5.6 kg/mm^2 pure tension, and 12.6 kg/mm^2 compression.

Epoxy resins have good dielectric properties; Araldite I has a dielectric constant at 20 °C of 6, a loss factor tg $\delta = 1.10^{-2}$ at 10^3 cycles, $tg\ \delta = 6.10^{-3}$ at 2.10^6 cycles, (Section 4.1).

These resins are known for their low vapour pressure. In vessels containing Araldite seals, pressures as low as 10^{-6} torr were obtained after reasonable pumping times (Saechtling[1092a]); the degassing of Araldite measured after 3hr pumping is at room temperature, about 1×10^{-3} lusec/cm^2 (see also Fig. 2.13). Recently Markley[806] published the results of an extensive study on the outgassing of epoxy resins. Figure 3.5 shows the results for a typical epoxy resin (Epon 828 cured with hardener A) used as a cylindrical sample (0.437 in. in diameter by 4 in. long). It can be seen that the amount of evolved gas is reduced after a degassing (at 100 °C for 23 hr) even if subsequently the sample is exposed to atmosphere (16 hr), indicating that the material has a low vapour pressure and contains absorbed gasses which slowly diffuse out. Figure 3.6 shows comparatively the gas evolution rates of several materials. Samples 1–5 and 9–11 were cylinders 0.5 in. in diameter and 4 in. long; samples 6, 7, 13

TABLE 3.3. EPOXY ADHESIVES

Resin	Hardener	Parts by weight of hardener to 100 pbw of resin	Recommended joint thickness (mm)	Curing time (min) at temp. (°C)						Recommended uses and remarks
				20	40	70	100	150	180	
Araldite 101	951	5–6	0.1–0.2	24hr	—	—	10	—	—	For bonding small metal surfaces
	930	6–7	max. 0.5	24hr	—	—	—	—	3–5	More heat resistant
	936	6.5	—	—	—	—	—	—	—	Curing only at 60 °C for 2.5 hr
Araldite 102	951	6–7	0.05–0.15	24hr	—	—	—	5	3	Bonding porous materials
	936	6	—	—	—	—	—	—	—	Curing only at 60 °C for 2.5 hr
Araldite 103	951	7–8	0.05–0.2	36hr	14hr	3hr	60	20	10	For bonding metals, ceramics, rubber; large surfaces. Hard joints
	930	6–10	0.2–0.5	24hr	14hr	2hr	60	20	10	
Araldite 105	960	6–10	3	24hr	—	—	—	10	—	Filling up cracks
Araldite 106	953 U	80	—	7hr	3hr	1hr	10	5	—	Resistant to vibration
Araldite 121	951	4–4.5	0.1–3	36hr	14hr	2hr	30	10	5	Bonding ceramics, synthetic resins to themselves and to metals; low expansion or heat sensitive materials
	930	2.5–5	0.1–0.5	36hr	14hr	2hr	30	10	5	
Araldite 123	951	5.5–6	0.1–3							
	930	3–6	0.1–0.5	36hr	14hr	1hr	15	5	3	Bonding large, metal ceramic surfaces
Araldite 1 (Natural)	one part, rods	—	0.05–0.2	—	—	—	—	3hr	55 10(200 °C)	Bonding non ferrous metals, ceramics

(Table 3.3 Continued)

Resin	Hardener	Parts by weight of hardener to 100 pbw of resin	Recommended joint thickness (mm)	Curing time (min) at temp. (°C)						Recommended uses and remarks
				20	40	70	100	150	180	
Araldite VIII (Natural)	paste	—	0.05—0.5	—	—	—	—	4hr	1hr 30(200 °C)	Bonding large badly fitting parts
Epon IX	one part paste	—	—	—	—	—	—	—	90	Bonding metal to metal; Max. service temp. 150 °C
Epon 901	B-1	23	—	24hr	—	—	1hr	—	—	Max. service temp. 120 °C Bonding, metal, glass, rubber plastics
	B-3	11	—	—	—	—	—	—	1hr	Max. service temp. 160 °C
Epon 907	B	80	—	24hr	—	1hr	—	—	—	Max. service temp. 80 °C ,,
Epon 929	one part paste	—	—	—	—	—	—	2hr	15(200 °C)	Max. service temp. 250 °C ,,
Gen Epoxy M 180	Versamid 140	100	—	—	—	3hr	—	20	—	Bonding metals
	Versamid 125	65—40	—	—	—	2hr	—	10	—	Bonding metals, rubber, PVC, Polyester
Gen Epoxy 190	Versamid 125	100—60	—	24hr	—	2hr	—	10	—	Bonding metals, rubber, PVC, Polyester
Torr Seal	Hardener	equal squeezed length	—	24hr	--	80	—	—	—	Bonding metal, ceramic, glass, or sealing up leaks; service up to 10^{-8} torr and 100 °C

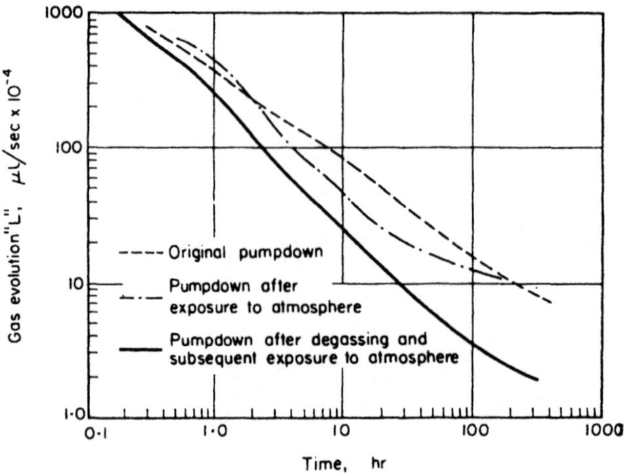

FIG. 3.5 Outgassing of Epon 828 cured with hardener A. Reproduced from Markley[806] *(Courtesy of Pergamon Press)*

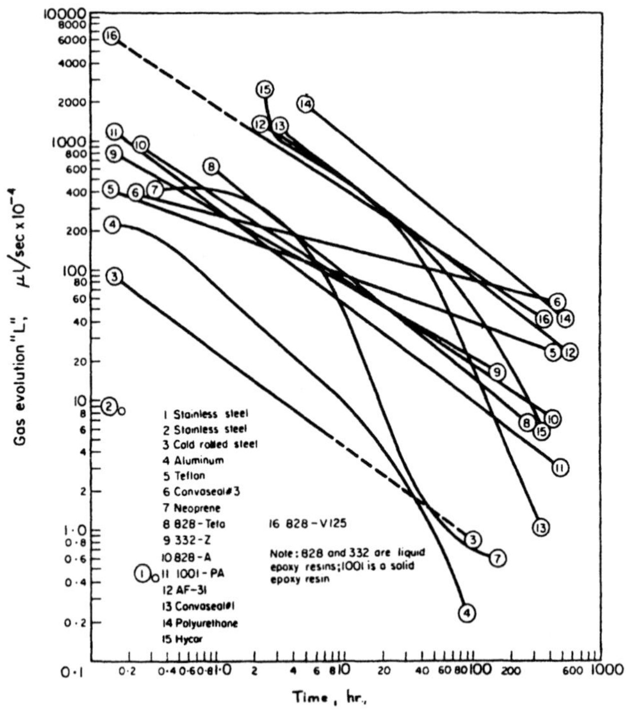

FIG. 3.6 Comparative outgassing rate of various materials. Reproduced from Markley[806] *(Courtesy of Pergamon Press)*

and 16 were rectangular 3/8 ×3/8 ×4.19/64 in.; sample 8 and 16 were cylinders 7/16 in. in diameter and 4 in. long; sample 12 was rectangular 1.37× ×0.8 ×9.18 cm; and sample 14 was cut from the wall of a larger cylinder and had a volume of 6.81 cm^3 and a surface area of 33.2 cm^2. Curves 4, 7, 8, 13 differ from the outgassing data generally measured (Section 21.13) probably because of particular contaminations in the measured samples.

33.2 Preparing the adhesives and completing the joints

In order to make an epoxy resin seal the procedure includes: the mixing of the resins (Section 33.21), the process of applying the resin on the surfaces (Section 33.22) and the curing (Section 33.23).

33.21 Mixing. The epoxy adhesives are usually supplied as a separate resin and hardener which should be mixed in order to obtain the product to be used. Some epoxy adhesives supplied as sticks, powder or paste have the resin already mixed with the hardener. When supplied separately the hardener should be added to the liquid or pasty resin at room temperature immediately before use. Hardeners include chemicals such as amines, polyamides and acid anhydrides and are supplied under various trade names and numbers (Table 3.3). The amines are suitable for cold curing, but they have less resistance to radiations; anhydrides are better for obtaining a good vacuum but they need longer curing times (Cross[249]).

Resin and hardener should be mixed carefully until the hardener is distributed homogenously throughout the resin. Small quantities are best mixed by hand; for larger quantities mechanical stirrers or vibromixers are suitable. To avoid partial curing during mixing (especially for adhesives with quick curing) it is advisable to cool the mixing containers (to 15–20 °C). Torr Seal (Varian[1257a]) is supplied in convenient tubes which eliminates weighing and allow the prepation of even very small quantities of adhesive without waste. From the two tubes equal lengths of resin and hardener are sqeezed on a flat surface, and the two are mixed so that the differently coloured hardener and resin result in an uniform coloured mixture. This resin should be used within a few minutes after being mixed.

33.22 Applying. Epoxy adhesives can be used to seal: metal-to-metal (see Lyons[790], Noggle[935], Boersch[143], Balain[74], McGuiness[830], White[1306]), glass-to-glass and-metal (see Quarrington[1020], Gale[419], Roberts[1070], Sayers[1100]), ceramic-to-glass and-metal (see Miranda[869], Kent[662]), mica-, sapphire-to-glass and-metal (see Roberts[1070], Martz[822], Schwartz[1112]).

Wheatley[1302b] has described a method of making vacuum seals to copper and Nylon by using epoxy resins (Epibond 100A; Furane Plastics Co.).

Immediately after mixing, the mixture should be applied to the surfaces to be bonded. To ensure perfect sealing it is essential to remove all dirt, oil, grease, oxides or other impurities from the surfaces to be joined, and is not enough only to remove the apparent dirt (e.g. wiping the surface with degreasing agents). For methods of cleaning see Sections 22.31 and 23.21 and Epstein[345]. To obtain a strong bond along with the cleaning, a rough surface is to be prepared. For metals, this surface is obtained by chemical or electrochemical etching processes (Section 22.31). For rubber, Epstein[345] recommends a treatment for 5 min (natural rubber) to 20–30 min (synthetic rubber) with concentrated sulphuric acid followed by washing with water, and rinsing with dilute (0.1–0.2 per cent) sodium hydroxide for neutralization. The resulting embrittled surface can then be flexed to produce a finely cracked surface, prior to the application of the adhesive.

The adhesive may be applied on the surfaces by painting, spraying or immersion. Usually it is sufficient to apply the adhesive on one of the surfaces to be bonded together. The adhesive should be applied on both surfaces if they are very rough or if one part should be inserted into the other (e.g. telescopic joint).

After drying slightly, the two surfaces to be sealed are joined together. A longer exposure to air produces absorption of moisture which will delay or even prevent the curing.

Although curing can take place without pressure, the parts to be joined should nevertheless be fixed in such a way so that the thickness of the resulting adhesive layer be the optimum for the given resin (Table 3.3).

If resin sticks are used (e.g. Araldite 1) the surface to be joined must be first heated to about 120–150 °C, and the stick rubbed over the surface.

If powder is used, this should be sprinkled on the cold surface and the part should then be heated to 120–150 °C so that the resin adheres to the surface and any air contained in the powder can escape before the parts are assembled. A quantity of adhesive (Araldite 1) of 140–160 g/m^2 is generally sufficient for normal surfaces; for rough surfaces the quantity needed is greater.

For liquid or pasty adhesives the amount to be applied is usually recommended by the supplier (Table 3.3) in terms of the dried coating prior to assembly. Sufficient adhesive must be applied to ensure complete coverage of the surfaces to be joined together and to provide a cured bond having a glue line thickness* (joint thickness) in the range of 0.1–0.2 mm.

The resin which is squeezed out of the joint can be easily removed with a spatula immediately after assembly of the joint or as long as the resin has not yet set.

Surfaces on which the resin should not adhere should be coated with a silicone grease (wax) or soap solution** or covered with oil paper. For

* The glue line or joint thickness is the thickness of the adhesive itself, determined as the difference between the overall bond and the sum of the thicknesses of the parts (Epstein[345]).

** Tools are best cleaned with hot soapy water before the adhesive starts to cure.

castings, the usual method is to employ as a releasing agent: silicone grease, Teflon, or chromium plating (Thompson[1222]).

33.23 Curing varies according to the adhesive employed and occurs at room temperature (cold curing), at slightly higher temperatures (heat curing at 80–100 °C) or at even higher temperatures (elevated temperature curing at 100–200 °C).

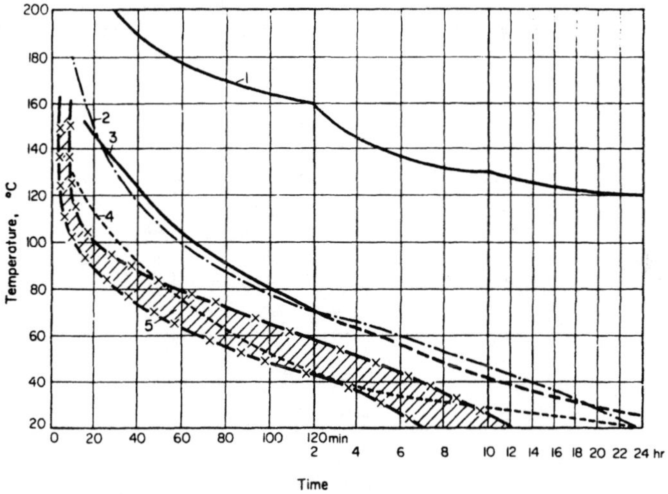

FIG. 3.7 Curing time of epoxy resins as a function of the curing temperature: (1) Araldite I; (2) Araldite 103—Hardener 930; (3) Epon 820—Versamid 140; (4) Araldite 101—Hardener 951; (5) Araldite 106—Hardener 953 U

The higher the temperature the shorter the curing time; this varies from minutes to days (Fig. 3.7 and Table 3.3). Curing temperatures over 100 °C may only be employed with thin layers of adhesive (max. about 0.5 mm) since otherwise bubbles may be formed which act as leaks in the seal.

The curing can be done by heating in an oven, with infrared lamps, induction heating, etc.

33.24 Dismantling. If an epoxy joint has to be taken apart, the most suitable method is to immerse the assembly for several days in thrichlorethylene or in warm dimethyl-formamide. Another method is to heat the joint to 130–150 °C and to pull the components apart at that temperature. Then the adhesive residue left on the surfaces may be scraped off after softening by an immersion for a few hours in dimethyl–formamide, nitrobenzene, phenol or cresol.

To remove *Torr Seal*, baking (500 °C) and soaking overnight in solvents such as *De Solv 292* (R.A.M. Chemical, Gardena, Calif.), *Stripper S.A.* or *Vis Strip* (Oakite, New York) is recommended.

33.25 Strength. The mechanical strength of epoxy seals depends on the design of the components and the type of adhesive used.

Epoxy seals and generally adhesive bonding require that the components be designed in such a manner, that the joint be strained as far as possible in the direction of the adhering surfaces.

The mechanical strength of joints made with heat curing adhesives are generally higher than that of those bonded with cold curing adhesives. The hot

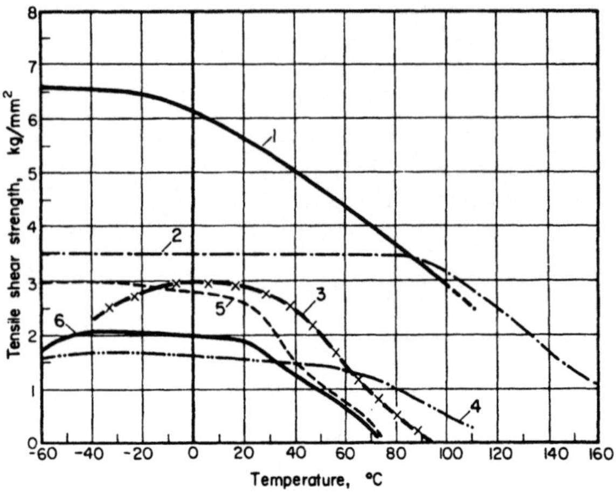

FIG. 3.8 Shear strength of epoxy resins as a function of the test temperature: (1) Araldite I; (2) heat-cured epoxy (Hess[549]); (3) cold-cured epoxy (Hess[549]); (4) Araldite 121—Hardener 951 (100 : 45); (5) Araldite 106—Hardener 953 U; (6) Araldite 103—Hardener 951 (100 : 8); (1, 4, 5, 6) test strip 170×25×1 mm (light metal) overlapped on 10 mm, tested after 10 min exposure to test temperature; (1) cured 1 hr at 200 °C; (4, 6) cured 1 hr at 100 °C; (5) cured 10 min at 150 °C; (2) cured 1 hr at 160 °C; (3) cured 7 days at 30 °C

strength of heat curing adhesives is also much higher than that of resins cured at room temperature (McGuiness[830], Hess[549]). Figure 3.8 shows the shear strength of some epoxy resin adhesives as a function of the temperature.

The cured adhesive in the joint is infusible, practically insoluble (Section 33.24) and resistant to ageing.

33.3 Vacuum sealing with epoxy resins

In establishing the design of an adhesive bonded joint, it must be remembered that these joints are *best in shear* and relatively *poor in peel.*

33.31 Butt joints placed in tension or compression have the stresses concentrated at the edges of the bond. The stress can be uniformly distributed under a normal load only if the ratio μ/E (Poisson's ratio to modulus of elasticity) is equal for the adhesive and joined parts or if this ratio for the adhesive is zero. Neither of these conditions is ever met in practice. Thus in practice an uneven stress distribution places a concentrated shearing stress at the edges of the joint, when the joint is subjected to tension or compression. The thermal expansion difference between the joined part gives similarly very pronounced stresses at the edges. Thus epoxy joints subjected to tension (or compression) should not be designed as butt joints; instead of these, lap joints are to be used. When two coaxial cylinders of similar diameters are joined end to end and subjected to torsion, a very uniform stress distribution results; thus for such applications butt seals can be successfully used.

33.32 Lap joints have the two parts overlapping each other, and are bonded by an adhesive film placed between them. The stress under load is not uniformly distributed over the bonding surface. Maximum stresses exist at the ends of the overlap. The best design which has a uniform stress distribution is obtained when: (1) the length of the overlap is small; (2) the adhesive is flexible; and (3) the joined parts are rigid. Tubular lap joints may be constructed with a telescopic arrangement (Fig. 3.9a), a step construction (Fig. 3.9b) or with outside or inside sleeve (Fig. 3.9c, d). The best way to join a plate to the end of a tube is illustrated in Fig. 3.9e.

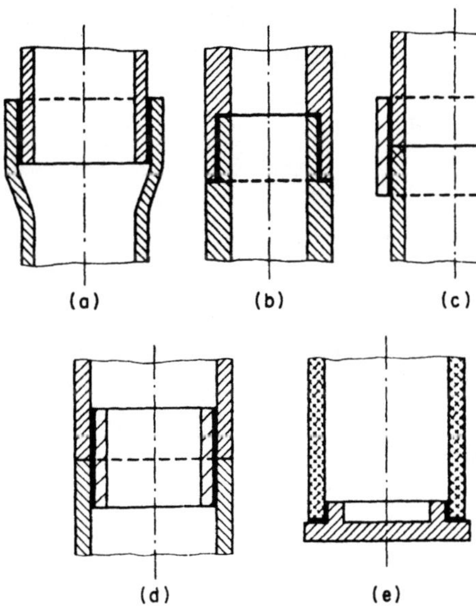

FIG. 3.9 Adhesive joining of pipes

Kent[662] used Araldite 1 (Table 3.3) to seal together glass and ceramic tubes, applying a thin layer of adhesive (from sticks) to the warmed ends of the two parts. The edges are placed together and temporarily secured by means of springs to hold them together throughout the curing (180 °C for 1 hr).

Lyons[790] sealed threaded pipes using epoxy resins, and Quarrington[1020] sealed thin corrugated copper discs to Pyrex tubes using Araldite.

Window seals using epoxy resins are described by Roberts[1070], Martz[822], Sayers[1100], Nelson[919] and other authors (Table 7.6).

Electrodes or thermocouple wires sealed by epoxy resins are described by Gale[419], Miranda[869], Balain[73, 74], Boersch[143], Noggle[935], Beil[137], Lipson[770b], Ehrenberg[333]. The description of such seals can be found in Section 43.1.

3.4 SILVER CHLORIDE SEALS

Silver chloride may be used for vacuum seals that must sustain higher temperatures than those to which wax or adhesive seal can resist. Silver chloride seals can be used generally up to 300 °C. The silver chloride has a low vapour pressure (Martin[812a]). At 300 °C the vapour pressure of silver chloride is about 10^{-7} torr, at 400 °C it is 10^{-5} torr and at 995 °C it is 10 torr (Espe[354]).

Silver chloride melts at 457 °C (Kohl[706]) forming a flowing fluid, which changes to a horn-like stuff which wets glass, quartz and metals (Angerer[34]). For the fusion of the silver chloride, porcelain or quartz crucibles should be used. Graphite is not suitable as some carbon will be dispersed in the molten silver chloride. Iron cannot be used because of the reactivity of the molten salt (Fugassi[411]).

The molten silver chloride can be cast in rods or sheets (Haynes[521]) from which gaskets, windows or other shapes can be cut (Simard[1138], Axilrod[60], Fuoss[415], Frank[398]).

Silver chloride is insoluble in water, alcohol, benzol, and acids. It is soluble in a solution of sodium thiosulphate ($Na_2S_2O_3$). Gold, platinum and silver are chemically entirely inert to silver chloride, and stainless steels, Monel, or nichrome are resistant to its chemical action (Kremers[717]).

The technique of sealing with silver chloride consists in heating the assembled parts to about 500 °C and melting the silver chloride by placing it on the hot joint, or by placing the silver chloride on the cold assembly and heating it. On cooling, the silver chloride expands; the design of the seal must allow for this effect.

The silver chloride may be applied to the seal as a powder or as pieces (Fugassi[411], Palmer[964], Weber[1289]), as a foil or strip (Greenblatt[479]), as preformed gaskets (Simard[1138], Frank[398]), or as a layer obtained by immersion of the parts in molten silver chloride (Benson[111]).

Silver chloride can be used to seal metal, glass, mica and salt windows (Section 72.6).

Molten silver chloride wets glass with difficulty at temperatures in the vicinity of its melting point. To obtain a satisfactory coating on glass (ground surfaces) it is essential to heat the glass strongly or to cover it first with a platinum film (Fugassi[411]). Wildy[1315] remarks that silver chloride does not readily wet either ground and polished, or freshly cleaved lithium fluoride. However the smallest trace of metallic silver or platinum on the surface makes it wet readily (Frank[398], Benson[111], Lord[783]). Silver chloride window seals are described in Section 72.6 (Table 7.6).

Silver chloride windows can be sealed with silver chloride. For this the ground end of a glass tube is coated with a thin layer of silver chloride. Then this coated end is reheated until the coating is molten and is pressed against the silver chloride window (Fugassi[411]). In order to retard the solarization (Section 71.1), it is recommended to paint the the exposed silver chloride surfaces with a coloured lacquer (Greenblatt[479]).

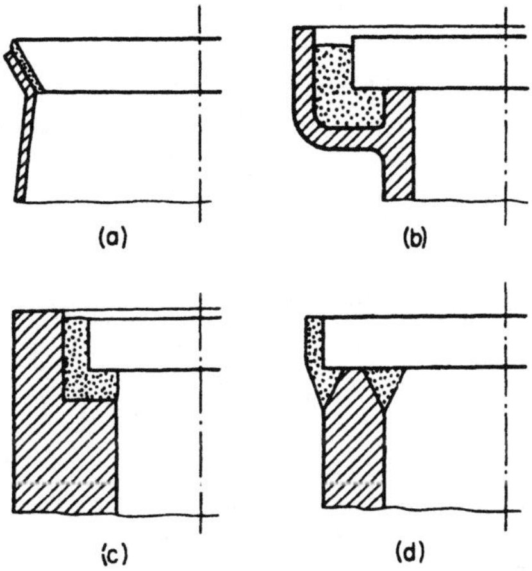

FIG. 3.10 Silver chloride seals; (a) with thin wall; (b) with channel; (c) with step; (d) with taper

The window seals should be made using a thin (thinned) wall metal part (Fig. 3.10a) as connexion between the window and the vessel (Lord[783], Wildy[1315]), providing a channel (Fig. 3.10b) around the window (Palmer[964], Weber[1289]) in a step (Fig. 3.10c) (Greenblatt[479]) or on a tapered edge (Fig. 3.10d) of the walls (Benson[111]).

Silver chloride was also used as the closing part (Section 61.32) of valves. Rapsberger[1027] describes a valve which closes by immersion of the outlet pipe in the silver chloride contained in a cup (Fig. 6.30a). A valve with glass bellows closing on a silver chloride seat is described by Kirslis[679] (see Fig. 6.30b).

3.5 SOFT SOLDERING

35.1 *Vacuum sealing with soft soldering techniques*

Soldered joints are sometimes considered as permanent connexions, but since they are used also to seal regular access ports on (experimental) vacuum systems (especially on systems connected with cryogenic work, see Scott[1114]), they are listed here as "semi-permanent seals".

35.11 Soft soldering alloys. A solder is a fusible metal or alloy, which adheres in the molten state to the parts to be joined, bonding them when it solidifies, (Belser[104, 105a], Clauser[317], Mebs[836]). The solders having melting points below 375 °C (or 400 °C) ar known as *soft solders*, and those melting above this temperature are the *hard solders*. Table 3.4 enumerates the mainly used soft solders, but also contains the hard solders having components with too high a vapour pressure to be used in bakeable systems. The hard solders having low vapour pressure components are enumerated in Table 2.18 (brazing).

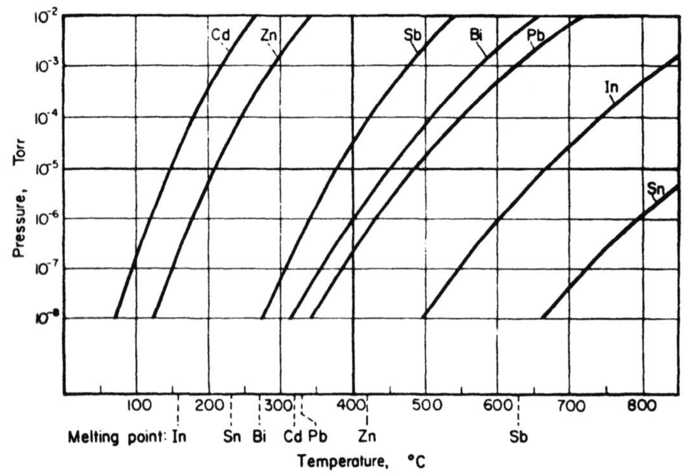

FIG. 3.11 Vapour pressure of metals used in soft soldering alloys

Figure 3.11 shows the vapour pressures of the metals commonly used in soft soldering alloys. Among these metals, cadmium and zinc have high vapour pressures; thus the alloys containing these metals are not recommended for use inside the vacuum system i.e. when they are used, their surface exposed to the vacuum should be kept to a minimum. Other components of soft solders (lead, tin, indium) have vapour pressures low enough to be used in high vacuum, but some of them (indium, bismuth) have their melting point so low that practically no heating of the system sealed with them is possible. In this respect they are equivalent to rubber seals; compared with rubber seals the soft soldered seals have the advantage of a much lower gas evolution rate at room temperature.

An interesting metal for soft soldering is indium (Cadenhead[190]). Indium and its alloys with silver, lead, tin, etc. are soft solders used in the temperature range 118–230 °C (Table 3.4). Kleen[685] recommends sealing by coating one of the parts to be joined with gold, and the other with indium. The joined parts are then heated to 200–250 °C.

The lead–tin alloys form the other group of solders, in the range of 180–280 °C; other alloys contain besides lead and tin, also antimony or silver (Table 3.4). When lead–tin alloys are used for vacuum seals in low temperature applications, the alloys should contain more than 50 per cent lead (Kaufman[658]). For cryogenic work, lead–tin alloys with about 70 per cent lead may be used. Although tin is brittle at low temperatures (below −27 °C) a tin–antimony (95/5) alloy remains stable in the cryogenic range.

Alloys of lead and tin containing bismuth and/or cadmium melt between 47 and 140 °C. Like pure bismuth, the alloys containing bismuth (more than 50 per cent) expand on cooling. With these solders it is advisable (White[1306]) to pre-tin the metal surfaces which are to be joined.

The solders prepared by many components, are very often inhomogenous and thus result in porous seals. Catalano[197b] recommends that the alloy (Rose metal) be melted and stirred, then thin samples be withdrawn from the melt by sucking the molten metal into 2–3 mm i.d. Pyrex tubing. When this alloy freezes, it expands, and the pressure shatters the glass, resulting in clean sticks of alloy.

A disadvantage of sealing with soft solders is the fact that most of these solders need the use of fluxes which have high vapour pressures. *Fluxes* are liquid or solid materials used to facilitate the wetting action of the soldering alloys. The fluxes dissolve the impurities (oxides) on the surfaces to be sealed, decrease the surface tension and the contact angle (Fig. 2.103) and provide an electrolytic ion exchange medium whereby one metal can plate out of the solder onto the metal to be joined. Table 3.5 gives some of the fluxes used in soft soldering. Ready made fluxes are available generally from the suppliers of the soldering alloys.

After the soldered seal is completed, it is recommended to *wash* off the remaining fluxes with a solvent (e.g. carbon tetrachloride, benzene), using brushes eventually. If fluxes containing rosin (Table 3.5) were used, the washing (brushing) should be done first with xylol and then with benzene (Laporte[741]).

35.12 Soft soldering of metal parts. Semi-permanent (dismountable) soldered joints are generally made by using a channel or an annular support for the solder (Scott[1114], Archer[46]). The channel (Fig. 3.12a, c) must have its outer

TABLE 3.4. SOLDERING ALLOYS

Melting range (°C) lower	Melting range (°C) upper	Alloy	Composition (%)	Remarks	Reference, Supplier*
47	47	Cerrolow 117 (Eutectic)	Bi (44.7), Pb (22.6), Sn (8.3), Cd (5.3), In (19.1)	The parts should be pre-tinned; after pre-tinning they are assembled and reheated to melt the alloy. These alloys shrink slightly on cooling	CP
58	58	Cerrolow 136 (Eutectic)	Bi (49), Pb (8.3), Sn (12), In (21)		CP
61	—	Wood's metal	Bi (50), Cd (12.5), Pb (25), Sn (12.5)	Super conducting at low temperatures	—
69	—	Lipowitz metal	Bi (50), Cd (10), Pb (27), Sn (13)		Mönch[878]
70	—	Cerrobend	Bi (50), Cd (10), Pb (26.5), Sn (13.5)	Expands on cooling	CP
94 (96)	100	Rose metal	Bi (50), Pb (25), Sn (25)		Mönch[878], Espe[350]
96	100	Lichtenberg metal	Bi (50), Pb (31), Sn (19)		Ardenne[47]
—	103	BiSnCd	Bi (50), Sn (37.5), Cd (12.5)		Ardenne[47]
—	—	Thermal-free solder	Cd (70.4), Sn (29.6)	Very low thermoelectric force with respect to copper	White[1306]
115	127	Cerroseal 35	In (50), Sn (50)	Low vapour pressure; for glass, metal, ceramics; no flux; pre-tinning	CP
115	—	Indalloy 1	In (50), Sn (50)	Flux: indium chloride	IC
124	—	Cerrobase	Bi (55.5), Pb (44.5)	Shrinks slightly on cooling; pre-tinning necessary	CP
138	—	Cerrotru	Bi (58), Sn (42)	Expands on cooling; pre-tinning or tin plating recommended	CP
140	—	BiCd Eutectic	Bi (60), Cd (40)	Not superconducting at temperatures above 0.8 °K	White[1306]
144	—	Indalloy 3	In (90), Ag (10)	Flux: indium chloride	IC
145	—	Indalloy	In (95), Ag (5)	Flux: indium chloride	IC

145	145	SnPbCd	Sn (50), Pb (32), Cd (18)		White[1306], Espe[350]
147	—	SnPbCd	Sn (34), Pb (33), Cd (33)	For soldering thin wires	Laporte[741]
147	—	Indalloy	In (90), Ag (5), Pb (5)		IC
151	—	Indalloy	In (95), Al (5)		IC
153	—	Indalloy	In (98.5), Pb (1.5)		IC
153	—	Indalloy	In (99), Cu (1)		IC
156	—	Pure indium	In (100)		White[1306], Espe[354]
158	—	InPb	In (79.3), Pb (20.7)		IC
178	183	SnAgPb	Sn (60), Ag (2), Pb (38)	For soldering of silver coated glass and ceramic	ES
181	181	SnPb Eutectic	Sn (64), Pb (36)	High tensile strength	Strong[1207] Espe[350]
183	190	SnPb	Sn (60), Pb (40)		ES
183	216	SnPb	Sn (50), Pb (50)	General use	ES
183	219	L.Sn 90	Sn (90), Pb (8.7), Sb (1.3)		DIN 1707
183	222	SnPb	Sn (45), Pb (55)		ES
185	225	SnPbSb	Sn (37.5), Pb (60), Sb (2.5)		White[1306]
183	238	SnPb	Sn (40), Pb (60)	Ductile solder	
183	242	L Sn 33	Sn (33), Pb (67)		Ardenne[47]
183	257	L Sn 25	Sn (25), Pb (73.3), Sb (1.7)	Strong in compression and tension	DIN 1707
183	305	L Sn 8	Sn (8), Pb (91.5), Sb (0.5)		DIN 1707
199	210	AS 10	—	Soldering alloy for aluminium	ES
199	267	AS 20	—		
183	280	SnPb	Sn (20), Pb (80)		ES
221	300	S X 10	Ag alloy, lead free		SS
221	228	Plumbsol	Sn-Ag		JM
221	305	Spec BTL	Sn (89), Ag (11)		Espe[350]
232	232	Pure tin	Sn (100)	Solder paint "Fryolux" may be used	FM
234	—	Indalloy 10	In (25), Pb (75)		IC

(Table 3.4 Continued)

Melting range (°C)		Alloy	Composition (%)	Remarks	Reference, Supplier*
lower	upper				
236	242	SnSb	Sn (95), Sb (5)	Recommended for low temperature -Kaufman[658]	ES
240	282	SnPbSb	Sn (5), Pb (91), Sb (4)		ES
250	315	TEC-Z	Ag (5), Zn (16.6), Cd (78.4)		HH
265	265	ZnCd Eutectic	Zn (17.5), Cd (82.5)	Useful at liquid helium temperature, as it is not superconducting	White[1306]
280	320	AgCdZn (LM 15)	Ag (5), Cd (80), Zn (15)		JM; Ardenne[47]
296	—	Comsol	SnAgPb	High strength; very small solvent action on copper	JM
309	310	SnAgPb	Sn (1), Ag (1.5), Pb (97.5)		ES
338	390	LM 5	Ag-Cd		JM
340	395	TEC	Ag (5), Cd (95)		HH
419	550	L Zn 98	Zn (98), Cu (2), Zn (98), Ni (2)		Ardenne[47]
605	620	Easy-Flo 45	Ag (45), Cu (15), Zn (16), Cd (24)	HH Flux or protective atmosphere (e.g. hydrogen)	
605	700	Easy-Flo 35	Ag (35), Cu (26), Zn (21), Cd (18)		
625	635	Easy-Flo	Ag (50), Cu (15.5), Zn (16.5), Cd (18)		HH
630	690	Easy-Flo 3	Ag (50), Cu (15.5), Zn (15.5), Cd (16), Ni (3)		
640	705	Silfos	Ag (15), Cu (80), P (5)		
640	705	Silfos 5	Ag (5), Cu (89), P (6)		HH

* CP — Cerro de Pasco Corp. 300 Park Avenue, New York 22 N.Y., U.S.A.; IC — The Indium Corp. of America, 1676 Lincoln Avenue, Utica, N.Y., U.S.A.; ES — Enthoven Solder Ltd., Upper Ordnance Wharf, Rotherhithe Street, London S.E.16, England; HH — Handy & Harman, 850 Third Ave., New York 22, N.Y., U.S.A.; JM — Johnson, Matthey Co., 78 Hatton Garden, London E.C.1, England; SS — Sheffield Smelting Co., Royds Mill Street, Sheffield 4, Yorks., England; FM — Fry's Metal Foundries Ltd., Tandem Works, Merton Abbey, London S.W.19, England; DIN — German Standard.

TABLE 3.5. FLUXES FOR SOFT SOLDERING

Flux	Remarks
Zinc chloride solution in water or preferably in alcohol (solution of about 1%)	acid flux
Solution of 2 pbw zinc chloride and 1 pbw ammonium chloride in 2 pbw water; the solution is diluted with alcohol (to about 5%)	acid flux
Solution of 15 pbw zinc chloride (and eventually 2 pbw ammonium chloride) in 10 pbw glycerine (and eventually 3 pbw alcohol)	less acid
Solution of 10 pbw ammonium chloride in 90 pbw petroleum jelly	
Solution of: 84 pbw alcohol, 0.5 pbw hydrochloric acid, 1 pbw acetic acid, 1.5 pbw zinc chloride, 14 pbw amyl-acetate	acid flux
Phosphoric acid solution	for soldering steel (and stainless steel)
Solution of 2 pbw rosin in 1 pbw benzol	non-corrosive
Solution of 2 pbw rosin in 1 pbw alchol and 0.5 pbw benzol	non-corrosive

edge lower than the inner edge to avoid the flow of the alloy inside the pipe or vessel.

The soldering alloy is contained in a cylindrical channel (Fig. 3.12a); sometimes the outer rim of the channel is omitted and the solder adheres only to a flat flange (Fig. 3.12b). Cones can be used for better alignment (Fig. 3.12c),

If the joints are not to be dismantled often, the soft soldered seals may be made in shapes similar to those of brazed seals (Section 22.33).

In cryogenic work it is often necessary to connect aluminium to stainless steel (e.g. the aluminium inner shell of a vacuum insulated storage vessel to a stainless steel pipe). For such seals Scott[1114] recommends transition joints made by friction-tinning the inside of the end of an aluminium tube with tin–lead solder and then completing the joint by standard soft soldering techniques to stainless steel. There is some advantage in joining the stainless steel to a brass or copper sleeve with silver brazing and then connecting this sleeve to the aluminium with tin–lead solder. In aluminium-to-stainless steel seals the aluminium part should be placed outside if possible, ensuring that a compressive stress will be obtained in the joint on cooling.

Soft soldering is generally completed using the heating with a soldering iron or a torch. If a torch is used it should not be played directly on the area to be soldered, especially when indium solders are utilized.

An unusual, but simple technique for the soldering of aluminium, stainless steel as well as glass and ceramics, consists in loading the solder with the

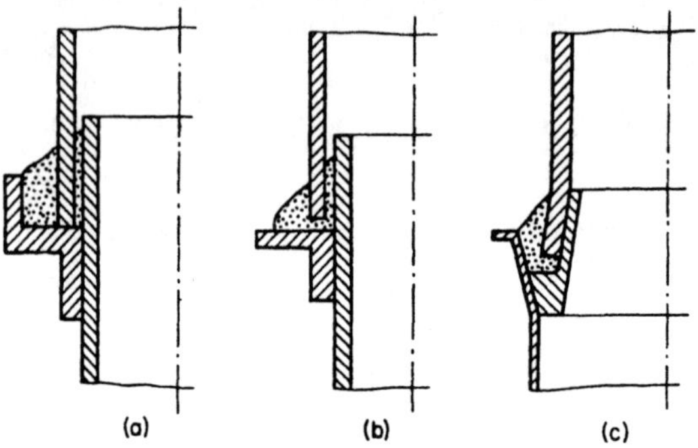

FIG. 3.12 Demountable soldered joints, with (a) channel; (b) flange; (c) channel and cone. After Archer[46] *(Courtesy of The Institute of Physics and The Physical Society, London)*

aid of a grinding wheel (Anon[39], McGuire[831]). The solder is first loaded on the grinder by operating the grinder with the abrasive head in contact with a stick of the solder. The heat of friction generated by the grinding melts the solder and spreads it on the wheel in an even layer. The loaded grinder head is then applied to the place to be tinned, the heat of friction again melts the solder which flows into close contact with the surface, where the layer was removed by the grinding.

35.13 Soft soldering of glass and ceramic parts. Glass and ceramic parts can be sealed to glass, ceramic or metal parts using soft solders (Bolton[149], Pearsall[981], Jamieson[628], Chandra[204], Archer[46], Haas[495], Ridyard[1057a]). For this purpose one of the following methods is used: (1) direct soldering with indium or its alloys; (2) soldering on a pre-metallized surface; (3) using cup seals (see Fig. 3.12).

When *direct soldering* with indium (or its alloys) is used, fluxes are not desirable; it appears that fluxes inhibit the wetting in this case (Belser[105a]). Cleanliness of the surfaces is essential; cleaning by scrubbing with detergents, and cleaning with chromic acid (Section 23.21) or positive ion bombardment are suitable for this purpose.

In accomplishing indium seals it is recommended (Bolton[149]) to use the minimum temperature necessary to make the joint. The indium solder does not flow over glass, quartz or ceramics, and must be applied by a rubbing action. If a soldering tool is used this must be very clean, and in particular it must be cleaned from any flux with which it was previously used.

The sealing process with indium solders includes the preparation of the metal and glass surfaces and the technique of assembling the metal–glass soldered seal (Cerro de Pasco[200]). *The metal part* should first be cleaned (Section 22.31) and coated with tin or tin–lead solder using a soluble flux such as zinc chloride (Table 3.5) or zinc-ammonium chloride, then wiped to leave a film of tin or solder of minimum thickness. This surface is then coated with the indium alloy (e.g. Cerroseal 35, see Table 3.4). Finally all traces of flux must be removed by a very thorough washing, thus resin-based fluxes are unsuitable for this application because of the difficulty of completely removing them. *The glass or ceramic parts* should be thoroughly degreased and washed with soap and water, and rinsed meticulously to remove all traces of soap or salts. Subsequent rinsing with alcohol or acetone is recommended. The heating of the cleaned glass to about 370 °C and followed by cooling to the working temperature has been found to improve the adhesion probably because the heating destroys the gas film on the surface. An additional cleaning by bombarding the surface with the discharge from a spark-coil has improved the adhesion of Cerroseal 35.

The cleaned glass (or ceramic) part should be warmed to a temperature some degrees above the lower melting point (Table 3.4) of the indium alloy (e.g. 122 °C for Cerroseal 35) and molten solder alloy applied with a wiping motion with a wad of cotton gauze. If either the alloy or the part is too hot the alloy will not adhere, but as the part cools to the correct temperature adhesion will occur. An efficient coating produces transparent parts on a perfect mirror when viewed from the opposite side.

In order to assemble the joint, the coated parts are warmed until the coating is just molten (the surfaces appear wet) and then they are brought together with a slight pressure; heavy pressure will squeeze out too much alloy. For glass-to-glass or-ceramic seals the remaining alloy film between the parts should be about 0.05–0.1 mm; for glass- or ceramic-to-metal seals the alloy film may be up to 0.15 mm.

Protection of the vacuum seals made with indium alloys is recommended, by covering them with a lacquer (e.g. Glyptal, see Table 3.2); this is required especially if the seal is of large dimensions.

Glass parts can be sealed to glass or ceramic parts by soft soldering their metallized surfaces. The metallization of glass can be done: by chemical silvering, using metallizing pastes of silver, platinum, gold, palladium (Monack[876] Heritage[543], Strong[1207], Umblia[1252], Angerer[34], Ballard[76], Oehme[952]), by vacuum evaporation or sputtering (Holland[581], Methfessel[850]) or by electroplating on a conductive layer e.g. graphite (Szymanowitz[1212], Rieck[1058], Jamieson[628]).

When soft soldering is used to seal metallized glass or ceramic parts to metal parts it is recommended to cover the metallized surface with a copper coating, which gives bonding strength to the soldered joint and is elastic enough to minimize the stresses. Figure 3.13 shows two examples of this practice, a glass to copper and a glass to aluminium seal. The seal shown in Fig. 3.13a represents that developed by Haas[495] which consists in: (1) Painting

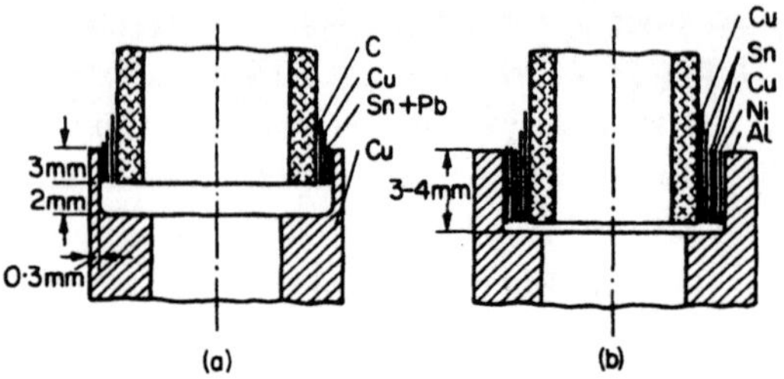

FIG. 3.13 Metallized glass-metal soldered seals: (a) glass-copper (after Haas[495]; *Courtesy of The American Institute of Physics)*; (b) glass–aluminium (after Chandra[204]; *Courtesy of Rudolf A. Lang Verlag, Esch/Taunus, Germany)*

the cleaned end of a (Pyrex) glass tube with a Du Pont conductive coating (Type F. No. 4666) allowing it to set for about 5 min and then heating it at 500 °C for 5 min (on Fig. 3.13a this layer is marked *C*); (2) Electro-plating of copper onto the conductive layer; (3) Machining on the end of the copper part a sleeve having thin walls (Fig. 3.13a) and annealing it; (4) Completing the seal of the copper plated glass tube to the copper part*, by soft soldering (with SnPb 50/50) them together. According to Haas[495] such seals were vacuum-tight after sixteen cycles between room temperature and 4.2 °K and more than twenty additional cycles between room temperature and 77 °K. The leak rate was less than 3.10^{-7} lusec.

Chandra[204] describes a technique of sealing glass tubes to aluminium, by using soft soldering. The technique consists in: (1) copper-plating the end of the glass tube by spraying; (2) tinning the copper-coated surface; (3) nickel-plating** the aluminium part (Fig. 3.13b); (4) copper-plating the nickel-coated surface; (5) completing the seal of the two copper layers by soft soldering the joint (Fig. 3.13b).

* The clearance of the copper sleeve around the tube was less than 0.05 mm.

** The nickel plating was done by immersion of the Al for about 30 min in a solution of 30 g/ nickel chloride, 10 g sodium hypophosphite, 100 g sodium citrate and 50 g ammonium chloride in 1000 ml water; the bath is used at 70 °C, drops of ammonia solution are added to obtain a pH of 5–8.

Archer[46] describes a glass to copper seal using soft solder placed in a channel between the concentric tubes (Fig. 3.14). In this seal the thin inner copper tube (1 Fig. 3.14) is driven through a truncated brass cone (2) which is soldered (6) to the coned outer copper tube (3). The copper tube (8) is attached to the assembly by the solder placed in the channel (6) so formed (see also Fig. 3.12). The upper, long channel formed between 1, 2 and 3 (Fig. 3.14) is filled with a low melting alloy (Archer[46] used an alloy consisting of 8 pbw Bi, 4 pbw Pb, 2 pbw Sn, 2 pbw Cd, 3 pbw Hg). The end of the glass tube (having rounded edges) is warmed and dipped into the low melting alloy (5), and when the alloy has set, some vacuum wax (7) is melted onto it.

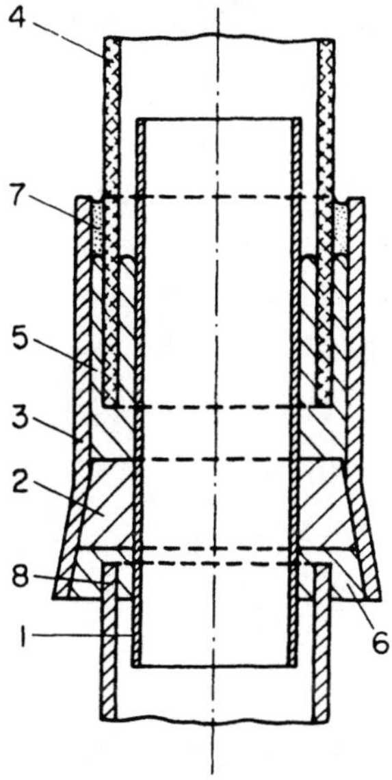

FIG. 3.14 Glass–metal seal with soldering alloy placed in a channel between concentric tubes. After Archer[46] *(Courtesy of The Institute of Physics and The Physical Society, London)*

35.2 Vacuum sealing with solder glasses

Solder glasses are used mainly for sealing parts which cannot be exposed to the high temperatures required in glass–glass (Section 2.3) or glass–metal (Section 2.4) seals and are successfully used for sealing mica (Section 72.4). Solder glasses were developed by various manufacturers (see Table 3.6) and are described by Donal[284], Dale[253–255], Dalton[256, 257], Gallup[422], Gleason[452], Flaschen[381], Alma[15], Haase[497], Knapp[694], Denton[274].

Solder glasses should meet the requirements listed on page 273.

TABLE 3.6. SOLDER GLASSES

Glass type	Main composition, weight (%)							Expansion $\alpha \times 10^{-7}$ (0–200 °C)	Softening point (°C)	Soldering temp. (°C)	Reference Supplier*
	PbO	B_2O_3	ZnO	Al_2O_3	SiO_2	NaO	Misc.				
Z 1	—	30.2	69.8	—	—	—	—	42	550	—	Dale[255]
T.H.G	30	59	3	1	7	—	—	50	—	650	Telefunken
8435	×	×	—	×	×	—	LiO	58 (300 °C)	549	—	Jena
PZ 2	65	35	—	—	—	—	—	66	415	—	
PZ 5	56	23	21	—	—	—	—	67	465	—	
PZ 8	54	18	28	—	—	—	—	68	400	—	BTH
PZ 1	64	29	7	—	—	—	—	71	450	—	
PZ 4	64	29	7	—	—	—	—	71.5	455	—	Dale[255]
PZ 3	67.5	32.5	—	—	—	—	—	72	440	—	
PZ 6	64	22	14	—	—	—	—	72.5	415	—	
PZ S7	62	12	21	—	5	—	—	73.5	380		
D 4	65	10	—	10	10	—	5 Bi_2O_3	74	496	—	Corning Dalton[256]
G 1	40	—	—	—	60	—	—	74 (300 °C)	568	—	Gallup[422]
PZ 9	63	16	21	—	—	—	—	74	380	—	
PZ S4	62	15	18	—	5	—	—	75	400	—	
PZ S8	62	9	24	—	5	—	—	75	370	—	BTH
P 3	71	29	—	—	—	—	—	76.5	430	—	
PZ S2	64	15	16	—	5	—	—	76.5	395	—	
PZ 11	67	15	18	—	—	—	—	77.5	360	—	Dale[255]
PZ S6	64	12	19	—	5	—	—	77.5	385	—	

D 5	70	10	—	10	10	—	—	78	472	—	Corning Dalton[257]
PZ S1	65	11	19	—	5	—	—	79	370	—	Dale[255]
GSS 34	68	29.5	2.5	—	—	—	—	80	360	525	GEC
PZ S5	64	12	19	—	5	—	—	81	375	—	Dale[255]
D 8	65	14	15	2.5	2.5	—	—	81	414	—	Corning
D 9	68	12	18	—	2	—	—	82	402	—	Dalton[256]
PZ 12	68	12	20	—	—	—	—	82.5	340	—	BTH
PZ 10	72	14	14	—	—	—	—	82.5	360	—	Dale[255]
D 6	75	11	—	11	3	—	—	83	440	—	Corning
D 7	72	18	5	2.5	2.5	—	—	83	428	—	Dalton[256]
8461	×	×	—	—	×	—	—	83	485	—	Jena
GSS 34A	57	24	2	—	—	—	17 CdO	84	—	525	GEC
7570	not known							84	440	560	Corning
D 2	44	11	—	11	—	28 PbF_2	—	84	434		Corning
D 3	73	11	—	11	—		(CdO)5	85	438		
D 1	75	11	—	11	3	—	—	85	440		Dalton[256]

(Table 3.6 Continued)

Glass type	Main composition, weight (%) PbO	B_2O_3	ZnO	Al_2O_3	SiO_2	NaO	Misc.	Expansion $\alpha \times 10^{-7}$ (0–200 °C)	Softening point (°C)	Soldering temp. (°C)	Reference Supplier*
G 6	43	26	—	—	31	—	—	85 (300 °C)	477	600	Gallup[422]
PZ 7	71	22	7	—	—	—	—	86	380	—	BTH, Dale[255]
T 209	78	15	—	1	5	1	—	90	365	410	Telefunken
F 3	74	11	—	—	14	1	—	90	400	540	Fischer
M 129		not	known					90	450	620	VB
PZ 13	72	13	15	—	—	—	—	91	335	—	BTH, Dale[255]
GSS 1	76	15	7	—	2	—	—	92	325	500	GEC
G 3	50	50	—	—	—	—	—	93 (300 °C)	435	—	Gallup[422]
P 2	79	21	—	—	—	—	—	95	370	—	Dale[255]
GSS 38	78	19	8	—	—	—	—	95	300	475	GEC
Ph	80	16	4	—	—	—	—	96	420	—	Philips
GSS 38A	65	16	2	—	—	—	17 CdO	97	—	475	GEC
G 2	60	—	—	—	40	—	—	108 (300 °C)	429	—	
G 4	60	40	—	—	—	—	—	108 (300 °C)	383	—	Gallup[422]
M 130		not	known					111	380	560	VB
G 7	62	18	—	—	20	—	—	117 (300 °C)	371	—	Gallup[422]
P 1	89	11	—	—	—	—	—	124	280	—	BTH, Dale[255]
8468	×	×	×	—	—	—	—	124 (300 °C)	330	—	Jena
G 5	70	30	—	—	—	—	—	124 (300 °C)	330	—	Gallup[422]

* Corning — Corning Glass Works, Corning, N.Y., U.S.A.; BTH — The British Thomson–Houston Co. Ltd., Rugby, England; Jena — Jenauer Glaswerk Schott u.Gen., Mainz, Germany; GEC — Osram-GEC Glass Works, East Lane, Wembley, Middlesex, England; Telefunken — Telefunken Works, Germany; Fischer — Glaswerk Fischer, Ilmenau, Germany; Philips — Philips, Eindhoven, Holland; VB — Verreries de Bagneaux, France.

(1) Should have softening points between 300–400 °C and working temperatures between 400–550 °C; the solder glasses used to seal mica to metal may have higher softening and working temperatures;

(2) Should have a viscosity of 10^4–10^6 poises at the temperatures where, the glass parts to be sealed, have a viscosity of 10^{12}–10^{13} poises (Section 23.11)

(3) Should have their expansion coefficients matching the expansion of the parts to be sealed;

(4) Should wet the surfaces to be sealed, and be able to form a strong bond;

(5) Should have a vitreous state, without any tendency to crystallization or to be decomposed due to the gases or vapours of the surrounding space; or as a result of the heating necessary during the sealing;

(6) Should have, at service temperature, a low rate of outgassing.

Solder glasses are commonly available in the form of *powder*, but sometimes also washers, sleeves or other shapes are used. If a solder glass is to be prepared from its components, these are melted together in a platinum or silver crucible and the resulting molten glass is poured into water. The solidified glass is ground into a powder (100–300 mesh according to the technique to be used in applying it to the surfaces). The grounding is done in a ball mill using water as the suspension liquid (with an addition of ammonia or lithium chloride) or in dry state. The dry powder is then mixed with a volatile liquid,* making up a paste which will be used for the sealing. For example a suitable suspension of Corning 7570 solder glass, consists of 50 g of mesh 325 powder with 25 cm^3 of methyl alcohol and a drop of saturated solution of Epşom salts; this suspension is recommended for spraying.

The glass solder suspensions are to be applied on the surface in thin (about 0.12 mm) layers.

Solder glasses can be used to seal glass to glass, glass to metal, metal or glass to mica, etc. Enamels were used by Eschbach[347] to lower the permeability of iron for hydrogen, and by this technique the permeability was lowered by a factor of about ten. Flaschen[381] quotes solder glasses consisting of binary and ternary mixtures of thallium, arsenic, and sulphur, fluid at 200–400 °C. Kallweit[649] proposed the use of a paste consisting of borax, glass solder and glass powder. This mixture should be smeared on the metal surfaces to be sealed, the assembly heated and oxidized at about 800 °C. To the film of oxide thus formed, the glass is then fastened. Dalton[256] sealed soda–lime glass parts with a "brazing glass" containing 60–85 per cent PbO, 5–15 per cent Al_2O_3, 0–40 per cent B_2O_3 and 0–20 per cent SiO_2. Jonas[643] used zinc

* Corning 7570 solder glass 100 mesh powder is usually mixed with a 1 per cent solution of 1000–1200 sec nitrocellulose in amyl acetate. This mixture may be applied to the surface with a funnel or other suitable orifice. When firing, sufficient time (1–15 min) should be allowed in the temperature range of 350–400 °C for the nitrocellulose to burn out completely, preventing blackening or bubbling.

borate ($\alpha = 36\text{–}60\times10^{7}$) and lead borate ($\alpha = 80\text{–}140\times10^{-7}$) to seal glass parts. He recommends the use of a metal ring (Cu 1–1.5 mm thick, Al 2–3 mm thick) coated on both sides with the solder glass, and included in the seal (Fig. 3.15b). When the ring is heated (by induction or by the direct passage of current) the solder glass melts and the preheated glass parts are then pressed to it from both sides. Embodying the heat source in the joint results in a favourable temperature distribution, the zone of highest temperature being restricted to the immediate vicinity of the joint (Fig. 3.15). As the metal ring remains in the joint, it can later again serve as a local heat source if the joint must be unsealed without damaging the parts.

Sloan[1149] describes a glass to Kovar seal using a glass frit containing 61 per cent lead oxide and approximately 25 per cent bismuth trioxide. The frit was applied by brushing or spraying the powder in an acetone solution, on the ground and cleaned glass and Kovar surfaces. The joint was completed at 300 °C in nitrogen and at 450 °C in hydrogen. Miller[863] used solder glasses to seal ceramic envelopes; the solder glass (code 7574) seal is completed in

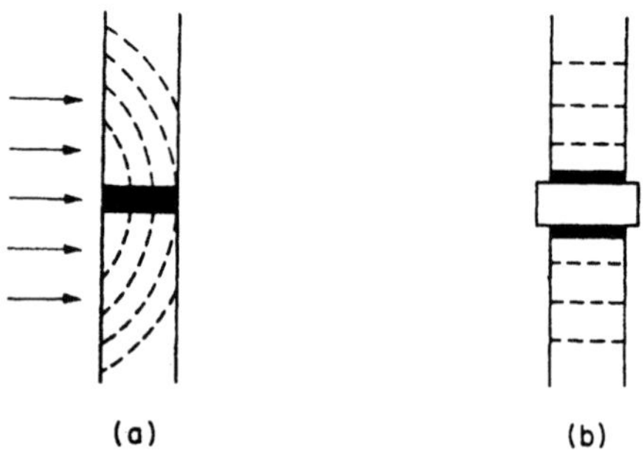

FIG. 3.15 Glazed butt seal; (a) direct seal; (b) seal with inserted glazed metal ring. After Jonas[649] *(By permission, of Philips Techn. Rev. 22, 53, 1960)*

nitrogen at 720 °C, maintaining a heating and cooling rate of 3 deg. C/min. Donal[284] and Wu[1335] used a low melting point lead borosilicate glass to attach mica windows on glass or metal (Section 72.4).

3.6 GROUND AND LAPPED SEALS

Ground glass (quartz) or lapped metal seals consist of two parts connected to each other on their ground or lapped surfaces. The surfaces may be placed one against the other without any other intermediate material, but usually for a vacuum-tight seal they must be greased (Section 36.4).

A ground joint is considered vacuum tight, if without grease it is tight to mercury at 20 °C, i.e. the mercury does not flows through the joint when exposed outside to atmospheric pressure and the inside of the joint is evacuated (Angerer[34]).

Glass ground surfaces are sealed against similar ones, against metals or sometimes against elastomers. Lapped metal surfaces can be sealed against glass or metal surfaces.

The ground surface may be flat (plane), conical, cylindrical or spherical.

36.1 Flat seals

Flat ground joints are used in applications where the joint must be closed and opened without moving the parts axially, but they are used also in some seals transmitting motion (Section 51.6), or in ground seals where the diameter is too large to use conical or spherical seals (bell jars, pipes, see Fig. 3.16).

In order to prepare a flat ground seal, the grinding compound is placed on a glass disc mounted on a turntable, which has its whole surface wed. The glass tube or vessel to be ground is pressed against this rotating disc, gripped closely to the end to be ground, and is moved from the centre to the

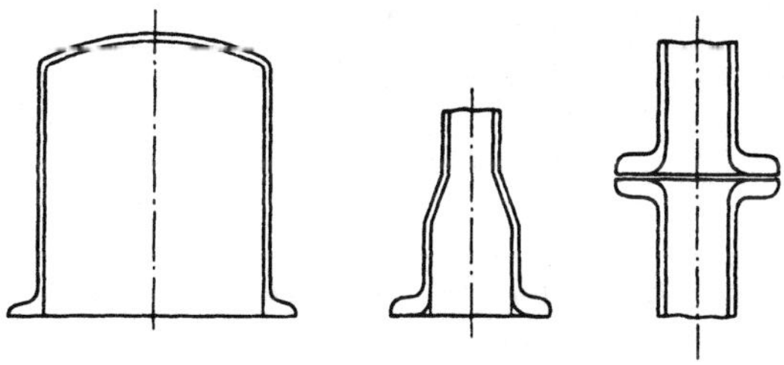

FIG. 3.16 Flat ground joints for (a) bell jar; (b) window; (c) pipe

outer edge of the disc on the side where the disc is moving away from the operator, and simultaneously rotated on its own axis. The procedure is repeated with progressively finer grinding powder, until if necessary the two parts to be connected are ground on each other. This last stage can be done only by using a guiding system which insures that the two parts are held parallel to each other.

Flat ground joints are sealed either by just greasing the surfaces or by using wax sealing (Section 31.2). When the ends of two pipes are sealed together by a flat ground joint (Fig. 3.16c) it is sometimes useful to place a thin gasket between them (Fig. 3.17). The pipe ends are clamped together fixing the clamp on the flared thickened walls (Fig. 3.16b, c) or on their conical ends (Fig. 3.17). Mancebo[798a] recommends that a film of aluminium (0.1 μ thick be evaporated on one of the optically flat ground flanges prior to mating them. The aluminium–glass bond is much weaker than the glass–glass bond, and so separation of such a joint is possible without damage to the ground surfaces. To re-use the flanges the old aluminium must be removed (dissolved) and a new film evaporated. For the same purpose, i.e. for ease of opening the seal, Fremlin[401] recommends the sticking of a ring, made of rubber sheet, to one of the surfaces; this method can be used only in vacuum systems, where the gases evolved from the rubber are admissible (Section 21.13).

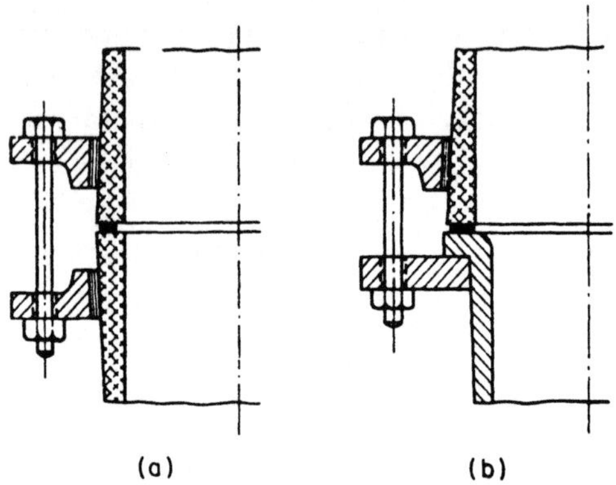

FIG. 3.17 Flat ground glass surface sealed with gasket (a) to pipe, (b) to flange. After Q.V.F. Ltd.[1021]

If the ground surface is optically polished the seal can be made without greasing the surfaces (Mancebo[798a]). Sears[1116] used the polished end of a Pyrex tube as the seat for an optically flat quartz plate, in a vacuum cut-off (see Fig. 6.32a). Forman[390] constructed a greaseless vacuum valve using the optically polished ends of two glass pipes inside the valve, as the closing system of the valve (see Fig. 6.32b). Another glass valve having the closing system based on ground glass joints is described by Lambe[730] (see Fig. 6.32c).

Metal lapped plane seals are used in the slide valves of the exhaust machines for the pumping of electric lamps, electron tubes, vacuum flasks, etc. (Matheson[825], Banki[79], Korolev[712], Ulmichek[1251], Roth[1083]). Such seals (Fig. 3.18) consist of two discs, one of which (usually the lower) is stationary, and the other rotating. The holes of the stationary disc are connected to the pumps, and those of the rotating part are connected to the vessels to be exhausted.

When the holes in the two lapped parts correspond to each other, a vacuum-tight connexion is established between vessel and pump. The seal is lubricated and tightened by the oil circulated in the concentric grooves on the plates (Fig. 3.18b). The lapping of such valves should be done very accurately, preferably mechanically on a system which ensures the movement of the two plates in respect to each other in such a way that they will never repeat exactly the same relative positions. This is done by rotating one of the plates,

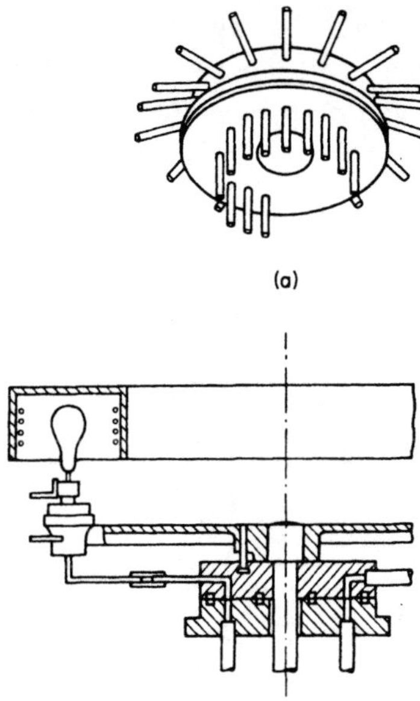

(a)

(b)

FIG. 3.18 Lapped seal used in slide valve: (a) side view of slide valve; (b) cross section through the slide valve of an exhaust machine

while the other one has a translational motion, but having the r.p.m. of the rotating plate indivisible by the number of strokes per minute of the other plate.

36.2 Conical and cylindrical ground seals

Conical ground seals consist of two parts: the inner and the outer member. The outer member can be the end of a pipe, the neck of a bottle, the port of a chamber or the cap for closing a pipe or port. The inner member is usually hollow as well, but sometimes it is just a plug. These seals are described by e.g. Mönch[879], Angerer[34], Holland-Merten[587], Jaeckel[623], Korolev[721], Anon[40], Eschbach[346], Urry[1253], Risz[1066].

Conical ground (glass) seals are made by forming the glass parts to the required dimensions and shapes and then grinding them on a suitable rotating chuck. For details on grinding techniques for tapered joints and stopcocks the references are e.g. Barr[81], Frost[409], Mönch[879], Willot[1322]. After grinding it is suggested (Skellett[1144]) that the surfaces be etched slightly to relieve the strain.

The dimensions of a conical ground seal are expressed by the diameters d and D at the small and large end of the taper and the length l of the ground zone (Fig. 3.19). The taper defined as $(D-d/l)$ used to be 1/10. Table 3.7 lists comparatively the conical ground joints according to the British, American and German standards for joints having 1/10 taper. The American standard joints are available with *full length, medium, and short length,** corresponding approximately to the *full length, three quarter, half and one quarter length* of the British standard, respectively to the *series 0, 1, 2, 3* of the German standard (Table 3.7).

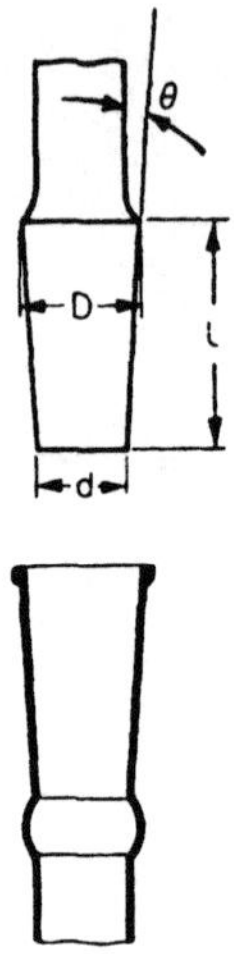

FIG. 3.19 Dimensions of tapered ground joints

The required precision of the conical joint is fairly high. The American standard requires that the taper should be 1 ± 0.006 mm/cm of length, and the German standard specifies a taper of $1 \pm 0.004/10$.

To eliminate the difficulty of seizure occurring in large diameter conical joints (Section 36.4) 1/5 taper is also used for them. Table 3.8 gives the dimensions of some such joints according to the German standard.

The pipes continuing the inner and outer members of conical ground joints have a diameter corresponding approximately to the outside diameter of the small ground end of the inner member of the joint. Large tubing cannot be sealed close to the joint because of the danger of warping the ground surface. It is advisable, however, (Barr[81]) to butt seal a larger tubing to the outer member at a point approximately 20 mm from the joint.

* For vacuum (greased) seals full or medium length are preferred; for waxed joints the short length can be used (see Fig. 3.2d).

TABLE 3.7. CONICAL GROUND GLASS JOINTS
(Taper 1/10; $\Theta = 2°51'45''$ see Fig. 3.19)

Dimensions (mm)			Standard designation*		
d	*D*	*l*	British	American	German
			full length	full length	Serie 0
3	5	20	—	—	NS 5/20
5	7.5	25	A 7	7/25	NS 7.5/25
7	10	30	A 10	10/30	NS 10/30
9.3	12.5	32	—	—	NS 12.5/32
9.25	12.5	32.5	A 12	—	—
9.5	12.5	30	—	12/30	—
11	14.5	35	A 14	14/35	NS 14.5/35
12.5	16	35	A 16	—	—
15	18.8	38	A 19	19/38	—
15.2	19	38	—	—	NS 19/38
20	24	40	A 24	24/40	NS 24/40
24.8	29	42	—	—	NS 29/42
25	29.2	42	A 29	29/42	—
30	34.5	45	A 34	34/45	NS 34.5/45
35	40	50	—	40/50	—
40	45	50	—	45/50	NS 45/50
45	50	50	—	50/50	—
50	55	50	—	55/50	—
54.5	60	55	—	—	NS 60/55
55	60	50	—	60/50	—
64	70	60	—	—	NS 70/60
65	71	60	—	71/60	—
78	85	70	—	—	NS 85/70
92	100	80	—	—	NS 100/80
97	103	60	—	103/60	—
			3/4 length	medium length	Serie 1
3.5	5	15	B 5	—	—
3.7	5	13	—	—	NS 5/13
3.8	5	12	—	5/12	—
5	7.5	16	—	—	NS 7.5/16
6	7.5	15	—	7/15	—
5.7	7.5	13	B 7	—	—
7.8	10	22	B 10	—	—
8.1	10	19	—	—	NS 10/19
10.1	12.5	24	B 12	—	—
10.4	12.5	21	—	—	NS 12.5/21
11.9	14.5	26	B 14	—	—
12.2	14.5	23	—	—	NS 14.5/23
13.4	16	26	B 16	—	—
16	18.8	28	B 19	—	—
16.4	19	26	—	—	NS 19/26
21	24	30	B 24	—	—
21.1	24	29	—	—	NS 24/29
25.8	29	32	—	—	NS 29/32

* See key on page 281.

(Table 3.7 Continued)

Dimensions (mm)			Standard designation*		
d	*D*	*l*	British	American	German
26	29.2	32	B 29	–	–
31	34.5	35	–	–	NS 34.5/35
31.1	34.5	34	B 34	–	–
36.3	40	37	B 40	–	–
41	45	40	–	–	NS 45/40
41.3	45	37	B 45	–	–
46.3	50	37	B 50	–	–
51.3	55	37	B 55	–	–
55.4	60	46	–	–	NS 60/46
65	70	50	–	–	NS 70/50
79.5	85	55	–	–	NS 85/55
94	100	60	–	–	NS 100/60
			½ length	medium length	Serie 2
4.1	5	9	–	–	NS 5/9
6.4	7.5	11	–	–	NS 7.5/11
8.2	10	18	–	10/18	–
8.7	10	13	–	–	NS 10/13
10.7	12.5	18	–	12/18	–
11.1	12.5	14	–	–	NS 12.5/14
12.5	14.5	20	–	14/20	–
12.8	14.5	17	C 14	–	–
13	14.5	15	–	–	NS 14.5/15
16.6	18.8	22	–	19/22	–
16.9	18.8	19	C 19	–	–
17.3	19	17	–	–	NS 19/17
21.5	24	25	–	24/25	–
22	24	20	C 24	–	NS 24/20
26.6	29.2	26	–	29/26	–
26.8	29	22	–	–	NS 29/22
31.7	34.5	28	–	34/28	–
32.1	34.5	24	–	–	NS 34.5/24
36.5	40	35	–	40/35	–
42.3	45	27	–	–	NS 45/27
56.9	60	31	–	–	NS 60/31
66.7	70	33	–	–	NS 70/33
81.3	85	37	–	–	NS 85/37
96	100	40	–	–	NS 100/40
			¼ length	short length	Serie 3
4.2	5	8	–	5/8	–
6.5	7.5	10	–	7/10	–
9.3	10	7	–	10/7	–
9	10	10	–	10/10	NS 10/10
11.3	12.5	12	–	–	NS 12.5/12
11.5	12.5	10	–	12/10	–
13.3	14.5	12	–	–	NS 14.5/12

(Table 3.7 Continued)

Dimensions (mm)			Standard designation*		
d	*D*	*l*	British	American	German
13.5	14.5	10	—	14/10	—
17.8	18.8	10	—	19/10	—
17.8	19	12	—	—	NS 19/12
22.8	24	12	—	24/12	NS 24/12
23	24	10	D 24	—	—
27.8	29	12	—	—	NS 29/12
28	29.2	12	—	29/12	—
28.1	29.2	11	D 29	—	—
33.3	34.5	12	—	34/12	NS 34.5/12
33.4	34.5	11	D 34	—	—
38.8	40	12	D 40	—	—
38.8	40	12	—	40/12	—
43.8	45	12	—	45/12	NS 45/12
48.8	50	12	—	50/12	—
53.8	55	12	—	55/12	—
58.8	60	12	—	60/12	NS 60/12
68.8	70	12	—	—	NS 70/12
69.5	71	15	—	71/15	—
83.8	85	12	—	—	NS 85/12

* Standard designation:
British Standard 572; supplied by Baird & Tatlock Ltd., Freshwater Rd. Chadwell Heath, Essex,, England; Griffin & George, Ealing Rd., Alperton, Wembley, England; The Thermal Sindicate Ltd., Wallsend, Northumberland, England (Silica joints).
American Standard, Natl. Bul. Stand. CS 21-39; supplied by Fischer Scientific Co. 633 Greenwich Str., New York 14; Eck & Krebs Inc., 27-09 40th Avenue, Long Island City N.Y., U.S.A.; Central Scientific Comp., 1700 Irving Park Road, Chicago 13, Ill., U.S.A.
German Standard DIN 12242/1954; supplied by E. Leybold's Nachfolger, Köln-Bayental, Germany; A. Pfeiffer, GmbH Wetzlar, Germany.

Conical ground joints are used also in shapes differing from the standard ones, i.e. the ground part is of standard dimensions but has supplementary or changed parts. Such seals are e.g. the *extended seals*, the *connexion seals*,

TABLE 3.8. CONICAL GROUND GLASS JOINTS (Taper 1/5)

Dimension (mm)			Designation DIN 12243
d	*D*	*l*	
50	60	50	NS 60/50
62	75	65	NS 75/65
75	90	75	NS 90/75

the *cooled seals* and the seals with a *guard vacuum*. Ground joints sealed with mercury are described in Section 37.2.

In standard conical joints the inner part ends at its smaller diameter and the outer part at its larger end. For the transmission of motion (Section 51.6) the inner part is *extended* inside the tube of the outer member (Fig. 3.20a).

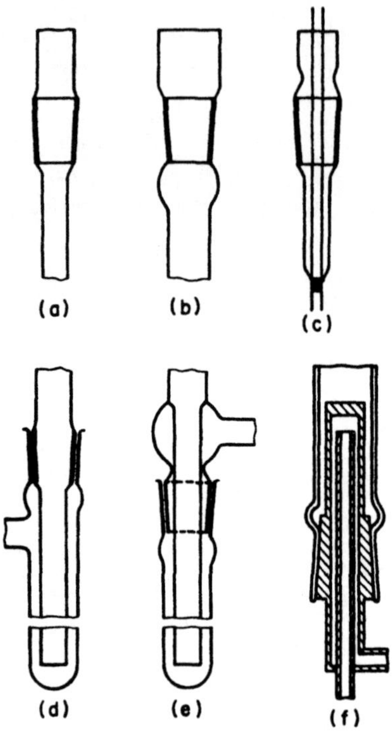

FIG. 3.20 Extended tapered ground joints; (a) extended inner part; (b) extended outer part; (c) extended inner part with lead-through; (d) cooling trap with exhaust pipe on outer part; (e) as (d) on inner part; (f) cooled metal electrode

Similar extended conical joints can be used to seal lead-throughs (Fig. 3.20 c, f) or cooling traps (Fig. 3.20d,e). Figure 3.20.f shows a conical seal having the inside member of metal, with a built-in electrode system and cooling jacket (Laporte[741]).

The outer member of the joint can also be extended forming a cup (Fig. 3.20.b) around the joint. This cup may be used for the cooling liquid or for sealing with mercury (Section 37.21).

For the dismountable connexion of conical joint members having different diameters, parts known as *connexion tapers* may be used (Fig. 3.21). The connexion tapers have two tapered ground surfaces either inside and outside the same plug (Fig. 3.21a) or on the two ends of a short connecting pipe (Fig. 3.21b). The connexions are used from large outside to smaller inside

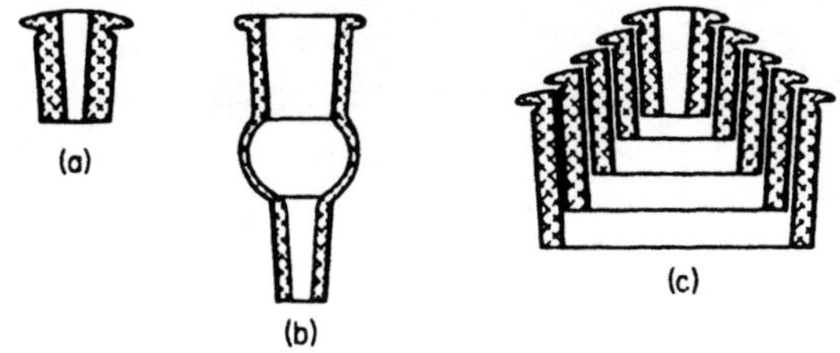

FIG. 3.21 Connexion tapers; (a) connexion plug; (b) connexion pipe; (c) connexion plug series

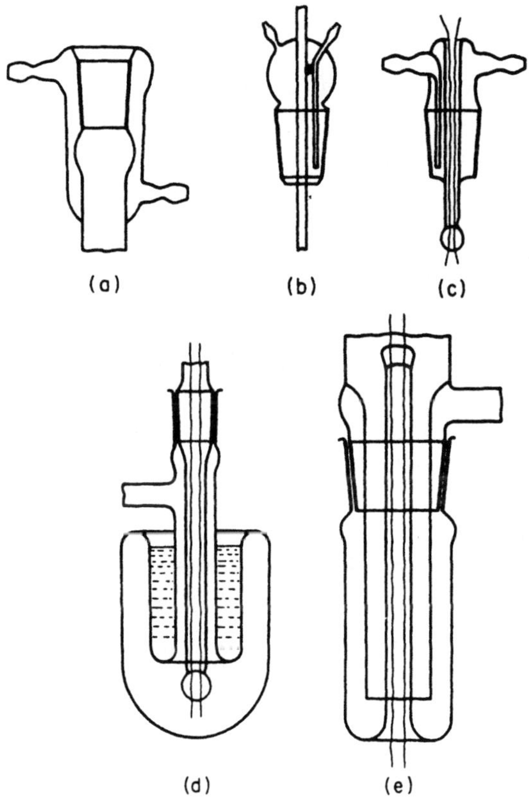

FIG. 3.22 Cooled tapered joints with (a) cooling jacket on outer part; (b) cooling jacket on inner part; (c) as (b) with lead-through; (d, e) seals with liquid air cooling

diameter (Fig. 3.21a) or from large inside to smaller outside diameter (Fig. 3.21b). Complete series of such connexions (Fig. 3.21c) are available.*

In order to avoid the penetration of grease vapours inside the vacuum system or to protect the grease from melting when the system is heated the

* Juffa & Co., Ilmenau, Germany.

ground joints may be constructed with cooling jackets (Fig. 3.22). The cooling jacket is provided around the outer part (Fig. 3.22a) or on the inner member (Fig. 3.22b). In this latter case the inlet pipe is extended inside the vessel. The cooled joints can be combined with extended ones. Figure 3.22c shows a cooled seal having an extension with lead-through wires, and Fig. 3.22d a seal having a cooling jacket around its electrode system. For better cooling, taper joints, provided with facilities for liquid air (nitrogen) may be constructed. The cooling of such parts is done, either by pouring the liquid air in the

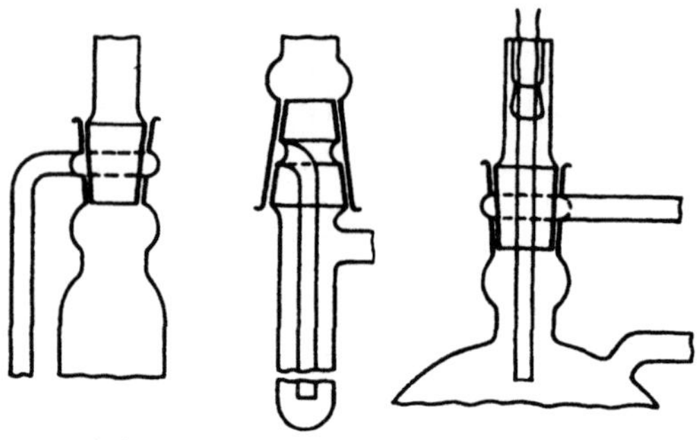

FIG. 3.23 Tapered joints with guard vacuum: (a) exhaust channel on the outer part; (b) exhaust channel in the inner part; (c) extended joint with guard vacuum

double-walled jackets provided around them (Fig. 3.22d) or by constructing the seal in the form of a cooling trap (Fig. 3.22e) which can be immersed in a container with liquid air.

Figure 3.23 shows examples of conical ground seals using a *guard vacuum* (see principle in Section 38.23). Such seals are used when the vapours of the grease should not reach the evacuated space of the system. The joint is greased only on the part outside the channel connected to the guard vacuum. The guard vacuum channel may be contained in the outside member (Fig. 3.23a) or in the inside part (Fig. 3.23b). Figure 3.23c shows a seal with guard vacuum and extension which carries lead-through wires.

Fine[376] constructed sliding conical ground glass joints, consisting of three parts (Fig. 3.24a). The connexion between parts 1–2 is a usual joint, but the second connexion (between 2 and 3) is a sort of extended joint which permits the opening and closing of the seal without the need of supplementary axial space, as is required by standard conical joints.

Using elbows with conical ground joints, dismountable chains can be built (Fig. 3.24b) allowing bending in any direction.

Cylindrical ground seals are rarely used in vacuum sealing due to the extreme difficulty in fitting the two parts together. Such seals can be used as shafts for the transmission of motion; they are placed vertically and sealed with oil contained in a cup (Fig. 3.20b) provided on the outer member. Decker[270] describes a cylindrical ground seal between a glass slug and a ground seat used to close an ultra-high vacuum valve (Fig. 6.35). Amariglio[25] used a cylindrical ground seal as a controlled leak (Section 61.45).

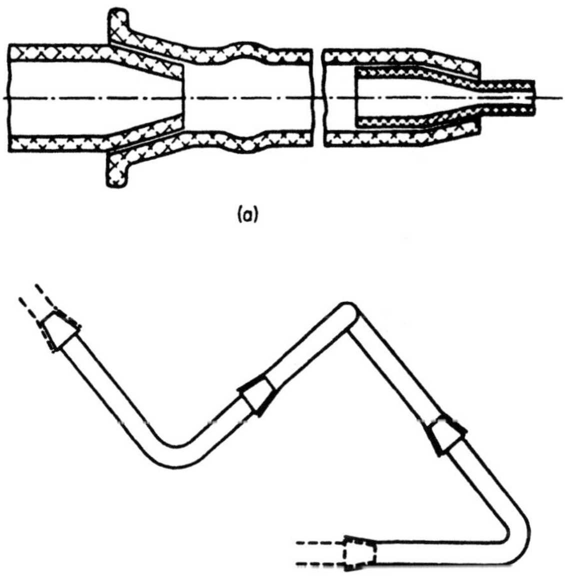

(a)

(b)

FIG. 3.24 Tapered joints allowing motion: (a) sliding, (b) bending

36.3 Spherical ground seals

Spherical or ball and socket ground joints were developed to be used in applications where the alignment of the parts to be joined is difficult or where motion of the parts with respect to each other is required.

A spherical joint consists of an inside part having the shape of a ball (or part of a ball) and an outer part (socket) ground to fit the ball (Fig. 3.25a). Two balls can be connected by a double socket (Fig. 3.25b).

The grinding of spherical joints is not difficult, since the spherical surfaces are the only ones which remain in continuous contact at every point when moved over each other in any direction. The grinding technique for these joints is described by Rubens[1090].

Spherical joints are designated by the ratio between the diameter of the ball and the inside diameter of the tubing (bore) both in millimetres (Table 3.9).

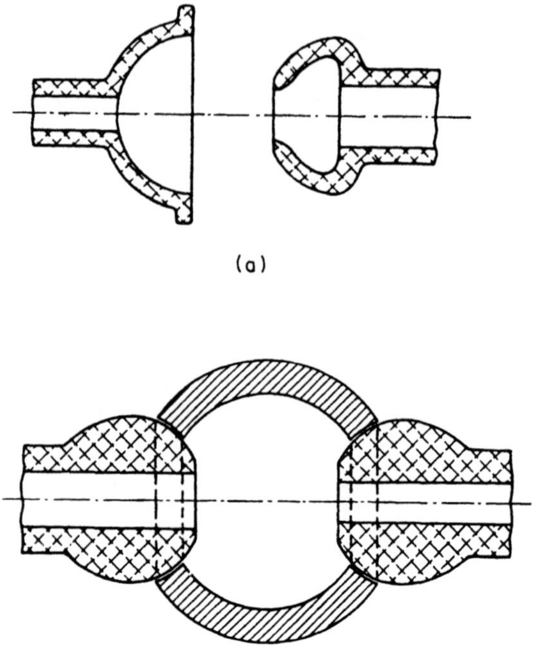

FIG. 3.25 Spherical ground joints: (a) ball and socket; (b) metal socket connecting two ball joints

TABLE 3.9. STANDARD SPHERICAL JOINTS
(Dimensions in mm)

Designation	Shank		Diameter of the ball	Remarks
	outside diameter	nominal bore		
7/1	–	1	7	
12/1	–	1	12	
12/1.5	–	1.5	12	capillary
12/2	7–9	2	12	joints
12/3	–	3	12	
12/5	7–9	5	12	
18/7	–	7	18	
18/9	12–13	9	18	
28/12	–	12	28	
28/15	19–21	15	28	normal
35/20	24–26	20	35	
35/25	–	25	35	bore
40/25	28–30	25	40	
50/30	38–40	30	50	joints
65/40	–	40	65	
75/50	–	50	75	
102/75	–	75	102	

Spherical joints are used for transmitting motion (Fig. 3.26a) especially for two-directional motion (Rubens[1090], Skidmore[1145], Section 51.6). They are also used in cut-offs (Section 61.13) with glass or metal balls (Fig 3.26b) as described by Neville[924], Metzler[852], or as the closing system of glass valves

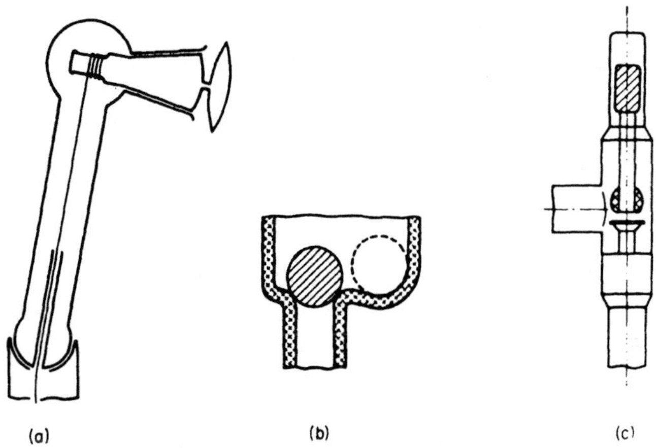

FIG. 3.26 Spherical joints for (a) motion transmission; (b) cut-off; (c) all glass valve seal

without grease (Klopfer[690], Vogl[1271], Decker[270], Yarwood[1341], Adam[3]). Such seals are shown in Fig. 3.26c and in Section 61.32e.

Finally spherical ground seals are used in break-off devices (Snow[1161], Norton[943]) as shown in Fig. 3.26a and Section 61.52.

36.4 Assembly and maintenance of ground seals

Ground seals can be used in vacuum technique only if they are properly *selected*, *assembled* and *maintained.*

The *selection* of the ground seal to be used must be based on the requirements which should be met by the particular joint. These requirements may be: the bore (diameter), the direction of opening of the joint, the space required for the opening, the maximum service temperature, the outgassing rate, and the various functional requirements (transmission of motion, electric current, etc.).

If small-bore ground seals are necessary, spherical joints should be used (see Table 3.9); for medium size (3–100 mm diameter) conical (Table 3.7) spherical or flat joints may be used, and for large diameters the flat joints are useful.

The direction in which the joint should be opened and closed and the space available for this, restricts the kind of joint which can be used in a particular application. For radial openings only flat joints are useful; if a small axial displacement is possible, spherical joints may be used. The opening of the standard conical joints (Table 3.7) require an axial displacement equal at least to the length of the joint. A conical joint which can be opened without requiring this condition is shown in Fig. 3.24a.

For room temperature the joints are to be greased with light greases (Table 3.10), but if the joint is to be used in a warm atmosphere, medium or heavy greases are necessary. At temperatures near to 100 °C special greases (Table 3.10) or cooled joints (see Fig. 3.22) must be used.

The influence of the outgassing rate of the grease can be kept to a minimum by using low vapour pressure greases (Table 3.10) as well as by limiting the penetration of the vapours to the evacuated space with correct greasing of the joint (Fig. 3.29b) and/or by providing a guard vaçuum in the joint itself (Fig. 3.23).

Ground joints may be *assembled* with their axes horizontal, vertical or in any other position. In some of these positions, the two parts of the joint can stand against each other without the need of clamping them together. However it is recommended that the parts be clamped together even if it does not appear imperative. One of the parts of the joint should be attached to the system, by using an elastic or flexible* clamping device.

The clamping of the members of a ground joint to each other can be done

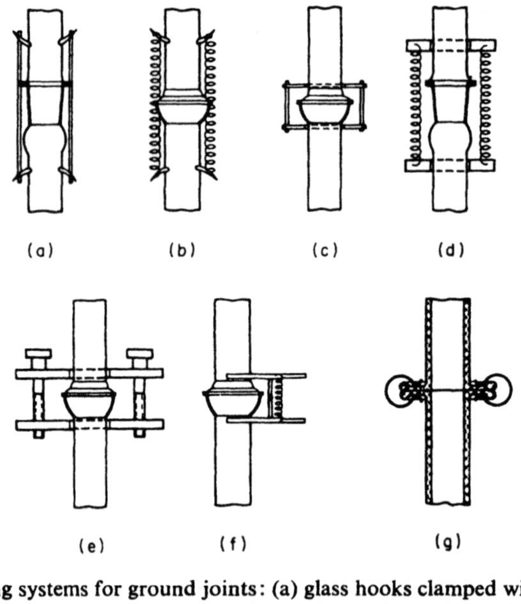

FIG. 3.27 Clamping systems for ground joints: (a) glass hooks clamped with rubber bands; (b) with springs; (c) jaws with rubber bands; (d) jaws with springs; (e) screwed clamp; (f) clamp for spherical joint; (g) clamp for flat joint

* Richard[1051] describes a technique to attach a heavy object to a glass system by means of a ball joint and spring suspension which compensates the weight of the object.

simply by providing rubber bands or springs between hooks on the parts (Fig. 3.27a, b) or between jaws placed on the parts (Fig. 3.27c, d). These hooks should not be simply made by attaching a glass rod and bending it (Fig. 3.28b) because it will then snap off when clamped, but the glass should flow smoothly into the body of the part (Fig. 3.28a). For more accurate clamping, the two parts of the clamp are fastened together by screws (Fig. 3.27e).

Spring-loaded clamps can be used with spherical seals (Fig. 2.27f) and simple clips (Fig. 3.27g) are used to clamp together flat seals. Experience in the laboratory showed that more spherical joints are broken by improper

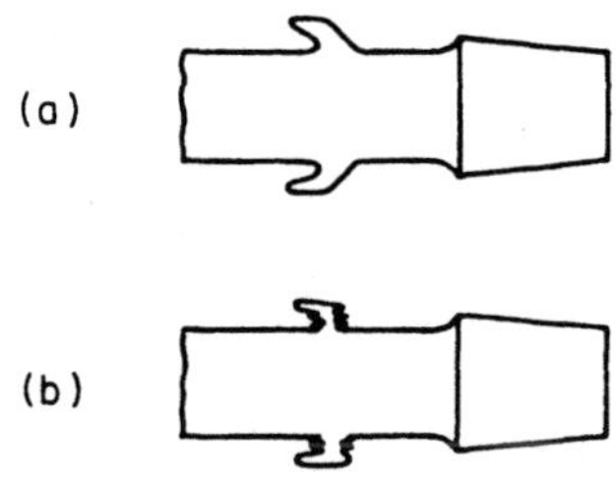

FIG. 3.28 Glass hooks: (a) correct shape; (b) incorrect shape

clamping than by any other cause (Barr[81]). If a spherical joint leaks, the cause is usually improper lubrication or small dirt particles which were not removed before the joint was assembled. Thus further tightening of the clamp of such a joint does not normally seal a leak, but usually leads to the joint breaking.

Careful handling is an important requirement for all ground seals (Kleinteich[687]). Ground joint parts should be stored and should be so handled as to avoid any scratching of the ground surfaces, avoiding any contact of these surfaces with hard or sharp edged objects. For the same reasons, the assembling or rotating of the ground members on each other should be avoided until they had been greased. During storage a thin sheet of soft material (plastic, paper, etc) should be placed between the two parts.

The *maintenance* of ground joints includes their *cleaning* and their *greasing*. Before use or re-use of a ground seal, the ground surfaces must be thoroughly cleaned from dust, dirt or any residual grease.

The washing-off of Ramsay and Apiezon greases (Table 3.10) may be done with benzine, benzol or carbon tetrachloride. The cleaning-off of silicone greases can be done (Roeben[1074], Edwards[327–330]) as follows:

(1) By rising the surface with paraffin or other suitable hydrocarbon solvent and subsequent cleaning with a warm chromic acid solution (Section 23.21) or subsequent cleaning with a warm solution of 5 g sodium perborate in 100 cm^3 of 10 per cent sodium hydroxide solution in distilled water.

TABLE 3.10. VACUUM GREASES

Designation	Max. service temp. (°C)	Dropping (melting) point (°C)	Vapour pressure (Torr)		Remarks	Supplier*
			at 25 °C	at higher temp.		
Vacuum grease P	25	55	10^{-5}	10^{-4} (100 °C)	10^{-8} (25 °C) degassed 2 hr at 90 °C	Leybold
Graise PB 1	–	–	–	–	light	CGR
Ramsay grease	25–30	56	10^{-7}	10^{-4} (30 °C)	–	Leybold
Apiezon L	30	–	10^{-9}	10^{-6} (135 °C)	only temporarily in well fitting joints	
Apiezon M	30	–	10^{-8}	10^{-6} (38 °C)	–	Edwards Shell
Apiezon N	30	–	10^{-7}	–	for tapered joints	
Vacuum grease R	30	65	5.10^{-6}	–	10^{-8} (25 °C) degassed	Leybold
Graise PB 2	(30)	–	–	–	for large joints	CGR
Graise PB 3	(30)	–	–	–	heavy	CGR
Lubriseal	30	40	–	–	for glass/metal	CENCO
Vacuseal light	50	–	10^{-5}	–	–	CENCO
Joint grease DD	58	120	–	–	for rotary seals	Leybold
Vacuseal heavy	60	–	10^{-5}	–	–	CENCO
Celvacene light	–	90	10^{-6}	–	–	CVC
Celloseal	–	100	10^{-6}	–	soluble in chloroform	Fischer

(Table 3.10 Continued)

Designation	Max. service temp. (°C)	Dropping (melting) point (°C)	Vapour pressure (Torr) at 25 °C	Vapour pressure (Torr) at higher temp.	Remarks	Supplier*
Apiezon T	110	–	10^{-8}	–	light m.p. grease	Edwards Shell
Celvacene medium	–	120	$<10^{-6}$	–	–	CVC
Cello grease	–	120	$<10^{-6}$	–	–	Fischer
Lithelen	150	210	very low	–	lithium soap	Leybold
Silicone stopcock grease	200	–	10^{-7}	10^{-5} (170 °C)	–	DCC Edwards
Silicone high vacuum grease	200	250	10^{-7}	10^{-5} (170 °C)	useful down to –40 °C	

* Leybold — Köln-Bayental, Germany; CGR — Comp. Générale de Radiologie (see ref. 224); Edwards High Vacuum Ltd. (see ref. 327–329); Shell Co. (see ref. 1129); CENCO (Central Scientific Comp. see ref. 199); Fischer Scientific Co. 633 Greenwich Street, New York 14; DCC — Dow Corning Corp. (see ref. 299); CVC — Consolidated Vacuum Corp. (see ref. 231).

(2) By rinsing initially with a hydrocarbon solvent and subsequent cleaning with a solution of 10 g sodium hydroxide and 5 g borax in 100 cm^3 of water.

(3) By cleaning the surface with a solution of 10–15 cm^3 of 50 per cent potassium hydroxide in 100 cm^3 of industrial methylated spirit, without allowing the ground surface to be in contact with the solution for more than *10 min.* With prolonged exposure to this solution, the glass may be etched.

For the greasing of ground joints the proper grease should be used, corresponding to the range of temperatures in which the seal is to be used as well as to the vapour pressure requirements of the system. Table 3.10 gives the designation (Trade Name) and characteristics of some greases supplied for this purpose. Generally the vapour pressure figures quoted are the "ultimate pressures" i.e. the vapour pressures of the grease in a degassed state. Thus a grease having in a fresh state a vapour pressure of 10^{-5} torr can reach 10^{-7} torr after a degassing of some hours, heated above its dropping temperature (Normand[941], Angerer[34]). In order to keep the grease clean, the grease supplied in tubes is preferred.

The grease is applied on the ground surface using rods (spatula) of soft materials (e.g. wood). Glass rods, or other hard or sharp tools which could scratch the glass should not be used, since internal scratches make the glass very prone to cracks. Obviously the best method is to apply the grease directly from a tube; this avoids the possibility of scratching the joint and minimizes the possibility of including dirt in the grease layer on the joint or in the remaining grease of the tube.

The grease is usually applied as strips on two diametrically opposed sides along the taper. The joint is then assembled and the two members are rotated* on each other until the lubricant is evenly distributed over the entire joint; under this treatment the joint should become transparent. On tapered joint sit is recommended that grease be applied only on the portion (about 1/3–1/2) which is on the atmospheric side of the joint (Fig. 3.29). If a heavy grease is used (see Table 3.10) in order to obtain a uniform spreading of the grease on the ground surface, the joint can be warmed slightly, without exposing it to a direct flame.

If while rotating the members to spread the grease over the ground surface a grating sound is heard, the rotation must be stopped, the joint opened,

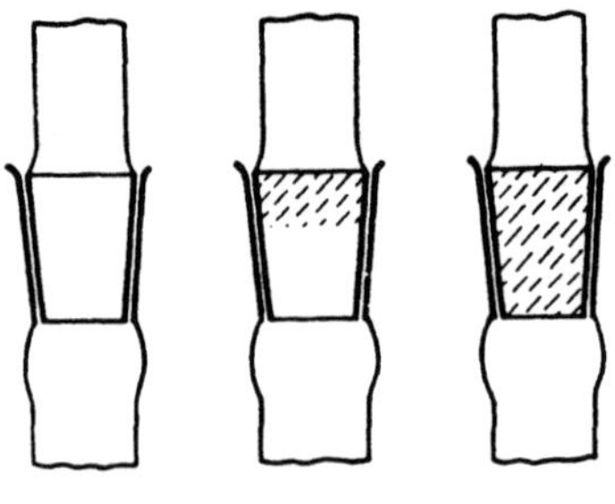

FIG. 3.29 Conical joint: (a) before greasing; (b) recommended greasing; (c) not recommended greasing (dotted part is greased)

cleaned and greased again. If after rotation, the surface of the joint presents any lines which do not disappear, cleaning and regreasing are also required. The same appearance can develop after some period of service, but in this case the lines are not of dirt (as in a newly greased joint) but consist of hardened grease. Neverthe]ess the joint should be cleaned and regreased in this case as well. If the grease flows down the inside of the joint (on the walls of the vessel) it should be opened, cleaned and regreased.

The combination of the lack of a lubricant (grease), the wedge action of the tapered members and the differences in expansion of the two member can

* Rotation in one direction only is recommended.

result in a condition in which, *it is not possible to open the joint*. This condition is also known as jamming, seizure, cemented joints, frozen joints, etc. If the seal is frozen, the use of brute force to open it is usually disastrous. The separation must be done skilfully by vibration, dissolving or melting the cement which keeps the members together.

Separation may be effected by tapping the edge of the outer member (socket) on wood or with wood, with a subsequent or simultaneous withdrawing of the inner member (cone).

A penetrating liquid solvent can often be used with success. The solution containing the detergent (preferably in alcohol) penetrates between the ground surfaces and loosens the cemented joint.

Heating can help in two ways. Melting the grease with hot water may be useful. Quick heating of the outer member can expand it and permit the withdrawal of the other member of the joint. This procedure needs some skill since the operation must be completed quickly (a hot brush flame should be used). Delay in removing the inner member allows it also to become heated, again tightening the joint. If the joint cannot be opened, the entire joint must be cooled before another attempt is made.

3.7 LIQUID SEALS

A *liquid seal* is a joint in which the gap between the connected parts is sealed by a material in the liquid phase.

If the liquid seal separates spaces having a pressure difference of an atmosphere, the sealing action may be based on: *hydrostatic pressure* (Section 37.1), on a high *impedance* to the flow of the liquid (Sections 37.2–37.4) or on the *surface tension* of the sealing liquid (Section 37.5). In order to minimize the height of the liquid column needed for the seal, in some seals a *guard vacuum* is used (Section 37.1) or the sealing *liquid is frozen* when the joint is exposed to pressure differences greater than what the liquid column can withstand (Section 37.6).

A general requirement for liquid seals is to use only liquids having at the service temperatures, vapour pressures lower than the pressure to be reached in the vacuum system (vessel), which they are sealing (see Table 3.11).

In seals where the sealing material is permanently liquid, mercury or oil is used. In frozen seals various low melting metals (or alloys) are used as the sealing medium.

Mercury has the danger of being toxic, of amalgamating with a series of metals (Au, Ag, Sn, Zn, etc.) and of attacking aluminium. Thus in systems containing these metals mercury cannot be used. Mercury has at room temperature a high rate of evaporation and gives an equilibrium vapour pressure of approximately 1.3 μ (1.3×10^{-3} torr) which corresponds to 10 mg mercury per m^3 of air. This is *toxic* according to the accepted standards in most countries (Biram[127]). Once handling of mercury has begun in a laboratory or factory it is difficult to keep the air burden below 75 microgrammes per m^3 (upper limit admitted by the standards). Mercury can be taken into the body through the skin as well as the mouth and the nose. However the most harmful way is considered to be by inhaling the mercury vapour. The mercury vapour in air can be kept low only with difficulty, due to the rapid rise of its vapour pressure with temperature (with a rise of 10 °C the vapour pressure is about doubled), the tendency of the mercury to form very small droplets (very large exposed surface), and the difficulty of cleaning it from surfaces since it does not wet the materials, and is insoluble in water. The handling of mercury has been extensively described by Biram[127].

37.1 Hydrostatic seals

A hydrostatic seal acts by balancing the pressure difference with a column of liquid. The most current hydrostatic seal used in vacuum systems is the open liquid column manometer. If, when sealing atmospheric pressure difference, the full column of mercury (760 mm) can be used (even if it is not easy), the use of a full column of oil (11–13 m) is out of question.

Mercury vapour may be prevented from penetrating into the vacuum system by floating a layer of a low vapour pressure oil (Table 3.11) on the mercury surface (recommended oil height is 1–3 cm). This can be done only on the mercury surface in contact with the evacuated space, but for reasons of accuracy it is better to float equal amounts of oil on both mercury surfaces. The oil dissolves gases, and during evacuation it boils (spills), thus slow exhausting is necessary to avoid this effect (Ayer[60a]).

The mercury sometimes sticks to glass. This can be prevented by placing the mercury in contact first with aluminium for a short time (Archard[45]).

Besides the full mercury column manometers (Fig. 3.30a), full column mercury seals may be used as rotary seals or seals allowing small axial movement (Fig. 3.30b). Chaudhri[209] describes a seal using a full column of mercury in an annular space between two concentric tubes, used as a safety valve to obtain constant pressure in a cell.

The height of the liquid column can be kept to a minimum if on a seal like that on Fig. 3.30b, the pressure on the outer level of the mercury is decreased (guard vacuum). Brueschke[176, 177] describes such a seal (Fig. 3.30c) in which a Wilson seal (Section 51.74) is placed (1) between the outer wall of the reservoir and the central (rotating) tube. The pressure is first reduced on the two sides of the mercury (or oil) column (3) by pumping through the connexion (2) with the valve (5) open. Thus the height of the liquid column can be kept small. The central shaft (4) holds a tube which dips into the liquid (3) permitting a limited vertical displacement as well. The shaft (4) can be rotated at 4000 r.p.m. maintaining at this time a pressure of 6.10^{-9} torr in the vessel (Brueschke[177]).

Liquid seals (hydrostatic seals) are also used in cut-offs (Section 61.13a).

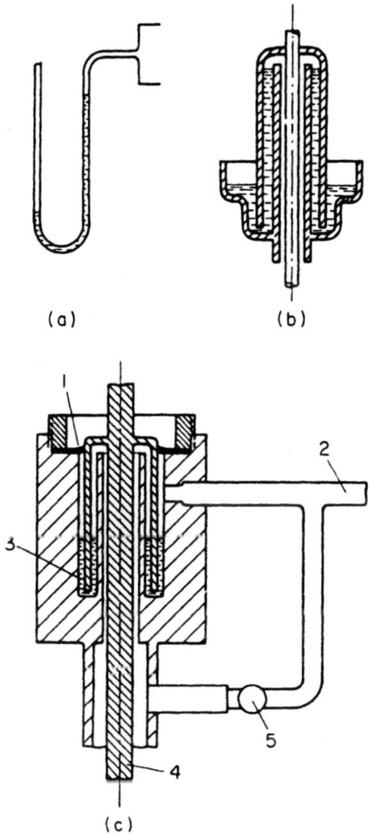

FIG. 3.30 Hydrostatic seals: (a) manometer tube; (b) full column mercury seal; (c) rotary seal with reduced column. After Brueschke[177] *(Courtesy of Pergamon Press)*

37.2 Mercury sealed ground and gasket joints

These seals are based on the fact that mercury placed over or near ground joints or gasket seals is able to close the small leaks remaining in such seals, without flowing through these leaks.

If ground joints are used without any lubrication, they will be cemented and it will be difficult, if not impossible, to open the joint. Thus in ground joints (especially tapered joints) a lubricant is necessary. When ground joints are to be sealed with mercury, the lubricant to be used is *graphite.* The graphite may be applied on the ground surfaces as a very thin layer by painting the surface with a diluted colloidal suspension of graphite (e.g. Aquadag) and allowing the liquid (water, alcohol) to evaporate. An alternative method is to rub a soft pencil (e.g. grade 4 B) on the ground surface. On conical joints the

TABLE 3.11. MATERIALS FOR LIQUID SEALS

Material	Vapour pressure p(torr)					
	Temp. (°C)	p	Temp. (°C)	p	Temp. (°C)	p
Oils						
Amoil	10	10^{-7}	25	10^{-5}	124	10^{-1}
Amoil S	8	10^{-7}	25	10^{-6}	146	10^{-1}
Apiezon A	—	—	20	10^{-6}	144	10^{-1}
Apiezon B	—	—	31	10^{-6}	162	10^{-1}
Octoil	15	10^{-7}	35	10^{-6}	155	10^{-1}
Octoil S	—	—	35	10^{-7}	174	10^{-1}
Silicon 702	−20	10^{-7}	41	10^{-6}	160	10^{-2}
Rotary pump oil (Shell CY 2)	—	—	25	10^{-1}	—	—
Metals						
Mercury	−5	10^{-4}	48	10^{-2}	126	1
Bismuth (M.P. 271 °C)	300	$<10^{-8}$	474	10^{-5}	802	10^{-1}
Gallium (M.P. 29.5 °C)	500	$<10^{-8}$	711	10^{-5}	—	—
Indium (M.P. 155 °C)	500	$<10^{-8}$	602	10^{-6}	—	—
Lead (M.P. 327 °C)	350	$<10^{-8}$	483	10^{-5}	—	—
Tin (M.P. 232 °C)	500	$<10^{-8}$	823	10^{-5}	—	—

graphite layer should be applied only on the ouside half (as in greasing, see Fig. 3.29). Nikolsky[932] prepared the lubricant by mixing *in vacuo* castor oil (preheated to 385 °C) and graphite (preheated to 750 °C), in order to obtain a gas free paste.

The mercury may be placed in these seals:

(1) In a cup, so that its level should cover the outside of the ground joint (cup seal, see Section 37.21):

(2) In a container in which the opening of the joint is immersed (immersion seal, see Section 37.22):

(3) In the space (groove) between two ground members or between two gaskets (guard seal, see Section 37.23).

37.21 The cup seal is used on conical joints or stopcocks (Fig. 3.31a, b) but can also be used on spherical or flat ground seals (Fig. 3.31c, e). All cup seals, having the mercury surface open, present two difficulties: the toxicity of the mercury and the fact that the joint cannot be tilted.

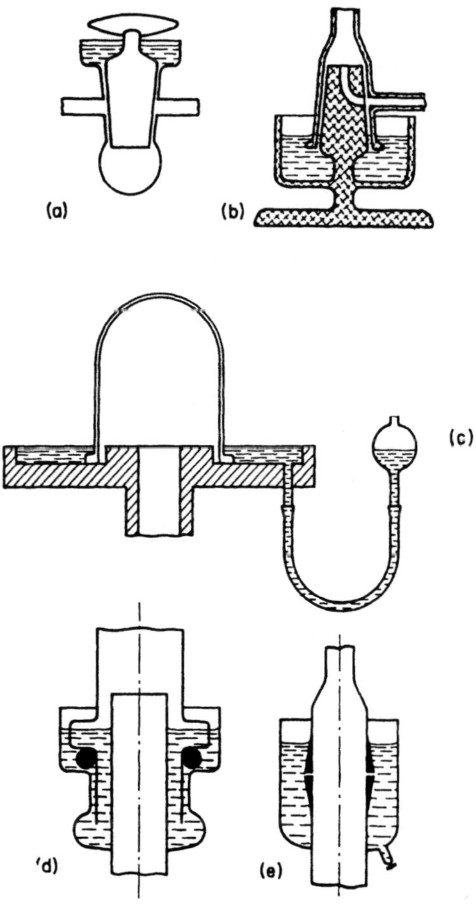

FIG. 3.31 Mercury-sealed ground and gasket joints: (a, b) stopcocks with mercury seal; (c) bell jar seal; (d) O-ring seal (after Gaunt[436]); (e) flat ground seal (after Lempicki[757]) *Courtesy of The Institute of Physics and The Physical Society, London)*

Cup seals are used mainly with the cup on the outer member of the taper joint or stopcock (Figs. 3.31a, 6.22) but if required for other purposes, the cup can also be on the inner member (Fig. 3.31b).

A bell jar can be sealed by mercury to its base plate (Fig. 3.31c); this seal should only be used if greased seals (Section 36.1), waxed seals (Section 31.2) or gasket seals (Sections 38.4, 38.5) cannot be used, since from such a seal, severe mercury poisoning may result, unless extremely effective ventilation is provided. The mercury may be poured around the bell jar after it is properly seated on the base plate, or introduced to the space around the bell jar by raising the level from a connected mercury reservoir (Fig. 3.31c).

A kind of mercury cup seal is used in quartz–metal seals (Fig. 2.86a) or direct platinum-to-Pyrex seals. Here a round platinum wire is sealed into the Pyrex tube, and the tube above the seal is partly filled with mercury, which assures the seal and serves as a conductor between the platinum and an electrode immersed in the mercury (Barr[81]).

Gaunt[436] describes a mercury-sealed O-ring joint, consisting of a cup provided on the inner pipe (Fig. 3.31d) in which the end of the outer pipe is immersed in mercury. The end of this pipe is provided with a shoulder on which sits the O-ring. When the apparatus is at atmospheric pressure the cap with the O-ring floats on the mercury. Evacuating the apparatus causes the cap to sink until the O-ring settles on its seat, thereby forming a seal. Gaunt[436] constructed such seals of about 50 mm in diameter without grinding the sealing surfaces. Smaller joints would possibly require light grinding of the seats. The seal has the disadvantage that the mercury is fully exposed to the evacuated space.

Lempicki[757] used a mercury-sealed flat ground seal (Fig. 3.31e), where the mercury remains outside the seal. The ground seal is first assembled, then the mercury is poured into the cup. Before opening the seal, the mercury is drained from the cup, through a pipe provided for this purpose.

37.22 The immersion seal contains the mercury in a vessel (part of the joint). When the inside part of the joint (2, Fig. 3.32) is rotated so that its opening

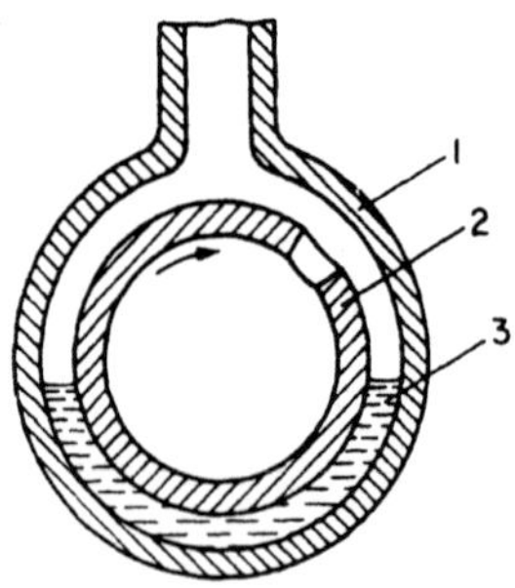

FIG. 3.32 Immersion seal

is immersed under the level (3) of the mercury the connexion between part (2) and part (1) is sealed. Stopcocks constructed on this principle may be used with their axes horizontal (Townes[1234]) or oblique (Stanier[1175]). See also Section 61.23.

37.23 In guard seals the mercury is placed at the two ends of a ground joint or in the space between two gasket seals (Fig. 3.33). Figure 3.33a shows a ground joint, where near to the ends of the joint two channels are provided

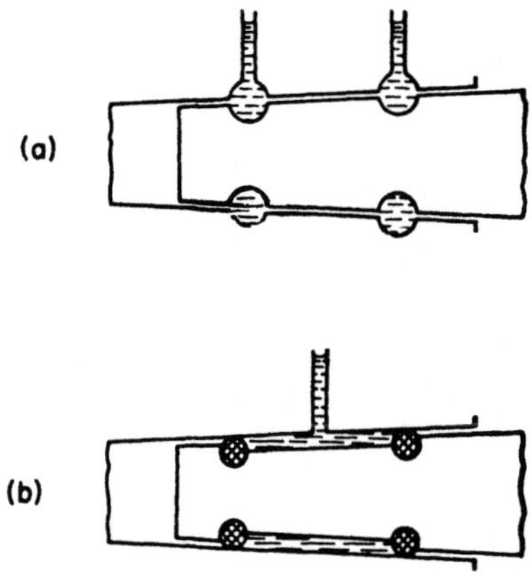

FIG. 3.33 Mercury guard seals: (a) ground seal; (b) O-ring seal

which are filled with mercury after the joint has been assembled. The same principle when applied to gasket (O-ring) seals, results in placing the plug provided with two O-ring seals and then filling the channel between the O-rings with mercury (Fig. 3.33b).

37.3 Mercury and gallium sealed glass frit

Due to its high surface tension (see Table 3.12) mercury does not flow through a dry glass frit having a porosity of less than 10 μ (about No. 4 frit*), even at pressure difference of an atmosphere. If the frit is wetted with another liquid (water, alcohol) the mercury can flow through (Espe[354]).

In these seals, the mercury may be placed as a layer (column) above the glass frit (porous sintered glass) as shown in Fig. 3.34a. The seal is tight for any reasonable pressure difference with the high pressure on the side of the

* Porosity number	1	2	3	4	5 and 3 in series
Pores (microns)	100–120	40–50	20–30	5–10	~2

mercury, and for a pressure difference equal to the height of the mercury layer, if the higher pressure is on the side of the frit. Wilson[1323] used this seal in a gas-circulating pump and Essig[358] in a valve (Sections 61.1 and 61.3).

Figure 3.34b shows a glass frit sealed with mercury from below. The seal can be opened or closed by changing the level of the mercury. This seal is used in cut-off (see Smith[1152], and Section 61.13c). An alternative construction of the seal is that shown on Fig. 3.34c, where the mercury closes the space between two glass frits.

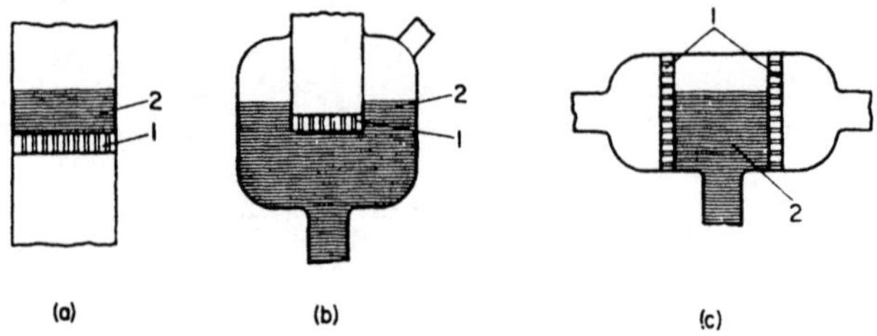

FIG. 3.34 Glass frits sealed with mercury: (a) with mercury layer above the seal; (b) with the mercury sealing below the frit; (c) with the mercury sealing between the frits; (1) glass frit; (2) mercury

In systems where the seal must be exposed to higher temperatures, mercury cannot be used because its vapour pressure is too high (see Table 3.11), and has to be replaced by a liquid of much higher boiling point. For most purposes, *gallium* is suitable as a replacement for the mercury. Gallium is liquid from 30 °C to very high temperatures (about 1900 °C) having a liquid range higher than any other metal. Gallium has a very low vapour pressure at the bakeout temperatures (400 °C–500 °C, see Table 3.11); it expands on solidification and may burst its container.

Gallium-covered glass sinters are more likely to become blocked than are mercury covered frits, and for this reason Beynon[120] recommends the use of several sinters in parallel. It is necessary to use a layer of gallium about 10 mm thick above the frit, to be sure of forming a leak-free seal.

37.4 Oil seals

Oil seals are used especially where the lubrication is needed simultaneously with the sealing action.

Flat lapped metal joints are sealed with oil in the sliding valves (Fig. 3.18) of the exhaust machines.

Cylindrical, rotating or sliding seals very often use oil as the sealing material. The most extensively used seal of this kind consists of two eccentric cylinders placed one inside the other, sealing on a line parallel to their axis (Fig. 3.35). This seal is used in one form or other in all rotary vacuum pumps (Barrett[86]).

For a reliable seal in this arrangement, the clearance between rotor and stator must be of the order of 2–3 μ. When the seal (pump) is new this requirement is easily met. If the two materials of the rotor and stator are not properly chosen, the friction (wear) increases the clearance until the oil cannot seal (Fig. 3.35b) the pressure difference existing across the two sides of the sealing line (Bacquet[69b]).

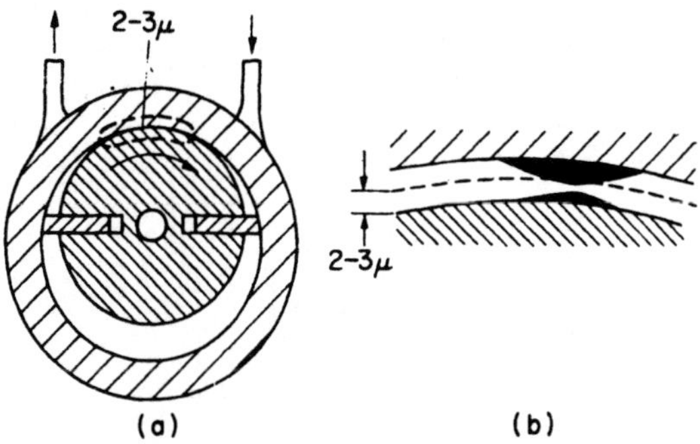

FIG. 3.35 Oil seal in rotary pumps

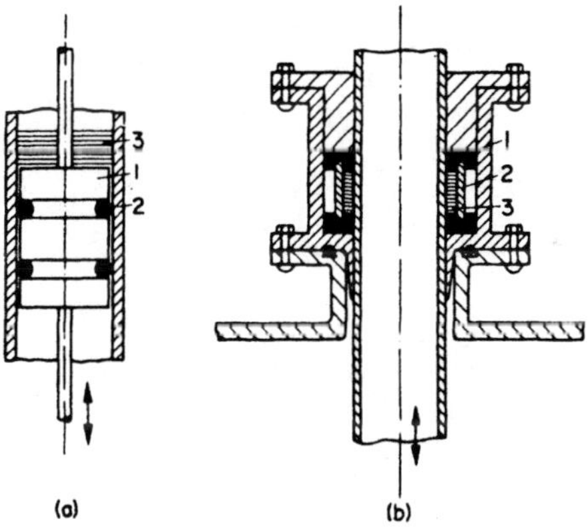

FIG. 3.36 Oil-sealed sliding joints: (a) piston seal; (b) seal for sliding pipe After Burgerjon[182a] *(Courtesy of North-Holland Publ. Co. Amsterdam)*

Oil-sealed cylindrical joints are constructed to seal sliding pistons (Fig. 3.36a) or sliding pipes (Fig. 3.36b). The piston (1) having rubber O-rings (2) slides in the cylinder, the seal being assured by the oil (3) above the piston. Lake[729a] used silicone oil in such a seal, and Gore[462] made a seal using dibutyl phtalate between two O-rings placed in the annular space formed by two concentric (sliding) glass tubes (Fig. 5.21a).

In order to seal a sliding pipe, Burgerjon[182a] used a U-packing (1) (Fig. 3.36b) closed on a cylindrical spacer (2); the cylindrical space (3) formed between the sliding pipe, the U-packing and the spacer are filled with oil.

Vertical rotary shafts may be sealed (and lubricated) with oil placed in a cup above or around the seal; a horizontal shaft must have the oil placed between rings, placed at some distance along the shaft. Trevoy[1237] described a rotary seal (Fig. 3.37a) with a glass capillary (1) of about 2 mm bore, in which a steel rod (2) is fitted (clearance about 2 μ). The joint is sealed by the oil placed in the cup (3). This seal was used at speeds up to 4000 r.p.m. Shaft

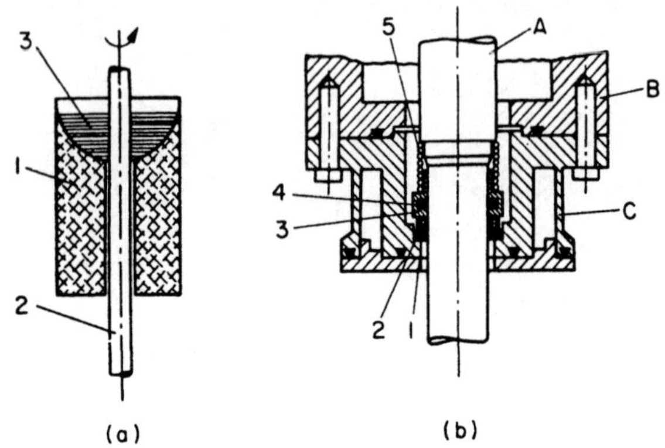

FIG. 3.37 Oil-sealed rotary joints: (a) glass capillary (after Trevoy[1237]); (b) ring seal. After Sikorsky[1136] *(Courtesy of the Institute of Physics and The Physical Society, London)*

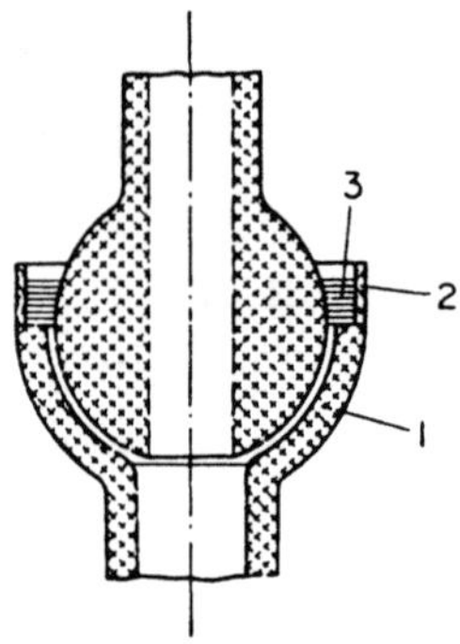

FIG. 3.38 Oil-sealed spherical ground joint (after Schrodt[1109])

seals sealed with oil and using elastomer or Teflon joints are described by Yarwood[1339], Marshall[809]. Other rotary shaft seals using oil as the sealing medium are described by Hill[562], Zobac[1356], Holland-Merten[587], Bode[141].

Sikorsky[1136] describes a rotary vacuum seal consisting of two rings (1 and 3, Fig. 3.37b) which seal the rotary joint. Ring (1) is made of compounded carbon and is held stationary sealed to the cover *C* (mounted on the body *B*) by a rubber O-ring (2). The other ring (3) is made of stainless steel and has a clearance fit on the rotating shaft *A*, to which it is sealed by the rubber O-ring (4). The stainless steel spring (5) has an interference fit on the shaft *A* on one side and on the neck of the ring (3) on the other side; it is wound so that a rotation of the shaft tends to increase the grip of the spring on the shaft and on the ring (3). The space between the shaft *A* and the cover *C* is filled with a suitable oil (e.g. Apiezon B). In order to cool the seal, the cover *C* includes a water channel. Sikorsky[1136] reported that the seal rotated at 600–900 r.p.m., maintains indefinitely a pressure of 2×10^{-4} torr in a 1.5 l. chamber while pumping with an oil diffusion pump of 30 l./min.

The oil seal can be used with *spherical ground joints* as well. Schrodt[1109] constructed a spherical joint (Fig. 3.38) of size 65/40 (see Table 3.9) and sealed it with oil poured in the annular space (3) between the glass ring (2) (about 6 mm high and 4 mm thick) fused to the rim of the socket (1).

37.5 Surface tension seals

Milleron[865, 865a] described the principles and techniques of constructing seals based on the surface tension of the liquids used. If two solids (Fig. 3.39a) are wet by a liquid which fill the gap between them, and on one side

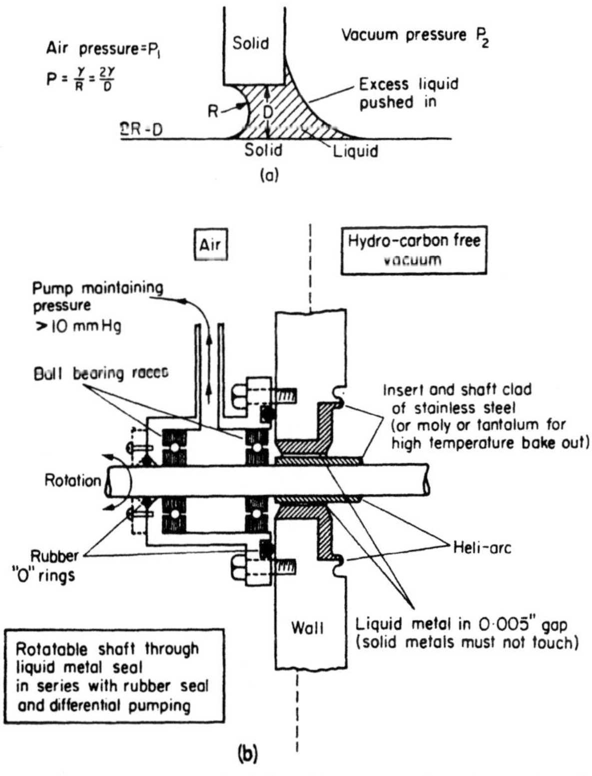

FIG. 3.39 Surface tension seal: (a) principle; (b) construction. Reproduced from Milleron[865] *(Courtesy of Pergamon Press)*

the pressure is atmospheric (P_1) and on the other it is vacuum (P_2) the pressure difference will tend to push the liquid towards the side with the lower pressure. The liquid will withstand the pressure difference if the gap D is small enough, and the surface tension of the liquid γ is high enough. Referring to Fig. 3.39a the equilibrium condition is given by:

$$P_1 - P_2 = \gamma\left(\frac{1}{R_1}+\frac{1}{R_2}\right),$$

where since $P_2 = 0$, and $R_1 \ll R_2$ and $2R_1 = D$ this gives:

$$D = \frac{2\gamma}{P_1},$$

e.g. a liquid metal with $\gamma = 500$ dyn/cm and $P_1 = 1$ atm $= 10^6$ dyn/cm^2 seals if the gap is less than $D = \frac{1000}{10^6} = 10^{-3}$ cm.

TABLE 3.12. SURFACE TENSION AND CLEARANCE FOR LIQUID SEALS

Liquid	Temp. (°C)	Surface tension dyn/cm	Maximum clearance (μ) for a pressure difference of 1 atm	100 torr
Gallium*	40	735	14.7	112
Tin	300	520	10.4	78
Mercury	15	487	9.5	72
Lead	350	420	8.4	64
Bismuth	300	370	7.4	56
Silver chloride	803	114	2.3	18
Water	20	73	1.4	11
Organic liquids	20	25–30	0.5–0.6	3.8–4.5

* Gallium produces embrittlement of aluminium and its alloys, thus its use with aluminium is to be avoided (Rostoker[1080a]).

Figure 3.39b shows a differentially pumped rotary shaft, sealed with a liquid metal film around the shaft (Milleron[865]). The clearance between the shaft and the "journal" in which the liquid was located was about 0.12 mm, for a shaft speed of maximum 10 r.p.m. The shaft must be cantilevered from ball bearings so that *no rubbing* can occur between the shaft and the "journal" where the liquid metal is located. Rubbing allows new substances to be formed and causes leakage through the seal.

37.6 *Molten metal seals*

Molten metals may be used as sealing media, assuring the seal only in *a frozen* state or withstanding some pressure differences in some constructions even when in the molten state.

Pakswer[963] constructed a window seal (Fig. 3.40a) in which a cavity was provided around the seal in order to trap the molten indium, thus permitting the seal to be baked at 350–400 °C. The cavity is closed on the tube side in

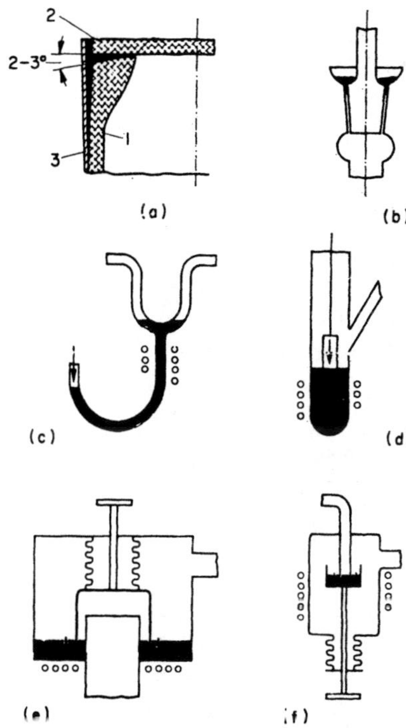

FIG. 3.40 Molten metal seals: (a) bakeable indium seal, (1) bevelled tube; (2) window; (3) shield (after Pakswer[963]); (b) stopcock sealed with Wood's metal; (c, d) cut-offs using indium; (e) cup and cap seal with moving cap; (f) cup and cap seal with moving cup

which the vacuum is made, by close matching of the ground surfaces of the window and its seat, avoiding (due to surface tension) the flow of indium in the evacuated space.

A taper joint (Fig. 3.40b) can be made vacuum-tight by filling the annular depression around it with Wood's metal (Hughes[606]). The Wood's metal (see Table 3.4) does not make a vacuum tight-joint with glass. Therefore it is necessary to platinize the glass surface locally and to tin the platinum locally (Section 35.13). A vacuum-tight seal can then be made by flowing Wood's metal between the tinned surfaces.

Molten indium (Paty[978], Axelrod[59]) or Wood's metal (Toby[1226]) were used to close cut-offs. These are constructed in a Y-shape (Fig. 3.40c) or as a T (Fig. 3.40d) and the metal is kept in the molten state during the sealing (Toby[1226]) or allowed to cool (Paty[978], Axelrod[59]).

Cup and cap seals (Fig. 3.40e, f) were made using Wood's metal (Hughes[606]), indium (Reynolds[1050]) or tin (Blanaru[130], Haaland[494]) as the sealing metal. To open the seal the metal is melted and the cap is raised (Fig. 3.40e, Reynolds[1050], Haaland[494]) or the cup is lowered (Fig. 3.40f, Hughes[606], Blanaru[130]). Valves using these seals are described in Section 61.32.

Solid mercury seals proved suitable in cryogenic use (Martina[811]), the metal being frozen in an O-ring groove by means of solid CO_2 or liquid air.

3.8 GASKET SEALS

38.1 Sealing mechanism of gaskets

Gasket seals are defined by the American Vacuum Society[26] as seals effected by *compressing* a gasket between the parts to be sealed.

Two flanges with good surface finishes joined together, make a good mechanical union, but a very narrow gap (channels) as shown by Fig. 3.41a always remains between them. Even if these channels are of micron sizes they constitute

(a)

(b)

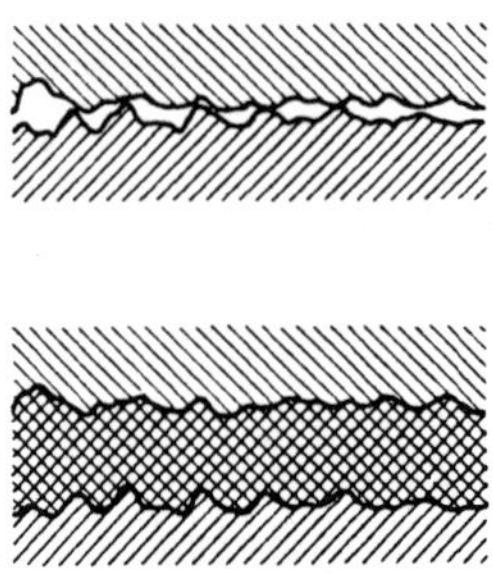

FIG. 3.41 Schematic cross section through a flange joint: (a) without gasket; (b) with inserted gasket

a considerable leak (Section 61.43 and Table 1.3) which cannot be permitted in high vacuum systems. To close these channels by pressing the two flanges on each other extremely high pressure would be required since the surfaces are very hard. Thus the sealing action is performed by interposing between the flanges a third material* (Fig. 3.41b) which fills up the small irregularities of the surfaces, requiring only reasonable compressive forces on the flanges.

*A gasket is a somewhat resilient part that is placed between two rigid surfaces to make a leak-proof seal (Fabian[364]).

The *pressure* required on the gasket in order to assure the seal, depends on:

(1) The material of the gasket;
(2) The surface finish of the flanges;
(3) The friction between the gasket and the flanges, i.e. the lubrication and the shape of the gasket and of the flanges.

The pressure exerted on the gasket deforms it. The *degree of deformation* required for a good seal depends on: (1) the material of the gasket, (2) the shape of the gasket seal, and (3) the kind of the effort on which the seal is based (compression, shear).

In order to fill out all the irregularities of the surfaces, the gasket should have some *plasticity*, and in order to maintain the pressure exerted on the seal the gasket should have *elasticity*. Unfortunately the properties of the various materials do not permit the full and simultaneous use of both the elasticity and plasticity. Rubber is used because of its excellent elasticity, while plastics and metals seal because of their plasticity.

The pressures required for rubber seals are moderate, and the deformations are not too great (see Table 3.13 and Fig. 3.50). Plastics require higher pressures and metals (excepting indium) require considerable pressures on the gasket–flange contact. In order to reduce the required force, the gasket–flange contact area is reduced (e.g. Sections 38.44, 38.56) to the minimum possible.

While closing a gasket seal, the pressure on the contact surface is gradually increased. At first the deformation is elastic, but on a further increase a plastic deformation occurs, initially on the edge or wedge provided on the flange, expanding gradually to the rest of the gasket. If the elastic limit of the gasket is exceeded, a considerable hardening of the material will occur. To maintain the tightness of the joint, the bolts must be retightened at regular intervals or constant strain systems must be used (see Sections 38.48, 38.58). Hardened gaskets cannot be re-used after the opening of the seal.

The ideal gasket would probably consist of two layers of a non work-hardening material of high ductility on a core of material of high strength (Wheeler[1304]).

The plastic flow of the gasket as required for the seal is aided in some sealing techniques by the heating of the seal, needed in any case for degassing purposes. In this way aluminium-cemented seals (Section 38.46) remain vacuum-tight although the securing bolts become loose (Holden[578]) due to the bond between the aluminium gasket and flange.

The hardness of the gasket must be less than that of the flanges. Figure 3.43 shows a comparison between the hardness of various groups of materials and the various corresponding hardness scales. Hardness values of various materials are given in Appendix B.1.

TABLE 3.13. SEALING FORCES REQUIRED FOR VACUUM-TIGHT GASKET SEALS

Gasket			Sealing			
Material	Shape of cross section	Dimensions (mm)	Force kg/cm (of gasket)	Pressure kg/cm² (of gasket)	Remarks	Reference
Natural rubber (hardness 45–60)	circular	3.5 mm diam.	1–4	—	The sealing forces are for room temp. (higher values), for warm (lower values)	Guthrie[491]
	circular	6.5 mm diam.	5.5–11			
	square	3.5×3.5 mm	2.5–8.5			
	square	6.5×6.5	5.5–14			
	square	10×10	11–35			
Rubber	circular	in V-groove	1.6–2.8	—		Garrod[430]
PTFE Teflon	flat	3 mm thick	—	123		Merkel[846]
		2.5 mm thick	—	140		
		1.6 mm thick	—	260		
		0.8 mm thick	—	450		
Indium	wire	1.6 mm dia.	min 5.9	—	In wire in trapezium groove	Adam[5]
	wire	0.8 mm dia.	7	—	To reduce dia. to 0.2 mm	Turner[1242]
	wire	1.5 mm dia.	56	—		Boulloud[153]
Lead	wire	1.5 mm dia.	300	—		Boulloud[153]
Gold					Corner seal	
	wire	0.25 mm dia.	500–100	—	Section 38.43	Munday[·06]
		0.5 mm dia.	280	Corner seal	Compressed to 0.25 mm	
	wire	0.2 mm dia.	430	Section 38.43	Compressed to 0.20 mm	Mark[803]
		0.75 mm dia.	300		Compressed to 0.37 mm	Lange[738]
		0.75 mm dia.	470		Compressed to 0.30 mm	
					Plane seal	
	wire	1.0 mm dia.	300		Section 38.42	Fischhoff[378]

		0.8 mm dia.	350		—	
	wire	1.5 mm dia.	410		—	Boulloud[153]
Copper	diamond	3 mm diagonal	500		see Fig. 3.121d	Hoch[568]
		2 mm dia.			Cu tube asbestos filled.	
	tube	0.3–0.5 mm wall	150	—	In coated	Bridge[167a]
		see Table				
	coined	3.19	300	—		Bridge[167a]
	knife	a = 1 mm see	600			
	edge	a = 0.5 mm Fig.	445	(annealed 290)		
	$\alpha = 30°$	a = 0.3 mm 3.143	375	(200 kg/cm)	Mongodin[879a]	
	knife					
	edge	round (Fig. 3.143)	300	(annealed 180)		Boulloud[153]
	shear	see Fig. 3.137	470	(annealed 295)		
	thin foil	0.04 mm	330		see Fig. 3.155a	
Copper OFHC	shear	see Fig. 3.137	375			Wheeler[1304]
	knife					
	edge	see Table 3.39	446			
	coined	see Fig. 3.121b	340			
	conflat	see Fig. 3.149	358			
Aluminium			730–	Cemented seal (section 38.46) to reduce dia.		Holden[578]
	wire	approx. 1 mm dia.	890	to 0.28 mm		Elsworth[839]
	wire	2.2 mm dia.	660	To reduce dia. to 1.5 mm		Redman[1036a]
	knife	see				
	edge	a = 1 mm Fig.	400			
	$\alpha = 30°$	a = 0.5 mm 3.143	260			Boulloud[153]
	knife					
	edge	round (Fig. 3.143)	210			
	shear	see Fig. 3.137	340			
	thin foil	0.2 mm	330	see Fig. 3.155a		
Iron (pure)	wire	1.5 mm dia.	1070	To reduce dia. to 1 mm		Peters[986a]

To obtain the deformation of the gasket it is subjected to a compression or shear force or to a mixture of them. In *compression* joints the effort is perpendicular to the supporting area (Fig. 3.42a). Compression seals using rub-

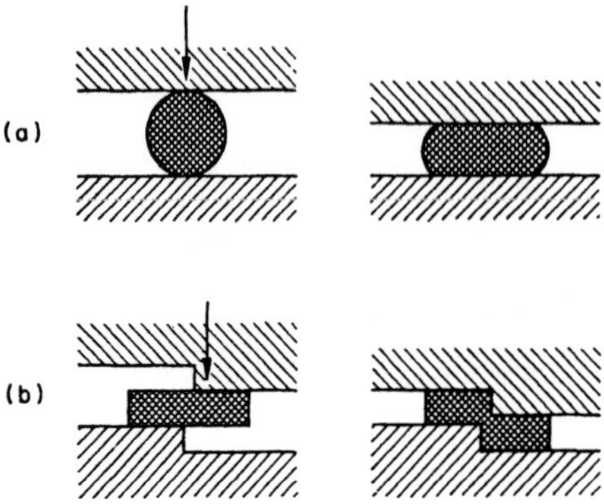

FIG. 3.42 Gasket under (a) compression; (b) shear

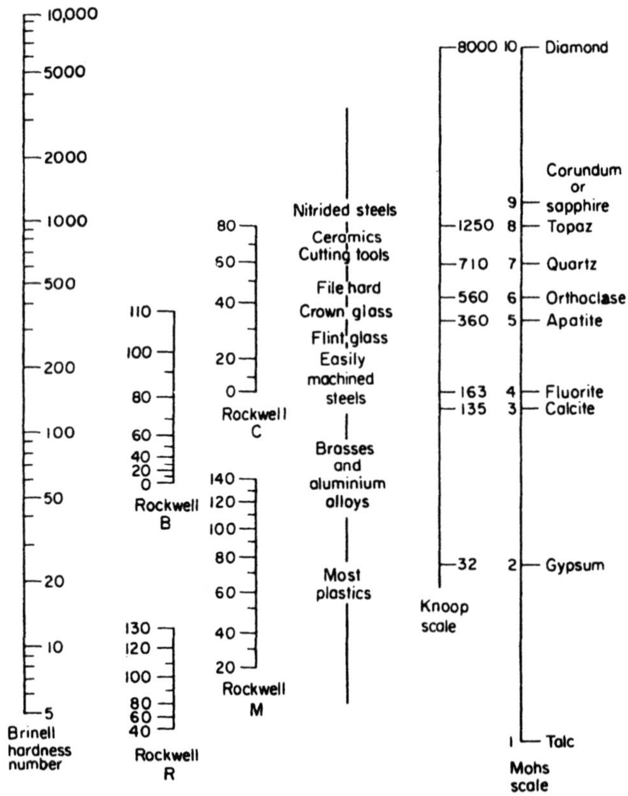

FIG. 3.43 Hardness scales for materials used in sealing techniques. *Adapted with permission from Kinney – Engineering Properties and Applications of Plastics, John Wiley & Sons, Inc. New York, 1957*

ber gaskets must be designed so that the total volume of the gasket remains constant, since *rubber is an incompressible material* i.e. the deformation of the rubber gasket in one direction is always compensated by an equal deformation in another direction (Fig. 3.59). When plastic or metal gaskets are used, wedge shaped flanges are used (Section 38.56) or the gaskets are of "coined" (Section 38.51) type.

In *shear* joints the effort is parallel to the supporting area (Fig. 3.42b); shear joints require theoretically less effort (than compression joints) for equal vacuum tightness. The possible deformation in shear seals, is smaller than under crushing. After a certain deformation any shearing joint becomes a compression seal, thus when a large deformation is required, the advantage of shear seals over compression seals is insignificant (Pierre[989]).

Gaskets should be *designed for sealing purposes only* and should not bear a mechanical load in order to align the parts of the seal. The mechanical alignment of the sealing parts should be made by separate stops provided on one or both parts (Guthrie[491]).

*38.2 Leaking mechanism in gasket seals**

Theoretically a gasket seal presents two different paths where a penetration of gas can occur from the outside into the evacuated system. These two ways are: *path a* (Fig. 3.44) through the space between the surface of the gasket and that of its seat (flange) and *path b* (Fig. 3.44) through the gasket material

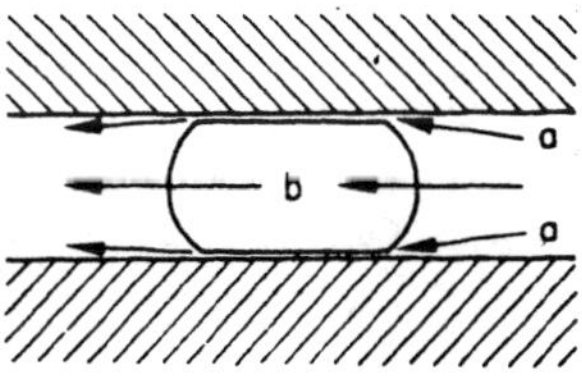

FIG. 3.44 Possible leak paths in gasket seals: (a) between the sealing surfaces; (b) through the gasket

* In a recently presented model of this mechanism, the sealing process is expressed by:

$$C = 1.93 \times 10^4 \sqrt{\frac{T}{M}} \frac{2\pi}{\ln (r_0/r_i)} \frac{A^2}{8.12} Ke^{-3\frac{P}{R}}$$

where C is the conductance (path a, Fig. 3.44) in cm^3/sec, T and M are the temperature (°K) and molecular weight of the gas, r_0 and r_i the outside and inside radii of the seal, A the peak-to-valley value of the surface roughness (cm), K a factor expressing the influence of imperfections (waviness, scratch, etc.), P the tightening pressure (kg/cm^2) and R the sealing factor (kg/cm^2). R—a new measure which expresses the sealing properties of the gasket material—is determined by using the Sealometer, a measuring apparatus constructed for this purpose. *See* A. ROTH and A. AMILANI, Sealing factors, their measurement and use in the design of vacuum gasket seals, *Trans. 3rd International Vac. Congress, Vol. 2*, Pergamon Press, 1966, and A. ROTH, Nomographic design of vacuum gasket seals, *Vacuum* **16**, 113 (1966).

itself (permeation). The existence of a leak through a *hole* inside the gasket is also possible, but its eventuality is small and must be considered as a rough fault.

Jordan[646] stated that the leak rates of satisfactory elastomer seals have been found to range between 6×10^{-5} and 3×10^{-8} lusec per centimetre of seal, depending on the type of elastomer and the temperature. Weitzel[1293] affirms that vacuum seals for cryogenic work should have a leak rate of less than 8×10^{-6} lusec/cm.

38.21 Leak through the seal. The leak rate through the space between the surface of the gasket and that of the flange depends — for a given gasket — on the pressure on the gasket and the surface finish of the flange. If the pressure is not high enough to force the gasket material into the scratches on the surface, the leak produced by these scratches may be important. A scratch of a quarter of a micron produces a leak of about 10^{-6} lusec (Henry[540]).

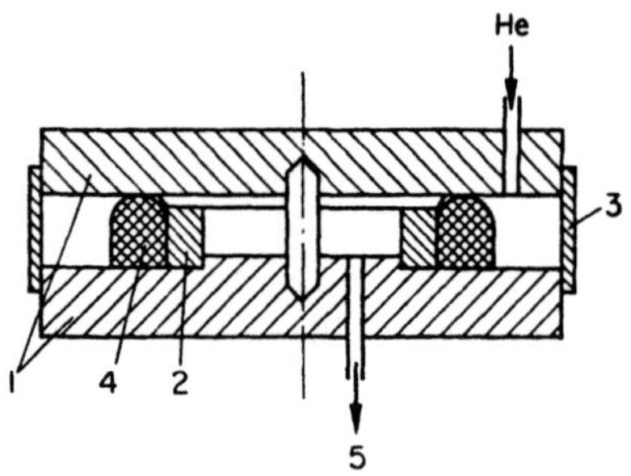

FIG. 3.45 Testing apparatus for the leak rate through a gasket seal: (1) gasket seats; (2) spacers; (3) sealing tape; (4) sample gasket; (5) connexion to leak detector. Reproduced from Kobayashi[703] *(Courtesy of Pergamon Press)*

Armand[49d] found that the conductance through the interface of two metallic surfaces which are in contact, is proportional to the square of the mean roughness of the surfaces, and varies as an exponential function of the force applied for pressing the sealing surfaces onto each other.

Kobayashi[703] published interesting results regarding the leaking mechanism of rubber gasket seals. The tests were done in an apparatus, shown schematically in Fig. 3.45. Two gasket seats (1) of mild steel were pressed together. For the various measurements, steel spacers (2) of different heights were used, allowing compression ratios (see Fig. 3.51) of the rubber gaskets from 5 to 25 per cent. Gaskets (4) of various kinds of rubber, having annular forms with circular, semicircular and square cross sections (72 mm diameter of the O-ring, 4 mm diameter or height of the cross section). By setting the sample gasket around the spacer and pressing the two seats together, the gasket was compressed to the height of the spacer used. The space inside the gasket was evacuated and connected (5) to a helium leak detector, and the space outside the gasket was filled with helium. This space was sealed outside with a tape (3). Exchanging the spacers, the leak rates for various compression ratios were measured. The compression ratios, where the leak rate becomes less than 10^{-6} lusec for various rubber hardnesses and gasket cross sections are plotted in Fig. 3.46. It was found that the relation between

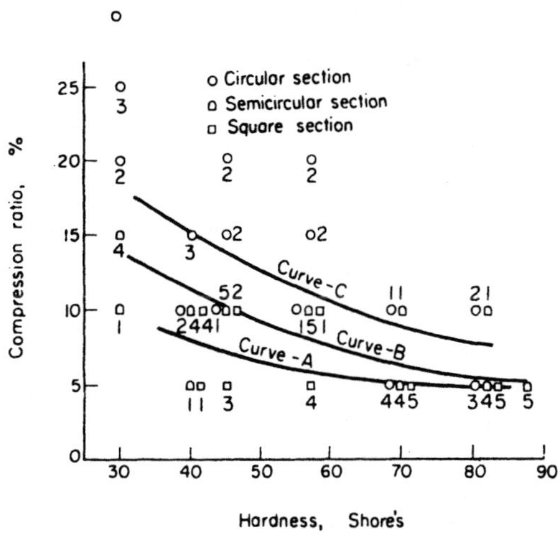

FIG. 3.46 Compression ratio vs. hardness for round, semi-circular, and square cross section rubber gaskets, required to get the leak rate smaller than 1×10^{-9} torr l./sec. Reproduced from Kobayashi[703] *(Courtesy of Pergamon Press)*

the compression ratio and the leak rate (through the gasket–flange surface) depends on both the *hardness* and the *shape* of the gasket and *does not depend* on the *kind* of rubber used. If the hardness of the rubber is greater than 50 (Shore) and the surface has no scratches in a radial direction, the leak rate becomes smaller than 10^{-6} lusec (tolerable) at a compression ratio of about 15 per cent, for any gasket shape (Fig. 3.46).

Among the various gasket shapes (Fig. 3.46), the square cross section gasket needs the minimum compression ratio and the circular one the maximum.

In seals using metal gaskets, the leak rate through the contact surface drops with the increase of the load (Armand[49]).

38.22 Leak through the gasket. It was found (Farkass[368]) that in the range of 10^{-9} torr the main gas load is the result of gas permeation through the elastomer sealants. The permeation being a function of the temperature (Section 12.2) and it was found that by reducing the temperature of the elastomer gasket to about -20 °C the pressure in the system could be reduced by a factor of ten.

Kobayashi[703] estlisabhed that with the usual rubber gaskets the leakage through the rubber itself (permeation) presents a great time-lag. With silicone rubber the permeation occurs more quickly. The results of permeation measurements, on square cross section gaskets (4×4 mm) are shown in Fig. 3.47. With the silicone rubber gasket, the permeation occurred within 1 min

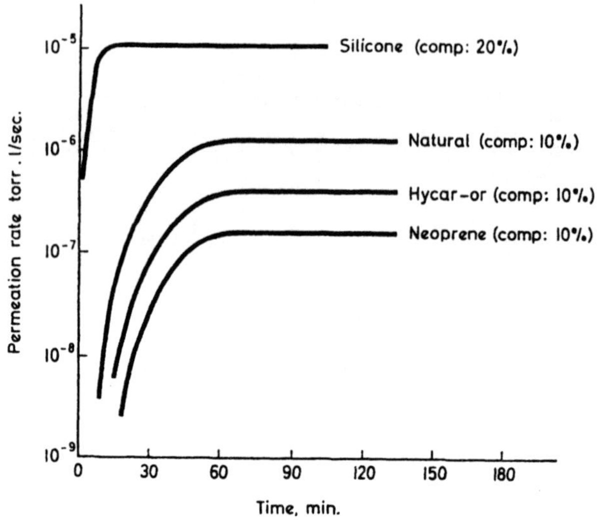

FIG. 3.47 Permeation rate as a function of time, through 4×4 mm square cross section rubber gaskets. Reproduced from Kobayashi[703] *(Courtesy of Pergamon Press)*

and saturated within 10 min; with other usual rubber gaskets the permeation occurs between 10–20 min and saturates between 70–120 min. For the values of gas permeation through gasket materials see Fig. 2.11.

38.23 Guard vacuum in the seals. The leak rate of gasket seals can be reduced using the technique known as *guard vacuum.* This technique consists of providing an enclosed evacuated space intermediate between the vacuum system (gasket) and the atmosphere (American Vacuum Society[26]). The guard vacuum is generally built as a double gasket system, but also double chambers have been constructed for this purpose.

Double gasket seals are useful not only to reduce the leak rate by guard vacuum, but also to quickly test the seal without evacuating the whole system, that is only the volume between the two gaskets. If the system is evacuated, the leak testing can be done easily by introducing a test gas in the volume between the gaskets (Kronberger[721], Peters[986a], Garrod[434], Avery[58]).

Sometimes the space between the two gaskets is filled with argon or some other inert gas (Willens[1316a]) or it may be used as a channel for a cooling liquid for the gasket (Farkass[367, 369]).

The leak rate through the seal depends on the conductance (Section 12.1) of the joint and the pressure difference across the seal. Thus *the guard vacuum reduces the leak rate by reducing the pressure difference across the seal.*

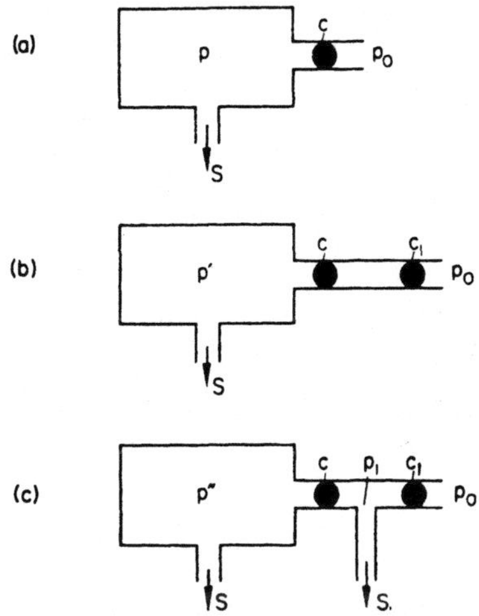

FIG. 3.48 Sealing principle of pumping systems: (a) with single gasket seal; (b) with double gasket seal; (c) with double gasket and guard vacuum

A gasket seal (Fig. 3.48a) with a conductance C, or leak rate $L = C(p_0-p)$ seals against the external pressure (usually atmospheric pressure). If a pumping speed S is used, in the vessel sealed with this gasket the lowest pressure which can be reached is p, the various values being related by the formula:

$$S \cdot p = L_C = C(p_0-p),$$

hence

$$p = \frac{C}{C+S}\, p_0.$$

If this gasket seal is doubled by a second one of conductance C_1 (Fig. 3.48b) the limiting pressure obtained with the same pumping speed S, will be

$$p' = \frac{C \cdot C_1}{(C+C_1)S+C \cdot C_1}\, p_0;$$

or with $C_1 = C$

$$p' = \frac{C}{2S+C}\, p_0 = \frac{S+C}{2S+C}\, p,$$

i.e. always $p > p' > p/2$. Thus the gain by this double gasket is not of great importance.

The limit pressure can be lowered much more if the space between the two gaskets is evacuated, i.e. a guard vacuum is used (Fig. 3.48c). If the pumping speed in the guard vacuum space is S_1 and the pressure p_1, the limiting pressure in the main vessel will be

$$p'' = \frac{C}{C+S} p_1 = \frac{C}{C+S} \cdot \frac{C_1}{C_1+S_1} p_0 = \frac{C_1}{C_1+S_1} p,$$

or

$$p'' = \left(\frac{p_1}{p_0}\right) p,$$

i.e. the pressure p'' in the vessel is reduced as compared with the pressure p (obtained with the simple gasket seal), by the same ratio as the pressure p_1 (in the guard vacuum space) is reduced compared with the outside pressure p_0. Considering the outside pressure $p_0 = 760$ torr and $p_1 = 10^{-1}$ torr (easily obtained with a rotary pump) the pressure inside the vessel will be:

$$p'' = 1.3\ 10^{-4} p.$$

On this basis ultra-high vacuum chambers with double walls were constructed (Ehlers[331], Kienel[672], Rivera[1067], Moll[871], Metcalfe[849]) consisting of an inner ultra-high vacuum chamber, surrounded by a second vacuum space with a guard vacuum. The limiting pressure which can be obtained in such a chamber is given by:

$$p'' = \frac{L_C + Q_d}{S} = \frac{C \cdot p_1 + Q_0 \cdot A}{C+S},$$

where L_C is the leak rate through the seal, Q_d is the gas load due to the outgassing, C the total conductance of the seal, p_1 the pressure in the guard vacuum space, Q_0 the specific outgassing rate (per unit of surface area), A the outgassing area, and S the pumping speed. Values of these factors are listed in Table 3.14 as obtained from the descriptions by various designers of double-walled ultra-high vacuum chambers.

It can be seen that for C, Q_d and S constant the relationship between p'' and p_1 is linear (Metcalfe[849]). Plotting p'' against p_1, a line with intercept a and slope b is obtained, knowing the values of a and b and the pumping speed S the total conductance of the seal C and the degassing rate Q_d can be calculated from

$$C = \frac{b}{1-b} S, \quad Q_d = \frac{a}{1-b} S.$$

TABLE 3.14. WORKING CONDITIONS IN VARIOUS DOUBLE-WALLED CHAMBERS

Outer chamber	Inner chamber	p_1 torr	A cm²	Q_0 torr l./sec /cm²	S lit./sec.	C lit./sec.	p'' torr	Reference
20 in. dia. 40 in. height	12 in. dia. 18 in. height	5×10^{-5}	6800	10^{-11}	200	7.5 10^{-5}	3.5 10^{-11}	Rivera[1067]
—	—	5×10^{-6}	—	—	100	2.0 10^{-3}	1.0 10^{-10}	Ehlers[331]
—	400 mm dia. 300 mm length	1×10^{-6}	—	—	100	1.0 10^{-1}	1.0 10^{-9}	Kienel[672]
48 in. dia.	32 in. dia. 16 ft³	5×10^{-5}	3.2 10^4	$2\times$ $\times10^{-12}$	750	1.0 10^{-4}	1.0 10^{-10}	Metcalfe[849]

The guard vacuum may be used between O-rings of square, circular or semi-circular cross section (Fig. 3.49a, b, c). A pump-out connexion is provided in one of the flanges, having one or more connexions to the space between the two gaskets (Guthrie[491], Scott[1114], Holland-Merten[587], Bridge[167a], Heywood[554]). A conical double O-ring seal with guard vacuum is shown in Fig. 3.90b (Schriever[1108]).

For case of assembly, the double O-ring can be made as a single moulding (Fig. 3.49d), the cross section of the O ring being in the form of a dumb-bell (Section 38.54) and the space between the two sealing parts being evacuated (Kronberger[721], Thomas[1220]). Seals with guard vacuum were also made by evacuating the space between the two edges (Fig. 3.49e) of a copper gasket (Heathcote[526]) or between concentric knife edges (Fig. 3.146b, van Heerden[528]).

Guard vacuum is used also in sliding or rotating seals. Brueschke[177] described such a seal (Fig. 3.30c) using a liquid column and Lake[729a] shows a double piston sliding seal using oil sealing and a guard vacuum between the pistons (Fig. 5.21b). Motion seals using guard vacuum are described in Section 51.8.

Hickman[557a] describes a guard vacuum seal using an inner gold wire (0.5 mm) gasket and an outer rubber gasket (1.5 mm thick sheet). When closing the seal the rubber gasket seals before the gold gasket. Further compression of the rubber gasket permits the making of the high vacuum, gold gasket seal.

Rivera[1067] constructed a double ultra-high vacuum chamber using differential pumping (guard vacuum). The *outer* chamber was constructed of stainless steel (20 in. diameter and 40 in. high). The chamber was provided with internal nichrome heaters and stainless steel radiation shields, and wrapped outside with (copper) water-cooling coils. Two (3 in.) sight ports, one for the top and one for the side, were provided and sealed with Viton A (Section 38.3) gaskets. This outer chamber was hoisted over the inner chamber. The *inner* chamber was 12 in. in diameter and 18 in. high, also constructed of stainless

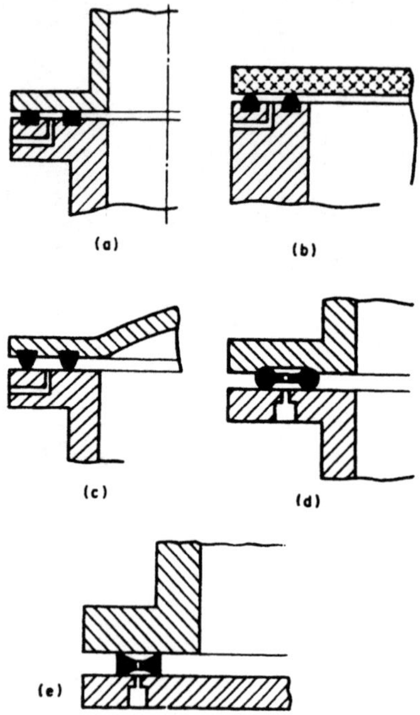

FIG. 3.49 Gasket seals using guard vacuum: (a) with square cross section gaskets; (b) with circular cross section gasket; (c) with semi-circular cross section gaskets; (d) with dumb-bell gasket; (e) with specially shaped copper ring (Fig. 3.136)

steel, wrapped with water-cooling coils and similarly provided with 3 in. sight ports on corresponding portions as on the outer chamber. The inner chamber was provided with a modified knife edge seal (Section 38.56b) of 13 in. diameter having a land 0.03 in. wide, held by clamps on an annealed copper gasket, leaving a gap of about 0.01 mils (0.25 μ). The pumping conditions are given in Table 3.14; the system allows a bakeout to about 490 °C.

Ehlers[331] and Adam[3] describe a system consisting of a thin-walled container built inside a high vacuum tank. The inside container is evacuated to very low pressures (Table 3.14) without having high vacuum seals, with the aid of the guard vacuum maintained in the tank (Table 3.14). The mechanical stress on the inner container is very low due to the low pressure gradient between the outer and inner chambers. Thus this chamber can have a very thin wall, which permits easy heating.

Metcalfe[849] describes a 455 l. double-walled ultra-high vacuum chamber. For performance data, see Table 3.14. Kienel[672] constructed a double-walled chamber where the outer chamber was sealed with rubber gaskets, but *no* gaskets were used for the inner chamber. The seals on this chamber were assured simply by the impedance of the closure (Table 3.14). The thin-walled inner chamber may be baked to 450 °C by the direct flow of current through the wall.

From the review of the double chamber systems, some conclusions can be summarized as follows:

(1) The double chamber with guard vacuum is an easy way to make the seals for ultra-high vacuum, especially in the applications where the chamber must often be opened and resealed.

(2) The inside chamber may be thin-walled since it is not exposed to great pressure differences, hence this chamber can be easily heated.

(3) The inner chamber can be sealed simply by impedance seals e.g. flat flanges (8 μ in finish) mating on 25–30 mm width (Metcalfe[840], Kienel[672]).

(4) If thin slits are to be avoided, as their degassing is difficult, it is better (Moll[871]) to provide the inner chamber with normal metal gaskets, without deforming them as required for sealing against atmospheric pressure.

38.3 Gaskets; materials and shapes

The gaskets used in vacuum sealing are made of rubber, plastics or metals. The general properties of these materials with regard to their use in vacuum systems are discussed in Section 2.1. Here it is intended to discuss the properties of these materials which effect their sealing action, or preclude their use in particular applications. Tables 3.16 and 3.17 list these properties and their effect on the design of the various seals, as well as the references to the literature, where the use of various gaskets is quoted.

38.31 Rubbers for gaskets. The main property which determines the extensive use of rubbers as gaskets is their *elasticity.* Other properties as degassing rate (Table 2.7), permeation to gases (Fig. 2.11), the permanent set at higher temperatures (Fig. 3.52), or their brittleness at low temperatures (Table 3.15) limit the range in which the rubbers can be used as gasket materials.

Four main types of rubber are used as gasket material: natural rubber, nitril, Neoprene, and silicone rubber. Recently a rubber-like material known as Viton was added to this list.

Natural rubber is a very clean material but it has a high permeability, a narrow temperature range, and a chemical instability, especially when exposed to light. It adheres to the surfaces it is in contact with and is difficult to remove. Nitrile rubbers (Buna N, Perbunan, Hycar) and Neoprenes have a much lower permeability than natural or silicone rubbers, but their gas evolution and temperature range do not permit their use in systems requiring low or high temperatures. Silicone rubbers have a wide temperature range, but their permeability for gases is very high. Viton* (A and B) has a lower permeability than any other rubber, its gas evolution is also the lowest, and can be used up to about 200 °C.

To obtain a vacuum-tight seal, the gasket must be compressed to a given ratio of its initial height. This compression ratio varies according to the type of rubber, the degree of leak tightness desired, and the shape of the seal. Figure 3.50 shows the values of the compression ratio of rubber gaskets as a function

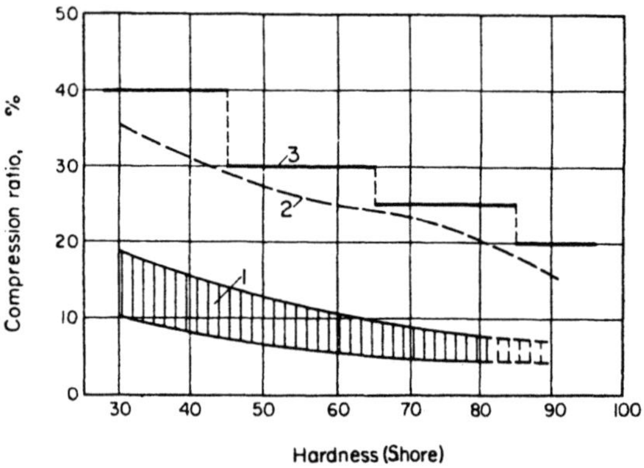

FIG. 3.50 Compression ratio of rubbers as a function of their durometer (Shore) hardness: (1) minimum compression ratio needed for a vacuum-tight seal (Fig. 3.46); (2) permissible compression ratio (Guthrie[491]); (3) maximum compression ratio for compression set test A.S.T.M. 395–49T (Fig. 3.51)

of the hardness of the rubber used. For Nygon (NGN[926]) a compression ratio of about 30 per cent is recommended. The compression of rubber gaskets results in temporary and/or permanent deformation, but according to Biram[128] the compression also increases the gas (sulphur) evolution from the gasket.

Rubber gaskets subjected to compressive deformation for some time, develop a residual deformation, known as *compression set*. While loading conditions and the resulting deformations of the gaskets vary from seal to seal, the quality of the rubbers in regard to their compression set is defined for standard loading and deformation conditions (A.S.T.M. 395–49 T). The test is made either with a *constant load* or a *constant deflection*, using cylindrical rubber probes having a 12 mm height and a 28 mm diameter.

* Viton (trademark of E. I. du Pont de Nemours Co.) is a copolymer of hexafluoropropene and vinylidene fluoride; its permeability for helium is equal to that of Neoprene, but for other gases it is lower by at least a factor of 4.

In the *constant load test* (Fig. 3.51a) the sample, having an initial height d_0 is compressed by a constant load p during 22 hr (or 70 hr) and kept during this time at a temperature of 70 or 180 °C. Due to the load, at the end of

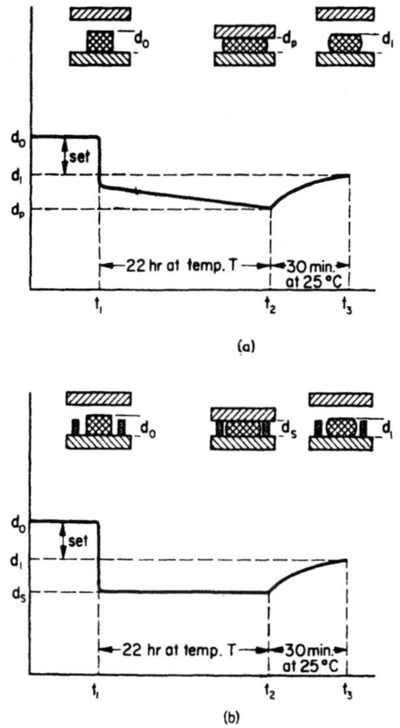

FIG. 3.51 Compression set test for rubbers (A.S.T.M. 395-49T): (a) with a constant load; (b) with a constant deflection

this period the height of the sample will be d_p. Then 30 min after the moment (t_2) when the load is removed and the probe cooled to 25 °C, its height will be d_1. The compression set of this rubber sample is then defined as

$$C_p = 100\,\frac{d_0 - d_1}{d_0}\ (\%).$$

In the *constant deflection test* (Fig. 3.51b) the sample is compressed to a given compression ratio $K = \dfrac{d_0 - d_s}{d_0}$ (not greater than the values given by curve 3, Fig. 3.50), and kept in this position for a period (22 hr) at a given temperature. Then, 30 min after the removal of the load (and cooling to 25 °C) the height of the sample will be d_1. The compression set is then given by:

$$C_K = \frac{C_p}{K} = 100\,\frac{d_0 - d_1}{d_0 - d_s}\ (\%).$$

The compression set C_K of various rubbers at various test temperatures varies according to the curves on Fig. 3.52. It can be seen that at room temperature the compression set of nitrile, butyle and silicone rubbers is less than 8 per cent. Nygon has a compression set of less than 7 per cent (NGN[926]). Chloroprene (Neoprene) rubbers have at room temperature a set of about 30 per cent. At low temperatures as well as at higher temperatures the compression set is higher (Fig. 3.52); silicone rubber (Gale[420]) has the largest range of temperatures with a low compression set.

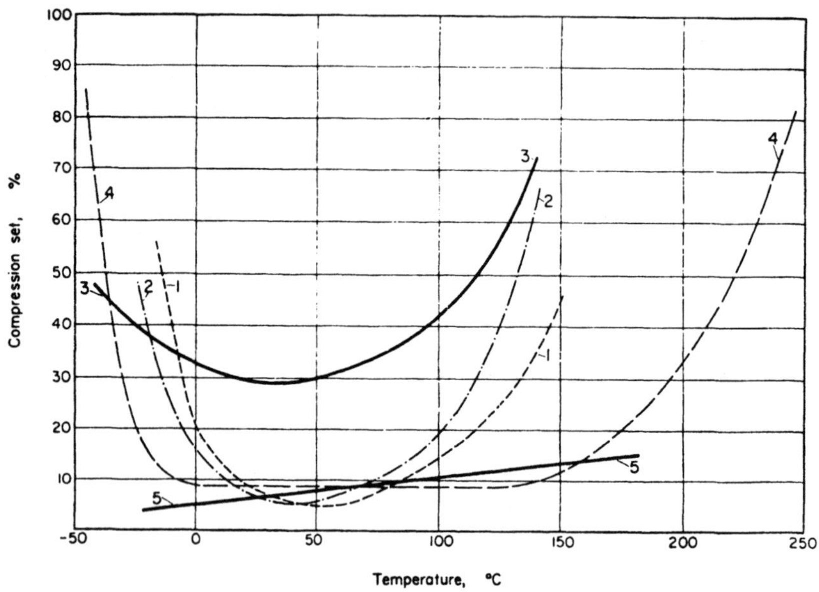

FIG. 3.52 Compression set of rubbers after 22 hr at various temperatures: (1) nitrile rubber (Wacker[1276]); (2) butyl rubber; (3) chloroprene rubber; (4) silicone rubber (Wick[1311]); (5) silicone rubber R 20 V (Wacker[1276])

If in a real seal the gasket is compressed between two surfaces (as in the compression set test) the stress set up will continue to change indefinitely owing to both the set and to gradual alterations in load caused by surface slippage (Guthrie[491]). The overall effect is called "creep". The greasing of the gaskets increases their tendency of slipping on the sealing surfaces, making it difficult to obtain the required sealing pressure (Section 38.48).

Rubber gaskets can be only heated up to moderate temperatures (Table 3.15) excepting silicone rubbers and Viton which permit heating to higher temperatures but still without reaching the temperatures required for bakeout (about 400 °C).

For thermal degradation of rubbers in vacuum see Strauss[1202].

TABLE 3.15. SERVICE TEMPERATURES OF ELASTOMER GASKETS

Gasket material	Service temperature range (°C)				Reference
	Permanent		Peak		
	min.	max.	min.	max.	
Natural rubber	−30	60	−65	75	Espe[350], Beske[118]
Styrene, Buna S	−25	75	−55	100	Fabian[364]
Butyl rubber, Isoprene	–	–	−55	150	Fabian[364]
Nitrile rubber, Buna N, Perbunan Hycar	−25	85	−50	150	Weitzel[1293]
Chloroprene, Neoprene	–	–	−50	120	Catton[198], Fabian[364]
Silicone rubber	–	–	−120	300	Wacker[1276], Nowak[948] Midland[856], Gondet[457a] Nitzsche[933]
Viton	–	150	–	260	Addis[9], Fabian[364]
Nygon	−50	(150)	–	–	NGN[926]
Teflon	−190	280	–	400	Riemerstma[1059]

38.32 Fluocarbons for gaskets. Fluocarbons are available under various trade names (PTFE, Teflon, Floun, Kel–F, Hostaflon (Table 2.14). The main physical properties of these materials are summarized in Appendix B; for further information see Cornell[238], Garlock[426], Horn[595], Schulz[1110], Kirby[677], Mehnert[841].

Teflon gaskets generally need higher sealing pressures (Table 3.13) than rubber gaskets. In order to obtain these pressures, the grooves are made narrower and on the flanges, ridges (Section 38.56) or small concentric or parallel grooves (Fig. 3.54a) are provided. According to Hearst[525] the compression ratio of Kel–F O-rings should not exceed 7 per cent.

An indentation on Teflon disapears after the load is removed. Riemerstma[1059] observed this effect after a bakeout to 330 °C, and affirms that this effect explains why in Teflon gasket seals it is not necessary to increase the closing torque over that required for the specified compression ratio.

Fluocarbon O-rings are not elastic and tend to become flat with use; to avoid this, the gasket is made elastic by building it as a Teflon clad over an

TABLE 3.16. REFERENCES ON THE USE OF GASKETS IN VACUUM SEALS

Gasket	Seal		
	O-ring Section 38.4	Thick gasket Section 38.5	Thin gasket Section 38.6
Rubber (Natural Nitrite, Neoprene)	Mooz[885], Elonka[338], Beske[118], Farkass[367], Tipton[1225], Evans[362], Weitzel[1293], Heywood[554], Holland[584], Garton[435], Stringer[1203], Gerow[445], Reilly[1044], Michelson[855], Juanigot[647a] Carson[195a]	Fabian[364] Strong[1206], Rose[1076]	–
Silicone rubber	Gondet[457a] Adams[6]	Mackinnon[793], Rose[1077, 1078], Gale[420]	–
Viton	Addis[9], Holland[584] Rivera[1067]	Beske[117]	–
PTFE Teflon	Hearst[525] Hart[517] de Vries[1273]	Nester[920], Cloud[218], Deruytter[275] Champeix[203], Nier[929] Giaimo[446], Billett[124] Riemerstma[1060]	Rueger[1091] Davies[262] Hintenberger[564]
Gold	Hoch[568], Johnston[642], Fischhoff[378], Ullman[1250] Klopfer[690] Bishop[129]	Munday[906] Mark[803]	–
Indium	Henry[540], Turner[1242], Horwitz[598], Bishop[129], Bride[167a], Fraser[400], Adam[5], Nelson[916], Fischhoff[378]	–	Martin[815] Reynolds[1049] Knudsen[700]
Aluminium	Holden[578], Henry[540], Hoch[568], Elsworth[339], Spees[1167], Heywood[555], Nelson[916]	Henry[540], Bleakney[132], Baldock[75], Higatsberger[559]	Henry[540] Hintenberger[564]
Copper	Henry[540], Hoch[568], Guthrie[491], van Heerden[527]	Neher[914], Mann[800], Hoch [568], van Heerden[527], Heathcote[526], Hees[529], Wheeler[1304], Lange[737], Pattee[975], Robinson[1073]	Paul[980] Hintenberger[564] Steckelmacher[1180] Brymner[179]
Nickel	Hoch[568], Heywood[555]	–	Brymner[179] Steckelmacher[1180]

(Table 3.16 Continued)

Gasket	Seal		
	O-ring Section 38.4	Thick gasket Section 38.5	Thin gasket Section 38.6
Monel	Hoch[568]	—	—
Silver	Hoch[568]	Drawin[303]	Paul[980] Hintenberger[564]
Iron (steel)	Hoch[568]	Carpenter[195]	Steckelmacher[1180] (stainless steel)
Lead (Tin–lead)	Bishop[129], Henry[540], Craig[243a], Green[478], Bridge[167a]	Elonka[338], Cloud[218]	—

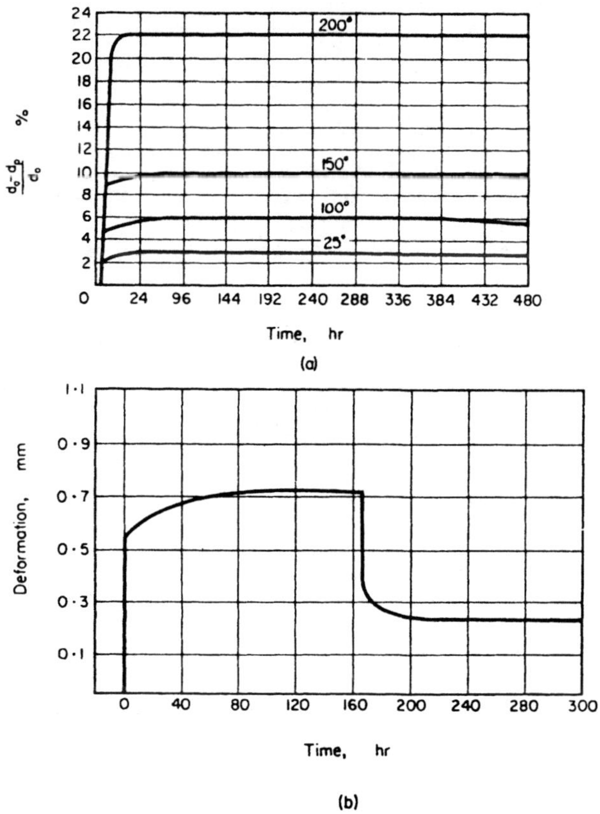

FIG. 3.53 Deformation of Teflon: (a) compression set by a load of 70 kg/cm^2; (b) deformation during and after loading by 70 kg/cm^2, for 160 hr at 25 °C (Wacker[1276])

TABLE 3.17. CHARACTERISTIC FEATURES OF GASKETS AND THEIR EFFECT FOR USE IN VACUUM SEALS

Feature	Rubber gaskets	Plastic gaskets	Metal gaskets
Compressibility	Incompressible; the seal should allow for its constant volume deformation.	Compressible; higher local pressures required; flanges with ridges indicated.	Compressible; Very high pressures required; sealing surfaces should be kept to a minimum.
Elasticity	Very elastic; the seal requires less pressure and maintains it.	Inelastic; the pressure on the seal must be often increased, bolts retightened or elastic (spring) loading used. Equal torque on all the bolts required.	
Permanent set	Small permanent set if the compression ratio is limited by flange–flange contact. Grooves or spacers required; gasket may be re-used.	Cold flow occurs under sealing load; small compression ratio used; to be used without flange–flange contact. Gaskets may be re-used after annealing.	Hardens under load. Gasket cannot be re-used many times.
Hardness	Very soft; the seal does not require very high surface finish of flanges.	Soft; The seal is based on parallel grooves and ridges on surface of the flanges.	Harder than other gaskets. Very good surface finish of flanges required.
Bonding	Easy to bond to itself or other materials; gaskets can be made from cord.	Teflon very difficult to bond to itself or other materials; gasket should be cut from a single piece.	Cold welding or welding and brazing may be used.
Outgassing	High; the surface exposed to vacuum should be kept minimum; heating limited.	Teflon has very low vapour pressure.	Very low vapour pressure even at high temperatures.
Permeability	Fair, excepting silicone rubber.	Teflon has a very low permeability.	Very low.
High temperature behaviour	Use limited to moderate temperatures; hardening accelerated by temperature.	Fair, especially Teflon. High expansion; the seal should be tightened at lowest service temperature.	Some metal gaskets may be used for bakeable seals. Seal should allow for differential expansion.

(Table 3.17 Continued)

Feature	Rubber gaskets	Plastic gaskets	Metal gaskets
Low temperature behaviour	Becomes brittle; cannot be used at low temperatures.	Teflon can be used at low temperatures; seal should be tightened at low temperature.	May be used at low temp. if differential expansion allowed for
Chemical behaviour	Resistant to oils and mercury.	Teflon is chemically inert.	Some oxidize, others are attacked by vapours (mercury).

elastic core* (rubber, spring), or the seal is constructed so as to maintain the pressure (e.g. with spring washers under the bolt heads). The deformation of Teflon is shown in Fig. 3.53.

The temperature range in which the fluocarbon gaskets may be used is wide (Table 3.15). Riemerstma[1059] established that the maximum temperature for systems containing Teflon is 425 °C, since at this temperature the material permanently changes from being relatively dense to porous in nature, and its volume grows by a large factor.

The gas permeation through fluocarbons is very small (Fig. 2.11).

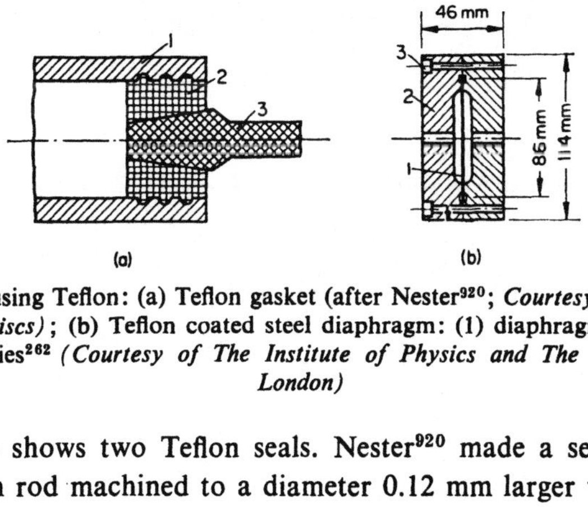

FIG. 3.54 Seals using Teflon: (a) Teflon gasket (after Nester[920]; *Courtesy of The American Institute of Physiscs)*; (b) Teflon coated steel diaphragm: (1) diaphragm; (2) flange; (3) bolts. After Davies[262] *(Courtesy of The Institute of Physics and The Physical Society, London)*

Figure 3.54 shows two Teflon seals. Nester[920] made a seal (Fig. 3.54a) using a Teflon rod machined to a diameter 0.12 mm larger than the inside diameter of the metal pipe (1) into which it is to be fitted. The Teflon rod (2) was then cooled in ice and inserted into the end of the metal pipe. The seal can be made even more reliable if the rod is machined to a larger diameter (about 0.25 mm depending on the diameter) and inserted in the metal pipe

* Such O-rings are supplied e.g. by Tanner Engineering Co. 22624 Avalon Boul., Wilmington, Calif., U.S.A.; M. Merkel Werke, Sanitasstrasse 17–21, Hamburg-Wilhelmsburg 1, Germany.

after cooling in dry ice. On warming the Teflon expands and gives a vacuum-tight joint, by filling up the grooves machined inside the metal pipe. In the Teflon plug, a tapered glass pipe 3 can be inserted. In order to dismantle the seal it must be cooled.

In order to avoid the difficulties arising in sealing thin Teflon diaphragms Davies[262] succeeded in making a seal (Fig. 3.54b) with a steel disc cleaned with abrasive, cathodically etched and washed coated with Teflon Dispersion 30 (from Lab. Apparatus and Glass Blowing Co., Manchester). After drying and sintering he obtained a tenacious Teflon coating of about 6 μ. The coated disc was then sealed between lapped flanges (Fig. 3.54b).

Fluocarbons are very difficult to bond to themselves (bonding method see Section 32.2). Thus if the opening to be sealed requires a gasket larger than can be cut from a single sheet, it is almost impossible to make a satisfactory vacuum-tight joint.

38.33 Metals for gaskets. The metals used for gasket seals are mainly gold, aluminium, copper, as well as indium, lead and rarely nickel, silver, iron, or other metals.

Gold has certain advantages as a gasket material. It does not oxidize when it is heated and its yield point is low (lower than e.g. that of aluminium). On the other hand, gold becomes much harder when it is deformed, the increased strength requires an increased force and the gasket is deformed and becomes useless. Gold gasket seals are quite expensive, but not unreasonable due to the high salvage value of the gold, which is about 75 per cent of the initial value (Caswell[197a]).

To obtain a vacuum-tight seal with gold gaskets, they have generally to be compressed to 40–50 per cent of their original cross section diameter (Lange[738], Mark[803]).

Gold gaskets are mainly made from wire. Ullman[1250] used a 0.5 mm diameter wire of 24 K gold (99.7 per cent). Caswell[197a] described the difficulties encountered with gold wire seals due to the imperfections (e.g. bits of included steel) in the drawn wire. Munday[906] recommends the use of gold gasket seals especially for non-circular openings.

The gold gasket can be made from wire by fusion butt welding the ends using a torch in air. Subsequent annealing to a red heat (for several seconds) is important (Ullman[1250], Caswell[197a]).

Aluminium is used as wire or as a thin sheet. Its advantage lies especially in its low cost.

A simple method of making an aluminium O-ring is to twist the ends of an appropriate length of wire and then to clip off the end of the twist. The short twisted end is covered with aluminium brazing flux and fused with a torch. A bead forms but a small amount of molten aluminium flows down between the wires by capillary action and forms a short fillet. The wire is then untwisted and the bead clipped off (Spees[1167]).

Holland[582] developed a sealing technique which relies on the diffusion of the aluminium into the flange through cracks in the oxide layer on the surface of the aluminium (Section 38.46). Aluminium gasket seals can be used up to about 400 °C.

Copper gasket seals may be used up to about 800 °C. A disadvantage of copper as a gasket material consists in the fact that it oxidizes readily when heated in air. The copper oxide evaporates and condenses on the faces of the flanges (near the seal) which them periodically have to be cleaned.

Copper is used as an O-rings made of wire, or as thick (Section 38.5) or thin (Section 38.6) gaskets of various shapes. Van Heerden[527] constructed copper O-rings from 1.5 mm diameter OFHC copper wire. The wire was bent in a circle on a jig with the ends sticking up. The ends were then welded electrically in a hydrogen atmosphere and the bead thus formed filed down and smoothed with emery paper. Finally the O-ring was annealed in a hydrogen atmosphere at 950 °C. In order to obtain a good seal, the use of OFHC copper and the treatment in hydrogen are essential.

Bridge[167a] described a seal using an O-ring made from a copper tube coated with indium and filled with asbestos, which prevented the tube from collapsing (tube diameter 2 mm, wall thickness 0.3–0.5 mm, O-ring diameter approximately 35 mm). He used this seal, because the force needed to make a tight seal with such a gasket is much less than when using copper wire (Table 3.13).

In shear or ridge seals (Sections 38.55, 38.56) copper gaskets made of OFHC copper (about 1 mm thick plate) and annealed in hydrogen at 950 °C are recommended (van Heerden[527], Lange[737]).

Indium is used as a gasket, generally in wire form (0.7–1.5 mm diameter), flattened between flanges to a thickness equal to 15–25 per cent of its initial

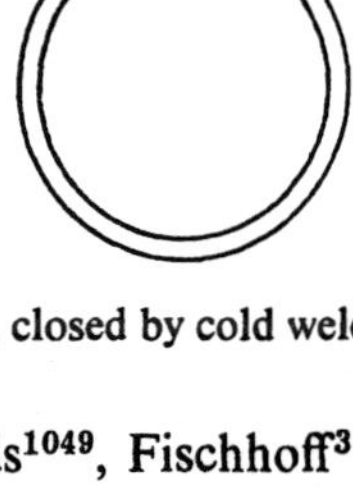

FIG. 3.55 Indium gasket, closed by cold welding the overlapping ends

diameter (Turner[1242], Reynolds[1049], Fischhoff[378]). The gasket is not a proper O-ring, but usually just an indium wire with overlapped ends (Fig. 3.55), which is sealed in the joint by cold welding of the indium, when the gasket is compressed between the flanges. Proper indium O-rings can be made by

TABLE 3.18. GASKET CROSS SECTION SHAPES

No.	Shape	Gasket (material)	Dimensions and reference	Reference
1.		O-ring Elastomers, metals	Table 3.19A Section 38.4	
2.		Metals, Teflon (with core)		Advanced[9a] Bridge[167a]
3.		Elastomer	Table 3.19B	
4.		Elastomer, with elastomer or metal web	Section 38.54 dumb-bell	
5.		Elastomer	Table 3.19B	
6.		Copper	0.015 × 0.015 × × 0.004	Lindsay[770]
7.		Copper	See Fig. 3.121a	Milleron[865]
8.		Diamond seal Copper	3 mm apex-to-apex See Fig. 3.121c	Hoch[568] Turnbull[1240a]
9.		Copper	10 × 4 mm Section 72.9	Bridge[167a]
10.		Copper	Table 3.39 and Fig. 3.137	
11.		Copper (indium coated)	Fig. 3.145a	Reynolds[1049]
12.		Trapezium Elastomer	Table 3.19C	
13.		Elastomer (metal protection)	Fig. 3.122	
14.		L-gasket Elastomer	Section 38.51c	

(Table 3.18 Continued)

No.	Shape	Gasket (material)	Dimensions and reference	Reference
15.		Copper	Fig. 3.144e	Mann[800]
16.		Copper	8 × 1.5 mm	Heathcote[526]
17.		Conflat seal Copper	Fig.3.149	
18.		Coined gasket Copper	Fig. 3.121b	

cutting the ends of an indium wire square and joining them (Fraser[400]) in a small flame with the aid of a suitable flux (Adam[5]).

Indium gaskets can be used up to 150 °C and for ultimate pressures not lower than 10^{-8} torr.

Knudsen[700] and Martin[815] mentioned the use of indium gaskets, cut from thin (0.18 mm) sheet. These gaskets make an efficient seal between metal and other surfaces but in opening the seal the gasket is destroyed. Since the tightness of the seal depends on the indium being made to flow slightly during the compression, it is important to cut the gasket narrow enough (e.g. 1.5 mm is a reasonable width). Reynolds[1049] used indium plated on the flanges, and Bridge[167a] coated his copper O-ring (Table 3.13) with indium. Power[1008] made a cemented seal (Section 38.46) using indium-coated aluminium wire-gaskets.

Bridge[167a] points out the advantage of using indium gaskets for seals at low temperature. Indium retains a high plasticity at this temperatures, so that it does not require such a high sealing pressure as lead gaskets.

Lead is rarely used as gasket material for vacuum seals. Green[478] describes a vacuum joint using a lead gasket made of 1.6 mm wire compressed between glass flanges. Dushman[314] mentions a Knudsen gauge (constructed by Lockenvitz*) sealed with lead washers. Craig[243a] used a lead–tin (50/50) alloy O-ring of about 1.5 mm cross section diameter to seal thin metal windows (see also Section 72.9).

Silver gaskets were used by Drawin[303].

Stainless steel is used by Advanced Prod. Co.[9a] to construct metal O-rings made of thin walled tube; their sealing action is based on the elasticity of the construction (Wittrick[1327a]).

* *Rev. Sci. Instr.* **9**, 417 (1938).

TABLE 3.19A. STANDARD AND COMMON DIMENSIONS OF ELASTOMER O-RINGS

Cross section diameter d (mm)	Inside diameter D (mm)	Nominal designation (inches) i.d. o.d. cross section diameter			Designation or code*										
		i.d.	o.d.	cross section	B.S.	G	E	A.S.	P.R.	G.S.	L	B	P	F.S.	JS
1.02 ± 0.07	0.74 ± 0.1	1/32	3/32	1/32	—	—	—	—	001	—	—	—	—	—	—
1.27 ± 0.07	1.07 ± 0.1	3/64	9/64	3/64	—	—	—	—	002	—	—	—	—	—	—
1.52 ± 0.07	1.42 ± 0.1	1/16	3/16	1/16	—	—	—	—	003	—	—	—	—	—	—
1.78 ± 0.07	1.78 ± 0.13	5/64	13/64	1/16	—	—	—	—	004	—	—	—	—	—	—
	2.54 ± 0.13	3/32	7/32	1/16	—	—	—	—	005	—	—	—	—	—	—
	2.90 ± 0.12	1/8	1/4	1/16	OS 1	R 101	VOR 101	1	006	—	—	—	—		—
	3.70	5/32	9/32	1/16	2	102	—	2	007	—	—	—	—	—	—
	4.47	3/16	5/16	1/16	3	103	—	3	008	—	—	—	—	—	—
	5.28	7/32	11/32	1/16	4	104	104	4	009	—	—	—	—	—	—
	6.07	1/4	3/8	1/16	5	105	105	5	010	—	—	—	—	—	—
	6.75	—	—	—	—	106	—	—	—	—	—	—	—	—	—
	7.65	5/16	7/16	1/16	6	107	107	6	011	—	—	—	—	—	—
	8.73	—	—	—	—	108	—	—	—	—	—	—	—	—	—

	9.25	3/8	1/2	1/16	7	110	110	7	012	—	—	—	—	—	—
	11.10	—	—	—	—	114	114	—	—	—	—	—	—	—	—
2.62 ± 0.07	4.74 ± 0.12	—	—	—	—	—	2A	—	—	—	—	—	—	—	—
	7.92	—	—	—	—	—	4A	—	—	—	—	—	—	—	—
	9.13	—	—	—	—	109	—	—	—	—	—	—	—	—	—
	9.20	3/8	9/16	3/32	8	111	—	8	110	—	—	—	—	—	—
	9.92	—	—	—	—	112	—	—	—	—	—	—	—	—	—
	10.78	7/16	5/8	3/32	9	113	113	9	111	—	—	—	—	—	—
	11.91	—	—	—	—	115	—	—	—	—	—	—	—	—	—
	12.37	1/2	11/16	3/32	10	116	116	10	112	—	—	—	—	—	—
	13.10	—	—	—	—	117	—	—	—	—	—	—	—	—	—
	13.95	9/16	3/4	3/32	11	118	118	11	113	—	—	—	—	—	—
	15.08	—	—	—	—	119	—	—	—	—	—	—	—	12	—
	15.55	5/8	13/16	3/32	12	120	120	12	114	—	—	—	—	—	—
	15.88	—	—	—	—	121	121	—	—	—	—	—	—	—	—
	17.12	11/16	7/8	3/32	13	122	122	13	115	—	—	—	—	—	—
	17.86	—	—	—	—	123	—	—	—	—	—	—	—	—	—
	18.75	3/4	15/16	3/32	14	124	124	14	116	—	—	—	—	—	—
	20.24	13/16	1	3/32	—	127	—	—	117	—	—	—	—	—	—
	20.63	—	—	—	—	128	—	—	—	—	—	—	—	—	—
	22.22	—	—	—	—	130	130	—	—	—	—	—	—	—	—
	23.81	—	—	—	—	132	132	—	—	—	—	—	—	—	—
3.0	7.0		—		—	—	—	—	—	—	—	11	—	—	—
	10.0		—		—	—	—	—	—	—	10/3	12	—	—	—
	14.0		—		—	—	—	—	—	—	14/3	13	—	—	—
	19.0		—		—	—	—	—	—	—	19/3	—	—	—	—
	20.0		—		—	—	—	—	—	—	20/3	—	—	—	—
	29.0		—		—	—	—	—	—	—	29/3	—	—	—	—
	38.0		—		—	—	—	—	—	—	38/3	—	—	—	—

(Table 3.19A Continued)

Cross section diameter d (mm)	Inside diameter D (mm)	Nominal designation (inches) i.d. o.d. cross section diameter			Designation or code*										
					B.S.	G	E	A.S.	P.R.	G.S.	L	B	P	F.S.	JS
3.0	55.0		—		—	—	—	—	—	—	55/3	—	—	—	—
	59.0		—		—	—	—	—	—	—	59/3	—	—	—	—
	84.0		—		—	—	—	—	—	—	84/3	—	—	—	—
	100.0		—		—	—	—	—	—	—	100/3	—	—	—	—
3.53 ±0.1	18.68 ±0.15	3/4	1	1/8	OS15	R125	—	15	210	—	—	—	—	—	—
	19.80	—	—	—	—	—	—	—	—	—	—	—	—	16	—
	20.22	13/16	1.1/16	1/8	16	126	VOR	16	211	—	—	—	—	—	—
	21.82	7/8	1.1/8	1/8	17	129	129	17	212	—	—	—	—	—	—
	23.40	15/16	1.3/16	1/8	18	131	131	18	213	—	—	—	—	—	—
	25.00	1	1.1/4	1/8	19	133	133	19	214	—	—	—	—	—	—
	25.80	—	—	—	—	134	—	—	—	—	—	—	—	—	—
	26.58	1.1/16	1.5/16	1/8	20	135	135	20	215	—	—	—	—	—	—
	27.8	—	—	—	—	—	—	—	—	—	—	—	—	21	—
	28.15	1.1/8	1.3/8	1/8	21	136	136	21	216	—	—	—	—	—	—
	29.75	1.3/16	1.7/16	1/8	22	137	137	22	217	—	—	—	—	—	—
	31.35	1.1/4	1.1/2	1/8	23	138	138	23	218	—	—	—	—	—	—
	32.93	1.5/16	1.9/16	1/8	24	139	—	24	219	—	—	—	—	—	—
	34.52	1.3/8	1.5/8	1/8	25	140	140	25	220	—	—	—	—	—	—
	36.09	1.7/16	1.11/16	1/8	26	141	—	26	221	—	—	—	—	—	—
	37.69 ±0.25	1.1/2	1.3/4	1/8	27	142	142	27	222	—	—	—	—	—	—
	39.69	—	—	—	—	144	—	—	—	—	—	—	—	—	—
	41.28		—		—	146	146	—	—	—	—	—	—	25	—

	42.86	—	—	147	—	—	—	—	—	—	—	—	—
	44.45	—	—	149	149	—	—	—	—	—	—	—	—
	46.04	—	—	150	—	—	—	—	—	—	—	—	—
	47.63	—	—	152	—	—	—	—	—	—	—	—	—
	49.21	—	—	153	—	—	—	—	—	—	—	—	—
	50.80	—	—	155	155	—	—	—	—	—	—	—	—
	52.39	—	—	156	—	—	—	—	—	—	—	—	—
	53.98	—	—	158	—	—	—	—	—	—	—	—	—
	55.54	—	—	159	159	—	—	—	—	—	—	—	—
	57.15	—	—	161	161	—	—	—	—	—	—	—	—
	58.74	—	—	162	—	—	—	—	—	—	—	—	—
	60.33	—	—	164	164	—	—	—	—	—	—	—	—
	61.91	—	—	165	—	—	—	—	—	—	—	—	—
	63.50	—	—	167	—	—	—	—	—	—	—	—	—
	65.09	—	—	168	—	—	—	—	—	—	—	—	—
	66.68	—	—	170	—	—	—	—	—	—	—	—	—
	68.26	—	—	171	171	—	—	—	—	—	—	—	—
	69.85	—	—	173	—	—	—	—	—	—	—	—	—
	71.44												
	±0.38	—	—	174	—	—	—	—	—	—	—	—	—
	73.03	—	—	176	—	—	—	—	—	—	—	—	—
	74.61	—	—	177	—	—	—	—	—	—	—	—	—
4.0	10	—	—	—	—	—	—	—	10/4	—	—	—	—
±0.1	18	—	—	—	—	—	—	—	—	14	—	—	—
	22	—	—	—	—	—	—	—	—	15	—	—	—
	23.5												
	±0.15	—	—	—	—	—	—	—	—	—	—	—	24/4
	25	—	—	—	—	—	—	—	—	16	—	—	—
	30	—	—	—	—	—	—	—	—	17	—	—	—
	33.5												
	±0.15	—	—	—	—	—	—	—	—	—	—	—	34/4

(Table 3.19A Continued)

Cross section diameter *d* (mm)	Inside diameter *D* (mm)	Nominal designation (inches) i.d. o.d. cross section diameter	Designation or code*										
			B.S.	G	E	A.S.	P.R.	G.S.	L	B	P	F.S.	JS
4.0	34	—	—	—	—	—	—	—	—	18	—	—	—
	38	—	—	—	—	—	—	—	—	19	—	—	—
	39.5 ±0.15	—	—	—	—	—	—	—	—	—	—	—	40/4
	42	—	—	—	—	—	—	—	—	20	—	—	—
	48	—	—	—	—	—	—	—	—	21	—	—	—
	54	—	—	—	—	—	—	—	—	22	—	—	—
	54.5 ±0.25	—	—	—	—	—	—	—	—	—	—	—	55/4
	60	—	—	—	—	—	—	—	—	23	—	—	—
	68	—	—	—	—	—	—	—	—	24	—	—	—
	69 ± 0.25	—	—	—	—	—	—	—	—	—	—	—	70/4
	76	—	—	—	—	—	—	—	—	25	—	—	—
	84 ± 0.4	—	—	—	—	—	—	—	—	—	—	—	85/4
	99	—	—	—	—	—	—	—	—	—	—	—	100/4
	119	—	—	—	—	—	—	—	—	—	—	—	120/4
	148.5 ±0.6	—	—	—	—	—	—	—	—	—	—	—	150/4
	173	—	—	—	—	—	—	—	—	—	—	—	175/4
5.0	15	—	—	—	—	—	—	—	15/5	—	—	—	—
	18	—	—	—	—	—	—	—	—	—	10	—	—
	20	—	—	—	—	—	—	—	20/5	—	—	—	—
	25	—	—	—	—	—	—	NW10	25/5	—	—	—	—
	28	—	—	—	—	—	—	—	—	—	15/20	—	—

30	—	—	—	—	—	—	—	30/5	—	—	—	—
35	—	—	—	—	—	—	20	35/5	—	—	—	—
38	—	—	—	—	—	—	—	38/5	—	—	—	—
40	—	—	—	—	—	—	—	40/5	—	25	—	—
45	—	—	—	—	—	—	—	45/5	—	—	—	—
50	—	—	—	—	—	—	—	50/5	—	—	—	—
53	—	—	—	—	—	—	—	53/5	—	—	—	—
55	—	—	—	—	—	—	32	55/5	—	30/40	—	—
57	—	—	—	—	—	—	—	57/5	—	—	—	—
60	—	—	—	—	—	—	—	60/5	—	—	—	—
68	—	—	—	—	—	—	—	68/5	—	—	—	—
73	—	—	—	—	—	—	—	—	—	50	—	—
75	—	—	—	—	—	—	—	75/5	—	—	—	—
80	—	—	—	—	—	—	50	80/5	—	—	—	—
85	—	—	—	—	—	—	—	85/5	—	—	—	—
87	—	—	—	—	—	—	—	—	26	—	—	—
88	—	—	—	—	—	—	—	—	—	65/80	—	—
90	—	—	—	—	—	—	70	—	—	—	—	—
97	—	—	—	—	—	—	—	97/5	27	—	—	—
100	—	—	—	—	—	—	—	100/5	—	*	—	—
105	—	—	—	—	—	—	—	105/5	—	*	—	—
110	—	—	—	—	—	—	—	110/5	—	—	—	—
120	—	—	—	—	—	—	—	120/5	—	—	—	—
125	—	—	—	—	—	—	100	125/5	—	100	—	—
135	—	—	—	—	—	—	—	135/5	—	—	—	—
144	—	—	—	—	—	—	—	144/5	—	—	—	—
148	—	—	—	—	—	—	—	—	—	125	—	—
160	—	—	—	—	—	—	—	—	—	150	—	—
165	—	—	—	—	—	—	—	165/5	—	—	—	—
170	—	—	—	—	—	—	150	—	—	—	—	—
195	—	—	—	—	—	—	—	195/5	—	—	—	—
205	—	—	—	—	—	—	—	—	—	*	—	—
230	—	—	—	—	—	—	—	230/5	—	200	—	—

(Table 3.19A Continued)

Cross section diameter d (mm)	Inside diameter D (mm)	Nominal designation (inches) i.d. o.d. cross section diameter			Designation or code*										
					B.S.	G	E	A.S.	P.R.	G.S.	L	B	P	F.S.	JS
5.0	265	—			—	—	—	—	—	—	265/5	—	—	—	—
	280	—			—	—	—	—	—	—	—	—	280	—	—
	305	—			—	—	—	—	—	—	305/5	—	—	—	—
	355	—			—	—	—	—	—	—	355/5	—	350	—	—
5.33 ±0.12	37.47 ±0.25	1.1/2	1.7/8	3/16	OS28	R143	—	28	325	—	—	—	—	—	—
	40.65	1.5/8	2	3/16	29	145	VOR	29	326	—	—	—	—	29	—
	43.82	1.3/4	2.1/8	3/16	30	148	148	30	327	—	—	—	—	—	—
	47.00	1.7/8	2.1/4	3/16	31	151	151	31	328	—	—	—	—	—	—
	49.70	—	—	—	—	—	—	—	—	—	—	—	—	32	—
	50.16	2	2.3/8	3/16	32	154	154	32	329	—	—	—	—	—	—
	53.35	2.1/8	2.1/2	3/16	33	157	—	33	330	—	—	—	—	—	—
	56.52	2.1/4	2.5/8	3/16	34	160	160	34	331	—	—	—	—	—	—
	59.70	2.3/8	2.3/4	3/16	35	163	—	35	332	—	—	—	—	—	—
	62.40	—	—	—	—	—	—	—	—	—	—	—	—	36	—
	62.87	2.1/2	2.7/8	3/16	36	166	166	36	333	—	—	—	—	—	—
	66.04	2.5/8	3	3/16	37	169	—	37	334	—	—	—	—	—	—
	69.21 ±0.38	2.3/4	3.1/8	3/16	38	172	172	38	335	—	—	—	—	—	—
	72.39	2.7/8	3.1/4	3/16	39	175	—	39	336	—	—	—	—	—	—
	74.63	—	—	—	—	178	—	—	—	—	—	—	—	—	—
	75.57	3	3 3/8	3/16	40	179	179	40	337	—	—	—	—	—	—
	78.75	3.1/8	3.1/2	3/16	41	180	—	41	338	—	—	—	—	—	—
	79.77	—	—	—	—	181	—	—	—	—	—	—	—	—	—

	81.92	3.1/4	3.5/8	3/16	42	182	182	42	339	—	—	—	—	(42)	—
	85.10	3.3/8	3.3/4	3/16	43	183	183	43	340	—	—	—	—	—	—
	88.27	3.1/2	3.7/8	3/16	44	184	184	44	341	—	—	—	—	—	—
	89.69	—	—	—	—	185	—	—	—	—	—	—	—	—	—
	91.45	3.5/8	4	3/16	45	186	186	45	342	—	—	—	—	—	—
	94.62	3.3/4	4.1/8	3/16	46	187	187	46	343	—	—	—	—	—	—
	97.79	3.7/8	4.1/4	3/16	47	188	—	47	344	—	—	—	—	—	—
	100.0	—	—	—	—	189	—	—	—	—	—	—	—	—	—
	101.0	4	4.3/8	3/16	48	190	190	48	345	—	—	—	—	—	—
	104.15	4.1/8	4.1/2	3/16	49	191	—	49	346	—	—	—	—	—	—
	107.3	4.1/4	4.5/8	3/16	50	192	—	50	347	—	—	—	—	—	—
	109.5	—	—	—	—	193	—	—	—	—	—	—	—	—	—
	110.5	4.3/8	4.3/4	3/16	51	194	194	51	348	—	—	—	—	—	—
	113.7	4.1/2	4.7/8	3/16	52	195	—	52	349	—	—	—	—	52	—
	117.5	—	—	—	—	199	—	—	—	—	—	—	—	—	—
	120.7	—	—	—	—	201	—	—	—	—	—	—	—	—	—
	123.8	—	—	—	—	203	—	—	—	—	—	—	—	—	—
	127.0	—	—	—	—	206	—	—	—	—	—	—	—	—	—
	130.2 ±0.58	—	—	—	—	208	208	—	—	—	—	—	—	—	—
	133.4	—	—	—	—	210	—	—	—	—	—	—	—	—	—
	136.5	—	—	—	—	213	213	—	—	—	—	—	—	—	—
	139.7	—	—	—	—	215	—	—	—	—	—	—	—	—	—
	142.9	—	—	—	—	217	217	—	—	—	—	—	—	—	—
	146.1	—	—	—	—	219	—	—	—	—	—	—	—	—	—
	149.2	—	—	—	—	221	221	—	—	—	—	—	—	—	—
6.0 ±0.15	114		—		—	—	—	—	—	—	—	28	—	—	—
	128		—		—	—	—	—	—	—	—	29	—	—	—
	142		—		—	—	—	—	—	—	—	30	—	—	—
	166		—		—	—	—	—	—	—	—	31	—	—	—
	185		—		—	—	—	—	—	—	—	32	—	—	—

(Table 3.19A Continued)

Cross section diameter *d* (mm)	Inside diameter *D* (mm)	Nominal designation (inches) i.d. o.d. cross section diameter			Designation or code*										
					B.S.	G	E	A.S.	P.R.	G.S.	L	B	P	FS	JS
6.0	222.5 ±0.8	—			—	—	—	—	—	—	—	—	—	—	225/6
	272	—			—	—	—	—	—	—	—	—	—	—	275/6
	321.5	—			—	—	—	—	—	—	—	—	—	—	325/6
	376	—			—	—	—	—	—	—	—	—	—	—	380/6
	425.5	—			—	—	—	—	—	—	—	—	—	—	430/6
7.0 ±0.15	113.7 ±0.38	4.1/2	5	1/4	OS53	R196	196	88	425	—	—	—	—	—	—
	114.7	—	—	—	—	197	—	—	—	—	—	—	—	—	—
	116.8	4.5/8	5.1/8	1/4	54	198	—	53	426	—	—	—	—	—	—
	120.2	4.3/4	5.1/4	1/4	55	200	—	54	427	—	—	—	—	—	—
	123.2	4.7/8	5.3/8	1/4	56	202	—	55	428	—	—	—	—	—	—
	126.4	5	5 1/2	1/4	57	205	205	56	429	—	—	—	—	—	—
	129.5 ±0.58	5.1/8	5.5/8	1/4	58	207	207	57	430	—	—	—	—	—	—
	132.7	5.1/4	5.3/4	1/4	59	209	—	58	431	—	—	—	—	—	—
	134.5	—	—	—	—	211	—	—	—	—	—	—	—	—	—
	135.9	5.3/8	5.7/8	1/4	60	212	—	59	432	—	—	—	—	—	—
	139.1	5.1/2	6	1/4	61	214	214	60	433	—	—	—	—	—	—
	142.2	5.5/8	6.1/8	1/4	62	216	—	61	434	—	—	—	—	—	—
	145.4	5.3/4	6.1/4	1/4	63	218	—	62	435	—	—	—	—	—	—
	148.6	5.7/8	6.3/8	1/4	64	220	—	63	436	—	—	—	—	—	—
	151.8	6	6.1/2	1/4	65	222	222	64	437	—	—	—	—	—	—

158.1	6.1/4	6.3/4	1/4	66	224	—	65	438	—	—	—	—	—	—
159.5	—	—	—	—	225	—	—	—	—	—	—	—	—	—
164.5	6.1/2	7	1/4	67	227	227	66	439	—	—	—	—	—	—
170.8	6.3/4	7.1/4	1/4	68	230	—	67	440	—	—	—	—	68	—
174.6	—	—	—	—	231	231	—	—	—	—	—	—	—	—
176.2	7	7.1/2	1/4	69	—	—	68	441	—	—	—	—	—	—
177.2	—	—	—	—	232	232	—	—	—	—	—	—	—	—
183.5 ± 0.75	7.1/4	7.3/4	1/4	70	234	—	69	442	—	—	—	—	—	—
189.9	7.1/2	8	1/4	71	236	236	70	443	—	—	—	—	—	—
193.7	—	—	—	—	237	237	—	—	—	—	—	—	—	—
196.2	7.3/4	8.1/4	1/4	72	238	—	71	444	—	—	—	—	—	—
200.0	—	—	—	—	239	—	—	—	—	—	—	—	—	—
202.6	8	8.1/2	1/4	73	240	240	72	445	—	—	—	—	—	—
208.9	—	—	—	74	241	—	—	—	—	—	—	—	—	—
215.3	8.1/2	9	1/4	75	242	242	73	446	—	—	—	—	—	—
221.6	—	—	—	76	243	—	—	—	—	—	—	—	—	—
228.0	9	9.1/2	1/4	77	244	244	74	447	—	—	—	—	75	—
234.3	—	—	—	78	245	—	—	—	—	—	—	—	—	—
240.7	9.1/2	10	1/4	79	246	246	75	448	—	—	—	—	—	—
247.0	—	—	—	80	247	—	—	—	—	—	—	—	—	—
253.4	10	10.1/2	1/4	81	248	248	76	449	—	—	—	—	—	—
259.7	—	—	—	82	249	249	—	—	—	—	—	—	—	—
266.1	10.1/2	11	1/4	83	250	—	77	450	—	—	—	—	—	—
272.4	—	—	—	84	251	—	—	—	—	—	—	—	—	—
278.7	11	11.1/2	1/4	85	252	252	78	451	—	—	—	—	—	—
285.1	—	—	—	86	253	—	—	—	—	—	—	—	—	—
291.5	11.1/2	12	1/4	87	254	—	79	452	—	—	—	—	80	—
297.8	—	—	—	88	255	—	—	—	—	—	—	—	—	—
304.2	12	12.1/2	1/4	89	256	256	80	453	—	—	—	—	—	—
316.9	12.1/2	13	1/4	90	257	257	81	454	—	—	—	—	—	—
329.6	13	13.1/2	1/4	91	258	258	82	455	—	—	—	—	—	—
341.5	—	—	—	—	—	—	—	—	—	—	—	—	84	—

(Table 3.19A Continued)

Cross section diameter *d* (mm)	Inside diameter *D* (mm)	Nominal designation (inches) i.d. o.d. cross section diameter	Designation or code*										
			B.S.	G	E	A.S.	P.R.	G.S.	L	B	P	FS	JS
7.0	342.3	13.1/2 14 1/4	92	259	—	83	456	—	—	—	—	—	—
	355	14 14.1/2 1/4	93	260	260	84	457	—	—	—	—	—	—
	367.7	14.1/2 15 1/4	94	261	—	85	458	—	—	—	—	—	—
	380.4	15 15.1/2 1/4	95	262	262	86	459	—	—	—	—	—	—
	393.1	15.1/2 16 1/4	96	263	—	87	460	—	—	—	—	—	—
8.0	208	—	—	—	—	—	—	—	—	33	—	—	—
	230	—	—	—	—	—	—	—	—	34	—	—	—
	250	—	—	—	—	—	—	—	250/8	—	—	—	—
	259	—	—	—	—	—	—	—	259/8	—	—	—	—
	260	—	—	—	—	—	—	—	—	35	—	—	—
	262	—	—	—	—	—	—	250	—	—	—	—	—
	280	—	—	—	—	—	—	—	280/8	—	—	—	—
	290	—	—	—	—	—	—	—	—	36	—	—	—
	330	—	—	—	—	—	—	—	—	37	—	—	—
	540	—	—	—	—	—	—	—	—	—	500	—	—
	622	—	—	—	—	—	—	—	—	—	600	—	—
10.0 ± 0.3	350	—	—	—	—	—	—	—	350/10	—	—	—	—
	370	—	—	—	—	—	—	—	—	38	—	—	—
	420	—	—	—	—	—	—	—	—	39	—	—	—
	448	—	—	—	—	—	—	—	448/10	—	—	—	—
	470	—	—	—	—	—	—	—	—	40	—	—	—
	475 ± 0.8	—	—	—	—	—	—	—	—	—	—	—	480/10

	524.5 ± 1.0	—	—	—	—	—	—	—	—	—	—	—	530/10
	550	—	—	—	—	—	—	—	—	41	—	—	—
	579	—	—	—	—	—	—	—	—	—	—	—	585/10
	580	—	—	—	—	—	—	—	—	42	—	—	—
	610	—	—	—	—	—	—	—	—	44	—	—	—
	633	—	—	—	—	—	—	—	—	—	—	—	640/10
	683	—	—	—	—	—	—	—	—	—	—	—	690/10
	690	—	—	—	—	—	—	—	—	46	—	—	—
	732.5 ± 1.2	—	—	—	—	—	—	—	—	—	—	—	740/10
	760	—	—	—	—	—	—	—	—	48	—	—	—
	782	—	—	—	—	—	—	—	—	—	—	—	790/10
	836.5	—	—	—	—	—	—	—	—	—	—	—	845/10
	840	—	—	—	—	—	—	—	—	—	800	—	—
	940.5	—	—	—	—	—	—	—	—	—	—	—	950/10
	1044 ± 1.4	—	—	—	—	—	—	—	—	—	—	—	1055/10
15.0	595	—	—	—	—	—	—	—	—	43	—	—	—
	625	—	—	—	—	—	—	—	—	45	—	—	—
	725	—	—	—	—	—	—	—	—	47	—	—	—
	780	—	—	—	—	—	—	—	—	49	—	—	—
	840	—	—	—	—	—	—	—	—	51	—	—	—
	1000	—	—	—	—	—	—	—	—	52	—	—	—
	1160	—	—	—	—	—	—	—	—	53	—	—	—
	1240	—	—	—	—	—	—	—	—	54	—	—	—
20.0	800	—	—	—	—	—	—	—	—	50	—	—	—
	1275	—	—	—	—	—	—	—	—	55	—	—	—

* B.S. - British Standard[169]; G - Angus[34a]; E - Edwards[329]; G.S. - German Standard DIN 2572/1940; L - Leybold[762]; B - Balzers[444]. P — Pfeiffer[987]; A.S. - American Standard A.N. 6227; P.R. - Precision Rubber Prod.[1004]; Plastic & Rubber Prod.[993]; F.S. - French Standard, see Tarbes[1214, 1215], Henry[539]; J.S. - Japanese Standard, see Kiyoshi[683].

TABLE 3.19B. ELASTOMER GASKETS WITH HALF ROUND (FIG. 3.56c) AND SQUARE (FIG. 3.56b) CROSS SECTIONS (Kiyoshi[683])*

(Dimensions in mm)

Nominal size	Actual size	Type I				Type II			
		Square	Half round			Square	Half round		
		(a)	(d)	(h)	(R)	(a)	(d)	(h)	(R)
24	23.5 ±0.15								
34	33.5								
40	39.5								
55	54.5 ±0.25	4	4	4	2	5	5	5	2.5
70	69.0	±0.1	±0.1	±0.1		±0.15	±0.15	±0.15	
85	84.0 ±0.4								
100	99.0								
120	119.0								
150	148.5 ±0.6								
175	173.0								
225	222.5 ±0.8								
275	272.0								
325	321.5	6	6	6	3	8	8	8	4
380	376.0	±0.15	±0.15	±0.15		±0.2	±0.2	±0.2	
430	425.5								
480	475.0 ±0.8								
530	524.5 ±1.0								
585	579.0								
640	633.5								
690	683.0								
740	732.5 ±1.2	8	8	10	4	12	12	12	6
790	782.0	±0.2	±0.2	±0.3		±0.35	±0.35	±0.35	
845	836.5								
950	940.5								
1055	1044.0 ±1.4								

*For the shape of the grooves see Figs. 3.114–3.118; the dimensions of the required flanges are listed in Table 3.35.

TABLE 3.19C. GASKETS WITH TRAPEZIUM CROSS SECTION*

(Fig. 3.56d) Dimensions in mm. (Leybold[762])

D	a	h	D	a	h	D	a	h
12	4	4.8	95.5	9.5	10	360	10	12
22	4	4.8	105.5	9.5	10	370	10	12
28	4	4.8	130.5	9.5	10	406	10	12
36	6	7.2	155.5	9.5	10	510	10	12
43	7	9.0	210	10	12	540	10	12
53	7	9.0	260	10	12	630	10	12
77	8	9.6	320	10	12	808	12	14.5

* For the shape of the grooves see Fig. 3.119; the dimensions of the corresponding flanges are listed in Table 3.38.

38.34 Gasket shapes. The shape of a gasket is described by its outline and its cross section. The most common outline of gaskets for vacuum sealing is circular (Sections 38.4–38.6), but sometimes gaskets with rectangular (Section 38.41e) or other outlines must also be used. The shape of the cross section of the gaskets is more varied as their outline. The circular cross section is desired in many seals and is preferred wherever possible. Table 3.18 summarizes the various cross section shapes of gaskets used in vacuum seals.

The gasket is described by: (1) the material, (2) the shape, (3) the dimensions. Figure 3.56 shows the dimensions to be indicated in order to define the

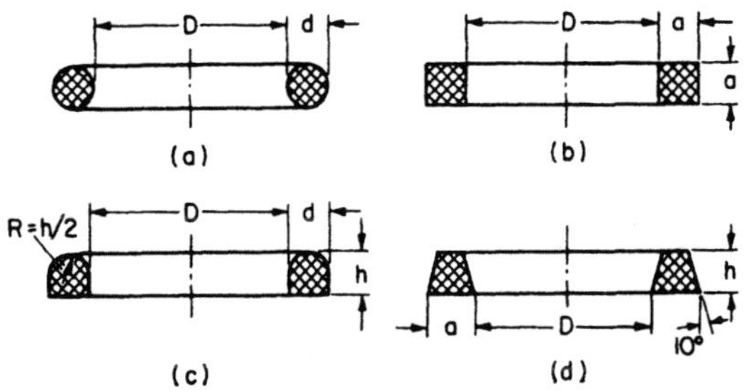

FIG. 3.56 Dimensions of gaskets (Tables 3.18, 3.19)

various gaskets having the commonly used cross sections. These gasket shapes are extensively used in elastomer gasket seals, some of them being officially standardized in various countries. Tables 3.19A–C give the list of the standard and usual elastomer gaskets.

The plastic and metal gaskets used in vacuum sealing are of a great variety of shapes (Table 3.18); their dimensions are not yet standardized and their usual dimensions are given in connexion with each particular kind of seal (Sections 38.4–38.6).

38.4 O-ring seals

An O-ring seal is a dismountable vacuum connexion which uses a circular gasket with circular cross section (Table 3.19A). The O-ring, made of an elastomer or a metal (Table 3.16), is compressed (or sometimes sheared) between the sealing parts. If the main compression force is exerted *axially* on the O-ring, the seal is known as a *flange seal;* if the force works *radially* the connexion is a *shaft seal.*

When the joint closes, the sealing parts touch the cross section of the O-ring in two, three or more points depending on the shape of the sealing parts (plane flanges, grooves, spacers).

Figure 3.57 shows some possible sealing arrangements of O-ring seals in which the gasket is compressed between two, three or four sealing parts.

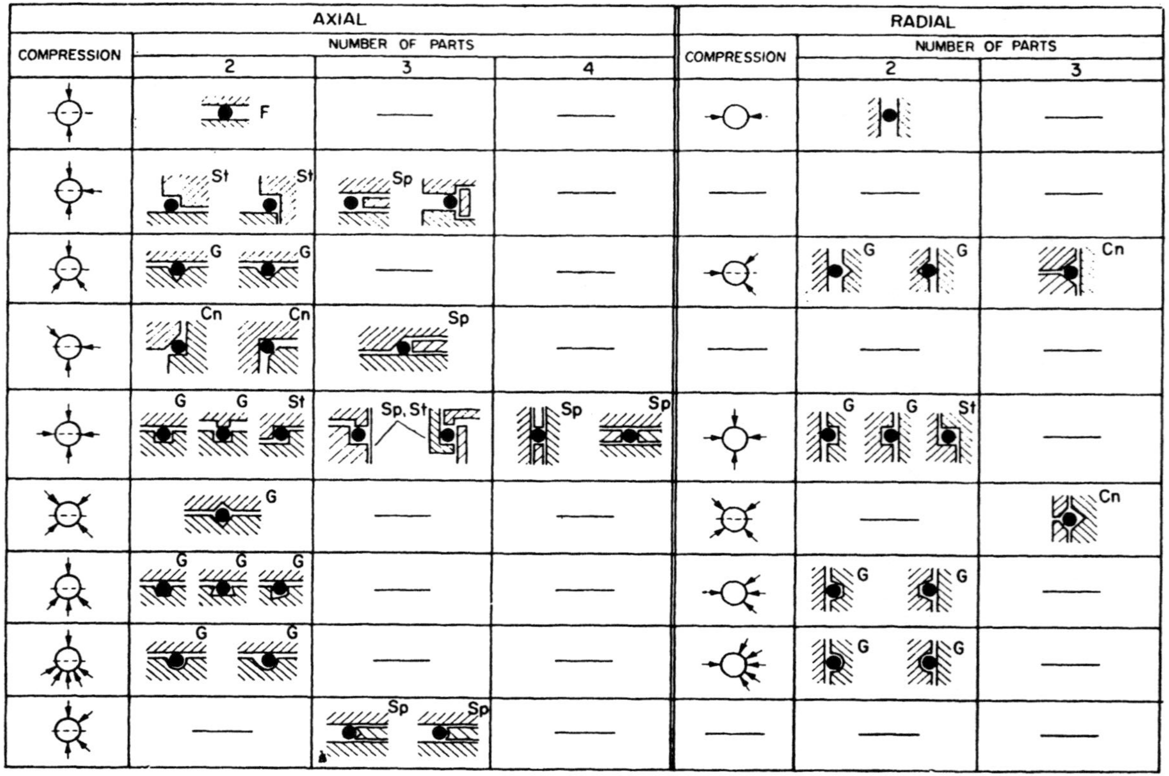

Fig. 3.57 Classification of O-ring seals. The axis of the seal is assumed to be vertical and on the right of the seal. *F* = flange seal; *G* = groove seal; *Sp* = spacer seal; *Cn* = conical seal; *St* = step seal

The various shapes presented in Fig. 3.57 are obtained by: (1) placing the O-ring in a groove machined in one of the sealing parts (Section 38.41); (2) closing the O-ring between the sealing parts and keeping it in place with spacers (Section 38.42); (3) compressing the O-ring in a conical seal (Section 38.43); (4) crushing the O-ring between the corners provided on the sealing parts (Section 38.44); (5) keeping the O-ring on a step and compressing it between the flanges (Section 38.45) or (6) compressing the O-ring between flat flanges (Sections 38.46 and 38.51).

38.41 O-ring groove seals. In *groove seals* one of the sealing parts (flange, coupling, shaft) has a machined recess designed to receive the O-ring. Basically a groove seal may be designed either for constant deflection of the O-ring or for constant load on the O-ring. The *constant deflection seals* (known also as *seals* with *limited compression*) are designed to limit the compression of the gasket to the compression ratio (Fig. 3.50) required for a vacuum-tight seal.

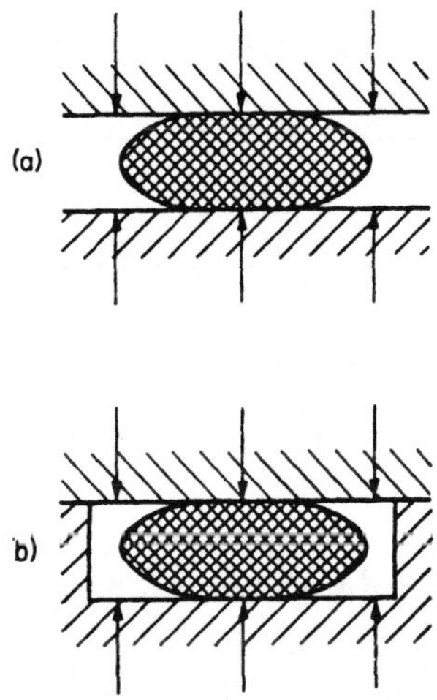

FIG. 3.58 Seals with (a) unlimited and (b) limited compression of the gasket

When this compression ratio is reached the sealing parts meet in a *metal-to-metal* contact (Brown[172]) enclosing the O-ring in the groove (Fig. 3.58b). The limited compression seals are preferred for elastomer gaskets (Table 3.17), while the arrangements with unlimited compression (Fig. 3.58a) are used with fluocarbon or metal gaskets, where the full load must always be kept on the gasket.

The groove should meet the following requirements:

a. *To receive the gasket,* i.e. to allow, when closed, enough cross sectional area to accommodate all the gasket cross section (in seals with limited compression) or the required part of the cross section (in seals with unlimited compression). For rubber gaskets the cross section of the groove should be *2–5 per cent greater* than the cross section of the O-ring, to allow for the uncompressibility of the rubber and the dead spaces in the corners (not filled up by the rubber). To fill up these dead spaces a very high local deflection of the gasket would be necessary producing a compression set (hardening).

b. *To ensure the compression ratio.* In order to obtain a vacuum-tight seal the compression ratio of rubber gaskets should be 20–40 per cent depending on the hardness of the rubber (Fig. 3.50). This means (related to Fig. 3.59) that the ratio B/d should be 0.8 to 0.6. The ideal groove should assure the same deflection in every part of the gasket cross section; any sharp corners in the cross section of the groove strongly deform the gasket, thus preventing it being re-used.

c. *To provide positive retention* of the gasket, i.e. to be able to keep the gasket in position when the seal is not closed.

d. *To allow easy removal* of the gasket.

e. *To allow the minimum trapped volume* (dead volume discussed at point *a* or volumes between the sealing parts, separated from the evacuated space by a thin gap of high impedance).

f. *To allow the minimum exposed surface* of the gasket to the evacuated space; this requirement is essential when common elastomer gaskets are used to seal high vacuum systems.

g. *To allow the re-use of the gasket* after the seal is opened. In this respect the grooves forcing the gasket to small angles or cutting it are of poor design (Table 3.20).

h. *To permit easy machining* and wide dimensional tolerances.

In order to meet some of these requirements and trying to meet them all various groove shapes were developed and used with O-rings. Table 3.20 lists these groove shapes with an estimate regarding their ability to meet the various requirements (a–h).

a. *Grooves with rectangular cross section.* The simplest and most commonly used grooves with rectangular cross sections (Fig. 3.59a) have their dimensions determined by:

(i) the incompressibility of the elastomer,
(ii) the required compression ratio (K, see Fig. 3.50)
(iii) the factor k allowing for the dead volume.

Thus the depth B and the width A of the groove (Fig. 3.59a) are given by:

$$A \times B = k \cdot \pi \cdot d^2/4$$
$$B/d = 1 - K$$

For rubber gaskets generally having a Shore hardness of between 40–60, a mean value of $K = 0.28$ may be used; this leads to a ratio $B/d = 0.72$ (Ardenne[47]). With a dead volume of 5 per cent ($k = 1.05$), a ratio $A/d = 1.15$ results.

TABLE 3.20. ABILITY OF GROOVES TO MEET THE VARIOUS REQUIREMENTS (G—GOOD; F—FAIR; P—POOR) FOR A PERFECT O-RING SEAL

Shape of groove cross section	Fig.	Requirement							
		a.	b.	c.	d.	e.	f.	g.	h.
Rectangular	3.59a	G	G	G	F	F	G	G	G
Triangular	3.65	P	P	P	G	G	F	P	G
Trapezoidal	3.66b	G	G	P	G	G	F	G	G
Dovetail	3.66a	G	G	G	P	P	G	P	P
Trapezium (asymmetrical)	3.66c	G	G	G	F	G	G	G	F
Semi-circular	3.70	F	P	F	G	G	F	P	F
Rounded-triangular	3.65	P	P	F	G	G	F	P	F
Ovoidal	3.71	F	G	F	G	G	F	F	P

The dimensions A and B of rectangular groves for various common O-ring cross-section diameters (see Table 3.19A) are given in Table 3.21. The groove should have a radius R (Fig. 3.59a) of

$$0.15d \leqslant R \leqslant 0.22d.$$

The edges of the groove should not be radiused to more than 0.25 mm, as a larger radius of chamfer would assist extrusion of the O-ring. In some cases it may assist machining, by providing a 5° (maximum) taper at the sides of the groove, but on no account should this angle be exceeded (Angus[34a], Everett[363a]) if the groove is still treated as rectangular.

For better retention of the O-ring, the width of a groove with rectangular cross section is sometimes made *equal* to the cross sectional diameter of the O-ring. To receive the compressed gasket, the groove must include a step (Fig. 3.59b). Bishop[129] quotes the dimensions of such grooves, leading to the formulae (Fig. 3.59b)

$$B = 0.72d,$$

$$C = 2E = 0.25d.$$

These formulae do not allow for a dead space in the groove. It is more realistic to allow for a dead space of about 5 per cent, hence to use the formula

$$C = 2E = 0.32d$$

TABLE 3.21. DIMENSIONS OF RECTANGULAR CROSS SECTION GROOVES (mm)**

O-ring cross sectional diameter* d			Flange seal (Fig. 3.59a)			Shaft seal (Fig. 3.63)				
nominal (mm)	nominal (inches)	minimum maximum (mm)	A	B	R[*3]	B–A[*2]	C[*2]	G[*2]	R_1	R_2
1.78	0.070	1.71 1.85	1.97 2.14 2.28[*4]	1.23 1.33 1.19[*4]	0.4	2.80 2.90	1.92 2.02	0.13	0.5	0.10
2.62	0.103	2.55 2.69	2.93 3.10 3.48[*4]	1.84 1.92 1.70[*4]	0.4	4.10 4.25	2.76 2.93	0.13	0.5	0.10
3.0	–	2.93 3.07	3.36 3.52	2.10 2.20	0.4	4.75	3.30	0.13	0.5	0.10
3.53	0.139	3.43 3.63	3.94 4.18 4.57[*4]	2.48 2.62 2.30[*4]	0.8	5.60 5.80	3.70 3.90	0.15	0.8	0.20
4.0	–	3.90 4.10	4.50 4.70	2.80 2.95	0.8	6.40	4.35	0.15	0.8	0.20
5.0	–	4.90 5.10	6[*1]	3.6[*1]	0.8	8.0	5.45	0.18	0.8	0.20
5.33	0.210	5.21 5.45	6.0 6.30 6.91[*4]	3.76 3.90 3.05[*4]	0.8	8.50 8.70	5.50 5.80	0.18	0.8	0.20
6.0	–	5.85 6.15	6.75 7.05	4.20 4.42	0.8	9.70	6.30	0.18	0.8	0.25
7.0	0.275	6.85 7.15	7.90 8.20 9.07[*4]	4.92 5.15 4.65[*4]	1.6	11.2 11.5	6.90 7.25	0.20	0.8	0.25
8.0	(0.312)	7.74 8.14	9.5[*1]	5.90[*1]	1.6	12.8	8.40	0.20	1.4	0.25
10.0	–	9.70 10.30	11.2 11.9	7.0 7.4	2.0	16.0	10.5	0.22	1.8	0.25
12.7	0.500	12.44 13.0	14.3 15.0	9.0 9.4	2.5	–	–	–	–	–

* See Table 3.19A.
** Flange dimensions, see Sec. 38.47.
*1 — Diels[280]; *2 — Edwards[327]; *3 — Precision Rubber Prod.[1004]; *4 — Genevac[442].

Kiyoshi[683] quotes the groove dimensions proposed in the Japanese Standard. This consists of a rectangular groove (Fig. 3.59c), having the sides bevelled at 45° on a width of 1 mm, for all the O-ring sizes used (Table 3.19A). The construction of the groove (Fig. 3.59c) is based on the ratios

$$B/d = 0.70 \quad \text{to} \quad 0.75,$$
$$A/d = 1.20 \quad \text{to} \quad 1.34,$$

which include a big (8–12 per cent) dead volume.

If the rectangular cross section groove is intended to be used with metal O-rings (e.g. indium) or without limiting the compression by a metal-to-

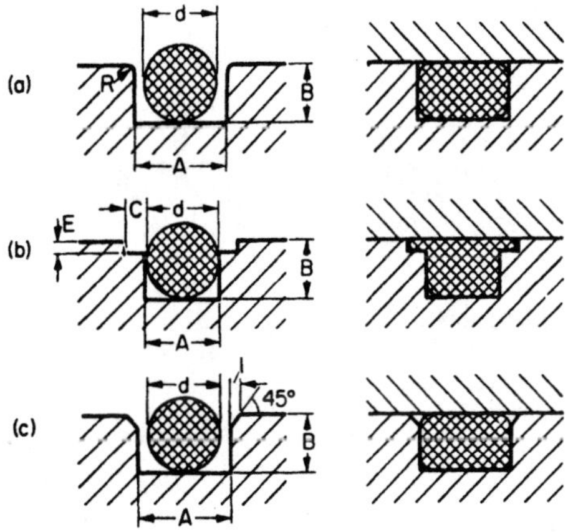

FIG. 3.59 Grooves with rectangular cross section, for O-ring seals (limited compression): (a) simple groove; (b) groove with step; (c) groove with bevel

metal contact of the sealing flanges, the groove should be made with its width *equal* to the cross-sectional diameter of the O-ring, and having a depth 15–30 per cent greater than the width. The gasket, placed in the groove, is compressed by the lip of the opposite flange (Fig. 3.60a). This seal has the drawback that the removal of the O-ring from the groove is not easy. In order to overcome this difficulty it is advisable to construct such seals on some demountable parts (Fig. 3.60b).

Richards[1051a] used this type of seal (Fig. 3.60c) as an end-seal on a tube without flanges. The seal consisted of a metal plate and retaining ring holding an O-ring, the section of which is constrained on three sides by the metal and presents the fourth side to the square-cut, flame-polished end of a glass tube.

In order to seal plugs closing holes in the wall of vacuum chambers, the O-ring may be placed in a groove provided in the head of the plug (Fig. 3.61). To avoid the friction of the O-ring while the seal is being closed, it is recommended to use bolts tightened with nuts (Fig. 3.61a). When the other side of the wall is not accessible, the screw must be cut in the wall, but in this case it is recommended (Carson[195a]) to cut a slot (Fig. 3.61b) across the threads in order to allow all the thread grooves to be readily evacuated.

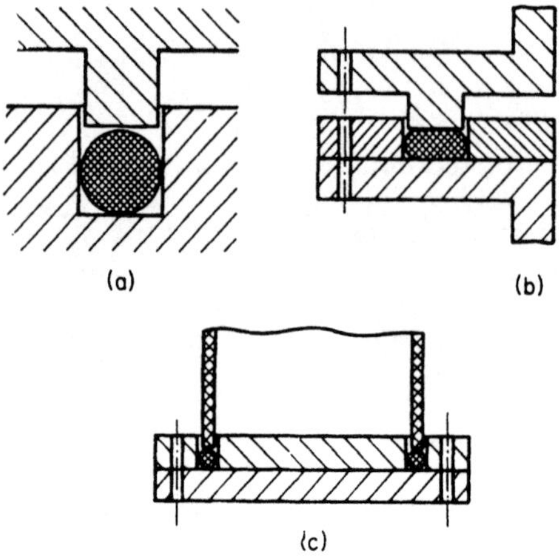

FIG. 3.60 Grooves with rectangular cross section, for O-ring seals (unlimited compression): (a) closed groove; (b) demountable groove; (c) as (b) for pipe edge seal

To allow for the possibility of evacuating the space inside the seal for leak detection purposes (Section 13.3) or guard vacuum purposes (Section 38.23), two concentric grooves are machined (Fig. 3.49b) and the space between them is connected to an exhaust pipe. Farkass[367] used a similar double O-ring seal having a concentric channel between the two grooves in which water was

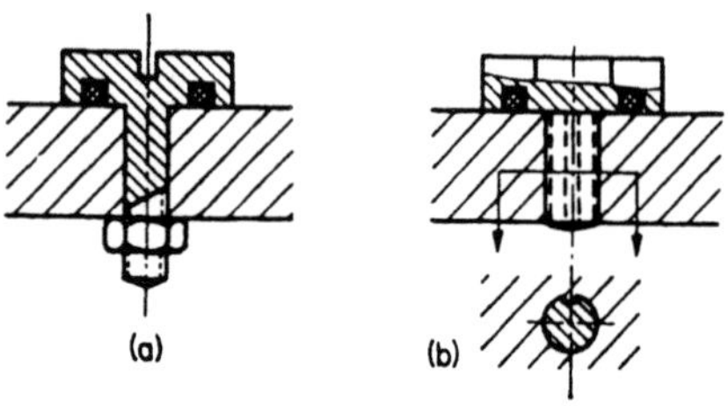

FIG. 3.61 Screws, sealed with O-rings in rectangular grooves: (a) bolt and nut; (b) screw with slot to evacuate the threads

circulated in order to cool the O-rings during the bakeout of the system. In this way ultra-high vacuum (10^{-8}–10^{-10} torr) was obtained in the chamber despite the rubber (Neoprene, Viton) O-rings. Farkass[369] mentions that during the experiments it was found that the O-ring sealed best when a high hydrostatic pressure in the channel (up to 10 kg/cm^2) pressed the O-rings against the flanges.

Grooved O-ring seals can be also used in joints where one of the sealing flanges has to be shifted with regard to other flange and the O-ring. Sliding plates sealed by O-rings are used in gate valves (Section 61.32h). In these applications the seal deviates from the standard design (Wahl[1281]), to avoid the

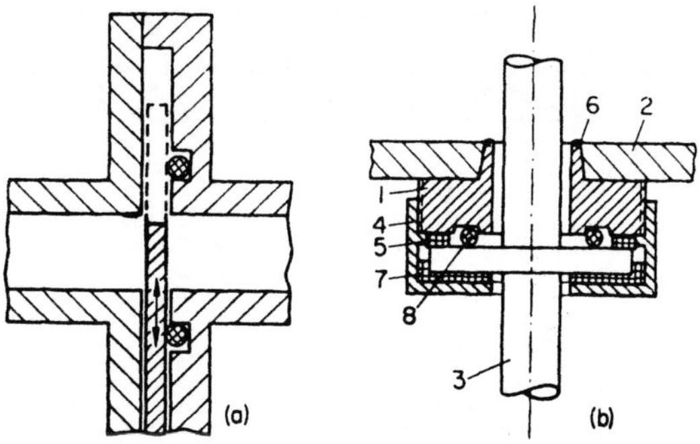

FIG. 3.62 O-ring and groove seals for special purposes: (a) sliding plate (after Wahl[1281]); (b) insulated lead-in (after Edwards[325])

danger of the O-ring being pushed out of its groove. To overcome this difficulty an O-ring one size smaller in diameter (*D*, see Table 3.19A) than that normally required for the groove, is to be used (flange dimensions, see Section 38.47).

Figure 3.62a shows a sliding gate sealed by an O-ring stretched into its groove, and Fig. 3.62b an O-ring seal used by Edwards[325] to join an electric

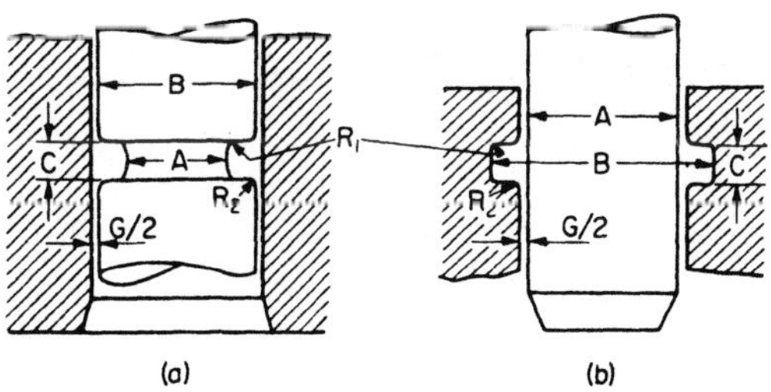

FIG. 3.63 Dimensions of rectangular groove O-ring shaft seals (piston O-ring grooves): (a) external; (b) internal (Table 3.21). *Courtesy of* Edwards[329]

lead-through or a shaft requiring a limited amount of rotation to a vacuum chamber. The O-ring (8) is located in a shallow groove (depth about 0.4 of the cross-sectional diameter of the O-ring) machined in the seat (1) brazed at (6) to the wall (2) of the vacuum chamber. The compression of the O-ring is limited to the specified value by the insulating ring (5), and the insulating ring (7) centres the lead (3) within the locking ring (4).

Sliding or rotating shafts may be sealed with O-rings, by placing them in grooves with rectangular cross section; such seals are known also as "piston O-ring seals". Figure 3.63 shows the construction details of such seals (Edwards[327]). The groove may be machined on the shaft (known as external seal, see Fig. 3.63a) or in the cylinder bore (internal seal, see Fig. 3.63b). For easier

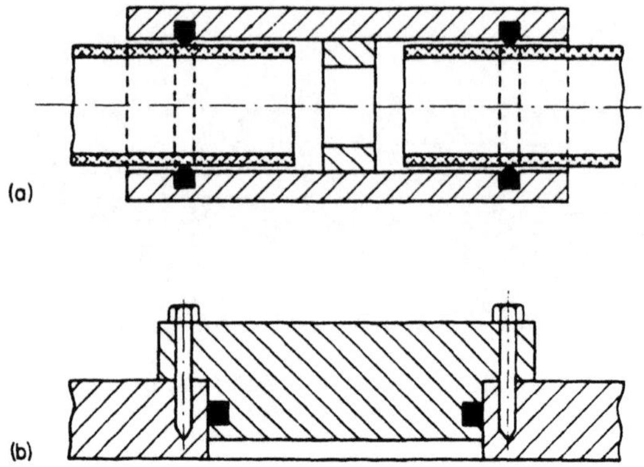

FIG. 3.64 O-ring seal in rectangular groove: (a) joint for glass tubes (after Yonts[1344]); (b) cover seal. After Kurie[723] *(Courtesy of The American Institute of Physics)*

machining the external type is to be preferred. The dimensions of these grooves corresponding to the various O-ring sizes are given in Table 3.21.

The O-ring may be placed as in an external shaft seal also when a cover plate of a port is to be sealed (Fig. 3.64b) having the smallest possible diameter (Kurie[723]). Yonts[1344] describes a joint for glass tubes, where the O-rings are placed as for an internal shaft seal into grooves machined in the inside of the connecting cylinder (Fig. 3.64a).

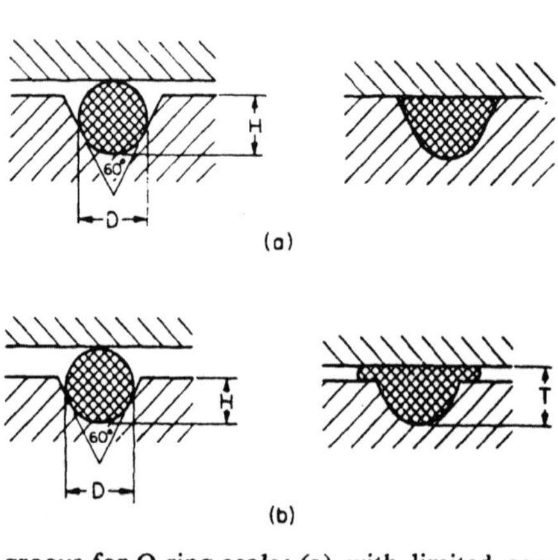

FIG. 3.65 Rounded V-groove for O-ring seals: (a) with limited compression; (b) with unlimited compression

b. *Grooves with triangular cross section.* If a V-shaped groove should receive an O-ring with metal-to-metal contact of the flanges assured at a compression ratio of 25 per cent, this groove would have an angle of about 112°. Such a groove cannot retain the O-ring, thus this kind of groove is not of general use.

By radiusing the apex of the groove, the so-called *rounded V-groove* is obtained, which is used either for seals with limited compression (Fig. 3.65a) or for seals, where just a part of the O-ring can be accommodated in the groove (Fig. 3.65b) the balance being squeezed between the flanges. The first kind of groove (Fig. 3.65a) is usually machined to a depth H of about 0.8 of the O-ring cross-sectional diameter. The second type (Fig. 3.65b) is made for a ratio $H/D \sim 0.6$, and a compression ratio $T/D \sim 0.8$ (Ardenne[47]). The di-

TABLE 3.22. DIMENSIONS OF ROUNDED V AND TRAPEZOIDAL (DOVETAIL)** GROOVES FOR ELASTOMER O-RINGS (mm)

O-ring* cross section nominal diameter		Rounded V-groove			Trapezoidal (dovetail) groove				
		limited	unlimited		40° (Fig. 3.66a, b)			30° (Fig. 3.67)	
		compression							
(mm)	(in.)	H (Fig. 3.65a)	H (Fig. 3.65b)	T*2 (Fig. 3.65b)	A*1	B*1	C*1	W*2	H*2
1.78	0.070	1.41	1.07	1.43	1.50	2.92	1.24	1.60	1.34
2.62	0.103	2.07	1.58	2.10	2.24	3.60	1.88	2.36	1.93
3.0	—	2.36	1.80	2.4	2.56	4.15	2.15	2.70	2.25
3.53	0.139	2.80	2.12	2.84	3.06	4.92	2.54	3.17	2.65
4.0	—	3.16	2.40	3.2	3.50	5.52	2.85	3.60	3.00
5.0	—	3.95	3.0	4.0	4.38	6.90	3.58	4.50	3.75
5.33	0.210	4.22	3.3	4.3	4.70	7.40	3.78	4.80	4.0
6.0	—	4.75	3.6	4.8	5.30	8.40	4.35	5.40	4.5
7.0	0.275	5.55	4.2	5.6	6.15	10.0	5.15	6.30	5.25
8.0	(0.312)	6.30	4.8	6.4	7.00	10.9	5.82	7.2	6.00
10.0	—	7.90	6.0	8.0	8.90	14.0	7.40	9.0	7.50
12.7	0.500	10.0	7.6	10.0	11.30	18.20	9.45	11.4	9.55

* See Table 3.19A. ** For dimensions of flanges see Section 38.47.
*1 — Edwards[327] (with ±0.05 mm). *2 — Ardenne[47].

mensions of the V-grooves corresponding to various O-rings are given in Table 3.22.

Garrod[430] used O-rings placed in V-shaped grooves (Fig. 5.19a) to seal the joint of two flanges permitting relative translation movement.

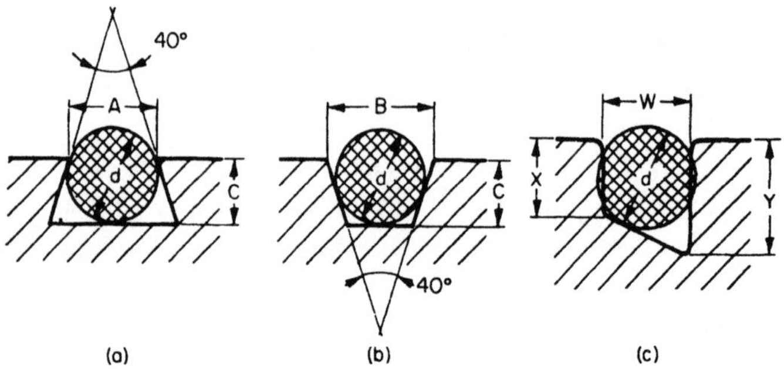

FIG. 3.66 Trapezium grooves: (a) dovetail; (b) open trapezium; (c) trapezium with parallel walls

c. *Grooves with trapezium cross section.* Three types of trapezium grooves are used to receive O-rings: the *open trapezium* groove (Fig. 3.66b), the closed trapezium or *devetail* groove (Fig. 3.66a) and the trapezium groove with *parallel* side walls (Fig. 3.66c).

The open trapezium groove may be preferred due to easy machining, but has the drawback of having a poor retention of the O-ring (see Table 3.20). The dovetail groove is difficult to machine and always has some trapped volume in the seal, but has a very good retention of the gasket. It is recommended for seals of large diameter (Camack[191]) or seals where the groove should hold the gasket when the flange is raised (Fig. 3.67). For this groove $H/D = 0.75–0.8$; $W/D \approx 0.9$.

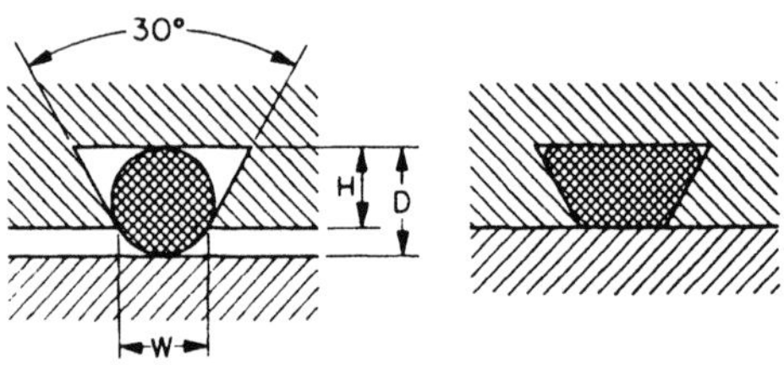

FIG. 3.67 Dimensions of dovetail grooves (Table 3.22)

Trapezium or dovetail grooves have to be machined with an apex angle of 30–40°. These grooves are almost always used in seals with metal-to-metal contact of the flanges. For a compression ratio of the gasket $C/d =$

= 0.72 and an apex angle θ the opening A of the groove (Fig. 3.66) is given by

$$A = \left(1.13 - 0.68 \operatorname{tg} \frac{\theta}{2}\right) \cdot d.$$

The dimensions of these grooves for various O-rings are listed in Table 3.22.

To avoid the difficult machining needed for the dovetail groove, as well as to eliminate the trapped volumes formed in these grooves, the trapezium groove with parallel sides was designed (Barton[92], Edwards[329]). This groove (Fig. 3.66c, 3.68) was obtained by reducing the width of a rectangular cross section groove (to obtain retention of the O-ring). However, in order to pro-

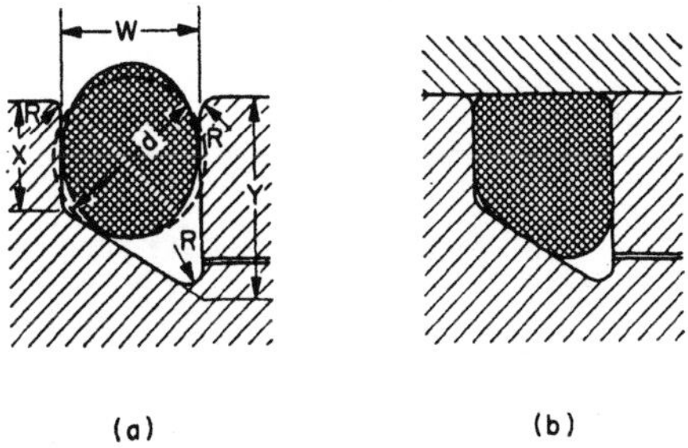

FIG. 3.68 Trapezium grooves for O-rings (dimensions see Table 3.23). *Courtesy of* Edwards[329]

vide sufficient cross-sectional area to completely accommodate the O-ring, and to obtain the required compression ratio for a good seal, the groove base is constructed with a slope.

For O-rings with cross-sectional diameter d, and tolerances as specified in Table 3.19A the dimensions of the groove (Fig. 3.68) are given by

$$W_{max} = d_{min} - 0.001 \quad \text{(in.)} \quad = d_{min} - 0.025 \text{ (mm)}$$
$$W_{min} = W_{max} - 0.03d \qquad X_{min} = A_{max}/W_{min} - 0.312 \cdot W_{min}$$
$$X_{max} = X_{min} + 0.03d \qquad Y_{min} = A_{max}/W_{min} + 0.312 \cdot W_{min}$$
$$Y_{max} = Y_{min} + 0.03d \qquad A_{max} = \pi \cdot d_{max}^2/4.$$

Table 3.23 lists the groove dimensions for various O-rings.

In this groove the O-ring completely fills the vacuum side (Fig. 3.68), a very small volume of gas being trapped on the atmospheric side, which if required can be evacuated through a small connexion provided in the flange. This connexion can be made axially (T_1, Fig. 3.69a) or radially (T_2, Fig. 3.69a). The seal can be constructed with double grooves for concentric O-rings (Fig. 3.69b), adding the advantages of the seals with guard vacuum (Section 38.23) to those of the trapezium groove O-ring seal.

TABLE 3.23. DIMENSIONS OF TRAPEZIUM GROOVES WITH PARALLEL SIDE WALLS (mm)
Edwards[329], Barton[92]

O-ring* cross section nominal diameter		Minim.	Maxim.	Groove (Fig. 3.68)						
				W		*X*		*Y*		*R*
(mm)	(in.)	diameter (mm)		max	min	max	min	max	min	
1.78	0.070	1.71	1.85	1.68	1.63	1.19	1.14	2.21	2.16	0.12
2.62	0.103	2.55	2.69	2.52	2.44	1.65	1.58	3.16	3.10	0.12
3.0	–	2.93	3.07	2.90	2.81	1.85	1.76	3.63	3.52	0.12
3.53	0.139	3.43	3.63	3.40	3.30	2.18	2.08	4.27	4.17	0.12
4.0	–	3.90	4.10	3.87	3.75	2.44	2.32	4.45	4.33	0.20
5.0	–	4.90	5.10	4.87	4.72	2.69	2.54	5.61	5.46	0.20
5.33	0.210	5.21	5.45	5.18	5.03	3.23	3.04	6.38	6.22	0.20
6.0	–	5.85	6.15	5.82	5.64	3.28	3.10	6.78	6.60	0.25
7.0	0.275	6.85	7.15	6.81	6.60	4.19	3.99	8.31	8.10	0.25
8.0	(0.312)	7.74	8.14	7.70	7.47	4.83	4.60	9.50	9.27	0.25
10.0	–	9.70	10.30	9.67	9.37	5.23	4.93	11.07	10.77	0.25
12.7	0.500	12.44	13.00	12.42	12.04	7.57	7.19	15.09	14.71	0.25

* See Table 3.19A.

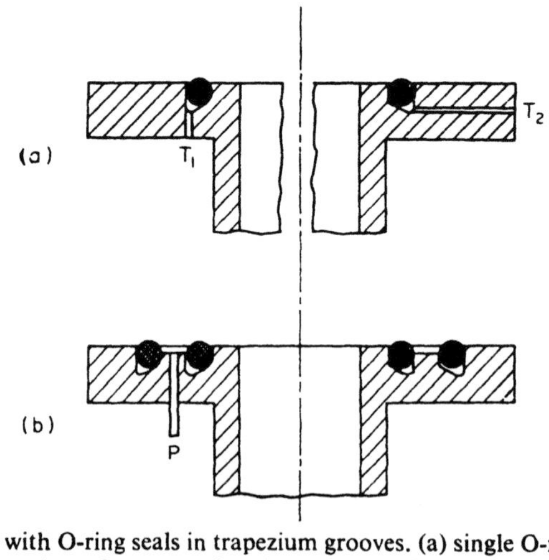

FIG. 3.69 Flanges with O-ring seals in trapezium grooves. (a) single O-ring; T_1 and T_2, axial and radial pumpout ducts for evacuating the trapped volume; (b) double groove seal with guard vacuum; P = pumpout duct

Adam[5] used a trapezium cross section groove (1.50 mm width; 1.37 mm depth on the deep side, 0.46 mm depth on the shallow side) to receive a 1.59 mm diameter indium wire (see also Table 3.13).

d. *Grooves with semi-circular and elliptical cross section.* Grooves with semi-circular cross sections do not form trapped volumes, but cannot receive the entire volume of the O-ring, thus these grooves are not used for elastomer O-rings, where the seal should be with limited compression.

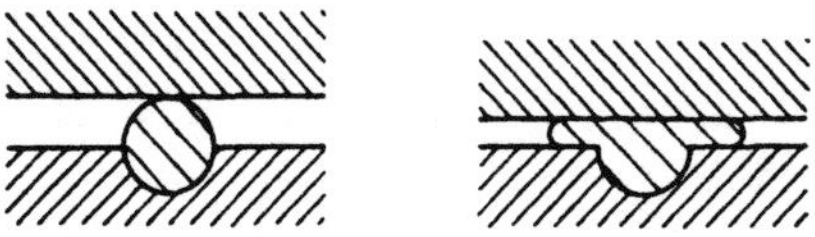

FIG. 3.70 Semi-circular cross section groove for seal with unlimited compression

Semi-circular cross-sectional grooves are used with metal O-rings (Fig. 3.70). Heywood[555] quotes vacuum seals with grooves of semi-circular cross section using single or double aluminium or nickel wire rings. Fraser[400] describes a seal for cryogenics, using an indium wire O-ring seated in a semi-circular cross-sectional groove, having its radius equal to the radius of the indium wire. Bishop[129] used a lead wire (3.1 mm diameter) O-ring sealed in a circular cross-sectional groove with the same width as the wire but less deep (2.4 mm) as its diameter.

The use of elastomer O-rings in grooves with semi-circular cross sections is mentioned by Doty[295], Branson[162]. These were used to seal glass tubes,

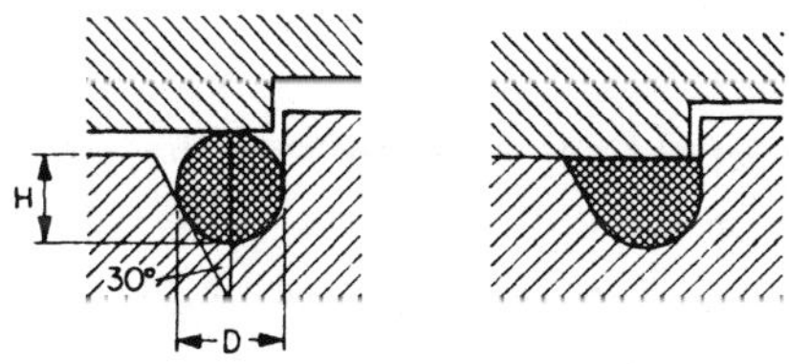

FIG. 3.71 O-ring seal, in ovoidal cross section groove

their edges were thickened, and on the edge a groove was formed, having the cross section a half circle with the radius equal to that of the standard O-rings (see Table 3.19A).

Elastomer O-rings sealed in semi-circular grooves may be used in sliding seals. A seal consisting of concentric glass tubes sliding in each other, using two O-rings placed in semi-circular grooves are described by Gore[462] (Fig. 5.21a). Kalmus[650] describes a similar seal for concentric metal pipes (Fig. 5.43).

Butterfly valves (Section 61.32h) have the O-rings seated in grooves cut in the edge of the valve plate (Holland[585], Edwards[328]).

Dorsten[204] describes a seal where the O-ring is placed in a groove having its cross section as *half an ovoid* (Fig. 3.71). This shape tends to correct the drawbacks of semi-circular cross section grooves i.e. to receive and to hold the O-ring. The dimensions recommended for this groove are determined by the ratio $H/D = 0.82$.

e. *Grooves for rectangular flanges.* Rectangular flanges may be equipped with O-ring gaskets (of circular cross section) by cutting the groove of one of the cross sections used for circular O-ring seals, but following the rectangular outline of the flange. The straight parts of the groove generally present no problem, but the corners are always a troublesome detail both for the machining and the fitting of the gasket in the groove.

Guthrie[491] describes a method of cutting straight grooves through the edge of the flange (Fig. 3.72a) which are then sealed with metal fillets or pieces of

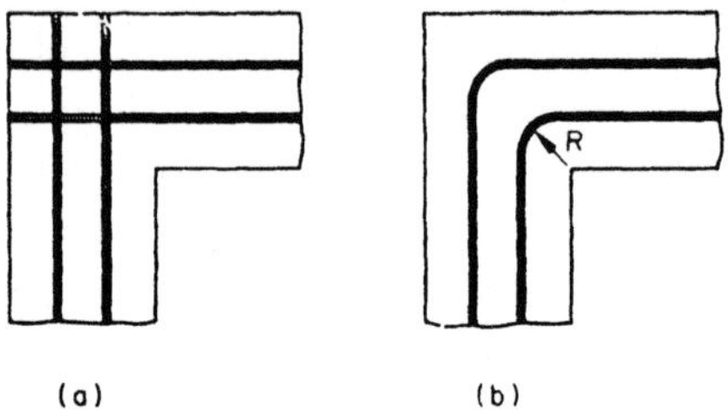

FIG. 3.72 Constructions for the corners of rectangular outline grooves: (a) with fillet; (b) with rounded corners

cord introduced in the groove outside the outline of the required groove (Fig. 3.72a). It is very important however that the grooves make no right angle bends (Kurie[723]), especially for flanges using very small gaskets (less than 3 mm cross section diameter), where small compression distances are allowed (Guthrie[491]). It is recommended that the perpendicular sections of the groove should be connected by a circular bend (Fig. 3.72b) whose radius is not less than that of the O-ring with corresponding cross section (Table 3.19A). The curve in Fig. 3.73 gives the smallest radii to be used at the rounded corners of rectangular outline grooves as a function of the cross-sectional diameter of the O-ring.

In vacuum seals grooves having even more difficult outlines than those required on rectangular flanges are sometimes necessary. The rectangular groove on a cylindrical surface is an example of such a difficult construction. Smotrich[1159] describes a seal using an O-ring seated in a rectangular groove machined on the (lateral) surface of a cylinder (see Fig. 5.46), and Louckes[785] presents a seal having a flexible strip sliding over the O-ring placed in a groove on the surface of, a cylinder (see Fig. 5.48).

A method of cutting grooves on the side of cylinders is described by Stark[1179a].

38.42 Spacer seals. The simplest O-ring seal consists of two flat flanges compressing the gasket between them. Unfortunately this seal does not provide for centring, retention and the other conditions required for any O-ring

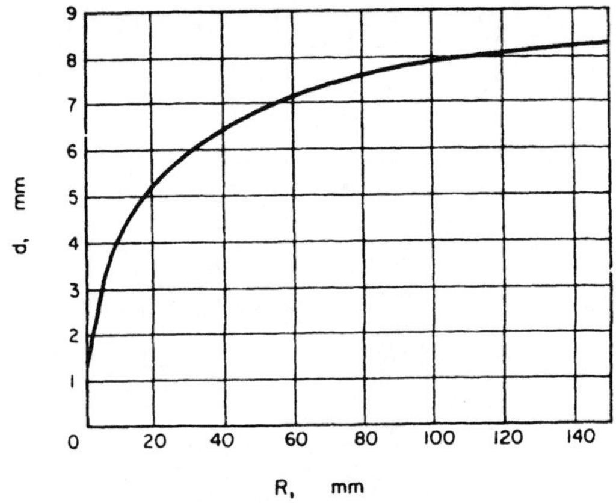

FIG. 3.73 Minimum radius of curvature (*R*) for gaskets, as a function of the cross section diameter (*d*) of the gasket

seal (Table 3.20), and because of the lack of a limit for compression it cannot be used with elastomer O-rings. Nevertheless simple seals of this kind (Fig. 3.74) were used with metal gaskets. Green[478] describes a seal using an O-ring of lead wire compressed between flat (glass) flanges. The lead wire of 1.6 mm diameter (1, Fig. 3.75) was compressed between two sections of a glass pipeline joint (supplied by Quickfit Visible Flow Ltd., see Q.V.F.[1021]), by pushing

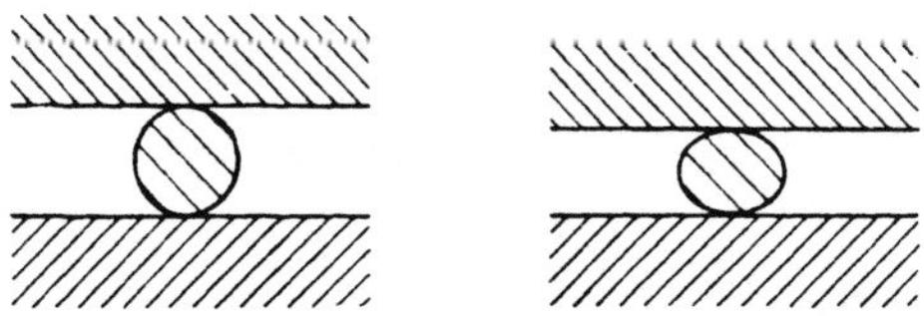

FIG. 3.74 O-ring (wire) sealed between flat flanges

together the two parts with the backing flanges and rings (3,4) fastened to the tapered (ground) surfaces of the pipes by inserts of graphite-impregnated asbestos (2, Fig. 3.75). A joint of this type (50 mm diameter) was baked at 300 °C for long periods (up to 64 hr) achieving in the system pressures of the order of 10^{-7} torr. A similar seal using indium gaskets has been utilized by Nelson[916] in dismountable electrodes (Fig. 4.10), and by Turner[1242] in a connexion of an electrode ring with radially displaced electrodes (Fig. 4.34).

Peters[986a] used high purity iron wire, compressed (Table 3.13) between tin-plated flanges with good surface finishes (Table 3.37). Redman[1036a] made leak-tight seals using a 99 per cent aluminium alloy wire gasket compressed (Table 3.13) between flat flanges.

Van Heerden[527] describes an O-ring seal, using OFHC copper O-rings* sealed between flat flanges, without any seat for the gasket. This is justified by the designer by the fact that only a flat surface allows for a perfect

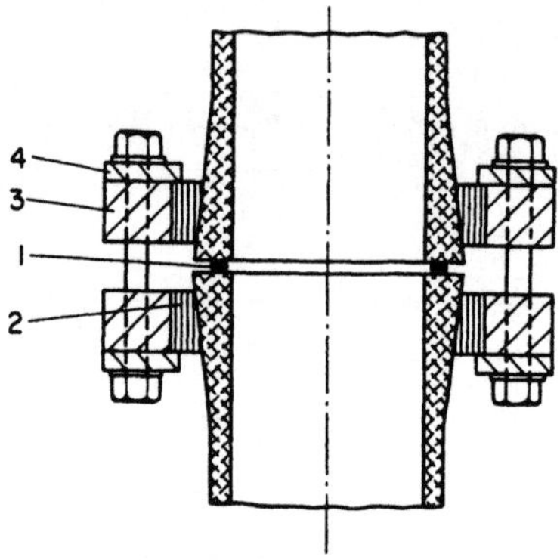

FIG. 3.75 Seal with lead O-ring between glass flanges. After Green[478] *(Courtesy of The Institute of Physics and The Physical Society, London)*

polishing. The position of the O-ring was nevertheless fixed by a flat ring having a thickness of half the thickness of the gasket, and screwed down on one of the flanges. This seal required retightening (Section 38.48) during the bakeout at 400 °C, since the copper O-ring expands more than the stainless steel flanges and the seal tends to open up during the cooling period. A disadvantage of this seal is that during bakeout the exposed edge of the O-ring oxidizes, the copper oxide thus formed evaporates and condenses on the flanges forming an irregular layer, which must always be removed by repolishing before each resealing.

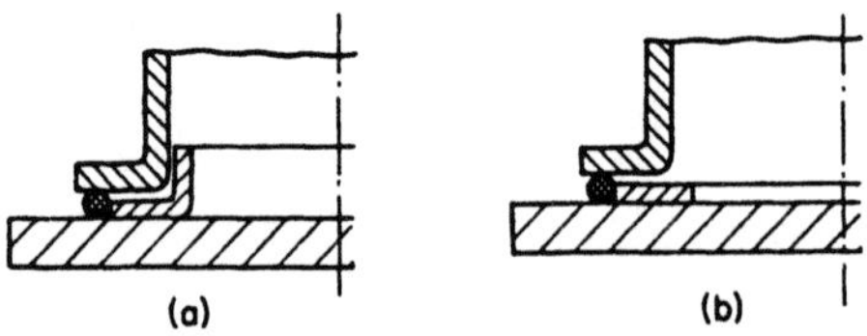

FIG. 3.76 Bell jar seals with O-rings, using (a) retaining plate, (b) retaining ring. After Williams[1318] *(Courtesy of The American Institute of Physics)*

* For the construction and treatment of this O-ring see Section 38.33.

If rubber O-rings are used between flat flanges and especially if they are greased, the danger exists that the O-ring could be sucked in. This may be avoided by placing a washer inside the O-ring (Lempicki[757]). If the thickness of this washer is about 0.7 of the cross section diameter of the O-ring, the washer serves also as *spacer*, which limits the compression of the O-ring.

This technique is used (Williams[1318]) in bell jar seals, having either a retaining plate (Fig. 3.76a) or a retaining ring (Fig. 3.76b).

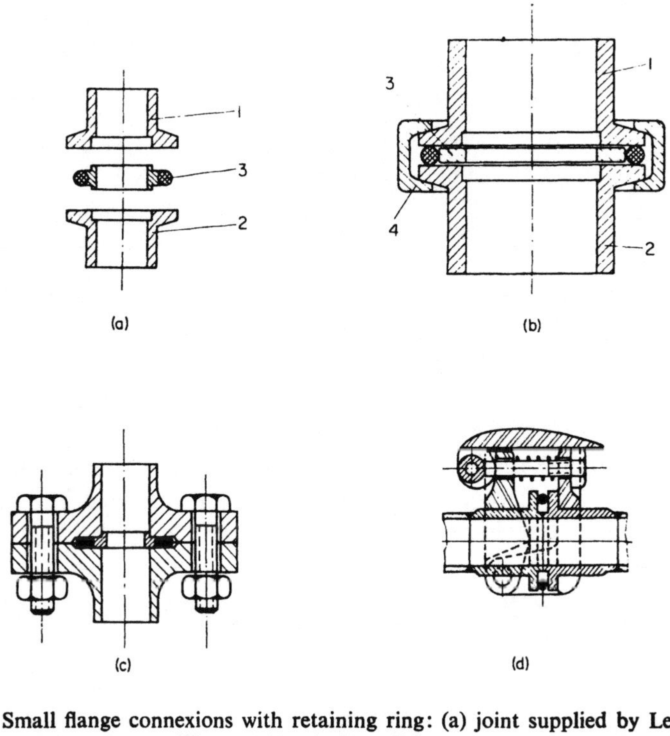

FIG. 3.77 Small flange connexions with retaining ring: (a) joint supplied by Leybold[762]; (b) joint supplied by Heraeus[542]; (1, 2) identical small flanges; (3) retaining ring; (4) clamping ring; (c) unit flange connexion supplied by Leybold[762]; (d) Mecavide connexion supplied by Laboratoires des Basses Pressions[728]

In order to hold the O-ring the retaining rings used in flange seals have generally a groove machined on their circumference. The groove has a V shape (or rounded V) or a semi-circular shape (Dunkel[309]). Small vacuum connexions using this type of seal are commercially available (Leybold[762], Heraeus[542], Lab. des Basses Pressions[728]) for pipes up to about 40 mm diameter. Figure 3.77 shows these seals in the closed and open positions. The parts of the Heraeus[542] seal can be seen on Plate 3 and its cross section on Plate 4.

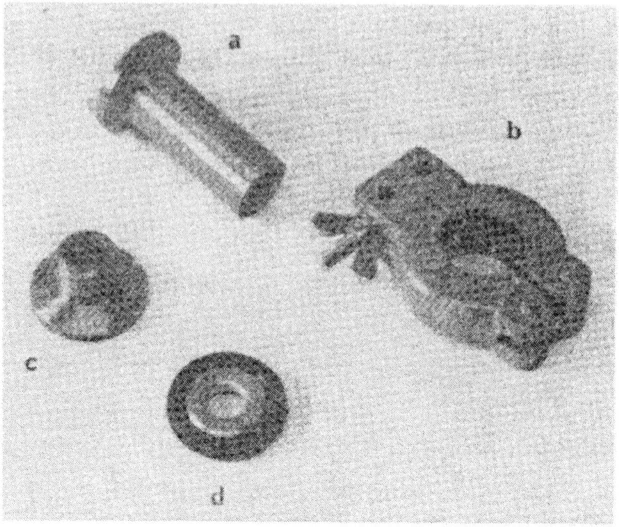

PLATE 3. Component parts of O-ring seals *(Courtesy of Heraeus[542])*

The Leybold[762] seal (Fig. 3.77a) has a retaining ring provided with rims to centre it in the recess of the flanges. In the Leybold[762] unit flange connexions (Fig. 3.77c) available for pipes up to 250 mm diameter, the retaining ring has its centring rim just on one side and the sealing surfaces are recessed, thus protecting them from damage during the handling of the flanges.

A similar joint (for connexions of 10–30 mm diameter) which utilizes a different clamping device (Fig. 3.77d) is available from Lab. des Basses Pressions[728].

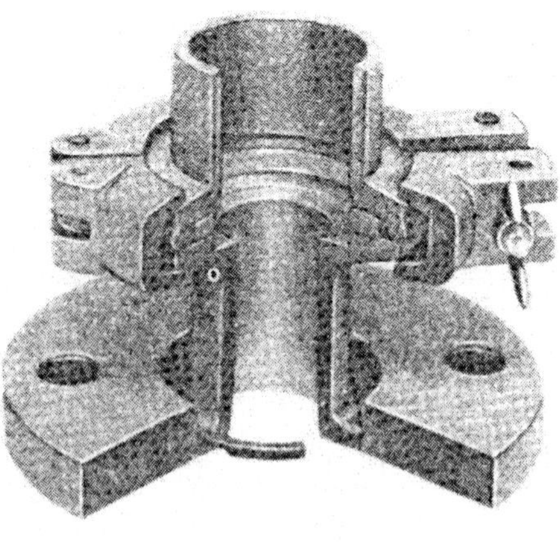

PLATE 4. Cross section through an O-ring seal *(Courtesy of Heraeus[542])*

By using concentric spacer rings placed inside and outside the O-ring, this can be locked as in a groove. Burrows[185] proposed a seal in which the elastomer O-ring is retained between two concentric metal rings, grooved on the surfaces facing the O-ring. The grooves are V-shaped (Fig. 3.78). In this seal, the relation between the O-ring cross-sectional diameter *(d)* and the dimensions of the groove should be:

$$x = d/2 \cos \theta,$$
$$\pi \cdot d^2/4 = 2 \cdot x \cdot y - y^2 \cdot \operatorname{tg} \theta/2.$$

For an angle $\theta = 20°$ (i.e. the angle of the V-groove is 140°) the dimensions on the grooves on the spacer rings are given by

$$x = 0.53 \cdot d,$$
$$y = 0.87 \cdot d.$$

Burrows[185] recommends that the faces of the retaining ring be slightly tapered (see dotted line on Fig. 3.78b) in order to counteract any out-of-parallel position of the flange surfaces during clamping.

Seals using O-rings kept between concentric retaining rings are available from Heraeus[542] and Geraetebau-Anstalt Balzers[444]. Figure 3.79 and Plate 6 show these seals and Tables 3.24 and 3.25 give their dimensions. Plate 5 shows a series of spacer rings holding the O-rings (Heraeus[542]).

Stiff[1190] described a spacer on which the O-ring is enclosed between two concentric cylindrical rings (1, 2, Fig. 3.80a) and the ends of the pipes (3, 4, Fig. 3.80a). To stop the rings from sliding a counterbore is machined in the pipe ends (the wall must be thick); the counterbore serves also as a stop for

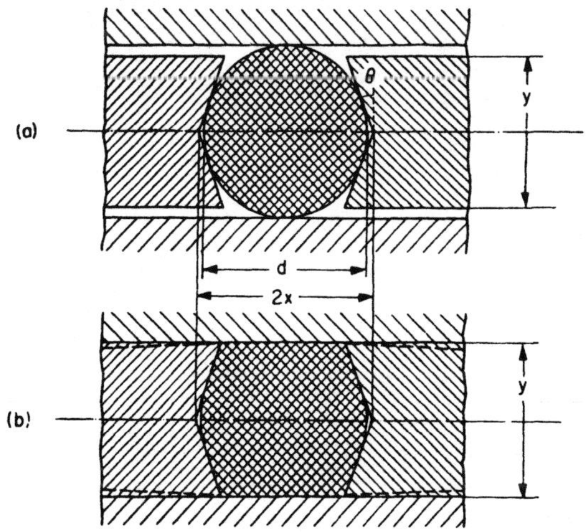

FIG. 3.78 Spacer seal with V-shaped grooves: (a) open; (b) closed (after Burrows[185])

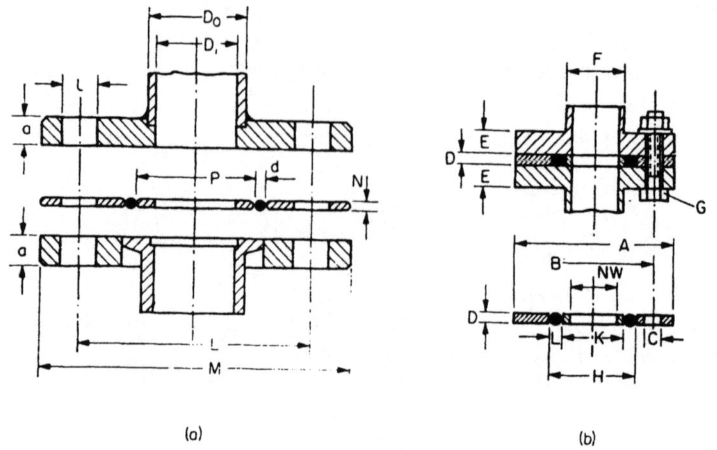

FIG. 3.79 Flanged unions with spacer discs: (a) union supplied by Heraeus[542]; (b) union supplied by Balzers[444]

TABLE 3.24. DIMENSIONS (mm) OF FLANGED UNIONS WITH SPACER DISCS (Fig. 3.79a) PRODUCED BY Heraeus[542]

NW	O-ring cross-sectional diameter	Spacer disc				Pipe and flange				Number of holes	Bolts
		P	*N*	*L*	*M*	D_i	D_0	*a*	*l*		
10	5	20	4.0	50	75	10	15	10	11.5	4	M 10×30
25	5	40	4.0	75	100	25	30	10	11.5	4	M 10×30
32	5	50	4.0	90	120	32	38	12	14	4	M 12×30
50	5	70	4.0	110	140	50	55	12	14	4	M 12×35
65	5	85	4.0	130	160	65	70	12	14	4	M 12×35
80	5	100	4.0	150	190	80	(85)	(14)	18	4	M 12×40
100	6	120	4.8	170	210	100	108	14	18	4	M 12×40
125	6	150	4.8	200	240	125	(133)	14	18	8	M 16×45
150	6	175	4.8	225	265	150	159	14	18	8	M 16×45
200	8	230	6.4	280	320	200	(209)	14	18	8	M 16×45
250	8	280	6.4	335	375	250	256	14	18	12	M 16×46

TABLE 3.25. DIMENSIONS (mm) OF FLANGED UNIONS WITH SPACER DISCS (Fig. 3.79b) PRODUCED BY Balzers[444]

NW	O-ring cross-sectional diameter	Space disc					Pipe and flange			Number of holes	Bolts
		K	*D*	*B*	*A*	H_{min}	F_{max}	*E*	*C*		
16	4	22	3	42	55	34	28	8	5.8	4	M 5×25
27	4	34	3	52	65	44	40	8	5.8	4	M 5×25
36	4	42	3	64	78	52	50	8	7.0	4	M 6×30
52	4	60	3	86	105	70	–	10	9.5	4	M 8×35
76	4	80	3	112	136	94	–	12	11.5	4	M 10×40
95	5	105	4	135	158	118	–	12	11.5	4	M 10×40
138	6	148	4.5	190	220	168	–	14	14.0	4	M 12×50
205	8	218	6	270	300	245	–	16	14.0	8	M 12×55
330	8	330	6	390	420	368	–	16	14.0	8	M 12×55
690	10	690	8	750	780	730	–	20	14.0	12	M 12×65

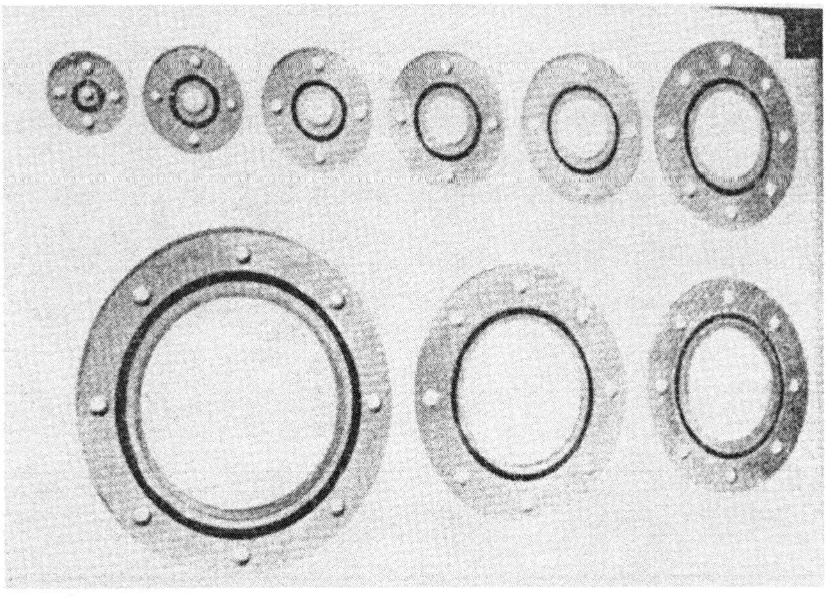

PLATE 5. Sealing discs with O-rings *(Courtesy of Heraeus[542])*

PLATE 6. Spacer seal *(Courtesy of Heraeus[542])*

the compression of the O-ring. The components are pulled together by the clamping flanges (5) pulling against the external spring steel retaining rings (6).

Edwards[329] recently produced a series of quick couplings based on a similar principle (Fig. 3.80b) using a single spacer cylindrical ring which holds the O-ring.

Bachman[65] sealed two pipes, one fitting inside the other, using the inside pipe as the inside cylindrical ring which holds the O-ring, and sealing the O-ring in the counterbore (step) cut in the inside of the larger pipe.

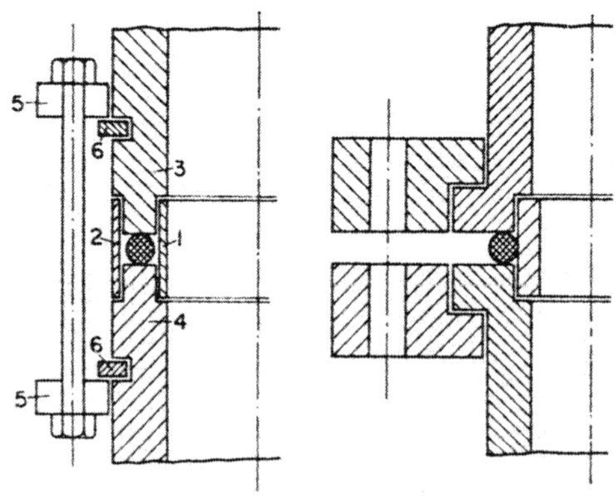

FIG. 3.80 O-ring seals with spacers: (a) with concentric tubular spacers (after Stiff[1190]); (b) with single spacer (after Edwards[329])

Richards[1053a] made a seal by placing two O-rings in a tapered joint and compressing them in the taper by means of a combination of tapered brass spacers. Feldman[361b] sealed a gold O-ring between flat flanges, by holding the O-ring from inside on a cylindrical spacer ring.

38.43 Conical seals. A pipe 1 (Fig. 3.81) can be tightly joined to a flange (2) of a vacuum vessel or pipe, by surrounding it with an O-ring, and compressing the O-ring towards the pipe and the flange with the *conically shaped* sealing

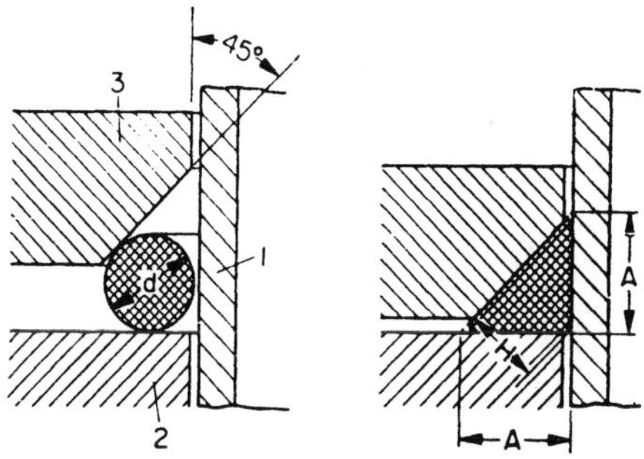

FIG. 3.81 Principle of conical seals using O-rings

ring or washer (3). The conical surface should form an angle of 45° with the surfaces of the pipe (1) and the flange (2) (Cassel[196], Pollard[996], Moore[882], Pollerman[998]). In this seal the cross section of the O-ring can be *correctly enclosed* in the triangle formed by the three sealing surfaces,

$$A \approx 1.32 \cdot d.$$

This corresponds to a ratio $H/d = 0.92$ (Fig. 3.81) just enough to assure a tight seal (see Fig. 3.50). If the dimension A of the washer is smaller than the above mentioned *optimum*, a seal with limited compression of the gasket is *not any more possible* (Fig. 3.82a). With values of A greater than the optimum, the compression ratio of the O-ring risks being insufficient (Fig. 3.82b) for a tight seal and may enclose large trapped volumes.

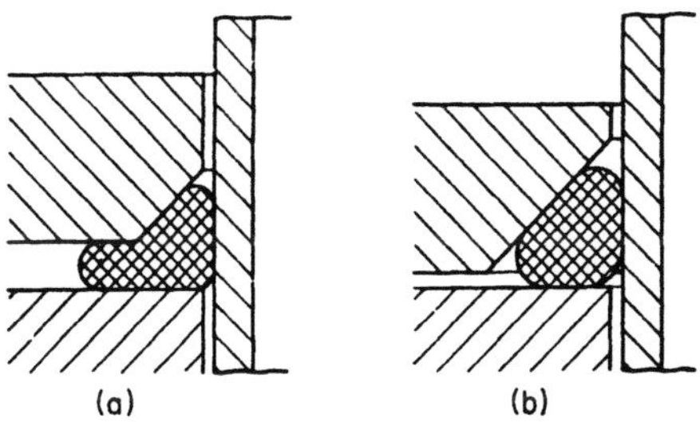

FIG. 3.82 Conical seals: (a) with dimension A (Fig. 3.81) less than optimum; (b) with dimension A more than optimum

The conical O-ring seal is used in three basic arrangements (Fig. 3.83). The arrangements differ in the relative position of the conical surface with respect to the evacuated space as well as in the number of sealing components. In the arrangement shown in Fig. 3.83a, the conical surface is placed

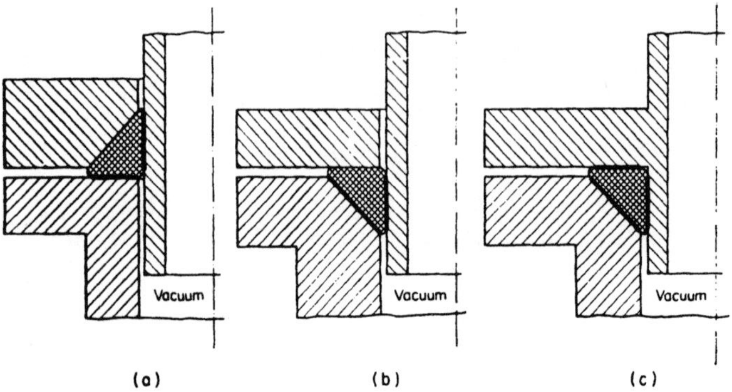

FIG. 3.83 Basic arrangements for conical O-ring seals: (a, b) seals consisting of three parts; (a) conical washer on atmospheric side; (b) conically shaped flange on vacuum side; (c) seal consisting of two parts with conically shaped flange on vacuum side

on the atmospheric side of the seal, and the vacuum seal is on the plane flange and the cylindrical surface of the pipe. This seal is the easiest to be constructed, especially when the pipe to be sealed is made of glass since it has a smooth surface. Here only the surface of the flange should have a good surface finish (Table 3.37) while the finish of the conical surface has no importance.

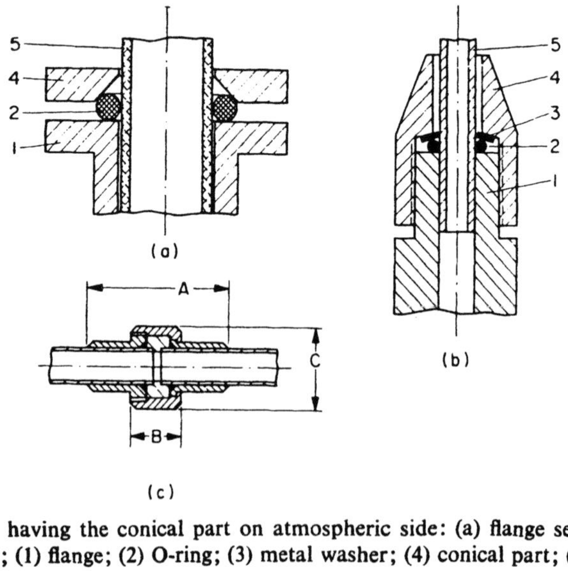

FIG. 3.84 Seals having the conical part on atmospheric side: (a) flange seal; (b) union; (c) double joint; (1) flange; (2) O-ring; (3) metal washer; (4) conical part; (5) joined tube

In the seal shown in Fig. 3.83b, the two sealing surfaces are the conical and the cylindrical, and generally it is more difficult to obtain a good surface finish on a conical than on a plane surface.

For better centring of the parts, the seal on Fig. 3.83c may be used. This seal is made between two parts instead of three, but does not permit the sealing directly on the surface of one of the pipes.

Figure 3.84 shows some O-ring seals of the type presented by Fig. 3.83a, with the conical surface on the atmospheric side. Pollerman[998], Ardenne[47], Espe[354], and Blanc[131] quote seals as shown in Fig. 3.84a, especially for joining glass or ceramic pipes to metal vacuum systems. Heinrich[532] describes a conical seal (Fig. 3.84b) obtained by machining the conical surface of the male portion of a standard flare fitting flat. He used this joint on tubings of 4–8 mm diameter, obtaining leak tightness (at the helium leak detector).

Vacuum unions based on the arrangement of Fig. 3.83a are available from N.G.N. Electrical[926] in the form of couplings using two O-rings (Fig. 3.84c). Plate 7 shows the parts of such an union, and Table 3.26 lists their dimensions.

PLATE 7. Vacuum union and its component parts *(Courtesy of N.G.N. Electrical[926])*

A conical seal having the conical part on the vacuum side (as in Fig. 3.83b) may be used to seal screwed plugs (Fig. 3.85a), pipes to vacuum vessels (Fig. 3.85b), pipes to pipes (Fig. 3.85c, d) or end plates to pipes (Fig. 3.85e).

For the plug seal (Fig. 3.85a) the side of the hole is bevelled at 45°, and a washer is used to avoid drag on the O-ring during the tightening of the seal. If the bevel is difficult to machine to a good surface finish, the plug may be sealed by placing the O-ring in a groove machined in the head of the plug (see Fig. 3.61b) or a step seal may be (see Fig. 3.95b).

The couplings for pipe-to-vessel connexion (Fig. 3.85b) are available from Genevac[442] (Style C_1 solderless connexion). In these connexions the pipe

TABLE 3.26. TUBE TO TUBE UNIONS, PRODUCED BY N.G.N. Electrical[926] (Fig. 3.84c; dimensions in inches)

Nominal bore	Length A	Length B	Diameter C	Length of the tube in the union	O-ring size B.S. (see Table 3.19A)
1/4	1 11/16	11/16	7/8	13/16	6
1/2	2 5/16	13/16	1 1/4	1 1/8	11
3/4	2 9/16	15/16	1 9/16	1 1/4	16
1	2 13/16	1 1/16	1 7/8	1 3/8	21
1 1/4	3 5/16	1 1/16	2 3/16	1 5/8	25
1 1/2	3 13/16	1 5/16	2 5/8	1 7/8	29
2	4 15/16	1 11/16	3 1/4	2 7/16	33

(1) is connected to the port (5) of the vacuum vessel, by pressing the washer (2) over the O-ring (4) with the aid of the locking ring (3).

Pollard[996] used seals with conical bevel (Fig. 3.85c). A cylindrical body (1) of bore greater by 0.1–0.25 mm than the outside diameter of the pipes to be joined, has 45° bevels on each end (2). An O-ring (3) is slipped over each

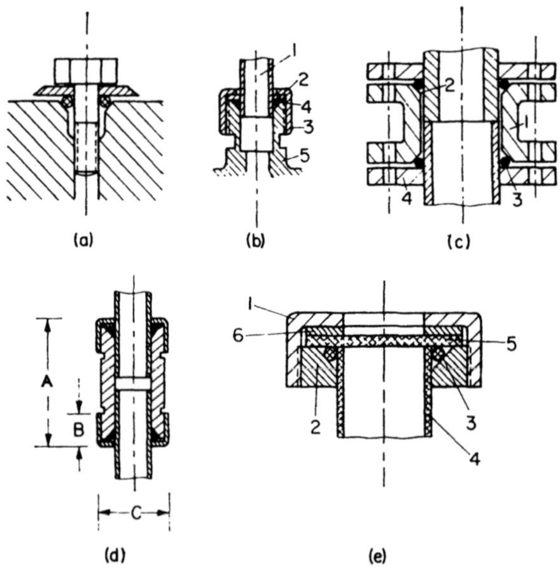

FIG. 3.85 Seals having the conical part on vacuum side: (a) plug seal; (b) pipe-to-vessel seal; (c, d) pipe-to-pipe seal; (e) end plate seal

pipe, and compressed into the bevel by the clamping flange (4). This type of seal is used in the couplings available from N.G.N. Electrical[926] (Fig. 3.85d). The components of these couplings are shown on Plate 8, and their dimensions are listed in Table 3.26A.

TABLE 3.26A. DIMENSIONS (inches) OF VACUUM COUPLINGS SUPPLIED BY N.G.N. Electrical[926] (see Fig. 3.85d)

Nominal bore	*A*	*B*	*C*	O-ring
1/4	1 3/8	3/8	1 1/16	VR 6
1/2	1 7/8	1/2	1 1/4	VR 11
3/4	2 5/16	5/8	1 9/16	VR 16
1	2 13/16	11/16	1 15/16	VR 21
1 1/4	3 1/4	15/16	2 1/8	VR 25
1 1/2	3 13/16	1	2 5/8	VR 29
2	4 5/8	1 1/8	3 1/4	VR 33

A demountable vacuum seal for attaching an end-plate to a glass tube is described by Moore[882]. In this seal (Fig. 3.85e) the O-ring (3) is compressed between the end plate, or window (5), the pipe (4) and the conically shaped collar (2) by the lock ring (1) and washer (6).

The seal of Fig. 3.86a (Leblanc[750]) is tightened by the nut (1) which applies a compressing force on the O-ring (3) by means of the ring (2). The O-ring

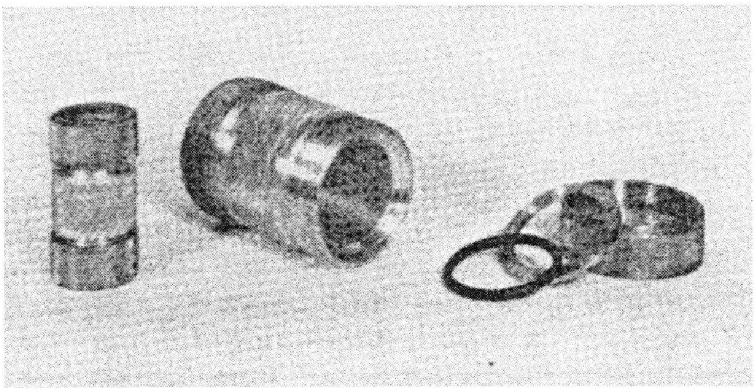

PLATE 8. Vacuum coupling *(Courtesy of N.G.N. Electrical[926])*

is thus sealed towards the glass pipe (4) and the polished tapered shoulder on which the O-ring is seated.

Horwitz[598] provided a collar on the end of a Pyrex pipe (Fig. 3.86b) and lapped a 15° bevel (4) around its outer edge. An indium O-ring (1) was pressed between this bevel and a cap (2).

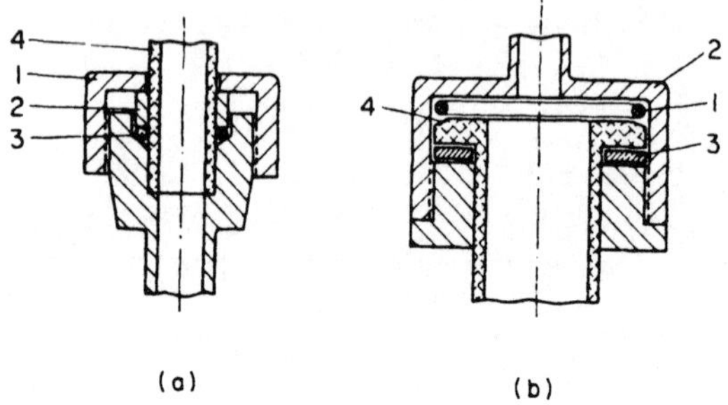

FIG. 3.86 Special conical O-ring seals: (a) glass-to-metal connector (after Leblanc[750]); (b) indium O-ring seal (after Horwitz[598]); (1) indium O-ring; (2) cap; (3) washer; (4) collar with edge ground to 15° bevel *(Courtesy of The American Institute of Physics)*

Garrod[432] describes a joint system which is readily adaptable to various sizes of glass or metal pipes (Fig. 3.87), using conical, O-ring seals, with a 26.5° taper. To make a vacuum-tight seal between pipes (1 and 2, Fig. 3.87a) the disc (3) of an appropriate size is chosen from the series (Fig. 3.87b). The

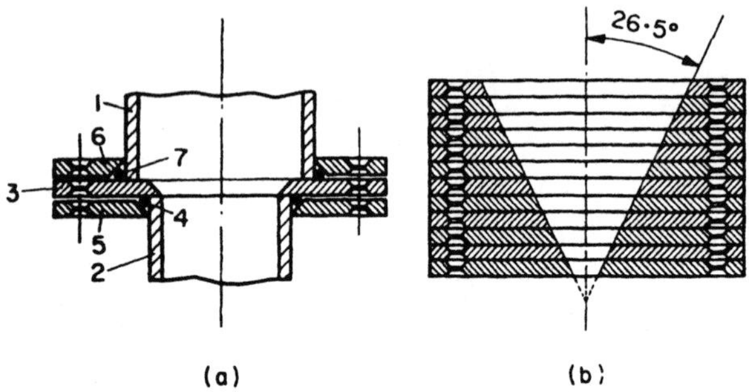

FIG. 3.87 Joints with ready adaptability to different sizes of pipes: (a) joint; (b) set of sealing plates. After Garrod[432] *(Courtesy of The Institute of Physics and The Physical Society, London)*

disc is chosen so that the smaller pipe just sits on it. The O-ring (4) is held in position by the disc (5) and a similar joint is made for the pipe (1) by a third disc (6) and O-ring (7). Due to the countersinking of the screw holes, the first disc (5) is sealed to (3) by 4 screws and after the second seal is made by another 4 screws.

Figure 3.88 shows the upper end of a vacuum chamber constructed by the author for heating objects with radio frequency currents. The lower end has the same kind of seals, and has a shaft seal (Section 51.71) on its axis which enables the object in the heating space of the r.f. coil to be raised.

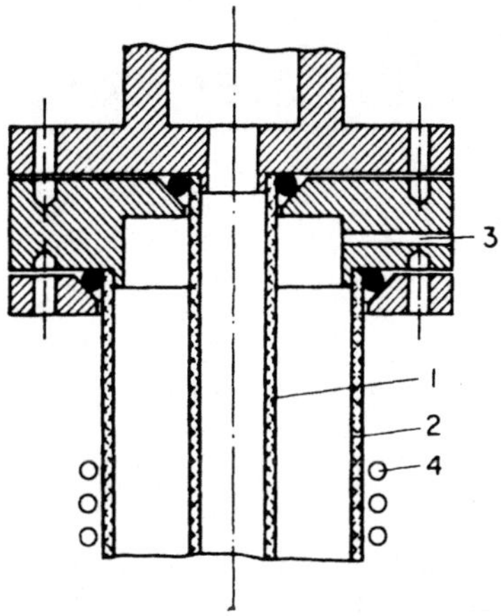

FIG. 3.88 Conical O-ring seals on a water-cooled double walled vacuum chamber: (1) quartz tube; (2) Pyrex tube; (3) water inlet; (4) r.f. coil

Centred conical connexions (of the type shown on Fig. 3.83c) are available from Edwards[329] (Fig. 3.89 and Plate 9). The union members are tightened by a securing nut, to form a rigid metal to metal joint, fully trapping the O-

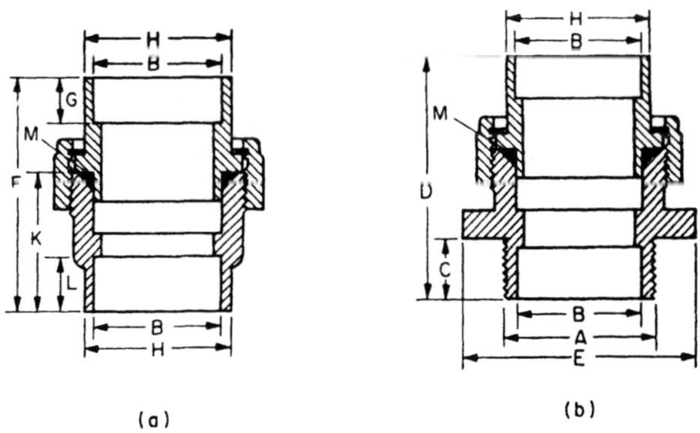

FIG. 3.89 Vacuum unions, supplied by Edwards[329]: (a) with plane ends; (b) with one end screwed

ring in a seal with limited compression. The knurled nut is fitted after the two halves of the union have been brazed in position, and secured by a retaining ring sprung into a recess in the knurled nut. The dimensions of these unions are given in Table 3.27.

Hodder[570] described a conical seal, modified from a refrigeration type fitting (Fig. 3.90a) by machining a groove of a special shape (Fig. 3.90a) to retain the O-ring. The dimensions of these seals are given in Table 3.28.

TABLE 3.27. VACUUM UNIONS SUPPLIED BY Edwards[329] (Fig. 3.89)

Size (in)		1/16	1/8	1/4	1/2	3/4	1	1 1/2	2
A	(in.)	1/8 BSP	1/8 BSP	1/4 BSP	1/2 BSP	3/4 BSP	1 BSP	—	—
B	(in.)	0.125	0.205	0.346	0.596	0.846	1.112	1.612	2.128
	(mm)	3.18	5.42	9.22	15.1	21.5	28.2	40.8	54.2
C	(in.)	3/8	3/8	3/8	3/8	1/2	1/2	—	—
	(mm)	9.52	9.52	9.52	9.52	12.7	12.7	—	—
D	(in.)	1 5/16	1 5/16	1 5/16	1 7/16	1 31/32	1 31/32	—	—
	(mm)	33.2	33.2	33.2	36.5	50.0	50.0	—	—
E	(in.)	0.520	0.705	0.915	1.195	1.475	1.850	—	—
	(mm)	13.4	17.9	23.2	30.3	37.4	48.0	—	—
F	(in.)	1	1	1	1 3/8	1 29/32	1 29/32	2 5/8	2 5/8
	(mm)	25.4	25.4	25.4	34.9	48.5	48.5	66.7	66.7
G	(in.)	1/4	1/4	1/4	3/8	3/8	3/8	3/4	3/4
	(mm)	6.35	6.35	6.35	9.52	9.52	9.52	19.1	19.1
H	(in.)	1/4	3/8	1/2	3/4	1	1 1/4	1 3/4	2 1/4
	(mm)	6.35	9.52	12.7	19.1	25.4	31.7	44.4	57.1
K	(in.)	1/2	1/2	1/2	3/4	1 1/8	1 1/8	1 1/2	1 1/2
	(mm)	12.7	12.7	12.7	19.3	28.6	28.6	38.1	38.1
L	(in.)	3/16	3/16	3/16	1/4	3/8	3/8	3/8	3/8
	(mm)	4.76	4.76	4.76	6.35	9.52	9.52	9.52	9.52
M*	VOR	101	2A	4A	121	130	136	146	159

* See Table 3.19A.

Schriever[1108] described a joint which uses a double O-ring seal on conical surfaces (Fig. 3.90b), having a guard vacuum between the two seals (see also Section 38.23). Edwards[324] made a simple seal called "rolling O-ring seal" (Fig. 3.90c), where the O-ring is placed over the end of the tube and the cover is pushed over it. This causes the O-ring to roll, and after one or more complete rotations the O-ring is in its equilibrium position and will hold the cover in position. This seal is useful only in low vacuum systems.

Conical joints with O-rings are very reliable and their applications are so varied that it is possible to build a complete vacuum system using only

Plate 9. Vacuum unions; one dismantled and showing retaining ring *(Courtesy of Edwards*[328]*)*

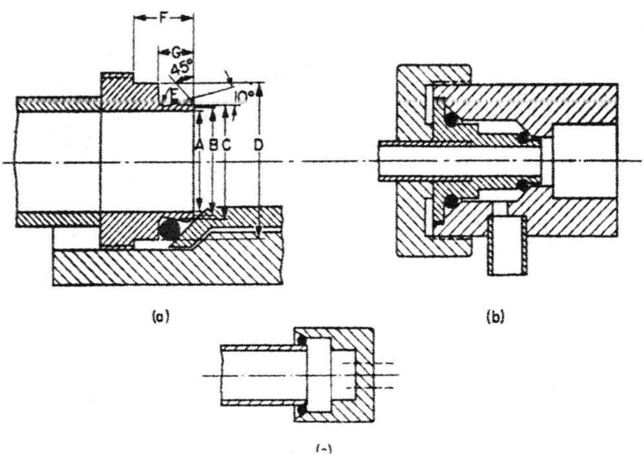

Fig. 3.90 Special conical O-ring seals: (a) modified from refrigeration type fitting (after Hodder[570]); (b) double conical O-ring seal (after Schriever[1108]); (c) rolling O-ring seal. After Edwards[324] *(Courtesy of The Institute of Physics, London, and of The American Institute of Physics)*

TABLE 3.28. MODIFIED REFRIGERATION FITTINGS TO CONICAL O-RING JOINTS (Hodder[570])
(Fig. 3.90a; dimensions in inches)

Size	*A*	*B*	*C*	*D*	*E*	*F*	*G*
1/4	3/16	7/32	0.236	0.344	1/32	5/32	0.100
3/8	9/32	5/16	0.387	0.531	1/32	7/32	0.130
1/2	13/32	7/16	0.455	0.641	3/64	1/4	0.157
5/8	1/2	17/32	0.572	0.750	3/64	9/32	0.168

this kind of joint. Figure 3.91 shows some typical applications of such joints on a vacuum system (Genevac[442]).

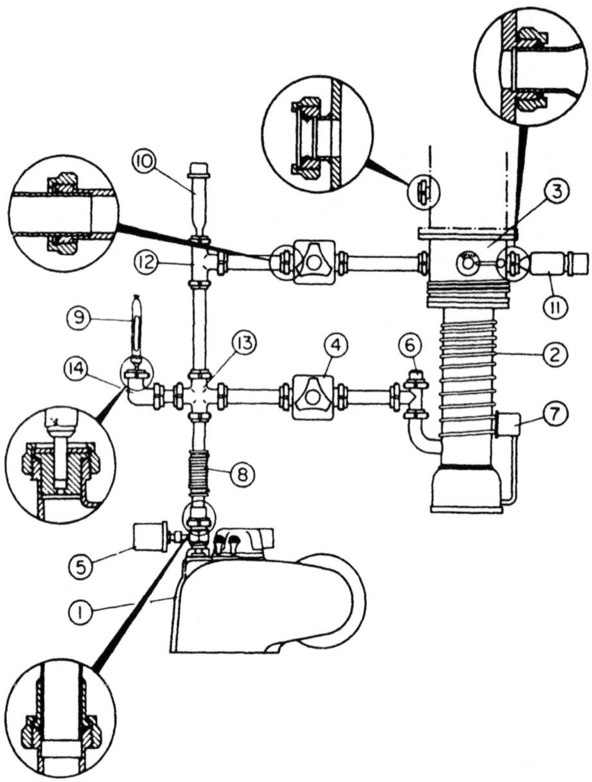

FIG. 3.91 Typical applications of conical O-ring seals on a vacuum system: (1) rotary pump; (2) diffusion pump; (3) baffle valve; (4) hand valve; (5) air admittance valve; (6) oil economizer; (7) thermal switch; (8) flexible connector; (9) discharge tube; (10) Pirani gauge head; (11) Penning gauge head; (12) solderless tee; (13) solderless cross tee; (14) solderless elbow *(Courtesy of Genevac[442])*

Weisbeck[1292a] describes a seal (Fig. 3.92) using 30° bevels in both the mating flanges, so that the O-ring is sealed in an equilateral triangle. It is noted that an O-ring (Perbunan) having a 5 mm cross-sectional diameter was sealed

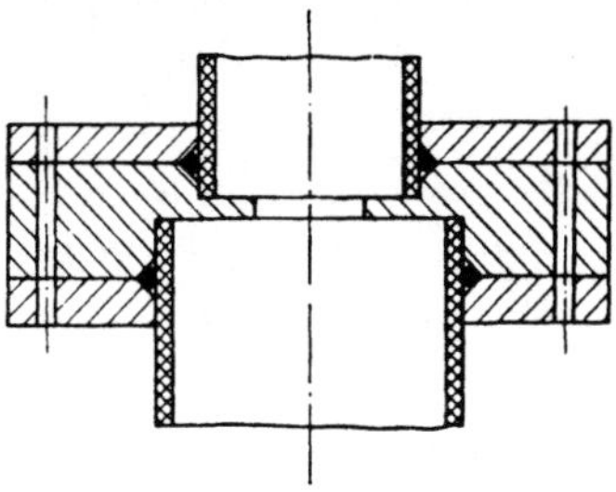

FIG. 3.92 Conical seal with the O-ring embedded in an equilateral triangle. After Weisbeck[1292a] *(Courtesy of Rudolf A. Lang Verlag, Esch/Taunus, Germany)*

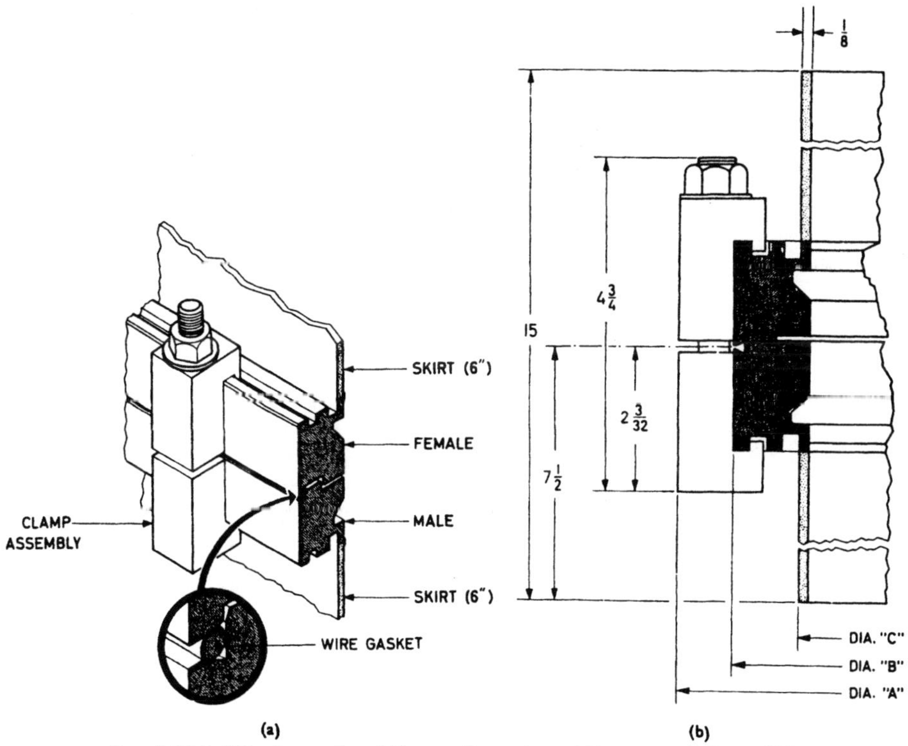

FIG. 3.92.A Wheeler seal and flange dimensions *(Courtesy of Varian[1257a])*

in an equilateral triangle having 6.7 mm sides. This corresponds to the theoretical equality between the O-ring cross-sectional area and the area of the equilateral triangle (without including any dead space). Allowing for a dead space of 4 per cent the sides of the equilateral triangle should be 1.37 times the cross-sectional diameter of the O-ring.

A conical seal with O-rings was used by Amoignon[29] in a leak valve (Fig. 6.122).

Wheeler[1304a] constructed a bakeable seal (Fig. 3.92A) in which an OFHC copper wire gasket is captured between the flanges. The sealing surfaces of the flanges are at 20° to the horizontal and the wire is supported by a vertical face. To ensure the location of the gasket against this vertical face, the gasket is made somewhat undersized and is slightly stretched as it is snapped into place. The sealing force is applied directly over the gasket by means of clamps.

According to Wheeler[1304a], this *capturing seal* geometry limits the distortion of the gasket material away from the sealing surface, and maintains a leaktight seal in cold state and during bakeout.

This type of seal is supplied by Varian[1257a]. The dimensions of flanges with this type of seal are shown on Fig. 3.92A and listed in Table 3.28A.

TABLE 3.28A. DIMENSIONS (inches) OF WHEELER LARGE VACUUM FLANGES, SUPPLIED BY Varian[1257a] (see Fig. 3.92A)

Size	12	14	16	18
A	15 5/8	17 5/8	19 5/8	21 5/8
B	14	16	18	20
C	12	14	16	18

38.44 Corner seals. These seals are based on the *crushing* of a metal O-ring in the corner of a step provided on the sealing flanges. The gasket used is generally of gold (Table 3.29), and is made by cutting a length of wire, annealing it, butting its ends and fusing them with a torch (Caswell[197]). Mark[803] recommends an annealing in air at 600 °C for 1 hr. To ensure the correct location of the gasket in the corner of the seal, the ring should be made slightly under-

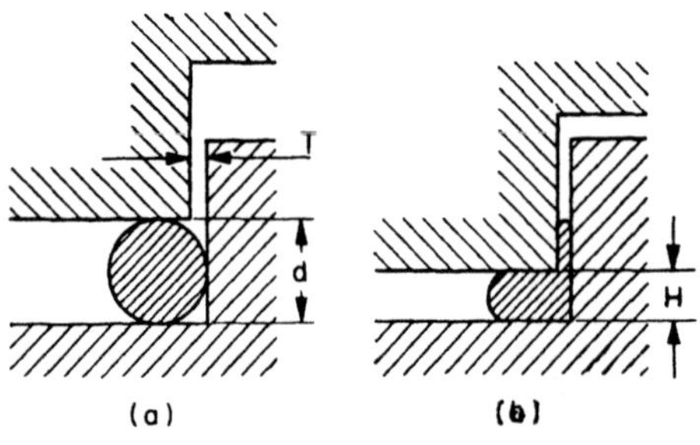

FIG. 3.93 Corner seal: (a) before and (b) after compression; T = radial clearance; H = compression limit

sized and it should be shaped to the exact diameter by pressing it over a conical aluminium form (Caswell[197]).

The location of the gasket in the joint (Fig. 3.93) should permit the crushing of the gasket in the shape shown by Fig. 3.93b. For a reliable vacuum-tight joint the ratio H/d should be 0.5–0.4 (Table 3.29). The gasket wire may be located in the seal by means of spring-loaded pins which retract as the wire is compressed (Connor[229]).

In order to crush the gasket correctly, the radial clearance in the step (T, Fig. 3.93) must be made with extreme accuracy (Table 3.29).

TABLE 3.29. CORNER SEALS WITH GOLD WIRE GASKETS

(Dimensions in mm)

Gasket		Flange		Reference
Wire diameter	Compressed to	Radial clearance T (Fig. 3.93)	Diameter range	
0.5	0.25	0.025–0.050	12–200	Grove[489]
1.0	0.5*	—	75	
0.5	0.25	—	75	Caswell[197]
0.5	0.25–0.20**	0.025–0.1	—	Mark[803]
0.75	0.37–0.30	0.050–0.1	—	
0.75	0.37	0.125	—	Dreyer[304]

* The same gasket may be used twice (Caswell[197]) if first compressed to 0.75 mm, then to 0.5 mm. ** Compression force see Table 3.13.

McDowell[828a] recommends a bakeable seal, which utilizes a gold gasket (0.5 to 1 mm diameter wire) which is compressed to 0.5–0.25 of its original diameter, in the corner of a large groove of rectangular cross section. The gold ring is best made slightly smaller than the inside diameter of the groove and stretched on mounting. The bottom of the groove has a slight slope (about 5°) towards the corner where the gold wire is placed. The gasket is compressed by the rectangular cross section ridge on the opposite flange, which fits into the groove.

Finally the surface finish of the sealing surfaces is important for the vacuum tightness of the seal. Grove[489] recommends a surface finish of 16 μ in r.m.s. (see Table 3.37).

Figures 3.94a, b show the general layout of a corner seal (Grove[489]). The arrangement on Fig. 3.94b has the advantage of well-defined sealing surfaces, the clamping flanges being separately applied to the sealing flanges. A corner seal which may be used also as a step or friction seal is shown in Fig. 3.96 (Marton[821]).

The corner seals can be baked and subjected to temperature cycling. Grove[489] mentions 40–50 cycles up to 450 °C.

Hickman[557a] used a double seal with an inside gold wire corner seal and outside rubber O-ring seal, separated by a guard vacuum (Section 38.23).

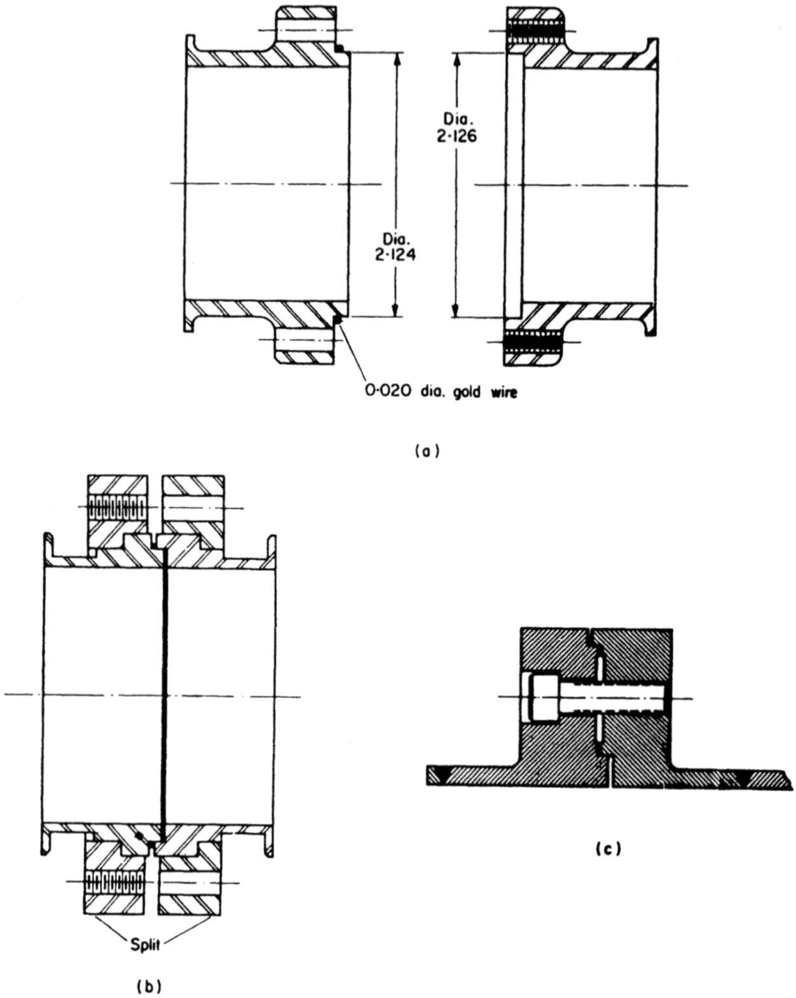

FIG. 3.94 Corner seals: (a), (b) simple seal (after Grove[489]); (c) double seal (after Mark[803]). *(Courtesy of Pergamon Press)*

Mark[803] and Dreyer[304] described a double gasket seal (Fig. 3.94c). The outer seal serves only as a balance, to minimize the radial distortion of the flange due to the compressive forces in the bolts.

38.45 Step seals. In these seals, the O-ring is placed in the step machined in one of the sealing parts, and is compressed against the other (flat) flange

(Fig. 3.95a) or against a second step machined on the other sealing part (Fig. 3.95b). The diameter of the shoulder should be made equal to the inside diameter of the O-ring. The cross-sectional diameter of the O-ring is small compared with the radial clearance, so that the elastic O-ring is simply flattened as in rectangular groove seals (Fig. 3.59).

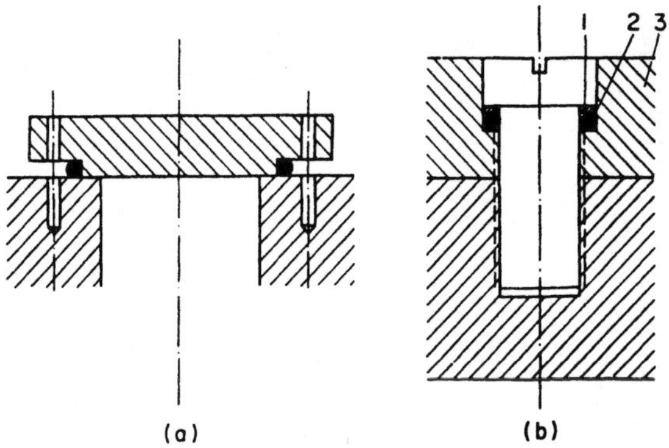

FIG. 3.95 Step seals with O-rings: (a) single step; (b) double step; (1) washer; (2) O-ring; (3) flange

Figure 3.95 shows a step seal (Kurie[723]) in which the O-ring has a limited compression. A seal with unlimited compression is shown in Fig. 3.95b. This technique may be used to seal a plug or a shaft to a vacuum vessel, if the wall is thick enough to permit the countersinking of the head of the plug (Fig. 3.95b) the washer and the O-ring. Other techniques for sealing plugs are shown in Figs. 3.61 and 3.85a.

Marton[821] described a flange design (Fig. 3.96) which could be used as a step seal with elastomer (Viton) O-rings (Fig. 3.96a), as a corner seal with gold wire gasket (Fig. 3.96b) as well as a friction seal with copper gaskets (Fig. 3.96c). The arrangement permits the system to be tested using elastomer O-rings and then allows it to be switched to one of the actual ultra-high vacuum seals.

A step seal with O-rings using a mercury seal is shown in Fig. 3.31d.

If the step cannot be machined on a single sealing flange, the step seal may be made up also with three parts (Fig. 3.97). The O-ring is placed over the pipe or rod (Fig. 3.97a) and compressed by a spacer ring, or by the edge of the cap which has to be joined to the pipe (Fig. 3.97b). The seal on Fig. 3.97b was used by Seki[1118] at liquid helium temperatures. The copper tube (4) is sealed to the cap (2) using an indium wire (1.5 mm) gasket compressed between the pipe (4), the cap (2) and the ring (3) screwed onto the part (1).

The flared end of a glass tube may be sealed to a metal flange, using the arrangement of Fig. 3.97c (Evans[361]). A similar seal was used by Randolph[1025] to join bellows to flanges (see Fig. 5.6b).

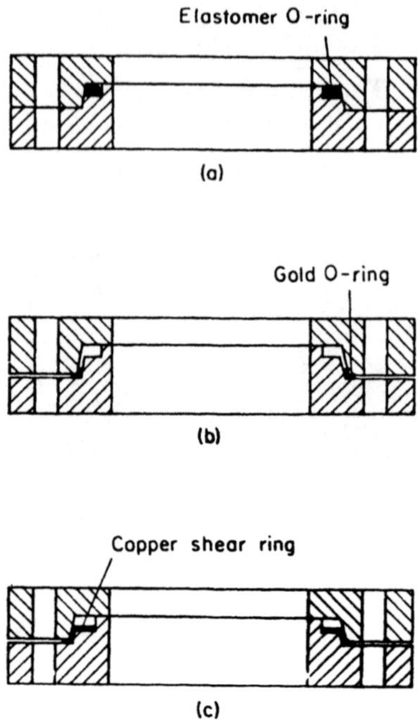

FIG. 3.96 Cross section of a pair of mating convertible flanges: (a) used as a step seal with elastomer O-ring; (b) used as a corner seal with gold wire; (c) used as a friction seal with copper gasket. Reproduced from Marton[821] *(Courtesy of Pergamon Press)*

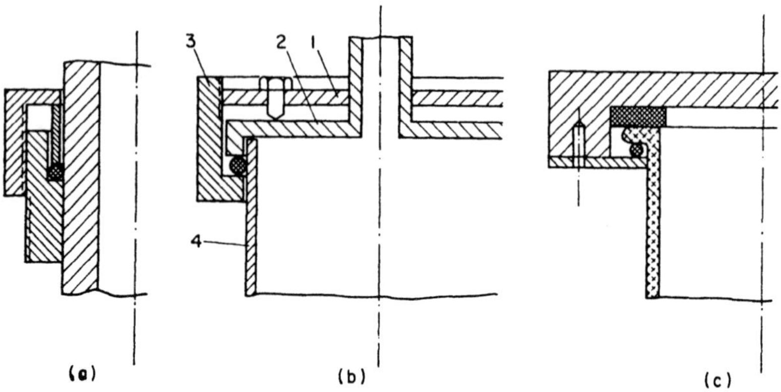

FIG. 3.97 Step seal consisting of three parts: (a) shaft seal; (b) straight end seal (after Seki[1118]); (c) "flared-end" seal (after Evans[361]) *Courtesy of The American Institute of Physics)*

38.46 Cemented seals. The cemented seals are based on the compression of an aluminium wire between flat flanges, combined with heating to a temperature high enough to form a good adherence between the parts (Holden[578]).

According to Holland[582], the gasket is made by fusing together the ends of a 20 s.w.g. (about 0.8 mm) aluminium wire to form a ring, which is then cleaned with caustic soda and water. When the wire is clamped between stainless steel flanges it does not cold weld, not even at pressures of 4 tons/in. of wire. However when the seal is baked at 250 °C or higher it adheres to the mating flanges even at more moderate pressures (5000 lb/in.). Gaskets which have not been baked above about 370 °C can be peeled off from the flanges. The gasket does not indent the flanges, but traces of aluminium may be left on the surface. Holland[582] explains the formation of the cemented joint as

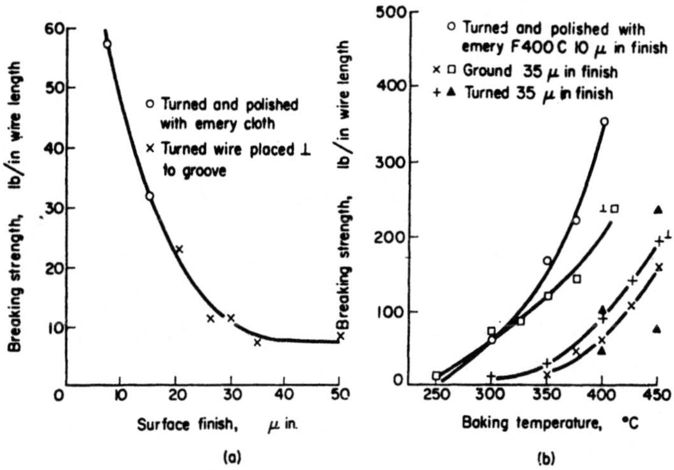

FIG. 3.98 Breaking strength of cemented seals when made with a forming pressure of 5000 lb/in. (893 kg/cm): (a) effect of the surface finish when baked at constant temperature of 300 °C for 10 min; (b) effect of baking temperatures with constant surface finishes and aluminium wire arranged parallel and perpendicular to the direction of the machining marks. Reproduced from Holland[582] *(Courtesy of Pergamon Press)*

follows. When the compressed gasket is baked, the aluminium flows and frictionally rubs against the flanges, thereby fracturing its surface oxide and adhering to the steel. Adhesion only occurs where the air has been previously expelled from between the junction of aluminium and flange, and the leading edges of the flowing aluminium contact the degassed metal of the flanges.

The strength of the cemented joint is directly proportional to the baking temperature (Fig. 3.98b) above a forming temperature (Elsworth[339]), since the breaking strength of the joint depends upon the surface finish of the flanges and is greater when the finish is smoother (Fig. 3.98a). Joints made with turned or ground flanges, with finishes rougher than 10 μin. have a breaking strength which depends on the position of the gasket in relation to the direction of machining (Fig. 3.98b). Greatest breaking strengths are obtained with the wire axis *normal* to the direction of the machining marks. The possible explanation (Holland[582]) of this uncommon behaviour is that when the wire is parallel to the marks its sideways expansion is impeded at the interface by the machining marks, so that frictional rubbing occurs mainly on the top of the asperities, while when the wire axis is at right angles to the direction of the marks it can expand sideways and rub over the whole of the flange surface.

In recent developments, with larger (15 in. diameter) cemented seals, it has been found (Holland[583]) that failure may occur due to the bolt tension which is relieved before adequate fractional adhesion has occurred, due to differences in the compression force of the various bolts or to dust particles settled between the mating flanges. Most of the difficulties in ensuring continuous and uniform compression during baking can be overcome by using Belville* or conical washers under the head of the clamping bolts.

Here the use of these washers is different from that generally required for spring washers in dismountable seals, where the washers have to deal with the differential expansion (Section 38.48). The difference is caused by the fact that in a cemented seal there is actual adhesion, so that after baking, the bolts could be slackened and the seal will still remain tight (Steckelmacher[1180]).

An incidental advantage of using a Belville conical spring washer was the reduction of the force required to make the seal, since the typical pressure required without spring washers (see Table 3.13) could be reduced by a factor of 2–3. The spring washer would also be able to take up any long term thermal creep of bolt materials after temperature cycling, so dispensing with the need for special creep resisting bolt materials (Steckelmacher[1180]).

Power[1003] proposes the use of stainless steel shims (0.25–0.10 mm thick) placed in the seal to control the compression of the wire gasket, and reports on attempts to avoid mercury attack on cemented gasket seals, by using aluminium containing silicon* or applying indium as a protective surface coating. Indium applied as a surface coating on an aluminium wire gasket has various interesting effects (Power[1003]). It acts as a surface lubricant between gasket and flange, reducing the clamping forces necessary for a particular compression (Fig. 3.99) or permitting extra compression with the original clamping forces. During compression the indium is mainly displaced to the inside and outside edges of the gasket; thin interfacial layers do, however, remain between flange and gasket. The indium coating helps to fill up surface imperfections on the flanges.

Indium coated aluminium gaskets do not need to be end welded; the ends of a cut length of wire may just be crossed, the indium producing a satisfactory bond on compression. Plate 10 shows a dismountable joint which has been made of short lengths of indium coated aluminium wire with ends just

* For the spring washers see Section 38.48 and Pierre[989], Haring[510a], Groupe de Techn. Nucléaire[488].

* Aluminium with 3–5 per cent silicon resist cold mercury attack and has considerable resistance to hot mercury vapours.

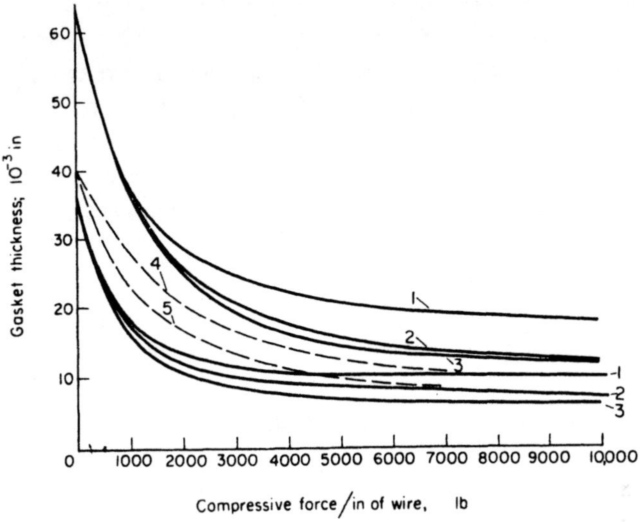

FIG. 3.99 Gasket thickness as a function of compressive force per unit length, for gaskets made from wires (cemented seals): (1) aluminium wire on plain flanges; (2) aluminium wire on indium-coated flanges; (3) indium-coated aluminium wire on plain flanges. Upper 1, 2, 3 curves for 0.064 in. diameter; lower 1, 2, 3 curves for 0.036 in. diameter wire. (4) aluminium–silicon (5 per cent) on plain flanges; (5) aluminium–silicon (5 per cent) on indium-coated flanges. Broken curves for 0.040 in. diameter wire. Reproduced from Power[1003] *(Courtesy of Pergamon Press)*

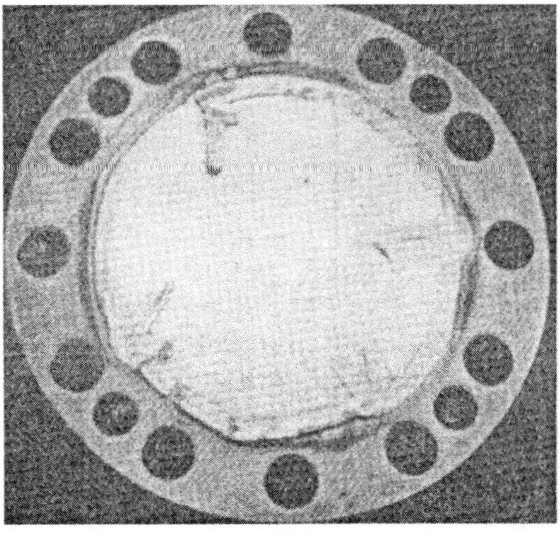

PLATE 10. Gasket made from four separate lengths of indiumized aluminium wire, with their ends simply crossed prior to clamping. After Power[1003] *(Courtesy of Pergamon Press)*

crossed, so that there were four unwelded junctions in the 90 mm diameter gasket. This joint remained vacuum-tight after four baking cycles to 350 °C.

In order to avoid the heavy flanges needed for the compression (Table 3.13) of the aluminium gasket, Comsa[227] proposes that separate clamping flanges be used only to compress the seal, during the first heating to 500 °C. In this way thin (1–2 mm) flanges can be used on the system itself, and the strong clamping flanges can be removed after the seal has been made.

38.47 Standard flanges and unions for O-ring seals. The previous Sections described the various O-ring seals but in those Sections (and especially in Section 38.41) the recommended dimensions of the flanges were not always given.

The dimensions of the flanges (unions) to be used with various O-ring seal arrangements (groove, conical) have been established by the various national standards and/or the standards used by various firms. Summarizing the recommendations of these standards, Table 3.30 gives the general rules for the design of flanges for grooved O-ring seals. In order to give more detailed examples, some standard dimensions are listed in Tables 3.31–3.35.

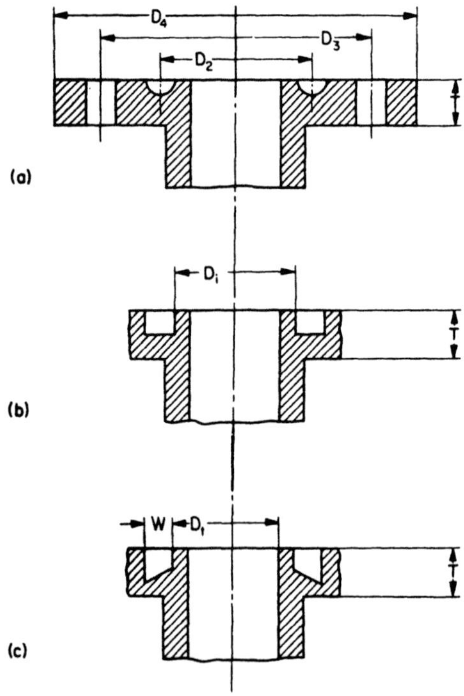

FIG. 3.100 Dimensions of flanges (Table 3.30) with (a) groove of semi-circular cross section; (b) groove of rectangular cross section; (c) groove of trapezium cross section

TABLE 3.30. GENERAL RULES FOR THE DESIGN OF GROOVED FLANGES FOR O-RING SEALS
(see Fig. 3.100)
(Dimensions in mm)

Dimension	Minimum	Maximum
D_2*1	$D_2 = 1.03 \cdot D_1 + 7$	$D_2 = 1.12 \cdot D_1 + 14$
D_i*2	$D_i = D - 0.15 \cdot d$	
D_t*3	$D_t = D + d - W$	
D_3*4	$D_3 = 1.12 \cdot D_1 + 34$	$D_3 = 1.18 \cdot D_1 + 40$
D_4	$D_4 = 1.14 \cdot D_1 + 46$	$D_4 = 1.30 \cdot D_1 + 56$
T	$T = 0.027 \cdot D_1 + 7.6$	$T = 0.027 \cdot D_1 + 10$

*1 For *V* or semi-circular grooves, see Fig. 3.65 and Table 3.22.
*2 For grooves with rectangular cross section, see Table 3.21; *D* is the i.d. of the O-ring, and *d* the cross-sectional diameter (see Fig. 3.56, and Table 3.19 A).
*3 For trapezium grooves, see Table 3.23.
*4 Bolt holes:

D_3	<100	100–300	300–450	450–550
Number of holes	4–6	8	12	16
Diameter of holes (mm)	9–14	9–18	14–20	14–20

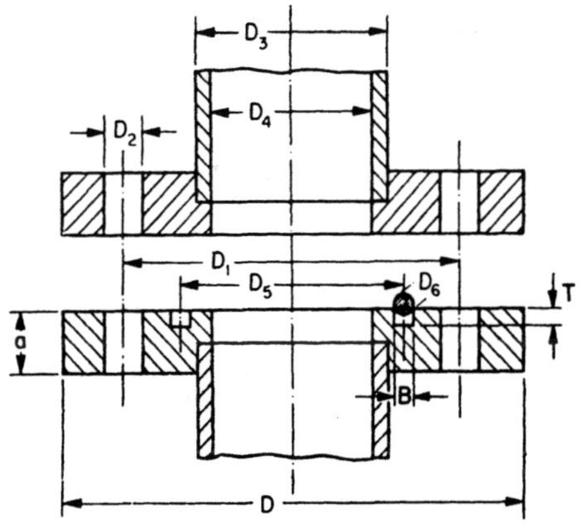

FIG. 3.101 Dimensions of flanges for O-ring seals (Table 3.31)

TABLE 3.31. FLANGES FOR O-RING SEALS CF. D.I.N. 2572–1940
(see Fig. 3.101; dimensions in mm)

Nominal size	D	D_1	D_2	Number of bolts	D_3	D_4	D_5**	D_6*	a	B	T	Bolts
N W 10	75	50	11.5	4	15	10	25	5	10	6	3.6	M 10×30
N W 20	90	65	11.5	4	25	20	35	5	10	6	3.6	M 10×30
N W 32	120	90	14	4	38	32	55	5	10	6	3.6	M 12×30
N W 50	140	110	14	4	55	50	80	5	12	6	3.6	M 12×35
N W 70	160	130	14	4	75	70	90	5	12	6	3.6	M 12×35
N W 100	210	170	18	8	108	100	125	5	14	6	3.6	M 16×40
N W 150	265	225	18	8	159	150	170	5	14	6	3.6	M 16×45
N W 250	375	335	18	12	256	250	270	8	15	9.5	5.9	M 16×45

* See Table 3.19A.
** See Tables 3.21, 3.22.

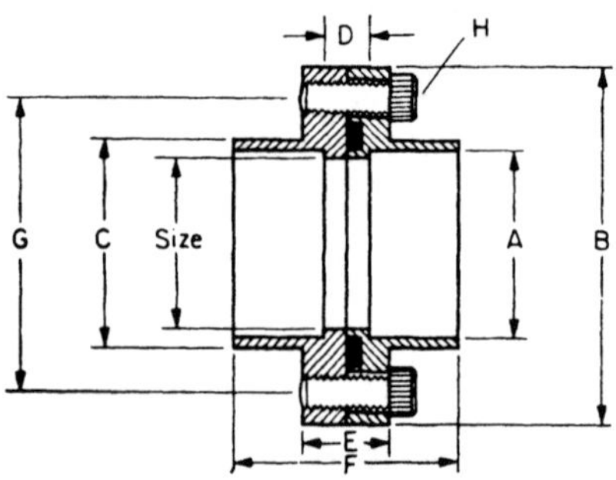

FIG. 3.102 Dimensions of flanges for O-ring seals, from Edwards[329] (Table 3.32)

TABLE 3.32. DISMOUNTABLE FLANGED JOINTS (Edwards[329])
(see Fig. 3.102)

Size	A	B	C	D	E	F	G	H	O-ring*
1/2 (in.) (mm)	0.596 15.1	1 1/2 38.1	3/4 19.1	3/8 9.52	1/2 12.7	1 1/4 31.7	1 1/8 28.6	4BA×1/2in	VOR 121
3/4 (in.) (mm)	0.846 21.5	1 5/8 41.3	1 25.4	3/8 9.52	1/2 12.7	1 1/4 31.7	1 5/16 33.6	4BA×1/2in	VOR 130

* See Table 3.19A; for groove dimensions see Table 3.21.

(Table 3.32 Continued)

Size	*A*	*B*	*C*	*D*	*E*	*F*	*G*	*H*	O-ring*
1 (in.) (mm)	1.112 28.2	2 1/8 54.0	1 1/4 31.7	3/8 9.52	1/2 12.7	1 1/4 31.7	1 13/16 46.3	4BA × 1/2 in	VOR 138
1 1/2 (in.) (mm)	1.612 40.8	4 101.6	1 3/4 44.4	3/8 9.52	3/4 19.1	1 1/2 38.1	3 1/4 82.5	1/4 BSF × × 3/4 in	VOR 146
2 (in.) (mm)	2.128 54.2	4 1/2 114.3	2 1/4 57.1	3/8 9.52	3/4 19.1	1 1/2 38.1	3 3/4 95.3	1/4 BSF × × 3/4 in	VOR 159
3 (in.) (mm)	3.148 79.9	7 1/4 184.1	3.312 84.1	3/4 19.1	7/8 22.2	1 1/2 38.1	5 3/4 146.0	1/4 BSF × × 3/4 in	VOR 184

* See Table 3.19A; for groove dimensions see Table 3.21.

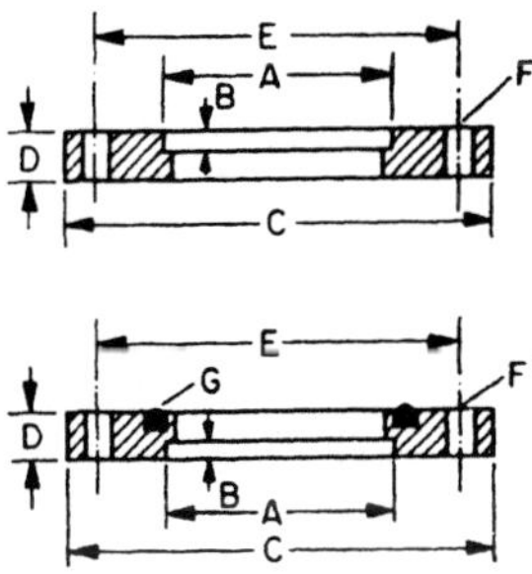

FIG. 3.103 Dimensions of flanges for O-ring seals, from Veeco[1262] (Table 3.33)

TABLE 3.33. O-RING FLANGES (Veeco[1262])
(Fig. 3.103; dimensions in inches)

Size	*A*	*B*	*C*	*D*	*E*	Number of holes *F*	*G**
1/4	1/4	1/16	1 1/8	3 16	7/8	3	8
3/8	3/8	1/16	1 1/8	3/16	7/8	3	8
1/2	1/2	3/32	1 3/8	1/4	1 1/8	3	12
5/8	5/8	3/32	1 3/8	1/4	1 1/8	3	12
3/4	3/4	3/32	1 3/4	1/4	1 13/32	3	17

* See Table 3.19A.

(Table 3.33 Continued)

Size	*A*	*B*	*C*	*D*	*E*	Number of holes *F*	*G**
7/8	7/8	3/32	1 3/4	1/4	1 13/32	3	17
1	1	1/8	2 1/4	5/16	1 3/4	3	21
1 1/8	1 1/8	1/8	2 1/4	5/16	1 3/4	3	21
1 1/2	1 1/2	1/8	3	5/16	2 9/16	4	(29)
1 5/8	1 5/8	1/8	3	5/16	2 9/16	4	(29)
2	2	5/32	3 1/2		3	4	(33)
2 1/8	2 1/8	5/32	3 1/2	3/8	3	4	(33)
2 1/2	2 1/2	5/32	4	3/8	3 1/2	4	(37)
2 5/8	2 5/8	5/32	4	3/8	3 1/2	4	(37)
3	3	3/16	5	7/16	4 1/4	6	(41)
3 1/8	3 1/8	3/16	5	7/16	4 1/4	6	(41)
4	4	1/4	6	1/2	5 1/4	6	(49)
	4 1/8	1/4	6		5 1/4	6	(49)

* See Table 3.19A.

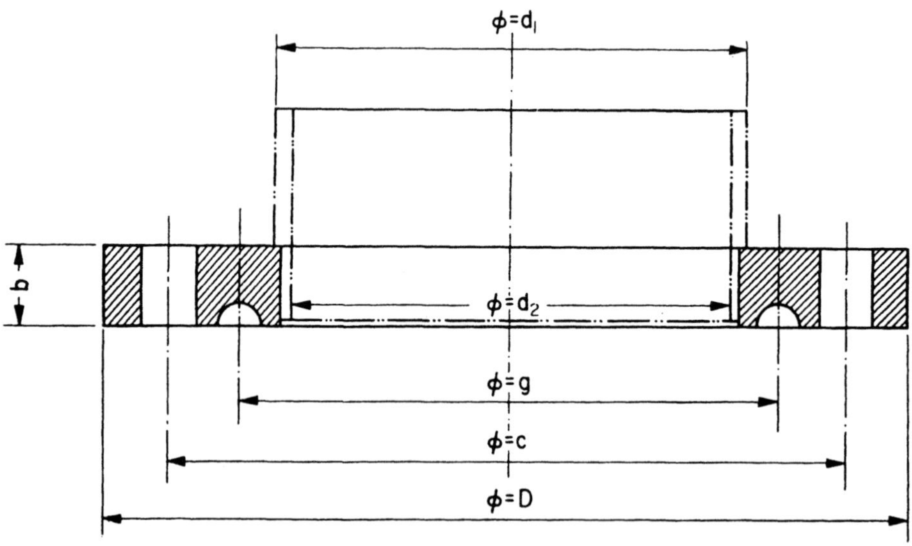

FIG. 3.104 Dimensions of flanges for O-ring seals (Table 3.34). After Henry[539]

TABLE 3.34. FLANGE DIMENSIONS (mm) OF FRENCH STANDARD PROPOSAL
Henry[539] (Fig. 3.104)

Nominal diameter	d_1	d_2	D	b	c	Bolts			g*
						Number	Dia.	Hole	
40	44.5	43	90	10	75	4	6	7	55
50	57	55.5	110	12	90	4	8	9	67.7
65	76	74	130	12	110	8	8	9	86.7
100	108	106	170	12	150	8	8	9	118.5
150	159	156	250	14	225	8	8	9	177.2
200	219.1	213	300	14	280	8	8	9	234.4
250	273.1	266	375	16	335	12	12	14	297.8
300	323.9	316	440	16	395	12	12	14	348.5
400	406.1	397	540	20	495	16	12	14	—

* The form of the groove cross section is at the free choice of the designer. The cross section area of the groove should be minimum equal to that of the O-ring, and the depth of the groove to 0.75 of the diameter (cross section) of the O-ring (Section 38.41).

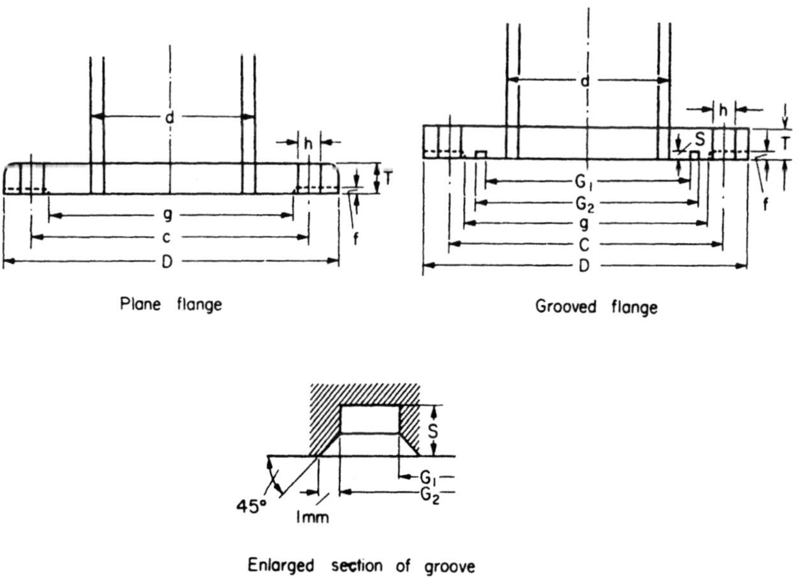

FIG. 3.105 Dimensions of flanges for O-ring seals (see Table 3.35). After Kiyoshi[683]

TABLE 3.35. JAPANESE STANDARD FLANGE DIMENSIONS* (mm)

Kiyoshi[683] (Fig. 3.105)

Nominal size	d	D	T cast flange	T other flange	f	g	C	Number of holes	h	G_1**	G_2	S
12	15	70	10	8	1	38	50	4	9.5	24	34	3
20	25	80	10	8	1	48	60	4	9.5	34	44	3
25	30	90	10	8	1	58	70	4	9.5	40	50	3
40	45	105	12	10	1	72	85	4	9.5	55	65	3
50	60	120	12	10	1	88	100	4	9.5	70	80	3
70	80	145	12	10	1	105	120	4	11	85	95	3
80	90	160	14	12	2	120	135	4	11	100	110	3
100	110	185	14	12	2	145	160	8	11	120	130	3
130	140	210	14	12	2	170	185	8	11	150	160	3
160	170	235	14	12	2	195	210	8	11	175	185	3
200	220	300	18	16	2	252	270	8	15	225	241	4.5
260	270	350	18	16	2	302	320	12	15	275	291	4.5
300	320	400	18	16	2	352	370	12	15	325	341	4.5
350	370	450	—	20	2	402	420	12	15	380	396	4.5
400	420	520	—	20	2	458	480	12	19	430	446	4.5
450	470	575	—	20	2	511	535	16	19	480	504	7
500	520	625	—	22	2	561	585	16	19	530	554	7
550	575	680	—	24	2	616	640	16	19	585	609	7
600	630	750	—	24	2	672	700	16	22	640	664	7
650	680	800	—	24	2	722	750	20	22	690	714	7
700	730	850	—	26	2	772	800	20	22	740	764	7
750	780	900	—	26	2	822	850	20	22	790	814	7
800	835	955	—	26	2	877	905	24	22	845	869	7
900	935	1065	—	28	2	983	1015	24	25	950	974	7
1000	1040	1170	—	28	2	1088	1120	24	25	1055	1079	7

* Japanese Industrial Standard J.I.S. B 2290–1957.

** O-ring dimensions see Table 3.19A; groove dimensions see Fig. 3.59c. It is recommended to use the grooved flange on the side facing the normal direction of flow.

Usual dimensions of flanges used with spacer seals are listed in Tables 3.24, 3.25. General dimensions of unions using conical seals are given in Tables 3.26, 3.27. More details on conical joints are presented in Table 3.36 and Fig. 3.106.

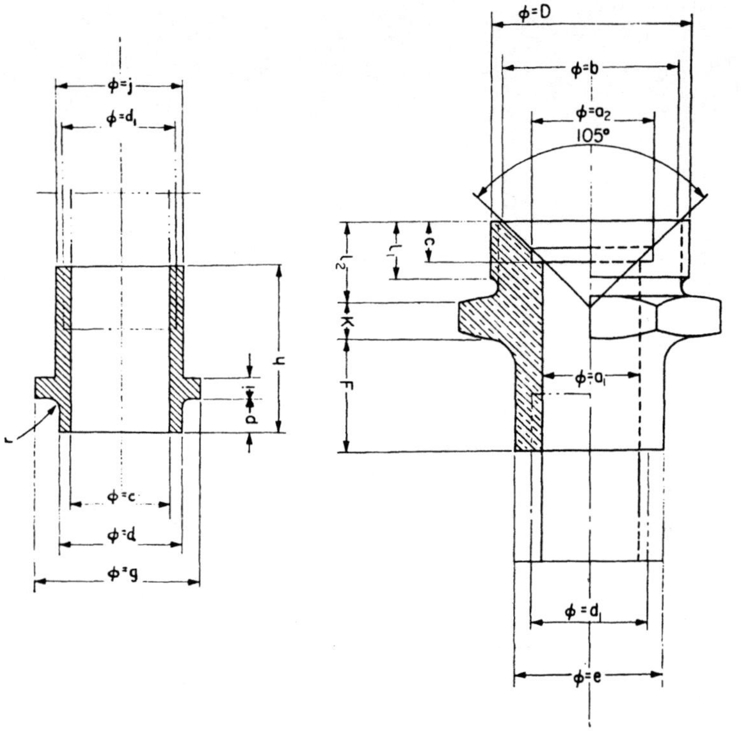

FIG. 3.106 Dimensions of conical joints (Table 3.36). After Henry[539]

TABLE 3.36. CONICAL JOINT. FRENCH STANDARD PROPOSAL* (Henry[539])
(Fig. 3.106; dimensions in mm)

Nominal diameter	10	(15)	20	25	32	40
d_1	14	20	25	30	38	44.5
D	26.44	33.25	41.91	47.80	59.61	69.30
a_1	13	17	24	29	35	43.5
a_2	15.3	20	28	34.3	40.5	50
b	22	29	37	44	54	64
c	3.8	4.8	5.0	5.9	6.8	7.0
k	6	7	8	9	11	12
e	19	25	31	38	45	55

* To be used with O-rings according to F.S. (see Table 3.19A).

(Table 3.36 Continued)

Nominal diameter	10	(15)	20	25	32	40
f	18	18	21	22	23	28
l_1	8	9.5	11	13	15	16.5
l_2	11	12	14	16	18	20
m	13	17	24	29	35	43.5
d	15.2	19.9	27.9	34.2	40.3	49.8
g	24	30	38.6	44.5	54	66
h	25	27	30	32	35	40
i	3	3	3	4	4	4
	18	22	30	38	45	53
p	4	5.8	5.8	6.5	8	9
r	0.5	1	1	1	1.5	1.5

* To be used with O-rings according to F.S. (see Table 3.19A).

38.48 Assembly and maintenance of O-ring seals. A seal may be well designed and constructed and nevertheless not be vacuum-tight, due to inadequate assembly or mistakes in handling or maintenance of the sealing parts. The main suggestions for the correct assembly, handling and maintenance may be grouped thus:

(a) regarding the O-ring itself,
(b) regarding the sealing surfaces, and,
(c) regarding the relative position in the seal (tightening of the seal).

Proper O-ring dimensions. The O-rings used are either purchased as ready-made products (available in a very wide range of sizes, see Table 3.19A), or for the larger sizes constructed from elastomer cord, or wire (Section 38.33). When ready made O-rings are used, the proper size should be selected (Smith[1153]), to exactly fit the seal in which it is to be used. The dimensions of the O-rings are specified in Table 3.19A and the O-ring sizes are listed in the various Tables giving the dimensions of the grooves, the unions and the flanges (Sections 38.41–38.47).

When using O-rings with *smaller cross-sectional diameters* than that provided in the design of the seal, the required compression ratio (see Figs. 3.46, 3.50) cannot be reached, and the seal may leak immediately or after a short time. With an O-ring having a *larger cross-sectional diameter* than designed for, the seal cannot be tightened until there is metal-to-metal contact of the flanges (see Figs. 3.64, 3.65) and/or the O-ring is sheared in the seal. The resulting seal can be leak-tight, but the alignment of the sealing parts is difficult, and a larger surface area of the O-ring is exposed to the evacuated space.

It is impossible to use O-rings with *larger diameters* (circumference) than provided for by the design and construction of the sealing parts, since the rubber is not compressible. The opposite solution, i.e. the use of O-rings with a *smaller diameter* than that required, *stretched* over the inside diameter of the groove, step or spacer is the most common *mistake* made in O-ring seals. Such a solution is equivalent to the use of an O-ring with a smaller cross-sectional diameter than that provided for the seal (see above) but has an additional danger. Due to the stretching, all the small irregularities on the surface of the O-ring are enlarged to real channels which make the seal leaky. Very often the O-ring is slightly damaged during the mounting operation itself by the cutting action of the edges over which it is stretched. This occurs especially if the edges of the groove, step or spacer are not radiused, or when the O-ring is stretched over the end of a glass pipe, which was not previously fire-polished.

When the O-ring is *made from elastomer cord*, three methods can be used for the joining of the ends: (1) bevel cutting and press fitting without cementing, (2) bevel cutting and cementing the ends together, and (3) vulcanizing the ends.

With some skill, the ends of the cord can be cut and the gasket can be fitted in the groove so that an adequate end-to-end seal is obtained. It is recommended that the ends of the cord be bevelled at an angle of about 30° (Ardenne[27]) in a direction such that when the gasket is compressed in the seal, this overlap will be closed. Asao[51] mentions a seal where the gasket was cut a little longer than the length of the groove, and by forcing the gasket in the groove the ends were brought so close to each other that a tight seal was obtained.

The cementing of the ends is an improvement over the previous techniques. Rubber cement is applied to each end surface of the cord and after drying, the surfaces are pressed together. The "self-vulcanizing" solution supplied by Edwards[329] contains two solutions which should be mixed in equal quantities just before use. A thin coat of the mixture should be applied on the roughened end surfaces of the cord, and allowed to dry for about 30 minutes. A second coat is then applied, and after about one minute the surfaces are pressed together into a firm contact and left under light pressure for 24 hours.

Vulcanizing is accomplished by first applying the proper cement and joining the ends together, and then by applying heat up to the temperature required for the particular rubber.

Clean O-ring surfaces. To make a tight seal the surface of the O-ring must be free of dust or any particles which would prevent the direct contact between the O-ring and the sealing surfaces of the flanges or parts of the union. It has been found (Young[1348]) that the cleaning technique used may also influence the gas evolution from the O-ring. O-rings (1 1/8 in. diameter) were cleaned in acetone, alcohol or trichlorethylene and dried in air at room temperature, and the gases remaining in the system, where these O-rings were used in seals, were analysed on a mass spectrometer. After obtaining a pressure of 10^{-5}–10^{-6} torr in the system, the mass spectrometer analysis indicated hydrocarbons and for the O-rings cleaned in trichlorethylene large peaks indicating hydrochloric acid. It was found that these contaminants can be reduced to negligible values by baking the O-ring in air at 100 °C for several hours (Young[1348]). No contaminants were observed when an O-ring was used, which had not been cleaned after unpacking, or greased with Apiezon N stopcock grease.

In order to avoid contamination of mercury in contact with rubber gaskets, Pike[990] recommends that the rubber should be boiled in a strong caustic soda solution for at least 1 hr, followed by boiling in distilled water.

Greasing of the O-rings. The grease may seal *for the moment* small scratches on the surface of the flange or of the O-ring itself, but it cannot be recommended as a sealing material. Greasing is particularly not recommended for seals where the gasket is not enclosed in a groove and may slide laterally.

Re-using the O-rings. O-rings can be re-used after the seal has been opened, only after a visual check on its integrity and state of permanent deformation has been made. The integrity of the surface may be checked by visual inspection of the surface in the slightly stretched state of the O-ring. If the surface is damaged or the cross section is not still circular (has edges), it is not advisable that the O-ring be re-used.

Sealing surfaces. The seal is restricted to some very small areas of the sealing parts (flanges, unions, etc., see Sections 38.1, 38.4). The surface finish of these sealing areas is a basic criterion for a reliable seal. The degree of the surface finish needed varies with the kind of gasket used (Table 3.37).

The required surface finish is obtained by machining, polishing, lapping etc. Conaboy[228] describes a method to form "precision" ends on silica tubes for O-ring seals, by straightening the silica with a graphite rod, on a glass blower lathe.

The small machining marks remaining on the sealing surface are incomparably less dangerous if they are *parallel* to the O-ring. Radial scratches, even if they are very small may often cause leakage. The cemented aluminium seal (Section 38.46) is an exception to this rule; here the strength of the seal is greater if the very small (Table 3.37) marks are normal to the sealing wire.

TABLE 3.37. SURFACE FINISH REQUIRED FOR SEALING AREAS

Gasket	Seal	Surface finish μ in (r.m.s.)	Remarks and References
Rubber O-ring	Spacer seal	min 63	Brown[172]
Gold wire	Plane seal Corner seal	4–8 16	see Fig. 3.74 Caswell[197a] see Fig. 3.94 Grove[489]
Iron wire (high purity)	Plane seal	16–32	Direction of marks parallel to wire Peters[986a]
Aluminium wire	Cemented seal between stainless steel flanges	10	Breaking strength greatest if marks normal to wire Holland[582]
Aluminium wire	Plane between glass and metal	8	Nelson[916]
Copper	Plane seal Gasket with beads Coined, knife edge, Conflat	50 32	see Fig. 3.121a Milleron[865] see Fig. 3.121b, Table 3.39, Fig. 3.149 Wheeler[1304]
Without gasket	Inside seal of double chamber with guard vacuum	8	see Section 38.23 Metcalfe[849]

Sharp edges on the sealing surfaces (grooves, steps) are to be avoided. Corners should be always radiused (see Tables 3.21, 3.23).

Metallic surfaces. Oxidized surfaces are equivalent to a rough surface finish. The oxidation of the surfaces occur especially during their heating for welding or brazing. In this respect it is recommended to machine the surface after the brazing, or to protect the sealing surface during brazing (protective atmosphere see Section 22.2).

Assembling the seal. The O-ring seals are assembled by placing the O-ring in the groove, or on the spacer or step and compressing it by closing (tightening) the sealing parts. When the seal is tightened with bolts (dimensions see Tables 3.24, 3.25, 3.31–3.35) they are placed symmetrically around the seal. The tightening of the seal should be done in steps, beginning with one of the bolts and tightening it slightly, then tightening to the same degree the diametrically opposed bolt, continuing with one next to the first, and then the one diametrically opposite to it, after the first cycle of tightening has been completed, the cycle is repeated, further tightening the bolts in the same sequence, etc.

The final tightening force (torque) required for a reliable seal, depends on the shapes and dimensions of the seal and the kind of gasket used. The torque on the bolts must produce the sealing forces (pressures) on the gaskets as listed in Table 3.13.

Seals (even standard ones) are constructed by the various firms with three or four bolts, although for a reliable seal it is advisable to provide a minimum of *six* bolts. Heywood[554] states that when there are only four or six bolts around the periphery of the seal, the leak rate is less if a smaller tightening force is used, since by strong torques the bending of the flange has a negative influence on the seal. As the number of bolts is increased the leak rate decreases with increasing torque.

In order to reduce the diameter of the flanges, required by the spacing of the bolts, Rollinger[1074b] suggests the use of special clamps instead of bolts. As these clamps apply the sealing force closer to the gaskets, they reduce the danger of bending the flange.

When seals are subjected to temperature cycling, the thermal expansion difference between the sealing parts and the bolts tends to open up the seal, thus

PLATE 11. Demountable flanged joints (dimensions in Table 3.32) *(Courtesy of Edwards[328])*

making retightening necessary. This can be prevented by spring-loading the bolts. Van Heerden[527] used a stainless steel ring with a shim between every two bolts, and Power[1003] recommends stainless steel shims of about 0.1–0.25 mm placed in the seal. Holland[583] and Steckelmacher[1180] recommend the use of "Belville" or conical spring washers placed under the head of the clamping bolts. The working principles of such washers are described by Haringx[510a].

PLATE 12. Quick release coupling *(Courtesy of Pfeiffer[987])*

PLATE 13. Quick release "Mecavide" coupling *(Courtesy of Laboratoires des Basses Pressions[728])*

Tasman[1215a] compensates for the expansion difference between flanges and bolts by placing under the head of the bolts, washers (cylinders) of a low expansion alloy (e.g. Nilo–40).

Small flanges are often closed with the aid of hand-tightened screws or of various systems of "quick release" couplings. Such clamping systems are used with the seals shown on Plates 3 and 4 and those on Plates 11, 12 and 13.

Figure 3.107 shows the correct steps to be used in assembling *conical* O-ring seals. The most important fact in assembling such seals is to prevent and to

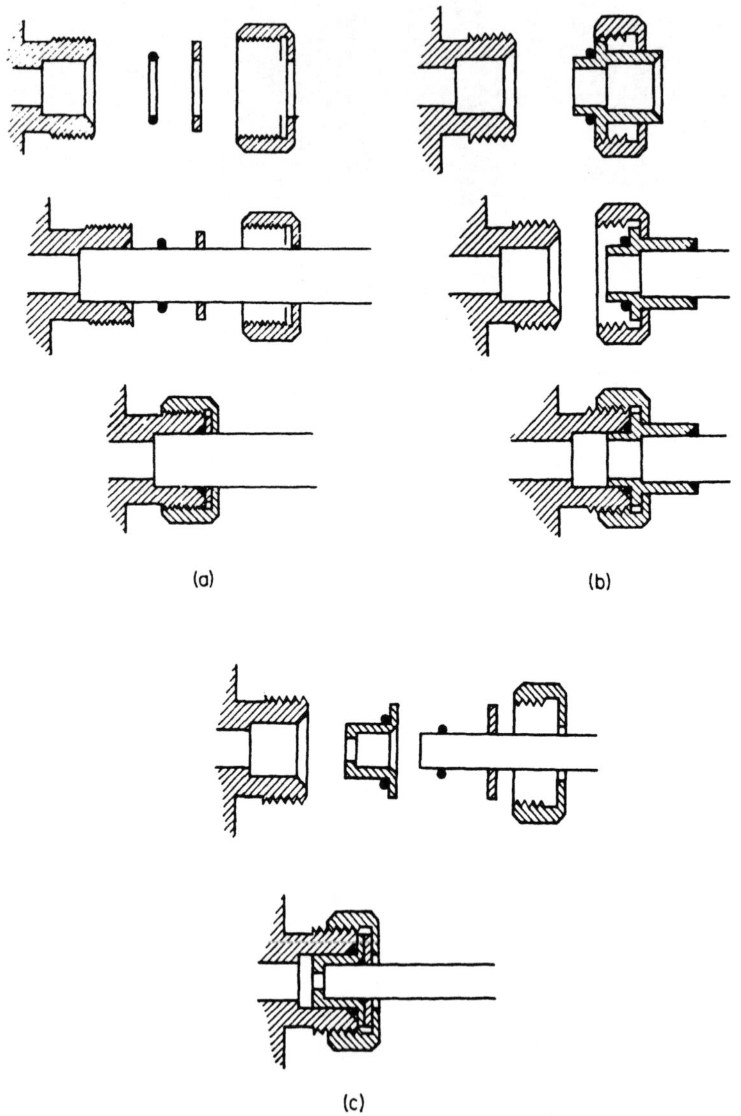

FIG. 3.107 Systems of closing conical O-ring seals *(Courtesy of Genevac[442])*

avoid drag on the O-ring during tightening. This is done either by keeping the two sealing parts without rotation (Fig. 3.107b) during the tightening of the locking ring or by the use of compression washers (Fig. 3.107a, c).

38.5 Thick gasket seals

Thick gasket seals are somewhat similar to the O-ring seals but they use gaskets of various cross-sectional shapes (see Table 3.18) having the width and height of their cross section of the same order of magnitude. The seals using thin gaskets are discussed in Section 38.6.

The thick gasket seals are grouped in the following paragraphs according to the relative position of the gasket and sealing parts (flanges) as well as the shape of the gasket and flange. Thus they are grouped in:

1. Plane seals (Section 38.51)
2. Conical seals (Section 38.52)
3. Cylindrical seals (Section 38.53)
4. Dumbbell seals (Section 38.54)
5. Shear seals (Section 38.55)
6. Ridge seals (Section 38.56)
7. Inflatable gasket seals (Section 38.57).

38.51 Plane seals. The most simple plane seal consists of a flat rubber gasket tightened between two *plane flanges.* For centring and to control the compression ratio, rectangular, square or trapezium cross-sectional gaskets are used placed in the appropriate *grooves.* In some applications *specially shaped* gaskets (elastomer or metal) are sealed between plane flanges.

a) *Plane flange seals.* Rectangular or square cross section elastomer gaskets may be sealed between plane flanges. It is common practice to slightly grease the sealing surfaces of the gaskets used in such seals. This procedure enables the seal to be tightened with light loadings (5–10 kg/cm^2 of gasket face before clamping) but results in frequent loosening of the seal initially clamped tightly. A gasket of square cross section (Fig. 3.108a) if compressed between two non-greased surfaces will keep the contact surfaces in their initial position (Fig. 3.108b) whereas if the surfaces are lubricated, a considerable slippage occurs (Fig. 3.108c). The slippage (migration of the contact surface) will not generally produce leaks, but the required subsequent tightening of the seal usually results in the gasket tearing.

If the gasket is sealed dry (non-greased) it requires higher loadings (15–30 kg/cm^2) depending on the hardness of the gasket and the roughness of the sealing surfaces.

Flat gaskets sealed between plane flanges should be provided with a possibility for centring. Foote[389] provided the gaskets with centring ears (Fig.

3.109). The seal was made directly on the end of the pipes, which were shaped to give sealing surfaces free of radial scratches. Foote[389] used an aluminium gasket, but the same seal may be used with rubber, Teflon or gold gaskets. Flat Teflon gaskets were used by Nier[929] to seal a mass spectrometer tube.

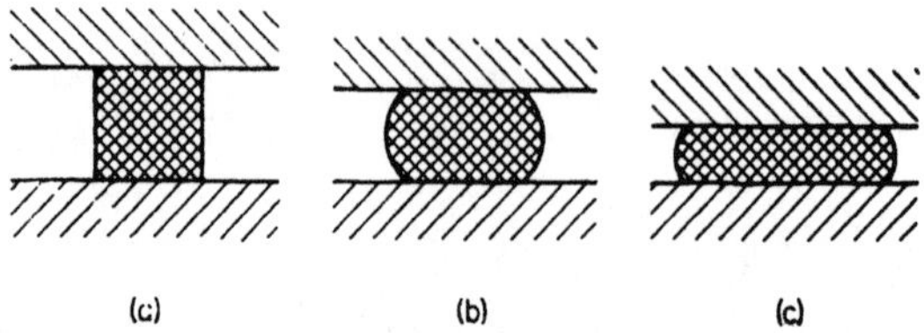

FIG. 3.108 Rubber gasket with square cross section: (a) before compression between flat flanges; (b) after compression between greaseless surfaces; (c) after compression between greased surfaces

Fremlin[401] stuck a rubber sheet to one of the surfaces of the ground seal of a vacuum desiccator in order to allow for quick release of the lid.

Plane seals are successfully used with bell jars, where either the atmospheric pressure (Fig. 3.110a) or mechanical clamping (Fig. 3.110b) provides the required loading of the gasket. The minimum bell jar diameter D (cm) at which a reliable seal can be made, based on the atmospheric pressure is given by

$$D \geqslant 4 \cdot p \cdot h,$$

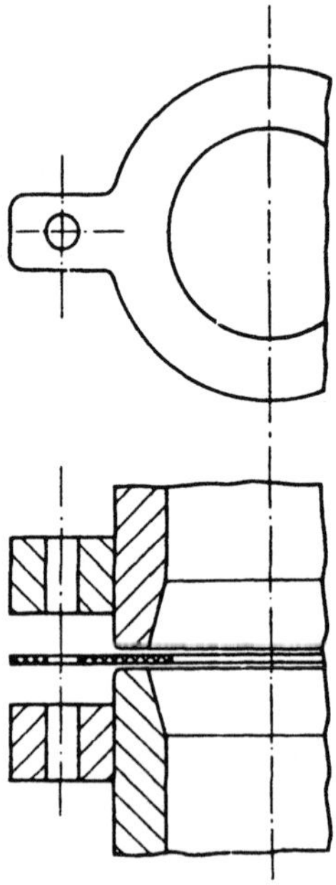

FIG. 3.109 Flat gasket seal. After Foote[389] *(Courtesy of The American Institute of Physics)*

where p is the load required on the gasket (kg/cm^2) and h the thickness of the bell jar wall seating on the gasket (cm). Experience shows that reliable seals based on atmospheric pressure (without mechanical clamping) may be made only with bell jars of about 15 cm (6 in.) or larger diameters. According to the formula given above, this corresponds, for a wall thickness of 4 mm, to a load of about 10 kg/cm^2.

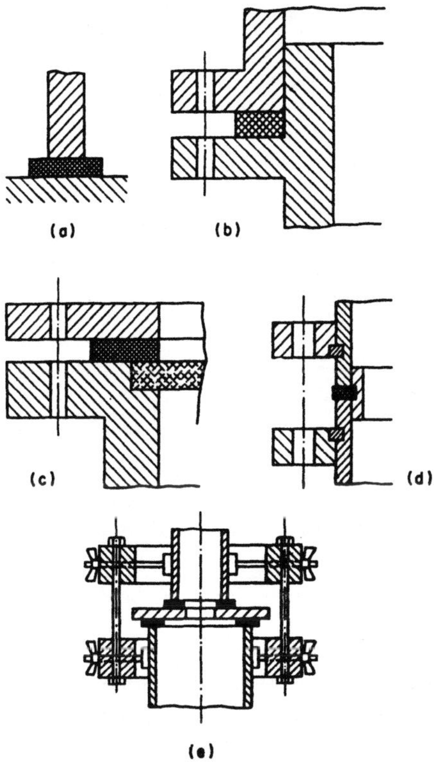

FIG. 3.110 Plane seals: (a) without clamping; (b) corner clamping; (c) window seal; (d) tubing seal with lock ring (after Lauritsen[746]); (e) double gasket seal for joining pipes with dissimilar diameters (after Weintroub[1292])

Windows or end plates can be sealed with flat gaskets and flat flanges, as shown in Fig. 3.110c.

Two pipes of the same diameter may be butt sealed, by placing a flat gasket between their ends. The inner ring (Fig. 3.110d), which serves to align the pipes, is retained by the gasket which fits into a shallow groove of the inner ring. The flanges assuring the tightening of the seal may be retained by lock rings (Lauritsen[746]).

Two pipes with dissimilar diameters may be joined using the arrangement of Fig. 3.110e (Weintroub[1292]). A metal plate is placed between the ends of the pipes, to which the two pipes are joined by means of plane rubber gaskets. A double clamp holds the two pipes and pushes them together.

Plane flanges have been sealed face to face by using a flat copper gasket (Goerz[455]) or aluminium gasket coated with a plastic material (Dobke[283]).

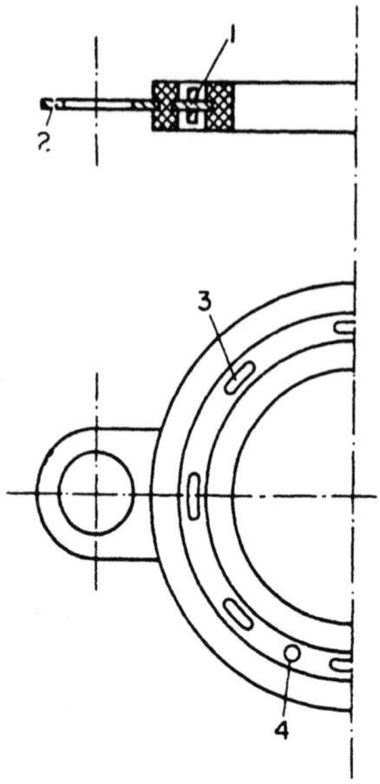

FIG. 3.111 Double ring gasket

Guard vacuum seals (Section 38.23) may be constructed using double rectangular cross-sectional gaskets (Fig. 3.111), the two concentric gaskets being connected by a metal ring (1) and provided with holding washers (2). The rubber parts of the gasket are usually vulcanized to the metal parts. Spacers (3) mounted onto the metal ring (1) act as mechanical stops when

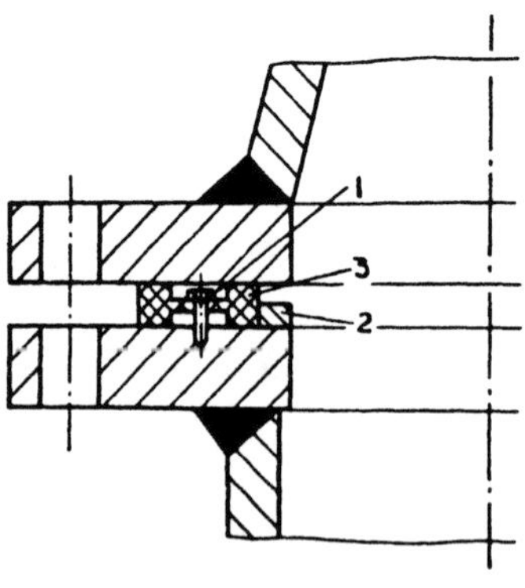

FIG. 3.112 Double ring gasket screwed to the flange. After Alvarez[24] *(Courtesy of The American Institute of Physics)*

the required compression ratio of the gasket is reached. The hole (4) provides the connexion between the two sides of the metal ring (1).

Alvarez[24] used a moulded rubber gasket (Fig. 3.112) held to the flange with screws, (1) and retained by a steel tack (2) welded to the flange along the vacuum side. This seal was used on a tank (accelerator) having a 40 foot long flange. Double gaskets are available as "dumbbell" seals (see Fig. 3.135).

b) *Groove seals.* In order to prevent sliding of the gasket during the tightening of the seal (Fig. 3.108) it is recommended that the gasket be placed

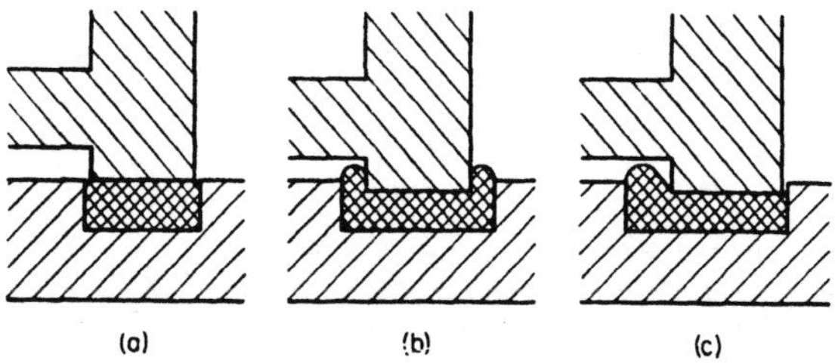

FIG. 3.113 Principles in designing flat gasket seals: (a) bad design; (b, c) good designs; (b) uncentred; (c) centred

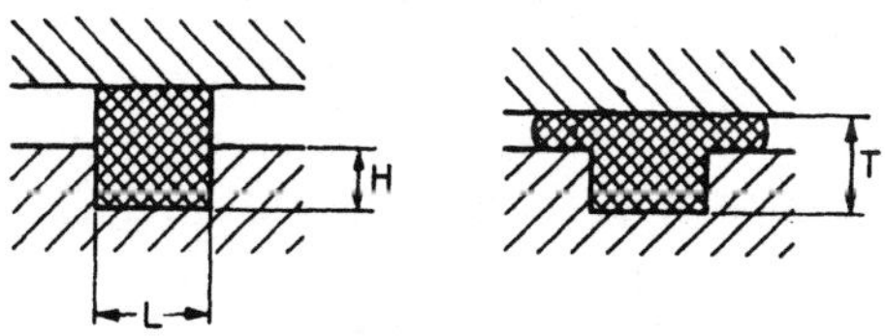

FIG. 3.114 Seal with gasket of square cross section in rectangular cross section groove of the same width

in a groove cut on one of the flanges. The groove may be designed for permanent loading of the gasket or for limited loading. In the first case the cross section area of the groove should be smaller than that of the gasket, and in the second case the groove should be able to accommodate the whole compressed gasket.

The groove should limit the compression ratio of the gasket to the required value (Fig. 3.50) and allow space for the deformation of the rubber gasket in directions perpendicular to the sealing force. Thus the design shown in Fig. 3.113a should be avoided. Arrangements as shown in Fig. 3.113b, c may be used, and the arrangement in Fig. 3.113c is preferred since it also ensures the centring of the sealing flanges.

When seals with rectangular cross-sectional gaskets are used (Figs. 3.114, 3.115) the construction from Fig. 3.115 is preferred. For the construction of Fig. 3.114 ratios of $H/L = 0.5$ and $T/L = 0.8$ are recommended (Ardenne[47], Guthrie[491]). In this design permanent sets of drastic proportions generally occur, leading sometimes even to the shearing of the gasket; the gaskets should usually be replaced in such seals after each opening.

The design shown in Fig. 3.115, in which the corners of the groove are tapered to provide an expansion space for the gasket, allows for a reasonable permanent set, permitting the re-use of the gasket after the seal has been opened. The recommended design ratios (Ardenne[47]) are in this case:

$$H/L = 0.6, \quad W/L = 1.5 \quad \text{and} \quad T/L = 0.8.$$

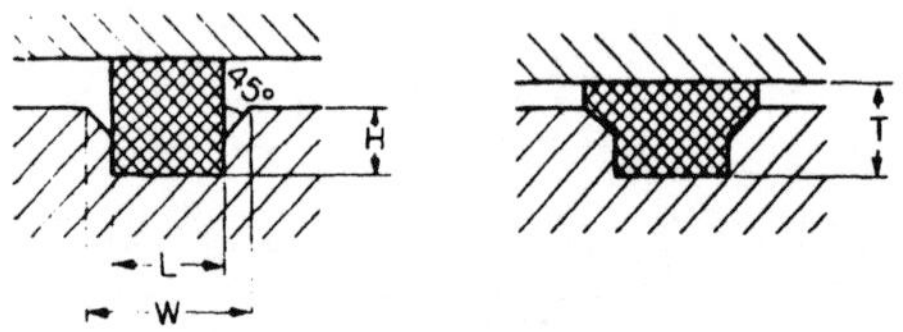

FIG. 3.115 Seal with gasket of square cross section in rectangular cross section groove with bevelled edges

The groove seal shown in Fig. 3.115 is generally used in joints between two flanges, but it can be adapted to seal more complicated joints as well. Figure 3.116 shows such a seal as it is used to close covers, ports, windows, etc.

In groove seals with limited compression of the gasket (see also Section 38.41) the groove should have a cross-sectional area 2–5 per cent greater than that of the gasket. In this way the gasket can be kept fully in the groove when the flanges are tightened (metal-to-metal contact).

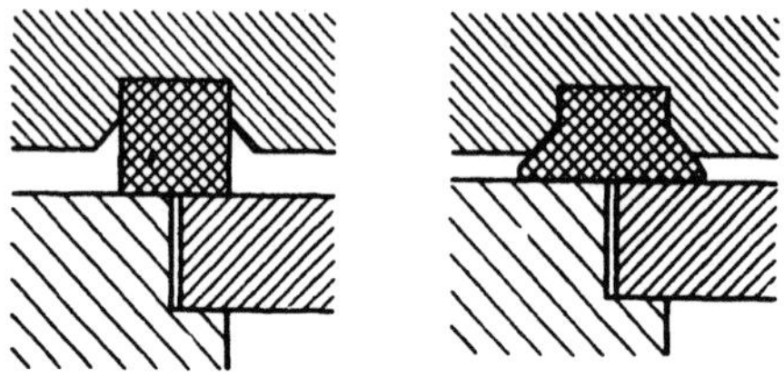

FIG. 3.116 The seal from Fig. 3.115 as used to join covers, ports or windows

A seal with metal-to-metal contact of the flanges, using square cross section rubber gaskets, may be designed as shown in Fig. 3.117 or Fig. 3.118. The seal on Fig. 3.117 (Ardenne[47], Pollermann[998]) should have the gasket placed on the vacuum side of the groove. The design ratios recommended for the design are $T/L = 0.75$ and $W/L = 1.4$.

The seal (Fig. 3.118) quoted by Smith[1151] as a good system to minimize the surface of the gasket exposed to the evacuated space, should have the following dimension ratios: $T/L = 0.75$, $B/L = 0.5$ and $W/L = 2.2$. Seals of this type are recommended by Guthrie[490a] who gives the suggested dimensions for such seals using double gaskets of square section (3/16, 1/4 and 3/8 in. sides).

If the gasket should be held firmly in the groove, seals with *trapezium* cross section gaskets (see Table 3.19C) may be used. These gaskets are placed

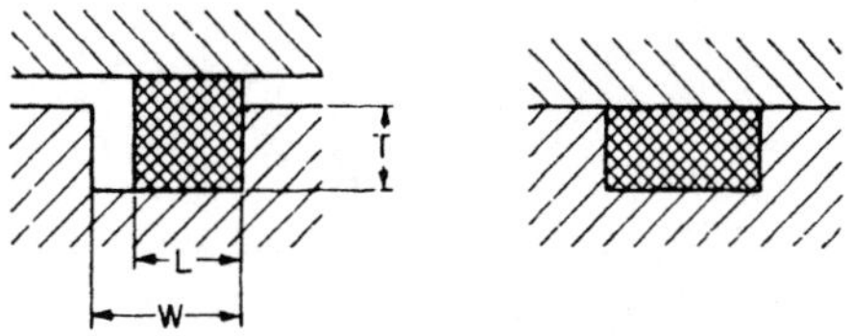

FIG. 3.117 Gasket of square cross section sealed in rectangular cross section groove

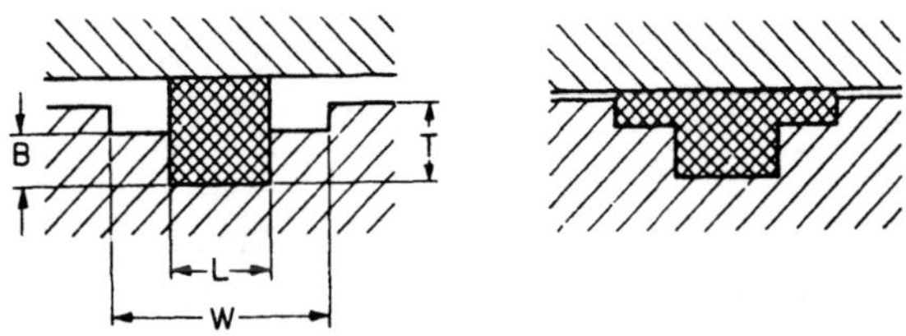

FIG. 3.118 Square cross section gasket sealed in rectangular cross section step groove

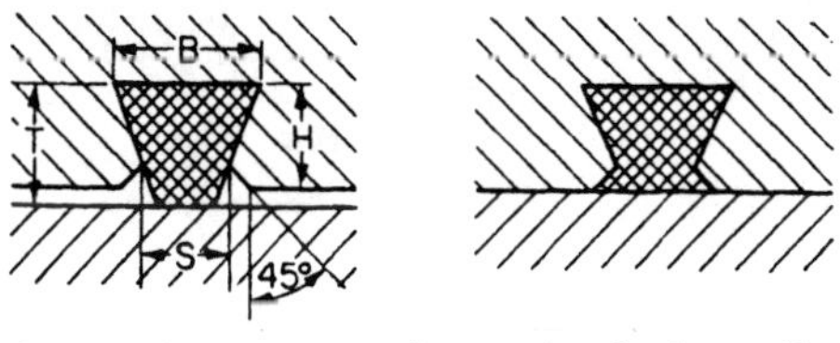

FIG. 3.119 Seal with trapezium cross section gasket in dovetail cross section groove

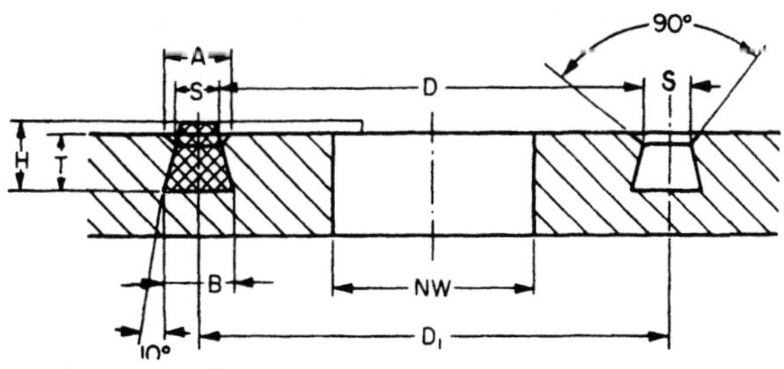

FIG. 3.120 Dimensions for trapezium cross section gasket seals (Table 3.38)

in grooves having the appropriate cross-sectional shapes (Fig. 3.119). The design ratios for the design of such seals are $T/H = 0.83$ and $S/B = 0.75$. Figure 3.120 and Table 3.38 give the detailed dimensions for seals using trapezium cross-sectional gaskets.

TABLE 3.38. STANDARD DIMENSIONS (mm) FOR SEALS WITH TRAPEZIUM CROSS SECTION GASKETS (DIN 2570; Diels[280]) (see Fig. 3.120 and Table 3.19C)

NW	D_1	H	B	A	D	S	T
10	16	4.8	4	5	13	3	4
20	26	4.8	4	5	23	3	4
25	32	4.8	4	5	29	3	4
32	42	7.2	6	7.5	37.6	4.4	6
40	50	9.0	7	9.5	44.7	5.3	7
50	60	9.0	7	9.5	54.7	5.3	7
65	85	9.6	8	10	79	6	8
90	105	10	9.5	11	97.7	7.3	8
100	115	10	9.5	11	107.7	7.3	8
125	140	10	9.5	11	132.7	7.3	8
150	165	10	9.5	11	157.7	7.3	8
200	220	12	10	13	212.5	7.5	10
250	270	12	10	13	262.5	7.5	10
300	330	12	10	13	322.5	7.5	10
350	380	12	10	13	372.5	7.5	10
400	416	12	10	13	408.5	7.5	10
500	550	12	10	13	542.5	7.5	10
600	640	12	10	13	632.5	7.5	10
800	820	14.5	12	14.5	811.2	8.8	12

c) *Seals with gaskets of special shapes.* Plane surfaces have been sealed by using various gaskets of special shapes (cross section).

One of the most commonly used gaskets with a special cross-sectional shape (see Table 3.18) is the *L-gasket*, used to seal bell jars or ports. L-gaskets are available commercially in various dimensions fitting the bell jars supplied, and are made from various elastomers (Neoprene, Viton, etc.). L-gasket seals have the drawback that the gasket surface exposed to the evacuated space is large; the seals are used only on dynamic vacuum systems equipped with strong pumps (e.g. evaporators).

Metal gaskets have been designed in a large variety of cross-sectional shapes, in the attempts of the various designers to solve the difficulties presented by the seals with metal gaskets (high-tightening forces, lack of elasticity). Figure 3.121 shows seals using the various shapes of gaskets, with the dimensions suggested for such seals.

Milleron[865] constructed bakeable seals using copper gaskets (Fig. 3.121a) between plane flanges. His copper gasket was provided with small *beads* raised above the bulk of the gasket. These small beads are forced into the broad bulk of the gasket to ensure that the seal will be maintained throughout the bakeout cycle. According to Milleron[865] the surface finish requirements are not excessively high (see Table 3.37), but the dimensions *B* and *C* of the gasket (Fig. 3.121a) are rather critical. The dimension *C* results from two contradictory requirements: that the bead should be as far away from the atmospheric side as possible in order to avoid oxidation in the vicinity of the bead, and that the bead should be as near as possible to the bolts in order to minimize the influence of the warping of the flanges, during the tightening of

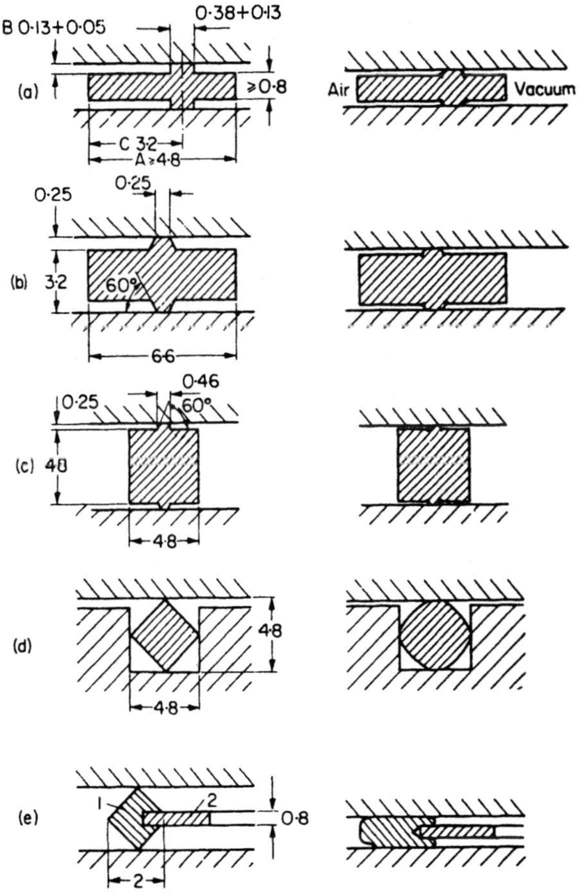

FIG. 3.121 Plane seals with gaskets of special shapes: (a) gasket with beads (after Milleron[865]); (b) coined gasket (after Wheeler[1304]); (c) gasket with ridge (after Goertz[455]) *(Courtesy of Pergamon Press)*; (d) Diamond cross section gasket in groove (after Hoch[568]); (e) diamond cross section gasket between flat flanges (after Kienel[671]) *(Courtesy of Rudolf A. Lang Verlag, Esch/Taunus, Germany)*

the seal. A seal using the gasket dimensions given in Fig. 3.121a has been successful.

Wheeler[1304] describes seals (1.5 and 6 in. diameter) in which *coined gaskets* were used. This gasket (Fig. 3.121b) has a sealing ridge, in the form of a truncated equilateral triangle. Seals using this gasket (OFHC copper) may be opened and remade with the same gasket. Wheeler[1304] reports on 35 successive sealing cycles, made with the same coined gasket, by increasing the pressure slightly with each retightening, to erase the impressions produced by the previous sealing. For good performance he recommends that these seals should only be rated for about *four* re-openings. The ability of the gasket to reseal is explained by the fact that the copper ridge is forced into the body of the gasket, without the development of high lateral forces (as in knife edge seals, see Section 38.56b) in the body of the gasket. It is probable that at some point in the body of the coined gasket, the elastic limit of the copper is not exceeded, providing the resilience which gives the ability to reseal after having been fully compressed. The required sealing forces are listed in Table 3.13, and the surface finish in Table 3.37.

It has been found that the coined gasket seal is sensitive to the degree of flatness of the sealing flanges. A deviation of 0.025 mm (0.001 in.) to 0.05 mm (0.002 in.) out of flat over approximately 1.9 mm (0.75 in.) of seal circumference was sufficient to prevent the making of good seals (Wheeler[1304]). These seals may be baked to 400 °C.

Goertz[455] used an OFHC copper gasket of rectangular cross section with a raised ridge on each side (Fig. 3.121c). This gasket was used either with a circular outline or a rectangular one, to seal openings of 1 to 6 in. diameter, obtaining leak rates of less than 7×10^{-8} lusec (see also Table 1.5) after temperature cycling (8–56 hr at 600 °C).

Hoch[568] and Turnbull[1240a] describe seals using copper gaskets of *diamond* cross sections (Fig. 3.121d) compressed between stainless steel flanges. It is recommended that the gasket be placed in a rectangular groove machined in one of the flanges. The groove should have a width equal to the apex-to-apex distance of the gasket (cross section) and a depth equal to 85 per cent of this distance, limiting the compression of the gasket to 85 per cent. The seal may be baked to 450 °C. The force required for making the seal is given in Table 3.13.

Kienel[671] reports on a gold gasket (1, Fig. 3.121e) of diamond cross section, used as a sealing bead on a stainless steel ring (2).

In order to limit the exposed surface of rubber gaskets, they may be shielded by a sheet metal ring shaped into a V (Fig. 3.122). The rubber gasket fits loosely into this V, being shielded from the vacuum side when the seal is tightened. This sheet-metal ring can be made from soft copper or steel (for systems containing mercury vapour).

Figure 3.123 shows another possibility of shielding the rubber gasket from any exposure to the evacuated space. It consists of enclosing the rectangular cross section gasket in a thin (0.05 mm) metal foil (Ardenne[17]).

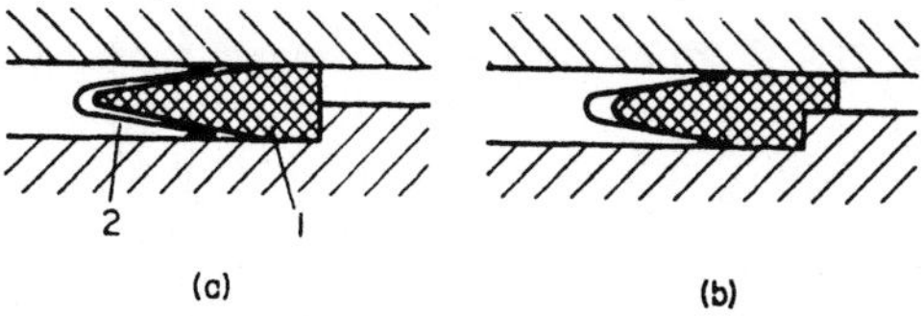

FIG. 3.122 Rubber gasket protected by sheet metal ring (vacuum on left side): (a) uncompressed; (b) compressed: (1) rubber; (2) sheet metal

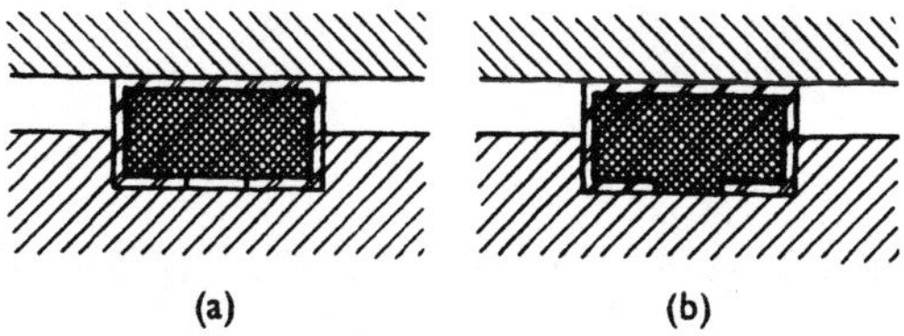

FIG. 3.123 Rubber gasket, completely enclosed in metal foil: (a) uncompressed; (b) compressed

38.52 Conical seals. These seals are based on gaskets previously shaped to conical forms or just brought to this shape by compressing them between the sealing parts. The most simple seal of this type is a *rubber plug* (Fig. 3.124a), which has the form a bung (Fig. 3.124b) when the seal should contain pipes or rods leading from the outside to the evacuated space.

In order to ensure the compression ratio required for a reliable seal, the conical seal is constructed as shown in Fig. 3.124c, where the conical gasket is compressed by a second plate provided with the required hole (or holes) for the pipe or rod crossing the gasket. This type of seal may be also used to seal wires (e.g. thermocouple) leading into the vacuum vessel (Stamper[1173]). The opposite construction (Fig. 3.124d) is also possible; the tapered end of a pipe may be sealed into a gasket placed inside a cylindrical opening. This type of seal was used by Nester[920] with a Teflon gasket (Fig. 3.54a) in low temperature work. If a Teflon gasket is used, some parallel grooves should be provided on the inside surface of the cylindrical opening (Fig. 3.124d and Table 3.17).

A pair of flanges with tapered sealing surfaces may compress the gasket either between parallel surfaces (Fig. 3.125a, b) or between these surfaces placed to form an angle (Fig. 3.125c). Goertz[455] describes a seal (Fig. 3.125a) where the gasket (1) used is originally flat, but is deformed to a conical shape

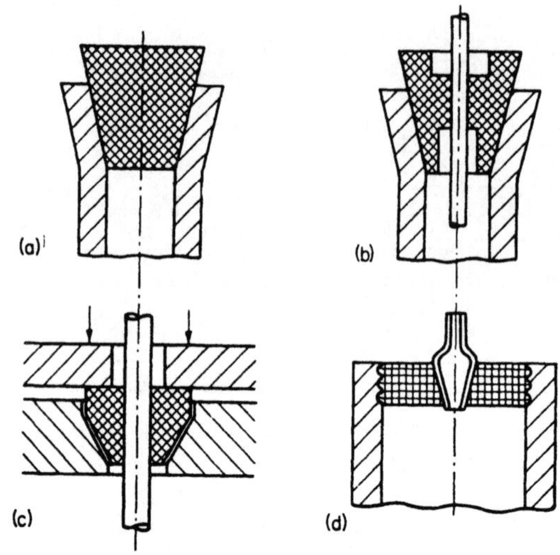

FIG. 3.124 Conical gasket seals: (a) simple rubber plug; (b) rubber bung; (c) conical gasket sealed by compression flange; (d) conical glass sealed in metal pipe with Teflon gasket

when the union is tightened, by the aid of the split rings (3) and an appropriately screwed joint. Medicus[839] connected conical vacuum joints (Fig. 3.125b) by placing a narrow rubber band (2) between the cones (1). This joint cannot be used in applications where the positioning of the joint must be

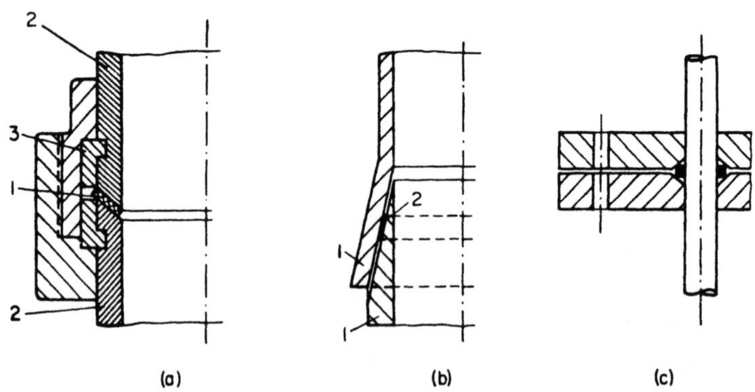

FIG. 3.125 Conical gasket seals, with (a) flat gasket deformed to conical shape: (1) gasket; (2) tube; (3) split ring (after Goertz[455]); (b) rubber band (after Medicus[839]); (c) rectangular cross section gasket (after Cowie[241]). *(Courtesy of The American Institute of Physics)*

very exact or where minute movements produced by the changes in pressure difference cannot be tolerated.

Conical seals having their tapered surfaces placed as to form an angle (usually 90°) can be used with gaskets of square cross section (Fig. 3.125c) in a similar manner (Cowie[241]) to the O-ring seals of this kind (Figs. 3.86 and 3.92).

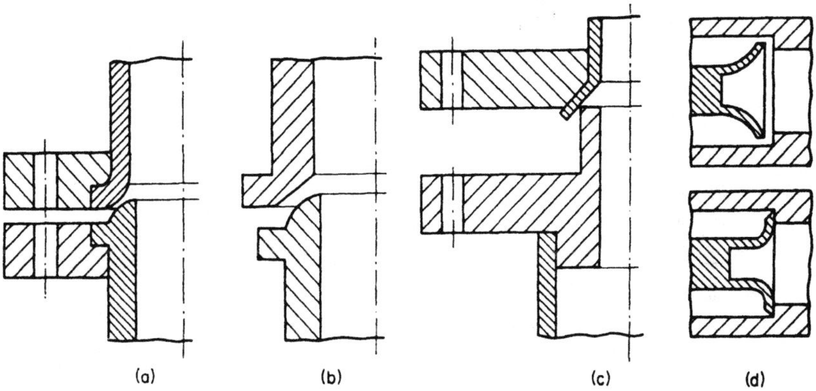

FIG. 3.126 Conical seals without gaskets: (a) between convex and conical surfaces; (b) between two convex surfaces (after Hall[503]); (c) flare seal (after Lichtman[764]); (d) seal with conical spring washer (after Wishart[1327]) *(Courtesy of Pergamon Press)*

Conical seals may also be used with metal gaskets or "without gasket". Lyubimov[791] quoted seals using aluminium gaskets, Hall[503], Lichtman[764] and Wishart[1327] described seals, where the flanges are sealing on each other without using a third part as a gasket.

In the seals described by Hall[503], the joint is made between a curved (convex) surface and a conical one (Fig. 3.126b) or between two convex surfaces (Fig. 3.126a). These seals may be used with 0.25 to 1 mm thick OFHC copper, copper–nickel, Monel or stainless steel gaskets, as well as without any gasket at all, and may be subjected to a long (420 hr) bakeout at 500 °C. Hall[503] assumed that the success in sealing without gasket is due to the very high compressive stress provided by the initial *line contact* of the sealing surfaces, which permits effective resealing even if the mating halves are not reseated in exactly the same position as in the previous sealing cycle.

Lichtman[764] constructed vacuum seals (Fig. 3.126c) by flaring the end of his copper tubing (slipped through the flange provided with a bevel). The opposite flange is provided with a prominence which seals against the flare of the tubing.

Wishart[1327] observed that a conical spring washer (Fig. 3.126d) may be used to make a seal without a gasket, since when compressed to the flat position, it increases in diameter and seals against the cylindrical wall. The conical washer was fabricated from a spring steel and the body was made of 304 stainless steel (spring disc considered harder than the body), and the diameter of the spring disc was kept within a few thousandths of an inch of the bore diameter in which it should seal. When the sealing force is removed, the spring of the disc contracts to its original conical form, leaving the walls and permitting it to be retracted. This seal was proposed for use as the closing member of valves or controlled leaks.

38.53 Cylindrical seals. In cylindrical seals the gasket is pressed radially on a cylindrical sealing surface. The cylindrical seals include: *rubber tubing joints*, *compressed gasket seals* and *lip seals.*

a) *Rubber tubing joints* are perhaps the most simple vacuum seals, although they should conform to some basic requirements:

(1) The rubber tubing should have the wall thickness needed to prevent collapse as shown by Fig. 2.17. Large diameter tubes can be strengthened by inserting helical wire springs (see Fig. 5.2d; Fremlin[401]).

(2) The rubber tubing should have an inside diameter smaller than the outside diameter of the pipe on which it is to be slipped over. It is difficult to give recommended values for the ratio of the two diameters, but on Fig. 3.127 it can be seen how a tubing with too large diameter fits on the end of a

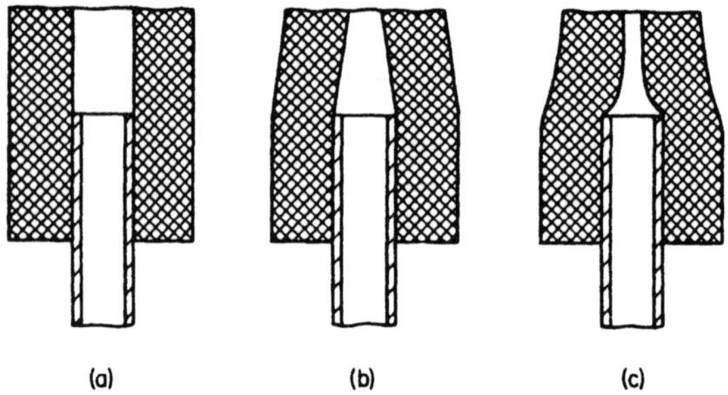

FIG. 3.127 Rubber tubing connexions; (a) the i.d. of rubber tubing too large; (b) good joint; (c) i.d. of rubber too small

pipe (Fig. 3.127a) compared with a joint having the correct diameter ratio (Fig. 3.127b) and one in which a rubber tube having too small a bore is fitted on the end of a pipe (Fig. 3.127c).

(3) The end of the pipe, which should receive the rubber tubing, should have radiused (metal pipes) or fire polished (glass pipes) edges. Cutting edges damages the inside of the rubber tubing, leading to leaking joints.

(4) It is recommended that the ends of the pipes on which the rubber tubings are to be slipped over be shaped to have ridges to enable easier gripping of the tubing (Fig. 3.128).

(5) When two pipes are butt joined with a rubber tubing connexion, it is recommended that the ends of the two pipes should touch inside the rubber (Fig. 3.129a) in order to minimize the area of rubber exposed to the evacuated space. Glass pipes should be brought in contact with care, and if the joint is to be bent, it is better to leave some distance between the ends of the butt joined glass pipes.

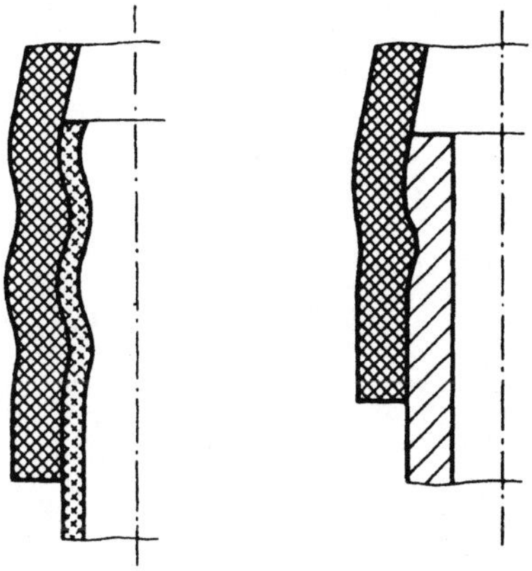

FIG. 3.128 Shape of pipe ends to be connected with rubber tubings: (a) glass tube; (b) metal pipe

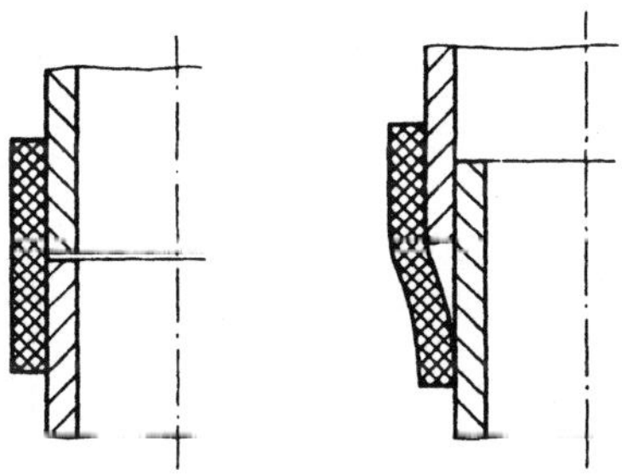

FIG. 3.129 Butt joints with rubber tubing: (a) for pipes with similar diameters; (b) for pipes with dissimilar but fitting diameters (not recommended because of the trapped volumes)

(6) Pipes with not too thick walls and having a good fit inside one another (Fig. 3.129b) may be joined with a short length of rubber tubing. If the fit of the two pipes is not perfect, it is better to provide the seal by means of a third connecting pipe of appropriate shape.

(7) In order to facilitate the sliding on of the rubber tubing over the end of the pipe, the inside of the rubber may be slightly greased; the use of water for this purpose is to be avoided.

(8) Usually vacuum joints made with rubber tubing (according to Figs. 3.127b and 3.128) are vacuum-tight without any external clamping. If clamp-

ing seems to be necessary, the use of wires wrapped over the seal should be avoided, because of the danger of cutting the rubber. Hose clamps (Fig. 3.130) may be used.

(9) The opening of rubber tube joints very often leads to breakage of the glass pipes. Thus it is recommended (especially when old joints are to be opened) *to cut* the rubber (longitudinally) instead of trying to withdraw it

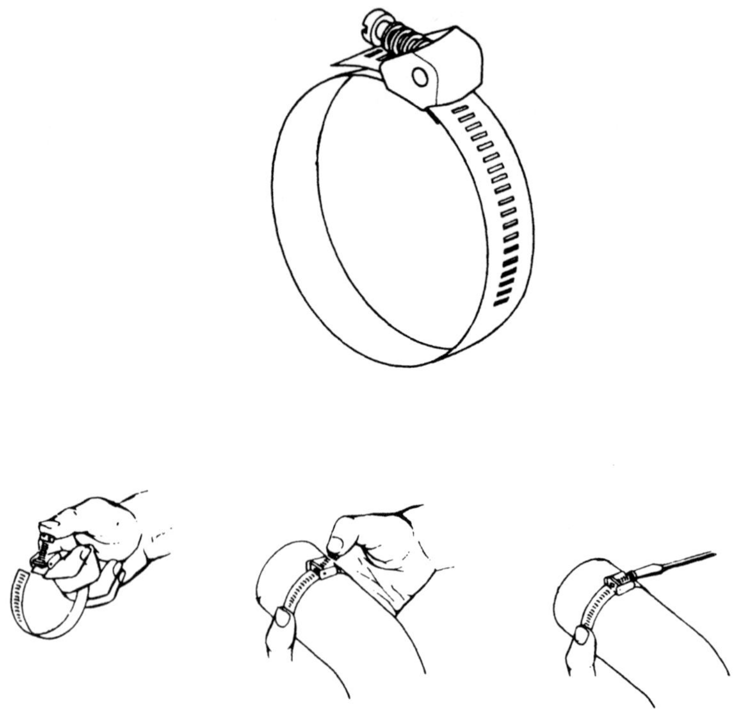

FIG. 3.130 Hose clamp and its closure over a rubber tube *(Courtesy of Veeco*[1262]*)*

from the pipe. However, such joints can be opened with some skill using a simple tool consisting of a 2–3 mm diameter wire bent in L-shape which is greased, inserted axially between the rubber and the pipe and moved around the joint in order to separate the rubber from the pipe.

For connexions which are to be *flexible* it is better to use special flexible connexions supplied commercially (e.g. Genevac[441], N.G.N. Electrical[926]) for such purposes (Section 51.3).

b) *Compression gasket seals.* Gaskets placed as shown in Fig. 3.131 are sealing axially on a cylindrical surface, due to the compression which they recieve in axial direction.

Kronberger[721] used this type of seal (Fig. 3.132a) to butt join two pipes with a sealing sleeve. The drawback of this seal lies in the fact that the alignment of the pipes and of the flanges with respect to the pipes is not assured. It can happen that the ends of the pipes are too close to one of the gaskets (dotted line, Fig. 3.132a). The joint has the advantage that the sleeve can slide over one of the pipes; thus the pipe can be cut and joined in position.

Pipes with dissimilar diameters may be joined with the expansion joint of Quickfit Visible Flow[1021], where the gasket (1, Fig. 3.132b) acts as a cylindrical seal for the glass pipe with smaller diameter (2) and as a plane seal for

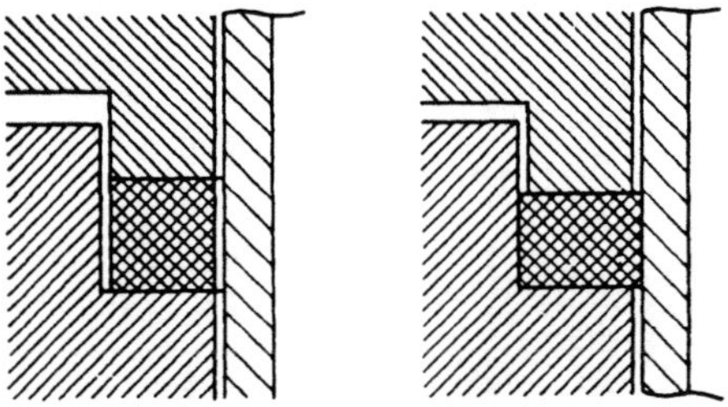

FIG. 3.131 Cylindrical seal; (a) uncompressed; (b) compressed

the larger pipe (3). The clamping system (4) holds the pipes together by pushing on the external conical surface of the larger pipe (3) and on the gasket (1). The seal is useful for pipes up to 100 mm (4 in.) diameter.

Two pipes with large difference between their diameters may be joined by cylindrical seals of the type shown in Fig. 3.133a (Guthrie[491], Roth[1083], Champeix[202], Banki[79]). In this seal, a cylindrical rubber gasket (1) fitted with a small clearance to the end of the larger pipe (2) is compressed by the washer (3), as the compression nut (4) is tightened. By this compression, the gasket (1) seals radially on the outside of the smaller pipe (5) previously inserted. In the construction of such seals there are two important requirements:

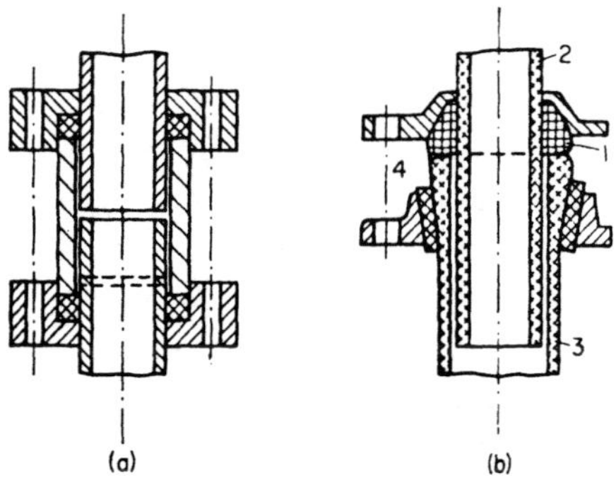

FIG. 3.132 Pipe-to-pipe joints; (a) with two cylindrical compression seals; (b) with a single gasket having a cylindrical seal towards one of the pipes and a plane seal towards the other pipe (after Q.V.F.[1021])

A close clearance between gasket (1) and pipe (2) as well as between the small pipe (5) gasket (1) and washer (3) must be maintained;

The seal must be tightened without rotating the washer; for this purpose ball bearings (6) are placed between the compression nut (4) and washer (3).

The seal on Fig. 3.133a is used to connect the exhaust tube of electric lamps and electronic tubes to the rotating part of the slide valves (see Fig. 3.18) of the exhaust machines.

Figure 3.133b shows a cylindrical seal, used to connect diffusion pumps to the vacuum systems (Champeix[202]). A cylindrical seal using Teflon is described by Guildner[489a].

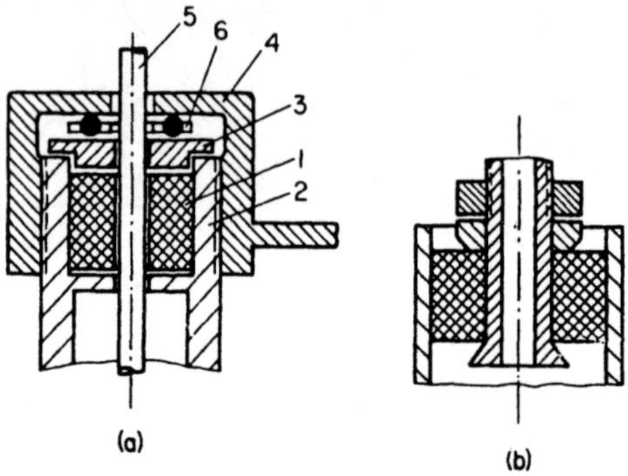

FIG. 3.133 Compression cylindrical seals; (a) for easy actuation; (b) semi-permanent sea

Cylindrical seals are also used on shaft seals (Section 51.73), electrical ead-throughs (Section 43.2) and vacuum valves (Section 61.32).

c) *Lip seals*, consist of gaskets sealing with their *lips* on cylindrical surfaces (Fig. 3.134).

In the construction described by Giaimo[446], the gasket consists of a Teflon ring (0.226 in. thick, 1.475 in. o.d. and 1 in. i.d.) machined to a U-shaped cross section (1 Fig. 3.134a) with thin lips (0.02 in.). A closely-wound helical spring (3) with a coil diameter of 0.185 in. wound from piano wire (0.01 in. diameter) was laid in the U. The sealing is accomplished by the lips of the U-gasket compressed between two surfaces (2) inclined at 35° to the normal. This arrangement produces a radial as well as an axial compressive force.

A lip seal using copper gaskets is supplied by Varian[1257a]. In this seal the conical washer (4. Fig. 3.134b) is placed *with the small diameter up* into the appropriately shaped mouth of the port (larger tube). The copper tube (1) is then pushed down through the washer until it makes firm contact with the bottom of the seating groove (5 Fig. 3.134b). The split bushing (3) is placed into position in the body of the port, and the lock nut (2) is tightened. Varian[1257a] recommends that the lock nut should be tightened only 5/6 to 1 turn, since if the nut is tightened excessively, the seal will not act as a lip seal and leaking will occur.

Lip seals using rubber gaskets (Wilson seals) are discussed in Section 51.74.

38.54 Dumbbell seals. Plane flanges (without grooves) can be sealed using dumbbell gaskets, moulded in the shape of two concentric gaskets connected

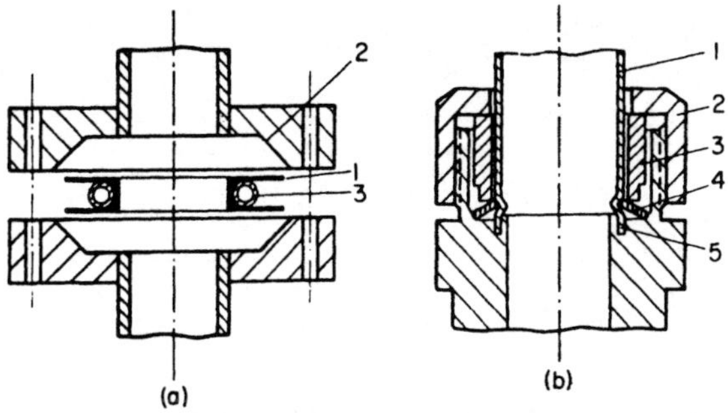

FIG. 3.134 Lip seals; (a) with Teflon gasket, after Giaimo[446] *(Courtesy of The American Institute of Physics and of R.C.A. Laboratories)*; (b) with metal washer *(Courtesy of Varian[1257a])*

by a web (Guthrie[491], Jacobus[621], Thomas[1220]). The web is usually perforated in order to connect its two sides; this makes the use of dumbbell seals with a guard vacuum possible (Section 38.23). To hold the gasket in place during assembly, pins held in the holes in one of the flanges are placed through the web, or in some construction (Fig. 3.112) the web is screwed to the flange.

In order to limit the compression ratio of the gasket as well as to minimize the area of the gasket surface exposed to the evacuated space, it is advisable that an inside metal spacer ring be provided in the seal. A gasket of this type,

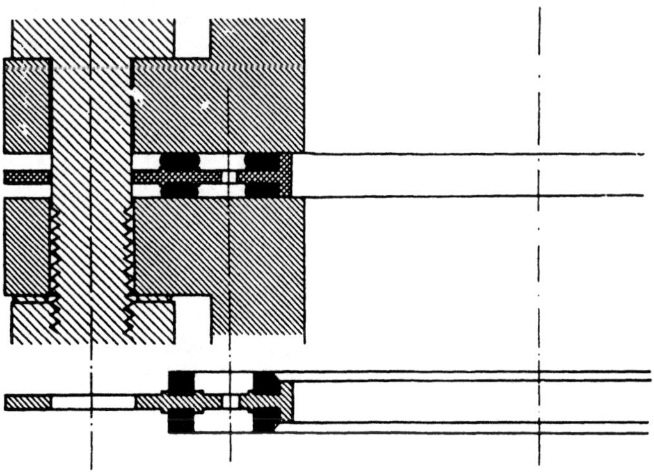

FIG. 3.135 Dumbbell gasket seal *(Courtesy of Vacuum Research Company[1253a])*

with the elastomer rings moulded onto a supporting metal ring which also provides the spacer ring is supplied by Vacuum Research Comp.[1253a] as shown in Fig. 3.135.

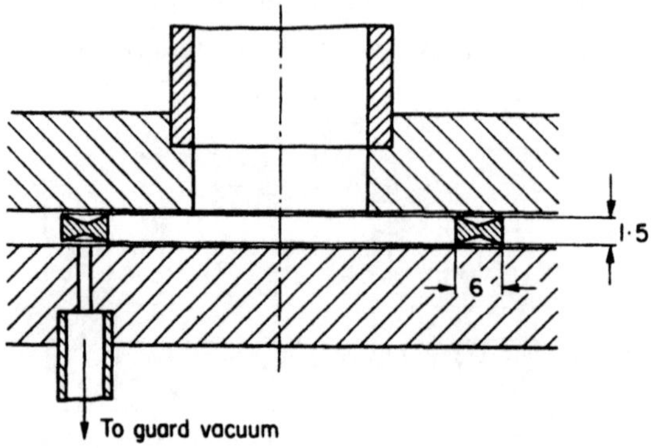

FIG. 3.136 Guard vacuum seal with specially shaped copper gasket. After Heathcote[526] *(Courtesy of The Institute of Physics and The Physical Society, London)*

Heathcote[526] describes a dumbbell type seal, which uses a metal (copper) gasket of X-shape cross section (Fig. 3.136). The initially 1.5 mm thick (6 mm width of the cross section) gasket was compressed to about 0.8 mm, between two stainless steel (or Monel) flanges and the spaces between the contact edges were evacuated. The seals (0.5–6 in. diameter) were leak-tight after baking to

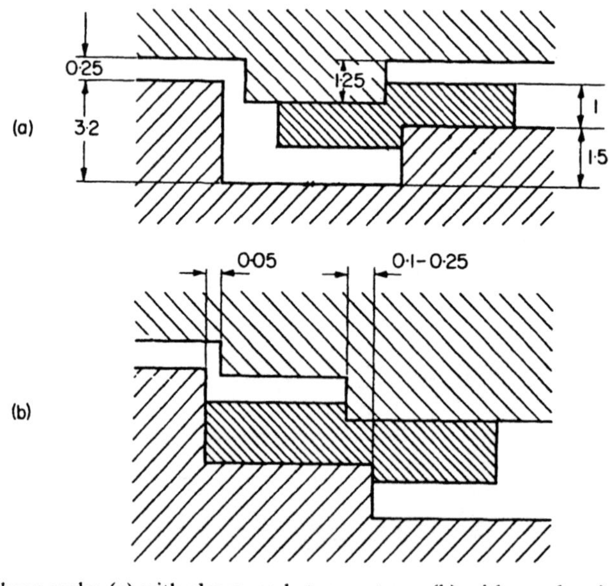

FIG. 3.137 Shear seals; (a) with clearance between steps; (b) with overlapping steps (after Carmichael[194])

450 °C. When remaking this joint, the light film of oxide formed between the copper contact edges is to be removed (with emery paper).

38.55 Shear seals. These seals are based on the shearing instead of the compression of the gasket (see Fig. 3.42b). The shear seal was developed by Lange[737] for ultra-high vacuum, and its various forms and requirements are discussed by Alpert[20], Carmichael[194], Henry[540] and Wheeler[1304]. The seal is known also under the name of "*step seal*".

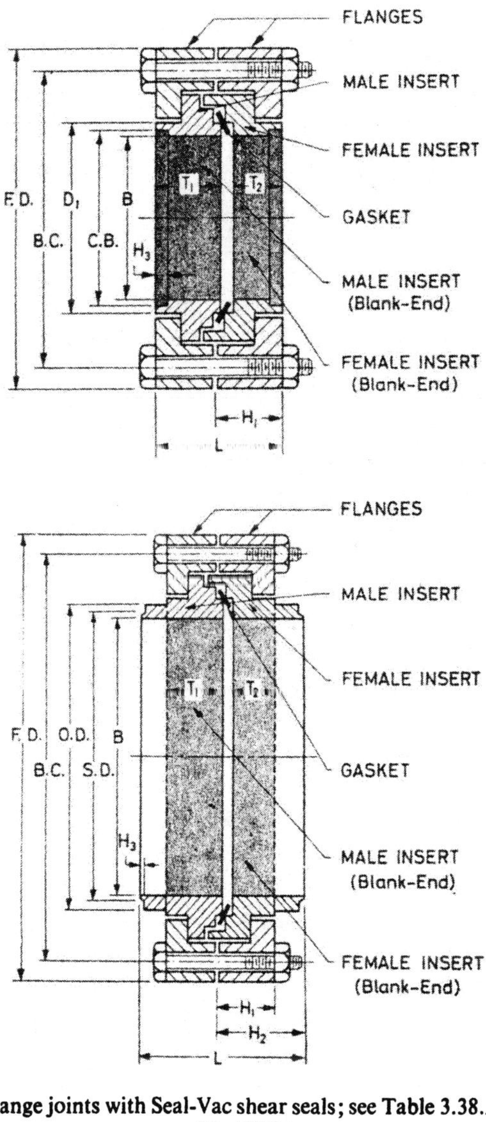

FIG. 3.137.A Flange joints with Seal-Vac shear seals; see Table 3.38.A *(Courtesy of Ultek*[1251a]*)*

Shear seals may be designed either with a clearance between the opposite shearing steps (Fig. 3.137a) or with overlapping steps (Fig. 3.137b). According to Lange[737] the gasket used in shear seals (OFHC copper, 1 mm thick) should be hydrogen annealed at 950 °C after cutting. The seal is made by tightening the bolts (12 bolts were used on a 6 in. flange, 3/8 in. thick) until the copper gasket is roughly one-half of its original thickness between the sealing edges (steps) The main advantages of these seals are: the easy machining of the flanges, ease of repair if damaged, and the fact that they allow vertical seals to be made (since the gasket is rigid enough). The seals were tested in vacuum at 450 °C (about 70 heat cycles) without failure (Lange[737]).

Wheeler[1304] compared the (step) shear seals with other seals (Knife edge, coined gasket, Conflat) using metal gaskets. He quotes the dimensions of the seal used for the tests. This seal was designed for a nominal step overlap of 0.010 in. (0.25 mm) and used an OFHC copper gasket, 0.040 in. (1 mm) thick. The seal had a 0.010±0.002 in. bite into the gasket on each side. The flange thickness was 0.920 in. for the 1.5 in. tubing, and 1.665 in. for the 6 in. tubing.

Wheeler[1304] recommends that the gasket used in shear seals be rated for use only for a *single* seal. However, he was able to use successfully the same

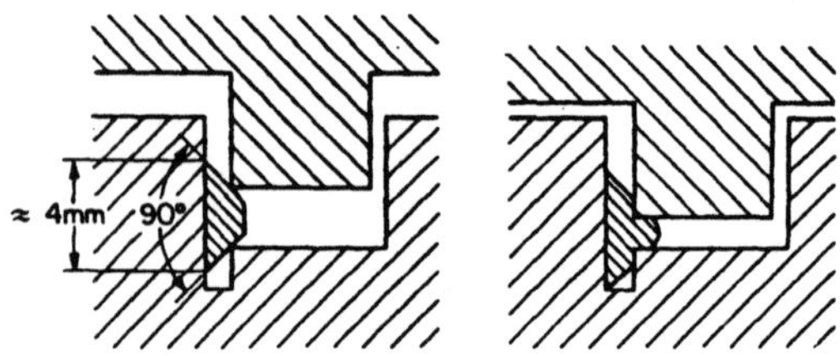

FIG. 3.138 Shear corner seal

gasket 5–7 times by increasing the pressure slightly each time. The forces required to tighten the seal are listed in Table 3.13.

Flange joints with step shear seals are available from Varian[1257a], Granville–Phillips[469], Ultek[1251a]. The dimensions of such flange joints manufactured by Ultek[1251a] are shown in Fig. 3.137A and listed in Table 3.38A.

A *shear corner seal* (see also Section 38.44) is described by Guthrie[491], Pollermann[998], Ardenne[47]. This seal uses a copper gasket with a trapezium cross section (Fig. 3.138) placed in a specially shaped groove in order to assure the shearing of the gasket. This seal was tested on flanges up to 50 mm diameter, with temperature cycling to 250 °C. Its drawback is the volume trapped by the gasket.

TABLE 3.38A. DIMENSIONS OF FLANGE JOINTS WITH SHEAR SEALS (INCHES) "SEAL VAC" CONNECTORS (Ultek[1251a])

(see Fig. 3.137A)

Tube size	1/4	3/8	1/2	3/4	1	1 1/2	2	3	4	4 1/2	5	6
B	0.190	0.315	0.440	0.690	0.87	1.375	1.875	2.812	3.772	4.272	4.687	5.656
S.D.	—	—	—	—	0.92	1.420	1.920	2.860	3.820	4.324	4.740	5.740
O.D.	—	—	—	—	1.0	1.50	2.00	3.00	4.00	4.50	5.00	6.00
C.B.	0.260	0.385	0.510	0.760	—	—	—	—	—	—	—	—
D_1	7/16	7/16	13/16	13/16	—	—	—	—	—	—	—	—
B.C.	15/16	15/16	13/8	13/8	$1\frac{5}{8}$	$2\frac{5}{16}$	$2\frac{3}{4}$	$3\frac{15}{16}$	$4\frac{15}{16}$	$5\frac{13}{32}$	$5\frac{31}{32}$	$7\frac{3}{16}$
F.D.	$1\frac{1}{4}$	$1\frac{1}{4}$	$1\frac{7}{8}$	$1\frac{7}{8}$	$1\frac{1}{2}$	$1\frac{3}{4}$	$1\frac{3}{4}$	$1\frac{3}{4}$	$1\frac{3}{4}$	2	2	$2\frac{1}{2}$
T_1	0.560	0.560	0.560	0.560	0.54	0.540	0.540	0.540	0.540	0.660	0.660	0.790
T_2	0.420	0.420	0.420	0.420	0.38	0.380	0.380	0.380	0.380	0.500	0.500	0.630
H_1	1/2	1/2	1/2	1/2	1/2	1/2	1/2	1/2	1/2	5/8	5/8	3/4
H_2	—	—	—	—	3/4	7/8	7/8	7/8	7/8	1	1	$1\frac{1}{4}$
H_3	1/8	1/8	1/8	1/8	1/16	1/16	1/16	1/16	1/16	1/16	1/16	1/16
L	1	1	1	1	$1\frac{1}{2}$	$1\frac{3}{4}$	$1\frac{3}{4}$	$1\frac{3}{4}$	$1\frac{3}{4}$	2	2	$2\frac{1}{2}$
Number of bolts	4	4	4	4	4	6	6	8	8	8	12	12
Bolt size	3/16	3/16	1/4	1/4	1/4	1/4	1/4	1/4	5/16	5/16	5/16	5/16

38.56 Ridge seals. In order to raise the local pressure on the surface where the flange contacts the gasket, *ridges* are provide on one or both flanges. The shape of these ridges varies from *tongues* or *rims* used with rubber gaskets, to simple or *multiple ridges* used with Teflon gaskets or the various *knife edges* used with metal gaskets.

a) *Tongue seals.* If the loading of a rubber gasket must be kept to a minimum, a tongue left on the othervise flat flange guarantles a seal without a great deflection of the gasket (Guthrie[491]). Figure 3.139 shows such a seal.

If the gasket is not of rubber but Teflon or a soft metal (e.g. lead), the seal should be provided with ridges on one of the flanges and small grooves of the same form on the other flange, machined inside the large groove holding the gasket. Figure 3.140 shows such a seal with a rounded double ridge, but these seals may have more concentric ridges and/or ridges of other cross sections

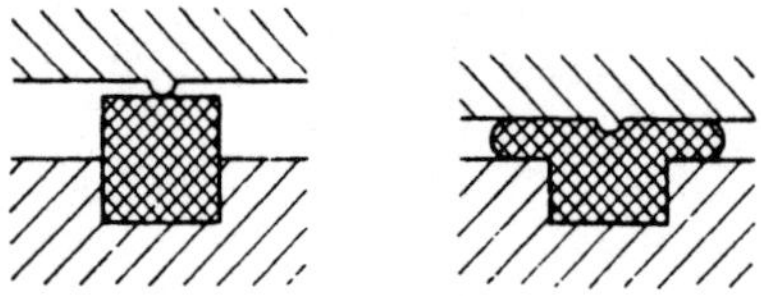

FIG. 3.139 Tongue seal with rubber gasket

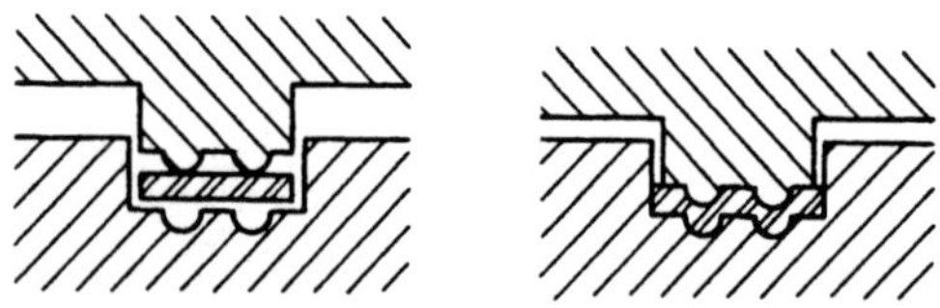

FIG. 3.140 Double ridge and groove seal

(e.g. trapezium). The seal on Fig. 3.140 may be used with a lead gasket (1–1.5 mm thick; Ardenne[47]). Gale[420] describes the possibilities of constructing silicone rubber gasket seals which can be used continuously at higher temperatures (about 150 °C). Examples of shapes and dimensions for such seals are shown in Fig. 3.141. These seals were constructed with an inside

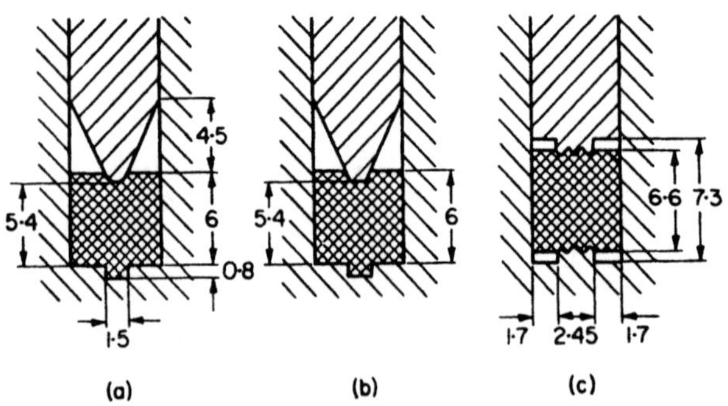

FIG. 3.141 Silicone rubber gasket in tongue seals: (a) with V-shaped tongue; (b) with trapezoidal tongue; (c) with small multiple ridges. After Gale[420] *(Courtesy of The Institute of Physics and The Physical Society, London)*

diameter of 110 mm (Gale[420]). For seals with a V-tongue (Fig. 3.141a) with a trapezoidal tongue (Fig. 3.141b) as well as for those with multiple concentric ridges (Fig. 3.141c) a compression ratio of 90 per cent is advisable.

Champeix[203] describes a shaft seal using a Teflon gasket compressed with a V-tongue (Fig. 5.26). A rotary vacuum seal using a rim which seals against a flat Neoprene gasket is described by Hayward[523] (Fig. 5.23a).

b) *Knife edge seals* are dismountable gasket joints made by forcing the knife edges provided on the flanges to bite into a metal gasket (van Heerden[527], Hees[530]). The clamping of a flat, soft, metal gasket between the knife edges (Fig. 3.142) makes the soft gasket material flow and fill up the small machining marks on the surface of the knife edge. Microscopic inspection of the gaskets used in such seals revealed that *the seal is made at the side of the knife edge* where a strong shearing of the gasket material occurs (van Heerden[527]).

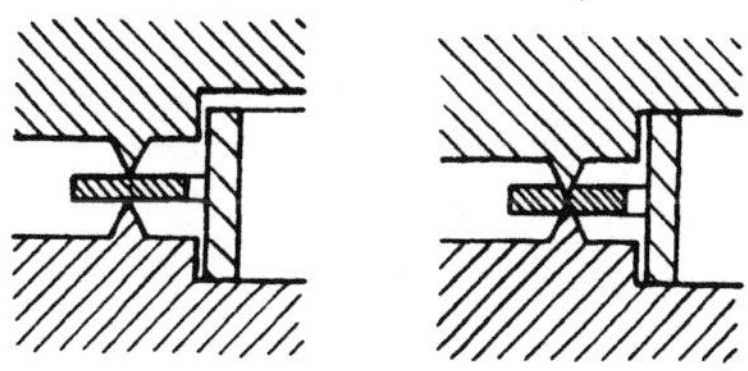

(a) (b)

FIG. 3.142 General layout of a knife-edge seal: (a) open; (b) closed

Wheeler[1304] made tight seals with knife edges which had been gouged with many 0.1 mm deep nicks across the edge.

The gaskets for knife-edge seals are usually of OFHC copper (Table 3.39) but gaskets of other metals* like mild steel (Carpenter[195]), nickel (Hees[530]), silver (Drawin[303]), silver-plated copper (Lichtman[764]) or indium-plated copper (Reynolds[1049]) have also been reported. A knife-edge seal using solid indium placed in a copper cup is described by Nicollian[927].

The copper gaskets to be used in knife-edge seals should be annealed in hydrogen at 950 °C (van Heerden[527]) or just air-fired to anneal followed by subsequent cleaning to remove the resulting oxides and reduce any scratches (Hees[530]). Circular scratches are of no importance but radial ones may cause leaks. Contrary to the preference of most designers for fully soft annealed gaskets, a test made by Wheeler[1304] led to the conclusion that the copper should be quite hard for a good, reliable seal.

The gaskets used in knife-edge seals should be flat, cut out from plates. The thickness of the gasket is usually 1–2 mm (see Table 3.39) but in special

*Aluminium was found to be *not suitable* as gasket for knife-edge seals (Carmichael[194]).

cases very thin (Robinson[1073]) or very thick (Pattee[975]) gaskets are also used.

If the gasket is used to locate the knife edges, it is machined from a heavier stock and is provided with the appropriate shoulders, which serve as locating devices (Hees[530]).

The knife edge must be of a harder material than the gasket. The harder and tougher the knife edge the better, since this minimizes the danger of nicking or burring. The knife edge is usually machined from stainless steel, but the carburizing of other steels to increase the hardness can also be considered. A knife-edge seal using flanges, gaskets and bolts of the same material,

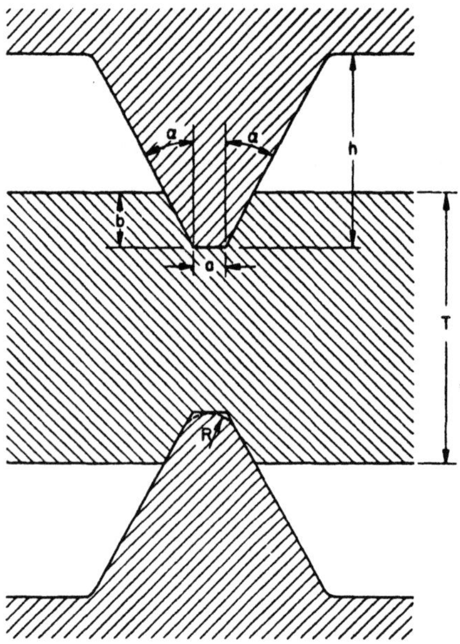

FIG. 3.143 Dimensions of knife edges (Table 3.39)

and hardening the surface of the knife edge by depositing (welding) on it a hard material (e.g. Stellite 6) is described by Carpenter[195].

The knife edge is in principle a ridge having a V-shape cross section (Fig. 3.142) with the sides at 30–45° to the normal (Fig. 3.143, Table 3.39). The apex of the knife edge may have a flat edge (Fig. 3.143) or a rounded one (dotted line, Fig. 3.143). The flat land on the apex is needed in order to generate a planar surface of contact between the knife edge and the gasket (Hees[530]) and also to facilitate alignment of the flanges, but its width must be kept to a minimum, since with its increase, the required sealing force is increased as

TABLE 3.39. DIMENSIONS (mm) OF KNIFE-EDGE SEALS
(See Fig. 3.143)

Gasket		Knife edge				Remarks	Reference
Material	Thickness T	α	h	a	b		
OFHC Copper	1	30°	1.587	0.125	–	Gasket annealed in H_2 at 950 °C.	van Heerden[527]
Copper	–	30°	0.63–0.76	0.075–0.150	–	Double ridge seal	Marker[805]
OFHC Copper	1.6	–	0.70	0.25	0.25	Knife edge located at 1.9 mm from outer side of the gasket	Wheeler[1304]
OFHC Copper	1.58	30 ± 5°	0.60–1.38	0.075–0.177	0.05–0.25	Profile and dimensions the same for seals with diameters up to 300 mm	Hees[530] Hees[529]
Copper or Aluminium	–	30°	–	0.3 1.0	–	Difference in sealing forces see Table 3.13	Boulloud[153]
OFHC Copper	3	30°	1.7	–	0.4	Double knife edge spaced 16 mm	Peters[986a]
Copper	–	45°	1.6	0.1	–		Papirov[966b]
OFHC Copper	10	45°	–	radius 0.125	–	Pipe butt seal see Fig. 3.145b	Pattee[975, 976]
Copper	0.25	45	1.6	–	–	Knife-edge flange between two gaskets see Fig. 3.148a, b	Robinson[1073]
Copper	1.58	45	2.0	radius 2	–	Gasket with V-shape bead, see Fig. 3.144c	Mann[808]
Silver plated copper	–	35	1.6	radius 0.25	–	Diameter up to 300 mm	Lichtman[764]
Mild steel	–	45	1.58–3.17	–	–	Knife edge hardened with welded deposit (Stellite 6); diameters from 1.5 to 10 in.	Carpenter[195]

well (see Table 3.13). The dimensions recommended by various authors are listed in Table 3.39.

The knife-edge seal is used almost exclusively on circular seals, although it seems that apart from the machining problems there is no reason why other geometries could not be used. The profile of the knife edge is independent of the diameter of the seal (Hees[530]) or the same profile can be used for a large range of seal diameters (Table 3.39).

The depth of bite (b, Fig. 3.143) of the knife edges into the gasket is generally small. Baker[72] quotes 0.16 mm for this depth, other authors (see Table 3.39) give depths of 0.05 to 0.4 mm.

Knife edges can be used as machined; the sealing surfaces should be specified for a 32 micro-inch finish. This is obtained usually by polishing with aluminium oxide paper and rouge paper (Wheeler[1304]). According to Baker[72], the knife edge should be first lapped with a brass lap using a 15 μ silicon carbide, followed by polishing with a second brass lap using a 3 μ diamond powder. The polish of the knife edges is important, while the finish of the gasket is relatively unimportant. This is because the filling up of a small scratch with sharp edges by the material of the gasket is like the flow of a highly viscous liquid through a narrow slit; it takes a long time and will not be completed before the pressure is released sideways (van Heerden[527]). It has been found (Baker[72]), that the annealed copper will flow into scratches a few microns in depth and completely seal them, but scratches 10 μ or more in depth are only partially filled. A leak-free seal will only be achieved, if no scratches are visible when viewed with a microscope at a magnification of X50, since a capillary leak 1 mm in length and 1 μ in depth will result in a leak rate of approximately 1×10^{-4} lusec.

Thin tightly adherent oxide films on the flange surfaces were never observed to cause seal failures, but heavy oxides can degrade the seal (Wheeler[1304]).

Knife-edge seals are used especially in bakeable joints. The seal may have a knife edge only on one of the flanges or may be based on knife edges on both flanges facing each other. In this latter case single or double (concentric) knife edges may be used.

When only one of the flanges is provided with a knife edge (Fig. 4.144a–c) the other sealing part may be a flange or pipe, sealed by other techniques (brazed glass–metal seal) to the vacuum system, or a gasket which seals on its other side by means of a different technique.

Nicollian[927] describes a valve using a knife edge seal between a cylindrical stainless steel part (1) and a cup (2) containing solid indium (3. Fig. 3.144a). The knife-edge seal may as well be used in the arrangement shown by Fig. 3.144b, where the outside of the copper cylinder (1) (the end of a copper to Pyrex Housekeeper seal) is threaded and screwed into the flange (2). By means of this flange and of the bolts (3) the flat end of the copper pipe (1)

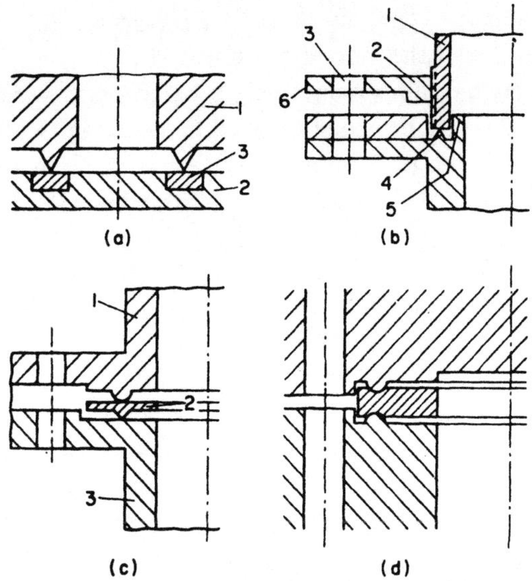

FIG. 3.144 Single knife-edge seals: (a) indium seal (after Nicollian[927]); (b) axial seal. After Drowart[305] *(Courtesy of The Institute of Physics and The Physical Society, London)*; (c) knife-edge, coined gasket seal. After Mann[800] *(Courtesy of The American Institute of Physics)*; (d) Curvac seal *(Courtesy of Ultek[1251a])*

is sealed against a knife edge (4). It is necessary to retain the end of the copper pipe (1) by the rings (5, 6) to avoid deformation of the copper (Drowart[305]).

In the seal described by Mann[800], one flange (1) has a cylindrical raised bead (of 1/12 in.) which presses into the gasket (Fig. 3.144c). The gasket is

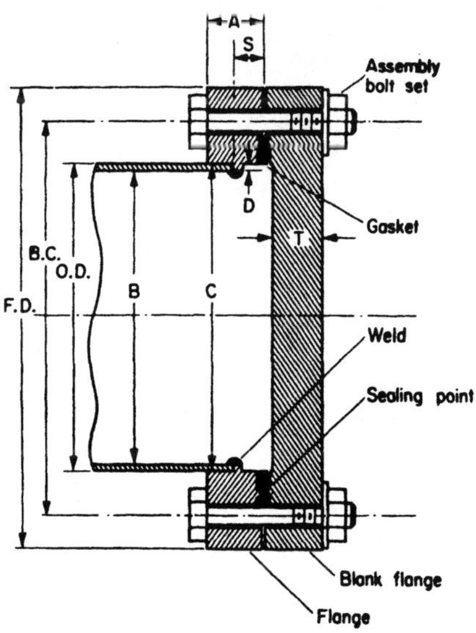

FIG. 3.144A Curvac flanges (Table 3.39A) supplied by Ultek[1251a]

flat on the side facing this bead, but on the other side it is coined (2) and seals against the flat surface of the second flange (3).

Ultek[1251a] supplies bakeable vacuum seals based on the sealing action of half toroidal shaped ridges provided on both flanges (Fig. 3.144d) against an OFHC copper gasket (about 2 mm thick); these are the seals known as "Curvac". Figure 3.144A and Table 3.39A give the dimensions of the various sizes of such seals.

TABLE 3.39A. DIMENSIONS (inches) OF CURVAC SEALS* (Ultek[1251a]) (see Fig. 3.144A)

Nominal size O.D.	1	1 1/2	2	2 1/2	3	4	5	6	8
F.D.	2 1/8	2 3/4	3 3/8	4 1/2	4 5/8	5 31/32	6 3/4	7 31/32	9 31/32
C	1.010	1.510	2.010	2.510	3.010	4.010	5.010	6.010	8.010
A	1 5/32	1/2	11/16	11/16	13/16	25/32	13/16	7/8	31/32
S	0.165	0.146	7/32	7/32	9/32	13/32	7/16	7/16	1/2
B	0.875	1.375	1.875	2.375	2.832	3.834	4.760	5.760	7.760
B.C.	1.625	2.312	2.850	3.628	4.030	5.128	5.969	7.128	9.128
Number of bolts	4	6	8	8	10	16	18	20	24
Bolt hole	17/64	17/64	11/32	11/32	11/32	11/32	11/32	11/32	11/32
T	0.160	0.160	0.310	0.310	0.433	21/32	11/16	3/4	27/32
D	0.067	0.067	0.067	0.067	0.088	0.088	0.125	0.125	0.125
Gasket** o.d.	1.290	1.895	2.425	3.243	3.598	4.743	5.567	6.743	8.743
Gasket i.d.	1.010	1.451	2.010	2.506	3.010	4.006	5.010	6.007	8.007

* Sizes 1–3 in. have rotatable flanges. The flange includes the bolt holes and is separate from and rotates with respect to the insert which incorporates the sealing ridge. Sizes 4–8 in. have the bolt holes and sealing ridges included in the flange.

** The gaskets are of OFHC copper, 0.080±0.0035 in. thick.

The knife-edge seals having a gasket between the two V-shaped ridges are shown in Fig. 3.142 and 3.143 and listed in Table 3.39. These seals should have flanges shaped to locate (centre) the gasket. Thin gaskets can be centred by locating them properly in one of the flanges, the second flange being centred with respect to the first one. If thicker gaskets are used they can be centred by both flanges. This later technique is used in the seal described by Reynolds[1049] who used an indium-plated copper gasket (Fig. 3.145a).

Pattee[975] described a seal (Fig. 3.145b) to connect two pipes, consisting of a thick copper cylinder (2) against which the ends of the two pipes (1) shaped into knife edges are pushed by the flanges (3). The dimensions of this seal are given in Table 3.39.

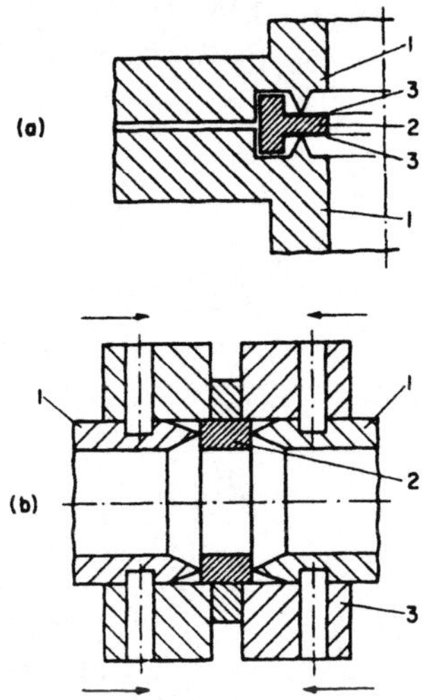

FIG. 3.145 Seals with knife edges facing each other: (a) seal with machined gasket: (1) flanges; (2) copper gasket; (3) indium coating (after Reynolds[1049]); (b) pipe butt seal: (1) heavy wall stainless steel tube; (2) copper gasket; (3) holding rings. After Pattee[975] *(Courtesy of The American Institute of Physics)*

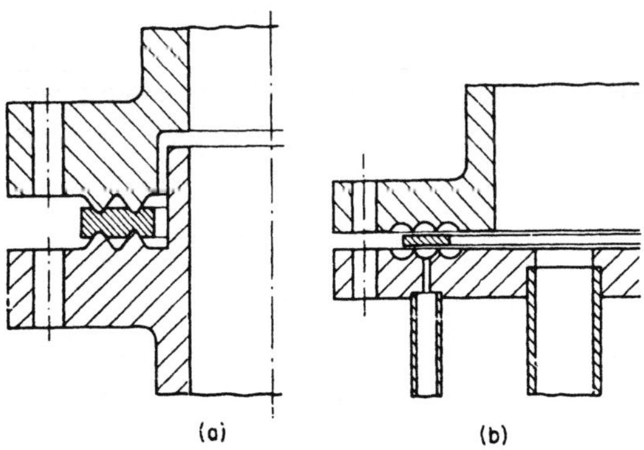

FIG. 3.146 Double knife-edge seals: (a) without guard vacuum. After Marker[865] *(Courtesy of Pergamon Press)*; (b) with guard vacuum. After van Heerden[528] *(Courtesy of The American Institute of Physics)*

The *double ridge knife-edge seals* (Fig. 3.146) are more difficult to machine but they have some advantages compared to single knife-edge seals. These seals with concentric knife edges may prevent the continued penetration of the knife edge into the gasket by trapping a volume of copper between the ridges (Marker[805]). The double ridge seals have proved successful in some other cases as well (Widmer[1312]), especially where the distortion of the flanges due to the bolting forces could cause difficulties. In this case one of the ridges is located outside of the bolt circle.

If the space between the two concentric knife edges is connected to a separate vacuum pump, any leak in the inner knife edge will have a leak rate reduced by a factor of 10^5–10^7 (Section 38.23) as compared to the single knife-edge seal. Rivera[1067] used a knife-edge seal on the high vacuum side of a double chamber with guard vacuum (Section 38.23).

When double knife-edge seals are used, both flanges must be heavy (Fig. 3.146) since if one of the flanges is thin, the forces exerted by the bolts and the gasket form a couple, which bends the flange and causes the inner knife edge to separate, resulting in leaks. The flanges should be at least 3/4 in. thick, unless they are pressed together with clamps with the point of application in the line between the two concentric knife edges (van Heerden[528]).

FIG. 3.147 Seals with the knife edges on inserted ring

Instead of making the knife edges on the flanges, the seal can also be made by using a hard ring (Fig. 3.147) having its cross section shaped to form knife edges on both sides (Ardenne[47]). By tightening the seal the knife edge of the ring will cut into the softer material of the flanges. The drawback of such a seal is the fact that it cannot be used in applications, where the seal should be opened and remade many times.

Robinson[1073] describes a seal made between two copper washers (0.010 in. thick) (2, Fig. 3.148a) clamped between the flanges and a steel ring (1) shaped with knife edges on both sides. The ring (1) has a groove cut peripherically to a depth which permits a cantilever action to the knife edges when the seal is tightened. The two flat copper washers may be replaced (in an alternative construction) by one U-shaped washer (Fig. 3.148b). This arrangement reduces the number of sealing surfaces. Bannock[80] sucessfully used the seal shown on Fig. 3.148a on a molecular sieve pump, maintaining a vacuum-tight seal over the temperature range from -196 °C to 500 °C.

Cohen[220] used a seal, where the knife edges were provided on the two sides of an alumina ring (Fig. 3.148c) sealing against two copper flanges. Such a seal provides electric insulation between the two flanges and is bakeable.

The *assembly* of knife-edge seals needs special care in order to meet the following requirements:

(1) The diameter of the mating knife edges must be absolutely the same. It appears that knife-edge diameters should be controlled to ± 0.05 mm (± 0.002 in.) and that the eccentricity should be held to within ± 0.12 mm (± 0.005 in.) (Wheeler[1304]).

(2) The two mating knife edges must be assembled in line, which requires a positive alignment of the flanges (Carmichael[194]). Hees[529] appreciates that for a good seal this alignment should be within 0.25 mm (0.010 in.).

(3) The mechanical layout of the flanges must locate the sealing edge as close as possible to the bolt circle. The flanges should be strong enough to prevent excessive distortion.

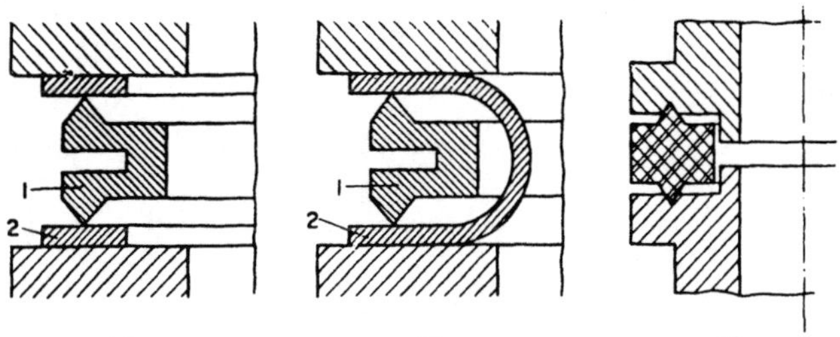

(a) (b) (c)

FIG. 3.148 Knife-edge seals with attached rings: (a) steel ring (1) sealing on two copper washers (2); (b) steel ring sealing on a single U-shaped copper washer. After Robinson[1073] *(Courtesy of The Institute of Physics and The Physical Society, London)*; (c) ceramic ring sealing on copper parts. After Cohen[220] *(Courtesy of the American Institute of Physics)*

(4) The seal should include enough bolts for uniform clamping and the bolts should be strong enough for ample clamping forces (Wheeler[1304]). The required forces are listed in Table 3.13.

(5) The sealing force should be applied normally to the surface of the gasket at the point where the knife-edge apex contacts it (Hees[529]).

(6) The tightening of the bolts should always be done gradually, going round and round. The bolts should be tightened just enough to give a good seal (van Heerden[527]).

Conflat seals have been recently produced by Varian[1257a] to be used in ultra-high vacuum systems, which are baked to high temperatures. The design has this name because it consists of a conical sealing surface and a flat copper gasket (Fig. 3.149). The sealing ridge is made with a vertical edge (normal to the gasket) and with the second side of the ridge inclined at an angle of 70°, although these angles have been altered in various sucessful tests (Wheeler[1304]). The edges are designed to bite 0.3–0.4 mm on each side of the 1.6–2.0 mm thick gasket (Fig. 3.149) Figure 3.150 and Table 3.40 give the dimensions of Conflat seals (Varian[1257a]).

Conflat seals have been resealed, using the same gasket, by increasing the pressure slightly with each seal, and by reversing the position of the gasket for each successive seal. It resulted (Wheeler[1304]) that the seal could be re-used 22 times. For good performance in practical situations, Wheeler[1304] recom-

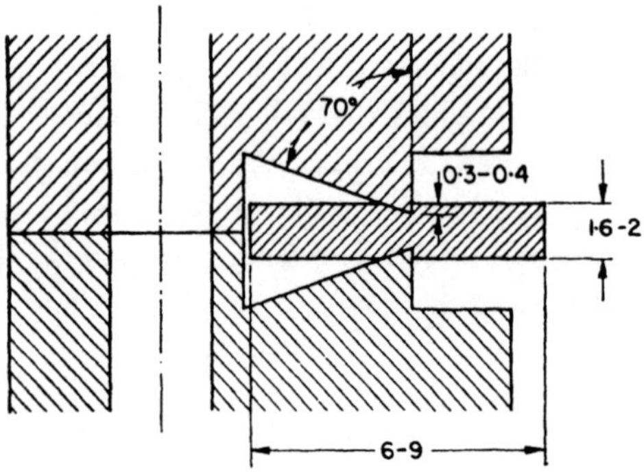

FIG. 3.149 Dimensions (mm) of a Conflat seal. After Wheeler[1304] *(Courtesy of Pergamon Press)*

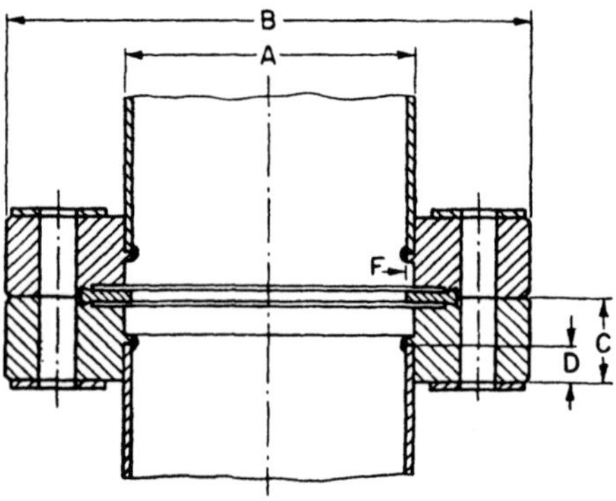

FIG. 3.150 Dimensions of Conflat flanges supplied by Varian[1257a] (Table 3.40)

TABLE 3.40. DIMENSIONS (inches) OF CONFLAT SEALS FROM Varian[1257a] (see Fig. 3.150)

Nominal size	*A* (±0.005)	*B*	*C*	*D*	Bolts			Gasket	
					Bolt circle diameter	Num-ber	Size	o.d.	i.d.
1½	1.510	2 3/4	1/2	0.209	2.312	6	1/4	1.895	1.451
2½	2.510	4 1/2	11/16	3/8	3.628	8	5/16	3.243	2.506
4	4.010	6	25/32	7/16	5.128	16	5/16	4.743	4.006
6	6.010	8	7/8	1/2	7.128	20	5/16	6.743	6.007

mends that the Conflat seals should be rated at 3 resealings with the same gasket. Sealing forces required are listed in Table 3.13.

Tobin[1225a] used Conflat seals to make thin (0.001 in.) stainless steel windows, by placing the stainless steel foil between the knife edge of the seal and copper gasket. To avoid the cutting of the foil it was necessary to dull (very slightly) the sharp portion of the ridge in contact with foil, by lapping it with fine silicon carbide paper. The resulting seal was baked at 400 °C, maintaining a pressure of 1×10^{-9} torr.

38.57 Inflatable gasket seals. Lock doors of vacuum vessels with pressures of 10^{-1}–10^{-2} torr may be sealed using inflatable gaskets (Ballif[77]). These seals have a nominal displacement which could seal clearance gaps up to 12 mm (about 1/2 in.) between the frame where the seal is placed and the movable door. Standard seals are commercially available with a square bead

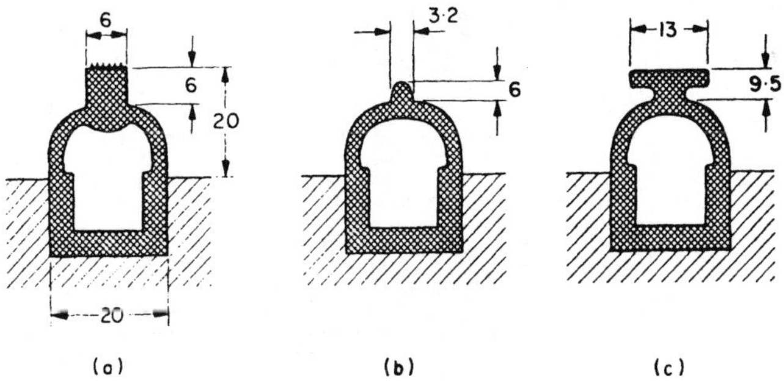

FIG. 3.151 Inflatable rubber gasket seals: gasket with (a) square bead; (b) round bead; (c) T-bead (after Ballif[77])

(Fig. 3.151a), a round bead (Fig. 3.151b) or a T-bead (Fig. 3.151c). According to Ballif[77] the T-bead results in most effective seals. He obtained seals with leak rates of 0.6 ft^3 per 24 hr for 12 linial foot of single gasket (about 0.4 lusec/cm) and 0.06 ft^3 per 24 hr (about 0.04 lusec/cm) for double seals with guard vacuum. For a reliable seal Ballif[77] recommends that:

(1) The seal inflation pressure should be 2.8–3.5 atm (40–50 p.s.i.g.), which is in excess of the design pressure for such inflatable gaskets.

(2) The seal should be designed so as to use the middle third of the nominal throw of the gasket, e.g. for gaskets with nominal throw of 0.75 in (Fig. 3.151) the working throw should be 0.25–0.50 in.

(3) The surface of the gasket should be free of any vendor or company identification mark, and the surface of the groove and door should have a finish of at least 200 micro inches r.m.s.

(4) The gaskets (from Fig. 3.151) should not be bent with radii less than 3–4 in.

Inflatable sheet metal gaskets are described by Rasor[1028]. Lucas[787] succeeded in making a vacuum-tight seal on a large oval-shaped window of 72 in. using the arrangement shown in Fig. 3.152. The glass window (1) is sealed with indium* gaskets (2) made from 0.150 in. wire, and the seal is compressed against the chamber wall by the inflatable stainless steel member (3) capable of 0.160 in. useful deflection. The seal was made at very low temperatures (liquid hydrogen). The inflation pressure of about 40 p.s.i. at these tempera-

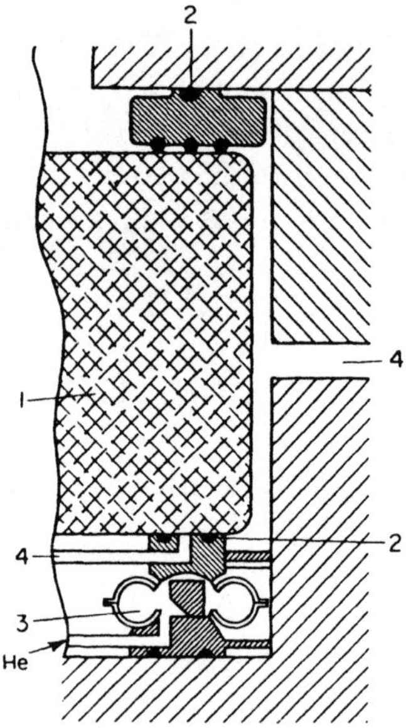

FIG. 3.152 Inflatable stainless steel member seal. After Lucas[787] *(Courtesy of The American Institute of Physics)*

tures (He filling) increases to 400–600 p.s.i. at 77 °K making a strong seal. The spaces between the gasket and window were equipped with pump-out connexions for guard vacuum (4).

38.58 Assembly and maintenance of gasket seals. To prevent repetition the reader is referred to Section 38.48 for the general rules regarding the assembly and maintenance of gasket seals.

When flat rubber gaskets are used (e.g. Fig. 3.110) they can be cut from rubber sheet. Cutting devices and methods are described by Heller[535], Allen[14] and Zobac[1356]. The sheet to be cut should always be backed by a wood plate, and covered by a circular metal plate having the diameter equal to that of the future gasket, or a diameter large enough to hold the rubber in place during the cutting.

* The seal using 99.999 per cent pure indium is very reliable; it was in use without any trouble for many years (H. P. Hernandez: Personal Communication).

It is recommended that the gaskets be utilized with the original smooth surfaces on the sealing flanges.

Medicus[837] describes a method of constructing gaskets by applying a rubber solution (para rubber in benzine or pure gasoline) on the surfaces to be sealed.

Some difficulty is often experienced during the inserting of rubber gaskets into dovetail grooves (Figs. 3.119, 3.120). The gasket (trapezoidal) should be placed obliquely in the groove (Fig. 3.153a) and using a tool (not sharp!)

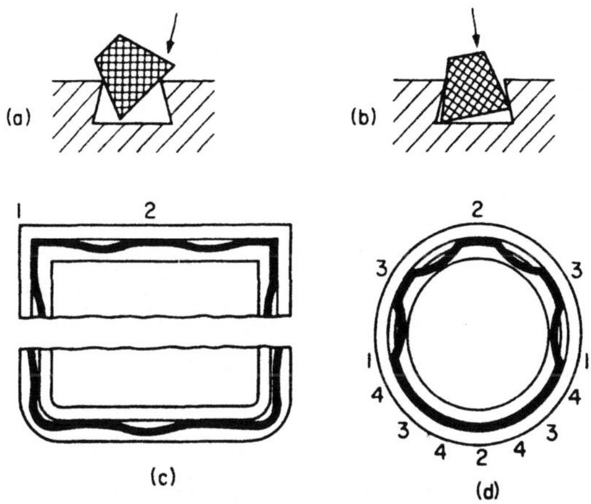

FIG. 3.153 Technique of placing the gasket in a dovetail groove: (a, b) pushing the gasket into the groove; (c, d) sequence of positioning the gasket

the edge of the gasket should be pushed over the side of the groove. This operation is done, part by part, over the whole length of the gasket. After the gasket is inside the groove (Fig. 3.153b) over its whole length, it should be pressed (heavy gaskets may have to be hammered) into its final position.

With large gaskets, it is advisable (Holland-Merten[587]) to follow the sequence 1, 2, 3, 4 suggested in Fig. 3.153c, d. If rectangular grooves are used it is essential that the gasket should not be stretched at the corners. For this, the sequence from Fig. 3.153c should be followed.

The assembly of a seal with a metal gasket should be done (Wheeler[1304]) in the following sequence:

(1) Clean the sealing surfaces with acetone and lint-free paper.

(2) Clean the gasket in the same manner (if necessary).

(3) Carefully assemble the two flanges and the gasket so as not to damage the gasket.

(4) Install the bolts and spin the nuts finger tight.

(5) Tighten the nuts in three to five steps, increasing the torque level by approximately 25 inch/pound (28.5 kg/cm) at each step.

In order to facilitate the opening of the seals it is advisable to place on one of the flanges 2–3 opening (grub) screws, screwed in this flange. After the screws of the seal are withdrawn, by tightening these screws, they push against the second flange, opening the seal.

38.6 Thin gasket seals

The seals grouped in this Section consist of a foil (usually metal) tightened between a plane surface and a surface provided with a ridge (Section 38.61), between two surfaces one holding a ridge or knife edge and the other a mating groove (Section 38.62) or between two tapered surfaces (Section 38.63).

38.61 Plane seals with thin gaskets. Thin layers of suitable metals (gold, silver, copper) interposed in the form of a foil of about 0.01 mm thickness and compressed between a plane flange and a second one having circular ridges, can make a vacuum-tight seal (Metro-Vickers[851]).

De Villiers[1267] described a seal (Fig. 4.33a) using a 0.25 mm thick aluminium foil as the gasket. Ruthberg[1092] has made a seal consisting of an aluminium foil (about 20 μ thick household aluminium foil) clamped between two Monel flanges, one of the flanges having an annular ridge (1/16 in. wide). The seal is bakeable at 400 °C. When disassembled, the aluminium foil adheres strongly to the surfaces (see also Section 38.46) and is removed with NaOH which does not affect the flange surface.

Strnad[1204] described a window seal based on the squeezing of a thin (0.05 mm) gold washer between the mica window and the rounded edge of a Kovar cylinder (Fig. 7.24a).

38.62 Groove and knife-edge seals. These seals, which utilize thin gaskets of silver, aluminium or copper, are based on the design of Paul[980] and Hintenberger[564]. The knife edge (55–90°) pushes the thin gasket into the mating V-groove (Fig. 3.154), or a rounded ridge pushes the gasket against a rounded V-groove (Jaeckel[623]).

The gaskets regularly used in these seals are 0.1–0.5 mm thick silver or copper or 0.2–0.5 mm thick aluminium (or Teflon) (Hintenberger[564], Papirov[966b]). Figure 3.155 and Table 3.41 give the dimensions of these seals, according to the construction of Hintenberger[564], which uses square outline flanges, with circular groove and knife edge, shaped at an angle of 60° (Fig. 3.155a); and according to the construction of Papirov[966b], which uses circular flanges with circular groove and knife edge shaped at 90°. Hintenberger[564] recommends that the foil (gasket) be preformed into a V-shape to a depth about 3/4 of the depth taken by the foil in the seal. The preforming avoids spoiling the foil by the sharp knife edge, or by placing the knife edge slightly

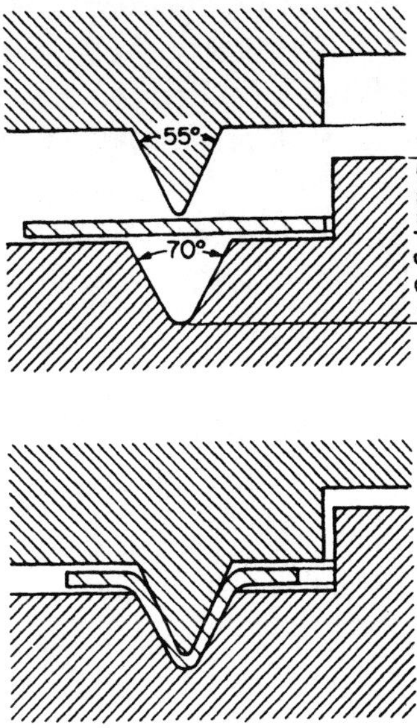

FIG. 3.154 Thin gasket, groove and knife-edge seals

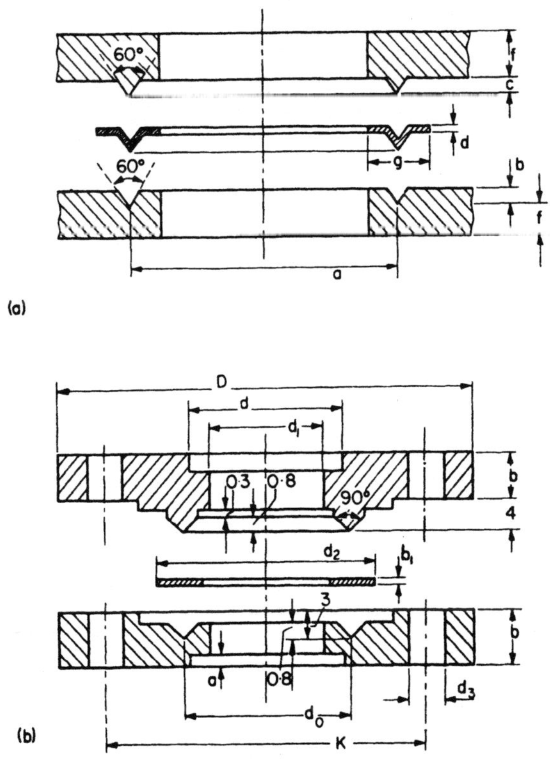

FIG. 3.155 Dimensions of thin gasket, groove and knife-edge seals (Table 3.41)

eccentrically with respect to the groove. In order to assure the sealing action, the knife edge is machined slightly higher than the depth of the groove (Fig. 3.155 and Table 3.41).

Knife-edge seals with thin gaskets are bakeable up to 400 °C.

TABLE 3.41. DIMENSIONS (mm) OF GROOVE AND KNIFE-EDGE SEALS (see Fig. 3.155)

a	b	c	d	f	g	Bolts Size	Bolts Number	Reference
15	0.5	0.8	0.2–	10	10	5	4	Hintenberger[564] see Fig. 3.155a
25	1.0	1.3	0.4	10	10	6	4	
35	1.0	1.3	0.2–	13	10	6	4	
50	1.0	1.3	0.4	14	10	8	4	
65	1.0	1.3	0.2–	14	10	8	4	
80	1.0	1.3	0.4	14	10	8	6–8	
100	1.0	1.3	0.2–	14	10	8	6–8	
120	1.0	1.3	0.4	14	10	8	6–8	

d_0	d_1	d	d_2	D	K	a	b	d_3	b_1	Number of bolts	Reference
15±0.1	10	14	20	58	40	3	10	9	0.4	4	Papirov[966b] Fig. 3.155b
20	15	19	25	65	45	3	10	9	0.4	4	
24	20	24	30	78	55	5	12	11	0.4	4	
45	40	45	50	110	80	5	14	13	0.4	8	
75	70	76	80	145	115	5	15	13	0.4	8	
160	150	156	170	230	195	8	22	13	0.5	12	
210	200	206	220	280	250	8	24	13	0.5	16	
310	300	306	320	380	350	10	28	13	0.5	24	

Warmoltz[1287] described a seal with a flange bearing a circular ridge immediately opposite a circular score on the other flange. The ridge has a cross section of an equilateral triangle (3 mm side) with the edge pressing on the gasket, which is a 0.3 mm thick copper sheet. The edge is radiused by 0.5 mm.

According to Hintenberger[564] and Henry[540] the surface finish of the knife edge is not critical; circular machining marks may be tolerated but radial ones should be avoided. Eccentricity of the knife edge relative to the groove may be tolerated up to about 0.3 mm.

38.63 Surface friction seals. Brymner[179] and Steckelmacher[1180] have constructed a seal (Fig. 3.156) where a thin (0.125–0.25 mm) metal gasket lies between the surfaces of a male cone on one flange and a mating female cone on the other flange. The seal depends on a surface friction effect instead of compression only. This fact is evident, because the seal remains perfectly good when after tightening the bolts are removed from the flanges (Armand[49a]). The reliability of these seals is based on the fact that the adhesive forces produced in sliding friction are much greater then those produced by pressures normal to the sealing surfaces, since the surface friction forces break up oxide and other surface contaminants, thus permitting local welding between pure metal surfaces.

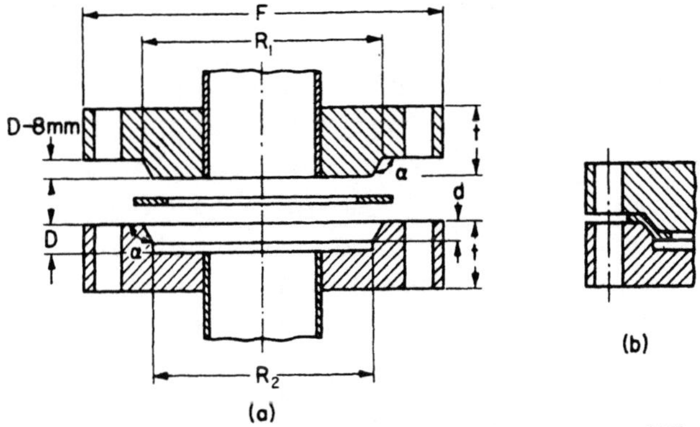

FIG. 3.156 Conical surface friction seal. After Brymner[179] *(Courtesy of The Institute of Physics and The Physical Society, London)*

TABLE 3.42. DIMENSIONS (inches) OF SURFACE FRICTION SEALS (Brymner[179]) (see Fig. 3.156)

Maximum tube o.d.	*F*	*t*	*D*	*d*	R_1	R_2	Bolts		
							PCD	Size	Number
1	2 1/2	1/2	1/8	1/16	1 5/8	1 19/32	2 1/8	2 BA	6
1 1/2	3	1/2	1/8	1/16	2 1/16	2 1/32	2 1/2	2BA–1/4 BSF	6
3 1/2	5	5/8	1/8	1/16	3 11/16	3 21/32	4 1/4	1/4–5/16 BSF	6–8
5 1/4	8	3/4	1/4	1/8	6	5 15/16	6 7/8	5/16–3/8 BSF	8–12
8 1/2	12	1	1/4	1/8	9 5/8	9 9/16	10 7/8	3/8–1/2 BSF	12–18
1/2	1/2	union	3/32	1/32	11/16	43/64	—	—	—

The gasket used in these seals may be of copper, soft iron, nickel, aluminium, or stainless steel. For soft iron and stainless steel gaskets the maximum number and size of bolts (see Table 3.42) should be used and the stainless steel gaskets should be used only with stainless steel flanges.

The flanges may be machined from mild steel or stainless steel. The angle α of the cones should be so chosen as to maintain the surface shear forces as large as possible. The component of the force produced by the bolts, which gives surface shear is equal to the product: bolt force $\times \cos(\alpha-90°)$. This component is a maximum when α is 90°, which is an impossible mechanical condition. An angle of 100° was found consistent with reasonable machining tolerances and resulted in satisfactory seals. The other dimensions of these seals are listed in Table 3.42 (Brymner[179]).

The surface friction seals, should be machined to such tolerances that with no gaskets the flanges mate on the conical portion of the seal, leaving reasonable gaps in the flat parts of the flanges both at the top and bottom of the gasket (Steckelmacher[1180]). The distance between bolts and conical shear surfaces should be a minimum to reduce bending moments. The nature of the forces forming the seal makes necessary the use of grub screws to force the flanges apart after the clamping bolts have been removed. Three or four such screws spaced around the circumference of one of the flanges are sufficient.

These seals may be baked up to 450° and can be used in ultra-high vacuum systems. As mating flanges are of the same material, they expand radially by equal amounts so that frictional forces across the seal are maintained during temperature cycling. Brymner[179] tested one of the seals at about 100 cycles up to 450 °C, during which period it has been re-made about 20 times using the same gasket, but it is affirmed that the seals may be used from –188 °C to 800 °C. Armand[49a] tested such seals (with 300 and 600 mm diameters) and found leak rates of 10^{-7}–10^{-8} lusec at room temperature after 10–16 cycles to 400 °C and leak rates of 10^{-4}–10^{-5} lusec at 400 °C.

A flange design, which can be used with elastomer O-rings, with a gold O-ring placed in a corner seal as well as with a surface friction seal, was constructed by Marton[821] (see Fig. 3.96).

CHAPTER 4

TRANSMISSION OF THE ELECTRIC CURRENT THROUGH SEALS

4.1 SELECTION CRITERIA FOR ELECTRICAL LEAD-THROUGHS

To transmit an electric current into a vacuum system or envelope, electrical leads should pass through the envelope.* These leads should be electrically insulated from the envelope (and other leads) and joined to the system by vacuum-tight seals. The electrical lead itself is selected usually according to the general rules for electrical connexions (see e.g. Mooradian[880]).

The selection of the materials forming the electrical lead-through and its vacuum seal is determined by the service requirements, i.e. the voltage, current, frequency, temperature, and obviously the conditions needed for an appropriate vacuum seal (Section 21.1). These requirements determine the insulation to be used (Section 41.1); the size and kind of the electric lead to be used is determined by the loading current (Section 41.2), the frequency and the temperature (Sections 41.3, 41.4).

According to the requirements, permanent (Section 4.2) or demountable (Section 4.3) electrical lead-throughs may be used.

41.1 Insulation

41.11 Bulk resistivity. A distinction must be made between bulk resistivity and surface resistivity of the materials (especially glasses and ceramics) used to insulate the electrical lead-throughs. The bulk resistivity is that of the mass of a body excluding the influence of various surface layers (gases, water). Glasses are known as insulating materials, but they are really electrolytic conductors, thus their resistivity is high at room temperature (up to 10^{19} ohm/cm) and falls with the increase of the temperature (down to 1 ohm/cm at 1200 °C). For most of the glasses the logarithm of the resistivity is linear with 1/T (°K). The resistivity of each particular kind of glass (or ceramic) is determined by the composition of the material. Lead glasses have higher resistivity than soda–lime or even boro-silicates (without lead). Thus it is a

* Bezold[121] describes a technique of transmitting electrical current without the use of lead-throughs crossing the wall of the vacuum vessel (see Fig. 5.58).

common practice to make the stems (see Section 42.2) of electric lamps and electron tubes from lead-glasses (see Table 2.10 and Fig. 4.1), while the rest of the enclosure (bulb) is made from soda–lime or other glass. Generally soda, potash or alumina decreases the bulk resistivity of the glass, while lime, magnezia or boron oxide increases it.

The bulk resistivity of the principal insulation materials used in vacuum technique is shown by Fig. 4.1 as a function of the temperature. In order to

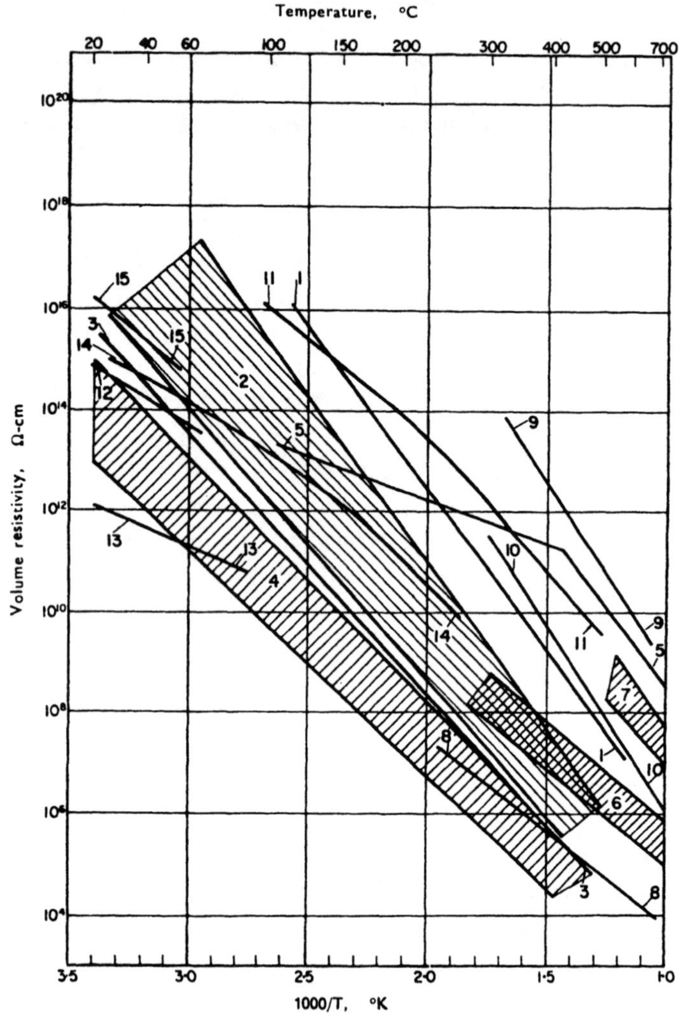

FIG. 4.1 Bulk resistivity of electrical insulation materials as a function of the temperature. Plotted after data from Shand[1124], Sedden[1117], Kohl[706], Ardenne[47], Espe[350, 354]. (1) Silica (quartz); (2) range of lead glasses; (3) Pyrex (Corning glass 7740); (4) range of soda–lime glasses; (5) Forsterite; (6) range of steatites; (7) range of special Steatites; (8) high voltage porcelain; (9) alumina; (10) zircon porcelain; (11) mica; (12) natural rubber; (13) Neoprene; (14) silicone rubber; (15) Teflon

compare various glasses, some glass manufacturers give for their glasses the value known as $T_{\varkappa 100}$ (see Appendix *B*.6), i.e. the temperature at which the resistivity of the given glass is 100 megohm.cm (Kirby[676], Stevels[1185], Sedden[1117], Hauffe[518], Flügge[387], Douglas[297]).

41.12 Surface resistivity. It was established (Edge[323]) that most glasses show much lower apparent bulk resistivities in normal atmospheres than when the measurements are carried out under vacuum. This decrease in resistivity is

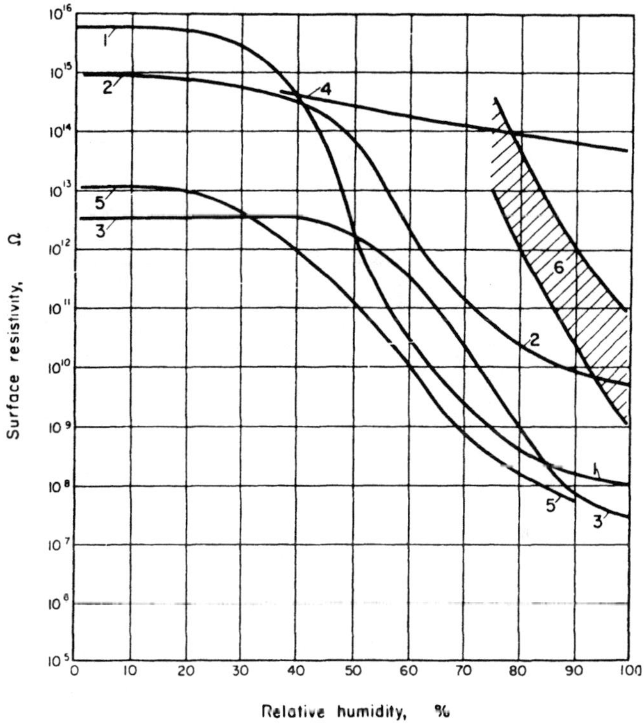

FIG. 4.2 Surface resistivity of glasses and ceramics: (1) silica (quartz); (2) Pyrex (Corning glass 7740); (3) Corning lime glass 0080; (4) Pyrex with surface treated by using silicone oil; (5) porcelain; (6) range of steatites. (Plotted after data from Kohl[706], Espe[350, 354])

caused by the relatively high *surface conductivity* of glass, due to adsorbed moisture and the products of atmospheric weathering (see Sections 21.3 and 23.21) and contamination. The surface conductivity of glasses (and ceramics) depends on the relative humidity of the surrounding atmosphere as well as on the kind of glass (or ceramic). Figure 4.2 shows the values of the surface resistivity at various glasses and ceramics.

In an electric device in which a glass (or ceramic) is used as the insulator, it is very important that it should retain its good insulating properties independently from the nature of the surrounding atmosphere. The surface *resisti-*

vity can be *increased* by outgassing at high temperature or by the application of non-hygroscopic films* on the surface of the insulating material.

The surface *conductivity* can be *increased* by coating the glass with a thin film of tin oxide applied on the surface by hydrolysis of stannic salt vapours at elevated temperatures. Gomer[456a] describes a technique with which surface resistivities of 10^{-3}–3 ohm/cm were reached.

The electrical resistivity of ceramics (alumina, zircon, beryllia) as well as that of mica and quartz is decreased by radiation damage (see Section 71.1) but generally these changes are reversible by annealing (Koenig[704]).

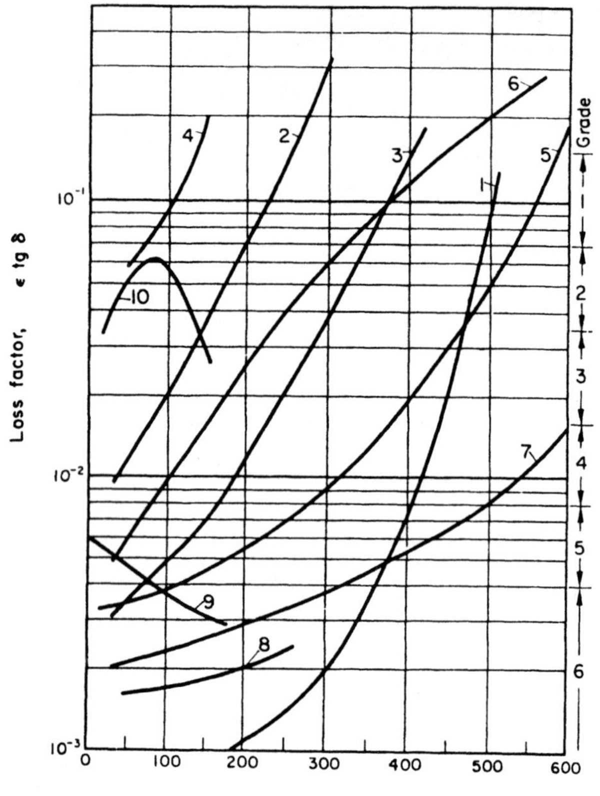

FIG. 4.3 Loss factor (at 10^5 cycles/sec) of insulating materials as a function of the temperature: (1) silica (quartz); (2) Nonex (Corning glass 7720); (3) Corning glass 7070 (low loss); (4) high voltage porcelain; (5) steatite; (6) zircon porcelain; (7) Forsterite; (8) mica; (9) silicone rubber (low loss factor type); (10) Teflon. (Plotted after data from Kohl[706], Steyskal[1189], Hippel[564a], Espe[350, 354], Nitzsche[933], Wick[1311])

* Silicone oils (e.g. Dow Corning 200 or Z 4141; Imperial Chemical Industries M 441; Midland Silicones D 4080, or MS 200; see Edge[323]) are used. The glass or ceramic is dipped into a 2 per cent solution by weight of DC-200 in thrichlor-ethylene or methylene-chloride. After removal it is air dried (30 min) at room temperature, and subsequently cured in air

41.13 Dielectric properties. The electric leads sealed into glass or ceramic form a condenser with the glass or ceramic. The power loss through the insulator is generally very small with regular low voltage seals. But in seals used with high voltages at high frequencies these losses may increase and heat up the insulator, ending in the melting of the glass or producing "break-down".*

The losses are proportional to the loss factor, i.e. the product of the dielectric constant ε** and the dissipation factor tg δ. Figures 4.3 and 4.4 give the

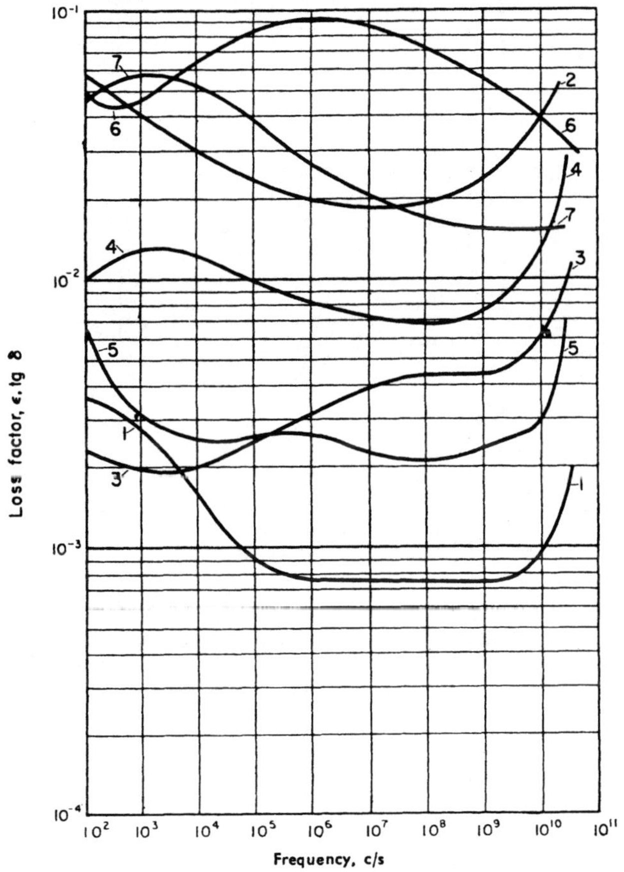

FIG. 4.4 Loss factor of insulating materials at room temperature as a function of the frequency: (1) silica (quartz); (2) Pyrex (Corning glass 7740); (3) Corning glass 7070 (low loss); (4) steatite; (5) Forsterite; (6) silicone rubber (medium loss factor type); (7) Teflon. (Plotted after data from Steyskal[1189], Kohl[706], Hippel[564a], Nitzsche[933], Wick[1311], Espe[350, 354])

for 30 min at 300 °C or 60 min at 275 °C. Other silicones (DC-2-4141, MS-D-3033) are applied in water solution usually by spraying.

*In vacuum devices the break-down may also occur due to a discharge in the residual gases. The voltage at which this kind of break-down occurs depends on the kind of the residual gas, the pressure, the distance between the electrodes and the material and shape of the electrodes.

**According to an empirical rule for glasses $\varepsilon \approx 2.2\,\gamma$ (γ is the specific gravity g/cm^3).

loss factors of some insulating materials as a function of the temperature and the frequency. From these figures it can be seen that high quality insulators are to be used especially in seals subjected to higher temperatures or where high frequencies are involved.

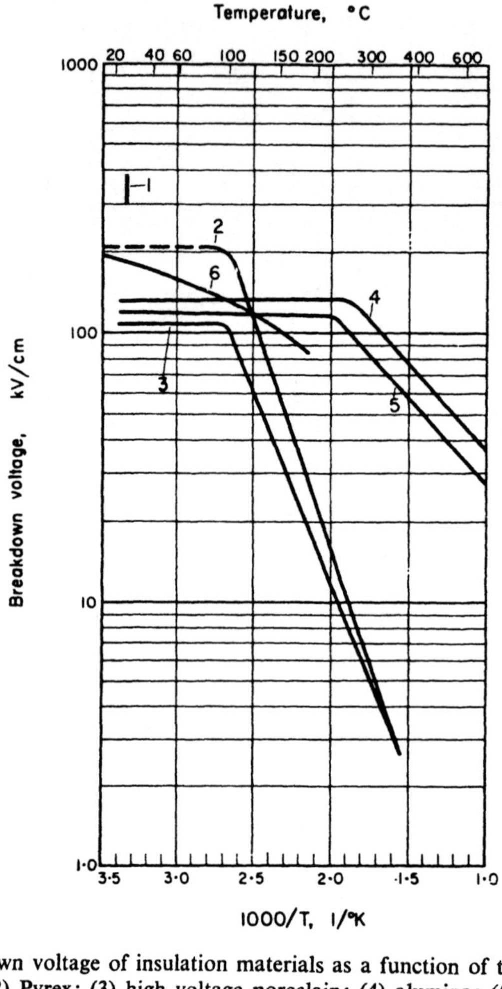

FIG. 4.5 Breakdown voltage of insulation materials as a function of the temperature: (1) silica (quartz); (2) Pyrex; (3) high voltage porcelain; (4) alumina; (5) zircon porcelain; (6) silicone rubber. (Plotted after data from Shand[1124], Kohl[706], Wick[1311], Espe[350, 354])

Figure 4.5 shows the break-down voltages for some insulating materials, again pointing out the influence of the temperature.

Break-down can also occur due to the accumulation of charges on the glass. The use of more conductive glasses as envelopes (bulbs) of electron devices may prevent the accumulation of charges due to electrons, avoiding the breakdown (puncture) of the envelope.

41.14 Electrolytic effects. At elevated temperatures voltage gradients may produce electrolysis of the glass (especially in glasses with low resistivity). Electrolysis is more probable when direct current is used, but it may also occur when alternating current is used since surface phenomena at the leads often cause a preferred conduction in one direction.

Electrolysis can occur between two electrodes sealed near to each other, but it may also occur between an electrode and the surface of the glass charged by electron bombardment.

The results of the electrolysis appear in different ways at the cathode (the more negative electrode) and the anode. At the *cathode*, metallic sodium is released and accumulated, and the space around the cathode develops a brown or blue colour due to the colloidal dispersion of sodium. If the glass contains lead oxides a black colour appears, because of the reduction of these oxides to metallic lead. Dumet leads (see Section 42.2) are discoloured or become light coloured because of the reduction of cuprous oxide to copper. At the *anode*, the glass is depleted in sodium, forming a silica rich (less conducting, see Fig. 4.1) glass. The oxygen released at the anode strongly oxidizes the metal of the leads, resulting in leaky seals. Dumet becomes darker in colour because of the formation of cupric (black) oxide.

As a result of the electrolysis, the pressure in the envelope is also affected by the evolution of gas from the seals, as by the lack of tightness of the seals. Gas bubbles appear on the electrodes, and some of this gas is released into the evacuated space. The reduction or oxidation of the lead wires results in leakage. The local composition change of the glass acts in the same manner as two very different glasses sealed together (Sections 23.12 and 24.12), producing stresses which may destroy the joint.

Düsing[317] proposes the testing of glass–metal seals for their behaviour as regards the electrolysis, by testing the service time of the seal at higher temperatures, since in these conditions electrolysis is favoured. If the obtained service time is considered insufficient, the temperature of the seal should be lowered by cooling, shielding or changing the shape of the seal, or the glass should be replaced by another glass (with higher $T_{\varkappa 100}$ point) (Stevels[1185], Kohl[706]). Düsing[316] states that the highest tolerable temperature of Kovar-to-glass seals is 200 °C. Above this limit, the glass electrolysis becomes troublesome (e.g. a seal at 200 °C having a lifetime of more than 7000 hr, had a lifetime of only 87.5 hr at 300 °C).

41.2 Loading current

The diameter of the leads should correspond to the current which must be carried by the electrode.

Figure 4.7 shows the maximum current which may be carried by leads of various materials at various diameters. On exceeding these values, the lead will be overheated, the vacuum seal will be destroyed and/or the pressure in the evacuated space will be increased due to gas evolution from the heated seal.

Because of the heating, the lead-throughs may be water-cooled (Sections 4.2 and 4.3), or just cooled by constructing the lead so as to present the largest surface for a given cross section area (more small diameter wires instead of one single large diameter wire or ribbon).

41.3 Frequency

At low frequencies the frequency of the current has no importance on the construction of the seal. At high frequencies the material of the lead and the insulation should be built according to special requirements.

At frequencies of 10^8 cycles/sec (and higher frequencies) the resistivity of some metals used as lead-throughs (Kovar) becomes intolerably high. In order to lower the resistivity of Kovar lead-throughs for high frequency use, they must be copper- or gold-plated (Düsing[316]).

The influence of the frequency on the loss factors of insulating materials is shown in Fig. 4.4.

41.4 Temperature

Lead-throughs as other parts of vacuum systems which have to be baked (e.g. for ultra-high vacuum) should withstand temperatures of up to about 450 °C.

A distinction should be made between lead-throughs which should withstand baking temperatures without carrying currents during the baking, and the lead-throughs which are heated while they are in service. If the lead-through is to be heated, the construction should exclude any material which does not withstand the required temperature; glass–metal or ceramic–metal seals may be used, but elastomers are excluded. If the electrode is to be heated simultaneously with its service period (e.g. furnace) the insulation material should have the required resistivity (Fig. 4.1), loss factor (Fig. 4.3), and breakdown voltage (Fig. 4.5) even at elevated temperatures. Such lead-throughs are available commercially in a variety of sizes and shapes (Sections 4.2 and 4.3).

4.2 PERMANENT LEAD-THROUGHS

Permanent lead-throughs consist of the electrical lead and the insulating material, sealed permanently to each other. Permanent lead-throughs are used in electric lamps, electron tubes and various other electrical vacuum devices (Roth[1083], Millner[867], Tzarew[1245], Köhler[708]). These permanent lead-throughs are based on glass–metal or glass–ceramic seals (Sections 2.4 and 2.5).

In this Section the various permanent lead-throughs are described according to their shapes, i.e. rod seals (Section 42.1), stem seals (Section 42.2), pin seals (Section 42.3), ribbon seals (Section 42.4), disc and cup seals (Section 42.5).

42.1 Rod seals

Metal rods (tungsten, molybdenum, FeNiCo, see Table 2.34) may be sealed into hard glasses, forming permanent lead-throughs. The technique used to build these seals is described in connexion with Fig. 2.74. Some usual rod seals are shown in Fig. 4.6. The rod is sealed into a part of the vacuum envelope having a specially designed shape (Fig. 4.6a) to avoid the build-up of dangerous stresses (Sections 24.14 and 24.5).

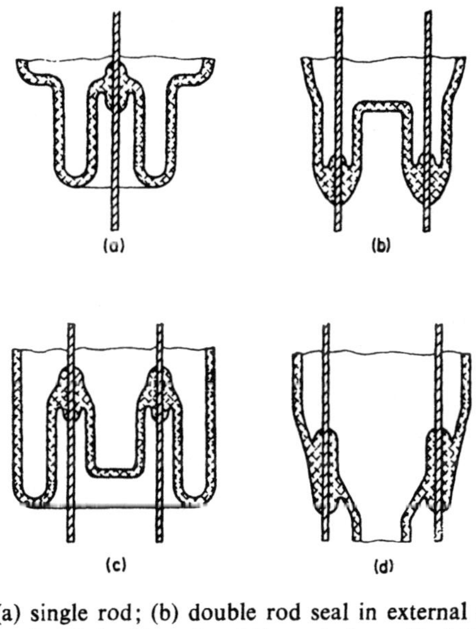

FIG. 4.6 Rod seals: (a) single rod; (b) double rod seal in external extensions; (c) as (b) but internal; (d) double rod seal through the wall of the device

The rod seals are usually used as double (or multiple) lead-throughs, placed on external extensions of the vacuum envelope (Fig. 4.6b), on parts protruding into the vacuum vessel (Fig. 4.6c) or crossing the wall of the system (Fig. 4.6d).

The rod (lead) should carry only currents which do not heat it sensibly. If the rod is loaded with currents which are too high, its rise in temperature produces rapid expansion differences, leading to stresses in the seal (Section 24.12). At the same time the resistivity of the glass is lowered (Fig. 4.1) and a break-down is possible. The rods (leads) should be designed in accordance with the admissible values plotted in Fig. 4.7 so as to load the lead-through of a given diameter made from a particular metal with a current lower than the maximum given by Fig. 4.7.

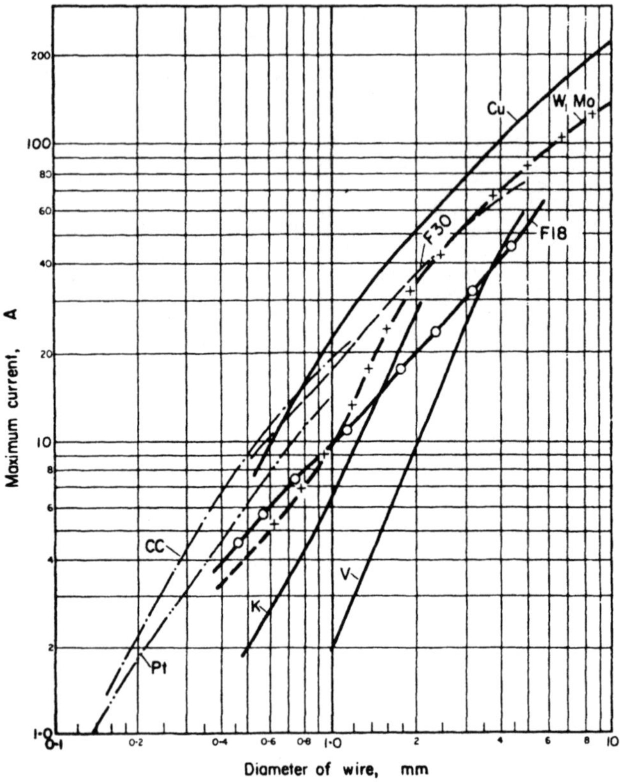

FIG. 4.7 Maximum admissible current as a function of the diameter of the sealed wire (rod). *CC* = copper clad wire; *Cu* = copper wire; *Pt* = platinum; *K* = Kovar; *V* = Vacon (see Tables 2.29 and 2.33); *W* = tungsten; *Mo* = molybdenum; *F18* = FeCr (18 per cent Cr); *F30* = FeCr (30 per cent Cr). (Plotted after data from Steiskal[1189], Mönch[879], Roth[1083], Ardenne[47], Espe[351])

In order to minimize the surface area of the rod exposed to the evacuated space, coaxial seals are used (Fig. 4.8). In a coaxial seal, the rod carrying the current may be insulated from the tube sealed to the envelope (Fig. 4.8a) or machined from the same piece with the tube (Fig. 4.8b). The arrangement shown on Fig. 4.8b minimizes the immediate heat transfer from the electrical lead to the glass–metal seal (Harrower[514]).

The lead-through is connected to the functional parts (anode, cathode, heater, etc.) by mechanical joints or by spot welding, brazing, etc. If the lead should be connected to thin metal layers deposited on the inside surface of the envelope, special techniques are to be used.

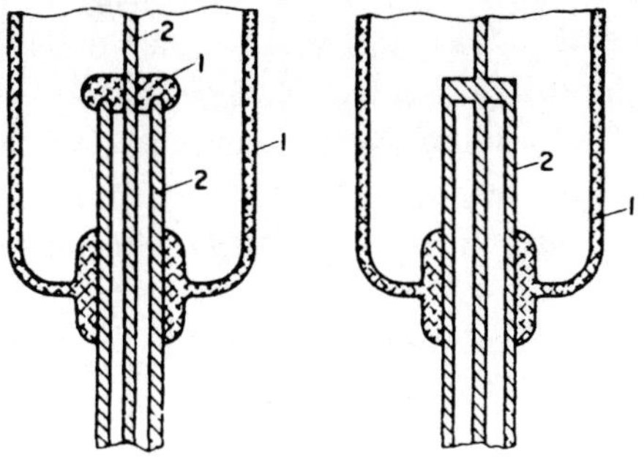

FIG. 4.8 Rod seals: (a) with coaxial tube; (b) with special electrode; (1) glass; (2) Kovar. After Harrower[514] *(Courtesy of The American Institute of Physics)*

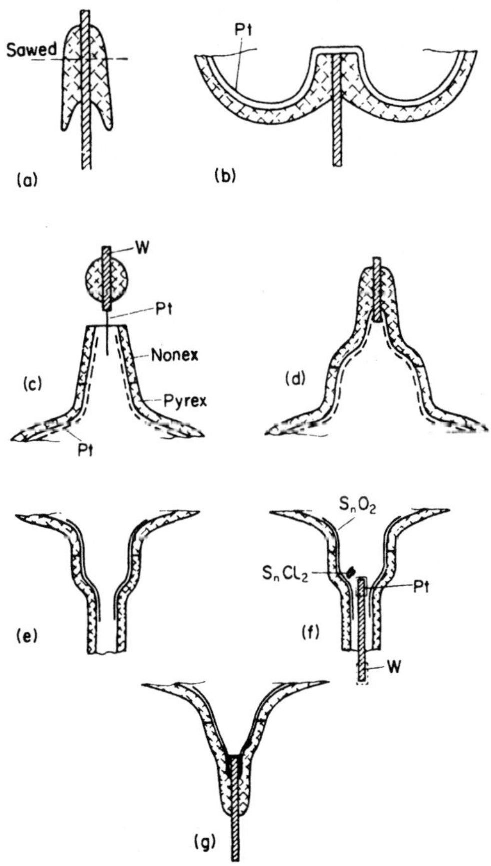

FIG. 4.9 Methods of connecting electrodes to metallization on the inside surface of the wall; (a, b) sawed electrode (after Wieder[1313]); (c, d) connexion with platinum stup (after Gomer[457]); (e, f, g) connexion with wrapped platinum foil. After Melmed[843a] *(Courtesy of The American Institute of Physics)*

Belser[104] recommends soldering with indium, which adheres to glass, quartz and ceramics, and is able to make electrical contact on layers having a thickness of only a few Ångstroms. Wieder[1313] suggests the method shown in Fig. 4.9a, b consisting in first beading the tungsten lead into an appropriate glass (see Table 2.34) and then sawing it off with a diamond saw near one end (Fig. 4.9a). After fire polishing to remove the sharp edges, the bead is sealed into the vacuum device (Fig. 4.9b) and after an electrolytic cleaning in KOH, the inside surface is platinized. Gomer[457] prepared the contact to the inside metallized surface (Fig. 4.9c) by using a tungsten rod beaded into Nonex glass, and having a platinum foil (1 mm wide, 5 mm long, 18 microns thick) spot welded to its end. A Nonex sleeve is collapsed over the tungsten bead so that the unplatinized section (Fig. 4.9c) makes the vacuum-tight seal, and the platinized section makes electrical contact with the thin platinum foil (Fig. 4.9d). Melmed[843a] coated the inside surface of the sidearm (Fig. 4.9e) with a conducting tin oxide, prebeaded and then sealed into this sidearm a tungsten rod, which was previously wrapped with a platinum foil, which was spot welded at both ends to the tungsten rod (Fig. 4.9f). A small crystal of stannous chloride (1 mm³) was then dropped into the sidearm, which was then heated until the glass collapsed on to the platinum foil. The resulting seal (Fig. 4.9g) can be used from liquid nitrogen temperatures up to 430 °C (Melmed[843a]).

Glass–metal lead-throughs are usually sealed to the glass envelope of the vacuum device or vacuum system, but if required these lead-throughs may be joined to the vessel by gasket seals. Figure 4.10 shows two examples of such joints.

Nelson[916] described a glass–metal seal, which is attached with a dismountable joint to the vacuum vessel (Fig. 4.10a). The electrical lead (1) is sealed

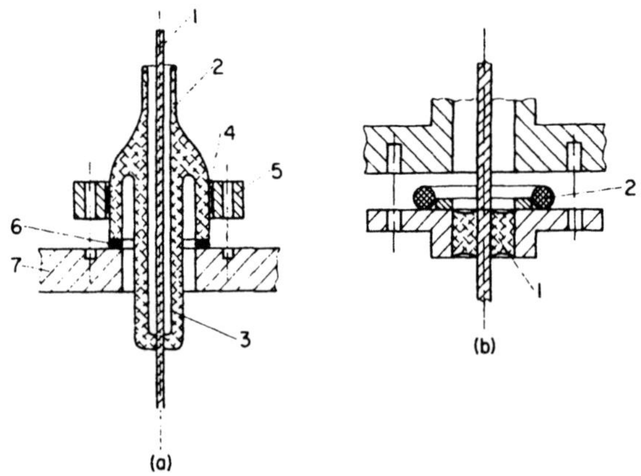

FIG. 4.10 Glass–metal seal used in demountable joints: (a) attached electrode; (1) copper electrode; (2) Pyrex glass; (3) W 1 glass; (4) asbestos; (5) backing flange; (6) gasket; (7) wall of the metal container. After Nelson[916] *(Courtesy of The Institute of Physics and The Physical Society, London)*; (b) compression seal (1) connected by gasket (2)

to the glass (2) by means of an intermediate glass (3). Nelson[916] used a standard copper to Wl glass stem seal joined to the Pyrex glass insulating support (at a distance of about 1/2 in. from the seal). The lead-through is sealed to the vessel (7) by an aluminium or indium gasket (6), the required compression force being given by the screws of the backing flange (5) applied on the asbestos insert (4) which sits on the tapered shoulders of (2). Nelson[916] used aluminium wire of 0.01 in. diameter (compressed in the seal to 0.002–0.003 in.) as the gasket for bakeable seals, and indium wire of 1/16 and 1/32 in. diameter in another seal. He points out the importance of a good surface finish of the sealing surfaces, especially when aluminium gaskets are used, rating the required surface finish to at least 8 micro inch.

If the lead-through is constructed as a glass–metal compression seal (Section 24.43) it may be connected to the vessel by elastomer or metal gasket seals as shown in Fig. 4.10b. Compression seals (for 5–10 A and 300–3000 V) of various shapes are commercially available (e.g. Standard Telephones[1174], Schott Jena). Adam[3] described such seals to be used at 35 A and 3000 V or 120 A and 3000 V, and which can be cycled up to 300 °C. Perott[983] described a procedure to construct such compression seal lead-throughs.

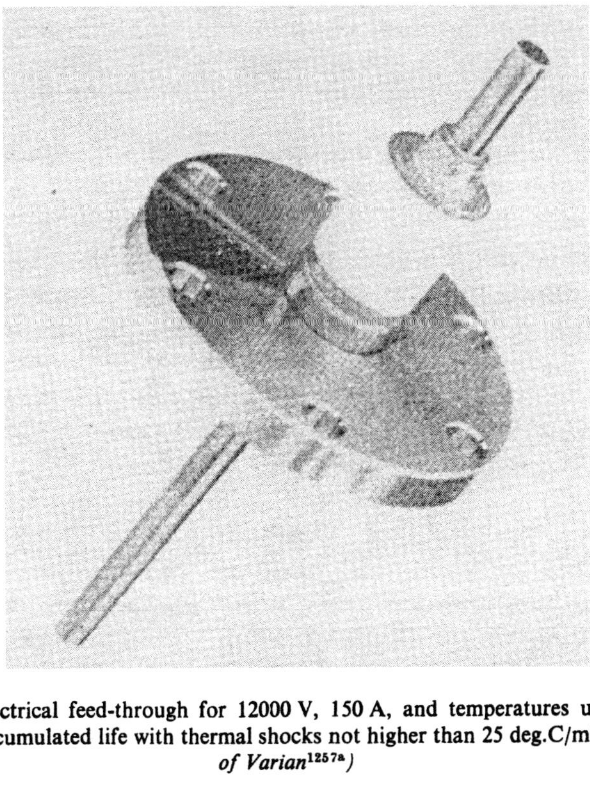

PLATE 14. Electrical feed-through for 12000 V, 150 A, and temperatures up to 450 °C. Has 500 hr accumulated life with thermal shocks not higher than 25 deg.C/min *(Courtesy of Varian[1257a])*

Metal rods sealed in ceramics are also extensively used as lead-throughs. Such seals are generally made* (see Section 2.5), connected to metal flanges which are then joined to the vacuum vessel or system by brazing (Section 22.3), by elastomer or metal gasket seals (Section 3.8). The ceramic–metal lead-throughs are used in various vacuum applications, from evaporation

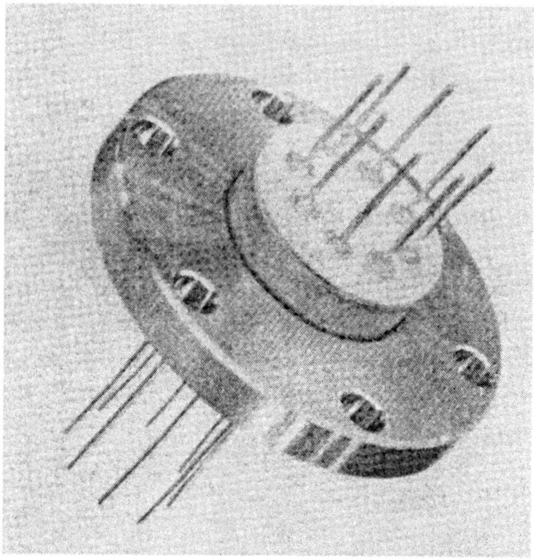

PLATE 15. Octal feed-through for 1000 V, 5 A per wire, and temperatures up to 300 °C. Life as feed-through from Plate 14 *(Courtesy of Varian*[1257a]*)*

plants (Hemmendinger[536]) to ionization chambers (General Radiological[440]). Beavis[101b] described an ultra-high vacuum radio-frequency lead-through based on a ceramic-to-metal seal (see moly-manganese process, Section 25.23) and joined to the vacuum system by an Ultek shear seal (Section 38.55).

Plates 14 and 15 show two electrical lead-throughs with ceramic-to-metal seal, supplied by Varian[1257a], equipped with stainless steel flanges to be used with Conflat gaskets (Section 38.56).

42.2 Stem seals

The "stems" are lead-throughs obtained by flattening a glass tube over the lead-wires to be sealed into the glass. The vertical stems have the glass collapsed (pressed) over the wires, in a direction perpendicular to the axis of the

* Ceramic feed-throughs for operating voltages up to 100 kV and currents up to 2000 A are available from e.g. Ceramaseal Inc., New Lebanon Center, New York.

wires in contrast to horizontal stems (see Section 42.3), where the glass is pressed parallel to the wires.

The usual (vertical) stems used in electric lamps are made by positioning the components of the stem in the jaws provided on the appropriate stem making machines (Roth[1084], Neumann[923], Ulmichek[1251], Banki[79]). These jaws (mounted on the periphery of a rotating table) carry the components of the stem (Fig. 4.11a) through a series of successively stronger cross flames. As a result

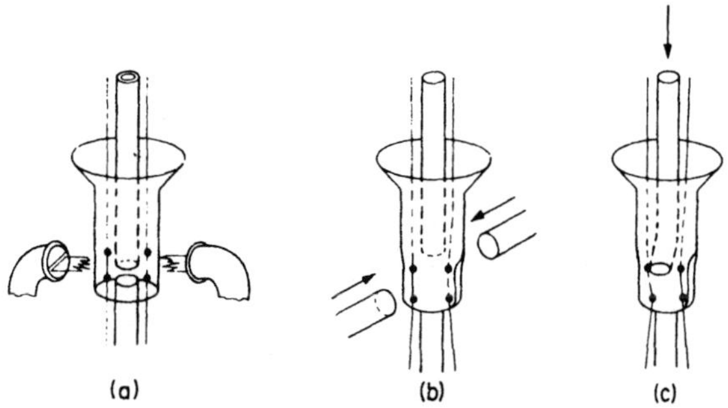

FIG. 4.11 Stages in forming a stem: (a) parts of the stem; (b) parts fused together; (c) stem, after compression and blowing-off of the exhaust tube connexion

of this heating the glass parts are softened; at this stage the pressing jaws are actuated (automatically) and the molten glass is pressed onto the wires (Fig. 4.11b). In a next step of the machine, air is blown through the exhaust tube, piercing a connecting opening through the wall of the stem (Fig. 4.11c). Subsequently the stem runs in some annealing fires and finally is transferred into an annealing oven (Sections 23.13 and 24.2).

The lead-in wires used in soft glass stems (see Table 2.32) consist usually of 2–4 butt welded parts. The part of the wire to be sealed into the glass is a portion of "Dumet" (copper-clad) wire. Towards the outside of the seal (on Fig. 4.11 upwards) the Dumet wire is continued by copper wire or sometimes is connected to the copper wire by a short part of Monel or Constantan wire serving as a "fuse". The Dumet is continued inside the vacuum device (downwards on Fig. 4.11) by a nickel (or Monel) wire.

The *Dumet wire* was developed (Fink[377], Eldred[336]) to replace the platinum wire used earlier in such seals. The Dumet is a copper clad iron–nickel (58/42) alloy core. It uses successfully the appropriate expansion of the iron–nickel core and the elasticity of the thin copper layer as well as the good adhesion to the glass of the oxidized copper (Sections 24.11 and 24.41). The sandwich structure of the wire results in quite an unusual expansion characteristic (Fig. 4.12). The axial expansion coefficient of the wire is $60–73 \times 10^{-7}$/deg.C and the radial expansion coefficient is $80–100 \times 10^{-7}$/deg. C. When this wire is sealed into a glass having a matching expansion with respect to the radial expansion of the wire, in an axial direction the stresses are taken up by the thin copper sheet. This principle was used later by Housekeeper in his unmatched seals (Section 24.41).

The Dumet wire is manufactured by fusing or welding the iron–nickel core (about 6 mm diameter and 40 cm length) into a tube of pure copper (with a wall thickness of about 0.5 mm) with an intermediate thin (50 μ) sheet of brass. This billet is swaged and then drawn into wire. After slight superficial

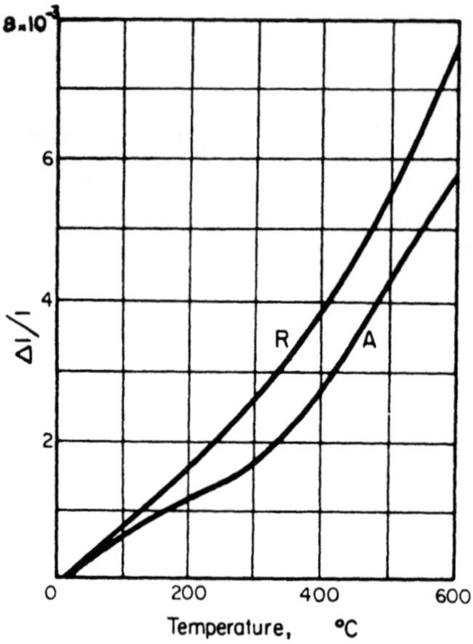

FIG. 4.12 Thermal expansion of Dumet wire; A = axial expansion; R = radial expansion

oxidation, the wire is passed through a solution of borax. Finally, the wire has a diameter of 0.25–0.75 mm. The copper layer forms about 25–35 per cent of the total weight, having a thickness of 15–50 μ, depending on the diameter of the wire. The borax layer on the wire is extremely thin (about 15 μ). The slightly oxidized copper (red, cuprous copper oxide, see Section 24.11) should assure a strong bond to the glass. The borax protects the oxide layer during the storage, and helps to create an intermediate glass layer rich in boron oxide during sealing, which facilitates the sealing.

Stems of hard glass are made using molybdenum (or tungsten) wires (Section 24.32) instead of Dumet.

The stem lead-throughs may have various shapes according to the number and kind of wires which are to be sealed into the glass. There are flat stems (Fig. 4.13a, b, c) with two or more (e.g. eight) wires (Fig. 4.13b). If the cross section of the wire required according to the current to be carried (see Fig. 4.7) is too large, multiple wires connected in parallel are to be used (Fig. 4.13c). To concentrate the wires on a small area and nevertheless to allow large enough distances between the adjacent wires, stems of the most various shapes (Fig. 4.13d–h) may be constructed.

Loughridge[786] discusses the faults of Dumet seals, and states that in order to detect these faults, first a visual examination should be made, using a binocular microscope with reflected light from a microscope illuminator. A common fault which can be detected in this way consists of a series of small parallel cracks in the glass at the surface of the wire. These cracks will run at right

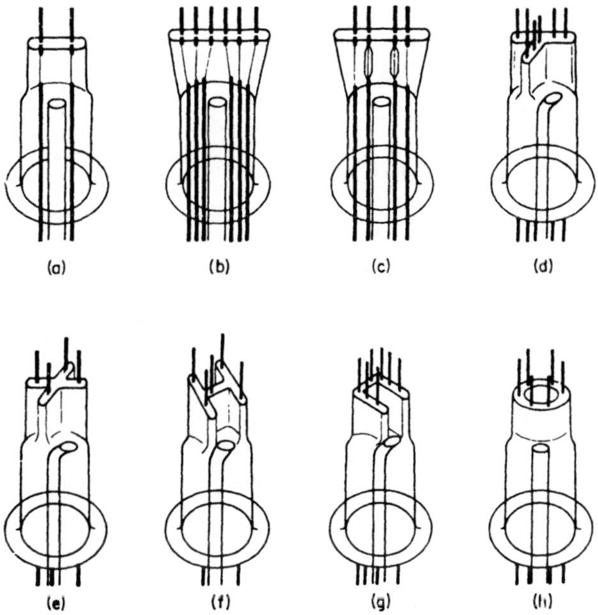

FIG. 4.13 Stem seal shapes: (a) flat stem with 2 wires; (b) flat stem with 6 wires; (c) flat stem with 2 single and 2 multiple wires; (d) T-stem; (e) X-stem; (f) H-stem; (g) U-stem; (h) O-stem

angles to the longitudinal direction of the wire, and are usually quite small. The cracks will generally be on one small segment of the wire and in an area where the seal was cold. Loughridge[786] believes that the presence of an oxide-borate layer at the glass-to-metal interface, which has not been heated sufficiently to diffuse these materials into the glass, will result in an even greater mismatch, and lead to glass fracture. In order to make the channels in the seal more visible, it is advisable to use a dye which penetrates the holes in the seal. Some of the faults of Dumet seals are produced by the faults of the Dumet itself, i.e. excessive oxidation, exposed core, irregular core, thin copper layer, distorted copper layer, etc.

42.3 Pin seals

A *pin seal* is a glass disc in which metal* pins are sealed perpendicular to the faces of the disc. These seals are also called *horizontal stems*, due to the position in which they are placed in the electron tubes, constituting their bottom. The two basic shapes of pin seals are shown in Fig. 4.14. The glass is in the form of a disc (Fig. 4.14a) or a cup (Fig. 4.14b) according to the

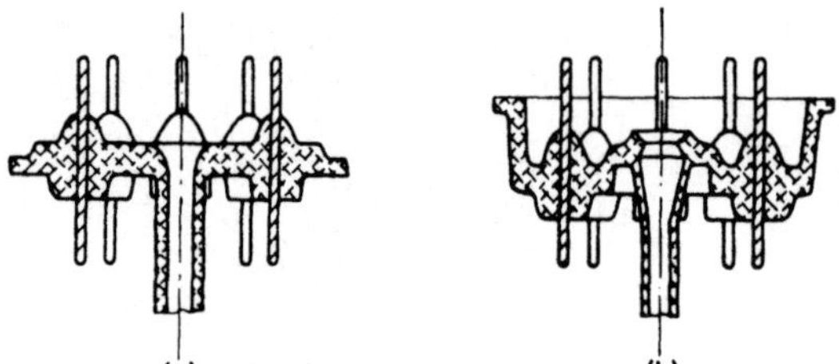

FIG. 4.14 Pin seals: (a) with disc; (b) with cup

requirements of the subsequent sealing into the vacuum device (Section 26.1). The pin seals are made with attached exhaust tubes (Fig. 4.14) or they are available also without this part (e.g. Siemens Ediswan[1135]).

The development of these seals is due to the advantages brought by them in the construction of electron tubes, compared with the vertical stems (Section 42.2) used earlier. The pin seals permit the construction of mechanically stronger tubes which are less high and have a circular, symmetrical arrangement of the electrical leads. The number of leads which can possibly be sealed is greatly increased (up to about 25).

Pin seals are made by pressing the glass (parallel to the axis of the pins) over the pins placed in suitable holders. The various techniques used for the pressing operation differ from each other mainly because of the shape in which the glass is brought into the seal, which determines the holders, the pressing technique, etc. The glass may be brought to the seal as:

(a) individual sleeves (Fig. 4.15a) placed around each pin (Violet[1269]);

(b) concentric glass rings (Fig. 4.15b) placed inside and outside the circle formed by the pins (Vamberi[1254]);

(c) drop of softened glass (Fig. 4.15c) placed between the pins (Vamberi[1254]);

(d) powdered glass (Fig. 4.15d) placed around and between the pins (Steyskal[1189]).

* The pins may be of Dumet, FeCr, Kovar (see Section 24.31). The diameter of the pins can be obtained from Fig. 4.7.

On laboratory scale a very convenient device for bringing electrical leads into a vacuum enclosure can be made by using the bottom (pin seal) of a discarded radio tube.

After placing the pins in their holder, and bringing the glass (by one of the techniques shown on Fig. 4.15) the pin seal is completed by combining adequate heating and pressing.

Figure 4.16 shows the technological steps used to complete a pin seal beginning from concentric glass rings.*

In the process using powdered glass, the pins are placed in their holder after being previously beaded in a thin sleeve of the appropriate glass. Using

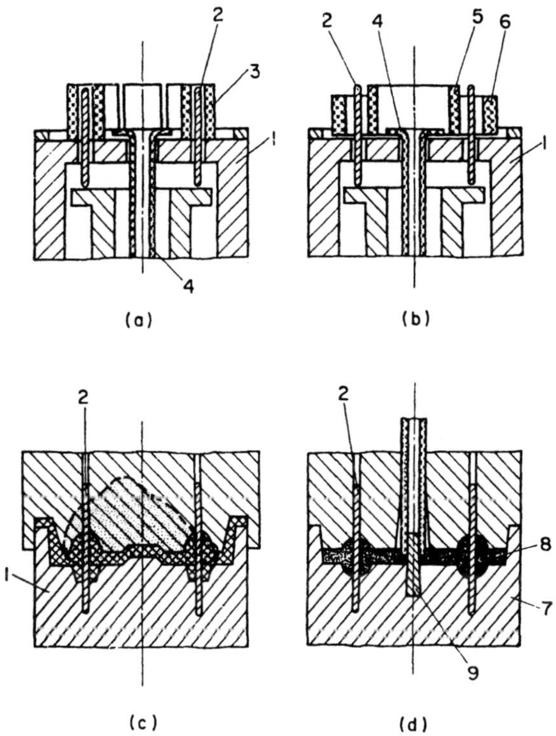

FIG. 4.15 Basic techniques for pressing pin seals, using (a) individual sleeves; (1) holder; (2) pin; (3) glass sleeves; (4) exhaust tube; (b) using concentric glass rings (5, 6); (c) using molten glass drop; (d) using powdered glass; (7) graphite holder; (8) powdered glass; (9) graphite holder for the exhaust tube

powdered glass of 350–750 μ particle size, a pressure of about 6–10 g/cm^2 is necessary. The sintering occurs usually in nitrogen.** Szalma[1211a] recommends for the sintering process of molybdenum pin seals (in matching glasses) a sintering temperature of 770–810 °C, with a temperature rise from the softening point of the glass (about 600 °C) to the sintering temperature at a rate of 10–80 deg.C/min, a sintering time of about 10 min and a cooling rate of 3–27 deg.C/min (for pin seals using 1 mm diameter pins).

* For more details see e.g. Espe[351, 354], Violet[1259], Millner[867].

** For more details see e.g. Espe[357], Dorgelo[289, 291], McKnight[833].

Pin seals may also be made by using regular glass–metal sealing techniques (Section 24.2). The pins are sealed into pre-pressed glass cups, profiled with bulges which, after being cut, leave the required holes for the sealing of the pins (Fig. 4.17).

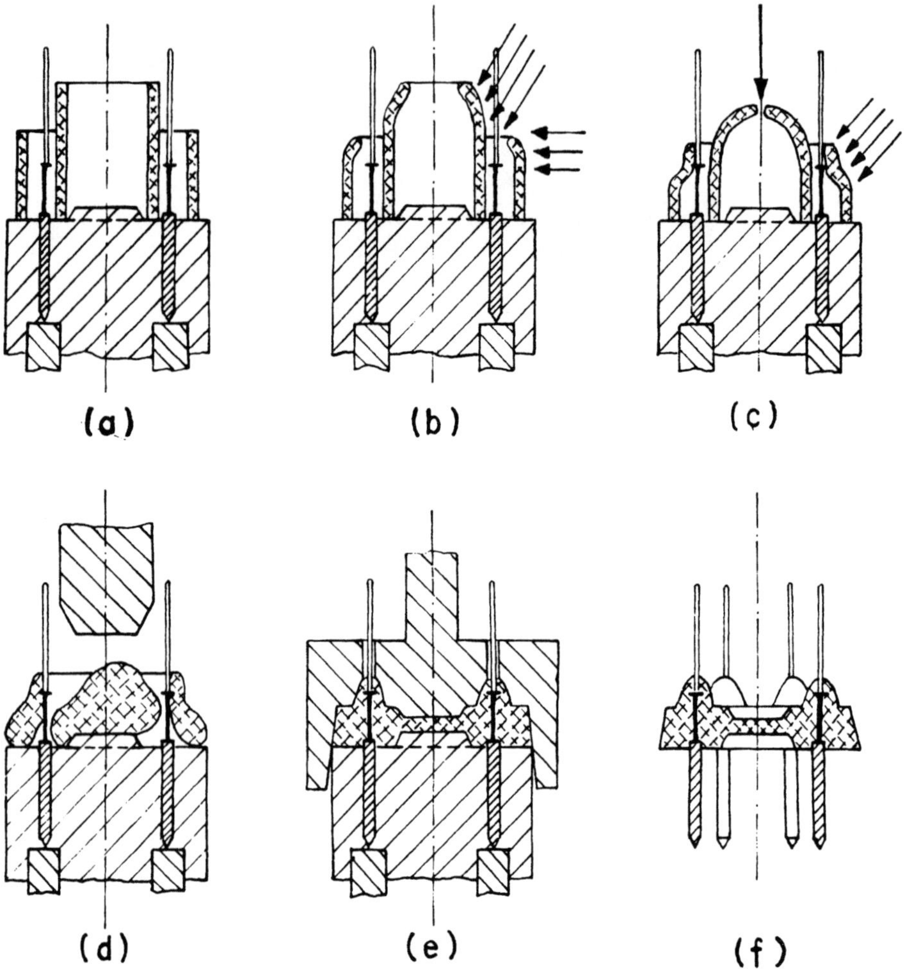

FIG. 4.16 Stages in completing a pin seal: (a) pins and glass rings placed on the holder; (b) preheating (arrows show flames); (c) strong heating (melting); (d) spacing the pins; (e) pressing; (f) finished pin seal

Kamphausen[652b] describes a method for making pin seals (with tungsten pins) for experimental purposes, by inserting the pins through a softened glass plate in such a way that the pins protrude on the other side but are covered with glass. Then, after cooling, the glass from the end of the pins is removed by etching in 10 per cent hydrofluoric acid.

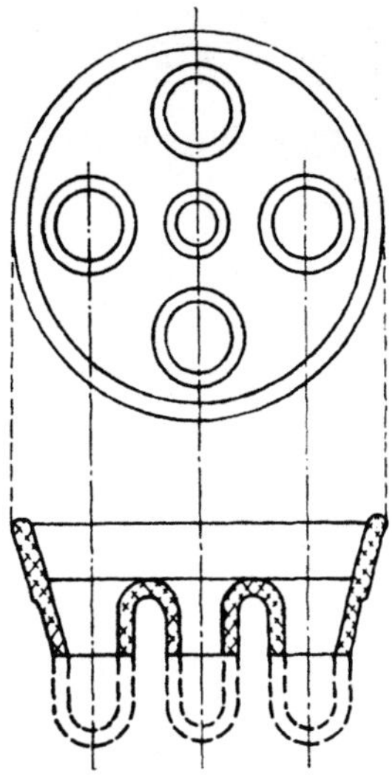

FIG. 4.17 Pre-pressed cup for pin seals

42.4 Ribbon seals

When the expansion of the metal does not match the glass (see Section 24.4), instead of wires or pins, thin ribbons are sealed through the glass. In such lead-throughs, copper or platinum ribbons are sometimes used (Table 2.35) but in the more current techniques, molybdenum ribbons are used. The techniques involved in the construction of such seals are discussed in Section 24.42.

The current carrying capacity of ribbon seals is generally very high compared with rod, pin or wire lead-throughs of the same cross-sectional area. Figure 4.18 shows the rated current for molybdenum ribbons (0.015 mm thick) sealed in quartz. A quick comparison with the maximum current permitted in wire (rod, or pin) seals (Fig. 4.7) shows that the cross-sectional area for the same current is about ten times less for the ribbon seals.

Figure 4.19 and Table 4.1 give the dimensions of some molybdenum ribbon seals.

Single ribbon seals are available only up to about 10 A. For higher currents twin or multiple ribbon seals should be used (Section 24.42).

The lead-throughs with ribbon seals present some rigidity problems. Some special techniques are used to ensure the rigidity of the very thin ribbon extending from the seal. It is possible to connect the extending ribbon to a leadwire which is held in a glass bead (Fig. 4.20a); to connect the internal end of the ribbon to a metal attachment held by a collar clamped to the glass tube holding the seal (Fig. 4.20b); or to ensure the rigidity of the ribbon by curling it into a semi-circle (Fig. 4.20c) (Reimann[1046]).

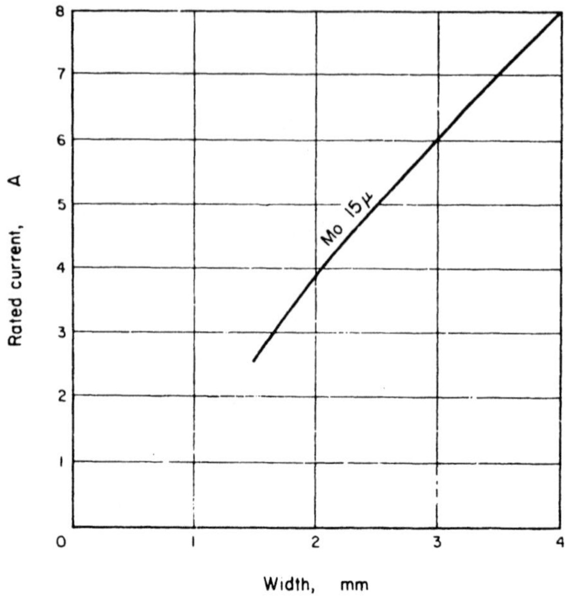

FIG. 4.18 Rated current for molybdenum ribbons (0.015 mm thick) sealed into quartz

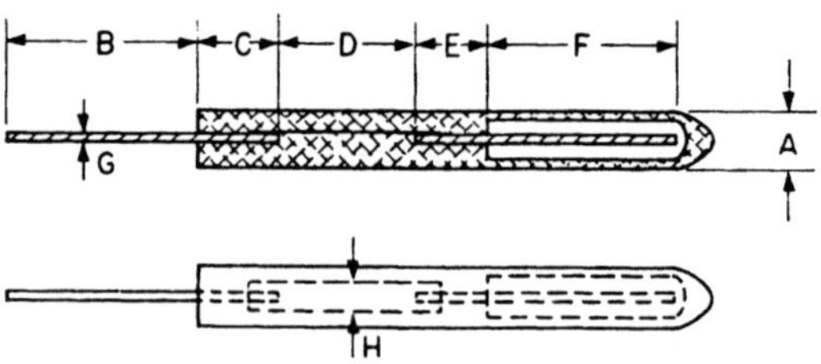

FIG. 4.19 Dimensions of molybdenum seals (Table 4.1)

TABLE 4.1. DIMENSIONS (mm) OF MOLYBDENUM RIBBON SEALS IN QUARTZ (Fig. 4.19) SUPPLIED BY Thermal Syndicate[1218a]

Rating	*A*	*B*	*C*	*D*	*E*	*F*	Molybdenum	
							wire *G*	ribbon *H*
2–3 A	4–5	25	8	12	8	25	0.3	1.5×0.0156
4 A	5–6	25	10	14	10	25	0.45	2×0.0156
6 A	6–7	25	10	18	10	25	0.7	3×0.0156
8 A	7–8	25	11	18	11	25	1.0	4×0.0156

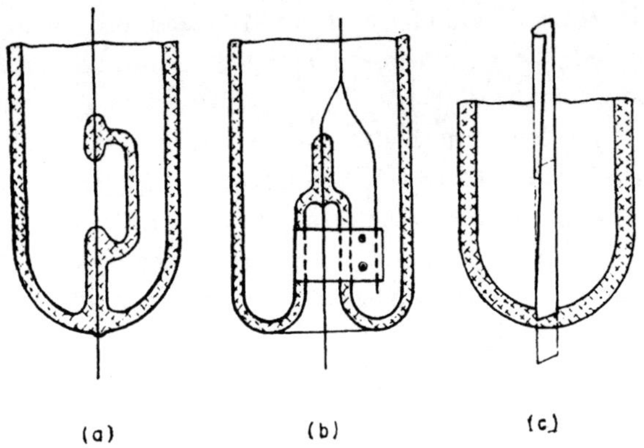

FIG. 4.20 Methods to give rigidity to ribbon seals: (a) with a second glass bead; (b) with metal attachment; (c) by curling the ribbon

42.5 *Disc and cup seals*

Electrical lead-throughs carrying heavy currents are generally constructed so as to separate the glass–metal (or ceramic–metal) seal from the electrical lead which carries the current. One of the methods used to achieve this (see also Fig. 4.8b) is by connecting the electrical lead through a metal disc or cup which is sealed (on its circumference) to the glass (or ceramic).

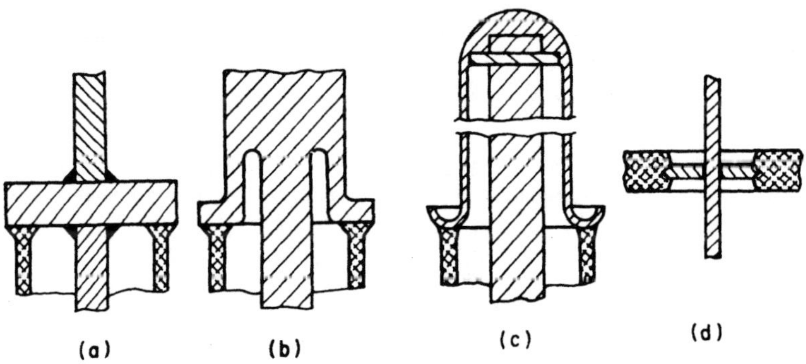

FIG. 4.21 Lead-throughs with disc seals: (a) metal disc with brazed connexion; (b) machined disc; (c) spinned disc; (d) molybdenum disc sealed in quartz

When a metal disc is used (Fig. 4.21) this may be sealed to the glass either by a matched seal (Section 24.3) or by an unmatched seal (Section 24.4).

The electrical lead is brazed to the two faces of the disc (Fig. 4.21a) or it is machined as one piece (Fig. 4.21b). When the electrical lead and the disc constitute a single piece, the disc which will be sealed to the glass should be separated from the rod carrying the current, by an annular cut (Fig. 4.21b) which assures the required temperature gradient between these two parts and avoids heating the glass–metal seal. The same result is obtained by using a metal cup (Fig. 4.21c) having the appropriate shape into which the electrical lead (rod) is brazed. By combining machined leads and cups or discs, a large variety of a shapes are available.

For lead-throughs sealed in quartz, a molybdenum foil (disc) may be used (Fig. 4.21d), which is sealed at the circumference (see Section 24.42) to the quartz and carries, in the middle, the lead wire brazed to it.

Instead of sealing metal discs or flared cups to the glass, metal cups may be used which are edge sealed in the glass (see Section 24.41) or onto ceramics (Section 25.4). Some arrangements for such seals are illustrated in Fig. 4.22.

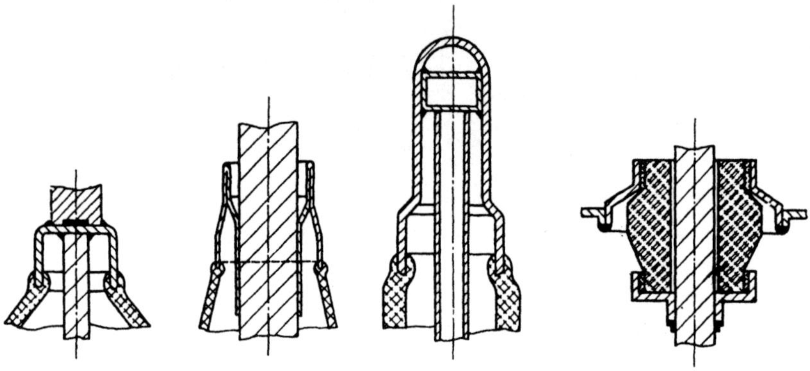

(a) (b) (c) (d)

FIG. 4.22 Lead-throughs with cup seals: (a) metal cup with brazed electrodes; (b) elastic seal; (c) spinned cup; (d) ceramic cup seal

The electrode may be brazed to the inside and outside of the cup (Fig. 4.22a) or just inside the cup (Fig. 4.22c). The electrical lead (rod) may pass through the cup, if this is made from two concentric, edge-welded (Section 22.32) rings, the inside ring has been brazed to the rod (Fig. 4.22b). Einspruch[334a] described a seal consisting of a short length of Pyrex tubing to which at each end, a length of stainless steel tubing is attached by a feather edge Housekeeper seal (see Fig. 2.83). The end of one of the stainless steel tubes is brazed to the metal vacuum chamber, and the other stainless steel tube carries the electrical lead.

When the electrical lead-through uses ceramics as insulation material, usually a cup is brazed or welded on the electrical lead (Fig. 4.22d), which is sealed to the ceramic part (see Section 25.2) and the ceramic part is sealed on the outside to a ring (elastic) brazed to the vacuum vessel.

Espe[351] describes a method to braze copper lead-throughs into iron–chromium cups to be sealed to glass. On the bottom of a FeCr cup, copper and borax is melted (Fig. 4.23a) and after cooling a hole is made to fit the copper rod (Fig. 4.23b). The copper rod, provided with an AgCu (72/28) alloy disc is introduced (Fig. 4.23c) through the hole, and the brazing is completed (Fig. 4.23d).

A method of constructing a cup seal (Kovar to glass) is described by Tasman[1215a]. A piece of glass (Corning 7052, or Philips 28) is inserted into the Kovar tube (part of the vacuum enclosure) which is profiled as shown in Fig. 4.24a. The glass is joined to the inner surface, bent around the edge, and then

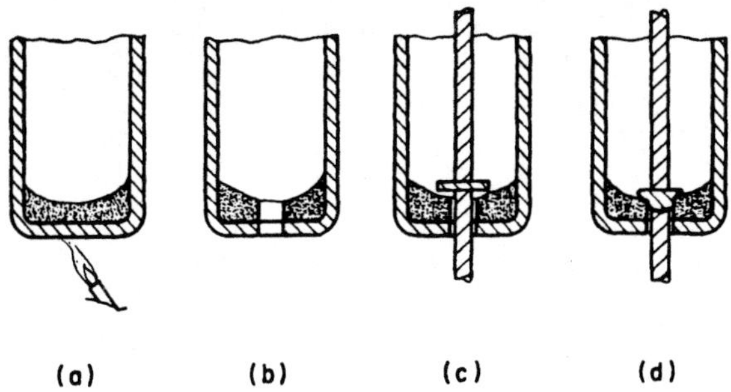

Fig. 4.23 Copper rod sealed through FeCr cup

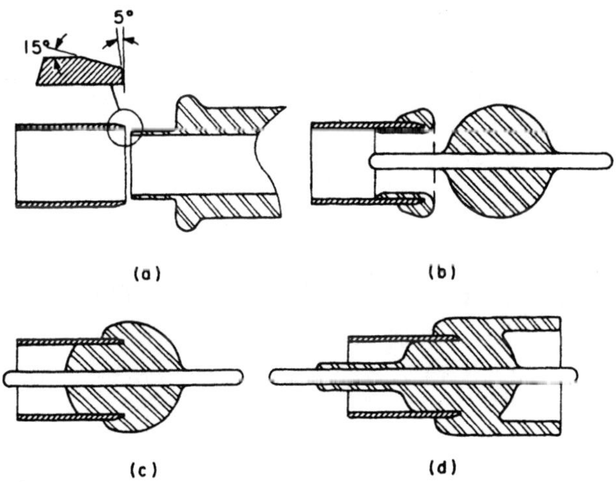

Fig. 4.24 Stages in completing a Kovar to glass lead-through. Reproduced from Tasman[1215a] *(Courtesy of Pergamon Press)*

melted onto the outer surface (Fig. 4.24b). A bead of the same glass is melted onto the Kovar rod and is inserted into the seal. The finished seal may have the shape shown on Fig. 4.24c for voltages up to 2000 V. For high voltages (up to 6000 V) the shape of Fig. 4.24d is recommended.

Tasman[1215a] also describes a lead-through using ceramic-to-metal seals (Fig. 4.25). The insert shows the details of a good way of brazing the lead to the cap, by means of a tapered copper plug and eutectic silver–copper alloy.

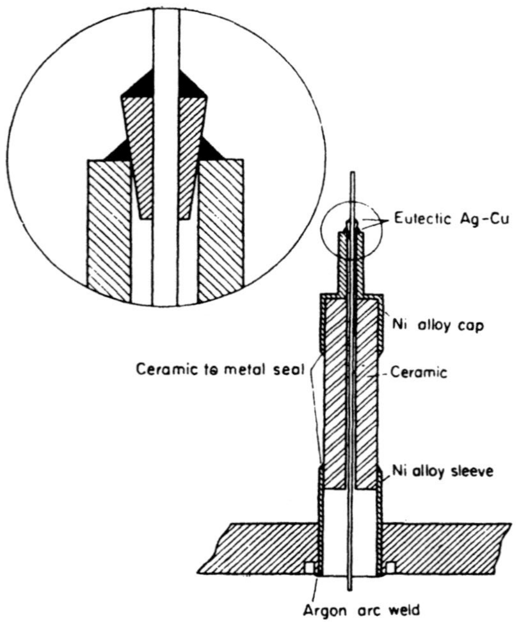

FIG. 4.25 Lead-through with ceramic-to-metal seal. Reproduced from Tasman[1215a] *(Courtesy of Pergamon Press)*

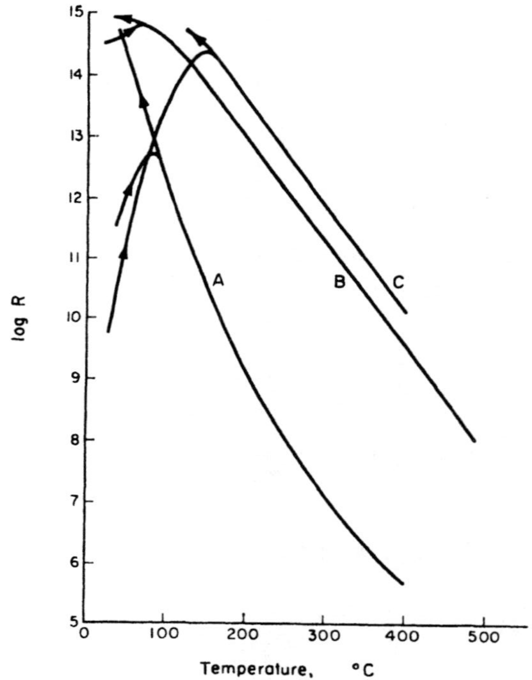

FIG. 4.26 Resistance of electrical lead-throughs as a function of the temperature. *A* Kovar–glass seal as in Fig. 4.24.c, d; *B* ceramic–metal seal as in Fig. 4.25; *C* seal with intermediate ring of high resistance glass. Reproduced from Tasman[1215a] *(Courtesy of Pergamon Press)*

The results of the measurements (Tasman[1215a]) of the resistances of these seals at various temperatures are plotted in Fig. 4.26. It can be seen that the resistance shows an initial rise with increasing temperature; this is due to the removal of an adsorbed moisture film (see Section 41.12). With ceramic insulation higher resistances were obtained than with a simple Kovar to glass seal. The curve *C* (Fig. 4.26) was obtained with Kovar to 28 (Philips) glass, using an intermediate ring of highly insulating (18 Philips) glass.

4.3 DEMOUNTABLE LEAD-THROUGHS

For various applications, demountable lead-throughs are required. In such cases, the electrical lead-through may be either dismountable by separating the electrical lead from the insulating (and sealing) part and this part from the vacuum vessel, or being able to separate the electrical lead-through as a whole from the vacuum vessel. In both cases the sealing and insulation, or just the sealing is done either by wax (resin) seals (Section 43.1) or by gasket seals (Section 43.2). As commercially available electrodes include the various insulation and sealing techniques, examples of such lead-throughs are presented separately in Section 43.3.

43.1 Waxed and resin sealed lead-throughs

For temporary use, electric leads may be sealed through vacuum enclosures, by using sealing waxes (Table 3.1). The wax may constitute only the vacuum seal or the seal and the insulation. Waxed seals have the drawbacks discussed in Section 3.1, but they are useful especially when continuous leads (thermocouple connexions) are to be taken through the walls of vacuum systems. A seal of this type is described by Scott[1114] (Fig. 4.27a) in which the wire (2) is placed to cross the wax (1) filling the groove provided on the end of the metal tube. In this arrangement the wax provides the insulation, as well as the vacuum sealing of both the electrical lead and the pipes (bell jar). Scott[1114] claims that wires insulated with enamel and silk can be used if the wax is kept fluid until it penetrates the silk insulation.

In some cases the wax provides the vacuum seal and only partly the insulation (Fig. 4.27b), the main insulation being provided by another material e.g. mica (Dufour[305a]).

Glass–metal seals connected to ground joints or simply tapered joints may be sealed with wax (Jouanigot[647a]).

Loctite (see Table 3.2) can be used to seal electrical leads through metal plates. This requires holes 0.1 mm larger in diameter than the wire passing through the hole. The wire is inserted through the hole, the space between the wire and the plate is filled with Loctite, and allowed to harden overnight (Reeber[1041]).

In order to seal electrical leads in Perspex, Lucite or Plexiglas, the lead is first immersed (10–15 min) in a hot acetic acid solution (50 per cent water) or in boiling glacial acetic acid, and then coated with a solution of Lucite in glacial acetic acid, and dried at 45° C. The coated wire may then be sealed into a hole in the Lucite plate (Giauque[447]).

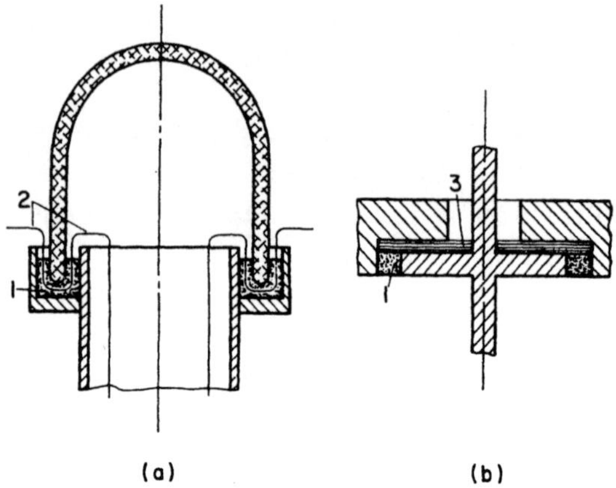

FIG. 4.27 Wax sealed lead-throughs: (a) continuous lead wire; (b) rod with machined or brazed disc; (1) wax; (2) lead-wire; (3) mica

Obviously, stronger and more reliable seals can be made using epoxy resins (Section 3.3). For this the electrical leads, previously thoroughly cleaned, are dipped in the epoxy resin (Balain[74], Lipson[770b]) and then sealed into the hole provided through the wall, or they are moulded in epoxy resin (Gale[419], Franke[398a], Noggle[935], Boersch[143]) and the assembly is sealed to the vacuum vessel by some other method (e.g. gasket seal, see Fig. 4.28c). When many wires are to be moulded into epoxy resin, they must be arranged to protrude through the bottom of the mould, spaced so that they cannot short circuit during the curing; for this purpose the wires are held in a split rubber stopper (Noggle[935]) closing the bottom of the pipe in which the mould is made. Balain[73] used a thin Lucite plate with proper holes, to space the wires.

Rogozinski[1074a] described a lead-through (Fig. 4.28a) having a central electrode (1) and a guard ring (2) sealed with Araldite into a cylindrical body (3). The shape of the parts is designed to generate compression stresses on the Araldite.

In some applications (as the accelerator tubes used in Van de Graaf generators) the resin is damaged on the inside of the seal if it protrudes or is exposed to the space with radiations. Miranda[869] suggests sealing the electrical leads for such applications, by applying a measured quantity of resin, and by providing the electrode with a groove (Fig. 4.28b). In this arrangement, only the part outside the groove is wetted, and the groove is of such dimensions as to take up the excess of resin squeezed inwards during the cementing process. Boersch[143] used lead-throughs insulated with Araldite (Fig. 4.28c) and sealed with rubber gaskets.

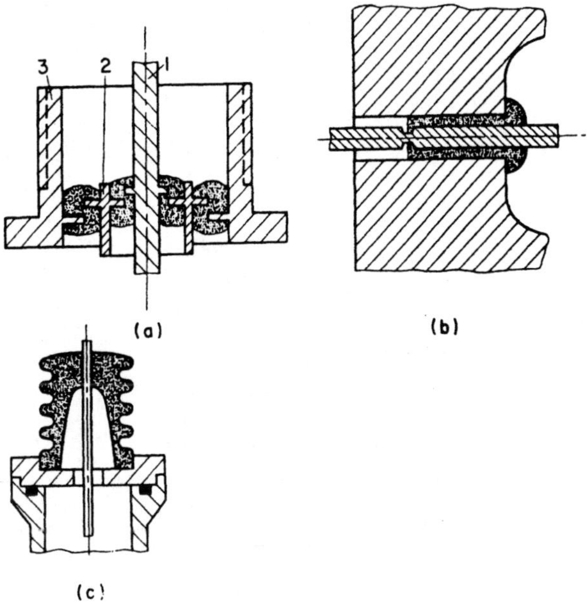

FIG. 4.28 Lead-through sealed with epoxy resin: (a) compression seal. After Rogozinski[1074] *(Courtesy of Société Française des Ingénieurs et Techniciens du Vide)*; (b) lead-through constructed to avoid radiation damage (after Miranda[809]); (c) Araldite insulator sealed with O-rings (after Boersch[143])

43.2 Gasket-sealed lead-throughs

Lead-throughs are sealed and insulated or only sealed, using the various forms of gasket seals (Section 3.8). In applications where gasket seals are used on pipes, ports, etc., there is no reason why the electrical lead-throughs should not be sealed in the same manner. When higher voltages or frequencies are used, the electrical insulation should be assured (even in gasket seals) by materials having appropriate insulating properties (see Figs. 4.1–4.5). To avoid excessive temperatures, the lead should have sufficient cross section (see Fig. 4.7).

Low-current electrical leads (wires up to about 0.5 mm diameter) may be sealed in various simple ways, using gasket seals (Fig. 4.29). If flat elastomer gaskets are to be used (Fig. 4.29a), the wires should be spaced so that they emerge radially between the bolts. By using vacuum grease on the gaskets this assembly can be made vacuum-tight (Scott[1114]). Without the grease, small channels are left along the sides of the wires. Fuschillo[417] suggests a technique of sealing lead-wires through rubber tubes (Fig. 4.29b, c). The wires (3) carefully separated, brought through the rubber tube (1) which is filled with vacuum grease (4). The seal is formed by compressing the rubber tube in a clamp (5). The system is especially useful for wires which must be

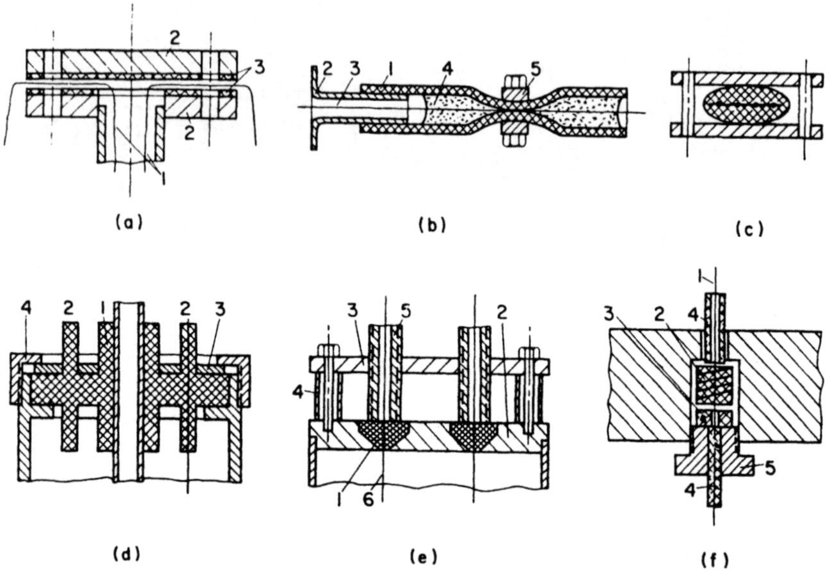

FIG. 4.29 Gasket seals of electric lead-through wires: (a) using a flat rubber gasket: (1) electric lead; (2) flanges; (3) rubber gasket; (b) with a clamped rubber tube (1); (2) glass or metal tube; (3) electric lead; (4) vacuum grease; (5) clamp. After Fuschillo[417] *(Courtesy of The Institute of Physics and The Physical Society, London)*; (c) cross section of seal b; (d) cylindrical seal, (1) rubber; (2) lead wire; (3) compression plates; (4) compression nut; (e) conical gasket seal, (1) gasket; (2) base plate; (3) support plate; (4) spacer; (5) compression screw; (6) lead wire. After Stamper[1173] *(Courtesy of The American Institute of Physics)*; (f) rubber spiral seal; (1) lead wire; (2) Neoprene spiral; (3) insulating spacer; (4) insulation sleeve; (5) compression screw. After Prior[1011] *(Courtesy of The Institute of Physics and The Physical Society, London)*

continuous (thermocouples) and it has been used (Fuschillo[417]) with uncovered, enamelled or glass insulated wires, in systems operating at 2×10^{-6} torr. Using a rubber pipe of 1/2 in. diameter, 0.12 in. wall thickness and 3.5 in. length, Fuschillo[417] was able to take out six wires of 0.09–0.3 mm diameter.

A very satisfactory seal developed (Guthrie[490a, 491]) for the entry of tubings and electrical leads into the vacuum system as shown in Fig. 4.29d. Riggs[1061] sealed (0.25 mm) gold wires by inserting them through holes in a fired soap-stone plug and between two semicylinders of fired soap-stone (steatite). The plug was then pressed into a brass fitting which was clamped against a Neoprene washer on the wall of the vacuum device. A steel ring held the two semi-cylinders so that the two Neoprene semi-circles between the semi-cylinders and the cylindrical plug bulged against the gold wires with just sufficient force to make a good seal.

In the seal described by Stamper[1173] (Fig. 4.29e), the wire passes through a flexible gasket* (silicone rubber) which fits snugly into a tapered hole in the plate. The wire passes axially through a hollow screw which allows pressure to be applied to the gasket. Prior[1011] uses a Neoprene gasket cut from a cord (about 6 mm long, and 6 mm diameter). To enable thin wires to be readily inserted through the axial hole of such a gasket, a helical cut (of a pitch

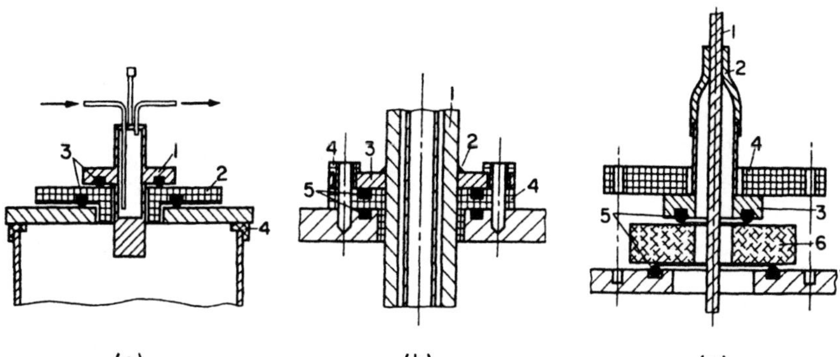

Fig. 4.30 Double O-ring (groove) seals for electric lead-throughs: (a) vertical lead-through, sealed by atmospheric pressure. After Stroud[1208] *(Courtesy of Pergamon Press)*; (b) bolted O-ring seal, (1) electrode; (2) brazed joint; (3) flange; (4) Teflon insulator; (5) O-rings (after Dickinson[279]); (c) flexible seal. After Hoch[569] *(Courtesy of The American Institute of Physics)*

approximately 2.5 mm) was made with a razor blade from the outside to the centre of the gasket, while the gasket was on a needle crossing it axially. It is then possible to unwind the Neoprene from the needle, and the wire can be threaded through the gasket by screwing the gasket onto the wire after making a right-angle bend on the wire. The helical cut (Fig. 4.29f) appears to have no detrimental effect on the vacuum-tightness of the gasket (Prior[1011]). After inserting the wire through the gasket, this is later compressed by a suitable screw, the wire being insulated by glass or ceramic sleeves and spacers (Fig. 4.29f).

O-ring seals are extensively used to seal electrical lead-throughs. Almost all the O-ring seals (see Section 38.4) may be applied to lead-throughs, the most commonly used being the grove seal (Section 38.41), with a single or double O-ring (Fig. 4.30).

* The rubber gasket can be formed in two steps. A rubber plug of about half the height of the hole is inserted first. With the leads placed through this plug, the space above it is filled with a rubber sealant (J. Stamper, Personal Communication).

A lead-through for low voltage and high current, where the seal is completely dismountable was described by Edwards[325] (see Fig. 3.62b). Some lead-throughs sealed with a double O-ring seal are illustrated on Fig. 4.30. Stroud[1208a] used a seal in which the water cooled electrode, is provided with a flange (1, Fig. 4.30a) holding an O-ring (3) in a groove; this O-ring seals against an insulator (2) made of Polythene or Teflon, which seals with a second O-ring onto the metal plate. In this case the metal plate is sealed to the vacuum vessel by means of an L gasket (4). Dickinson[279] used a lead-through of a similar construction (Fig. 4.30b) for use on a furnace to 2500 °C. Hoch[569]

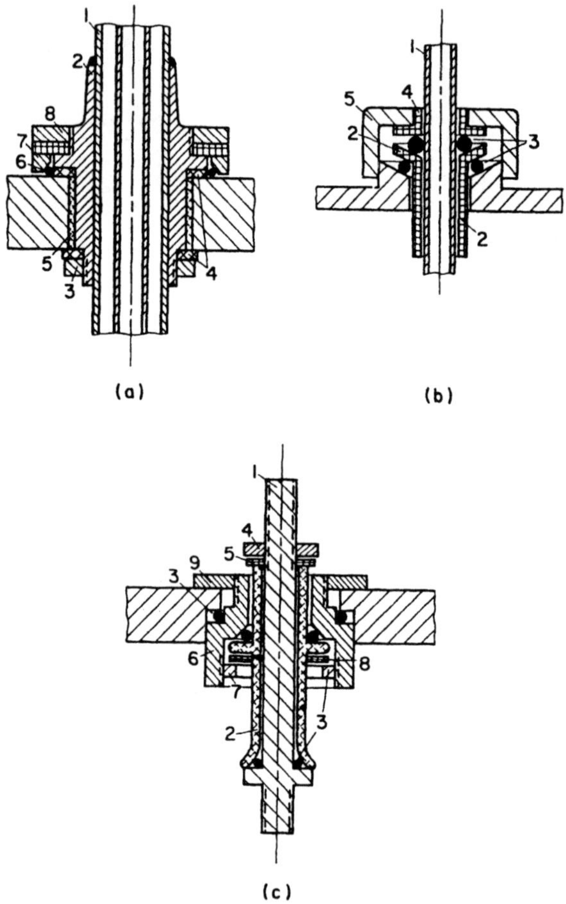

FIG. 4.31 Conical O-ring seals for lead-throughs: (a) seal with glass tube and insulation washer. After Moore[884] *(Courtesy of The Institute of Physics and the Physical Society, London)*; (b) seal with Teflon bushing, (1) electric lead; (2) Teflon bushing; (3) O-ring; (4) insulation washer; (5) closing nut. After Gandy[423] *(Courtesy of The American Institute of Physics)*; (c) seal with profiled glass tube, (1) electric lead; (2) glass tube; (3) O-rings; (4) nut, to close the lower O-ring; (5) washer; (6) intermediate body; (7) nut, to close the medium O-ring; (8) washer; (9) nut, to close the external O-ring *(Courtesy of* Edwards[329]*)*

describes a flexible lead-through seal (Fig. 4.30c) consisting of the electrical lead (1) sealed by an elastic connexion (2) to the ring (3). The ring (3) is sealed by an intermediate glass spacer (6) and the double O-ring seal (5) to the wall of the vacuum vessel, the assembly being held together by the clamping system (4).

Spacer seals (Section 38.42) or step seals (Section 38.45) can also be used to join lead-wires (rods) to vacuum systems (Smith[1153], Reilly[1044], Pesterfield[986], Kramers[716]). An electrode joined to the vacuum system by a seal using indium wire gasket is described by Holland[584].

Conical seals (Section 38.43) are commonly used especially in connexion with small size lead-throughs. Some constructions of lead-through sealed with conical seals using O-rings are illustrated in Fig. 4.31. These seals may be made using one (Fig. 4.31a), two (Fig. 4.31b) or three (Fig. 4.31c) O-rings, straight (Fig. 4.31a) or profiled electrodes, or insulators (Fig. 4.31c), or bushings (Fig. 4.31b). The construction shown on Fig. 4.31a (Moore[884]) has a

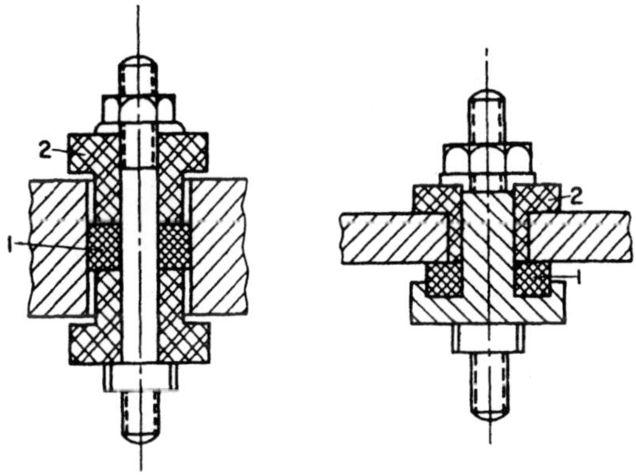

(a) (b)

FIG. 4.32 Lead-throughs, sealed with flat rubber gaskets: (a) seal with enclosed gasket, 1) rubber gasket; (2) ceramic parts; (b) seal having the gasket outside, (1) rubber gasket; (2) insulating part

profiled metal bush (2) brazed to the electrical lead (1) and joined to the wall of the vacuum vessel by the nut (3). Mica washers (4) and a glass tube (5) insulate the lead-through. The seal is made by means of the O-ring (6), by the insulated ring (7) and the nut (8). This seal may be applied on a straight hole in the wall of the vacuum chamber. The construction of Fig. 4.31b (Gandy[423]) is very simple, but needs a machined part on the wall of the vacuum chamber. The construction of Fig. 4.31c (Edwards[329]) needs a simple step in the wall of the vacuum vessel, but a profiled glass part which insulates the lead-through and seals outside and inside with O-rings in conical seals.

Where the electrical lead-through is insulated with ceramic spacers or spacers of other insulating materials (Ardenne[47], Zobac[1356]) it may be sealed using simple flat rubber gaskets (Fig. 4.32).

De Villiers[1267] joined a Kovar–glass terminal by screwing down the Kovar skirt (5) of the terminal onto an aluminium foil gasket (5 Fig. 4.33a) of about 0.25 mm thickness, by means of a specially shaped ring (4) and a tubular screw (2). The lead-through shown in Fig. 4.33b (Pockman[995]) is sealed

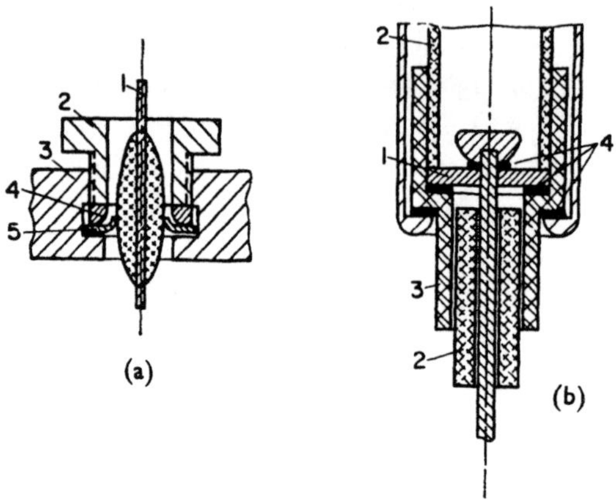

FIG. 4.33 Lead-throughs sealed with flat metal gaskets: (a) seal with aluminium foil, (1) Kovar; (2) tubular screw; (3) vacuum chamber; (4) sealing ring; (5) aluminium foil (after de Villiers[1267]); (b) seal with lead gaskets, (1) steel disc; (2) loose fitting glass tube; (3) porcelain or glass; (4) lead (Pb) gaskets. After Pockman[995] *(Courtesy of The American Institute of Physics)*

entirely by lead (Pb) gaskets, and insulated by glass or ceramic tubes and spacers.

The flat rubber gasket (Fig. 4.34a) having the lead-in wires sealed radially into it, was suggested by Taylor[1217]. Any metal may be used as the conductor, but according to Taylor[1217] silver, copper or nickel ribbons or tubings plated with brass have been found to give best adhesion to the rubber. The gasket is vulcanized in a mould with the metal inserts in place.

Greenland[480a] suggested using a metal ring (Fig. 4.34b) interposed between the rim of the bell jar and base-plate. The ring is provided with radial holes in which insulated electrodes may be sealed. The ring itself is sealed towards the bell jar and the base-plate with L-gaskets. Consolidated Vacuum Corp.[231] supplies feed-through rings based on this principle, for bell jars of 14, 18 and 24 in. diameter. These rings are provided with a grooved O-ring seal on their bottom, and the radial ports are shaped so as to be sealed with the CVC Con-O-Ring gasket seals (an O-ring held between two concentric aluminium rings, see also Section 38.42).

Turner[1242] used an electrode ring, in which commercially available glass-to-metal seals were sealed using indium wire gaskets; the electrode ring itself was sealed to the vacuum system by indium wire seals.

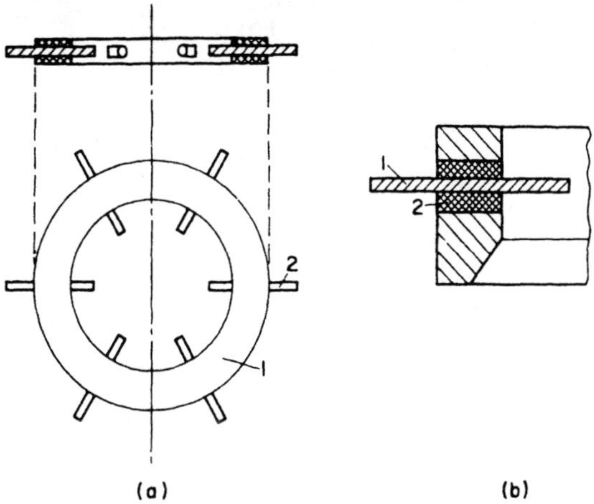

FIG. 4.34 Radial lead-throughs: (a) in rubber ring, (1) rubber; (2) metal insert. After Taylor[1217] *(Courtesy of The American Institute of Physics)*; (b) in metal ring. After Greenland[480a] *(Courtesy of The Institute of Physics and The Physical Society, London)*

Electrical lead-throughs may be sealed with simple cylindrical seals, in which the metal rod is just forced through a cylindrical rubber plug (Herne[544]), or using cylindrical rubber gaskets as in Fig. 4.35 (Edwards[329]).

Figure 4.36 illustrates some lead-throughs using conical seals (Section 38.52). If a conical rubber is used (Fig. 4.36a) it is recommended that the rubber plug (2) be compressed by a disc (3) instead of basing the seal just on the electrical lead (1) forced through the stopper (D'Eustachio[276a]).

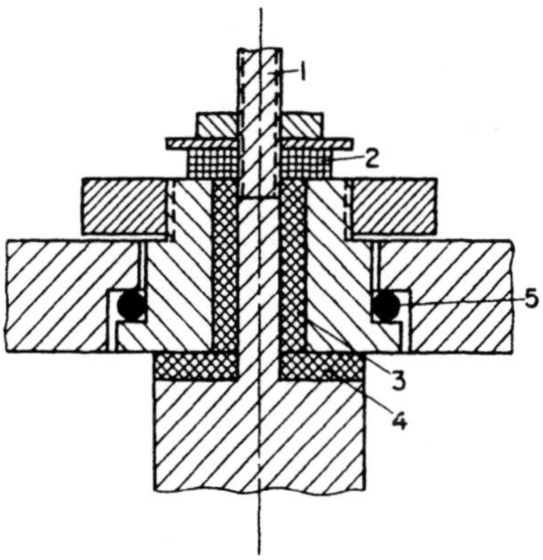

FIG. 4.35 Lead-through, sealed by a cylindrical gasket: (1) electric lead; (2) insulation washer; (3) cylindrical rubber; (4) flat rubber gasket; (5) O-ring (dimensions of the lead-through see Fig. 4.38)

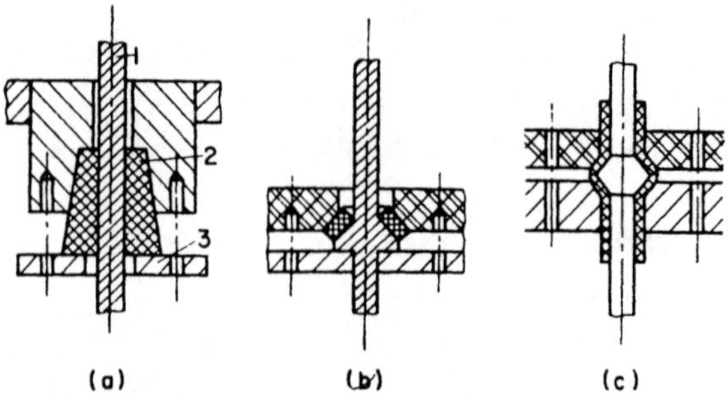

FIG. 4.36 Lead-through, sealed by conical seals: (a) with rubber stopper (after D'Eustachio[276a]); (b) with flat rubber gasket compressed to conical; (c) with rubber tubing, slipped over the lead-through

Figures 4.36b and c illustrate conical seals where the lead-throughs are provided with tapered parts which should ascertain the sealing, by pressing against a flat annular gasket (Fig. 4.36b) or using a rubber tubing slipped over the lead (Fig. 4.36c).

In order to minimize the possibility for arc-over when the seals are used in a poor vacuum, Bron[170a] and Knight[696a] designed lead-through seals using Teflon bushings which project into the vacuum chamber.

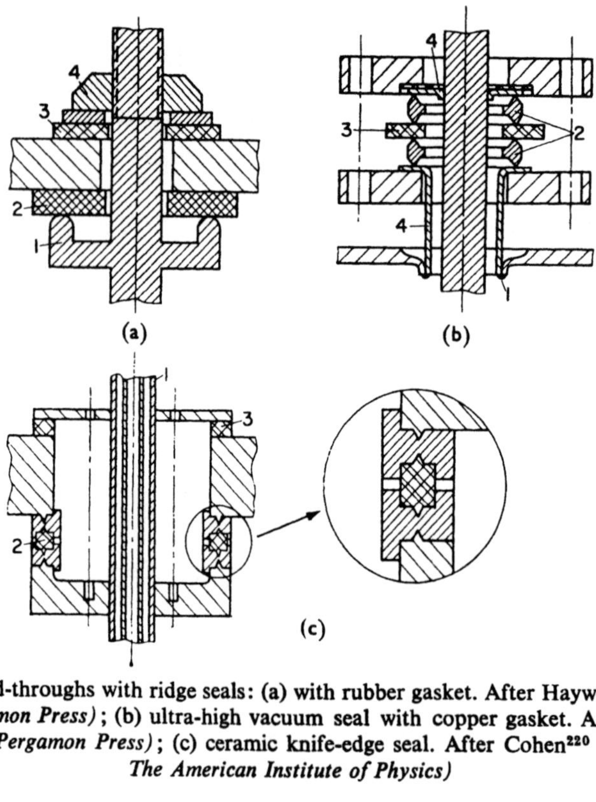

FIG. 4.37 Lead-throughs with ridge seals: (a) with rubber gasket. After Hayward[524] *(Courtesy of Pergamon Press)*; (b) ultra-high vacuum seal with copper gasket. After Ehlers[331] *(Courtesy of Pergamon Press)*; (c) ceramic knife-edge seal. After Cohen[220] *(Courtesy of The American Institute of Physics)*

Hayward[524] describes a simple ridge seal (Section 38.56) used in connexion with an electrical lead-through (Fig. 4.37a). The ridge (1) seals against a flat rubber gasket (2) tightened by the nut (4). On the opposite side, the insulation is ensured by the mica washer (3).

Dismountable lead-throughs for ultra-high vacuum systems are described by Moll[871]. Ehlers[332] shows a seal for lead-throughs (Fig. 4.37b) used in ultra-high vacuum systems with double-walled chambers and a guard vacuum (Section 38.23). The seal is welded at (1) on the ultra-high vacuum side, and at the guard vacuum side the copper gaskets (2) provided with ridges seal against the ceramic plate (3) and the sleeves (4). Cohen[220] described a water-cooled (copper) electrical lead (1 Fig. 4.37c) sealed with a double knife edge (2) (see also Fig. 3.148c) and insulated by the ceramic ring with the knife edge (2) and the ceramic ring (3).

43.3 Commercially available lead-throughs

Electrical lead-throughs are supplied by most of the manufacturers of vacuum equipment (Section 11.3). In order to provide data for the designer of vacuum seals requiring lead-throughs, the description of some commercially available lead-throughs is given.

Figure 4.38 shows the dimensions and the shapes of vacuum electrodes available from Edwards[329]. These electrodes are shown in Plate 16, their construction is illustrated in Figs. 4.31c and 4.35 respectively, and their dimensions are listed in Table 4.2.

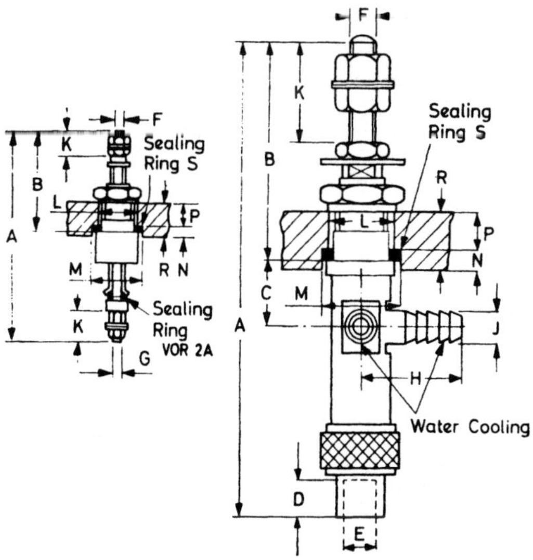

FIG. 4.38 Dimensions of vacuum sealed electrodes supplied by Edwards[327] (Table 4.2)

TABLE 4.2. DIMENSIONS OF VACUUM ELECTRODES FROM Edwards[329] (see Fig. 4.38)

Type	6*B*	7*B*	8	9
Rated voltage V	10.000	5.000	60	60
Rated current A continuous intermittent	 100 150	 15 20	 200 300	 300 400
A in. mm	6 1/2 165	3 1/4 83	6 5/8 168	8 203
B in. mm	3 1/4 83	1 1/2 37	3 5/8 92	3 3/4 95
C in. mm	– –	– –	– –	1 1/8 29
D in. mm	– –	– –	11/16 17.4	11/16 17.4
E in. mm	– –	– –	3/8 9.5	1/2 12.7
F in.	3/8 BSF	2BA	3/8 BSF	1/2 BSF
G in.	3/8 BSF	3BA	–	–
H in. mm	– –	– –	– –	1 3/4 45
J in. mm	– –	– –	– –	1/2 12.7
K in. mm	1 25.4	7/16 11.0	1 7/8 48	1 3/4 45
*L** in. mm	1.064 27.03	0.564 14.33	1.064 27.03	1.064 27.03
*M** in. mm	1.343 34.13	0.774 19.66	1.343 34.13	1.343 34.13
N_{min} in. mm	11/32 8.7	13/64 5.1	11/32 8.7	11/32 8.7
P_{max} in. mm	5/8 15.9	7/16 11.0	11/16 17.5	11/16 17.5
P_{min} in. mm	1/2 12.7	3/16 4.8	5/16 7.9	5/16 7.9
R_{min} in. mm	27/32 21.4	25/64 9.9	21/32 16.7	21/32 16.7
S (Table 3.19 A)	VOR 135	VOR 118	VOR 135	VOR 135

* Tolerance, plus 0.002 in. (0.05 mm), minus 0.

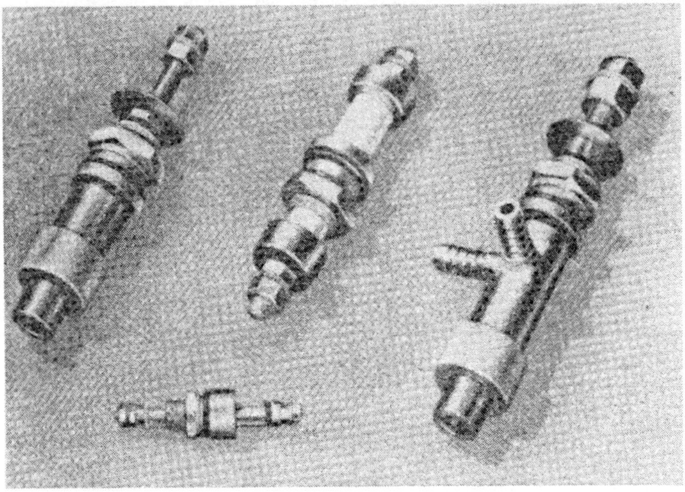

PLATE 16. Dismountable vacuum-sealed electrodes *(Courtesy of Edwards[328])*

The electrical lead-throughs supplied by Leybold[762] are illustrated in Fig. 4.39, and their characteristics are listed in Table 4.3.

TABLE 4.3. CHARACTERISTICS OF CURRENT LEAD-INS FROM Leybold[762] (Fig. 4.39)

Type (Fig. 4.39)		*a*	*b*	*c*	*d*	*e*	*f*	*g*
Insulation		Synthetic resin		Ceramic		Glass	Perbunan	
Poles		4	2	1	1	1	1	1
Maximum	Volts	220	220	3000	6000	3000	40	40
	Amps	1	15	15	30	120	250	2500
Remarks		*1	*2	*3	*4	–	water cooling	
Max. temperature (°C)		80 (120)	80 (120)	80 (150)	500	80 (200)	–	–

*1 The maximum temperatures in brackets are for Vitilan O-rings.
*2 The lead-in is tested for a leak rate of 10^{-5} torr. l./sec.
*3 Lead-in for Penning open-type gauge head.
*4 Electrical resistance at 20 °C is 5×10^{15} ohms at 400 °C is 5×10^{11} ohms.

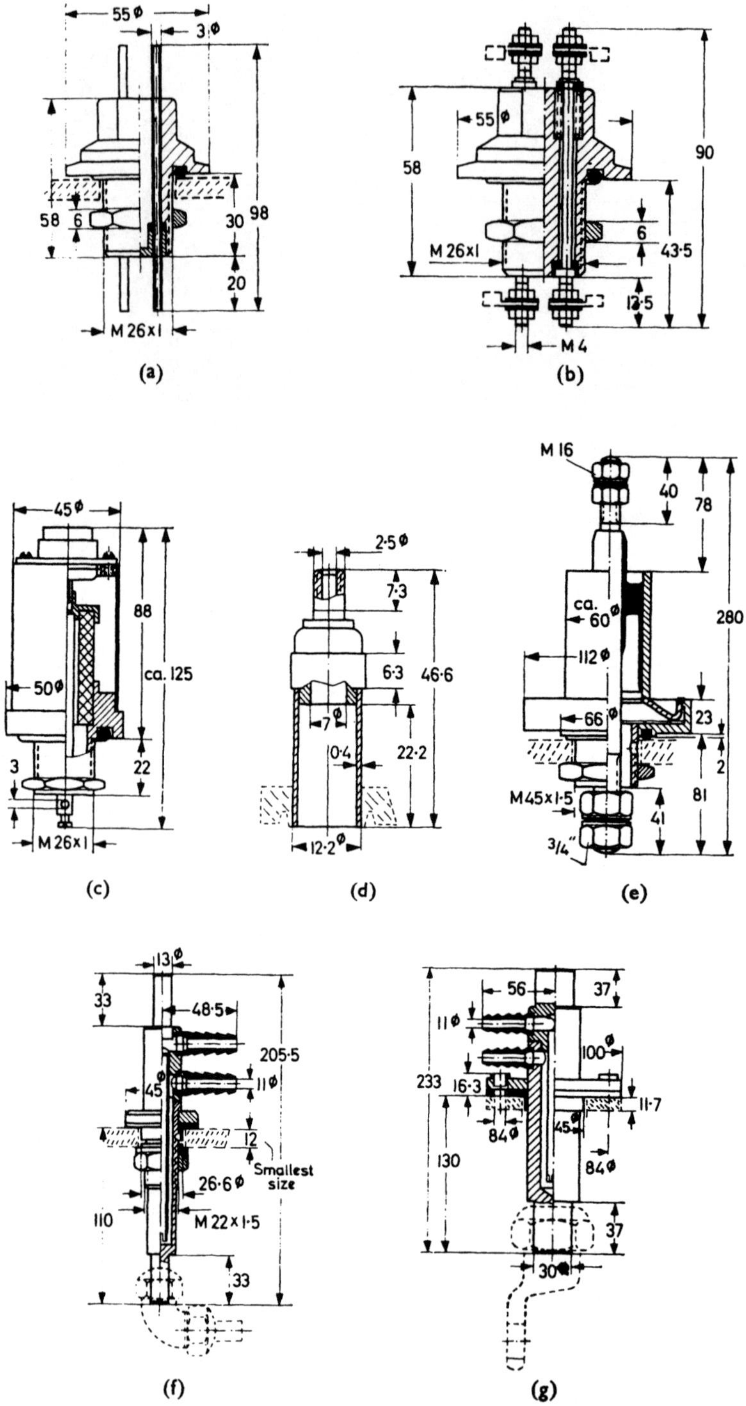

FIG. 4.39 Current lead-ins supplied by Leybold[762] (Table 4.3)

Figure 4.40a shows the electrical lead-through for the transmission of high-voltages supplied by Balzers[444] insulated with ceramic. The water-cooled high-current lead-through manufactured by Balzers[444] is illustrated in Fig. 4.40b and its dimensions are listed in Table 4.4.

The lead-throughs supplied by Pfeiffer[987] are shown in Fig. 4.41 and their dimensions are given in Table 4.5.

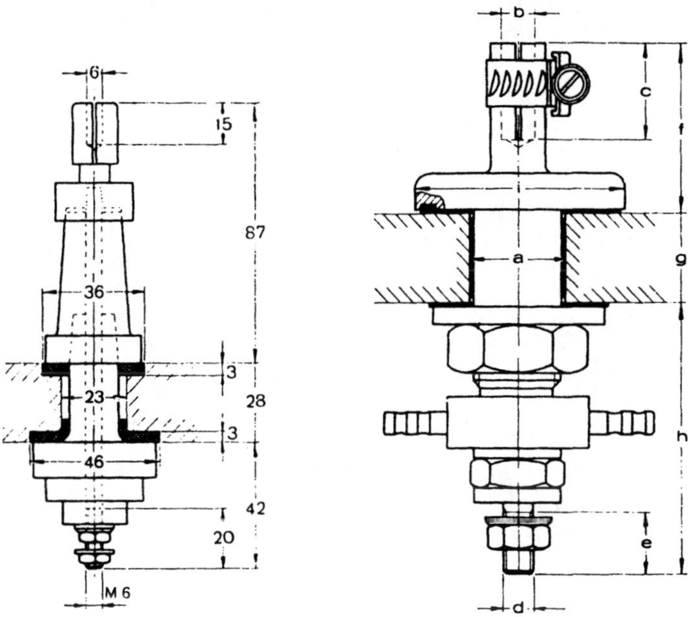

FIG. 4.40 Current lead-ins supplied by Balzers[444] (Table 4.4)

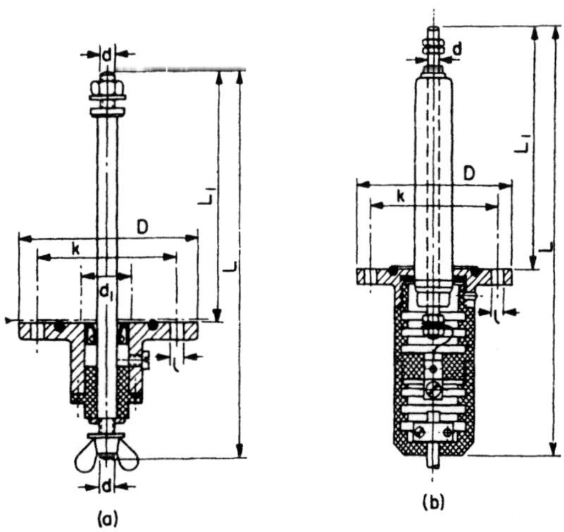

FIG. 4.41 Current lead-ins supplied by Pfeiffer[987] (Table 4.5)

TABLE 4.4. DIMENSIONS (MM) OF HIGH-CURRENT LEAD-THROUGHS FROM Balzers[444] (Fig. 4.40b)

Dimensions	*a*	*b*	*c*	*d*	*e*	*f*	*g*	*h*	*i*
Type 1	32	12	30	M12	20	55	28	87	70
Type 2	32	12	30	12	30	54	28	88	60

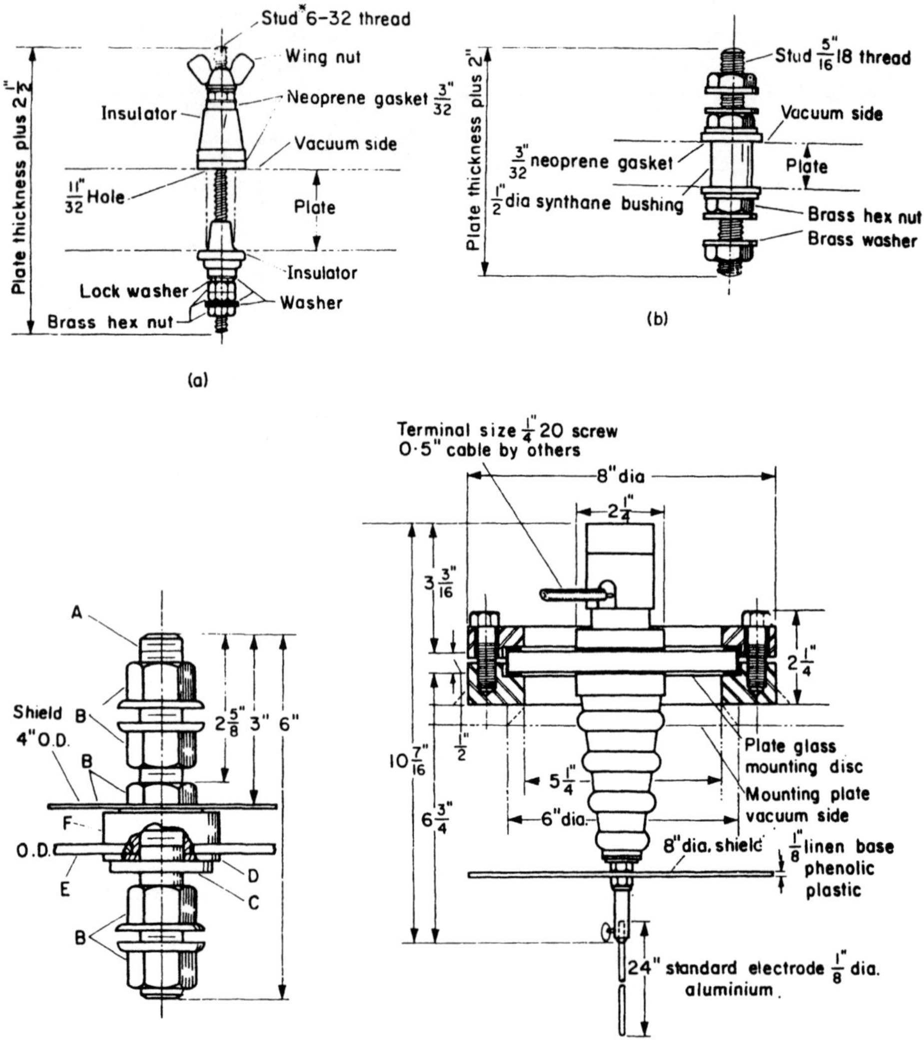

FIG. 4.42 Electric feed-throughs supplied by N.R.C.[912]: (a) power feed-through for 115 V, 10 A; (b) for 50 V, 100 A; (c) for 400 A; (d) for 7500 V, 0.2 A at 10^{-2} torr and 30000 V, 0.5 A (max) at atmospheric pressure

TABLE 4.5. DIMENSIONS OF ELECTRICAL LEAD-THROUGHS FROM Pfeiffer[987] (Fig. 4.41)

Type (Amps/Volts)	Dimensions (mm)							Remarks
	d	L	L_1	d_1	D	k	l	
200 A/ 220 V	M 8	185	120	30	88	70	7	Fig. 4.41a
1000 A/ 220 V	M20	330	155	40	130	100	14	Fig. 4.41a
500 A/ 40 V	M12	230	125	30	130	100	14	water cooled; Fig. 4.41a
1000 A/ 40 V	M16	345	155	32	120	90	14	Fig. 4.41a
2500 A/ 40 V	M20	360	160	40	130	100	14	Fig. 4.41a
1 A/10,000 V	M 6	220	125	30	88	70	7	Fig. 4.41b

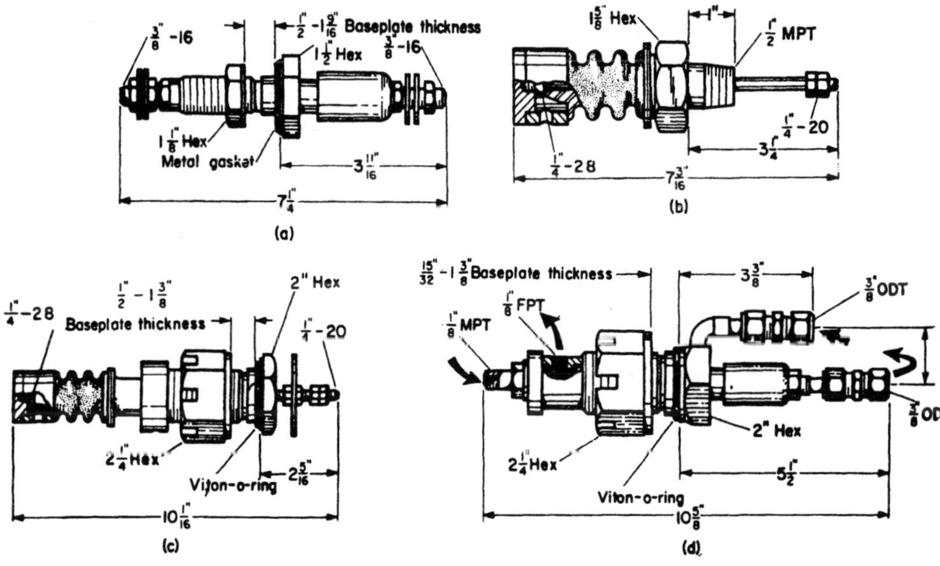

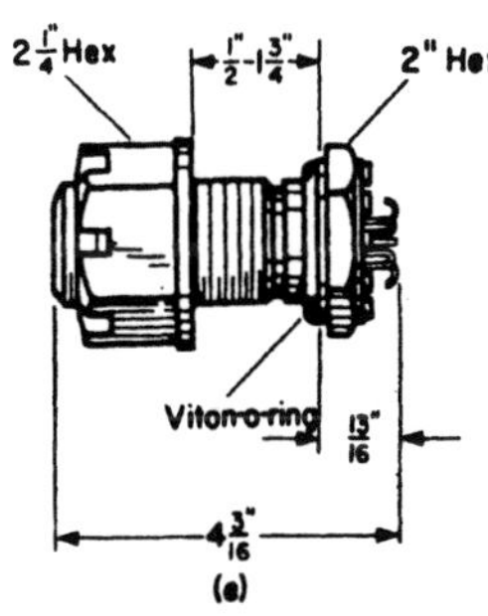

FIG. 4.43 Feed-through supplied by C.V.C.[231] rated (a) for 300 A continuous service or 400 A intermittent (20 per cent duty cycle, 3 min on, 12 min off) service; (b) for 25000 V, 10 A d.c.; (c) for 25000 V, 10 A d.c.; (d) water cooled h.f. electrode for 200 V, 10 kW, 465 kcycles; (e) multiple-pin electrode for thermocouple wire, supplied with two interchargeable headers; one having 8 wire feed-throughs rated at 5.5 A, 2500 V; the other having 8 sleeves with a minimum i.d. of 0.038 in. in which wire feed-throughs must be inserted

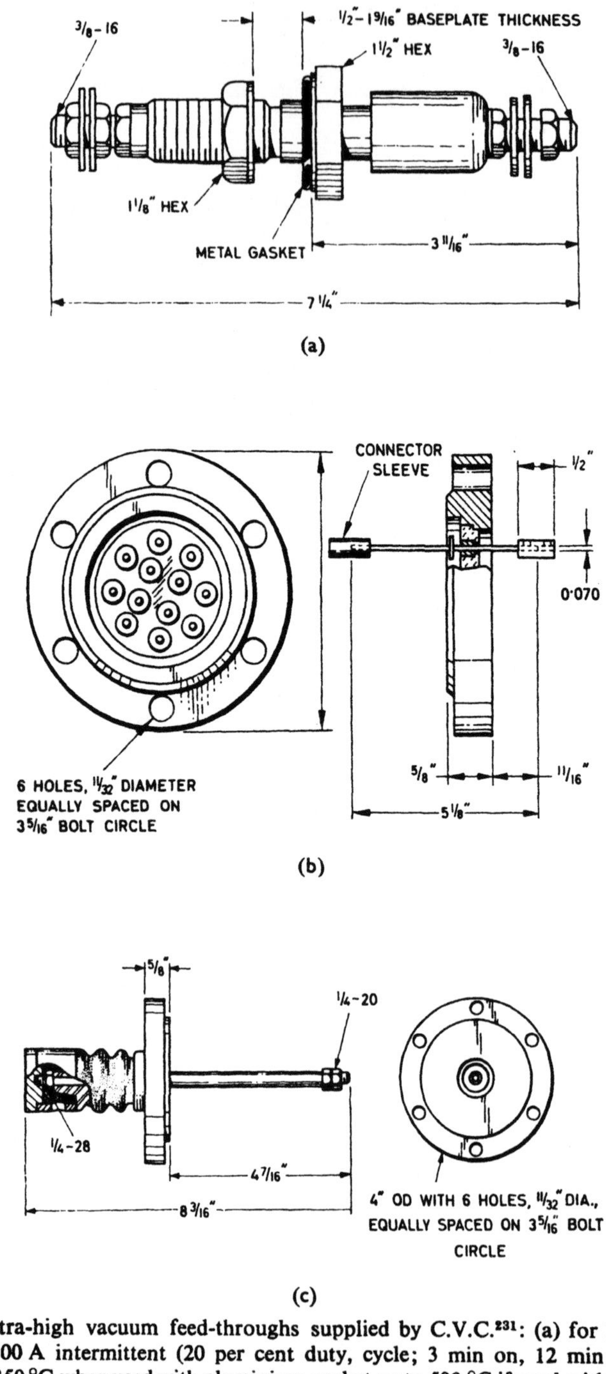

FIG. 4.44 Ultra-high vacuum feed-throughs supplied by C.V.C.[231]: (a) for 200 A continuous and 300 A intermittent (20 per cent duty, cycle; 3 min on, 12 min off) service; bakeable to 350 °C when used with aluminium gasket or to 500 °C if used with gold gasket; (b) multiple-pin, bakeable to 350 °C (aluminium gasket), or to 400 °C (gold gasket); (c) high voltage feed-through for 25000 V, 10 A d.c. bakeable as (a)

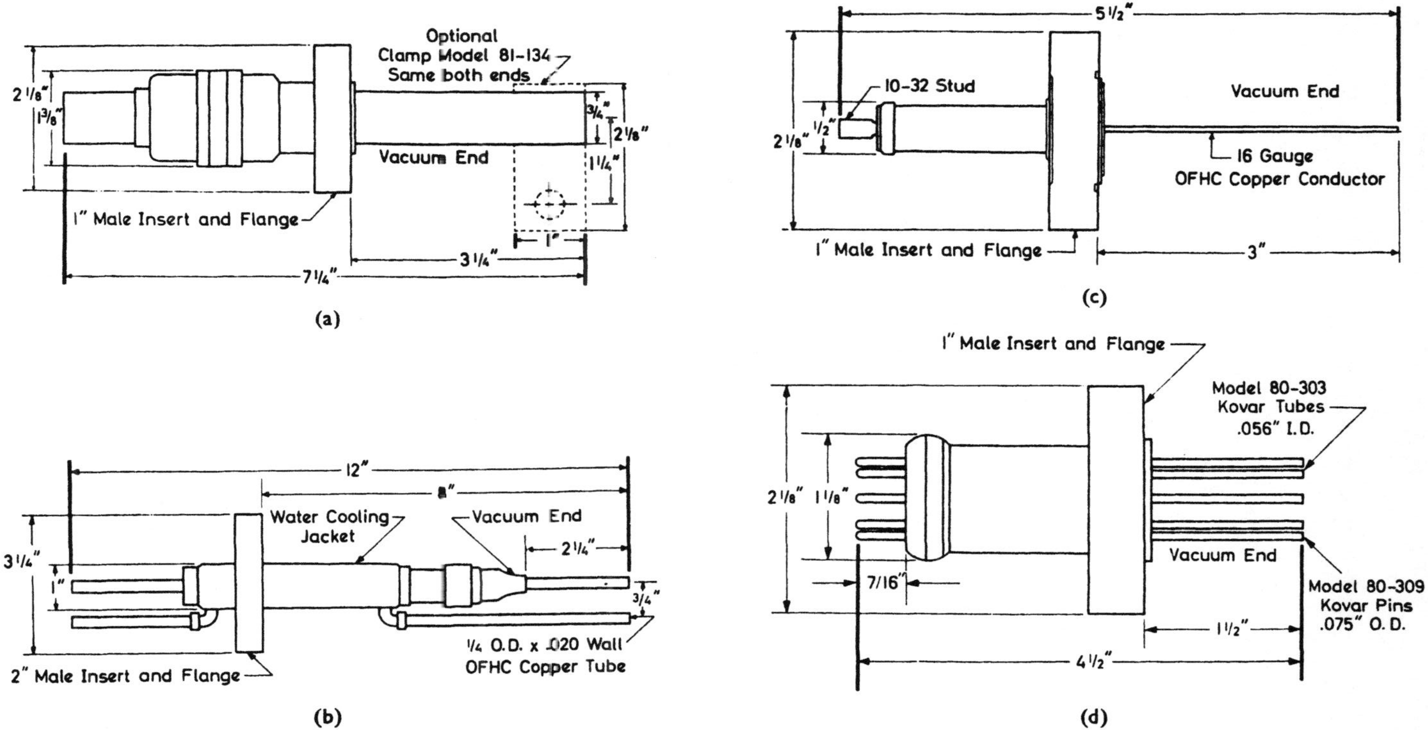

FIG. 4.45 Ultra-high vacuum feed-throughs supplied by Ultek[1251a]: (a) Kovar-to-glass seal for max. 500 A, max. 3000 V (below 10^{-4} torr), max. bakeout 400 °C, max. temperature change 25 °C/min; (b) r.f. feed-through with water cooling, using a Housekeeper Pyrex-to-copper seal, rated for max. 200 A, 3500 V RMS (400 Kc–1 Mc), bakeable to max. 450 °C, max. temperature change 25 °C/min; (c) high voltage feed-through with alumina insulator, for max. 5 A, 12000 V (below 10^{-4} torr), bakeable to max. 400 °C at a heating or cooling rate of 25 °C/min; (d) multiple-pin feed-through with Kovar pins or 0.056 in. i.d. Kovar tubes; for max. 1000 V (below 10^{-4} torr); max 5 A per pin; bakeable to 400 °C as (c)

The electrical lead-throughs supplied by NRC Equipment Corp.[912] are illustrated in Fig. 4.42.

Figure 4.43 shows the electrical lead-throughs supplied by Consolidated Vacuum Corp.[231] and Fig. 4.44 the electrical lead-throughs for ultra-high vacuum supplied by the same firm.

The electrical lead-throughs supplied by Ultek[1251a] are illustrated in Fig. 4.45.

CHAPTER 5

THE TRANSMISSION OF MOTION THROUGH SEALS

5.1 MECHANICAL TRANSMISSION

51.1 Classification

The transmission of motion from the outside into the vacuum chamber is a very frequent requirement in the various vacuum technologies. The most frequent purpose requiring the transmission of motion inside the evacuated space is obviously the actuation of the closing systems (Section 6.1) of stopcocks, cut-offs or valves. Other technological purposes include: the relative motion of parts inside the vacuum system (Smith[1154]), the motion of shutters (Gray[471], Shapiro[1125]), the rotary motion of targets (Croes[247], Fuschillo[416]) or the tilting of crucibles in vacuum (Matthews[826]).

The techniques used to transmit movement into the evacuated space may be classified according to the kind of energy utilized in mechanical transmission (Section 5.1), magnetic transmission (Section 5.2) and electrical (heat) transmission (Section 5.3).

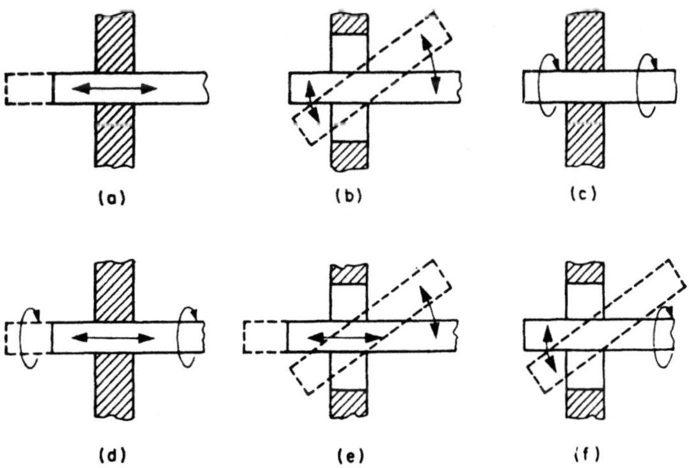

FIG. 5.1 Transmission of motion into evacuated chambers: (a) sliding; (b) tilting; (c) rotation; (d) sliding and rotation; (e) sliding and tilting; (f) tilting and rotation (wobbling)

It is required and possible to transmit movement by sliding, tilting, rotation, sliding and rotation, sliding and tilting, tilting and rotation or sliding, tilting and rotation. Figure 5.1 illustrates the various kinds of movement and Table 5.1 summarizes the cross references to the various examples of the techniques described in this Chapter.

The techniques used to transmit the movement into the evacuated space are: (a) the tilting of the vacuum device (Section 51.2); (b) the bending of elastic pipes (Section 51.3); (c) the deformation of bellows (Section 51.4); (d) the deformation of diaphragms (Section 51.5); (e) the relative motion of ground seals (Section 51.6); (f) the motion of gasket seals (Section 51.7); (g) the motion of guard vacuum seals (Section 51.8); (h) the use of magnetic fields (Section 5.2); (i) the use of heat transfer or electric current (Section 5.3).

TABLE 5.1. TECHNIQUES FOR TRANSMITTING MOVEMENT INTO AN EVACUATED SPACE

Technique	Kind of movement						
	Sliding	Tilting	Rotation	Sliding and tilting	Sliding and rotation	Rotation and tilting	Sliding, rotation and tilting
Elastic pipes	Fig. 5.2a	Fig. 5.2d, b 5.3	Fig. 5.2c 5.3A	Fig. 5.2b	Fig. 5.2a	Fig. 5.2b	Fig. 5.2b
Bellows	5.5 5.7	5.8 5.9	5.10	—	—	—	—
Diaphragms	5.13 5.14	5.15	5.15	—	—	5.15	—
Ground seals	5.17	5.16c 5.18a	5.16	5.18a	—	5.16c 5.18a, b	5.18b
Gasket seals	5.19 5.21	—	5.20, 5.22 5.23, 5.25 5.26, 5.30–-5.41	—	5.20 5.24	5.27 5.29	5.25a Table 5.13
Guard vacuum	5.49		5.26, 5.50 5.51	—	—	—	—
Magnetic	Table 5.14	—	5.54 5.55	—	—	—	—
Electric	5.56	5.57	—	—	—	—	—

A classification of the vacuum seals used to transmit rotation into the evacuated space is presented in Table 5.1A.

TABLE 5.1A SEALING TECHNIQUES FOR ROTATION IN VACUUM*

Group of seal	Principle	Type of seal	Kind of seal	Pressure range (torr)	Remarks
Shaft penetration seals (uninterrupted shaft; leakage occurs through slit between rotating and stationary elements)	Sliding contact (rubbing)	Solid-to-solid contact	O-ring	$760–10^{-5}$	Fig. 5.20 1*
			Chevron (lip)	$760–10^{-5}$	Fig. 5.30
			Face seal	$760–10^{-6}$	Fig. 5.33
		Solid-to-liquid contact**	Surface tension	$10–10^{-7}$	Fig. 3.39
			Fluid head (mercury)	$760–10^{-3}$	Fig. 5.49
			Slinger	$760–10^{-5}$	2*
			Visco	$760–10^{-7}$	3*
	Clearance	Solid-to-gas contact and controlled pressure difference**	Centrifugal pump	$760–10^{-11}$	
			Molecular pump	$10–10^{-11}$	4*
			Labyrinth seal	$760–10^{-11}$	5*
			Diffusion pump	$10^{-1}–10^{-8}$	6*
Hermetic seals (unbroken interface between vacuum and environment; leakage solely by diffusion through the solid material; require bearings in the vacuum)	Mechanical	Flexing element	Bellows	$760–10^{-7}$	Fig. 5.10
			Wobble plate	$760–10^{-6}$	Fig. 5.15
			Harmonic drive	$760–10^{-6}$	7*
	Field transmission	Stationary membrane	Electric fields	$760\ 10^{-9}$	limited to torque 2 kg. cm
			Permanent magnets	$760–10^{-9}$	Fig. 5.53, 5.54 8*
			Electromagnetic	$760–10^{-9}$	9*

* Classification reproduced from: W. Courtney, J. Lavelle, R. Britton and A. S. Denholm, Sealing techniques for rotation in vacuum, *Astronautics and Aeronautics*, *2* (No. 2), 40 (1964). *By Courtesy of Ion Physics Corporation.*

** The Slinger, Visco, Centrifugal and Molecular seals are speed dependent.

1* Limited to surface speeds up to 100 ft/sec.

2* Similar to the fluid seal; the gravitational pressure gradient is replaced by a pressure gradient associated with rotation of a fluid, which spins with the shaft. There is no sealing when the shaft is stationary.

3* The shaft has right-hand and left-hand threads cut on adjacent length. These cause pumping towards their junction as the shaft turns. Only one direction of rotation is permitted in a given seal.

[*Continued on page 494*

51.2 Transmission of motion by tilting the vacuum device

In certain cases it is possible to achieve movement of parts inside the vacuum chamber (device) by the physical manipulation of the sealed-off container, or of the container connected to the vacuum system by some flexible connexion (elastomer pipe, bellows). Transmitting motion in this manner into the evacuated space is an easy but inaccurate method; thus it has very limited use.

Shutters can be made to function in this way by tilting the entire container, and using the action of gravity on a counter-weight applied to the shutter (Bachman[65]).

The transmission of motion by tilting may also be applied to open glass gas containers under vacuum (Section 61.51).

51.3 Transmission of motion through elastic pipes

Elastic pipes were used in simple or sophisticated arrangements to transmit translation, tilting or rotation into vacuum chambers. Rubber pipes can be used if the gas evolution (Section 21.13) may be tolerated in the given application. Some simple techniques to transmit motion using elastomer pipes are illustrated in Fig. 5.2. Limited translational and rotational movement may by transmitted using a shaft (1, Fig. 5.2a) sealed by a rubber sleeve (2). The rubber tubing can be sealed to the body and the shaft just being slipped over (Fig. 5.2a) or joined by a wax to these parts (Fig. 5.2b).

Norbury[937] used the construction shown on Fig. 5.2b. A rotating member (1) passes through a metal tube (2) and rubber tubing 3 in. in length (3) with a wall thickness of 1/8 in. joined to the metal tube and to the shaft by wax joints (4, 5). The maximum torsion which may be obtained in such a construction is controlled by the length and the wall thickness of the rubber. Norbury[937] claims to obtain with the dimensions used by him a twist of 90°.

4* See e.g. C. E. Williams and J. W. Beams, *1961 Vac. Symp. Trans.*, Pergamon Press, Oxford, 1962, p. 295 and W. Becker, *Advances in Vacuum Science and Technology*, Pergamon Press, Oxford, 1960, p. 173.

5* Based on guard vacuum; see e.g. F. Arnold and W. K. Stair, *The Labyrinth Seal, Theory and Design*, Rep. M.E. 5-62-1, OTS.TID. 15986, March 1962.

6* A diffusion pump, where the body is steady and the jets are mounted on the shaft. This transfers the sealing problem from the high vacuum range to the backing pump range.

7* The harmonic drive utilizes the controlled elastic deflection of one or more parts for the transmission of mechanical motion through a sealed wall; see C. W. Musser, Harmonic Drive, *Machine Design*, **32** (April 14), 160 (1960), and Fig. 5.3A.

8* Radially aligned magnet design preferable for higher torques; a radial magnetic drive for 20 kg.m torque at 50 r.p.m. is described by D. A. Guerdon, Hermetic magnetic Couplings, *Product Engineering*, **29** (April 28), 96 (1958).

9* The rotor is placed in the evacuated chamber, the stator outside. They are limited to speeds up to 1000 r.p.m. by the high electrical losses occurring above these speeds; see. D. A. Guerdon, *Product Engineering*, 32 (Nov. 27), 58 (1961).

Ne vertheless rubber tubing may be used to obtain *unlimited* rotary motion, by using a crank (1 Fig. 5.2c), the handle of which is enclosed in the rubber tubing (2) in such a way that the rubber flexes on the crank handle. The end of the crank rotates freely in the stopcock (3) which seals the end of the rubber (Bachman[65]).

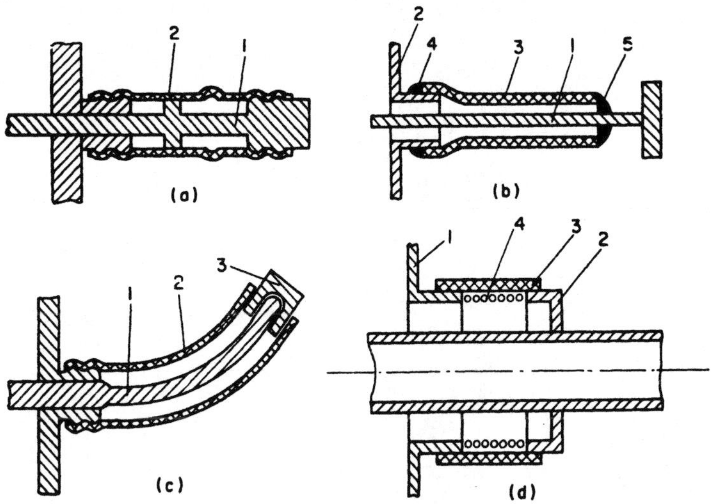

FIG. 5.2 Transmission of motion by means of elastic pipes: (a), (b) shaft with rubber tubing for limited translation and rotation (after Norbury[937]); (c) rubber tube seal for unlimited rotary motion; (d) rubber tube seal for tilting motion (after Fremlin[401])

In order to allow the tilting of the shaft, it may be connected as shown in Fig. 5.2d. A large diameter tube (1) connected to the vacuum chamber and a flange (2) on the moving shaft are joined by a length of rubber tube (3), which is prevented from collapsing due to the atmospheric pressure by a helical wire spring (4). Fremlin[401] recommends that the spacing between the turns of the helix be limited by "lacing" with a flexible plastic strip, to prevent the turns bunching during operation. The rubber tube should be clamped from outside (Section 38.53).

N.G.N. Electrical[926] supplies flexible couplings (Fig. 5.3 and Plate 17) consisting of a short length of vacuum rubber, having in the bore a set of moulded O-ring projections, which grip onto the tube when inserted into the coupling. A reduced bore section in the middle of the coupling acts as a stop for the tube ends. Having inserted the tubes the coupling may then be bent through an angle of 10 to 45° without the danger of losing the vacuum tightness.

Von Ubisch[1247] transmitted movement into a vacuum chamber by moving a steel rod placed with ample clearance inside a soft copper tube (about 1/4 in. diameter and 2–3 in. length) which can be twisted. Although such tubes may withstand an occasional twist of one or more revolutions, it is advisable to keep the rotation within 90°. Lange[738] found this method useful on ultra-high vacuum systems, provided that the rotation required is not more than about 30° and the number of cycles of operation is less than 25.

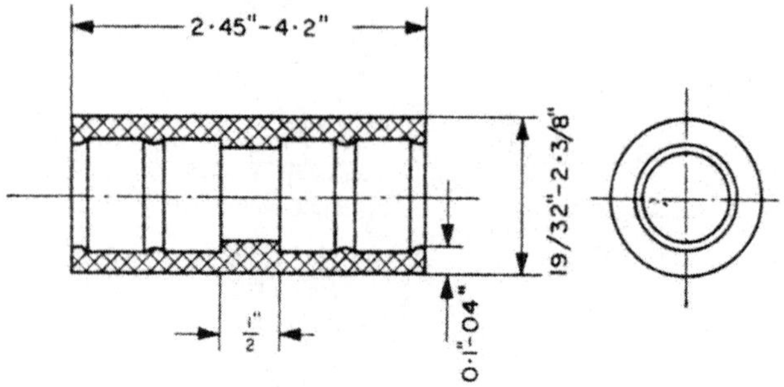

FIG. 5.3 Flexible "Nyvac" coupling supplied by N.G.N. Electrical Ltd.[926]

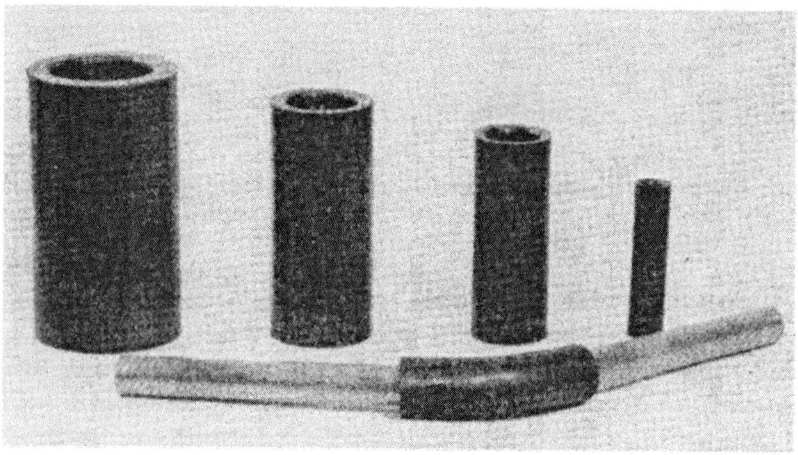

PLATE 17. Flexible couplings *(Courtesy of N.G.N. Electrical[926])*

Hunter[612a] describes a system for the transmission of rotary motion having high torques and high speeds using using all-metal seals. This motion seal cons sists of a thin-walled metal tube (3, Fig. 5.3A) which is closely fitted at it- centre (both inside and outside) with ball bearings (2, 4) that have races per- mitting some flexure. This assembly is then squeezed into an elliptical shape, and maintained by a shaft (1) with elliptical cross section, introduced inside the inner bearing and by a hollow elliptical driving shaft (5) surrounding the outer bearing. The thin-walled tube (3) is held fixed in position by attaching its end to the vacuum system. The outer shaft (5) can be rotated, and since the two shafts cannot be deformed, the inner shaft (1) must follow the outer one.

The binding of the system is prevented by the fact that both the tube and bearings are thin enough to flex. Figure 5.3A shows the relative positions of the components during a 90° rotation. The flexible tube (3) was (Hunter[612a]) of carbon steel (4 in. long, 2 in. diameter, and 0.010 in. wall thickness), and it was squeezed to an elliptical shape having a difference between the maximum and minimum diameters of 1/64 in. This device transmitted torques of more tann 125 ft.lb.

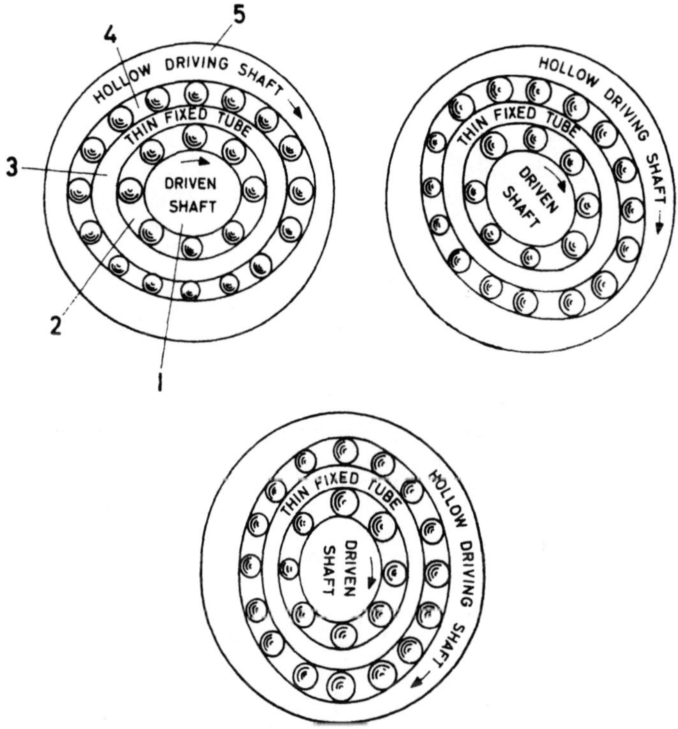

FIG. 5.3A Double bearing seal for rotary motion. Reproduced from Hunter[612a] *(Courtesy of Pergamon Press)*

The flexible connexions available from various suppliers (Balzers[444], Pfeiffer[987], N.G.N. Electrical[926], Heraeus[542]) are made usually from brass (Tombac) pipes reinforced by their undulated (spiral) surfaces. These connexions are available in the diameter range of 15 to 200 mm, and can be bent in a *static* service to a radius of about 1.5 times their external diameter, but for *dynamic* service (where the connexion is bent repeatedly, as in Fig. 5.2c) the allowed bending radius is about 7 times the outside diameter.

51.4 Transmission of motion using bellows

Bellows are pipes having their walls bent to form consecutive parallel rings. This corrugated shape allows limited axial compression or bending. The bellows are made from elastomers (rubber, Teflon), metals (bronze, Tombac, stainless steel) and in special cases from glass.

Rubber bellows may be used to connect pipes forming angles up to 180° (Plate 18) between them.

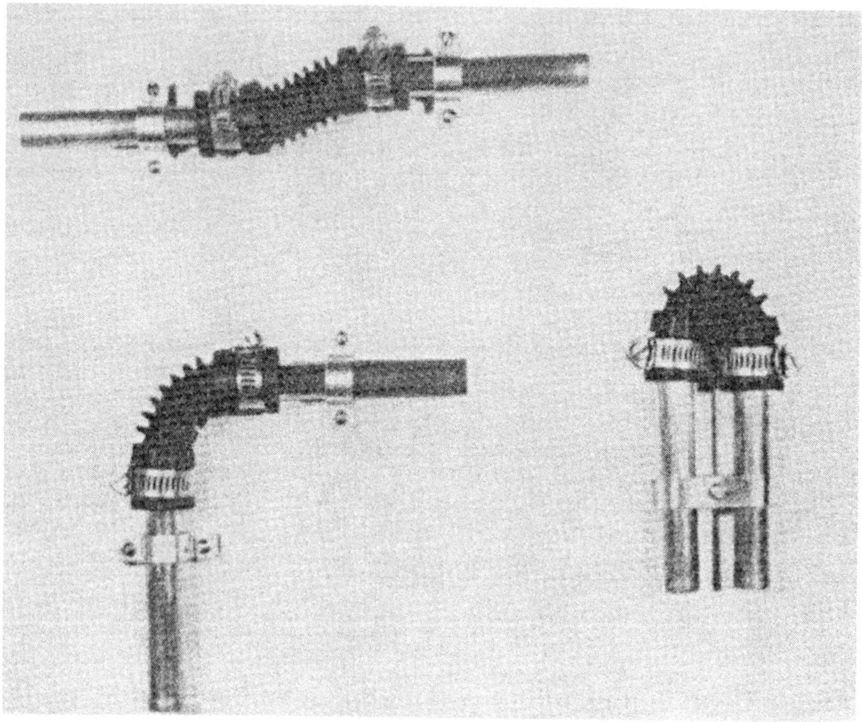

PLATE 18. Rubber bellows *(Courtesy of Genevac*[441]*)*

Metal bellows allow an axial compression of 20–30 per cent of their net length but most manufacturers advise design of the stroke to be a maximum of 10–15 per cent. These figures refer to slow motion (from 1 stroke/min as is required in closing or opening valves, shutters, etc., up to 60 strokes/min as required in dynamic use). For higher stroke frequencies the allowed compression is much less. In order to keep the deformation of the bellows in the allowed limits, the design should incorporate appropriate restraints. Metal bellows and flexible connexions are desirable for the isolation of vibrations from the mechanical pumps, and for the relative mechanical adjustment of the parts of the vacuum system.

Metal bellows designed for use as flexible vacuum connexions are available from most of the firms producing vacuum equipment* (Section 11.3).

Figure 5.4 and Table 5.2 give the dimensions of bronze flexible vacuum couplings supplied by Edwards[328].

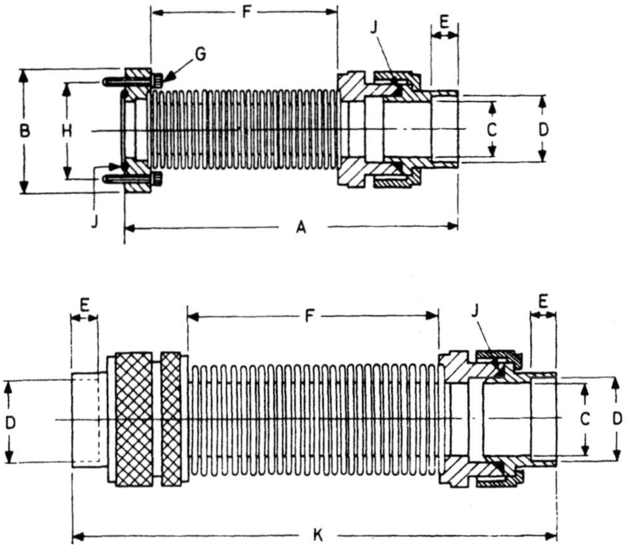

FIG. 5.4 Dimensions of flexible vacuum couplings supplied by Edwards[329] (see Table 5.2)

TABLE 5.2. FLEXIBLE VACUUM COUPLINGS (Fig. 5.4) FROM Edwards[328]

Size	1/2 in.	3/4 in.	1 in.	Size	1/2 in.	3/4 in.	1 in.
A in.	4 1/4	4 5/8	5 7/16	F in.	2 1/2	2 5/8	3 17/32
mm	108	117	138	mm	64	67	90
B in.	1 1/2	1 5/8	2 1/8	*G*	4 *BA* ×	4 *BA* ×	4 *BA* ×
mm	38	41	54	in.	5/8	5/8	1/2
C in.	1/2	3/4	1	*H* in.	1 1/8	1 5/16	1 13/16
mm	13	19	25	mm	29	33	46
D in.	0.596	0.846	1.112	*J**	VOR	VOR	VOR
mm	15.1	21.5	28.2		121	130	136
E in.	3/8	3/8	3/8	*K* in.	5 1/4	5 7/8	6 3/4
mm	9.5	9.5	9.5	mm	133	149	172

* Stainless steel bellows are available from e.g. B.O.A. Metallschlauchfabrik A.G., Kellerstrasse 45, Lucerne, Switzerland.

* O-ring dimensions see Table 3.19A.

Figure 5.5 and Table 5.3 give the dimensions of stainless steel flexible connexions supplied by National Research Corp.[912].

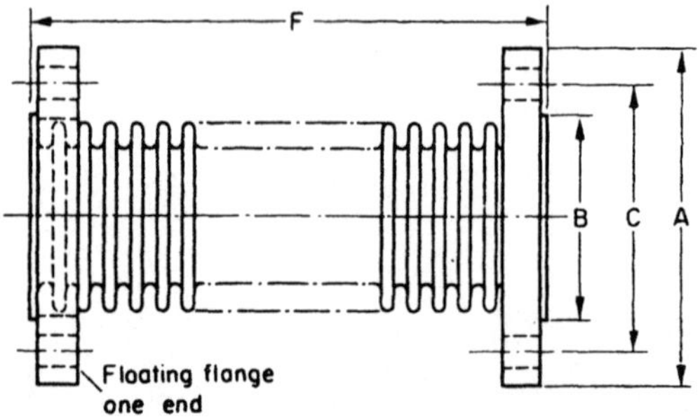

FIG. 5.5 Dimensions of flexible vacuum connectors supplied by N.R.C.[912] (Table 5.3)

TABLE 5.3. FLEXIBLE VACUUM CONNECTORS (Fig. 5.5) FROM National Research Corp.[912]
Dimensions in inches

Pipe size	*A*	*B*	*C*	*F*	Bolt holes	
					Number	Diameter
2	6	3 5/8	4 3/4	7	4	3/4
3	7 1/2	5	6	8	4	3/4
4	9	6 3/16	7 1/2	8	8	3/4
5	10	7 5/16	8 1/2	8	8	7/8
6	11	8 1/2	9 1/2	8	8	7/8

When the bellows are installed in a simple direct connexion, torques and strains can be introduced into the system because of the external pressure tending to collapse the bellows (Leffert[755]). To prevent the torque a second bellows may be installed opposite to the first, and the ends of the two bellows connected by an external frame.

The ends of metal bellows are sealed to vacuum couplings (Fig. 5.4) or flanges (Fig. 5.5) by brazing, or brazed directly to the body of the port. Some arrangements used to braze the ends of the bellows to the flanges or couplings are shown in Fig. 2.43. The ends of the bellows may be welded to the flanges using the techniques illustrated in Fig. 5.6a (Tasman[1215a]). The flange (1) is profiled to a thin collar and fits snugly over the end of the bellows (3). The snugly fitting protecting ring (2) is inserted into the bellows, and is welded, (argonarc, heliarc) restricting the weld to the extreme edge (4) and keeping it away from the corrugated surfaces. Randolph[1025] seated the flared ends (1 Fig. 5.6b) of the bellows on elastomer O-rings (2) and clamped them to flange plates (3) by the annular rings (4) that are split along a diameter, and have a lip fitting in the space between the corrugations.

The *axial motion* permitted by the bellows (Garrod[430]) is successfully used to seal all-metal valves (see Section 61.33b) where the bellows are sealed (brazed) at one end to the port and at the other end to the moving stem (Fig. 5.7a).

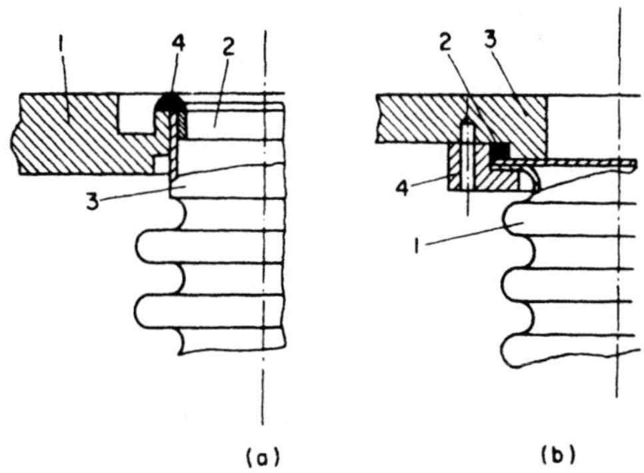

FIG. 5.6 Methods of sealing bellows to flanges: (a) edge welding. After Tasman[1215a] *(Courtesy of Pergamon Press)*; (b) O-ring seal. After Randolph[1025] *(Courtesy of The American Institute of Physics)*

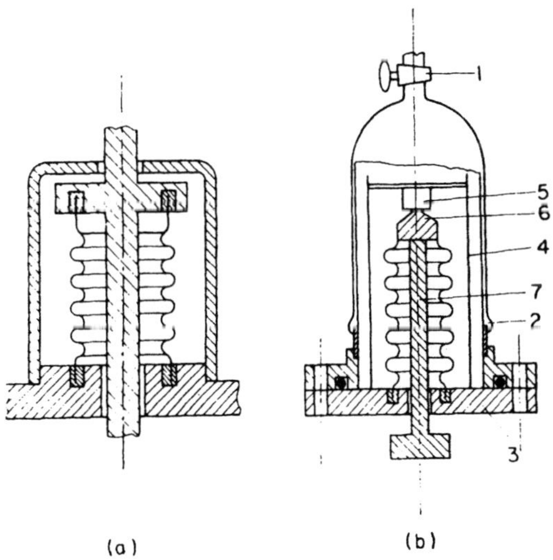

FIG. 5.7 Bellows used to seal axial translation motion: (a) in a sealing system of a valve; (b) in a device to open under vacuum a sealed container (after Schacher[1101])

Schacher[1101] used the bellows in a device for opening small sealed containers under vacuum (see also Section 61.5). The device consists of a Pyrex tube provided at one end with a stopcock (1 Fig. 5.7b) and at the other end with a glass–metal (FeNiCo) seal (2) to which the flanges on the bottom are attached.

The flange (3) holds two support rods (4) carrying the specimen (5). After evacuating the device and closing the stopcock (1), the object (5) is punctured by the steel point (6) by moving the rod (7) axially.

A bellows-sealed, linear-motion feedthrough is supplied e.g. by Varian[1257a]. This feedthrough (Plate 19) has a one inch maximum axial travel, and is actuated by the rotation of the hand drive control (moves 0.1 in. per revolu-

PLATE 19. Linear motion feed-through; maximum travel 1 in.; counter indicates travel to 0.001 in.; bakeable to 200 °C. *(Courtesy of Varian[1257a])*

tion) the travel being indicated by a counter to 0.001 in. with an accuracy of 1 per cent. The feedthrough can be loaded up to 3 lb. and can be baked to 200 °C (or for short periods to 400 °C) requiring the removal of the counter and external drive mechanism during baking.

The axial motion transmitted through bellows, may be used in connexion with the rotary motion obtained by other techniques (e.g. magnetical transmission). Gerber[444a] constructed "dual-motion" feedthroughs by combining the linear motion (limited to 2 in.) through bellows with an electromagnetic coupling for rotary motion.

Radial motion as required in centreing the joints used in electron or ion optics (Ardenne[47]) may also be transmitted by using bellows (another technique, see Fig. 5.19). In such applications the joint can be designed to permit radial motion around a spherical surface (Fig. 5.8a), on a conical surface (Fig. 5.8b), or on plane surfaces (Fig. 5.8c).

Limited *tilting motion* may be transmitted through bellows by using the techniques shown in Fig. 5.9. The construction shown on Fig. 5.9a was described by Brose[171] and consists of a long metal rod (1) which passes through the solid cap (2) into the interior of the vacuum chamber. The cap (2) is sealed to the bellows (3), which is brazed to the metal tube (4), which is the port of a metal chamber, or is sealed to the glass chamber (5).

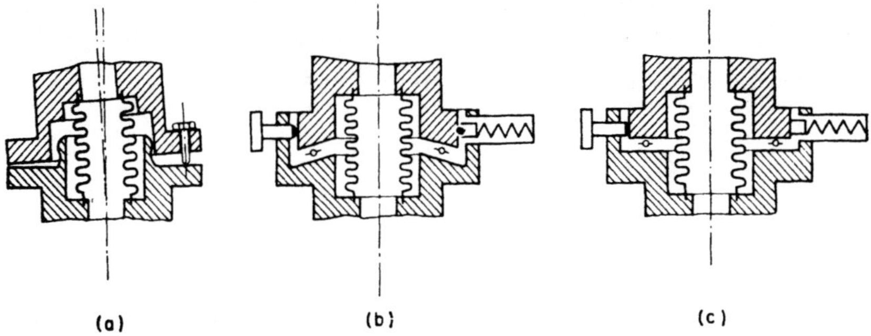

FIG. 5.8 Bellows used to seal joints for radial motion: (a) joint centred on a spherical surface; (b) joint centred on a conical surface; (c) joint centred on a plane surface

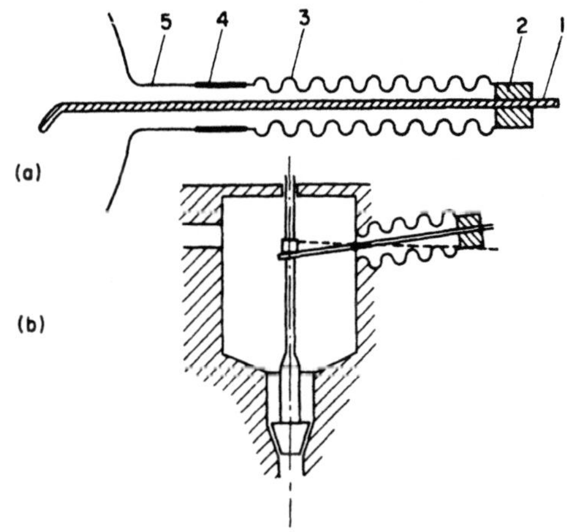

FIG. 5.9 Bellows used to seal joints for tilting motion: (a) vacuum lever. After Brose[171] *(Courtesy of The Institute of Physics and The Physical Society, London)*; (b) lever changing the tilting into translation motion. After Della Porta[272] *(Courtesy of Pergamon Press)*

Della Porta[272] used a system in which the tilting of a bellows-sealed handle (Fig. 5.9b) is transformed into a translational motion, which closes and opens a ground joint.

Rotary motion can be transmitted with bellows sealed devices as illustrated in Fig. 5.10. If continuous rotary motion is necessary the device should be designed to give minimum bending to the bellows. Thus an arrangement such as shown Fig. 5.10a (Guthrie[491], Tasman[1215a]) may be used. Where slow or limited rotary motion should be transmitted the bellows may be bent more and the construction may be designed in a more simple manner (Fig.

5.10b; Morse[894], Snyder[1162]). Since the life of these devices is determined by the flexed bellows, the angle of tilt of the wobble shaft should be kept to a minimum, a fact that reduces the torque which can be transmitted by such a system. A shaft seal using the design shown on Fig. 5.10b is commercially available from e.g. National Research Corp.[912].

For special applications *glass bellows* are required and constructed. Boll[147] describes a bakeable valve (Fig. 5.11a) sealed by glass bellows (3). This valve

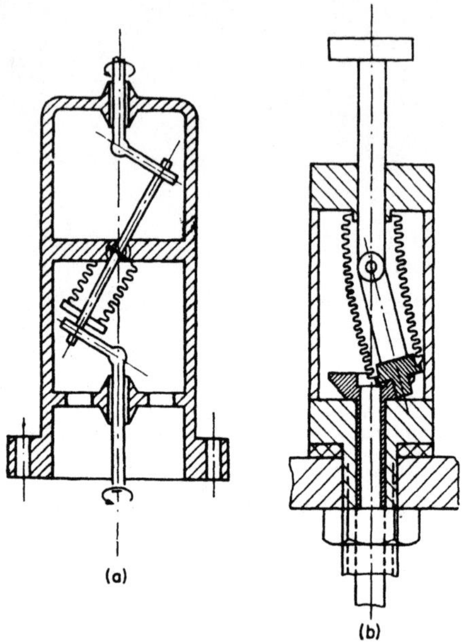

Fig. 5.10 Rotation transmitted through bellows: (a) with tilted bellows; (b) with bent bellows

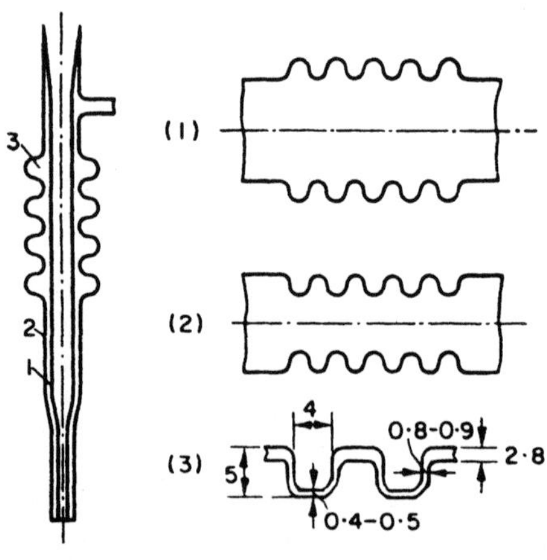

Fig. 5.11 Glass bellows: (a) bellows seal (after Boll[147]); (b) stages in constructing a glass bellows. After Pompeo[999] *(Courtesy of The American Institute of Physics)*

is closed by fusing together the ends of the pipes (1 and 2) by means of a torch (see also Section 61.32d). Pompeo[999] points out the techniques to be used in the construction of glass bellows. The technique used by some glass-blowers is to form the bellows on the tube by blowing the glass into a "doughnut" shape, (1, Fig. 5.11b). This has the disadvantage of placing the thin glass wall on the outside of the tube where it is subject to breakage. This can be overcome by placing the thin wall on the inside of the tube (2, Fig. 5.11b). This operation requires the forming on a glass-blower lathe. Pompeo[999] gives the dimensions of the corrugation (for a glass pipe of 60 mm o.d.) as shown in 3, Fig. 5.11b. A glass bellows having 5 convolutions with these dimensions permits a compression of 0.5–0.7 mm at a load of about 20 kg.

51.5 Transmission of motion using diaphragms

A diaphragm is an elastic membrane which is placed such that it separates the evacuated space from the surroundings. The diaphragm sealed around its lip has enough elasticity to permit limited motion of its central portion. If a motion-transmitting rod is sealed at the centre of the diaphragm (the rod perpendicular to it), limited translation motion or tilting can be transmitted inside the sealed space.

To transmit motion into an evacuated space, "flat diaphragms" are generally used; their name refers to their general shape, since even if they are flat initially, during work, convolutions or distortions of the surface occur. To avoid elongation during work, one or more convolutions are provided for at the moulding of the diaphragm. This allows flexing rather than stretching of the diaphragm. Cook[233] describes a technique to form a diaphragm from a metal foil by spinning.

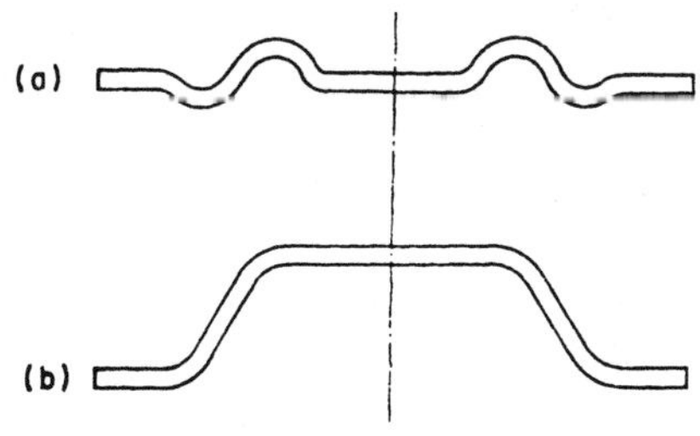

FIG. 5.12 Diaphragms: (a) convoluted; (b) dished

The travel available with a convoluted diaphragm (Fig. 5.12a) is approximately *twice* the height of the convolution. Convolution heights greater than the thickness of the diaphragm are generally difficult to make. Dished diaphragms (Fig. 5.12b) permit larger travel, but this height is also limited in the usual designs to about *one-quarter* of the diameter of the diaphragm. By stretching flat diaphragms the maximum available stroke is generally only 7–9 per cent of their diameter (Taplin[1213a]).

In order to ensure reliability in service, it is advisable to follow some basic requirements:

(1) The flexing annulus around the moving shaft (plate) should be kept wide enough to permit easy flexing, but not too wide for the stroke required, (Fig. 5.13a).

(2) The deflection of the diaphragm should be limited to 90 per cent of the maximum possible stroke, by incorporating a mechanical stop in the design (Fig. 5.13b).

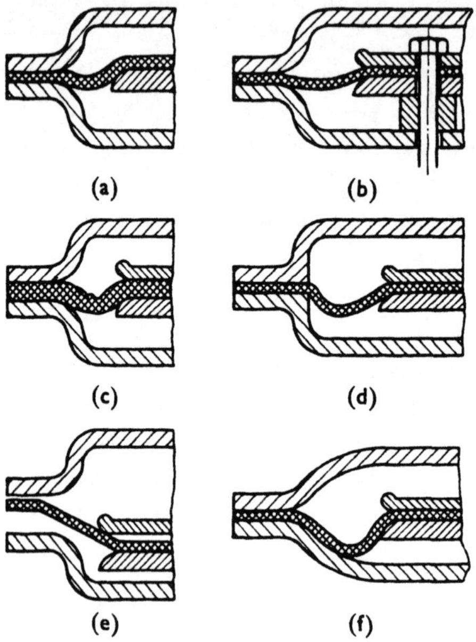

FIG. 5.13 Details for the design of a diaphragm seal

(3) The diaphragm should be constructed from a material thin enough to prevent wrinkling (Fig. 5.13c).

(4) Everywhere the diaphragm is sealed between flanges, sharp corners (Fig. 5.13d) should be avoided. It is recommended that every edge in contact with the diaphragm be radiused at least to radii twice the thickness of the diaphragm.

(5) If the diaphragm is moulded with convolutions or dished, the bends should be kept outside the working area. The design shown on Fig. 5.13e is not recommended.

(6) The case holding the diaphragm should be deep enough, or the diaphragm stretched enough so that contact with the case wall (Fig. 5.13f) is avoided.

Diaphragms are used extensively as motion seals in manometers (Jaeckel[623]) and in vacuum valves (see Sections 61.32g and 61.33b).

Diaphragms are generally circular in outline, but annular shapes may also be used if the application requires this. An annular diaphragm was described by Batzer[96] (Fig. 5.14), consisting of a ring of 0.032 in. OFHC copper sheet

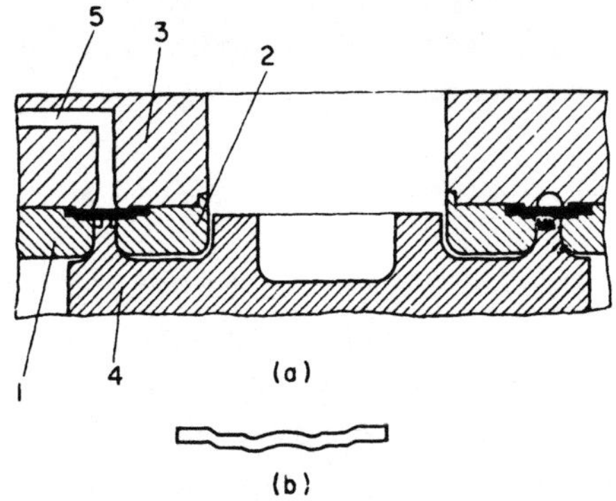

FIG. 5.14 Diaphragm seal in pneumatic bakeable valve: (a) the closing system of the valve; (b) shape of the diaphragm used. After Batzer[96] *(Courtesy of Pergamon Press)*

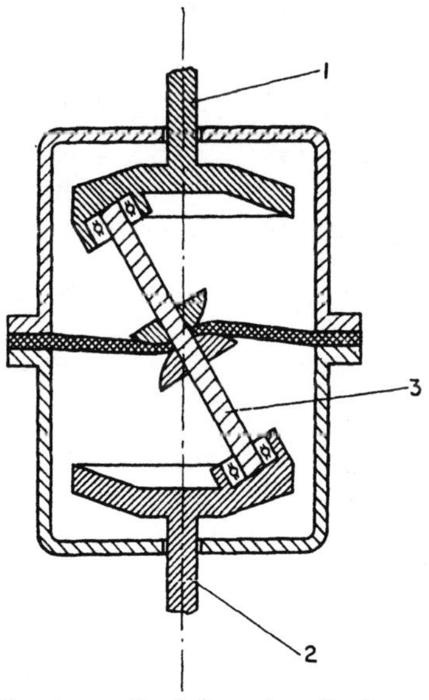

FIG. 5.15 Rotary motion transmitted through a diaphragm (after French[403])

shaped as in Fig. 5.14b. The diaphragm is clamped vacuum-tight by two stainless steel rings (1, 2, Fig. 5.14a) against a third (3). This arrangement was used by Batzer[96] as a closing system for a valve (Section 61.32g), the diaphragm being pressed uniformly against the valve disc (4) by the pressure exerted hydraulically (5).

French[403] describes a rotary seal using a diaphragm (Fig. 5.15) in an arrangement similar to that shown in Fig. 5.10. The driving shaft (1), in alignment with the driven shaft (2), rotates this latter by means of the wobble shaft (3), sealed through the centre of the diaphragm (4). The edge of the diaphragm is clamped between the upper and lower half of the housing. The centre of the diaphragm through which the wobble shaft passes is clamped between flanges, thus sealing the top half (open to atmosphere) from the bottom (vacuum). During the rotation the only motion of the diaphragm is a moderate flexing.

51.6 *Transmission of motion using ground seals*

Ground seals (Section 3.6) may be used to transmit motion into vacuum chambers, if the drawbacks of the ground seal itself (vapour pressure of the grease) may be tolerated in the system. The motion which can be transmitted through ground seals is restricted and the manipulation requires care, particularly in heavy apparatus (Brose[171]). This kind of motion transmission is used in connexion with glass vessels (Highhouse[560]) though it may be used on metal systems as well but consisting of lapped metal seals (Matheson[825], Banki[79], Roth[1083]) or Teflon motion joints.

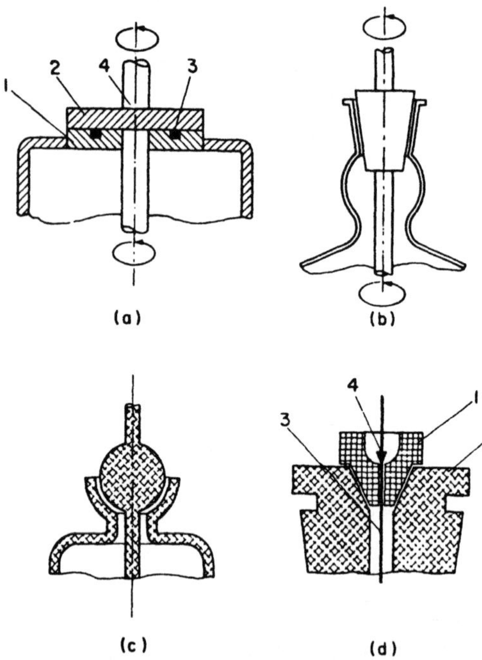

FIG. 5.16 Rotary motion transmitted through ground seals: (a) lapped flat seal; (b) conical ground seal; (c) spherical ground seal; (d) seal for a vacuum stirrer. After Nester[922] *(Courtesy of The American Institute of Physics)*

Flat ground glass joints or lapped metal seals (see Fig. 3.18) may be used to transmit *rotation* or very limited radial motion (centring). In such seals, one of the ground or lapped parts (1 Fig. 5.16a) is sealed to the vessel. The other part (2) is placed upon this and can be rotated, while the ground or lapped face seals are placed against the mating part. Ground seals are greased and lapped metal seals are lubricated with oil (e.g. castor oil), which flows through a circular channel (3). The shaft (4) which moves with the upper plate (2) transmits the motion into the chamber. Lapped seals are able to transmit rotation up to 3–4 r.p.m.

Conical ground seals or spherical ones (Fig. 5.16b, c) are also frequently used to transmit slow rotation. These seals are usually greased, but if the absence of grease is required, they can be constructed e.g. with a mercury seal. Howe[603] constructed a greaseless vacuum rotor seal using a standard ground glass joint, provided with Teflon ring gaskets (see Fig. 3.33b) and with mercury sealing.

Nester[922] transmitted rotary motion (to a vacuum stirrer) using a device consisting of a plug (1) of self lubricating plastic (Nylon, Teflon, polyethylene) fitting into the port of the vessel or into an adapter (2 Fig. 5.16d). The plastic plug should be of small size (e.g. 7/15, see Table 3.7) in order to minimize thermal expansion which might break the glass. A passage is provided through the plug, in which the shaft (3) is located. This passage is wide at the top and narrows down to form an inverted conical bearing seat with a 30° taper, and continues as a narrow passage to the bottom of the plug. The bearing is a conical, hardened, stainless steel bead (4) on the shaft.

A cylindrical ground seal with oil, which is able to transmit rotation up to 4000 r.p.m., is illustrated in Fig. 3.37a (Trevoy[1237]).

Axial translational motion may be transmitted in the evacuated space using one of the seals shown on Fig. 5.16b, c, by transforming the rotary motion into a linear displacement. Slow translational motion can be transmitted with the system illustrated* in Fig. 5.17a. For more rapid translational motion, the arrangement shown on Fig. 5.17b may be used. These techniques may only be used for motion on a vertical line, since they raise or lower a part suspended on the thin wire wound on the shaft (Fig. 5.17a) or on a pulley (Fig. 5.17b). For the transmission of limited translation (pulling) or of a motion in strokes the technique illustrated in Fig. 5.17c can be used. Accurately adjustable translational motion may be transmitted in the vacuum vessel by the system shown in Fig. 5.17d. This consists of a nut (1) rotating with the plug of the ground joint, and a guiding system (2) which avoids rotation of the shaft (3). By rotating the plug of the ground joint, the shaft screws into the nut receive a controlled translational motion.

* A spherical ground seal for the transmission of translational motion is shown in Fig. 3.26a (Rubens[1090]).

Tilting motion may be transmitted using spherical ground joints* (Section 36.3). Skidmore[1145] used a spherical joint to transmit a wobbling motion describing a conical surface (Fig. 5.18a). Lloyd[776] used the positive and negative components of an achromatic pair of lenses (Fig. 5.18b) to construct

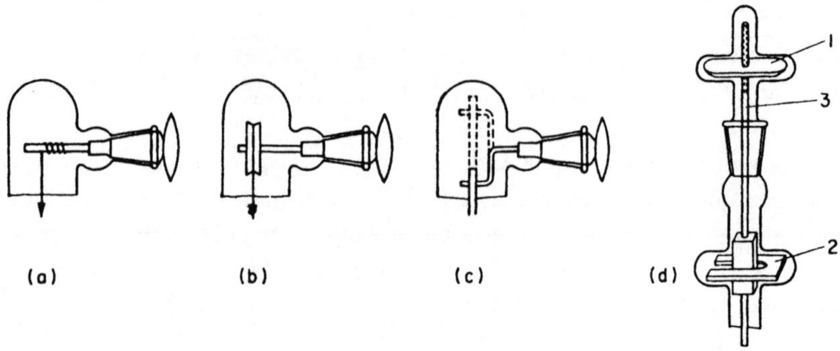

FIG. 5.17 Axial translation, transmitted by rotation of a conical ground seal: using (a) wire wound on shaft; (b) wire wound on pulley; (c) stroke motion; (d) screw

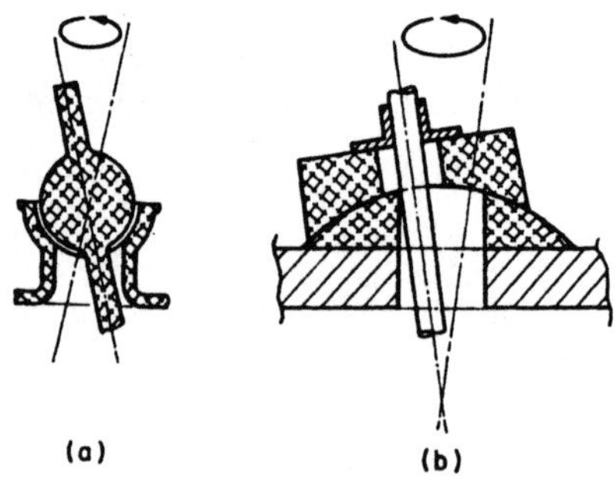

FIG. 5.18 Spherical ground seals used to transmit wobbling motion: (a) spherical joint (after Skidmore[1145]); (b) joint constructed from lenses. After Lloyd[776] *(Courtesy of the American Institute of Physics)*

a seal permitting motion in three directions. Spherical ground seals used to open gas samples under vacuum, by bending the sample tube, were described by Norton[943] and Snow[1161] (see Fig. 6.132c).

51.7 Transmission of motion using gasket seals

The transmission of the various kinds of motion may be achieved through vacuum-tight gasket seals, as O-ring seals (Section 51.71), rim seals (Section 51.72), cylindrical and conical seals (Secion 51.73), lip seals (Section 51.74) and spring-loaded lip seals (Section 51.75), and friction seals (Section 51.76).

* See footnote, p. 509.

The commercially available shaft seals are described in Section 51.77 and the seals constructed to allow angular displacement are discussed in Section 51.78.

51.71 O-ring seals for transmitting motion. Radial, axial (sliding), or rotary motion are frequently transmitted using appropriate O-ring seals. *Radial* motion may be obtained with an arrangement as that shown on Fig. 5.19a

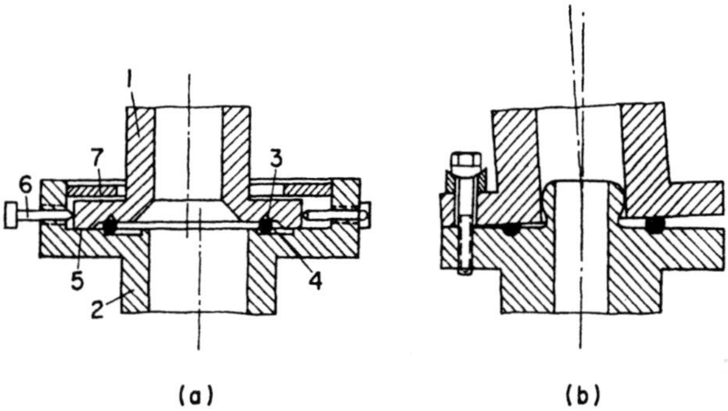

FIG. 5.19 O-ring seals to transmit radial motion: (a) with sliding O-ring. After Garrod[430] *(Courtesy of The Institute of Physics and The Physical Society)*; (b) by compression of an O-ring

consisting of the flange (1) sealed to the flange (2) by an O-ring (3) placed in a groove (Fig. 3.65). The flange (2) is provided with a shallow trough (4) on which the O-ring can move. To limit the compression of the O-ring the two flanges make metal-to-metal contact on the surface (5). The upper flange is displaced by the screws (6), the assembly being guided and kept free from dust by the cover (7). The trough (4) is lubricated with a small amount of silicone grease and graphite (1 p.b.w. graphite and 10–15 p.b.w. oil). According to Garrod[430] such a system may be used for loads of 15–25 kg (0.7–1.2 kg/cm of O-ring), i.e. for diameters of 50–100 mm (2–4 in.). For smaller diameters a spring loading of the seal is required; with larger diameters the load on the O-ring is too big to permit displacement.

Figure 5.19b illustrates an O-ring seal permitting a limited tilting. The upper flange rotates around the spherical surface provided on the lower flange, the motion being limited by the maximum compression of the O-ring on one side and the minimum compression required for a tight seal, on the diametrically opposed point.

Sliding and/or *rotary* motion can be transmitted in principle through shaft seals using *one* O-ring (Fig. 5.20a, b) but in current practice seals using *two* O-rings are preferred (Fig. 5.20c, d), since they ensure better centreing of the shaft. It is easier to machine the groove on the shaft (Fig. 5.20a) but when the shaft must have a long stroke the groove should be made in the wall of the port (Fig. 5.20b, c) (Kurie[723], Fox[394]). The recommended dimensions of the shafts, grooves and O-rings for such seals are given in Table 3.21 and Fig. 3.63.

A sliding or rotary shaft seal with a double O-ring can be constructed without grooves, by placing a cylindrical spacer (Fig. 5.20d) between the O-rings and compressing the assembly with a closing ring.

The O-rings used in motion seals should be lubricated with a low vapour pressure grease or oil (see Sections 36.4 and 37.4).

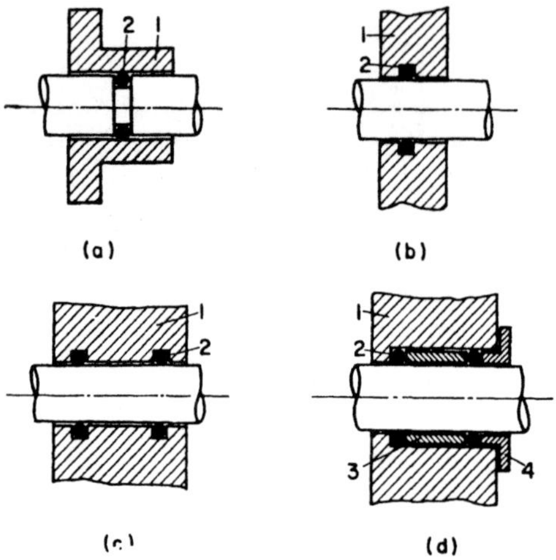

FIG. 5.20 Sliding and rotary O-ring shaft seals: (a) with single O-ring in a groove machined on the shaft; (b) with single O-ring in a groove machined on the chamber; (c) with double O-ring seal; (d) with two O-rings and a spacer; (1) chamber wall; (2) O-rings; (3) spacer; (4) sealing ring

Gore[462] describes a glass sliding joint (Fig. 5.21a) consisting of precision bore glass tubing (1), in which a smaller tube provided with grooves is inserted. The smaller tube has grooves holding O-rings (3). The space (2) between the O-ring is filled with dibutyl-phtalate (or oil) by using a hypodermic syringe. The smaller tube is able to slide inside the larger one, the seal being able to maintain presures of the order of 10^{-3} torr (Gore[462]).

Lake[729a] describes a sliding seal made in two stages (Fig. 5.21b). In each stage a piston (1) mounted on the rod (2) slides inside the precision bore tube (3). On each piston two O-rings placed in grooves assure the seal. To improve the seal and to provide lubrication a small amount of silicone oil (4) is maintained above the pistons. To minimize the leakage the space between the pistons is provided with a guard vacuum (Section 38.23) evacuating this space by the connexion (5). According to Lake[729a] such a seal is able to maintain pressures of the order of 10^{-6} torr and with continuous running at 100 strokes per minute the leakage of oil past the lower piston is not more than a few cubic centimetres per week. A small trough mounted near the bottom of the driving rod prevents this oil from dispersing in the vacuum chamber.

Sikorski[1136] described a rotary seal for 600–900 r.p.m. (Fig. 3.37b), and Gaunt[436] used a motion seal with O-rings and mercury sealing (Fig. 3.31d).

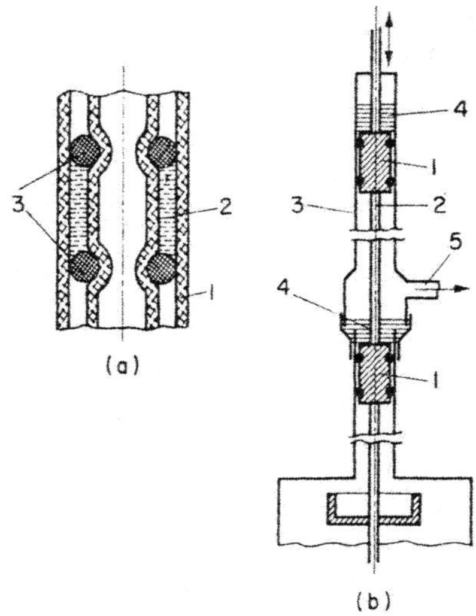

FIG. 5.21 Oil-lubricated O-ring seals for transmitting motion: (a) sliding joint between two concentric pipes (after Gore[462]); (b) double piston sliding seal. After Lake[729a] *(Courtesy of The Institute of Physics and The Physical Society, London)*

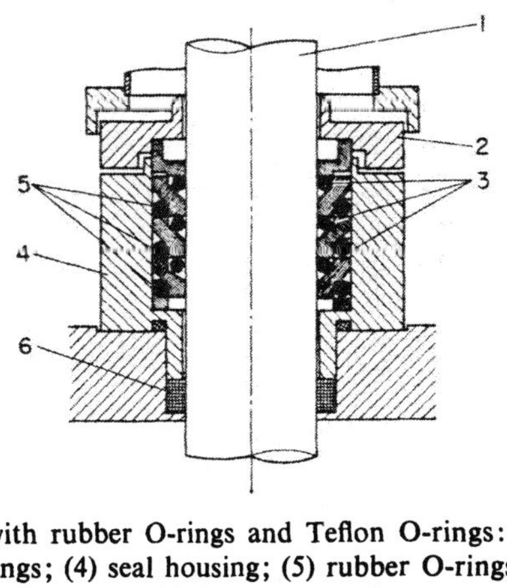

FIG. 5.22 Rotary seal with rubber O-rings and Teflon O-rings: (1) actuating shaft; (2) seal nut; (3) Teflon O-rings; (4) seal housing; (5) rubber O-rings; (6) Teflon shaft guide bushing. After Smith[1151] *(Courtesy of Pergamon Press)*

Smith[1151] described a greaseless seal (Fig. 5.22) in which the three inner sealing members are Teflon O-rings, which provide the sliding surface for the shaft. The three outer rings are rubber O-rings, which provide elasticity in the seal and make the static vacuum joint against the housing. Such a seal has considerably more frictional drag than lubricated O-ring, Wilson or Chevron seals, and sufficient additional force must be provided for its actuation. Great care should be taken to protect the shaft and seal from dust, dirt and oil.

A rotary seal using a gland with Teflon and Neoprene O-rings is quoted by Watson[1288b], and a motor sealed with Viton O-rings in order to be placed in high vacuum is described by Anastasio[29b].

51.72 Rim seals for transmitting motion. Hayward[523] described a rotary vacuum seal (Fig. 5.23a) consisting of a screwed base (1) upon which rests the rubber

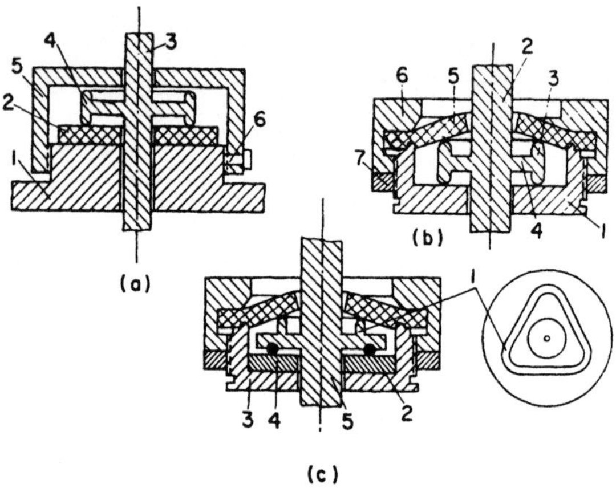

FIG. 5.23 Rotary rim seals: (a) with flat gasket (after Hayward[523]); (b) with elastic gasket; (c) shaft for continuous service with greased triangular rim and grease (after Hayward[524])

gasket (2), lightly lubricated with a vacuum grease. The shaft (3) passes through the central hole in (1) and carries a rim (4) bearing on the gasket (2). For a shaft with 1/2 in. diameter, the diameter of (4) is about 3/4 in., and the radius of the ridge is about 1/32 in. for a gasket of 1/16 in. thickness. The shaft is held in position by the hollow knurled nut (5) screwed onto the base (1) and locked by (6).

Hayward[524] described the modifications of the seal shown on Fig. 5.23a, which were made in order to better utilize the elasticity of the rubber gasket (Fig. 5.23b), and to improve the seal for continuous service (Fig. 5.23c). The rotary seal of Fig. 5.23b consists of a body (1) in which the shaft (2) having a rim (3), attached via the disc (4) rotates. The Neoprene gasket (5) is held in position by the screwed nut (6), which is provided with a locknut (7). The face of the body on which the Neoprene rests has a raised ring to ensure a better seal between the body and the gasket. The top of the rim is above the top of the body to ensure a good seal between the rim and the gasket. Such a seal has a leak rate less than 10^{-4} lusec, but the lubrication between rim and gasket is insufficient to be used for continuous (motor) operation.

The rotary seal shown on Fig. 5.23c consists of a body (3) containing a hardened steel washer (4), and the shaft (5) holding a disc which has on the upper side a triangular rim (1) (instead of the circular rim of Figs. 5.23a, b); and on the other side three holes carrying steel balls (2), which bear on the steel washer (4). The triangular rim is packed with vacuum grease and while running distributes the grease over the under-surface of the Neoprene gasket.

51.73 Cylindrical and conical seals for transmitting motion. The most simple shaft seal consists of a rubber plug through which a rod passes in the vacuum chamber. If the rod is greased such a seal permits limited sliding and rotary

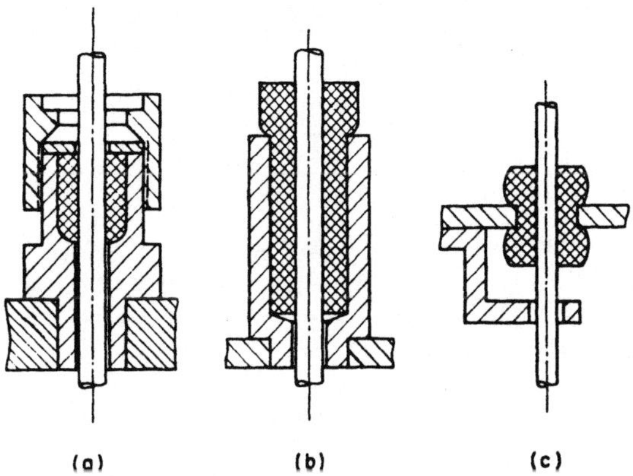

FIG. 5.24 Shaft seal with cylindrical gaskets: (a) adjustable; (b) long path; (c) with pilot bearing. After Lauritsen[746] *(Courtesy of The American Institute of Physics)*

motion. Figure 5.24 illustrates some simple methods using a length of rubber tubing (Fig. 2.17) to seal a shaft. The rubber is compressed in a screwed fitting (Fig. 5.24a) or forced into a longer, drilled holder (Fig. 5.24b). If the rubber pipe is just plugged into a hole provided in the wall of the vessel (Fig. 5.24c), a pilot bearing is necessary to prevent the tipping of the seal and leakage along the wall-rubber contact surface (Lauritsen[746]).

Cowie[241] described a shaft seal (Fig. 5.25a) using a rubber gasket with a square cross section, sealed between two tapered surfaces and the shaft. This seal permits sliding, rotation and tilting of the shaft. Ridenour[1057] used a shaft seal consisting of two Neoprene washers (1) (with an i.d. smaller than the diameter of the shaft) pressed on a metal plug (2) shaped conically at both ends, as shown in Fig. 5.25b. Billett[124] designed a rotary shaft (Fig. 5.25c) using thin (1.5 mm) PTFE (Teflon) washers interspaced with metal washers. The clearance between the metal washers and the shaft was about 0.25 mm providing space for flow of the Teflon, under the pressure applied axially on the assembly. This seal worked satisfactorily at temperatures between 22–200 °C, having a leak rate less than 10^{-5} lusec.

A rotary seal using conical Teflon gaskets was described by Howard[602a]. The seal consists of a Teflon gasket fitting on the cylindrical surface of the shaft, and being shaped on its outside surface to a taper of 14°. The cylindrical surface fitting onto the shaft is provided with parallel circular grooves (0.025 in. deep, 0.040 in. wide, and spaced 0.020 in.); the deformation of the protuberances between the grooves enable a tight fit. The tapered outside surface of

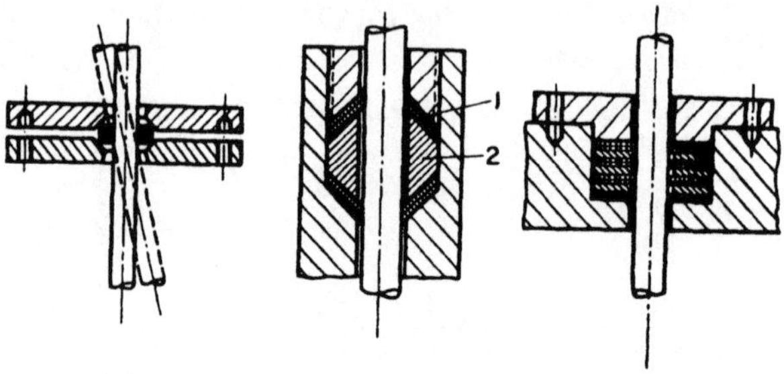

FIG. 5.25 Shaft seals: (a) with square cross section gasket (after Cowie[241]); (b) with conical gaskets (after Ridenour[1057]); (c) with flat gaskets (after Billett[124]). *(Courtesy of The American Institute of Physics and of The Institute of Physics, London)*

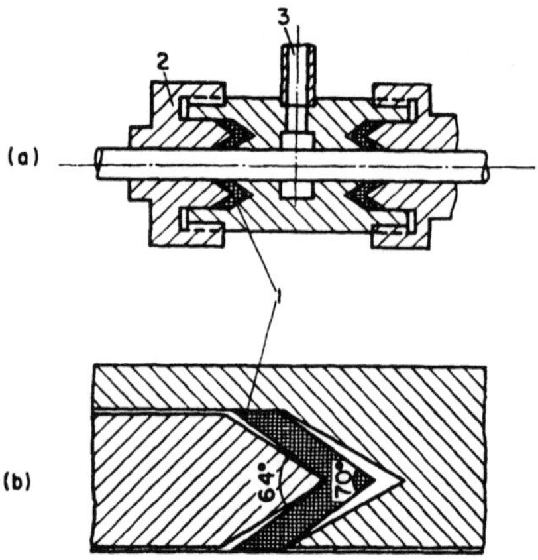

FIG. 5.26 Sliding shaft seal with V-shaped Teflon gasket and guard vacuum. After Champeix[203] *(Courtesy of Société Française des Ingénieurs et Techniciens du Vide)*

this Teflon gasket fits into the tapered inner surface of a second Teflon gasket, tightened to the body with small grooves as in the first gasket. The two Teflon gaskets are separated by a conically shaped brass part. When the Teflon gaskets are compressed axially, the small brass cone separating them causes them to tighten laterally against the rotating shaft and the outer housing. It was found that, while the shaft was rotated at 20 r.p.m., at about 5×10^{-5} torr, the leak rate was less than 2×10^{-9} cm^3/min helium at S.T.P. The seal was provided with a connexion for a guard vacuum.

Champeix[203] described a seal for a sliding and rotary motion (Fig. 5.26) using Teflon gaskets (1) with a V-shape cross section pressed between a mating groove (see Fig. 5.26b) and a knife edge, shaped part (2). The seal was provided with guard vacuum (3). The seal was tested at 10^{-5} torr, obtaining a leak rate of about 8×10^{-4} lusec without moving the shaft; each stroke of the shaft produced a throughput of 0.1 mm³ N.T.P.

51.74 Lip seals. The "lip seals" known as Wilson seals (Wilson[1325]) are based on the sealing action of a rubber sheet towards a shaft crossing it through a hole cut with a diameter considerably smaller than that of the rod.

The periphery of the rubber sheet is held tightly by a circular metal part (4, Fig. 5.27), and the rubber close to the shaft is distorted and bent out from the plane of the sheet, its lip sealing against the shaft. It is usually stated that atmospheric pressure forces the washer against the shaft (Wilson[1325]) helping to form the seal. Actually it is not the atmospheric pressure which is predominant in forming the seal, but rather the snap of the gasket material (Dawton[267], Turnbull[1238]). Thus the inner edge of the rubber is forced against the shaft, so that the seal remains vacuum-tight, even when the shaft is rotated or slides in and out rapidly.

For satisfactory service the design and construction of Wilson seals should meet the following requirements:

(1) The surface of the shaft must be smooth. Rough shaft surfaces increase the leak rate of the seal. This increase is particularly marked with reciprocating motion, when circumferential markings (scratches) give a higher leak rate than longitudinal ones (Dawton[267]). The sliding rod (shaft) does not necessarily have to be particularly straight (Wilson[1325]).

(2) The gasket should have a smooth surface and should not be too soft. Rubbers with a Shore (durometer) hardness of 50–60 are recommended for the purpose. Gaskets cut from sheet material (usually 1.4–1.6 mm thickness) are satisfactory, provided that the hole into which the shaft moves is smooth, round and has squarely cut edges. A sharp, circular cutter, lubricated with a soap solution, produces just such a smooth cut.

(3) The size of the hole is not critical, but the hole should be of a diameter so that the rubber is distorted out from the plane of the sheet by about 3 mm when the rod is inserted in the hole (Wilson[1325]). Thus the hole should be about 0.65–0.8 of the shaft diameter (see Table 5.4).

(4) In the assembly of the seal it is important that the rubber near the rod be free (see Fig. 5.27) so that the rod can rotate or slide freely without jamming.

(5) The sealing gasket should be flexed at the centre by a conical bevel (about 30°) on the metal plate supporting the rubber (2, Fig. 5.27). When the seals are used warm (100 °C), a much better performance is obtained if the

washer is given additional elastic properties by having a spring within it (Section 51.75) to keep it close to the shaft (Dawton[267]).

(6) The lubrication of the seal must be adequate. Dawton[267] recommends the use of a thin, high vacuum oil (as Apiezon B) rather than a high vacuum grease.

(7) The compression of the nut holding the rubber sheet should be just sufficient to seal. If it is closed too tightly, it will squeeze the rubber out of shape and generally permit leakage.

The basic dimensions of simple Wilson seals with various shaft diameters sare listed in Table 5.4 in connexion with Fig. 5.27.

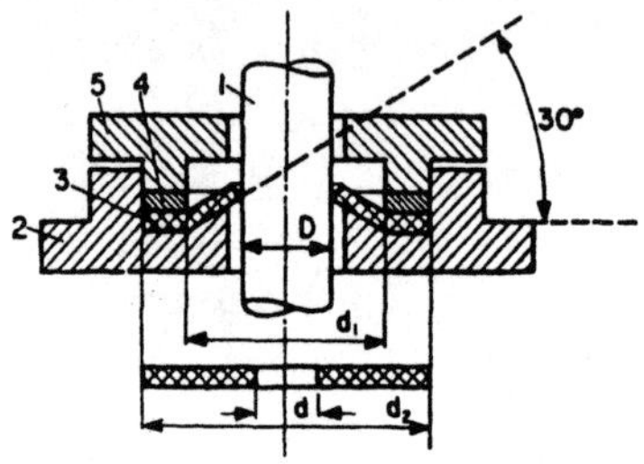

FIG. 5.27 Shape and dimensions of a Wilson seal: (1) shaft; (2) base plate; (3) rubber washer; (4) metal ring; (5) lock nut

TABLE 5.4. DIMENSIONS (mm) OF WILSON SEALS (Fig. 5.27)

D	d	d_1	d_2	h
4	2.5–3.0	14	18	
6	4.0–4.5	16	20	
10	6.5–8.0	20	24	
15	10.0–12	25	30	
20	13–16	30	34	1.4–1.6 (Gasket thickness)
30	20–24	40	50	
40	26–32	50	65	
50	33–40	60	82	
60	40–48	70	96	
70	45–56	80	110	

Wilson seals can be used with shafts from 1.5 mm up to very large diameters (e.g. 70 mm) but at larger sizes (over about 20 mm) it must be ensured that the pressure difference will not force the shaft into the chamber. McLean[834] quotes a frictional resistance of 250 g at a Wilson seal of 1 in., while the force due to the atmospheric pressure on the shaft of such a seal is about 5 kg. Thus with well-greased Wilson seals (of larger sizes) the frictional resistance is not important compared with the force due to the atmospheric pressure.

Wilson seals are very often constructed using a double gasket (Fig. 5.28). The space between the gaskets is evacuated (Snyder[1162]) or filled with grease (Holland[581]). If the Wilson seal is to be used on a glass plate, Smith[1153] proposed the appropriated arrangement (Fig. 5.28c).

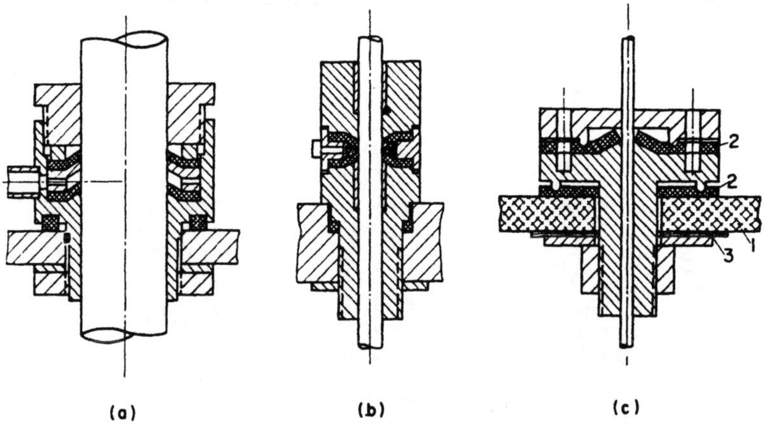

FIG. 5.28 Wilson seals with double gasket: (a) with parallel gaskets and guard vacuum; (b) with opposite gaskets (Edwards[328]); (c) Wilson seal on a glass plate; (1) glass; (2) Neoprene; (3) mica. After Smith[1153] *(Courtesy of The Institute of Physics and The Physical Society, London)*

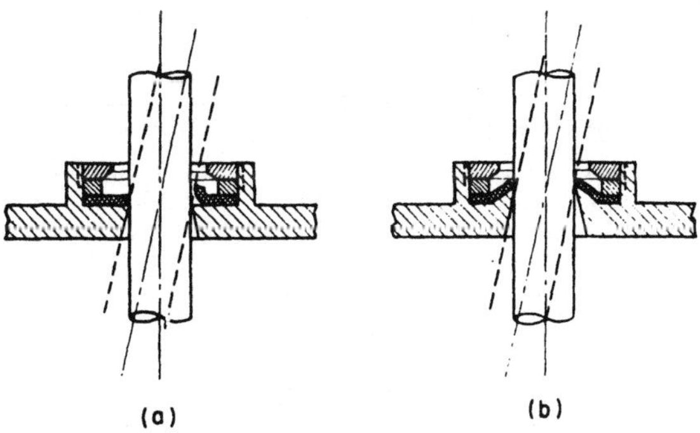

FIG. 5.29 Wilson seals permitting a tilting: (a) bad design; (b) good design

It is possible to construct a Wilson seal so as to permit a limited tilting of the shaft (Fig. 5.29). In this case it is recommended (Dawton[267]) that the metal support should be as close as possible to the washer, or be without the 30° lip (Fig. 5.29).

Dawton[267] quotes the leak rates of the Wilson seals (see also Apkarian[43, 44]). For a shaft maintained steady (without motion) the leak rate was

$1-2\times10^{-3}$ lusec after 8 hours of pumping (including the real and virtual leak; see Section 1.3). The same shaft when rotating slowly (1 rev/sec) had a leak rate of about 4×10^{-3} lusec. For reciprocating motion, the leak rate is proportional to the distance the shaft moves and is independent of the time or speed if this is below 4 cm/sec. In these conditions the leak rate is about 0.01 l./micron/cm of travel. At higher speeds the leak rate per unit motion of the shaft, is increased. The leak rate during the inward motion of the shaft is six times than that during the outward motion.

A variation of the Wilson seal, known as a "chevron seal" is a very reliable seal for large slowly rotating or sliding shafts. It consists of a gland similar to that of the Wilson seal, but containing several heavily greased gaskets in a pile without spacers (Fig. 5.30).

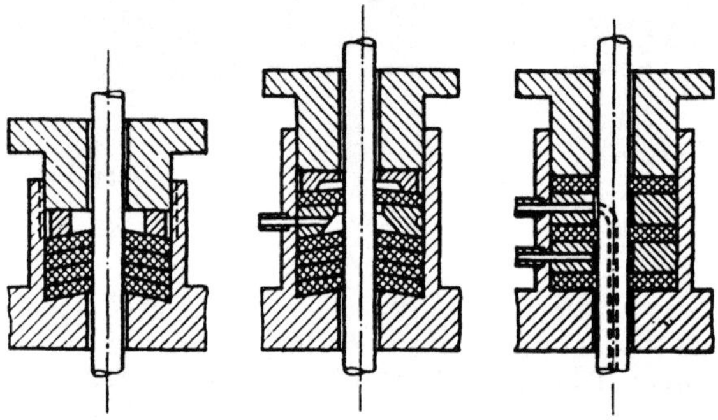

(a) (b) (c)

FIG. 5.30 Chevron seals: (a) standard; (b) with guard vacuum; (c) used as a stopcock. After Bauer[97] *(Courtesy of The American Institute of Physics)*

In order to allow the testing of the seal, in some constructions a spaced gasket is placed outside the chevron group, in an arrangement similar to that used in Wilson seals (Fig. 5.30b).

Bauer[97] describes a method of using the chevron seal as a stopcock. For this purpose the gaskets are alternated with thick metal washers, having radial holes (Fig. 5.30c). The shaft has a longitudinal hole connected to a radial one. By the translation of the shaft this hole can be connected to one of the various holes in the metal washers.

51.75 Spring-loaded lip seals. Lip seals using elastomer gaskets of V or U shape cross sections forced onto the shaft by a spring are available under the names of Gaco-seals*[1], Paulstra-seals*[2], Garlock-seals*[3] or Simmerring-seals*[4].

*[1] G. Angus & Co. Wallsend-on-Tyne, England.
*[2] Soc. Paulstra, 9 rue Hamelin, Paris, France.
*[3] Garlock Packing Co., Palmyra, New York, U.S.A.
*[4] Carl Freudenberg, Simrit-Werk, Weinheim/Bergstr., Germany.

These seals consist of profiled elastomer ring (1 Fig. 5.31) placed in a casing (2), and held in place by an L-shaped metal ring (3). The lip (4) of the gasket is forced onto the shaft by a spring (5), embedded in the gasket or mounted near the lip of the gasket.

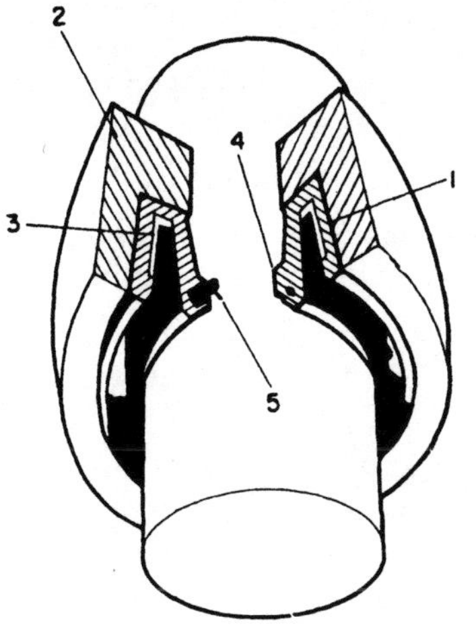

FIG. 5.31 Spring-loaded lip seal

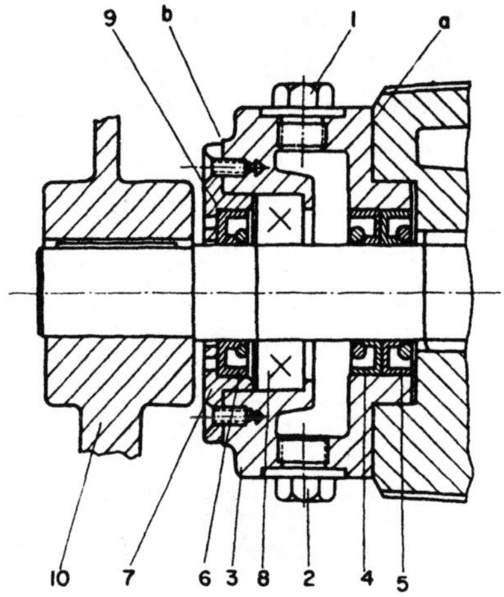

FIG. 5.32 Spring-loaded lip seals, used to seal the rotary shaft of a vacuum pump (*Courtesy of Leybold*[763])

These seals are very reliable if during their assembly provisions were made to follow some basic requirements (Chaillou[201]):

(1) The gasket should be mounted with the sealing lip on the high pressure side.

(2) Before assembly, the external surface and the lip of the gasket should be greased (e.g. silicone grease).

(3) To avoid any damage of the gasket (lip) during assembly, it is recommended that the end of the shaft be provided with a tapered part, or at least the edges be radiused.

(4) If the shaft has steps (Fig. 5.32) a conical sleeve should be slipped over the reduced part of the shaft so that the gasket can be slid along the shaft without the danger of cutting it when it passes over the step.

(5) Some force is required to introduce the shaft through the gasket; for larger shafts the use of a press is required.

Greenhouse[479a] recommends that for a given Garlock closure, an oversize shaft should be used, e.g. to use a 0.264 in. shaft with a 1/4 in. closure.

Figure 5.32 illustrates an arrangement (Leybold[763]) in which a double-lip seal gasket (4, 5) is used to seal the rotary shaft of a vacuum pump. The spring loaded gaskets are carried in the casing (3), which is attached to the body by covering the face (a) with a sealing paint (Section 3.2). This casing carries the ball bearings (8), and is sealed on the other side by the gasket (6), held in the hole (9) of the lid (7). The casing (3) is filled (1, 2) with oil.

A lip seal with a spring loaded Teflon gasket was described by Giaimo[446] (Fig. 3.134a).

The leak rate of the lip seals with spring loaded gaskets (as in Fig. 5.31) is reported (Chaillou[201]) to be less than 5×10^{-4} lusec. With such a seal, a rotary motion of speeds up to 18,000 r.p.m. was obtained, but in order to maintain a pressure of 10^{-5} torr in the chamber, a double gasket seal was used, providing a guard vacuum between the gaskets (Sitney[1142], Chaillou[201]).

51.76 Friction seals. It is possible to seal rotary shafts by lubricating the friction surface of the stationary and rotating parts. These seals are illustrated in Fig. 5.33, for a vertical and a horizontal shaft.

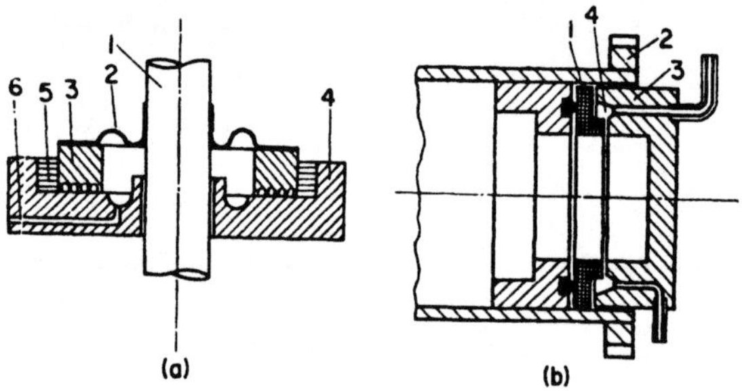

FIG. 5.33 Friction seals: (a) with diaphragm and oil lubrication; (b) with cellulose acetate butyrate on lubricated cast iron (after Toffer[1229])

The shaft protruding in the evacuated space (1, Fig. 5.33a) carries a membrane (2) which holds a friction ring (3) provided with small grooves on its lower surface. When the shaft and the ring (3) rotate, the surface of the ring (3) slides on the surface of the stationary casing (4). The lubrication of the surfaces (in friction) is made by the oil (5). The oil flowing through the seal is collected by (6).

Toffer[1229] used the friction of a Tenite II (Tennesee Eastman cellulose acetate butyrate) ring (1) which rotates with the gear (2, Fig. 5.33b), running against the lapped face of the cast-iron stationary back plate (3). The oil for the seal was supplied by a V-groove (4) surrounding the friction surface. Capillary forces retained the oil between the sealing surfaces, which were covered by a thin film of Molycote and oil mixture upon assembly. The Tenite ring is sealed to the rotating body by an O-ring seal. According to Toffer[1229] this seal was able to maintain a pressure of the order of 10^{-5} torr, operating it at 137 r.p.m. The use of a cooling fan is recommended.

The use of Teflon as a self-lubricating friction surface is discussed by Willens[1316]. For an oil-sealed rotary shaft see Fig. 3.37b.

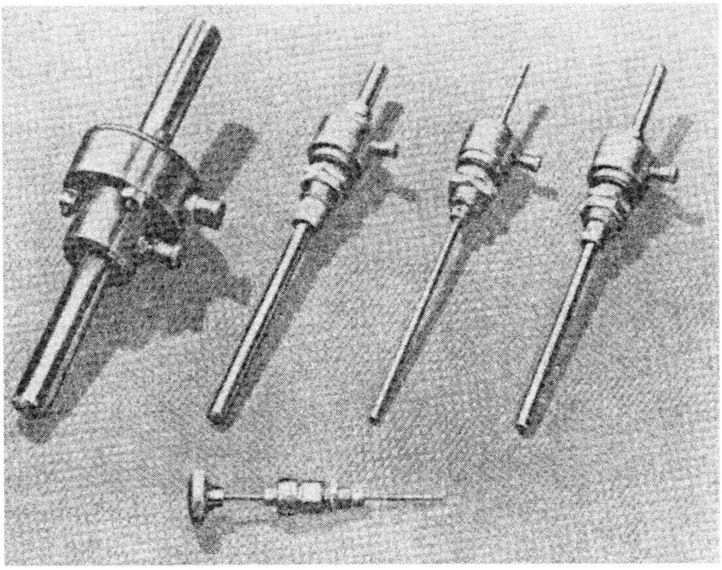

PLATE 20. Shaft seals *(Courtesy of Edwards[328])*

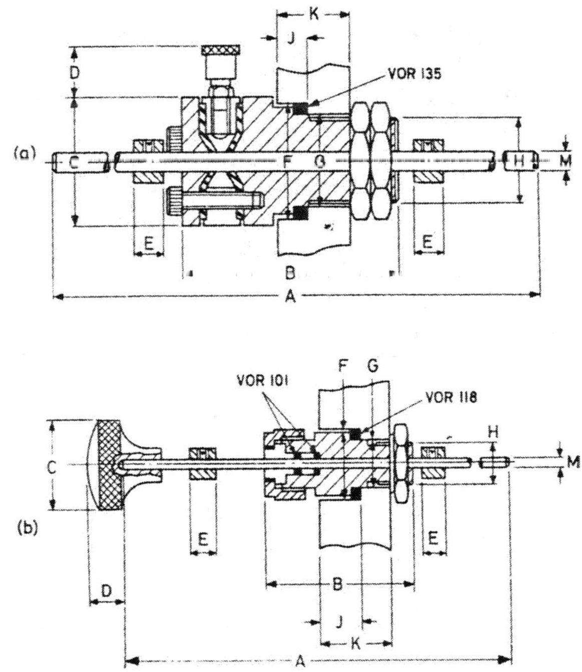

FIG. 5.34 Dimensions of shaft seals supplied by Edwards[328] (Table 5.5)

TABLE 5.5. SHAFT SEALS AVAILABLE FROM Edwards[328] (Fig. 5.34)

Size	1/8 in.	1/4 in.	3/8 in.	1/2 in.
A in. mm	6 152	12 305	12 305	12 305
B in. mm	1 13/16 46	2 9/16 65	2 9/16 65	2 11/16 68
C in. mm	1 1/4 32	1 1/2 38	1 5/8 41	1 11/16 43
D in. mm	3/8 9.5	13/16 21	13/16 21	13/16 21
E in. mm	5/16 8	3/8 9.6	5/8 16	3/4 19
*F** in. mm	0.774 19.66	1.343 34.13	1.343 34.13	1.343 34.13
*G** in. mm	0.564 14.33	1.064 27.03	1.064 27.03	1.064 27.03
H in.	1/2 BSF	1 BSF	1 BSF	1 BSF
J_{max} in. mm	0.562 14.29	0.342 8.71	0.350 8.89	0.342 8.71
J_{min} in. mm	0.187 4.76	0.340 8.66	0.345 8.76	0.340 8.66
K_{max} in. mm	1 25.4	1 25.4	1 25.4	1 25.4
K_{min} in. mm	7/16 11	9/16 14	9/16 14	9/16 14
M in. mm	1/8 3.2	1/4 6.3	3/8 9.5	1/2 12.7
Figure 5.34	b	a	a	a

* Tolerance, plus 0.002 in., plus 0.05 mm, minus 0.

51.77 Commercially available shaft seals. Shaft seals for the transmission of rotary and/or translational motion are manufactured by the various firms constructing vacuum equipment (Section 11.3). Some examples of shaft seals, available as ready made items are described in this Section.

Edwards[328] supply a small shaft seal (with 1/8 in. shaft) using O-rings (Fig. 5.34b), and larger shaft seals (Fig. 5.34a) using Wilson seals. These shaft seals are shown in Plate 20, and their dimensions are listed in Table 5.5.

Balzers[444] supply shaft seals for rotary motion (max. torque 0.2 kg/m; max. rotation 1500 r.p.m.) as shown in Fig. 5.35 and listed in Table 5.6.

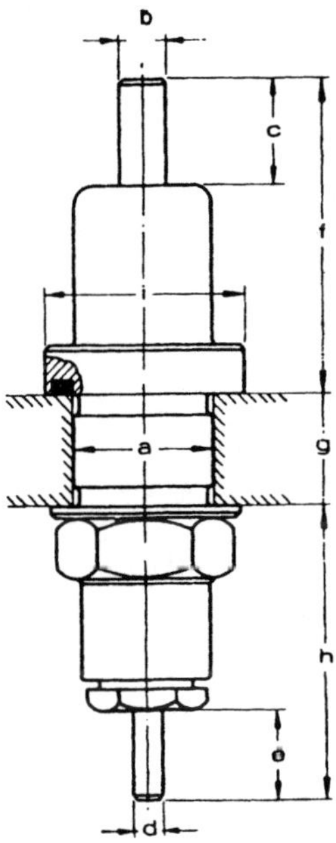

FIG. 5.35 Dimensions of shaft seals supplied by Balzers[444] (Table 5.6)

TABLE 5.6. DIMENSIONS (mm) OF ROTARY SEALS AVAILABLE FROM Balzers[444] (Fig. 5.35)

Size	*a*	*b*	*c*	*d*	*e*	*f*	*g*	*h*	*l*
1	36	12	36	9	30	87	28	80	52
2	36	12*	12*	9	30	23	28	80	52
3	45	10	22	10	22	80	28	80	65
4	45	10	27	10	12	42	28	60	69

* With internal hole.

The rotary seals available from Leybolds[762] are shown in Fig. 5.36 and their characteristics are listed in Table 5.7.

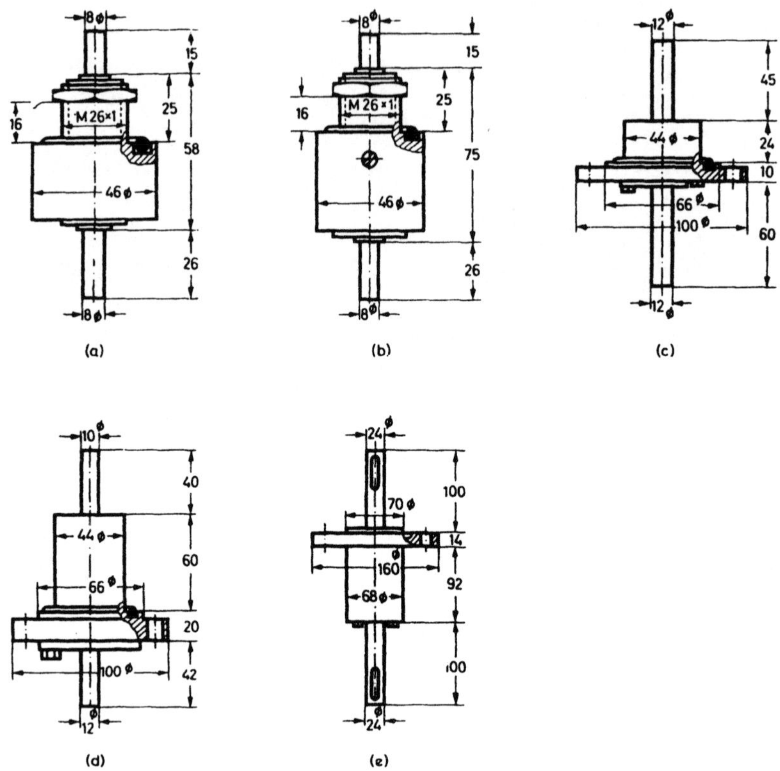

FIG. 5.36 Dimensions of shaft seals supplied by Leybold[762] (Table 5.7)

TABLE 5.7. HIGH VACUUM ROTARY TRANSMISSIONS* SUPPLIED BY Leybold[762] (Fig. 5.36)
(Dimensions in mm)

Type**	*G* 27	*G* 27 *K*	*F* 45	*F* 45 K	*F* 65 K
Shaft diameter	8	8	12	10	24
Admissible r.p.m.	100	3000	100	3000	3000
Admissible torque kg.m	0.2	0.2	1	0.5	10
Figure 5.36	(a)	(b)	(c)	(d)	(e)

* Nominal leak rate less than 10^{-5} torr. l./sec; oil seal.
** The symbol *K* means that the shaft seal is equipped with ball-bearings.

The rotary seals available from Pfeiffer[987] have the construction illustrated in Fig. 5.37 and the dimensions listed in Table 5.8.

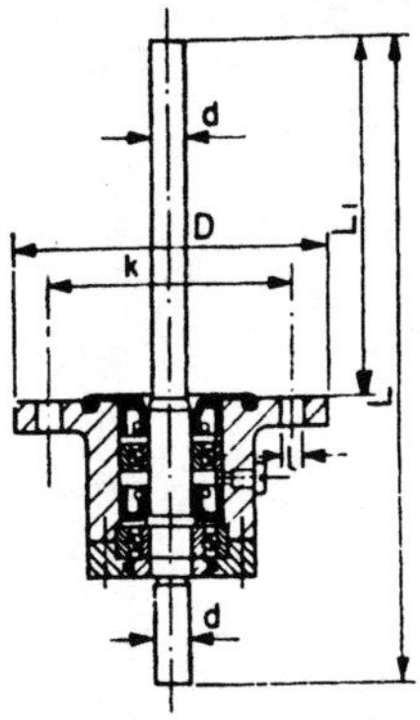

FIG. 5.37 Dimensions of shaft seals supplied by Pfeiffer[987] (Table 5.8)

TABLE 5.8. DIMENSIONS (mm) OF ROTARY* SEALS AVAILABLE FROM Pfeiffer[987] (see Fig. 5.37)

Type	d	L	L_1	d_1	D	k	l
10	10	180	100	30	88	70	7
20	20	270	120	32	120	90	14

* For up to 1000 r.p.m.; oil lubrication.

Heraeus[542] supply rotary shaft seals with shafts of 10 and 20 mm diameter. Figure 5.38 illustrates some details of the construction of these shaft seals and gives their dimensions. Table 5.9 lists the service characteristics of these seals.

TABLE 5.9. CHARACTERISTICS OF ROTARY SEALS AVAILABLE FROM Heraeus[542] (see Fig. 5.38)

Type	10	20
Figure	a	b
Admissible torque kg.m	0.5	5
Admissible r.p.m.	100	1000

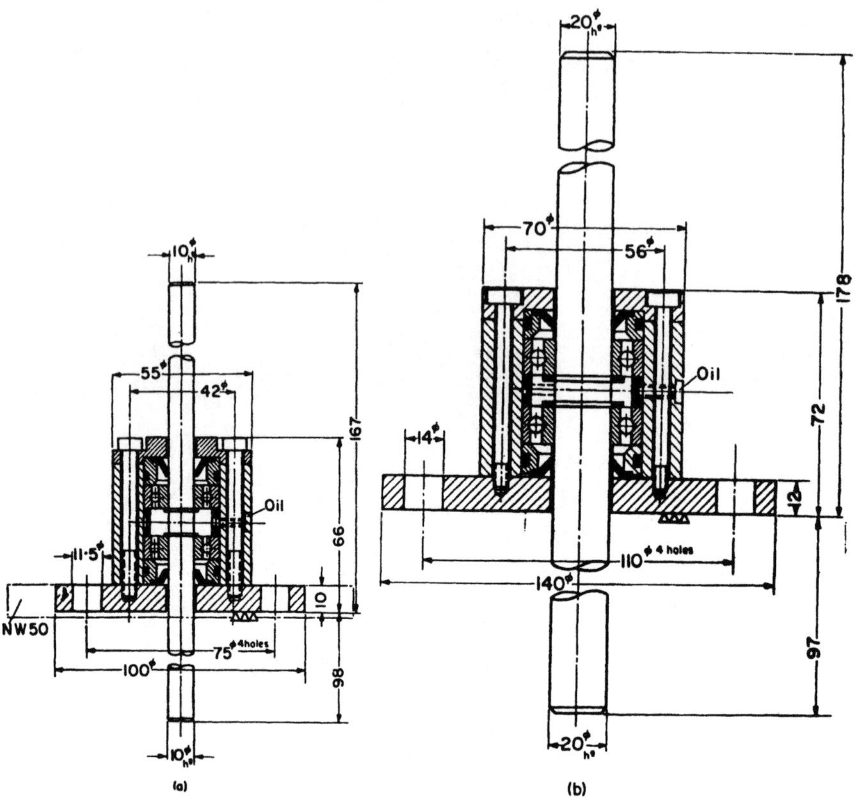

FIG. 5.38 Dimensions of shaft seals supplied by Heraeus[542] (Table 5.9)

Consolidated Vacuum Corp.[231] supply rotary seals as shown in Fig. 5.39 and listed in Table 5.10.

TABLE 5.10 CHARACTERISTICS OF ROTARY SEALS AVAILABLE FROM C.V.C.[231] (Fig. 5.39)

Type	SR-25	SR-37	SR-50	SR-75
Figure	a	b	c	d
Shaft diameter in.	1/4	3/8	1/2	3/4
Rotation r.p.m.	—	500	500	500
Motion	rotary and sliding		rotary	rotary
Lubrication	grease	oil	oil	oil

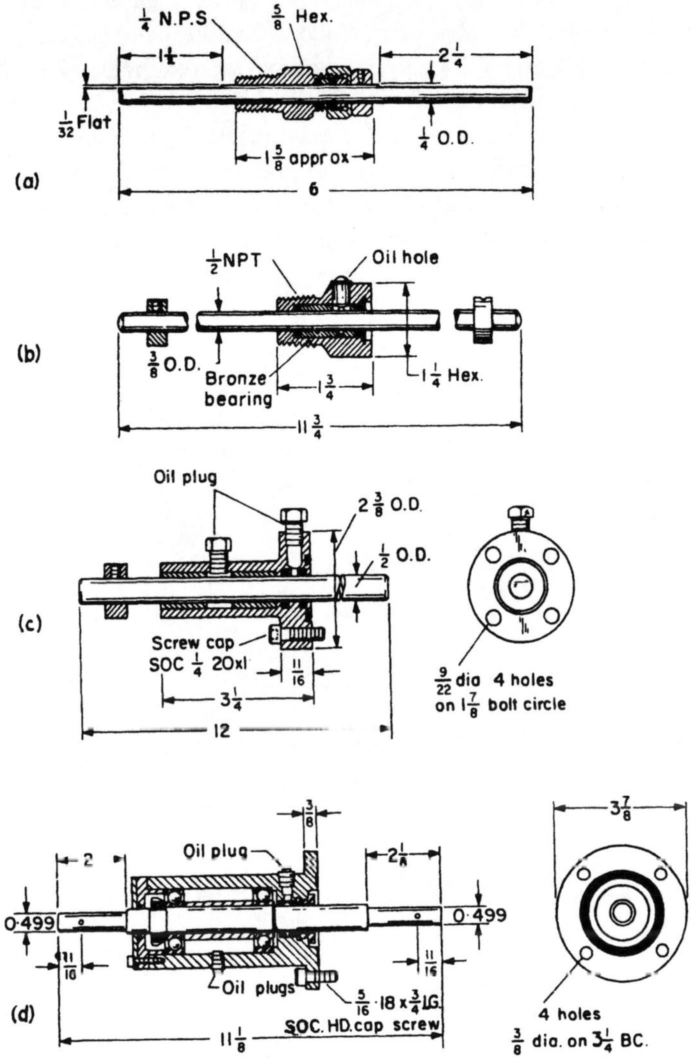

FIG. 5.39 Dimensions of shaft seals supplied by C.V.C.[231] (Table 5.10)

National Research Corp.[912] manufacture rotary seals (Fig. 5.40a) and push-pull vacuum seals (Fig. 5.40b). Table 5.11 list the dimensions of the rotary seals.

Officine Galileo[953] supply two rotary seals as shown in Fig. 5.41 and listed in Table 5.12.

51.78 Seals for angular displacement. The most difficult seals required in vacuum technique are those permitting tri-dimensional displacements of a given part. Such a displacement is composed (Fig. 5.42a) of an axial motion

(Ax), a radial (R) and an angular motion (An). The angular motion may be centric (Fig. 5.42b) or eccentric (Fig. 5.42c). Usually the construction of the seals with eccentric angular motion is much more simple than of those requiring centric angular displacement.

The design of the seals for angular displacements depends not only on the kind of this displacement, but also on the amount of motion required. The amount of motion is given usually as the *angular deflection* i.e. the angle between the middle and the extreme positions during deflection (see Fig. 5.42b, c). The maximum angular deflections obtainable with various seal designs are listed in Table 5.13.

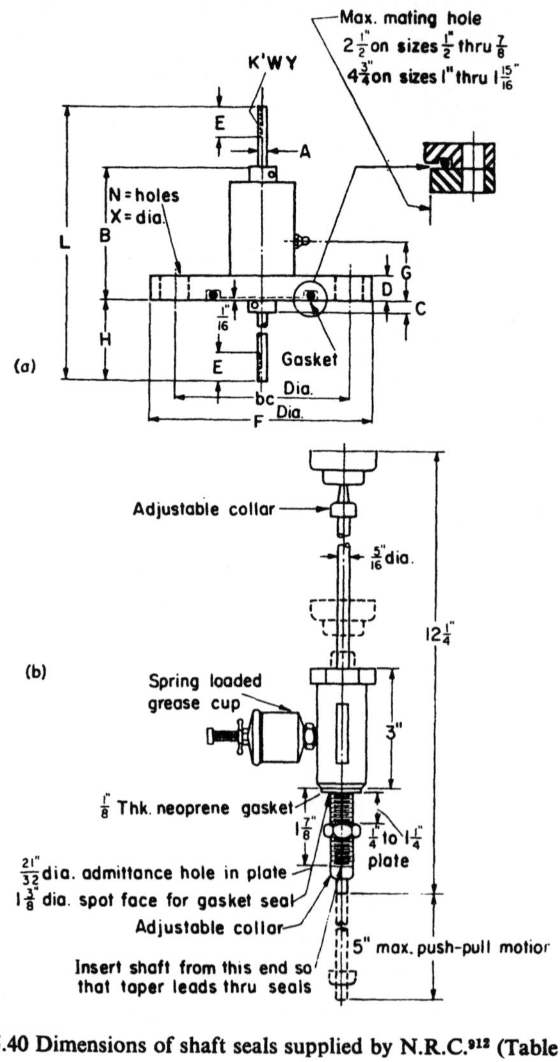

FIG. 5.40 Dimensions of shaft seals supplied by N.R.C.[912] (Table 5.11)

TABLE 5.11. DIMENSIONS (inches) OF ROTARY SEALS* SUPPLIED BY N.R.C.[912]
(Fig. 5.40a)

A	*B*	*C*	*D*	*E*	*F*	*G*	*KWY*	*bc*	*N*	*X*	*H*	*L*
1/2	3 23/32	9/32	5/8	3/4	6	1 7/8	$\frac{1}{16}\times\frac{1}{8}$	$4\frac{3}{4}$	4	3/4	$1\frac{11}{16}$	$6\frac{1}{2}$
5/8	3 25/32	11/32	5/8	15/16	6	1 7/8	$\frac{3}{32}\times\frac{3}{16}$	$4\frac{3}{4}$	4	3/4	$2\frac{1}{8}$	$7\frac{1}{2}$
3/4	3 13/16	3/8	5/8	$1\frac{1}{8}$	6	1 7/8	”	”	4	3/4	$2\frac{3}{8}$	$8\frac{1}{2}$
7/8	3 29/32	15/32	5/8	$1\frac{5}{16}$	6	”	”	”	4	3/4	$2\frac{9}{16}$	9
1	6 17/64	61/64	3/4	$1\frac{1}{2}$	9	2 3/4	$\frac{1}{8}\times\frac{1}{4}$	$7\frac{1}{2}$	8	3/4	$3\frac{7}{16}$	$11\frac{1}{2}$
1 3/16	6 5/16	1	3/4	$1\frac{7}{8}$	9	2 3/4	”	”	8	3/4	$3\frac{7}{8}$	$12\frac{1}{2}$
1 7/16	6 13/32	$1\frac{1}{8}$	3/4	$2\frac{3}{16}$	9	2 3/4	$\frac{3}{16}\times\frac{3}{8}$	”	8	3/4	$4\frac{9}{16}$	$14\frac{1}{2}$
1 11/16	$6\frac{1}{2}$	$1\frac{3}{16}$	3/4	$2\frac{5}{8}$	9	2 3/4	”	”	8	3/4	$5\frac{1}{8}$	16
1 15/16	$6\frac{3}{8}$	$1\frac{5}{16}$	3/4	3	9	2 3/4	$\frac{1}{4}\times\frac{1}{2}$	”	8	3/4	$5\frac{3}{4}$	$17\frac{1}{2}$

* O-rings A.S. 39 for 1/2 to 7/8 rotary seals (see Table 3.19A).
A.S. 57 for 1 to 1 15/16 rotary seals.

TABLE 5.12. CHARACTERISTICS OF ROTARY SEALS AVAILABLE FROM Officine Galileo[953] (see Fig. 5.41)

Size	6	12
Admissible torque kg.m	0.2	1.5
Admissible r.p.m.	1400	600

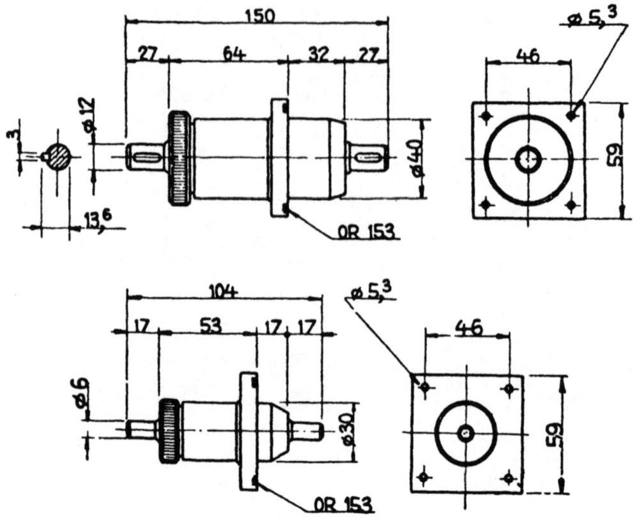

FIG. 5.41 Dimensions of shaft seals supplied by Officine Galileo[953] (Table 5.12)

Kalmus[650] described a seal (Fig. 5.43) which consists of a separate flange of special shape and a cylindrical tube. The seal between them is accomplished by means of a rubber O-ring, which also gives the coupling its flexibility. The O-ring fits round a polished section of the tube and is kept under compression in a groove which has an outside diameter 0.012 in. smaller than the natural outside diameter of the O-ring. The width of the groove is 0.005 in. less than the thickness of the ring, and its depth is sufficient to prevent the ring being pushed inwards along the tube by the pressure difference. The tube can be tilted relative to the flange (see Table 5.13) and can be kept at the desired angle by adjusting three screws in a collar, which itself can be located anywhere along the length of the tube.

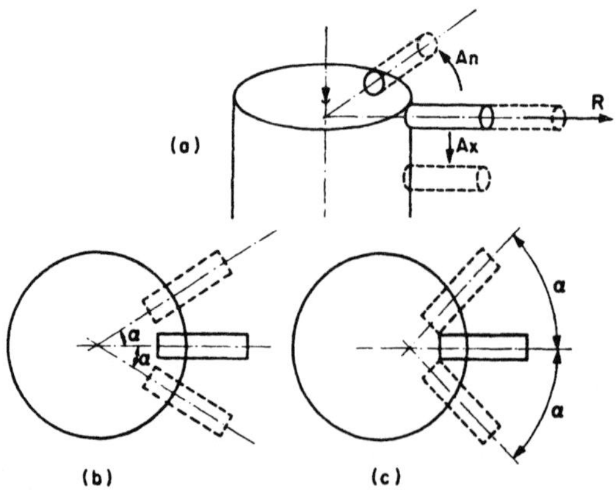

FIG. 5.42 Angular motion: (a) tridimensional motion; (b) centric axial motion; (c) eccentric axial motion

TABLE 5.13. ANGULAR MOTION SEALS

Maximum angular deflection α (deg)*	Centric (C) Eccentric (E)	Remarks	Reference
5	E	Bellows; see Fig. 5.9a	Brose[171]
5	E	O-ring; see Fig. 5.19b	Ardenne[47]
10	E	O-ring; Fig. 5.43	Kalmus[650]
10	E	Rubber pipe; Fig. 5.2d	Fremlin[401]
10	C	O-ring seal on cylindrical surface; Fig. 5.46	Smotrich[1159]
12	E	O-ring seal on spherical surface; Fig. 5.44a	Lee[754]
30	E	As before; Fig. 5.44b	Retzloff[1048]
50	E	Wilson seal on sphere; Fig. 5.45	Brannen[161]
85	C	O-ring seal on cylindrical surface; Fig. 5.47	Verba[1266]
90	C	As before; Fig. 5.48	Louckes[785]

* See Fig. 5.42.

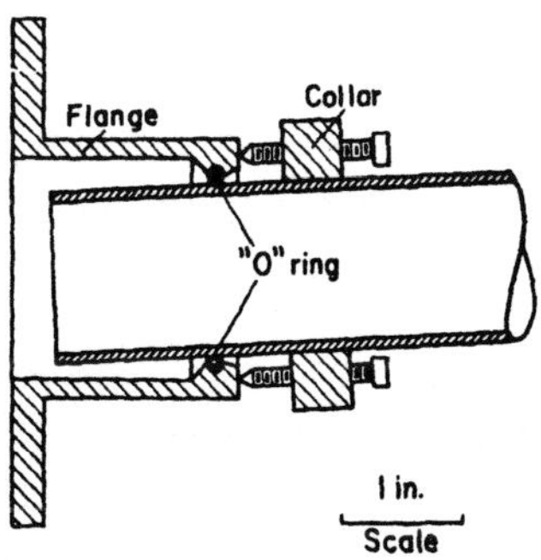

FIG. 5.43 Pipe-to-pipe, angular seal. Reproduced from Kalmus[650] *(Courtesy of Pergamon Press)*

Figure 5.44 shows two designs of tilting joints sealed with O-rings against spherical surfaces. Lee[754] designed the seal shown on Fig. 5.44a consisting of a part (1) having a spherical surface and being brazed to the flange (5), a rim (3) and the second flange (4). The rim (3) and the flange (4) are cut with concave spherical surfaces of the same radius as (1). The sealing O-ring is placed in a groove in the flange (4).

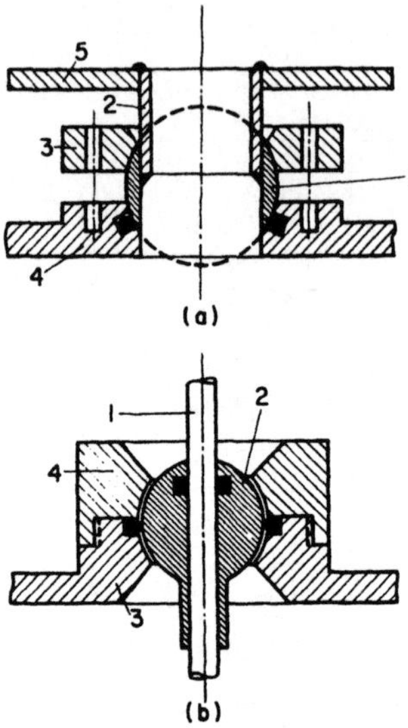

FIG. 5.44 Seals for angular motion, using O-rings sealing against spherical surfaces: (a) only tilting. After Lee[754] *(Courtesy of The Institute of Physics and The Physical Society, London)*; (b) with sliding motion. After Retzloff[1048] *(Courtesy of The American Institute of Physics)*

The seal described by Retzloff[1048] permits simultaneous tilting and sliding (Fig. 5.44b). It consists of a sphere (2) fitting a spherical socket (3, 4) and sealed by an O-ring. A rod (1) may slide along a diameter of the sphere being sealed by another O-ring. In the construction by Retzloff[1048], the sphere had 1 in. diameter, and the socket was made of two parts, machined to an accuracy of 0.001 in.

Brannen[161] described a tilting joint using a Wilson seal. The sphere (1) is pinned (2, Fig. 5.45) on a diameter to the holding ring (3) sealed to the vacuum chamber. A rubber ring (4) held in a frame as in Wilson seals (Section 51.74) and presses against the air-side surface of the sphere. This seal permits quite a large angular deflection (see Table 5.13). By pinning the sphere to the inner ring of a thrust bearing, whose outer ring is fixed to the chamber, the sphere can be rotated as well.

Smotrich[1150] described a seal permitting 1 in. vertical travel, 20° of rotation about a vertical axis and unlimited rotation about a horizontal axis. The rota-

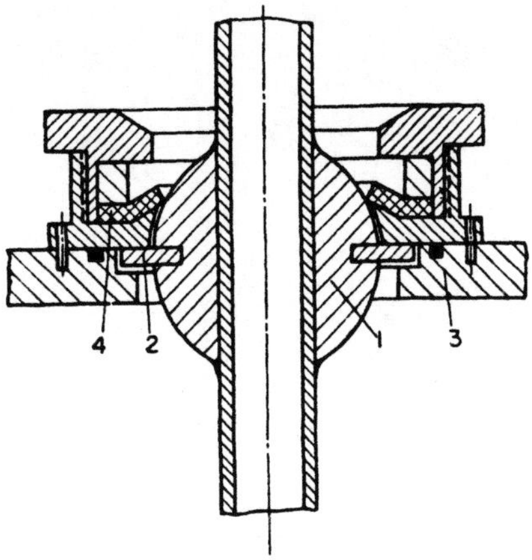

FIG. 5.45 Wilson seal on spherical surface. After Brannen[161] *(Courtesy of The American Institute of Physics)*

tional motion about a vertical axis and the vertical travel are obtained by moving the cylinder (1, Fig. 5.46) in the stationary housing (2) containing a gasket (3) sealing on a cylindrical surface, and seated in a groove machined in the surface of the cylinder. The rotation about a horizontal axis is sealed by a double O-ring seal (4).

Verba[1266] used a scattering chamber (Fig. 5.47) consisting of a stationary cylinder (1), upon which a cylindrical sleeve (2) slides in a helical motion.

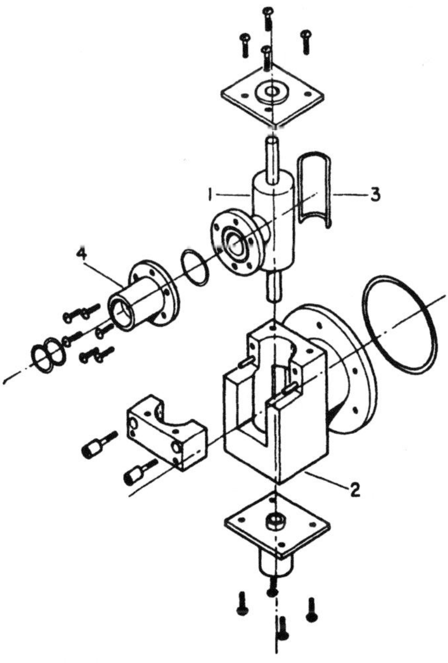

FIG. 5.46 O-ring seal for axial, radial and angular displacement. After Smotrich[1159] *(Courtesy of The American Institute of Physics)*

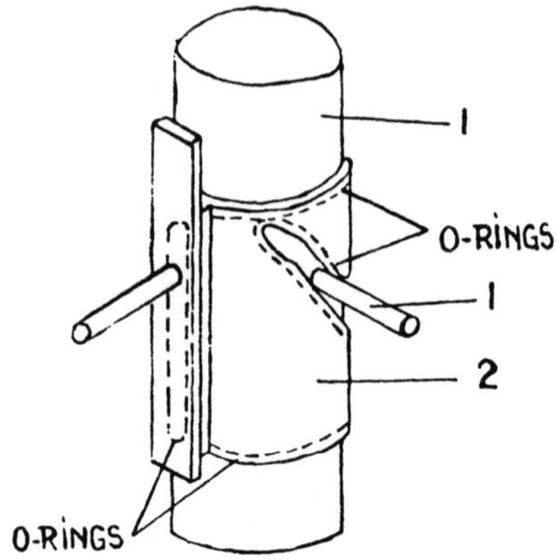

FIG. 5.47 O-ring seal for helical motion. After Verba[1266] *(Courtesy of The American Institute of Physics)*

This chamber is sealed by O-rings round the top and bottom of the sleeve, round the spiral slot, and round the vertical slot. This seal permits a continuous angular motion over a wide range (see Table 5.13). It is recommended to lubricate the O-rings with silicone stopcock grease.

Louckes[785] used a vacuum seal (Fig. 5.48) consisting of a flexible strip (*a*), sliding over an O-ring (*b*) of 0.139 in. cross-sectional diameter compressed by about 0.007 in. The strip was made of 0.005 in. stainless steel (304) with angle stiffeners (*c*) of 0.01 in. stainless steel, spot welded across the strip, and spaced at 5/16 in. The offset in the keeper piece (*d*) was about 0.007 in., giving an overall clearance for the strip of about 0.002 in.

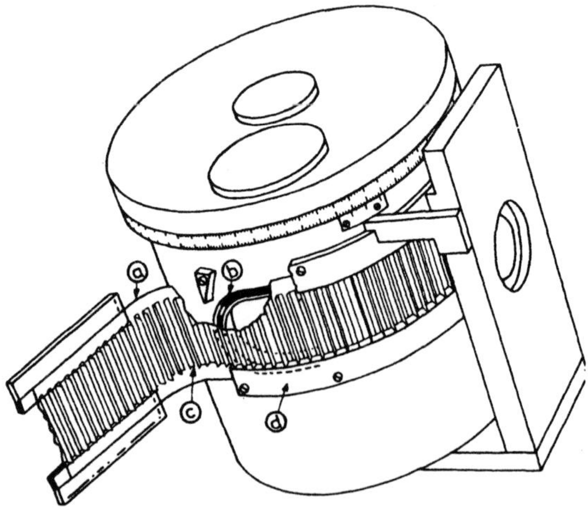

FIG. 5.48 O-ring seal for large angular motion. After Louckes[785] *(Courtesy of The American Institute of Physics)*

51.8 Motion seals using a guard vacuum

The principle of the guard vacuum seals is explained in Section 38.23. In sliding or rotary seals the guard vacuum is used to ensure good vacuum-tightness with liquid, gasket or double chamber seals.

A sliding seal using double piston, O-rings, oil and guard vacuum was described by Lake[729a] (see Fig. 5.21b).

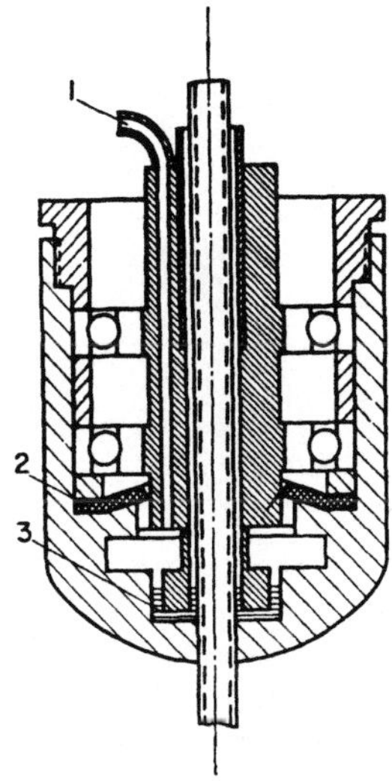

FIG. 5.49 Rotary seal with mercury and guard vacuum for stationary shaft and rotating housing. After Brueschke[176] *(Courtesy of The American Institute of Physics)*

The seal described by Brueschke[177, 178] using a mercury seal and guard vacuum and based on the principle shown in Fig. 3.30c, may be constructed having either the housing or the shaft stationary. The first case is shown in Fig. 3.30c and has the guard vacuum connexion on the housing. In the second case (Fig. 5.49) the guard vacuum connexion (1) is placed along a cylindrical stationary part around the shaft. The Wilson seal (2) is in this case placed at the bottom of the seal, above the small mercury bath (3). The lower portion of the stationary shaft (tube) is provided with vanes to keep most of the mercury from rotating with the housing, and thus the effects of the centrifugal force are greatly reduced. If the first pumpdown is not done carefully (see also Fig. 3.30c) or a sudden leakage through the Wilson seal occurs, the mercury is splashed in the seal.

A rotary seal using liquid metals (gallium, indium or tin) and guard vacuum was described by Milleron[865, 865a] (Fig. 3.39b).

Armstrong[49b] described a large rotary seal ($6\frac{1}{4}$ in. diameter) using a double O-ring seal (Teflon O-ring with rubber core, 1, Fig. 5.50) and a guard vacuum

(2). With a guard vacuum of 5×10^{-3} torr, it was possible to maintain an ultimate pressure of 3×10^{-7} torr in the container. The O-rings used were compressed from a cross-sectional diameter of 0.24 in. to 0.22 in.

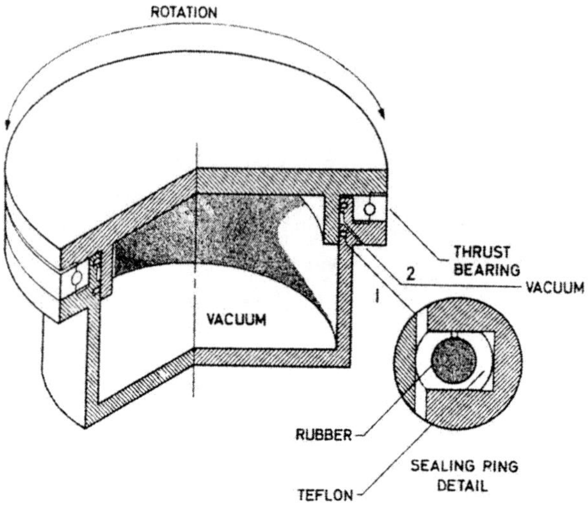

FIG. 5.50 Rotary vacuum seal using Teflon O-rings and guard vacuum. After Armstrong[496] *(Courtesy of The American Institute of Physics)*

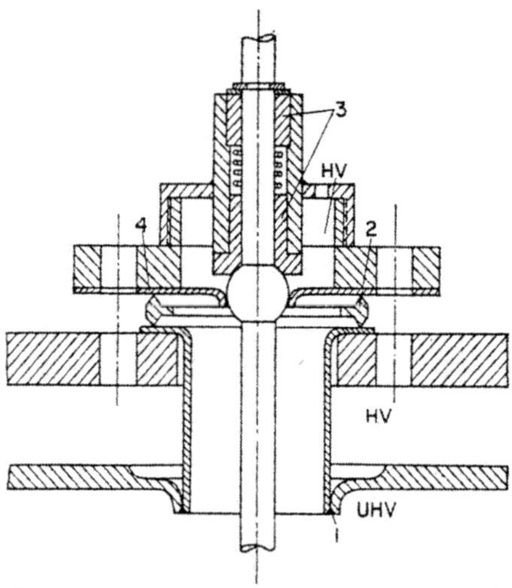

FIG. 5.51 Rotary seal for ultra-high vacuum double chamber. After Ehlers[331] *(Courtesy of Pergamon Press)*

A motion seal using a guard vacuum between V-shaped Teflon gaskets is shown in Fig. 5.26 (Champeix[203]). Sitney[1142] and Roberts[1069] used spring-loaded lip seals (see Section 51.74) forming between them a space where the guard vacuum is used. Sitney[1142] quotes speeds of 6000–18000 r.p.m. in spaces having a pressure of 8×10^{-6} torr. Roberts[1069] used his seal at 4200 r.p.m. and measured a pressure of 7×10^{-7} torr at 1000 r.p.m. and 1×10^{-7} torr at zero r.p.m. in the chamber.

The rotary seals used in chambers with double wall and guard vacuum (see Section 38.23) are sealed with regular (multiple) O-ring seals (Section 51.71), or lip seals (Sections 51.74, 51.75) at the point where the shaft crosses the outer wall, and a simple cylindrical or spherical clearance seal at the crossing of the inner wall. Figure 5.51 shows such a seal (Ehlers[332]). The joint is welded at (1) has a "coined" copper gasket at (2) and a rubber gasket seal at (3). The inner wall (4) has a conical seat in which the sphere on the shaft, seals by friction. The tightness is maintained by the high vacuum (HV) in the space between the walls, ensuring the possibility of holding an ultra-high vacuum (UHV) in the chamber.

5.2 MAGNETIC TRANSMISSION

52.1 Translation using magnetic fields

Magnetic fields may be used to transmit motion inside the evacuated space by placing a magnet (or electromagnet) outside the chamber and transmitting the field through the wall, which should be of a non-magnetic material. The electromagnet may be placed in the chamber as well and energized from the outside, but this solution presents the need of electrical lead-throughs and degassing or break-down difficulties.

The translational motion of iron slugs or rings is achieved using rod (Fig. 5.52a), or horseshoe (Fig. 5.52c) permanent magnets or electromagnet coils (Fig. 5.52b, d). With these methods simple motions are easily produced, but also complex motions may be obtained adding various mechanical arrangements. Table 5.14 lists some applications of this technique described elsewhere in this book.

To obtain a reliable and reproducible motion, using outside magnets, it is recommended that the following simple requirements be observed.

(1) The outside magnet should be placed as close to the wall as possible.

(2) The wall must be of nonmagnetic material.

(3) The shape of the iron slug should match both the shape of the magnet (Fig. 5.52a, c) and that of the walls or inside parts guiding the motion.

(4) In order to minimize degassing, friction and scratching, slugs embedded in glass should be preferred.

(5) The strength of the field and the distance between the magnet and iron slug, should be chosen in such a manner as to avoid strong shocks*

* Except when the purpose is to break-off ampules (Section 61.5).

TABLE 5.14. APPLICATIONS OF MOTION ACTUATED BY OUTSIDE MAGNETS*

Direction of motion	Shape of moved part	Position of moved part		Purpose	Remarks	Reference
		free	energized			
Vertical	Iron slug	floating on Hg, In	raised above Hg level	Change of Hg, In level	Fig. 6.8b	Paty[978] Hammond[505] Axelrod[59]
	Iron slug	same	same	Opening of sinter-glass Hg cut-off	Fig. 6.14.a	Glaister[451]
	Glass cap with Fe core	same	same	CO_2 admission	Fig. 6.14b	Köhler[708]
	Iron slug	floating on Hg	immersed in Hg	Change of Hg level	Fig. 6.8a	Thomson[1221]
	Iron slug	raised	immersed in Hg	same	Fig. 6.6	Haul[519]
	Iron cylinder	—	raised	Closes Sn valve	Fig. 6.29a	Blanaru[130]
	Iron cylinder in glass	—	raised	Opens ground glass valve	Fig. 6.33b	Yarwood[1341] Vogl[1271] Thiele[1218b]
	Iron slug	—	—	Breaks seals	Sec. 61.5	
Horizontal	Iron slug	Translation and rotation			Fig. 5.52	
	Iron tip	Translation		Ground glass valve	Fig. 6.32c	Lambe[730]
	Iron slug	same		same	Fig. 6.35	Decker[270]
	Iron slug in glass	same		Actuate ground ball on seat	Fig. 6.34b	Metzler[852]

* Magnetic valves see Figs. 6.85-6.89.

on the wall or the level of liquid (mercury, indium) placed inside the chamber. It is advisable to use shock-absorbers (springs) in the places where the slug may knock the wall (Fig. 5.52b).

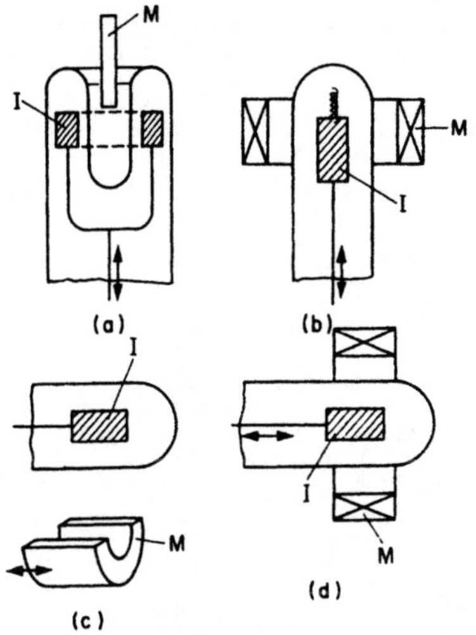

FIG. 5.52 Translation motion by outside magnet: (a) vertical motion with rod magnet and iron ring; (b) vertical motion with electromagnet and iron slug; (c) horizontal motion with horseshoe magnet and iron slug; (d) horizontal motion with electromagnet and iron slug

52.2 Rotary motion transmitted by magnetic fields

Magnetic transmission permits the rotation (slowly or at high speeds) of parts placed inside the evacuated space. The part to be rotated should be mounted on proper bearings and should be connected to an armature able to be rotated by the magnetic field. For slow rotation the arrangements shown in Fig. 5.53 may be used, with a magnetic rod (Fig. 5.53a) or a horseshoe

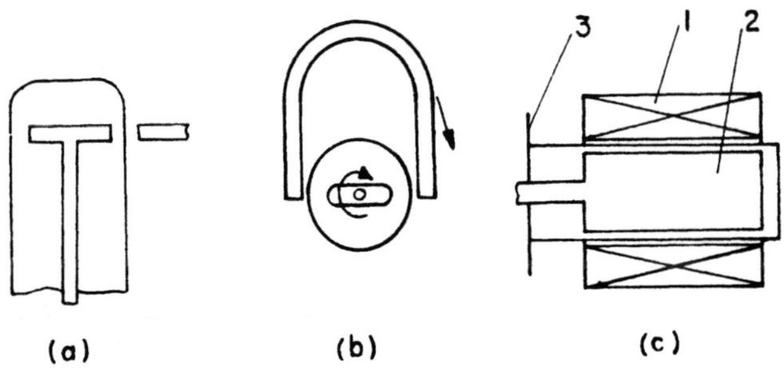

FIG. 5.53 Rotary motion by outside magnet: (a) magnet placed inside and rotated by outside magnet; (b) armature rotated by outside horseshoe magnet; (c) shaft rotated by outside magnet (see Plate 21)

magnet (Fig. 5.53b). Figure 5.53c and Plate 21 shows the rotary feed-through supplied by Varian[1257a]. In this item the rotary motion is transmitted by a magnetic coupling between a rotatable magnet (1) outside, and a rotatable shaft (2) inside the vacuum chamber (3). When the outside magnet is rotated (750–500 r.p.m. maximum), the shaft inside the vacuum is caused to

PLATE 21. Rotary magnetic feed-through *(Courtesy of Varian[1257a])*

rotate in the same way. This feedthrough is available in a standard size (about 2 9/16 in. diameter and 5 in. long, with a shaft of 5/16 in. being able to transmit a torque of 4 in.lb) and in a large size (5 11/16 in. diameter and about 7 in.long, being able to transmit a torque of 6 ft.lb). Both the feedthroughs are bakeable (with magnet removed) at about 400 °C since they are equipped with Conflat flanges (see Section 38.56b).

Coenraads[219a] described a high speed (10,000 r.p.m.) magnetic drive (Fig. 5.54) used with molybdenum-disulphide coated ball bearings, in a vacuum of 1×10^{-7} torr. A quartz window (1 Fig. 5.54) about 2 mm thick and 60 mm diameter was sealed with O-rings (Section 72.8). The magnetic coupling between magnets (2) and (3) is made leaving a gap of about 5 mm between them.

Translational motion can be combined with rotary motion. Gerber[444a] described a feedthrough for "dual-motion" with an electromagnetic coupling for rotary motion and a bellows seal for the linear motion (limited to about 2 in.).

Translational motion may be transmitted magnetically into the evacuated space (Fig. 5.55) and transformed there to a controlled rotation (Zobac[1356]).

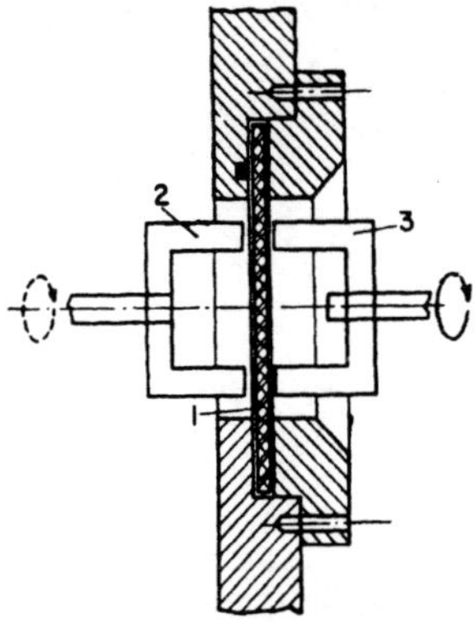

FIG. 5.54 Magnetic drive. After Coenraads[219a] *(Courtesy of The American Institute of Physics)*

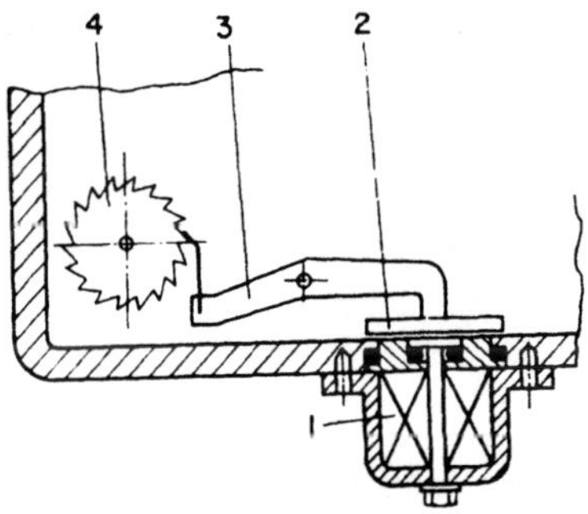

FIG. 5.55 Rotary motion from strokes given by an electromagnet

A stroke motion is exerted on the armature (2) by the electromagnet (1), and through the lever (3) it is transformed in a controlled rotary movement of the wheel (4).

5.3 ACTUATION BY HEAT TRANSFER OR ELECTRIC CURRENT

53.1 *Motion based on thermal expansion*

The thermal expansion of metal wires or bi-metal ribbons can be successfully used to transmit motion into vacuum systems. The heat required to actuate the motion is transferred through the walls of the vacuum system or is generated inside by an electric current.

Bachman[65] described a method to position a movable part (1 Fig. 5.56a) by using the expansion of a wire (3). If the wire (3) is heated by resistance heating, its expansion permits the spring (2) to move the element (1). This method was used by Flinta[384] to actuate a leak valve (Fig. 6.125b). The same principle may be used without the spring, by utilizing the force of the gravity.

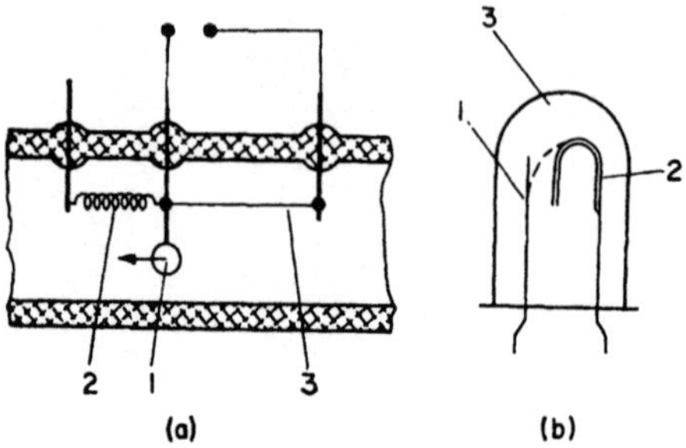

FIG. 5.56 Motion by expansion: (a) of wire; (b) of bi-metal

Bimetal strips or ribbons* may be successfully used to actuate motion in vacuum. The bimetal element (cantilever, U-shaped strip, helical coil) has a deflection proportional to the temperature rise. Heated from the outside through the wall, with a resistance placed inside or by an electric discharge the bi-metal can be used to move light parts or to connect or disconnect electrical circuits. Figure 5.56b shows the principle in which the bi-metal is used in the starters of discharge (fluorescent) lamps. Between electrodes (1 and 2) a glow discharge occurs (through the space 3 filled with neon, argon, neon-helium, etc.) which heats up the bi-metal (2). This expands, and short-circuits the starter. The discharge stops, the bi-metal cools, and re-opens the circuit. If the circuit is not changed by other switches, the described sequence continues in cycles.

53.2 Irreversible motion by burning-out

The most simple motion device (in vacuum) based on the "burning-out" consists of a part (1, Fig. 5.57a) hung on a wire (2). If the wire is burnt-out by passing a current through it, the part (1) will fall. Figure 5.57b shows the same method, but used with a spring (3) which moves the part (1) upwards, if the wire (2) is burnt out (Mönch[879]).

In the arrangement shown on Fig. 5.57c by burning out the limb (1), the weight (2) gives the required motion to the element (3). The same displacement can be achieved (Fig. 5.57d) by using a spring (2) (Bachman[65]).

*Available from e.g. H. Wiggin & Co., Birmingham 16, England, or W. M. Chace Co., Detroit 9, Mich., U.S.A.

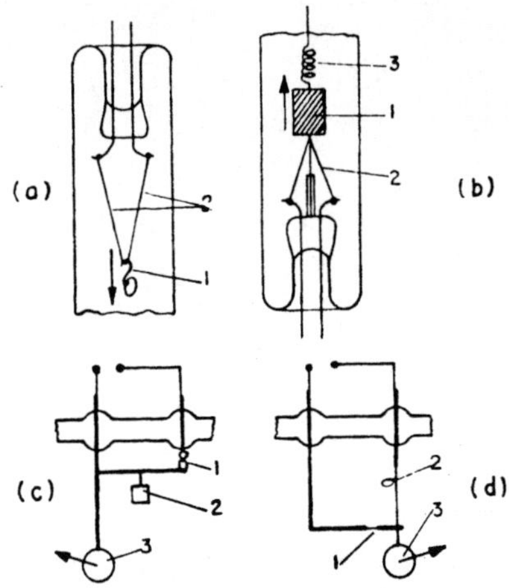

FIG. 5.57 Motion by burning-out: (a) with weight hung on wire; (b) same as (a) with spring; (c) with counter-weight; (d) with weight and spring

53.3 *Actuation with electric current*

The most obvious way to transmit motion into a vacuum vessel would be perhaps by using motors or servomotors placed inside the vacuum system, and energized through electrical connexions crossing the walls by lead-throughs. Some authors (Butler[188], Mc Nally[835]) quote such arrangements, but the method is not extensively used in vacuum applications, due to the large volume taken by the motors inside the vacuum, the difficulties of degassing, lubrication and the danger of break-down (see Section 41.1).

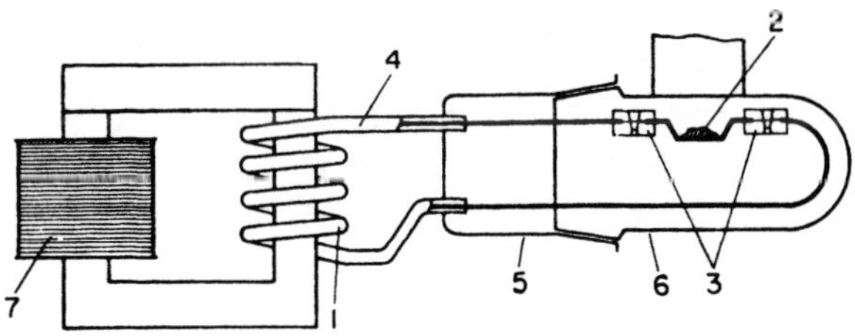

FIG. 5.58 Transmission of electric current by induction. After Bezold[121] *(Courtesy of Rudolf A. Lang Verlag, Esch/Taunus, Germany)*

An interesting solution is given by von Bezold[121] who used the induction transmission of high currents into the vacuum chamber (Fig. 5.58). The secondary coil (1, Fig. 5.58) is contained in a glass tube (4) bent into a spiral and connected to the vacuum system (6) by a conical ground joint (5). In the application described by von Bezold[121] the secondary circuit is closed through the heater (2), held in the clamps (3). The primary coil (7) is placed outside the vacuum system.

CHAPTER 6

SEALS USED IN THE TRANSFER OF MATERIALS

In every vacuum application, gases, liquids or solids are to be transferred from the vacuum system to the surrounding space either from outside into the vacuum system or from one part of the system to another. The seals used in connexion with material transfer must allow a port of adequate dimensions to be open during the transfer, while still keeping the rest of the system vacuum-tight. The port itself should be vacuum-tight when closed. The ports and their seals differ somewhat in design according to the state of the material to be transfered: gaseous (Section 6.1), liquid (Section 6.2) or solid (Section 6.3).

6.1 SEALS FOR THE TRANSFER OF GASES

During the exhaust of any vacuum system, gases are transferred from one part of the system to another, and it is always desirable to have means of shutting off various parts; cut-offs (Section 61.1) are used when the pressure difference is not too large, and stopcocks (Section 61.2) or valves (Section 61.3) for larger pressure differences.

When gases are to be introduced into the vacuum system reversible or irreversible seals may be used. Stopcocks or valves are used for large through puts, and leaks (Section 61.4) for small and very small throughputs (reversible seals). Various procedures are used for the opening of gas containers (Section 61.5) or their break-off (irreversible) seals.

61.1 Cut-offs

61.11 Principles. A cut-off is defined (British Standard[169a]) as a device in which a liquid surface is used to separate two parts of a vacuum system. The author prefers to restrict the meaning of the term (cut-off) to those liquid seals where the closing and opening action is due to the change in the *level* of the liquid (American Vacuum Society[26]). Thus liquid seals in which the closing action is achieved by immersing the edge of a cap in a liquid will not be considered as cut-offs, but as valves with a liquid seal (Section 61.32a).

Cut-offs are not available as separate units, but are generally incorporated in various systems (Kenty[664], Wyllie[1336, 1337], Slack[1147], Jenkins[632]) as devices for gas transfer, gas analysis, McLeod gauges etc.

A good cut-off should meet the following requirements:

(1) The cut-off should have the maximum conductance in the open position; thus a short length of large bore tubing should be included in the cut-off portion.

(2) It should close readily for as high as possible pressure differences. As the pressure difference increases, the arms of the cut-off have to be longer, in order to hold the required liquid column.

(3) The cut-off should give the lowest possible outgassing rate. For this purpose the liquid with the lowest vapour pressure should be used; the cut-off should be kept at the lowest possible temperature; and/or the exposed surface of the liquid should be kept to a minimum. Usually the liquid used is mercury, although its vapour pressure is not low. In special cases gallium or indium may also be used. To keep the surface to a minimum a small bore tube should be used, but this is contradictory to requirement (1). The mercury vapour can be prevented to some extent from penetrating into the evacuated space, by covering the mercury column with an oil having a low vapour pressure (Ayer[60a]).

(4) The cut-off should have the minimum volume and its operation should not appreciably change the volume of the system. This requirement is met by a concentric arrangement (Fig. 6.7d).

(5) The cut-off should be able to withstand sudden changes in pressure due to an accident or unskilled handling, without flooding the system with the liquid. To meet this requirement, cut-offs using floats are preferred.

(6) The cut-off should have a locking device to be used both in the open and closed position of the cut-off (Figs. 6.4–6.6).

A cut-off consists of a closing system (Section 61.13) and an actuating device for the liquid (Section 61.12).

61.12 Actuating devices in cut-offs. The actuation device of a reliable cut-off should allow the raising and lowering of the liquid, and the locking of the cut-off in the desired position.

(a) *Raising devices.* The liquid used in most of the cut-offs is mercury. When mercury is handled, it should be kept in mind that it is a *toxic* material. The precautions to be taken during the handling of mercury are described in detail by Biram[127] (see also Section 37.2).

Various systems may be used in order to raise the mercury in the cut-off (Stohr[1197a], Mönch[879], Holland-Merten[587], Roth[1083]). In all these systems the mercury is pushed into the cut-off from a container usually placed underneath the cut-off and connected to it through a pipe, joint, etc. The level of the mercury in the cut-off is changed either by changing its *level* in the container (Fig. 6.1, 6.2) or by changing the *volume* of the container (Fig. 6.3).

Figure 6.1 shows some raising devices, in which the level of the mercury in the cut-off is moved by changing the *position of the container* relative to the

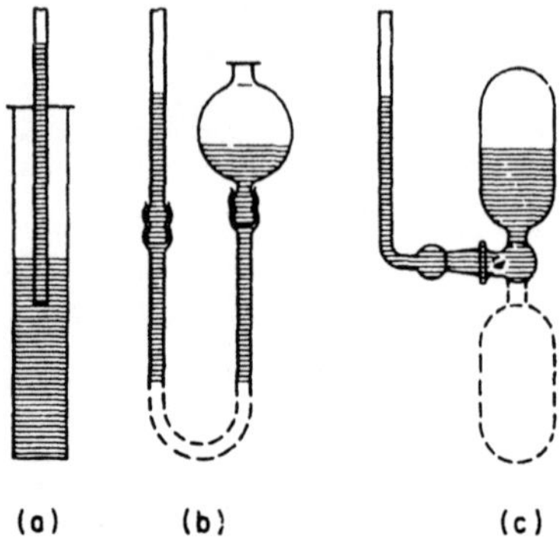

FIG. 6.1 Devices for raising the mercury, based on the change of the position of the container: (a) with concentric containers; (b) with a connexion of elastic pipe; (c) by reversing the position of the container

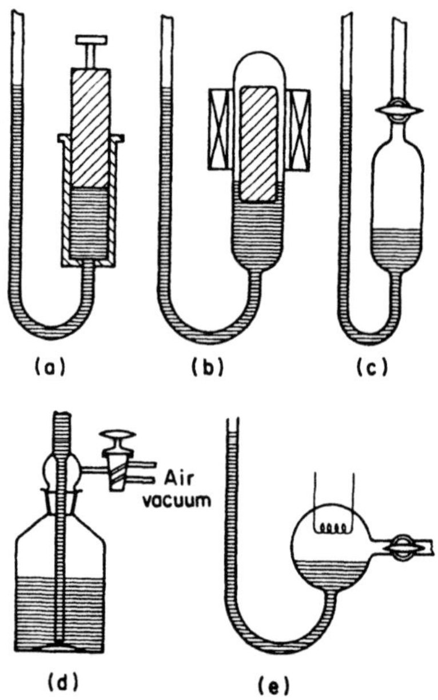

FIG. 6.2 Devices for raising the mercury, based on the change of its level in the container, using (a) piston; (b) plunger; (c, d) gas pressure; (e) heated filament

cut-off. This may be done by raising or lowering a concentric container (Fig. 6.1a), or a container connected through an elastic tube (Fig. 6.1b), or by reversing the container connected by a ground joint (Fig. 6.1c).

When a barometric column of mercury is used, it is advisable to provide a level difference of at least 80 mm between the "open" and the "closed" position of the cut-off, to avoid undesirable opening or closing of the cut-off due to changes in atmospheric pressure.

Figure 6.2 shows actuating devices based on the change of level in the container without moving the container itself. This may be done by pressing a piston on the level of the mercury (Fig. 6.2a), immersing (magnetically) a plunger into the mercury (Fig. 6.2b), or by changing the air pressure above the mercury (Fig. 6.2.c–e). The pressure above the mercury may be above atmospheric pressure (Fig. 6.2c), if the column on the cut-off side is at least 760 mm, or below the atmospheric pressure (Fig. 6.2d). In the latter case a two-way stopcock is used which may connect the container to a vacuum pump or atmosphere. The system shown on Fig. 6.2e is based on a rise of pressure due to the expansion of the gas introduced above the level of the mercury, when heated by the filament placed there.

The change in volume of the mercury container may be done by using containers having elastic walls, such as rubber balls (Fig. 6.3a), rubber tubes (Fig. 6.3b), diaphragms (Fig. 6.3c), or bellows (Fig. 6.3d).

(b) *Locking devices.* Cut-offs using the above mentioned raising devices, are often required to be locked in the open or closed position. This may be done by various methods which either guarantee the constancy of the system until a further operation is made or even guarantee this in the case of pressure variations in the cut-off.

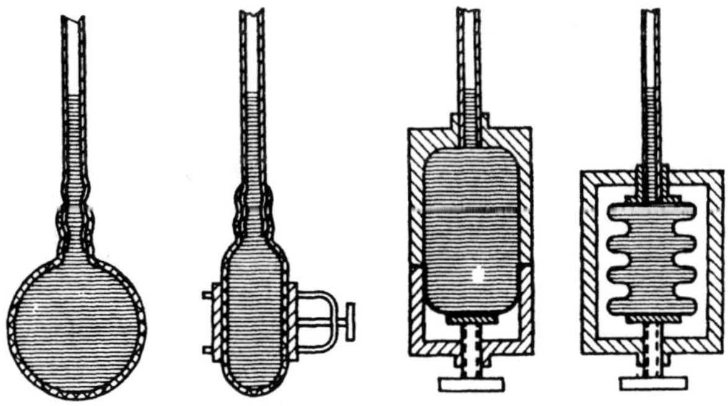

(a) (b) (c) (d)

FIG. 6.3 Devices for raising the mercury, based on the change in volume of the container, with (a) rubber ball; (b) rubber pipe; (c) diaphragm; (d) bellows

Figure 6.4 shows some locking devices which may be used on cut-off, but which do not guarantee the invariability of the cut-off during changes in pressure. These devices are to be used in connexion with a raising device using a magnetically displaced plunger (Fig. 6.2b). This plunger may be locked in its raised position by seating it on a magnetically actuated stop (1 Fig. 6.4a), by withdrawing it (if it is a ball) into a side arm (Fig. 6.4b), by hanging the hook connected to the plunger (Fig. 6.4c) or by rotating the plunger normal to a slot (Fig. 6.4d). In the cases shown in Fig. 6.4c, d the iron load of the plunger must be asymmetrical in order to permit its rotation by the external magnet.

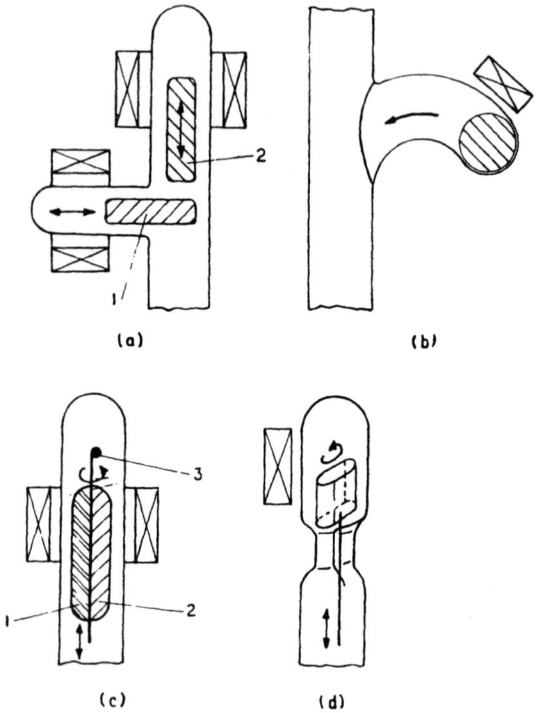

FIG. 6.4 Locking devices for plungers, or floats, using (a) a magnetically controlled side catch (1) to lock the plunger (2); (b) a ball, magnetically pulled in the side arm; (c) an asymmetric plunger, hung on holder (3), having the part (2) magnetic and the part (1) non-magnetic; (d) a plunger head, rotated to sit on the shoulders of a slit

Kenty[666] described a cut-off, which is locked in the closed position, by the steel ball resting on a ground surface (Fig. 6.5). To close the cut-off, the ball (1) floating on the mercury is drawn down by an outside magnet until it seats against the ground surface at the bottom of the bulb (2). Meanwhile, the mercury is forced over to close the cut-off, and surface tension causes it to withdraw into the constriction (4), forming the meniscus shown in the figure. When the magnet is withdrawn, hydrostatic pressure over the small area of (4) is unable to lift the ball (1) and surface tension prevents mercury from spreading under the ball. The device is thus locked until the ball is slightly raised by a magnet. In the position of the cut-off, the mercury seeks a level determined by hydrostatics and the capillary depression of the meniscus in the construction (3). The bulb (2) should have a bulge, to permit the flow of the mercury which entered above the ball (1).

Haul[519] locked his cut-off by freezing the mercury (Fig. 6.6). In order to close the cut-off, the stainless steel ring (1) is first pulled down into the mer-

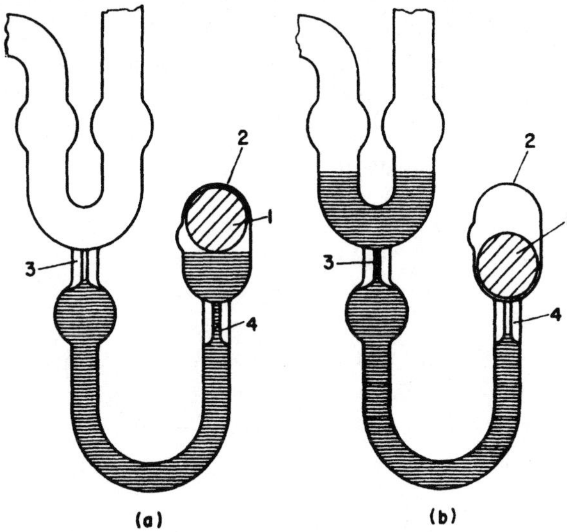

FIG. 6.5 Locking system based on the surface tension of the mercury in a capillary. After Kenty[000] *(Courtesy of The American Institute of Physics)*

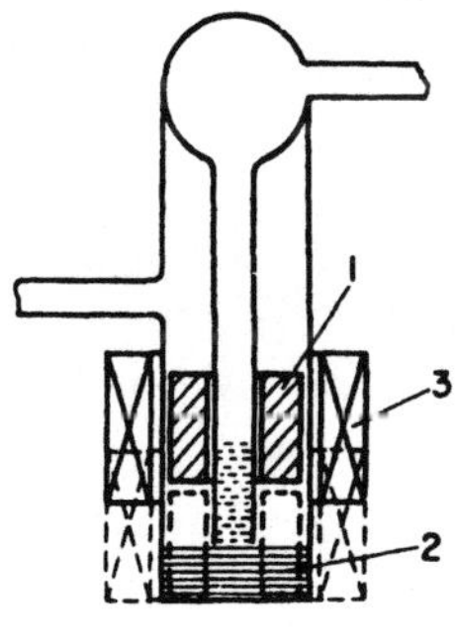

FIG. 6.6 Locking a cut-off by freezing the mercury. After Haul[519] *(Courtesy of The Institute of Physics and The Physical Society, London)*

cury reservoir (2) by means of the magnet (3). The mercury is then frozen in position by immersing the cut-off in a Dewar flask containing an alcohol–carbon dioxide (dry ice) freezing mixture. In this state the cut-off was able to maintain a pressure of 10^{-6} torr, against large pressure differences.

A difficulty that occurs very often in cut-offs, is the sticking of the mercury onto the walls. This happens especially in small-bore, glass tubes, and the cause is almost always on account of lack of purity of the mercury. To avoid the sticking of mercury to glass, Rosenberg[1079a] recommends grinding the bore of the capillary. The ground bore permits small pockets of air between the glass and the mercury to reduce the surface forces retarding the flow of the mercury. The method is also used industrially; the exhaust tubes of discharge lamps (e.g. fluorescent lamps) through which the small amount of mercury is introduced into the lamp, are chemically etched on their inside surface. Frank[1237a] described a technique for grinding capillaries. Archard[45] described a method of preventing mercury from sticking to glass, by first placing it in contact with aluminium for a short time. Apparently a thin surface compound is formed which changes the surface tension.

61.13 Closing systems in cut-offs. Cut-offs are closed and opened by changing the position of the level of the liquid. The proper closing action may be achieved by the liquid itself, or by floats moved on the level of the liquid. The liquid may close the connexion between two parts of the system, either by entering into a Y-shaped part of the connexion, or by sealing on a porous (sintered), glass surface. Hence the cut-offs may be closed by: (a) simple liquid seal, (b) floats, or (c) porous (sintered) glass.

For special reasons cut-offs with liquids other than mercury are also used (see paragraph d).

(a) *Cut-offs with simple liquid seals.* The simplest cut-off is a U-tube having at its base the connexion to the mercury reservoir (Fig. 6.7a). This shape has the drawback that the cut-off position is not well defined. In this respect the cut-off having a Y-shape is an improvement (Fig. 6.7b). For a well-defined cut-off position, a separating wall in a tube (Fig. 6.7c) or a concentric arrangement (Fig. 6.7d) may be used. A simple cut-off which is able to operate in various flow ranges for the gas is shown in Fig. 6.7e. The straight connexion between I and II is closed by a crossing wall, so that the gas must flow through the capillaries. By raising or lowering the level of the mercury one or more capillaries (of various bore) may be left open to the flow; by raising the mercury at the bottom of the last capillary the cut-off may be completely closed.

In order to avoid the need of atmospheric mercury columns, the cut-off may be constructed so as to allow simultaneous exhausting of the two sides of the cut-off. Such cut-offs are shown in Fig. 6.8. The cut-off constructed by Thompson[1221] has a solenoid (1 Fig. 6.8a) which when energized pulls the iron cylinder (2) down and the mercury closes the connexion (3) of the cut-off. The cut-off described by Hammond[505] (Fig. 6.8b) is closed when the steel slug (3) floats on the mercury. By raising it with the magnet (4) the level of the mercury changes leaving the connexion between (1) and (2) open.

(b) *Cut-offs with floats.* An adequate closing action in a cut-off may be achieved by a float carried on the surface of the liquid. This system has the advantage that when the cut-off is closed, the vapours of the liquid are excluded from the evacuated space. It permits a positive closing, being able to withstand higher pressure differences than cut-offs closed by the liquid itself.

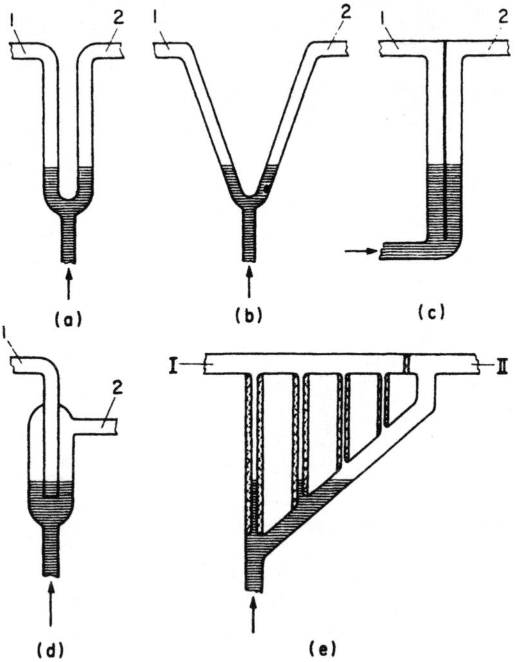

FIG. 6.7 Cut-offs with liquid seal: (a) U-shape; (b) V-shape; (c) with separating wall; (d) concentric pipes; (e) cut-off for variable flow

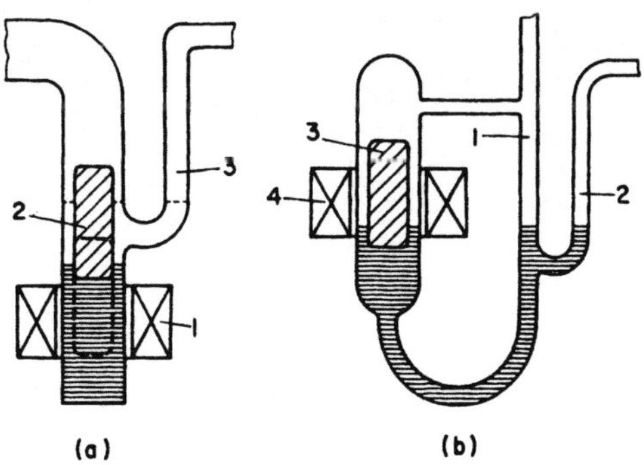

FIG. 6.8 Cut-offs, operated by magnetically controlled plunger: (a) with the plunger in the cut-off part (after Thomson[1221]); (b) with the plunger in the side tube. After Hammond[505] *(Courtesy of The Institute of Physics and the Physical Society, London)*

The floats have either *cylindrical* or *spherical* shapes. If cylindrical floats are used (Fig. 6.9), provision should be made to ensure that they will always rise and fall vertically. For this purpose, either bulges (1) are provided on the float (Fig. 6.9a) or indentations are made in the pipe (Fig. 6.9b) to guide the neck of the float. The ground part of the float is usually spherical, and the corresponding seat on the body is tapered (Ditchburn[282], Barr[81]). The taper (2, Fig. 6.9a) must be fairly sharp to avoid sticking. If the float should stick slightly and then fall suddenly it would be smashed on the bottom of the guiding tube. To avoid this, it is recommended that the connecting tube be led out laterally so as to leave some mercury in the bottom (3, Fig. 6.9a).

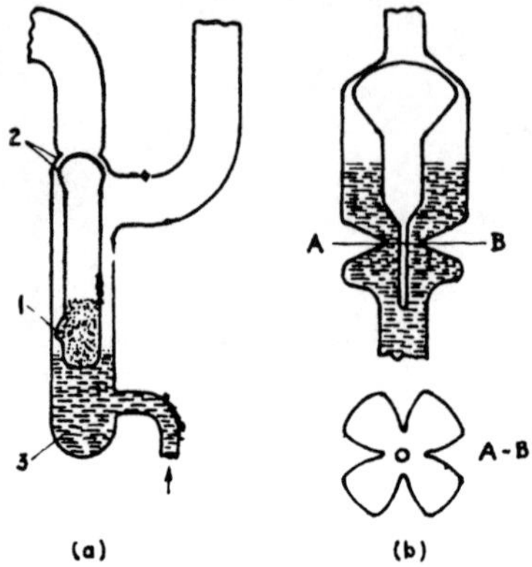

FIG. 6.9 Cut-offs with float: (a) cylindrical; (b) mushroom

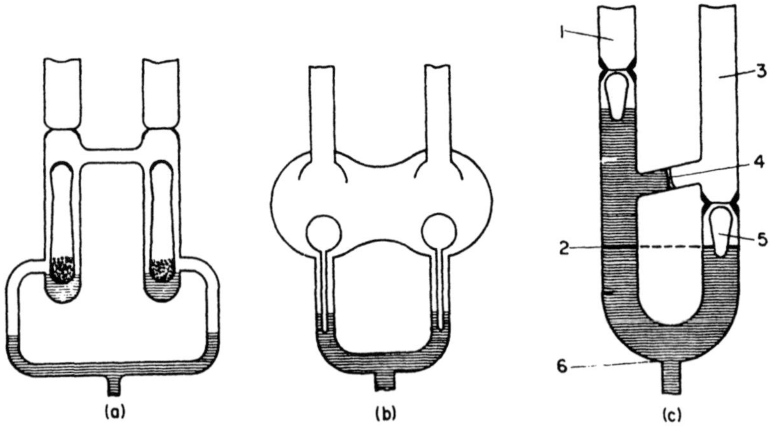

FIG. 6.10 Cut-offs with double float: (a) in separate pipes; (b) in the same chamber; (c) in a differential arrangement

The floats can either be empty or partially filled with mercury. Greene[480] described a float having the shape of a cup; this is filled with mercury so that the seal is made by the ground joint and this mercury.

To avoid contact with the mercury on both sides of the cut-off and have higher conductances, cut-offs with two floats are recommended (Wagner[1280], Klemenc[688], Ditchburn[282], Knor[699], Miller[860]). Some constructions of such cut-offs are illustrated in Fig. 6.10. The two floats may be placed in parallel, interconnected pipes (Fig. 6.10a, c), or in the same vessel (Fig. 6.10b). The cut-off shown in Fig. 6.10c operates only when the pressure in the arm (1) is higher than in (3). The mercury is slowly lowered to level (2), allowing the gas in the system to diffuse through the sintered glass disc (4), from arm (1) towards arm (3). During this step the float (5) is seated. When the pressures on sides (1) and (3) are equalized, the mercury can be lowered below the level (6), thus permitting rapid evacuation of the side (1) (Miller[860]).

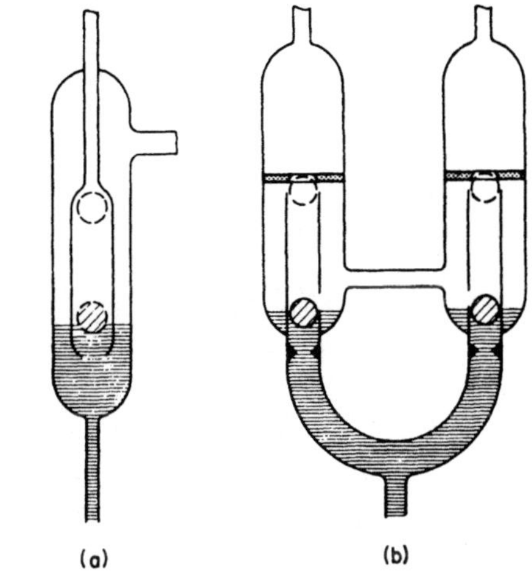

FIG. 6.11 Cut-offs with ball: (a) single (after Serfas[1120]); (b) double (after Neville[924])

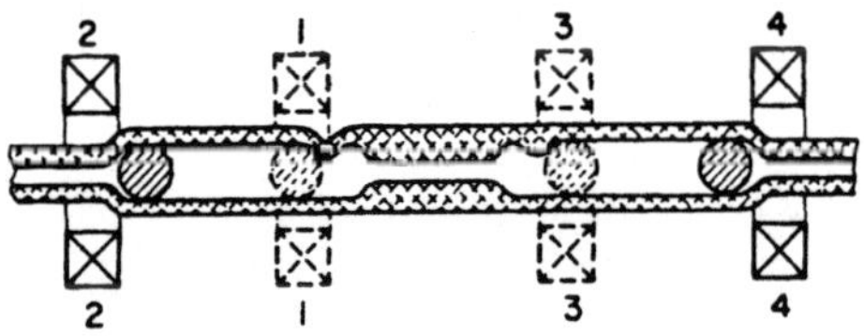

FIG. 6.12 Mercury stop and leak. After Bottomley[152] *(Courtesy of The Institute of Physics and The Physical Society)*

Steel *balls* may be successfully used as floats in mercury cut-offs (McFarland[829], Serfas[1120], Neville[924]). Figure 6.11a shows a cut-off with a single ball (Serfas[1120]). In the construction shown on Fig. 6.11b (Neville[924]) uses two balls, sealing against sintered discs drilled with appropriate central holes. When the cut-off is opened by lowering the mercury level, the ball on the side connected to the vacuum sticks at first, but the gas passes slowly through the sintered disc, and when the pressure is nearly equalized, this ball drops and opens an unrestricted path to the gas flow.

In order to control the flow of the mercury, Bottomley[152] used a device with two steel balls closing on ground surfaces at the extremities of the cylindrical portions of a glass bulb (Fig. 6.12). This device is capable of acting as a stop for the flow of mercury, or as a slow leak, depending on the position of the balls, controlled by powerful magnets (positions 1, 2 and 3, 4).

(c) *Cut-offs with sintered glass.* The cut-offs using sintered glass are based on the fact that, due to its surface tension, mercury does not pass through a glass frit. The cut-offs based on this principle can be constructed according

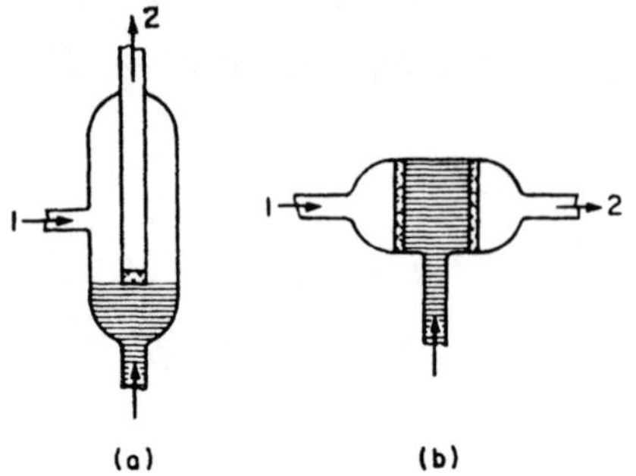

FIG. 6.13 Cut-offs with sintered glass: (a) with single glass frit (unlimited raising of the mercury); (b) with double glass frit (limited raising)

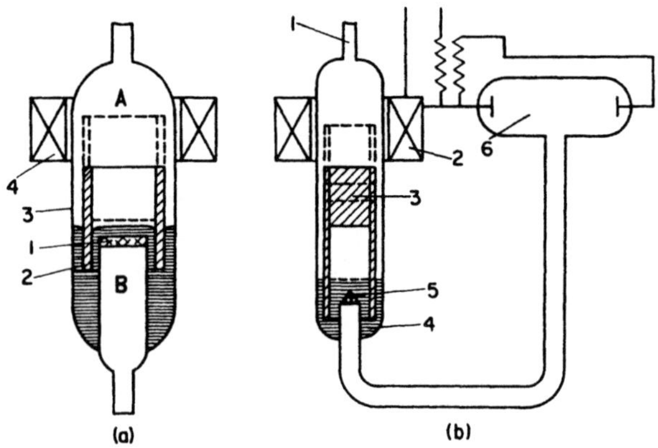

FIG. 6.14 Cut-offs with sintered glass covered by mercury: (a) cut-off. After Glaister[451] *(Courtesy of The Institute of Physics and The Physical Society, London)*; (b) cut-off for refilling discharge tubes (after Köhler[708])

to one of the sealing methods illustrated in Fig. 3.34. The glass frit may be placed at the end of the inside pipe (Fig. 6.13a); when the rising level of the mercury reaches this pipe, the connexion between (1) and (2) will be closed (Block[139], Almond[16], Honig[591], Gorman[463a], Parks[970]). The mercury may close the space between two parallel glass frits (Fig. 6.13b). In this cut-off, the pressure of the mercury determines the blow-off presure. When the pressure in the arm (1) exceeds that of the mercury, it will flow from the space between the frits, allowing the gas to flow through the cut-off (Smith[1152]).

Glaister[451] described a cut-off (Fig. 6.14a) consisting of a slightly convex glass frit (1) which is covered by a layer of mercury when the cylindrical iron

(2) floats on the mercury. The glass envelope (3) is surrounded by a solenoid (4), allowing the cylinder (2) to be lifted clear of the surface, thus allowing the mercury to run from the glas frit, and opening the cut-off. The cut-off is able to withstand a pressure difference of one atmosphere from *A* towards *B*, but only a few torr from *B* towards *A*.

When discharge tubes are to be operated at constant gas pressure and the gas must be refilled (e.g. Moore lamps), the cut-off illustrated in Fig. 6.14b may be used. The gas enters through (1) and if the solenoid (2) is energized, the glass cylinder provided with an iron core (3) is raised, thus permitting the mercury (4) to leave open the porous plug (5) so that the gas can flow to the discharge tube (6).

(d) *Cut-offs with molten metals.* Molten metals such as indium, gallium or various low melting alloys were used as sealing liquids for cut-offs. In the cut-off described by Paty[978], the height of the molten indium is controlled by a glass-enclosed iron slug, which is raised out of the indium by a magnet and is held in position by a stop as shown in Fig. 6.4a. Axelrod[59] construct-

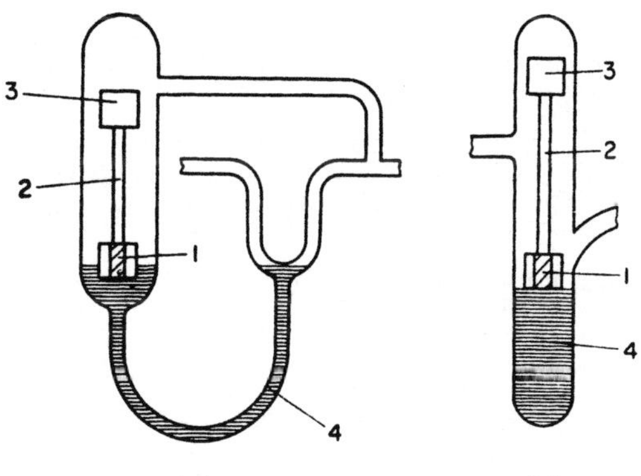

FIG. 6.15 Cut-offs with indium: (a) U-shape; (b) compact. After Axelrod[59] *(Courtesy of The American Institute of Physics)*

ed the cut-offs shown in Fig. 6.15a, b on the same principle, for use in ultra-high vacuum systems. The construction shown on Fig. 6.15a consists of a soft iron slug (1) enclosed in glass, connected through a stem (2) to a stabilizer (3), which helps to keep the float assembly vertical. In the open position of the cut-off, this assembly floats on the indium (4). After bakeout, the float and the stem are pulled into the indium, by an external magnet. It is held in this position until the indium solidifies, and keeps the float in position, keeping the cut-off closed by the increased height of the indium. The system shown in Fig. 6.15b operates on the same principle, having a more compact construction, and also a higher impedance in the open position. In order to decrease the impedance, it is advisable to construct this cut-off with a deep well for the indium, and a thin stem on the float.

Beynon[120] describes a cut-off using molten gallium (Fig. 6.16a). Lowering the soft iron (1) into one limb of the gallium (2) filled tube, with an external magnet (3), causes the level of the molten metal to rise in the other limp and close the cut-off. On the same device a gallium sealed glass frit (4) was used. Behrndt[103a] used a similar cut-off but with an alloy of 62.5 per cent gallium, 21.5 per cent indium and 16 per cent tin (m.p. 107 °C).

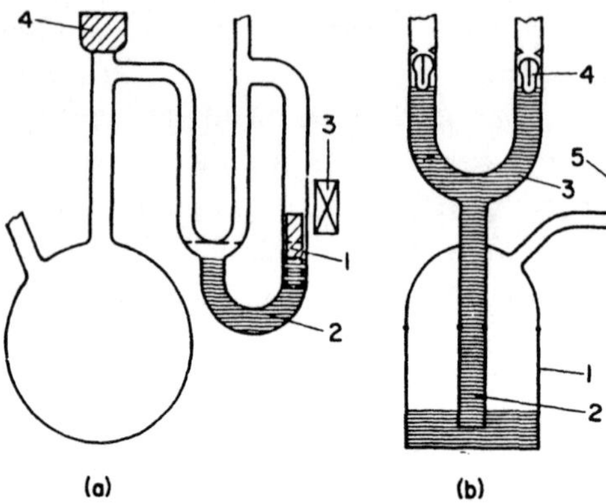

FIG. 6.16 Cut-offs with (a) gallium. After Beynon[120] *(Courtesy of Elsevier Publ. Co., Amsterdam)*; (b) Wood's alloy. After Toby[1226] *(Courtesy of The American Institute of Physics)*

Toby[1226] described a cut-off using Wood's alloy (see Table 3.4). This cut-off (Fig. 6.16b) consists of a Kovar cup (1) and an inner Kovar tube (2) sealed by a graded seal to the Pyrex cut-off (3). The molten Wood's metal is raised by admitting nitrogen through the double valve (5). Molten Wood's metal tends to oxidize in air. The rising molten metal closes the cut-off and is stopped by the floats (4). Since the alloy is not as dense as mercury, the floats should not be completely filled with iron. The Wood's metal should be allowed to *cool only in the Kovar* part since it cracks the glass when cooling.

61.2 Stopcocks

61.21 Shapes and dimensions. The American Vacuum Society[26] defines the stopcocks as small valves in which a *plug* can be rotated inside a *body* or barrel, which is usually tapered, while a hole in the plug registers with corresponding holes in the body in the open position of the stopcock.

Glass stopcocks are extensively used in glass vacuum systems. The plug (stopper or key) of the stopcocks consists of a conical ground part and a handle. The conical part is made on a solid (Fig. 6.17a) or hollow piece (Fig. 6.17b). If the plug is solid the connecting bore is drilled through it.

Hollow plugs are made from pipe, and are provided with a hole in the wall (Fig. 6.17b). Sometimes the hollow plugs are provided with an inside connecting pipe (Fig. 6.17c). The handle of the plug may be symmetrical (Fig. 6.17a, c) or asymmetrical (Fig. 6.17b). The asymmetrical handles are preferred for hollow plugs having a single hole and are generally arranged with the longer side facing the bore (Eschbach[346]). When symmetrical handles are used with such plugs, the side corresponding to the hole should be marked with a dark spot. Wilson[1324] points out the importance of well-marked stopcocks, i.e. to avoid faults in the handling of the system.

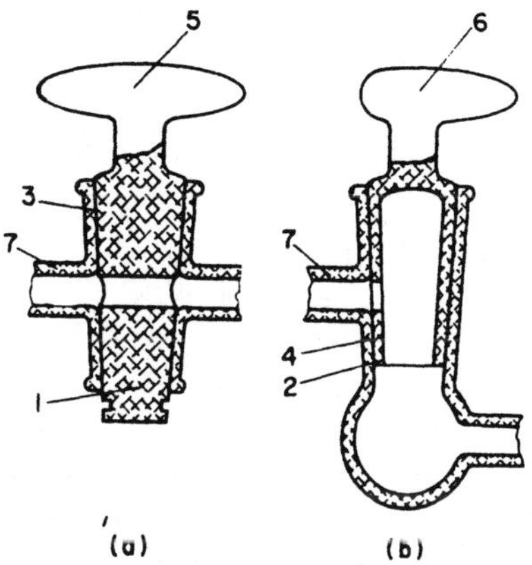

FIG. 6.17 The parts of the stopcock: (1) solid plug; (2) hollow plug; (3) open shell; (4) closed shell; (5, 6) handles; (7) connexions

The solid plug of the stopcock should protrude beyond the small end of the barrel, but no plug should extend over the large end of the barrel (Parr[971]). This is necessary in order to prevent a ridge forming on the plug during the grinding operation or later due to wear. Solid plugs sometimes have a groove on their small end (Fig. 6.17a) for retaining keys or rubber rings.

The shell or barrel carries the side connexion or outlet tubes, and is ground on its tapered inside surface. With solid plugs or hollow plugs having an inside connecting tube, shells opening on both ends should be used (Fig. 6.17a, c). The outlet tubes must have the same bore as the hole in the plug. Enlargements of the holes in the shell will form pockets for grease and dirt and will give rise to leaks.

In order to guarantee the leak tightness of the stopcock, the design with the shell closed at its small end (Fig. 6.18c) is recommended. This arrangement avoids the atmospheric pressure on the small end of the plug and thus ensures that the plug will be pressed into the shell, greatly reducing the leak rate of the stopcock. This construction can be used only with hollow plugs or with plugs having inside connecting tubes. If a solid plug is used, a possibility for evacuating the cup closing the shell should be provided (Fig. 6.18b).

A particular stopcock is *defined* by its *shape* and *size*. The shape of a stopcock includes that of the plug and shell as well as the number of ways, the number and position of the outlets, and the position of the bore. The combination of these various elements result in a large number of stopcock shapes. The more commonly used ones are shown in Fig. 6.18 and described in Table 6.1.

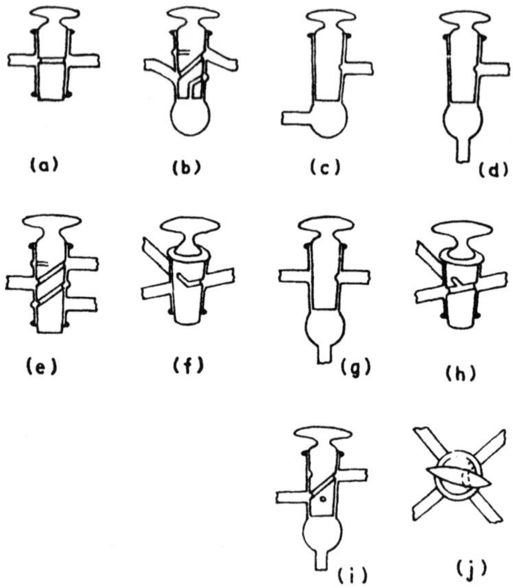

FIG. 6.18 Shapes of glass stopcocks (Table 6.1)

The main characteristic in defining a stopcock is the number of *ways*. Manufacturers and consumers often confuse the number of *ways* with the number of *outlets*. The number of ways is that of the connexion possibilities of the stopcock, and is not necessarily equal to the number of outlets. Thus the stopcocks in Fig. 6.18a–d have one way and two outlets; those in Fig. 6.18e–g have two ways and three outlets; those in Fig. 6.18h–i have three ways and three outlets.

The plug generally has the standard taper* of 1 : 10 (see Table 3.7) and may have a straight bore, single or double oblique bore, or connecting tube (see Table 6.1). The shell may be open at both sides, or closed at one side, and the outlets may be straight (in the same plane, perpendicular to the axis of the plug), offset (in various planes, perpendicular to the axis), or in L, T

* For details of making the taper see, e.g., Frost[490], Barr[81], Parr[971].

TABLE 6.1. GLASS STOPCOCKS*

Number of ways	Plug	Bore	Shell	Position of outlets	Number of outlets	Fig. 6.18
1	solid	straight	open	straight	2	a
1	solid	oblique	closed	Z (offset)	2	b
1	hollow	hole	closed	Z (offset)	2	c
1	hollow	hole	closed	L (right angle)	2	d
2	solid	double-oblique	open	forch	3	e
2	solid	L	open	straight	3	f
2	hollow	hole	closed	T	3	g
3	solid	T	open	straight	3	h
3	hollow	connecting tube and hole	closed	T	3	i
4	solid	L	open	straight	4	j

* Some suppliers of glass stopcocks are listed in connexion with Table 3.7

(one outlet in the axis and the others in planes perpendicular to the axis). The stopcock with offset outlets and an oblique bore, overcomes the difficulty occurring in those with straight bore, where there is a tendency to form a greaseless ring in the plane of the bore, thus leading to leakage.

The *size* of the stopcock is generally expressed simply with the aid of the *bore* in the plug. All the other dimensions of the stopcock are more or less connected to the bore diameter; the differences between various standards or dimensions utilized by various suppliers are not critical. Standard stopcocks are available with bore diameters of 1 to 35 mm. The largest effective bore for practical glass stopcocks is considered to be about 50 mm (Pirani[992]). The limitation of the bore limits the conductance of the stopcock (see Table 6.3). It is considered that the use of glass stopcocks is only practical for overall pumping speeds up to about 20 l/sec. Figure 6.19 and Table 6.2 give the dimensions of some standard stopcocks.

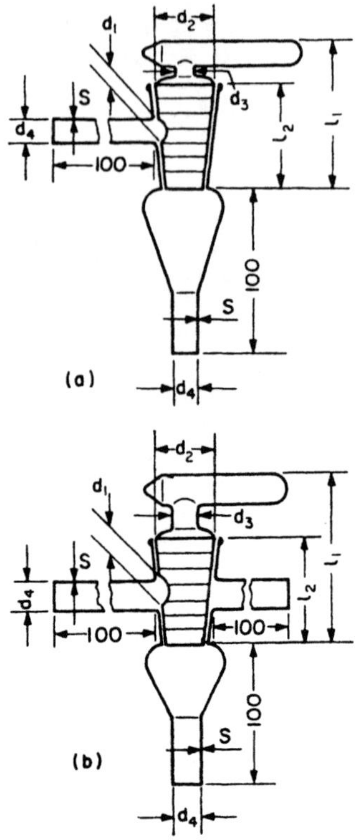

FIG. 6.19 Dimensions of a glass stopcock (Table 6.2) with (a) one way; (b) two ways

TABLE 6.2. DIMENSIONS (mm) OF GLASS STOPCOCKS
(after D.I.N. 12557; see Fig. 6.19)

d_1	6	8	12	20	35
Taper	1:10	1:10	1:5	1:5	1:5
l_1	60	80	100	125	165
l_2	40	50	70	80	100
d_2	20	30	40	50	75
d_3	10	14	20	25	35
d_4	9.25	11	15.2	25	40
Wall thickness	1.0–1.5	1.25–1.75	1.5–2.5	2.0–3.0	2.0–3.0

61.22 Greased stopcocks. The usual glass stopcocks must be lubricated using vacuum grease (see Section 36.4). This fact limits their use to vacuum systems where the pressure should not be lower than the vapour pressure of the grease (about 10^{-6} torr), and where the part containing the stopcock must not be baked. If lower pressures have to be reached, the degassing of the grease may be facilitated by using special constructions (Fig. 6.20). If the grease can be attacked by the vapours handled in the plant, greaseless stopcocks (Section 61.23) should be used.

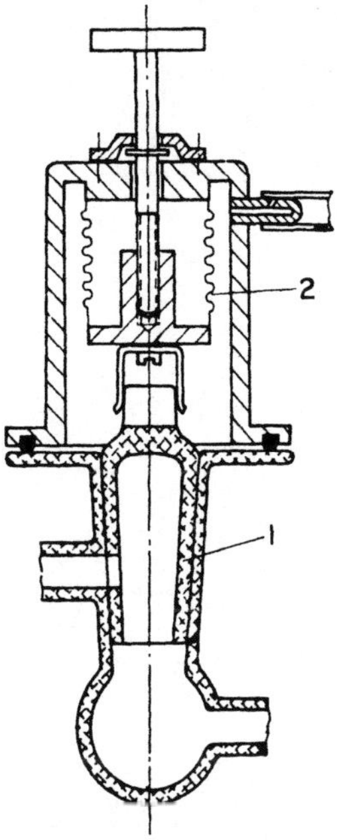

FIG. 6.20 Arrangement for the rapid degassing of a greased stopcock. After Horseling[597] *(Courtesy of Philips Techn. Rev.* **17**, *184, 1955)*

Stopcocks are best lubricated by applying the grease at both ends of the plug. No grease should be applied near to the bore. After the grease is applied on to the plug with a spatula, the plug is inserted in the shell and turned slowly *in one direction* until the lubricant is worked over the whole area. The stopcocks should be regreased at quite short intervals (sometimes daily). Before regreasing a stopcock it should be first thoroughly cleaned, removing the old grease (see also Section 36.4, and Maglio[794], Roeben[1074], Barr[81]). The difficulty most often encountered with stopcocks is leakage, due to hairlines that develop in the thinned-out grease. This must always be watched for and avoided. When the stopcock is "frozen" do not apply brute force to move it. The methods to be used are described in Section 36.4.

The gas evolution from the greased stopcock can be decreased by using low vapour pressure greases (see Table 3.10); by degassing the grease before use (Normand[641]); or after being applied onto the stopcock; or by maintaining the stopcock at lower temperatures.

Horseling[597] described a stopcock arrangement (Fig. 6.20) which permits raising the plug (1) in a sealing system using bellows (2), first exposing the whole greased surface to be degassed in vacuum, then inserting the plug into the shell and using the stopcock.

If necessary, either the plug and/or the shell may be cooled. Figure 6.21 shows an arrangement which can be used to cool the plug of a stopcock

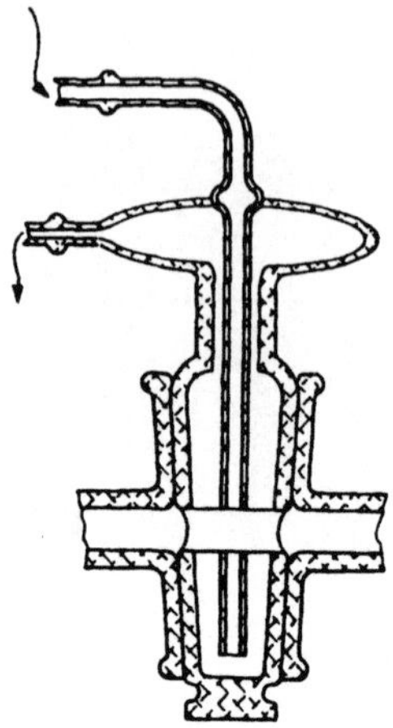

FIG. 6.21 Stopcock with cooled plug. After Brennan[165a] *(Courtesy of The Institute of Physics and The Physical Society, London)*

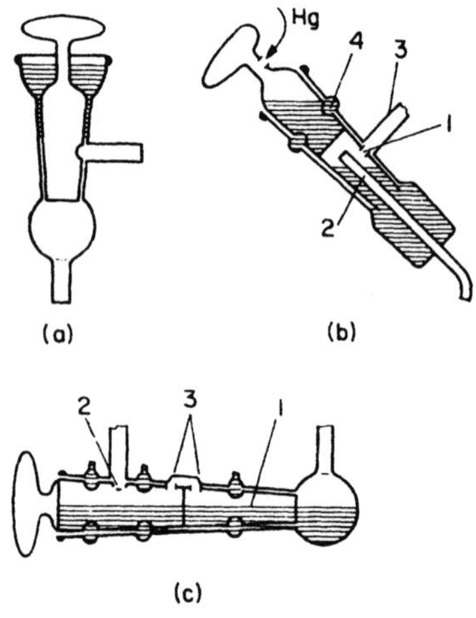

FIG. 6.22 Mercury sealed, graphite lubricated stopcocks: (a) vertical; (b) inclined (Stanier[1175]); (c) horizontal (Townes[1234])

(Brennan[165a]). Arrangements for the cooling of the shell or cooling with liquid air are illustrated in Fig. 3.22. According to Rapp[1026], a greased, ground-glass joint may be cooled to a temperature of -78 °C (acetone–CO_2 mixture) and remain leak-tight. When cooling to liquid nitrogen temperature cracking of the grease occurs. A preliminary quenching in acetone–CO_2 will minimize these difficulties.

61.23 Greaseless stopcocks. In vacuum systems where gases are handled which dissolve the grease, greaseless stopcocks must be used (Bachman[69a], Buchman[181]). The most common of these are lubricated with graphite and sealed with mercury (see also Section 37.2).

These stopcocks may be constructed to operate in vertical, oblique or horizontal positions. In the stopcock made for use in a vertical position (Fig. 6.22a), the mercury is placed in the cup provided on the large end of the shell. Stanier[1175] described a stopcock clamped with its axis inclined to the vertical (Fig. 6.22b) so that when the plug is rotated by 180° the hole (1) is below the surface of the mercury in the base of the stopcock. Thus the low pressure side (3) is completely closed from the side (2). The plug and shell fit together sufficiently well to prevent the mercury being forced between them, when the side (2) is at atmospheric pressure. Mercury, introduced into the upper portion of the plug through a small hole in the handle, flows into the groove (4) through the holes provided.

A horizontal greaseless stopcock has been described by Townes[1234] (Fig. 6.22c). In the open position, communication between the two chambers (1, 2) of the stopcock takes place through the holes (3) which open into a small trough in the outer wall. When the plug is turned 180° these holes are submerged in the mercury and the communication between chambers (1) and (2) is closed. The grooves provided around the stopcock are separately filled with mercury.

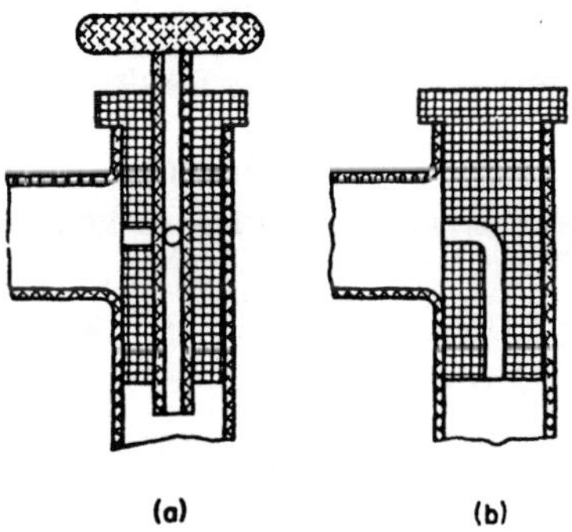

FIG. 6.22A Stopcocks with polyethylene plug: (a) with glass pipe inserted through the plug; (b) with channel in the plug. After Raats[1022] *(Courtesy of The Institute of Physics and The Physical Society, London)*

Raats[1022] described stopcocks using a glass shell and a polyethylene plug, with or without a glass pipe inserted through the plug (Fig. 6.22A).

Greaseless stopcocks using P.T.F.E plugs are supplied by the Springham Co.*

*G. Springham & Company Ltd., Harlow New Town, Essex, England.

61.3 Valves

A valve is defined (American Vacuum Society[26], Société Française des Ingénieurs et Techniciens du Vide[1122a], D.I.N.[281a]) as a device for adjusting the rate of flow of a fluid or for completely stopping the flow.

61.31 Principles and classification. A conventional valve consists of three major parts (Fig. 6.23):

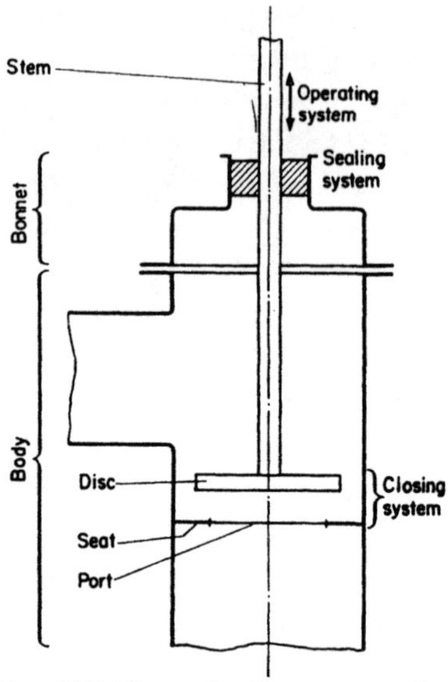

FIG. 6.23 The parts of a vacuum valve

(1) The *body* of the valve containing the external openings;

(2) The *bonnet* of the valve, i.e. the part through which the motion is transferred;

(3) The *stem* which transfers the motion and closes or opens the *port* of the valve, by pressing a *disc* against the *seat*.

The function of a valve (Krieger[719]) is to adjust or stop the flow and is achieved by the closing system. This system needs motion inside the valve. The motion should be transmitted from outside and the moving part should be sealed. Thus any valve consists of (Fig. 6.23): a *closing* system, a *sealing* system and an *operating* system.

The *closing system* of a valve may be based on: liquid seals, molten metal seals, silver chloride seals, fused glass seals, ground joints, diaphragms, or gasket seals (Fig. 6.24).

The *sealing system* is based either on packings or it may be packless, using bellows, diaphragms or magnetic motion transmissions.

The *operation system* transmits the motion using mechanical, pneumatic, magnetic or electrical means of motion transmission.

Since from the combination of these systems a very great number of valve designs is possible, the number of types constructed is really very large. In order to cover this large range, the following sections will describe the various designs of the closing (Section 61.32), sealing (Section 61.33) and operation (Section 61.34) systems, giving real valve constructions to illustrate these designs. The design of the valve should obviously choose its elements in

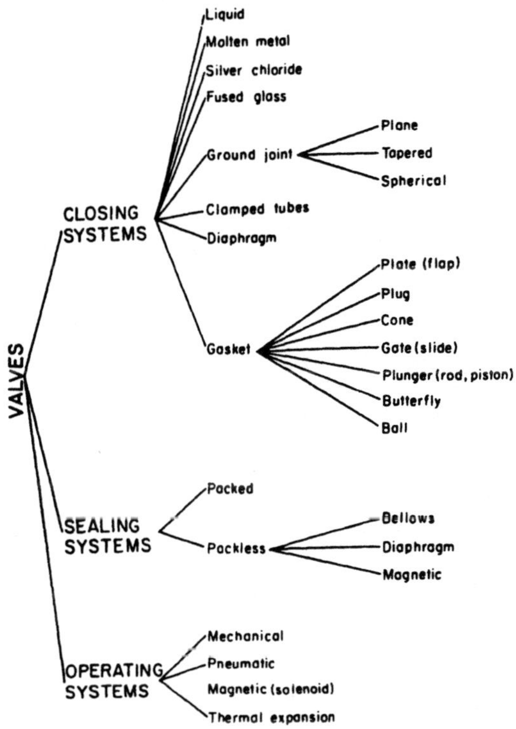

FIG. 6.24 Classification of the systems forming a valve

such a way that the resulting valve meets the service requirements. Section 61.35 describes the basic features of vacuum valves constructed for isolation, air-admittance, throttling, and other special purposes. The valve chosen for a particular vacuum system should also have the *shape* corresponding to its location on the plant. It is possible to choose from the valves available from the various suppliers the appropriate straight-through, angle or elbow valve.

An *ideal valve* should meet the following requirements (D'Eustachio[277], Yarwood[1340], Moreau[887]):

(1) The valve should be mechanically strong, simple to construct, dismantle, and reassemble. The valve should consist of a minimum number of parts, having the smallest surface area in order to decrease the outgassing rate and allow easy cleaning.

TABLE 6.3. CONDUCTANCE AND TIGHTNESS OF VALVES

Valve	Size	Open	Closed	Leak rate (lusec)	Reference
		conductance (l./sec)			
Glass stopcocks					
Hollow plug	1 mm bore	0.06	–	–	
	6 mm bore	6	–	–	
	30 mm bore	500	–	–	
Solid plug	1 mm bore	0.01	–	–	
	6 mm bore	2.5	–	–	
	30 mm bore	180	–	–	
Greaseless ball ground joint	10 mm dia.	–	4×10^{-4}	–	Decker[270]
	–	–	10^{-4}–10^{-5}	–	Klopfer[690]
					Adam[3]
Valve with silver chloride	Fig. 6.30b	–	–	$1\times5\ 10^{-4}$	Kirs-lis[679]
Right angle, plate valves with shaft seals	10 mm	8	–		
	20 mm	15	–		
	32 mm	27	–		
	50 mm	47	–	$<1\times10^{-2}$	Ley-bold[762]
	65 mm	160	–		
	100 mm	340	–		
Right angle, plate valves	1 in.	10.5	–	–	Varian[1257a]
	1.5 in.	35	–	–	
Flap valves, bellows sealed (see Plate 24)	32 mm	50	–		
	65 mm	330	–		
	100 mm	750	–		
	150 mm	1600	–	$<1\times10^{-4}$	Ley-bold[762]
	250 mm	4700	–		
	350 mm	9000	–		
	500 mm	18000	–		
Gate valves	2 in.	135	–		
	4 in.	2680	–	–	C.V.C.[231]
	6 in.	6400	–		

(Table 6.3 Continued)

Valve	Size	Open conductance (l/sec)	Closed conductance (l/sec)	Leak rate (lusec)	Reference
Gate valves	4 in. 6 in.	2750 6500	– –	–	N.R.C.[912]
Gate valves	6 in.	4200	–	$1\times6\ 10^{-7}$	Ultek[1251a]
Gate valve, Viton closing, bellows seal	1 cm²	–	–	$<1\times10^{-6}$	Beske[117]
Valve closed with knife edge on In	–	–	$<10^{-14}$	–	Esch-bach[349]
	–	35	–	$<2\times10^{-8}$	Drawin[302]
*UHV valve, closed on gold gas-ket, bellows sealed	Fig. 6.58	–	–	10^{-5}	Yates[1342]
UHV valves*	0.1 in² 1 in²	– –	10^{-10}–10^{-11} 10^{-8}–10^{-9}	– –	Ed-wards[328]
UHV valves* bellows sealed	3 cm² 20 cm²	3 40	7×10^{-8}	–	Atlas[53a]

* For ultra-high vacuum valves see Table 6.28.

(2) The valve should have the maximum possible conductance in the open position without unduly increasing the volume of the system. In most valves, the port has a smaller cross section area than the connecting tubing; thus it constitutes an impedance to the flow. In order to avoid a further increase of the impedance, the stem, disc or other parts should be placed or withdrawn outside the path of the gas flowing through the valve (see flap valves, gate valves). In order to permit a comparison between the various valves, Table 6.3 summarizes the order of magnitude of the conductances of various types of valves.

(3) The valve should have the minimum possible leak rate through the closing system, when it is in the closed position (see Table 6.3).

(4) The valve should have the minimum possible leak rate from outside. The body and the bonnet should be made from non-porous materials, and the seal between the body and the bonnet, as well as the shaft seal, should be leak-tight. The valve manufacturers generally guarantee that valves with packings have leak rates less than 1×10^{-2} lusec, and that the packless valves (bellows) have leak rates less than 1×10^{-6} lusec.

(5) The virtual leak created by the valve should be very small. The valve should not contain materials with high vapour pressure (Section 21.13), double weld (Section 22.32), or other trapped spaces, e.g. in the screws, gaskets, etc. The valve should not have a rough inside surface on which a large amount of water vapour may condense while the valve is open to air.

(6) The valve should have a positive action not influenced by pressure difference. It should be able to operate at a large range of pressures. It is not essentially necessary that the valve should be able to close having the vacuum or the pressure on any one of the sides.

(7) The valve should operate rapidly. The quick action of a valve is mostly required for closing the system, or to isolate the interested part. If fast ope ration is necessary, magnetic systems should be used (see Table 6.4).

TABLE 6.4. OPERATING-TIME OF VALVES

Operation	Opening or closing time (sec)	Remarks	Reference
Closing manually a screw operated valve	5–10	—	
Closing a lever operated valve	2–3	—	
Closing manually a stopcock	1–2	—	
Electro-pneumatic valve	2	see Fig. 6.83	
Electro-magnetic valves	0.05–0.010	—	
Solenoid operated plate valve	0.01	see Fig. 6.56	Knudsen[701]
Pneumatic gate valve	0.012	—	Round[1089a]
Electromagnetic gate valve	0.005	—	Lavrov[746a]
Pneumatic valve	0.003	see Fig. 6.82b	Meyer[853]

(8) The valve should be easy to operate or should allow of being operated by remote handling or interlock. Sometimes the valve has to operate at a predetermined pressure after a predetermined time interval, or various valves have to operate in a predetermined sequence.

(9) The valve should be capable of a great number of cycles. For quality valves the manufacturing firms quote figures up to 100,000 operating cycles.

(10) The valve should be chemically resistant. The materials forming the valve should not be corroded by the gases flowing through them. If the vapours can condense on the parts of the valve, its construction should allow for easy removal of the deposits.

(11) The valve should be bakeable. Bakeable valves are described in Section 61.35f.

Unfortunately no valve can meet all these requirements. Thus in choosing a valve for a particular purpose a compromise should be made, insisting on the most important requirements for that particular purpose (e.g. bakeability when the valve is to be used in an ultra-high vacuum system).

A review of the advantages and drawbacks of some vacuum valves was recently published by Blanc[131a].

61.32 Closing systems of valves. The closing system of a valve consists of the parts which close or open the port* of the valve (Fig. 6.23). The closing action is based on one of the basic features listed in Fig. 6.24.

(a) *Closing systems with liquid seals.* The sealing material commonly used in these systems is mercury, but other liquids may also be used if their physical or chemical properties correspond to the particular purpose (Section 3.7). Generally these valves cannot be used when the closing system of the valve should withstand large pressure differences, due to the danger of forcing out the liquid from its correct place.

The most simple valves using a liquid seal are those where a cap is utilized to close the valve (Fig. 6.25). The valve described by Schmitt[1104] (Fig. 6.25a) consists of a ring seal (1), a curved neck (2) and a side arm (3), the annular space formed by the ring seal being filled with mercury to about half its depth. The valve is closed by dropping a small steel cap (4) over the inner tube. The cap floats on the mercury and closes the connexion between the two sides of the valve. To open the valve, the cap is lifted by an external magnet and placed into the curved arm, where it lies until needed. The valve can also be used as a throttling (leak) element, by using one or more caps having the required passages (pin-holes), and withdrawing them into various arms.

A valve of this construction using gallium, is described by Neubert[922a].

The closing cap can be carried on a wire having a counterweight at the other end (Fig. 6.25b) or on rigid stems (Fig. 6.25c, d). The counterweight or the stem is moved by an external magnet (Section 52.1). The design shown on Fig. 6.25d (Townes[1234]) has a locking system consisting of an iron bar encased in glass and having a rectangular cross section. When rotated by

* The *port* of a valve is the variable opening inside the valve through which the fluid should pass from the entrance to the exit of the valve.

90° this bar rests on the sides of the slit, through which it may slide in its initial position.

Essig[358] described a valve (Fig. 6.26) consisting of a porous disc (1) carried by a sliding shaft, and two other porous discs (2) sealed on the end of the inlet and outlet pipes, and immersed in mercury (3). By moving the disc (1) over

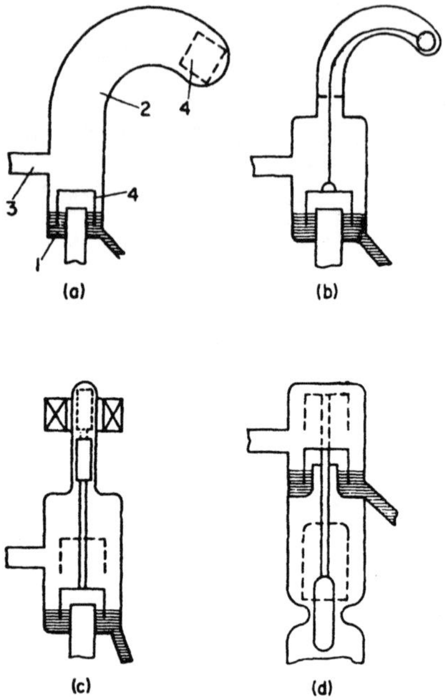

FIG. 6.25

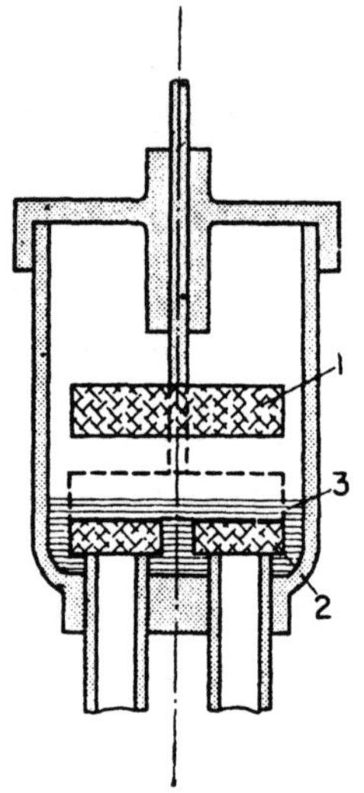

FIG. 6.26

FIG. 6.25 Closing systems of vacuum valves using a liquid (mercury) seal: (a) with dropping cap; (b) with cap suspended on wire with counterweight; (c) with cap held on stem, directed upwards; (d) as c with stem directed downwards

FIG. 6.26 Valve with mercury and porous discs. After Essig[358] *(Courtesy of The American Institute of Physics)*

the discs (2) the mercury is forced out from between them, and the gas can flow through the porous discs from the inlet to the outlet pipe.

(b) *Closing systems using molten metals.* Some valves use molten metals in a similar way to those utilizing liquid seals. However, the operating principle of these valves differs because of the fact that the sealing metal is kept liquid only during the closing or opening of the valve. The closing system using liquid metals is known also as "soldered" seal, since in the solidified state the used metals (tin, Wood's metal, indium) in fact form such a seal. This permits the valves with such closing systems to be used with larger pressure differences across them.

Hughes[606] describes a valve using Wood's metal (Section 35.11). To open the valve, the Wood's metal placed in the cup (1, Fig. 6.27) is melted by heating. The iron part (2) is unhooked by magnetic handling (Fig. 6.4) and the central assembly with the cup is allowed to rest on the bottom of the glass tube (3). Since the Wood's metal does not wet the glass well, the closing portion of the inside tube (4) is made of Kovar.

Indium is used in the valve (Fig. 6.28) constructed by Reynolds[1050]. The bottom of this valve is a copper block in which a groove is milled, and filled with indium. A second groove on the outside of the copper block, contains a NiCr heater. By means of this heater, the indium may be melted, and the

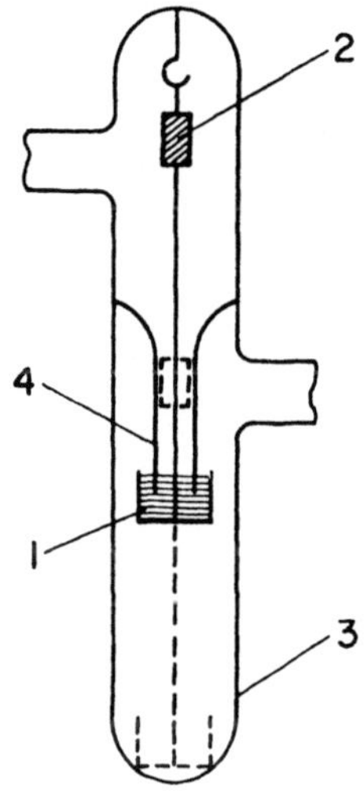

FIG. 6.27 Valve using Wood's metal in the closing system. After Hughes[606] *(Courtesy of The American Institute of Physics)*

valve plate can be raised, operating it with the bellows-sealed shaft. This valve can be used up to 150 °C, but if tin is utilized instead of indium, the valve can be heated up to about 200 °C. Since indium and tin have low vapour pressures (Fig. 3.11), such valves may be used in high vacuum systems.

Papirov[966b] describes a valve for ultra-high vacuum, which has the shape of a plate valve but is closed by melting some metals (Au, Ag, Sn, In, Cu) contained in a powdered refractory metal filler (Fig. 6.92).

Blanaru[130] constructed a valve (Fig. 6.29a) using tin as the molten metal of the closing system. The body of the valve is a (molybdenum) glass tube (1) which has a narrowed end (9). A glass cup (6) containing some tin (7) has attached to its bottom another glass tube (8) closed at both ends and containing an iron cylinder (11) placed between two non-magnetic, helical springs (10). To avoid mechanical shocks a helical spring (12) is placed in the guiding tube (9). A Kovar tube (3) is sealed on to the central glass tube (2).

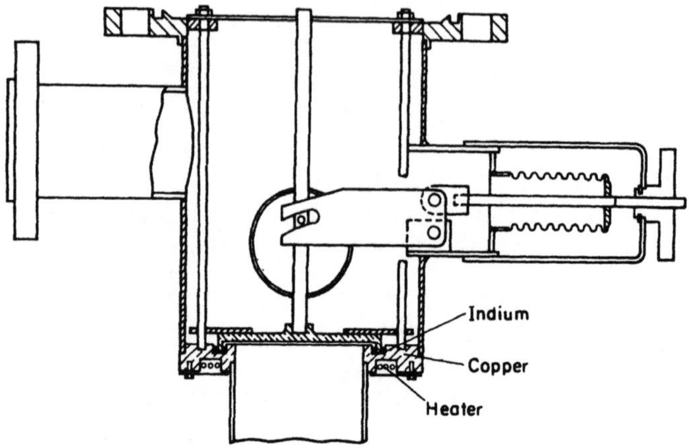

FIG. 6.28 Valve using indium seal in the closing system. After Reynolds[1050] *(Courtesy of Pergamon Press)*

A collar (5) made of tantalum foil is soldered to the Kovar tube, forming a circular groove which is filled with tin (4). The melting of the tin and the outgassing of the parts is done with the high frequency coil (13). The vertical movement of the cup is obtained by means of an external solenoid. To close the valve the cup is raised until the tantalum collar is entirely immersed in

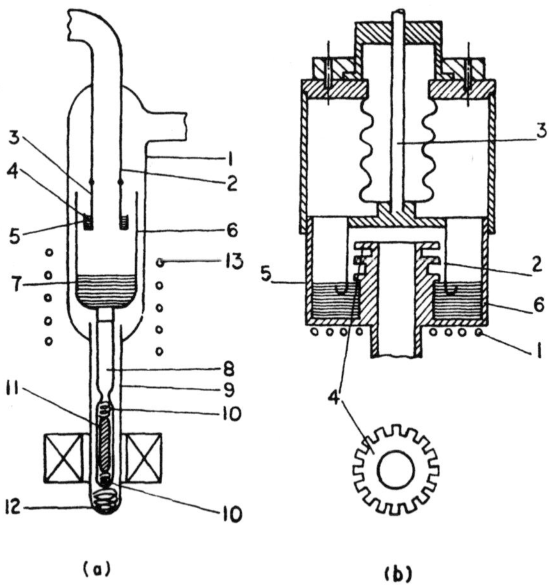

FIG. 6.29 Valves using tin in the closing systems: (a) with glass body. After Blanaru[130] *(Courtesy of The Institute of Physics and The Physical Society, London)*; (b) with metal body. After Haaland[494] *(Courtesy of The American Institute of Physics)*

the molten tin (7), and kept there until solidified. To open the valve, the tin is melted again and the cup is lowered.

Haaland[494] described a metal valve closed by molten tin (Fig. 6.29b). The tin (5) is placed in the cold drawn steel cup (6) and is melted by an external heater (1). The closing cap (2) is then raised by means of a bellows-sealed shaft (3). Since tin does not wet iron before about 500 °C, the tin must be wetted to the closing cap and cup before assembling, by heating in hydrogen. The cap must be provided with a tin-holding lip, in order to produce a reliable seal after bakeout. The splash guards (4) are necessary to prevent the loss of tin when the valve is re-opened after air has been admitted to the closed valve.

(c) *Closing systems using silver chloride.* Since silver chloride may be used for vacuum sealing (Section 3.4) valves using this material in their closing systems have also been constructed.

Ramsperger[1027] used silver bellows (1 Fig. 6.30a) coated with fused silver chloride and closed at the upper end with a cup containing fused silver chloride (2). The valve is closed by extending the bellows so that the end of the outlet (Pyrex) tube (3) is immersed in the silver chloride in the cup (2).

Kirslis[679] utilized a glass bellows (1 Fig. 6.30b) 18 mm i.d. and 43 mm o.d. enclosed in a glass tube (2). The silver chloride seat (3) closes against the inlet pipe and is moved by actuating the lever system (4). Vaughan[1260] described a similar glass, bellows-sealed valve, where a pointed glass rod closes against the silver chloride placed in a cup on the end of a capillary.

Green[478a] used a valve (Fig. 6.30c) in which the silver chloride (1) was transferred by melting and then solidified in the capillary (2), thus closing the

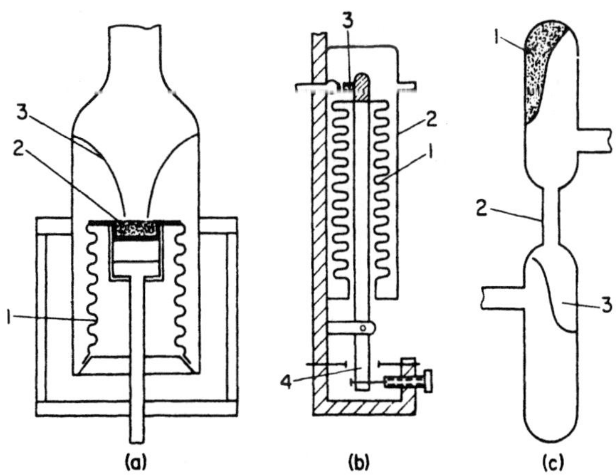

Fig. 6.30 Valves using silver chloride seals: (a) with metal bellows (after Ramsperger[1027]); (b) with glass bellows (after Kirslis[679]); (c) closing by melting the silver chloride (after Green[478a])

connexion between the inlet and outlet. To open the valve, the silver chloride is melted again and allowed to flow into the trap (3).

d) *Closing systems based on fusing glass joints.* Boll[147] constructed a valve, which is closed by fusing together the ends of two concentric capillary glass tubes (see Fig. 5.11a).

Sill[1137] described a valve (Fig. 6.31) where a tungsten coil (1) heats the central glass tube (2) until the glass softens. If the pressure on the inside exceeds that on the outside by an amount at least equal to $4\,\sigma/r$, where σ is

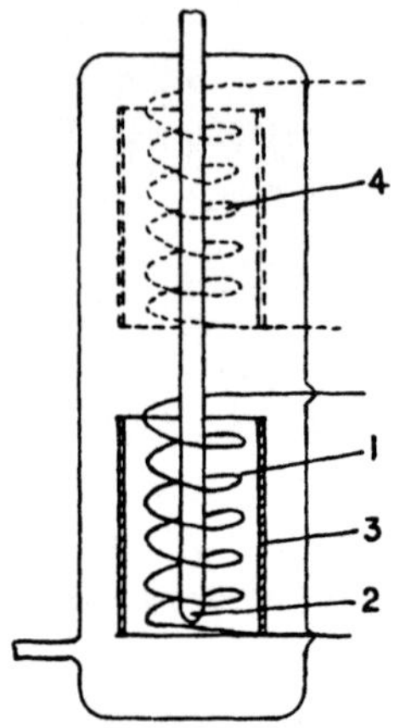

FIG. 6.31 Closing system based on fusing the glass. After Sill[1137] *(Courtesy of The American Institute of Physics)*

the surface tension of the glass (σ = 450 ergs/cm^2 Morey[888]) and r the radius of the hole, a hole will be blown in the central tube. With glass tubing 5 mm in diameter a minimum pressure difference of about 5 torr is necessary before the hole can be blown and the valve opened. The valve can be closed by heating the glass to the softening point, when no pressure difference exists across the hole. Surface tension forces then cause the hole to close. Using a nickel radiation shield (3), such a valve was operated successfully six times. When the pressure difference was about one atmosphere, the valve opened in 15 sec for a power dissipation of about 100 W. An additional coil and radiation shield (4) can be mounted so that when the central tube is excessively shortened after repeated operation, the next coil can be used. The valve can be baked, but some gas is always evolved when the glass is softened.

e) *Closing systems based on ground joints.* Flat, conical, cylindrical or spherical ground joints (see also Section 3.6) may be used as the closing system of vacuum valves. If these joints are used ungreased, the valve may be baked.

Klopfer[690] quotes that in the closed position a greaseless ground joint has leak rates of 7–70 lusec (10^{-5}–10^{-4} l./sec).

Valves of this category are sometimes improperly refered to as stopcocks. Really they differ from the stopcocks in the fact that the two ground surfaces are in *contact only in the closed position of the valve.*

Figure 6.32 illustrates the construction of some valves using *flat* ground joints as their closing system. In the design shown on Fig. 6.32a (Sears[1116]) a

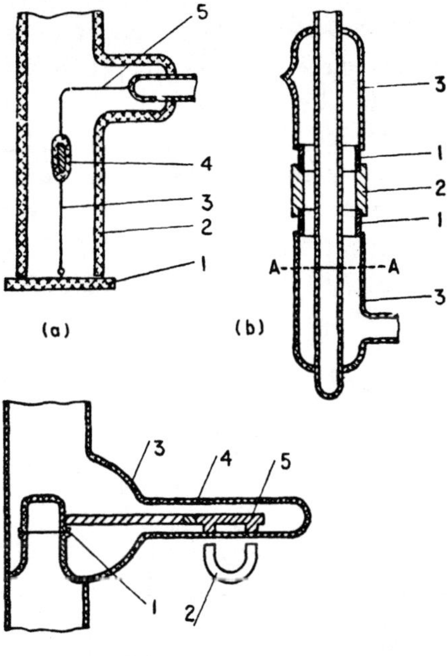

FIG. 6.32 Closing systems using plane ground seals: (a) polished plate on pipe end (after Sears[1116]); (b) butt pipes (Forman[390]); (c) cap, sliding on pipe end. After Lambe[730] *(Courtesy of The American Institute of Physics)*

plate (1) of optically flat quartz (25 mm diameter and 1 mm thick) seats against the polished end of an 18 mm Pyrex tube (2). A quartz eye sealed to the centre of the disc (1) is attached via a quartz fibre (3) to an iron slug encased in quartz (4). This assembly is suspended on a quartz cantilever spring (5), which serves to seat the disc against the opening of the tube. To open the valve the upwards force of the spring (5) is counterbalanced by the downward force of an external solenoid. In order to actuate this valve, the pressure difference should be small.

The valve from Fig. 6.32b (Forman[390]) closes at the butt joint of the optically polished ends of the inner tube. The sleeve of this valve consists of two Kovar cylinders (1) brazed to a nickel cylinder (2) and sealed to the glass envelope (3). After the glass–metal assembly is completed, it is cut on a glass saw at *A–A* (see Fig. 6.32b), and after annealing, the cut surfaces are optically polished. After this, they are placed in contact, the inside tube is evacuated, and then the outer envelope is sealed. Precautions should be taken to prevent the inner tube from deforming. The opening and closing of the valve is achieved by heating or cooling the outer sleeve. According to Forman[390] such a valve was capable, in the closed position, of maintaining a pressure of 10^{-5} torr for a week in a 0.6 litre vessel. The nickel sleeve (2) may be used up to about 200 °C; above this temperature it distorts.

Figure 6.32c shows a glass valve in which the two parts of the flat ground glass joint (1) move parallel to the ground surface (Lambe[730]). The closing cap is moved horizontally by the external magnet (2), which acts on the soft iron end (5), which moves in the side tube (4). The connexion (3) between the side arm and the body is made in such a way as to accommodate the cap when the valve is open. The valve may be opened only when no pressure difference exists across it. A valve based on this principle was used by Anderson[32a], who studied the influence of the out-of-flatness of such seals on their tightness.

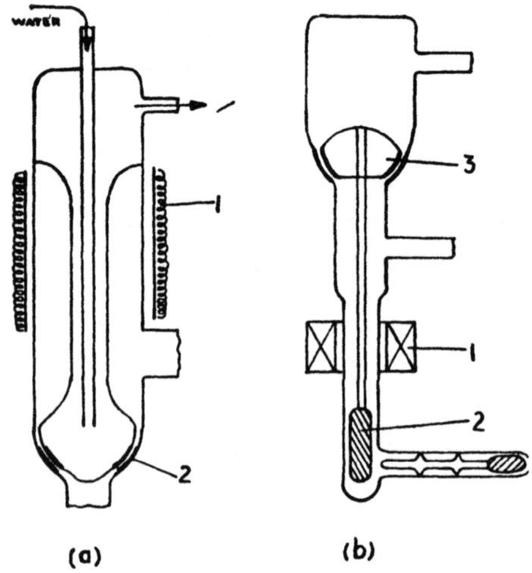

FIG. 6.33 Closing system using tapered or spherical ground joints: (a) actuated by thermal expansion. After Harrison[512] *(Courtesy of The Institute of Physics and The Physical Society, London)*; (b) magnetically operated, (1) electromagnet; (2) iron slug; (3) ball and socket joint. After Vogl[1271] *(Courtesy of The American Institute of Physics)*

Conical and ball, ground joints can be successfully used as the closing parts of valves, but the ball joints, having less probability of sticking, are preferred. The conical or ball joint parts may be separately supported on coaxial glass tubes (Fig. 6.33a). If the outer tube is heated by an external heater (1) and the inner tube is water-cooled (Harrison[512]) due to the expansion differences, the ground joint (2) will be slightly opened. The throughput of the valve may be controlled by adjusting the temperature difference between the two coaxial tubes. If the inner tube is silica the necessary expansion difference may be obtained without cooling.

Vogl[1271], Decker[270], Yarwood[1341], Thiele[1218b] described valves using a spherical ground joint, where one of the members can be raised to open the valve by the action of an external magnet or solenoid. Such a construction is shown in Fig. 6.33b.

Adam[3] describes a valve of a similar construction, using a greaseless spherical ground joint, and quotes a conductance in the closed position of 1×10^{-4} l./sec. This valve was used in an ultra-high vacuum system down to pressures of 10^{-10} torr.

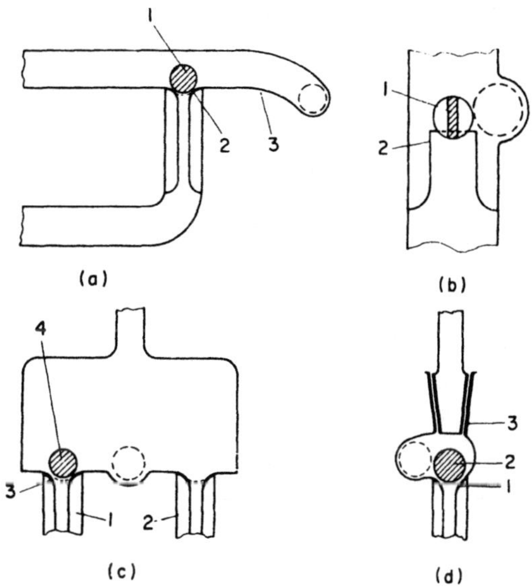

FIG. 6.34 Closing systems using a ball seated on a conical ground surface: (a) with steel ball and side arm (Bachman[65]); (b) with glass ball (Metzler[852]); (c) two way valve with glass ball (Thomas[1220a]); (d) demountable valve with tapered ground joint (Bach[64])

Figure 6.34 shows some valve designs using a steel ball sitting on a ground seat. In the construction shown in Fig. 6.34a (Bachman[65]), the steel ball (1) placed in the side tube (3) may be moved by an external magnet, to sit on the ground shoulders of the connexion (2).

Metzler[852] described a valve (Fig. 6.34b), which consists of a Pyrex ball (1) which sits on the polished Pyrex seat (2) sealed into the enlarged portion of the vacuum line. The glass ball used had a surface finish of 20–30 micro in. and was drilled to take a small magnetic slug, which is carefully sealed into the ball. The valve was used to close a space having a pressure of 10^{-9} torr from another with 10^{-5} torr.

Thomas[1220a] used a valve where various pipes (1, 2) were butting onto ground seats on a plate (3), on which a glass sphere (4) filled with iron may be moved from one seat to the other by means of an external magnet (see Fig. 6.34c).

In order to facilitate cleaning of the valve, Bach[64] constructed his valve (Fig. 6.34d), closed by a steel ball (2) resting on a ground seat (1), by using a standard conical ground joint (3).

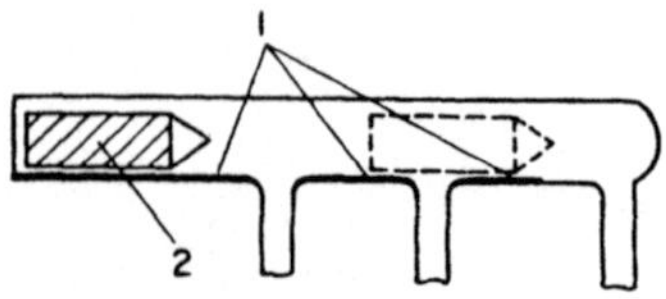

FIG. 6.35 Closing systems with cylindrical ground joint (after Decker[270])

A glass valve with a cylindrical ground joint (Fig. 6.35) is described by Decker[270]. This valve consists of a cylindrical ground seat (1) and a glass enclosed iron slug (2) to fit the seat. The valve is closed or opened by moving the ground slug in and out of the seat by an external magnet.

f) *Closing systems using clamped elastomer tubes*. The most simple valve consists of a clamp placed on the connecting elastomer tube. The method may be used on the low vacuum side of temporary vacuum systems, where the outgassing rate of the elastomer tube can be tolerated.

Figure 6.36 shows various clamping methods using mechanical devices. Although the operating system of all these clamps is different, the closing of the elastomer tube is achieved in all these clamps by pressing it between two transverse plates or rods.

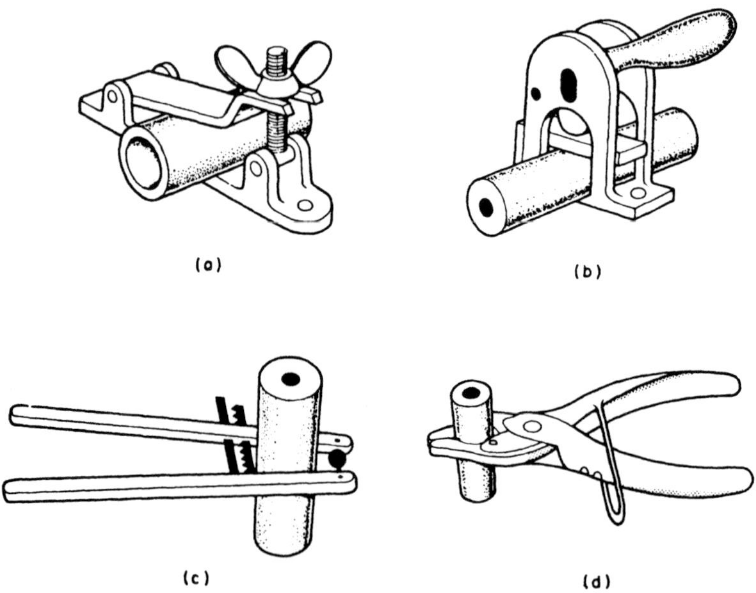

FIG. 6.36 Clamping systems for elastomer tubes: (a) with screw; (b) with lever and cam; (c, d) tongs or nippers

The systems using clamps have the advantage that no moving part has to be introduced through the wall, but their main drawback is the lack of reproducibility due to sticking.

Altaiskii[23] described a valve (Fig. 6.37) consisting of a metal rod (1), enclosed in a length of rubber tube (2), and a clamp unit (3). The metal rod has two central ducts, butting in a half-round groove (4). A clamping ball (groove) of 8 mm diameter, and ducts of 4–5 mm diameter may be used with

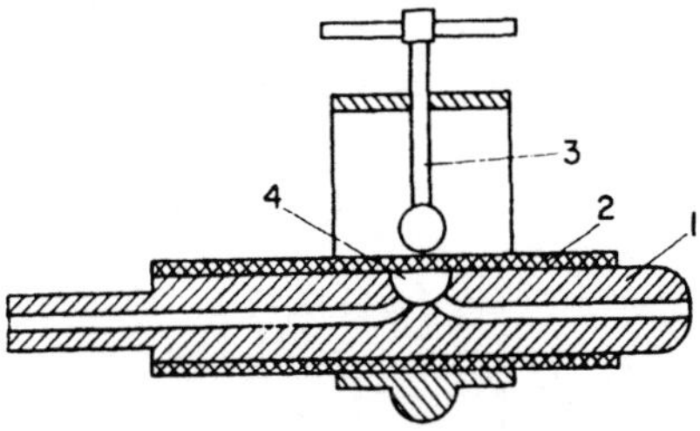

FIG. 6.37 Valve, consisting of an elastomer tube on a metal plug (after Altaiskii[23])

a rubber tubing having 22 mm o.d. and 4 mm wall thickness. The valve is closed by pressing the rubber in the groove, using the clamping ball.

g) *Closing systems using diaphragms.* In these valves a rubber diaphragm (1, Fig. 6.38) is forced against the seat (2) by the external shaft (3) operated from the head (4). The diaphragm functions here as both the closing and the sealing system of the valve (see Fig. 6.23).

Such valves are supplied by several manufacturers, generally using rubber diaphragms. A valve using an annular metal diaphragm was described by Batzer[96] (Fig. 5.14).

Figure 6.38A shows the construction of the diaphragm (Saunders) valves supplied by Edwards[328] and Table 6.5 lists their dimensions.

Figure 6.39 shows the handle-operated, right-angle and straight-through, diaphragm valves supplied by Leybold[762], and Table 6.6 lists their dimensions.

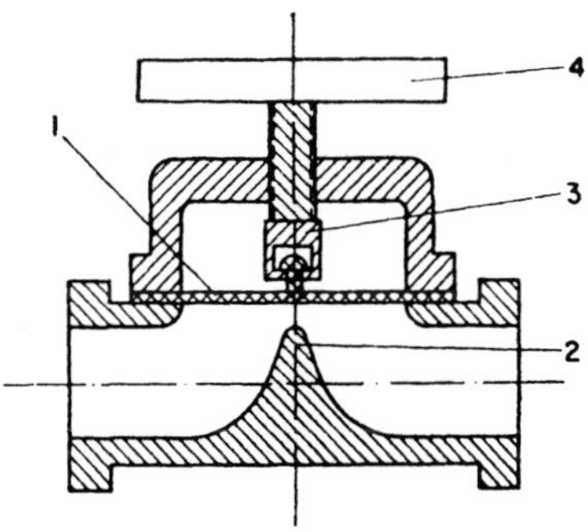

FIG. 6.38 Diaphragm valve

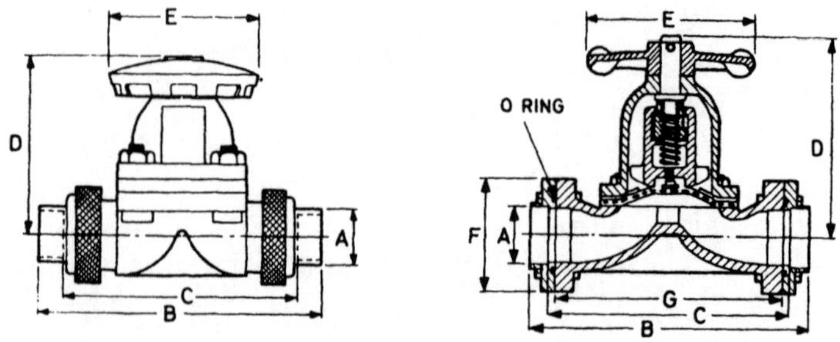

FIG. 6.38A Diaphragm valve supplied by Edwards[328] (Table 6.5)

TABLE 6.5. DIMENSIONS OF "SPEEDIVALVES" SUPPLIED BY Edwards[328] (see Fig. 6.38A)

Size		1/4 in.	1/2 in.	3/4 in.	1 in.	1 1/2 in.	2 in.	3 in.
A	in.	1/4×0.346	1/2×0.596	3/4×0.846	1×1.112	1½×1.612	2×2.128	3×3.128
	mm	6.35× 8.77	12.7× 15.25	19.05× 21.5	25.4× 28.2	38.1× 40.9	50.8× 54.1	77.5× 54.1
B	in.	3 3/8	4 1/16	5 5/16	5 13/16	7 5/8	8 7/8	10
	mm	85.5	100.3	135.0	147.8	193.6	225.5	254
C	in.	2 3/4	3 5/16	4 9/16	5 1/16	6 1/2	7 3/4	8
	mm	69.9	84.0	115.9	128.6	165.1	196.8	203.2
D	in.	2 1/4	2 3/8	2 15/16	3 5/8	5 3/8	6 5/8	9 1/8
	mm	57.2	60.3	74.6	92.1	136.5	168.2	231.8
E	in.	1 1/4	2 1/4	2 3/4	3 1/8	4 1/2	5 1/2	7 3/4
	mm	32	27.2	69.9	79.4	114.3	139.8	196.8
F	in.	–	–	–	–	4	4 1/2	7 1/4
	mm	–	–	–	–	101.6	114.3	184.1
G	in.	–	–	–	–	6 1/2	7 3/4	8 1/2
	mm	–	–	–	–	165.1	196.8	215.9
O-ring (see Table 3.19A)		VOR 4A	VOR 121	VOR 130	VOR 136	VOR 146	VOR 159	VOR 184

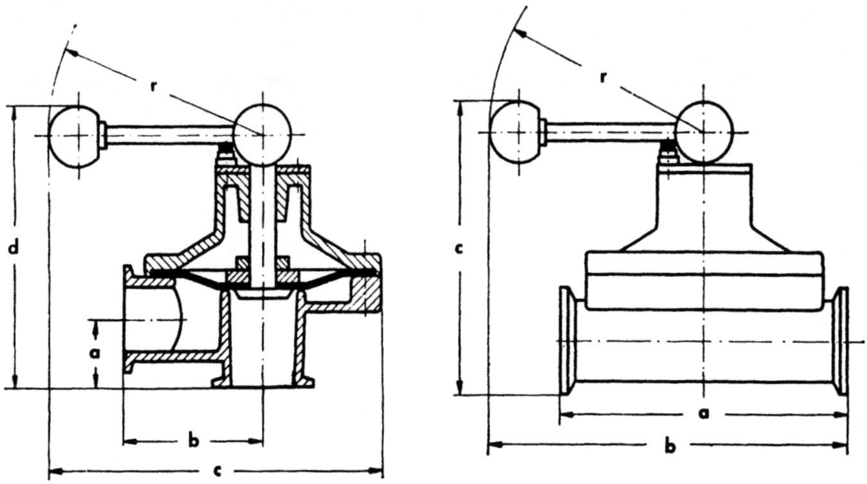

FIG. 6.39 Diaphragm valves supplied by Leybold[762] (Table 6.6)

TABLE 6.6. DIMENSIONS (mm) OF DIAPHRAGM VALVES SUPPLIED BY Leybold[762] (see Fig. 6.39)

Size	Right angle			Straight-through		
	10	20	32	10	20	32
a	20	25	35	100	120	150
b	50	60	75	103	152	190
c	115	142	179	95	120	150
d	110	115	146	—	—	—
r	73	92	115	73	92	115

Figure 6.40 and Plate 22 illustrate the shape and construction of the diaphragm valves supplied by Genevac[441] and Table 6.7 lists their dimensions.

h) *Closing systems using gasket seals.* In any vacuum system, where gasket seals are utilized, valves using gaskets may also be used. A large variety of such valves are available from various suppliers. The basic closing principles used in valves with gaskets are schematically illustrated in Fig. 6.41. Table 6.8 describes the characteristics of these closing systems. It must be mentioned that the nomenclature for these valves is not well established; thus sometimes the valves defined here as "cone valves" or those called here "plunger valves" are called "plug valves", some flap valves are called gate valves, etc.

Plate valves (see Table 6.8) are closed by pressing a gasket between the plate (disc) and the seat (Fig. 6.23). These valves are derived from those used for liquids, gas or steam, being modified to meet some of the requirements (Section 61.31) established for vacuum valves. The usual modifications are:

(i) the introduction of Neoprene gaskets to provide a vacuum seal in the closing system of the valve;

(ii) the introduction of a vacuum-tight, motion seal;

(iii) the construction of the body from non-porous material, and the machining of the internal surfaces to a good finish.

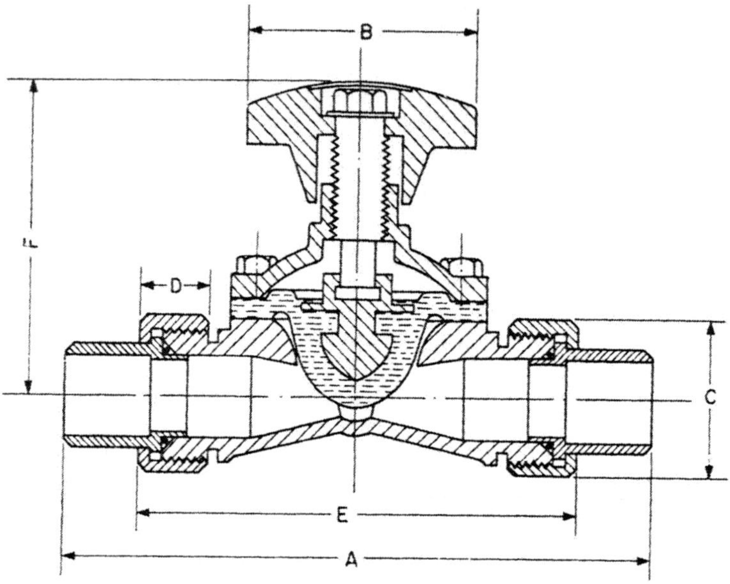

FIG. 6.40 Diaphragm valves supplied by Genevac[441] (Table 6.7)

PLATE 22. Diaphragm valve *(Courtesy of Genevac[441])*

TABLE 6.7. DIMENSIONS (inches) OF DIAPHRAGM VALVES SUPPLIED BY Genevac[441] (see Fig. 6.40)

Type	HV 15	HV 20	HV 25	HV 30	HV 35
A	3 1/2	5 1/16	6 1/8	7 1/8	8 1/4
B	1 3/8	2 3/4	3 1/4	3 1/4	3 1/4
C	7/8	1 7/16	1 3/4	2 1/8	2 3/8
D	7/16	5/8	11/16	3/4	3/4
E	3 1/8	4 1/16	4 5/8	5 3/8	6
F	2 1/2	3 3/16	3 1/8	4 1/4	5
Nominal bore	1/4	1/2	3/4	1	1 1/4
O-ring*	OS 6	OS 11	OS 16	OS 21	OS 24

* See Table 3.19A.

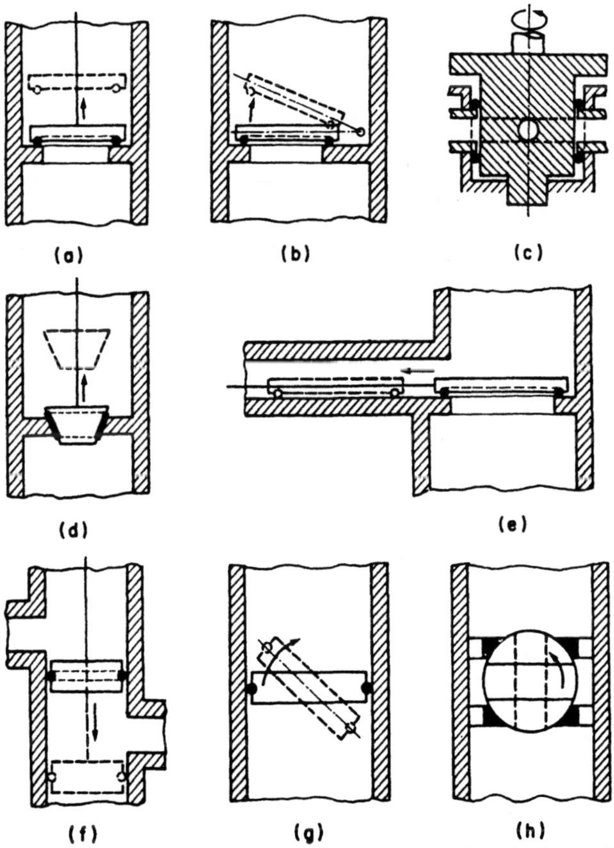

FIG. 6.41 Closing systems of vacuum valves, using gasket seals: (a) plate valve; (b) flap valve; (c) plug valve; (d) cone valve; (e) gate valve; (f) plunger valve; (g) butterfly valve; (h) ball valve

TABLE 6.8. CHARACTERISTICS OF THE VALVES CLOSED WITH GASKETS (see Fig. 6.41)

Designation	Closing part	Position of gasket	Gasket seals against	Motion	Figure
Plate	disc	Face of the disc	seat	Translation normal to port	a
Flap	disc	Face of the disc	seat	Translation and/ or rotation	b
Plug	plug	On plug or shell	plug and shell	Rotation	c
Cone (Nose)	cone	On tapered surface	seat	Translation normal to port	d
Gate	disc	Face of the disc	seat	Translation parallel to port	e
Plunger (Piston)	plunger	On cylindrical surface of plunger	wall	Translation axial	f
Butterfly	disc	Edge of the disc	wall	Rotation; axis in the plane of the disc	g
Ball	ball	On the seats	ball	Rotation	h

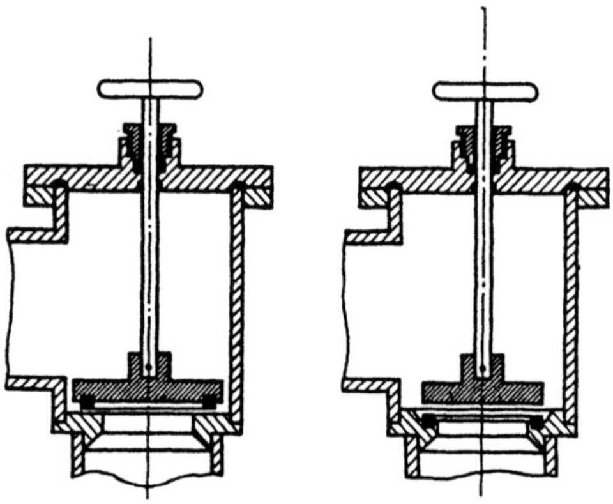

(a) (b)

FIG. 6.42 Stem operated plate valves, having a single closing system: (a) with gasket on the plate; (b) with gasket on the seat

Earlier, many workers modified (Moore[883]) or constructed by themselves (Cowie[242], Garrod[429]) such valves to fit their vacuum systems. Now a large variety of such valves are available from the suppliers of vacuum equipment (Section 11.3).

The gasket of the closing system may be of Neoprene, Viton, Teflon or a metal and may have rectangular, square, trapezium, or circular cross section (Section 3.8). It may be placed on the closing disc (Fig. 6.42a, 6.43) or on the

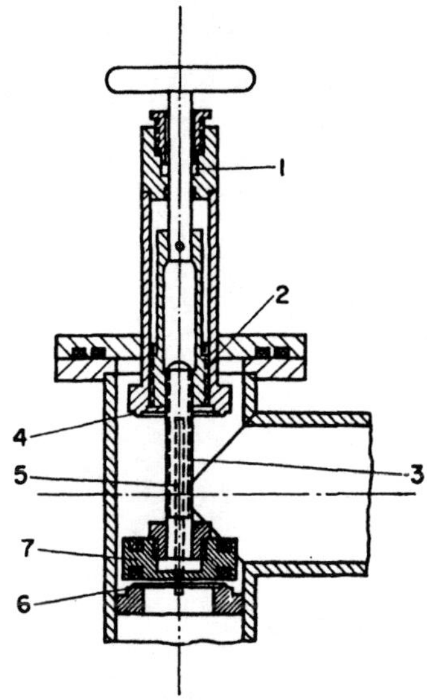

FIG. 6.43 Stem-operated plate valve, having a closing system both for closed and open positions: (1) Wilson seal; (2) standard pitch (right) screw; (3) large pitch (left) screw; (4) sealing seat; (5) guide; (6) closing seat; (7) plate

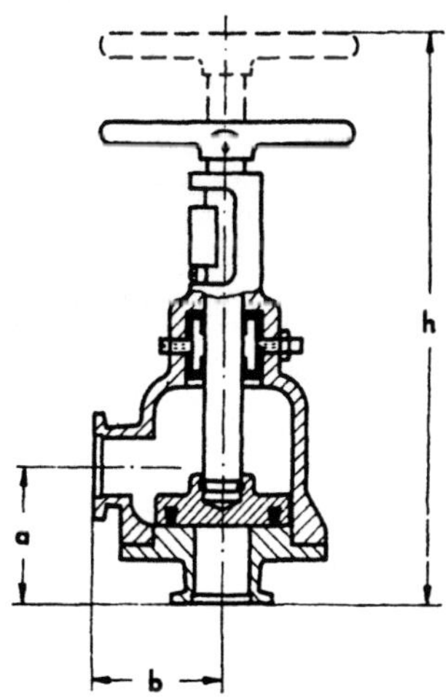

FIG. 6.44 Dimensions of right angle disc (plate) valves supplied by Leybold[762] (Table 6.9)

seat (Fig. 6.42b). To allow easy replacement of the gasket, it is generally preferable to place the gasket on the disc. In some valves, the closing disc is provided with a gasket on both sides (Fig. 6.43), the second gasket closing

TABLE 6.9. DIMENSIONS (mm) OF PLATE VALVES SUPPLIED BY Leybold[762] (see Fig. 6.44)

Size		a	b	h
Angle valves with small flange	NW 10	30	30	144
	NW 20	50	50	210
	NW 32	50	50	210
Angle valves with standard flanges	NW 10	50	50	150
	NW 20	60	60	190
	NW 32	77	77	250
	NW 50	85	85	275
	NW 65	105	105	319
	NW 100	130	130	390
	NW 150	170	170	460
	NW 500	370	270	720

PLATE 23. Disc valves *(Courtesy of Pfeiffer[987])*

against a second seat, achieving the sealing of the stem in the open position of the valve. The particular construction shown in Fig. 6.43 (Guthrie[490], Ardanne[47]) is provided with a differential screw system, which permits fast opening or closing of the valve.

Figure 6.44 and Table 6.9 show the plate valves supplied by Leybold[762] and lists their dimensions.

Figure 6.45 shows the valve supplied by Pfeiffer[987] and Table 6.10 list their main dimensions. Plate 23 illustrates two shapes of such valves.

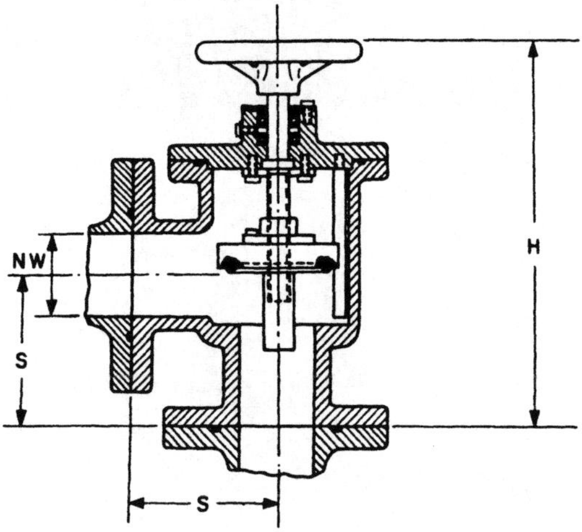

FIG. 6.45 Dimensions of right angle disc (plate) valves supplied by Pfeiffer[987] (Table 6.10)

TABLE 6.10. DIMENSIONS (mm) OF PLATE VALVES SUPPLIED BY Pfeiffer[987] (see Fig. 6.45)

NW	*S*	*H*	*NW*	*S*	*H*
10	55	170	50	110	270
15	55	170	65	120	300
20	75	230	80	150	360
25	75	230	100	150	360
32	95	250	125	180	410
40	110	270	150	200	440

Figure 6.46 shows the overall dimensions of right angle valves available from Genevac[441], listed in Table 6.11.

The lever-operated plate valves supplied by Balzers[444] are shown in Fig. 6.47 and their dimensions are listed in Table 6.12.

Figure 6.47A shows the bellows sealed plate valves available from Ultek[1251a]. The dimensions of these valves are listed in Table 6.12A.

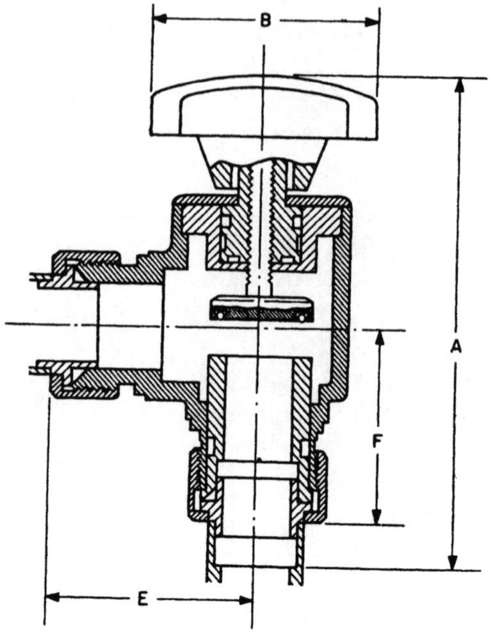

FIG. 6.46 Dimensions of right angle valves supplied by Genevac[441] (Table 6.11)

TABLE 6.11. DIMENSIONS (inches) OF THE RIGHT ANGLE VALVES FROM Genevac[441] (see Fig. 6.46)

Nominal bore	*A*	*B*	*E*	*F*
1/2	5	2 3/4	1 1/2	1 1/2
3/4	6	3 1/4	2 1/8	2 1/16
1	6 3/4	3 1/4	2 7/16	2 7/16
1 1/4	7 1/8	3 1/4	2 9/16	2 5/8

Plate valves are generally constructed of metal and have elastomer gaskets in their closing system. However such valves have been constructed from glass, and metal valves closed-on metal seals have also been made.

Doty[296] described a glass valve provided with glass bellows, which is closed with a Viton A flat gasket, glued to the glass with Epoxy resin.

Sinclair[1140a] removed the O-ring of a standard valve and substituted an indium seal. Nicollian[927] described a valve (Fig. 6.48) in which the closure is accomplished by a knife edge pressed into solid indium. The indium (1) is contained in a copper cup (2), which is completely wet by the indium when heated above the melting point of the indium, forming an indium rich eutectic alloy. This wetting prevents the formation of voids at the copper–indium

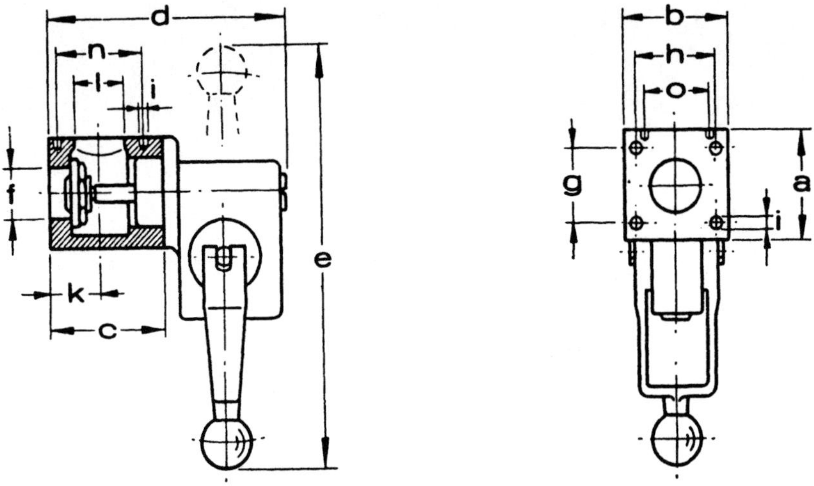

FIG. 6.47 Lever operated disc (plate) valves supplied by Balzers[444] (Table 6.12)

TABLE 6.12. DIMENSIONS (mm) OF THE VALVES SHOWN IN Fig. 6.47. SUPPLIED BY Balzers[444]

Type	*a*	*b*	*c*	*d*	*e*	*f*	*g*	*h*	*i*	*k*	*l*	*n*	*o*
36	70	70	78	164	276	32	44	58.3	M6	36	36	58.3	44
64	90	90	95	181	326	54	62	72	M8	50	64	72	62

interface, which could trap gas every time the valve is opened to air. Because of this alloying, the indium block, which has a tendency to stick to the knife edge (3) during closure, is restrained from parting from its container upon opening. This valve is bakeable at 350 °C.

Eschbach[349] described a dismountable ultra-high vacuum valve (Section 61.35) in which the sharp edge of the valve plate is pressed into the seating covered with indium, tin or silver placed in a groove.

The plate valves constructed for straight-through mounting on to the inlet of diffusion pumps (Fig. 1.1) should have large openings and should be operated from the side. In these valves the closing plate is actuated by a screw mechanism, placed parallel to the plane of the plate, combined with a lever or another mechanism which transmits the motion to the plate.*

Figures 6.49–6.51 illustrate the construction of three baffle valves commonly used in vacuum systems. The screw and lever operated valve shown in Fig.

* The plate should remain in line with the port in order to act as a baffle.

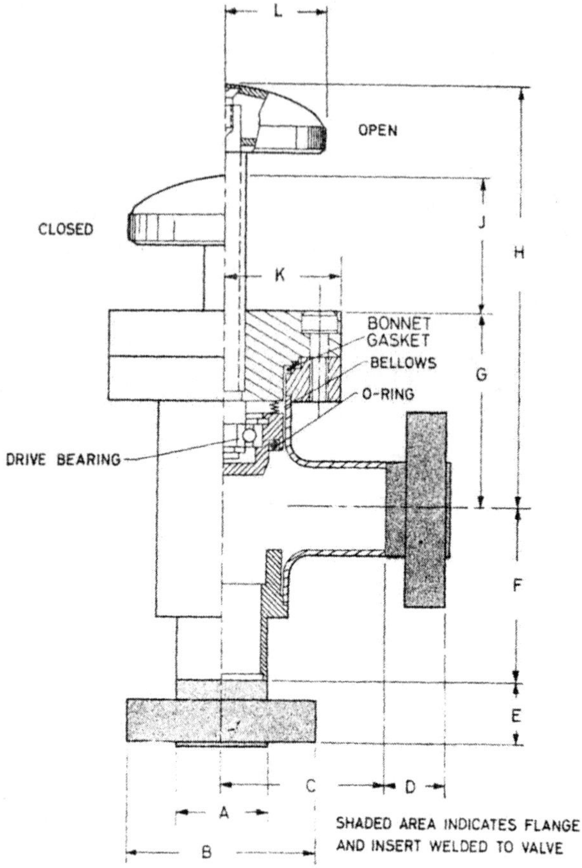

FIG. 6.47A Angle valves supplied by Ultek[1251a] (Table 6.12A)

TABLE 6.12A. DIMENSIONS (inches) OF THE PLATE VALVES* SUPPLIED BY Ultek[1251a] (see Fig. 6.47A)

Size	1″	2″	3″	Size	1″	2″	3″
A	1	2	3	G	2 1/8	2 7/8	2 7/8
B	2 1/2	3 1/4	4 1/2	H	4 3/4	6 3/8	6 1/2
C	1 13/16	2 5/16	3 3/8	J	1 1/2	1 1/2	1 1/2
D	3/4	7/8	7/8	K	1 5/16	1 7/8	2 7/8
E	3/4	7/8	7/8	L	1 1/8	1 5/16	1 5/16
F	1 13/16	2 5/16	3 3/8				

* Equipped with Viton gaskets; can be heated to 150 °C.

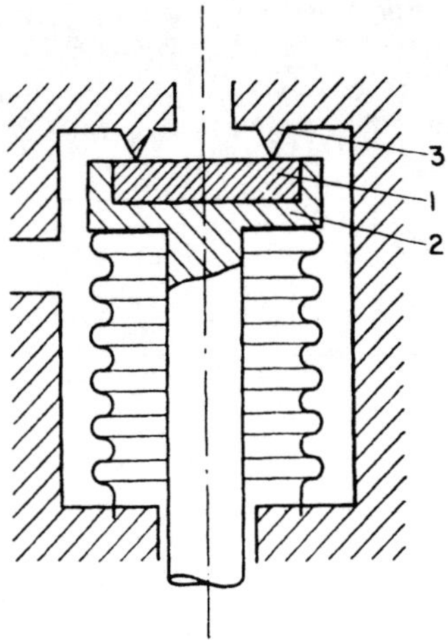

FIG. 6.48 Plate valve closed with knife edge seal on indium (after Nicollian[927])

6.49 is the type supplied by Leybold[762]. Their dimensions are listed in Table 6.13.

The baffle valve shown in Fig. 6.50 is constructed by Edwards[328] in the sizes listed in Table 6.14.

Figure 6.51 illustrates the design of a baffle valve, where the plate is actuated by a cam (Ardenne[47], see Boersch, *Z. Phys.* **130,** 513, 1951). Plate (baffle valves) having similar designs with that shown in Fig. 6.51 are available from Associated Electrical Ind.[52], Pfeiffer[987], Heraeus[542]. A cam operated valve having a square plate is described by D'Eustachio[277].

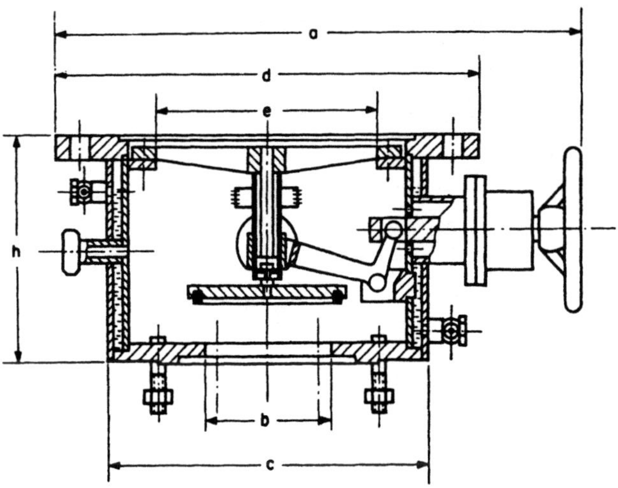

FIG. 6.49 Dimensions of the screw and lever operated plate valves supplied by Leybold[762] (Table 6.13)

TABLE 6.13. DIMENSIONS (mm) OF PLATE VALVES (Fig. 6.49) SUPPLIED BY Leybold[762]

Size	NW 65/100	NW 150/250	NW 250/350
a	330	480	610
b	70	125	250
c	210	280	400
d	210	375	490
e	100	250	350
h	150	200	260
Conductance (l./sec) at 10^{-3}–10^{-5} torr	500	1500	3800

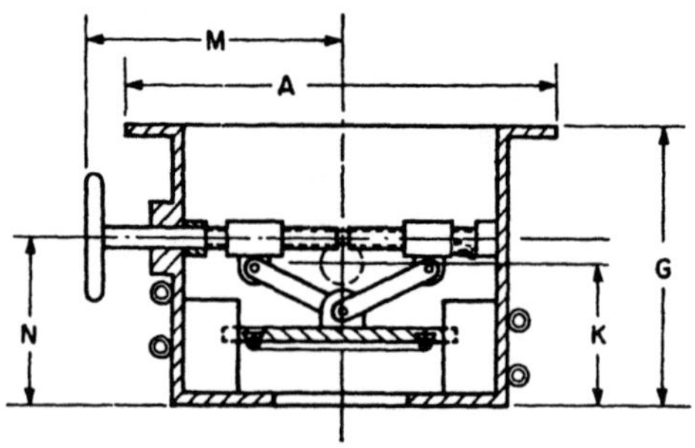

FIG. 6.50 Dimensions of screw and lever operated plate valves supplied by Edwards[328] (Table 6.14)

TABLE 6.14. DIMENSIONS OF THE BAFFLE VALVES (Fig. 6.50) SUPPLIED BY Edwards[328]

Size	2 in.		4 in.		6 in.		9 in.	
	(in.)	(mm)	(in.)	(mm)	(in.)	(mm)	(in.)	(mm)
A	7	178	9 3/4	248	12 1/4	311	15 1/2	394
G	3 13/16	97	6	152	6 15/16	176	8 7/16	214
K	2 1/16	52	1 1/2	38	2	51	2 1/4	57
M	4 9/16	116	6 1/4	159	8 1/2	216	13 1/16	332
N	2 5/16	59	3 9/16	91	3 1/2	89	4 1/2	114

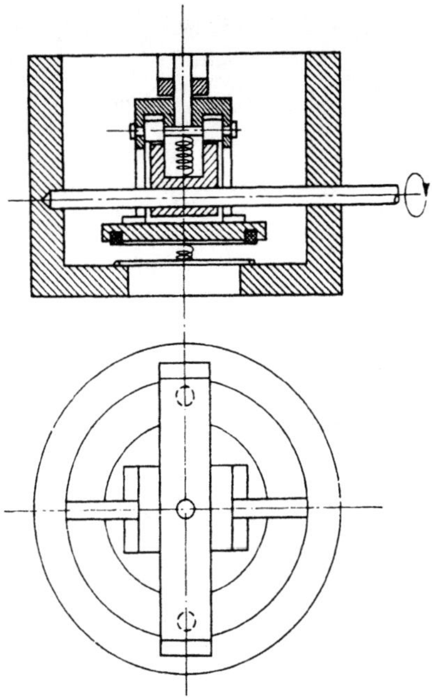

FIG. 6.51 Plate valve, operated by screw and cam

Flap valves (see Table 6.8) are provided with a plate, attached by an arm to a shaft which can rotate about an axis situated in the plane of the port outside its periphery (Fig. 6.52a). This elementary design has the drawback that considerable torque is required to open the plate against pressure.

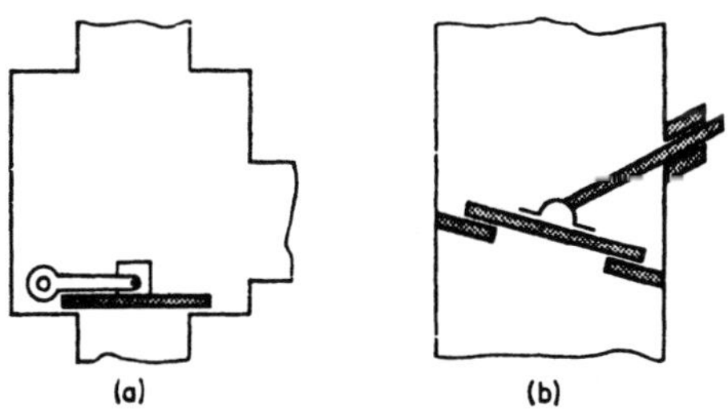

FIG. 6.52 Basic arrangements for flap valves; (a) with shaft; (b) with lever

Buck[180] described a valve (Fig. 6.53) constructed on this principle, but it was provided with a device to facilitate its opening. On rotation of the shaft (3), the toe of the cam (1) tilts and lifts the plate (2) from its seat, thus "cracking" the valve and equalizing the pressure on the faces of the plate. Further rotation of the shaft (3), first carries the edge of the plate with the arm (4), and then lifts the plate.

The port and the plate of flap valves are usually circular, but if required these may have other shapes as well. King[673a] described a flap valve having the port in the shape of a slot (1 by 10 in.) closing on an O-ring placed in a groove of rectangular outline.

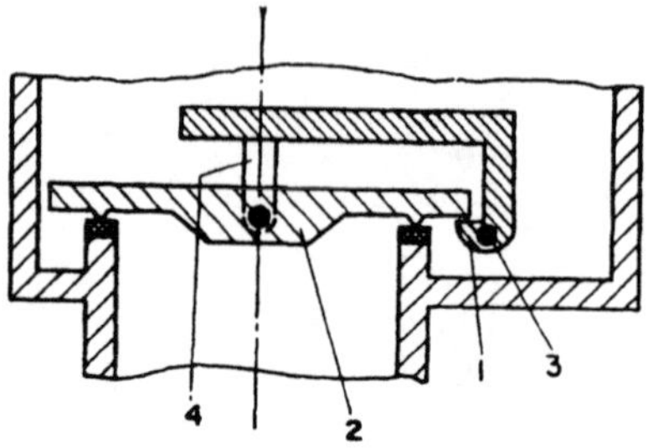

FIG. 6.53 Closing system of flap valve, permitting its opening against atmospheric pressure. After Buck[180] *(Courtesy of The Institute of Physics and The Physical Society, London)*

Instead of using a rotating shaft, lever systems have also been utilized to actuate the plate. Figure 6.52b illustrates the principle of such a flap valve. Jennings[634] described a flap valve (Fig. 6.54) operated by rotating shafts and locked by levers. To open the valve, the lower member of the compression unit is turned anti-clockwise, so that the toggle joint unlocks and the two

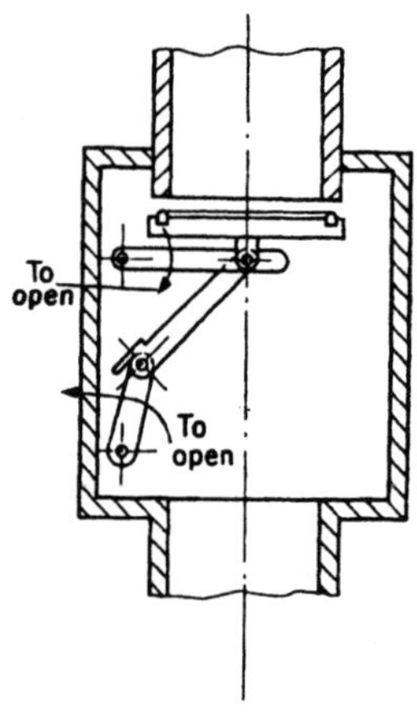

FIG. 6.54 Flap valve operated by radius bar and toggle joint. After Jennings[634] *(Courtesy of The American Institute of Physics)*

parts of the cantilever compression member fold together, allowing the valve plate to drop down, out of the way of the gas flow. Moreau[887] described a flap valve of a similar construction, where the plate enters — in the open position of the valve — into a recess on the body, thus completely removed from the valve opening.

Figure 6.55 shows the schematic arrangement of flap valves supplied by Balzers[444] in the open and closed positions. When the valve is open, the plate is tilted in the direction of the flow. These valves are operated either by a lever, or a servo-motor (dimensions see Table 6.15).

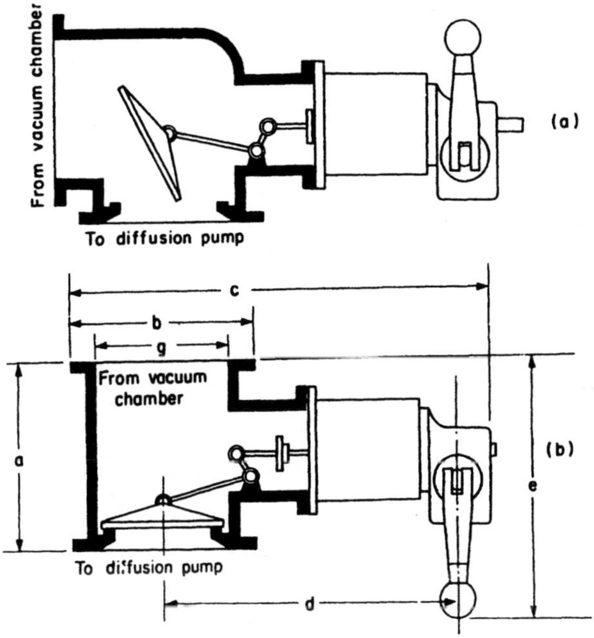

FIG. 6.55 Dimensions of flap valves supplied by Balzers[444]; (a) elbow valves; (b) straight-through valves (Table 6.15)

TABLE 6.15. DIMENSIONS OF VALVES (FIG. 6.55) FROM Balzers[444] (millimeters)

Type	*a*	*b*	*c*	*d*	*e*	*g*	
PVH 190	240	300	527	337	326	230	Lever operated
PVH 310	254	420	641	390	326	350	
PVE 190	240	300	695	—	415	230	Motor operated
PVE 310	254	420	810	—	430	350	

Plate 24 shows the sectional view of a Leybold[762] flap valve, where the lever system is easily visible. This valve is available with openings of 32 to 350 mm (conductance see Table 6.3).

To prevent damage in a vacuum system by a fast inrush of air, flap valves with electrical triggering mechanism may be used. Knudsen[701] described such

Plate 24. Sectional view of a hand-operated, bellows-sealed flap valve *(Courtesy of Leybold*[762]*)*

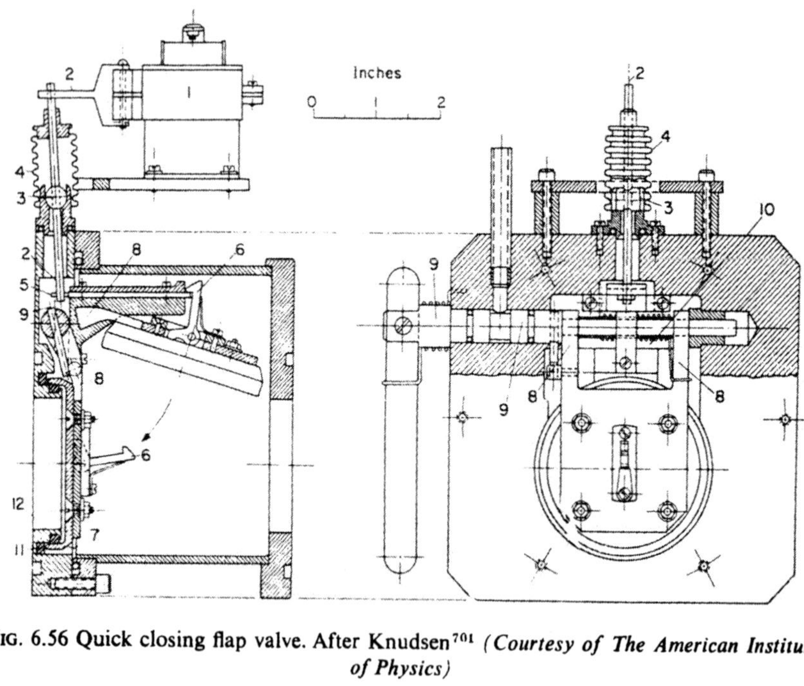

Fig. 6.56 Quick closing flap valve. After Knudsen[701] *(Courtesy of The American Institute of Physics)*

a valve which closes in less than 0.01 sec after the start cf an inrush of air (Fig. 6.56). A spark gap which holds off a high voltage in a good vacuum breaks down as soon as the pressure rises above a certain value. This energizes and closes solenoid (1), which swings lever (2) on its pivot axis (3). The motion of (2) is transmitted inside the valve through the bellows (4) to the sliding release tongue (5), thus releasing catch (6). The closure (7) mounted on the arms (8), swings on the pivot shaft (9) by the action of the coil (10). The closing of (7) is cushioned by the Neoprene washer (11) and the vacuum seal is made at the O-ring (12). When (7) is closed, the build-up of atmospheric pressure against it makes the O-ring seal vacuum-tight.

Plug valves consist of a plug, rotated inside the body so that a hole in the plug registers with corresponding openings in the body. The construction of these valves is similar to that of stopcocks, but their seals are based on gaskets.

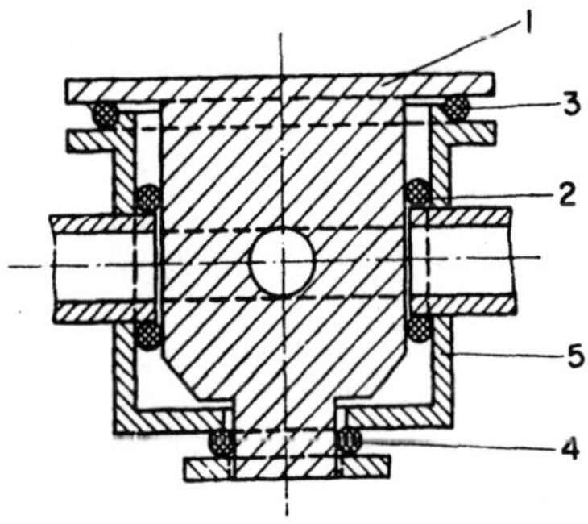

FIG. 6.57 Plug valve. After Green[475] *(Courtesy of The Institute of Physics and The Physica Society, London)*

A plug valve (Green[475]) is shown in Fig. 6.57. Here the plug (1) closes on the O-rings (2) placed around the ports and is sealed against the body (5) by the O-rings (3 and 4).

Plug valves (and metal stopcocks) are generally used only in rough vacuum systems (see e.g. Ministry of Agriculture[868]).

Cone or nose valves (Table 6.8) are closed by pressing the tapered or spherical end of a plug on a conical seat. This plug closes either against a gasket or directly on the seat.

Yergin[1343] modified a steam valve, by placing on the tapered plug two grooves in which rubber O-rings were seated. Jacobs[619a] described a cone valve which may be used for vacuum and high pressures. A glass valve using a polyethylene closing nose on a tapered seat is described by Raats[1022]. Brown[173] quotes valves fitted with Teflon seats. Q.V.F.[1021] manufactures glass valves closing on PTFE seats, and sealed by bellows.

Yates[1342] described a bakeable, bellows sealed ultra-high vacuum valve. The end of the closing stem (1), (3) is specially shaped (4, Fig. 6.58) so that the gold ring gasket (2) after having been deformed (Fig. 6.58b) while closing the valve, is brought back in its original shape when the valve is opened (Fig. 6.58c). After a number of operations the gold ring will be hardened. The baking of the valve helps to anneal it.

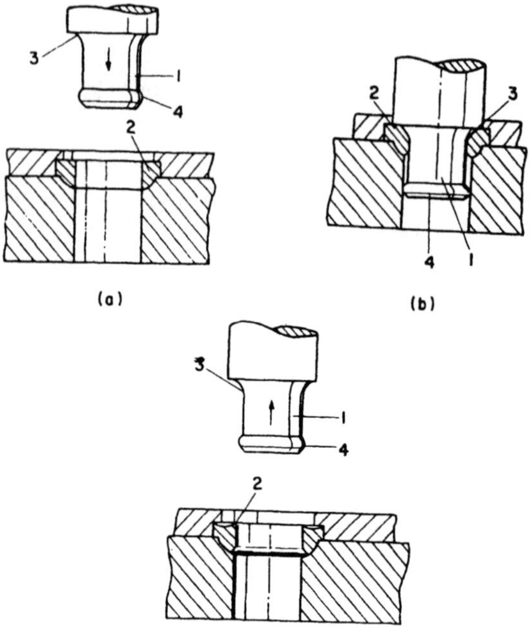

FIG. 6.58 Closing system for an all-metal valve, with gold gasket. After Yates[1342] *(Courtesy of Rudolf A. Lang Verlag, Esch/Taunus, Germany)*

Bakeable valves for ultra-high vacuum have their closing system based either on hard metal noses pressed against softer seats (see also Green[476]) or soft metal noses sealing against hard seats (Section 61.35f).

Friauf[404] described a valve, where the shallow taper (about 8° half-angle) of the valve stem, seals against the slightly rounded edge of the valve seat. Radabaugh[1023] described a valve closed by a ball (Fig. 6.59). In this valve, a section (1) of the stem is reduced in size to such an extent as to give some flexibility. This allows the ball (2) to align itself in the seat (3).

Gate valves (see Table 6.8) have their closing system based on a gasket seal between the seat around the port and the gate which moves parallel to the port. The advantage of these valves consists in the fact that in the open position the port is left completely free. The various gate valves differ in the exact motion of the gate during closing. There are gate valves which close just by forcing the gate to slide motion between the seats (e.g. Fig. 6.60). In other valves the gate slides in front of the port, and then it is closed by a

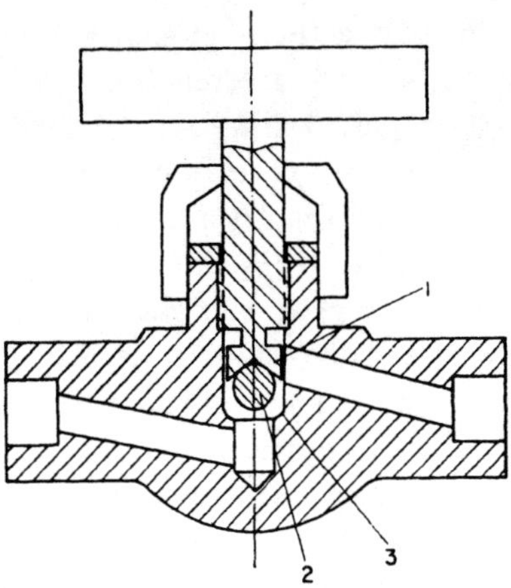

FIG. 6.59 Valve closed with a ball. After Radabaugh[1023] *(Courtesy of The American Institute of Physics)*

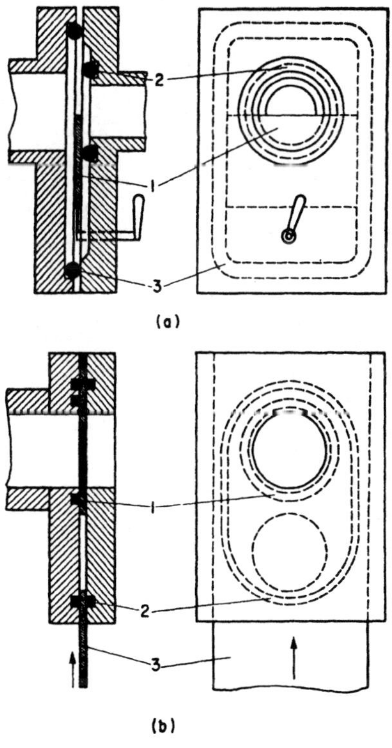

FIG. 6.60 Gate valves with vane: (a) sliding over the O-ring. After Wahl[1281] *(Courtesy of The American Institute of Physics)*; (b) seated on the O-ring. After Graziotti[472] *(Courtesy of Société Française des Ingénieurs et Techniciens du Vide)*

small motion normal to the port, actuated by a cam (Fig. 6.61), guided rollers (Fig. 6.63), sliding wedges (Fig. 6.65) or levers (Fig. 6.66). The sliding motion of the gate can be straight (Fig. 6.60–6.70) or circular (Fig. 6.71).

Valves with the gates sliding on O-rings placed in appropriate grooves machined in the body, have been described e.g. by Wahl[1281], Asao[51], Graziotti[472]. In the gate valve described by Wahl[1281] (Fig. 6.60a) a gate (1) slides over the O-ring (2) to close the port around which this O-ring is placed. The two plates forming the body of the valve are sealed by another O-ring (3) placed in a groove of rectangular outline. As the vane (1) does not sit on the whole surface of the O-ring, the O-ring is easily pushed out of its groove. To avoid this Wahl[1281] used an O-ring one size smaller in its outside diameter than that for which the groove was machined. Bondelid[150] added protective pins to hold the O-ring. Graziotti[472] avoided this difficulty by constructing the gate valve (Fig. 6.60b) in such a manner that the vane is seated on the whole surface of the O-ring both in the open and closed position of the valve. In this valve the vane (3) is provided with a hole. When the hole is placed in front of the port (sealed by the O-ring 1) the valve is open. The vane is sealed to the body by a second O-ring (2).

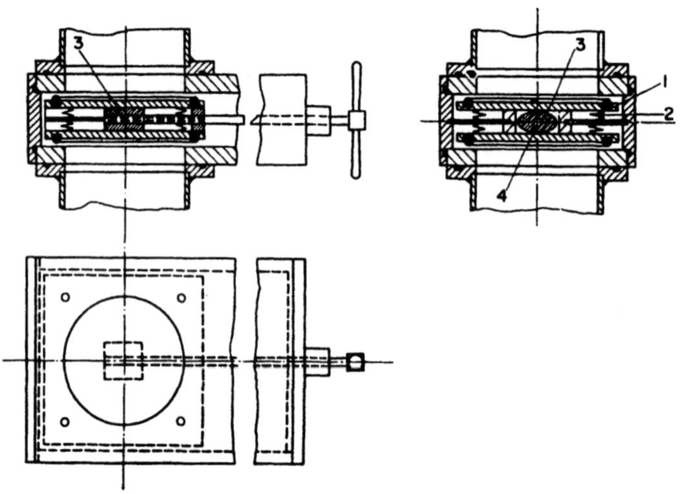

FIG. 6.61 Cam operated gate valve. After King[673] *(Courtesy of The American Institute of Physics)*

In the valve shown on Fig. 6.61 the gate comprises of two rigid plates (1) which are normally drawn together by the springs (2) to permit free motion in the supporting frame. In the closed position of the valve, the plates are forced apart by the action of a cam (3) fixed on the shaft (4). To close the valve the shaft is first pushed in, and then by rotating the handle the plates are forced apart, pressing the gasket against the seat around the port (King[673]).

Hartman[516] described a gate valve in which the gate is operated by a shaft extending through an O-ring seal (1, Fig. 6.62) placed in the edge of a thick plate forming the body of the valve. Near its end, the shaft (2) is turned to a smaller diameter (3) eccentric to the axis of the shaft. The gate (4) is a flat circular plate with a hole drilled through a stud turned on the back, to clear

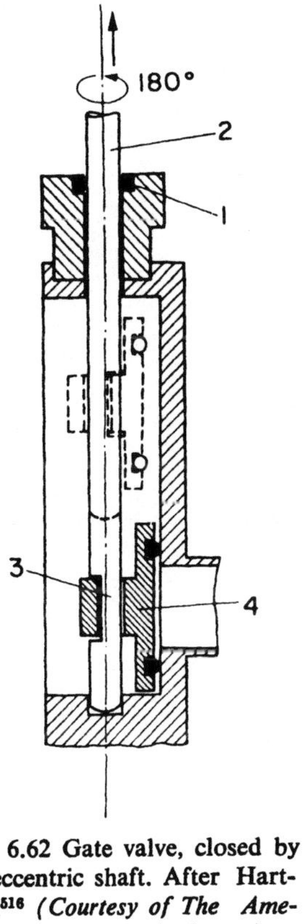

FIG. 6.62 Gate valve, closed by an eccentric shaft. After Hartman[516] *(Courtesy of The American Institute of Physics)*

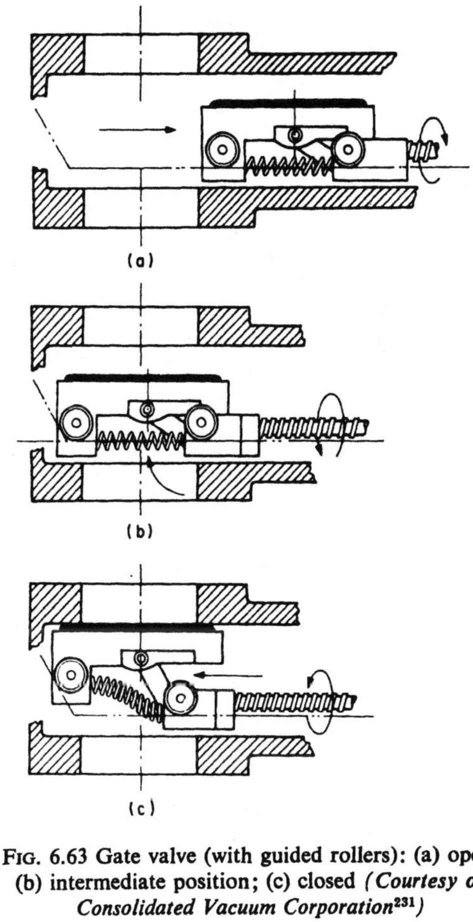

FIG. 6.63 Gate valve (with guided rollers): (a) open; (b) intermediate position; (c) closed *(Courtesy of Consolidated Vacuum Corporation[231])*

the shaft diameter. The gate slides into a slot milled in the thick plate and carries an O-ring in an edge recess. By sliding the shaft, the plate moves in front of the port; by rotating it, the O-ring on the plate is compressed to the seat.

Figure 6.63 illustrates the closing system of gate valves, where the plate is guided by rollers. This allows the valve to be closed by a single kind of motion of the shaft. By rotation of the handle the threaded actuating shaft pushes the carriage which holds the plate in front of the port, and here the guides

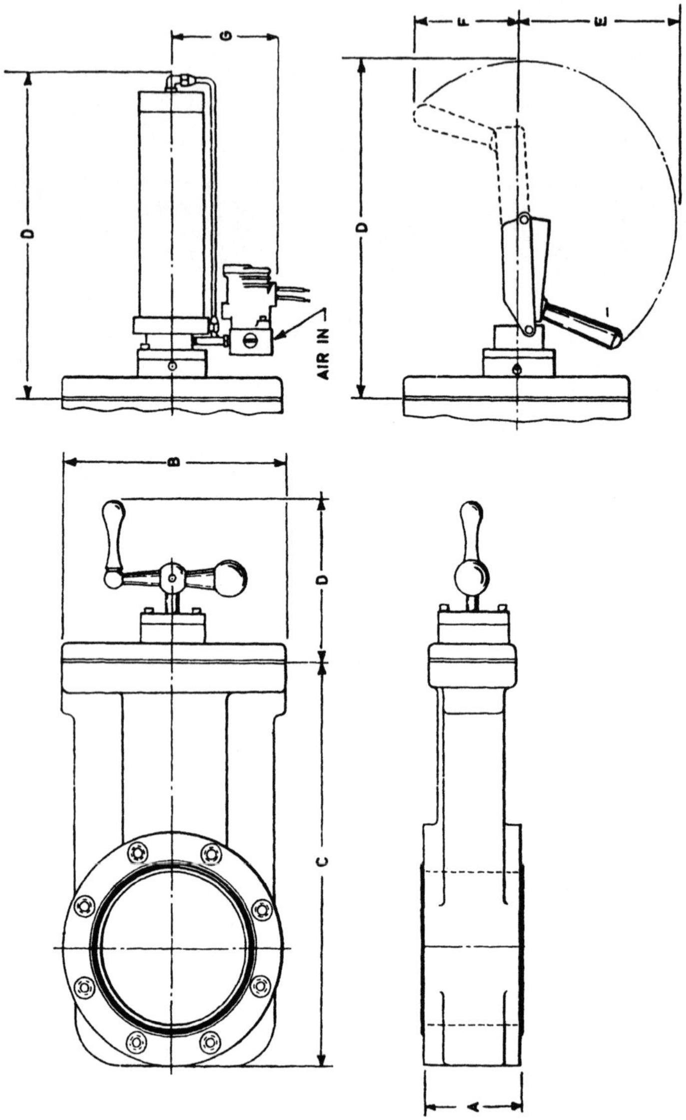

FIG. 6.64 Dimensions of gate valves supplied by C.V.C.[231] (Table 6.16)

force the carriage towards the port. This type of closing system is used in the gate valves supplied by Consolidated Vacuum Corporation[231] (Fig. 6.64). The dimensions of these valves are listed in Table 6.16.

TABLE 6.16. DIMENSIONS (inches) OF GATE VALVES (Fig. 6.64) SUPPLIED BY Consolidated Vacuum Corp.[231]

Nominal size	*A*	*B*	*C*	*D**		*E*	*F*	*G*
2	2 5/8	5	7 3/16	Throttle Quick Pneumatic	5 1/16 9 7/8 8 13/16	5 3/4	5 1/8	4 1/2
4	4 7/16	9	14 13/16	Throttle Quick Pneumatic	6 3/16 14 13/16 12 11/16	7 1/2	5	4 13/16
6	4 3/4	11	18 3/4	Throttle Quick Pneumatic	7 1/2 15 1/2 16 3/8	7 1/2	5	5 7/16

* The valves are available operated by (a) manual throttling with a threaded actuating shaft; (b) manual quick-acting which utilizes a double toggle for rapid actuation; and (c) pneumatic, which employs a double-acting air cylinder for remote control.

Horikoshi[594a] described a thin gate valve (50 mm) having a large aperture (100 mm), in which the sliding frame is provided with inclined slits. When the plate is stopped in front of the port, these force the plate against the seat.

Plate 25 shows a gate valve manufactured by Heraeus[542].

A gate valve closed by sliding wedges has been described by Boyd[156] (Fig. 6.65). In the open position, the wedge steps of the two movable plates (1, 2) mesh together. The plate (2) is supported on a leaf spring (4). When the valve is closed the two plates move forward to cover the port. Plate (2) rides forward on the leaf spring (4), while being pushed by the articulating spring (3). When plate (2) jams against the end of the housing, plate (1) continues to move forward for a small distance, pushing plate (2) at right angles by the action of the wedge steps. Plate (2) is thus pressed against the gasket fixed around the port.

Lever-operated, closing systems in gate valves have been described by Grivet[485], Morton[896], Ortel[956]. Figure 6.66a shows a gate valve (Morton[896]), where the plate (1), which closes onto the seat of the port (2), is linked to a second plate (3) forming a "parallel rule" unit. The plate (3) slides on guides (6). To close the valve, plate (3) is moved towards the port, and when plate

PLATE 25. Plate valve *(Courtesy of Heraeus[542])*

(1) comes into contact with the face plate (4) it is constrained to move upwards. When the valve is to be opened, by moving the plate (3) to the right, the plate (1) will first travel downwards until the step (5) is cleared; then both plates will move together back into the housing.

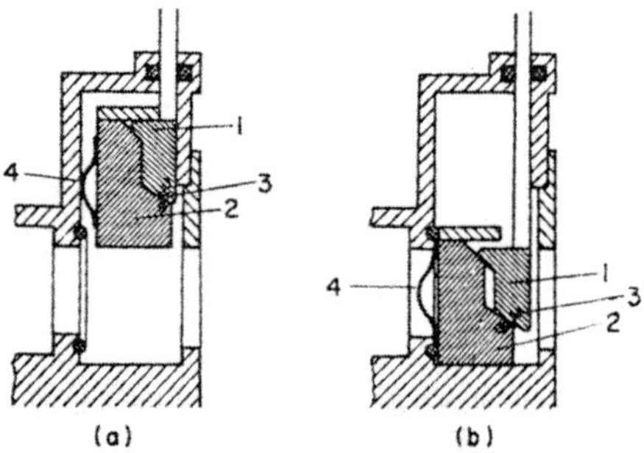

FIG. 6.65 Gate valve, with a plate sliding on wedge steps. After Boyd[156] *(Courtesy of The American Institute of Physics)*

Figure 6.66b illustrates the closing system of the gate valves manufactured by Vacuum Research Company[1253a]. In these valves the rotation of the shaft (1) connected by the part (2) to the plate (3), slides the plate (3) towards the

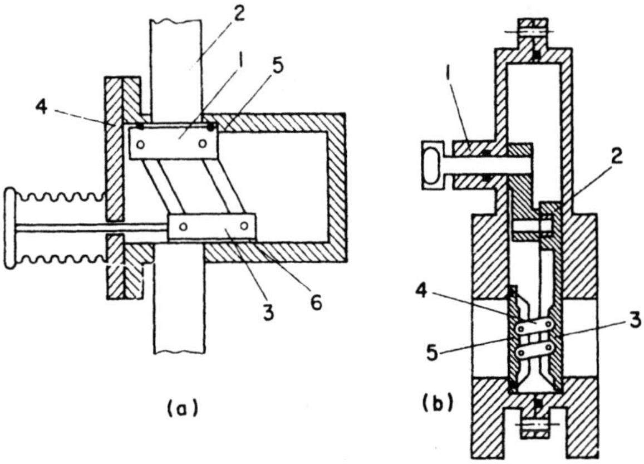

FIG. 6.66 Gate valves, closed by lever actuated gates: (a) construction of Morton[896] *(Courtesy of The Institute of Physics and The Physical Society, London)*; (b) construction of Vacuum Research Company[1253a]

port. When the plate (5) must stop, it is pushed towards the seat by the parallel levers (4). Figure 6.67 and Table 6.17 give the overall sizes of some such gate valves supplied by Vacuum Research Company[1253a].

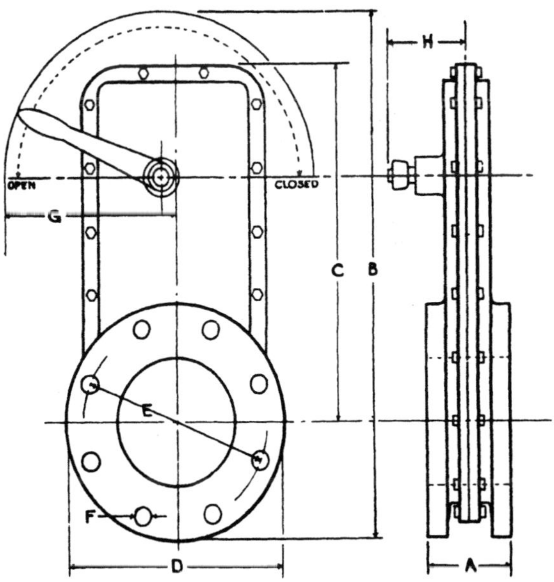

FIG. 6.67 Dimensions of gate valves supplied by Vacuum Research Company[1253a] (Table 6.17)

TABLE 6.17. DIMENSIONS (inches) OF GATE VALVES (Fig. 6.67) SUPPLIED BY Vacuum Research Co.

Size	*A*	*B*	*C*	*D*	*E*	*F*	*G*	*H*	Holes
1	3	9 5/16	5 9/16	4 1/4	3 1/8	1/2	4 5/16	2 3/4	4
2	3 1/4	12 3/8	7 3/4	6	4 3/4	5/8	5 1/2	3	4
4	3 3/4	19 5/8	12 1/2	7 1/4	5 1/4	3/8	8 5/8	4	6
6	4	24 1/8	17	9 1/4	8	3/8	8 5/8	4	8

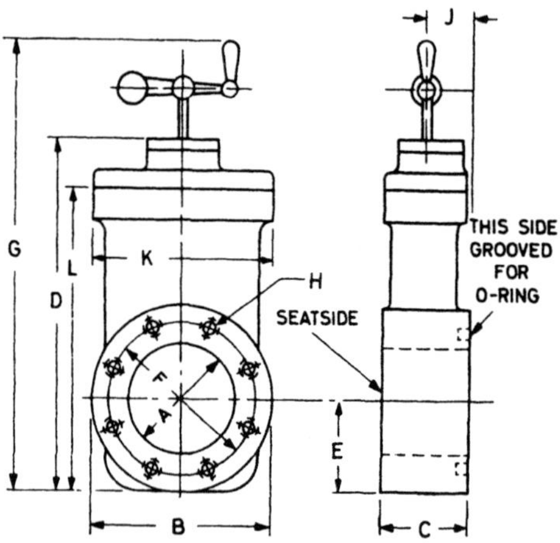

FIG. 6.68 Dimensions of gate valves supplied by N.R.C.[912] (Table 6.18)

TABLE 6.18. DIMENSIONS (inches) OF GATE VALVES (Fig. 6.68) SUPPLIED BY National Research Corp.[912]

Size	4	6	Size	4	6
A	5 1/4	7 1/16	G	21	25 1/2
B	9	11	H	8×5/8–11	8×3/4–10
C	4 7/16	4 3/4	J	2 1/16	2 1/8
D	17	21 1/8	K	9	11
E	4 1/2	5 1/2	L	14 13/16	18 3/4
F	7 1/2	9 1/2	Conductance (l./sec)	2750	6500

Figure 6.68 shows the gate valve manufactured by National Research Corp.[912] and Table 6.18 lists their dimensions and characteristics.

Figure 6.68A shows the dimensions and the operating system of the 6-inch gate valve, manufactured by Ultek[1251a], for flanges with Seal-Vac seals (see Fig. 3.137A).

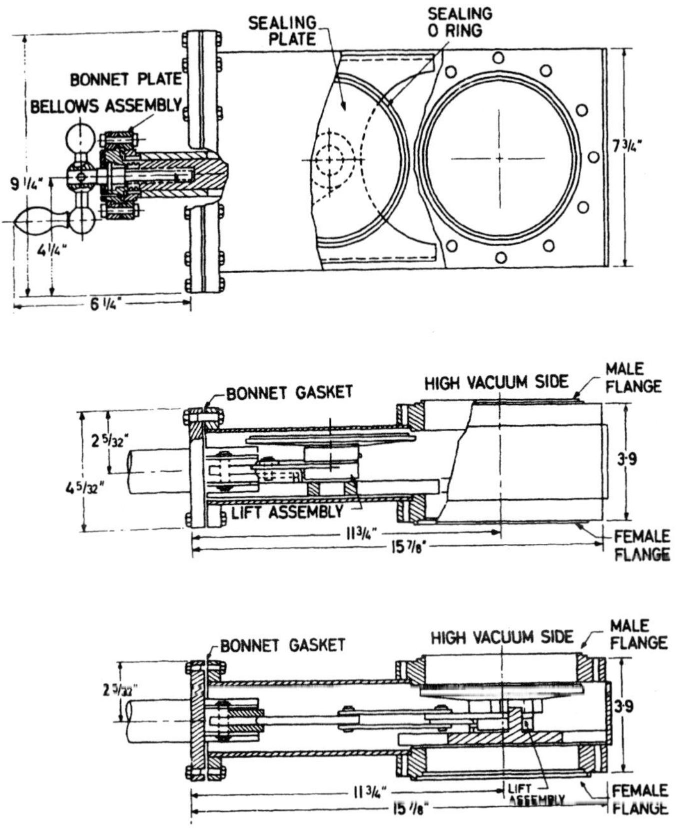

FIG. 6.68A Dimensions of gate valves supplied by Ultek[1251a]

It is possible to use rotating plates (instead of ones with translation motion) as the closing parts of gate valves. Smith[1155] has found that with proper design it is possible to slide the one plate over the other, without damaging the gasket (Fig. 6.69a). Lissberger[771] proposed the construction (Fig. 6.69b), where the two plates are concentric, the aperture in the lower plate being surrounded by an O-ring, and closed or opened according to the position of the aperture in the upper (rotating) plate. Munday[906a] described a gate valve, where the closing plate is enclosed between two flanges and is moved by the rotation of one flange with respect to the other.

Plunger valves (see Table 6.8) are closed by either a piston, plunger or rod which slides *axially* in a cylindrical body. Although the ports of the valve are usually placed on the cylindrical surface of the body, it is possible to place one

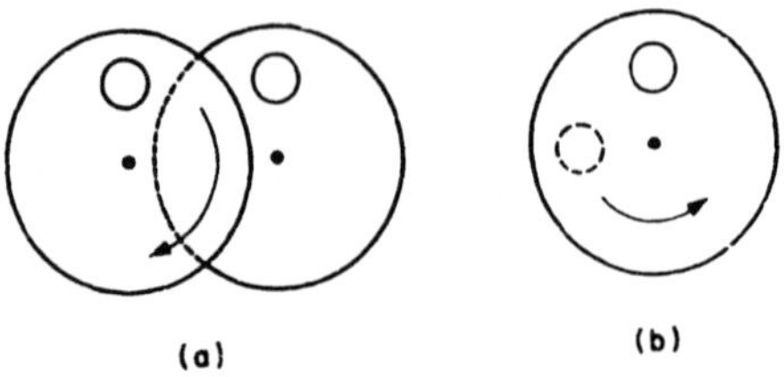

FIG. 6.69 Gate valves with rotating gates: (a) juxtaposed (Smith[1155]); (b) superposed (Lissberger[771])

of these ports normal to the axis (Fig. 6.70), or even connect it through the plunger itself (Fig. 6.72).

The valve may be closed or opened by the rod sliding in a plastic or elastomer stopper, provided with an axial and a radial channel as in Fig. 6.70a, (Raats[1022]), or a rod sliding through a chamber (Fig. 6.70b), sealed through

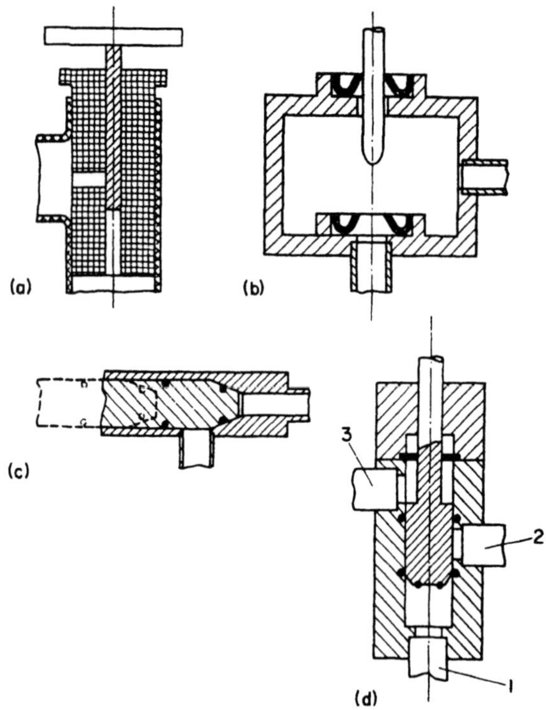

FIG. 6.70 Plunger valves: (a) with a metal rod, closing on polyethylene (after Raats[1022]); (b) metal rod closing on V-shape elastomer gasket (after Kinder[672a]); (c) metal plunger with O-rings on the plunger (after Fuller[412]); (d) plunger with O-rings on the plunger and the body (after Stein[1181])

the cover by a gasket with the appropriate shape (e.g. V-shape), and closing on the bottom of the chamber on a similar gasket. Kinder[672a] described a valve, having three such rods, each of which may close the connexion provided in front of it; the valve was used to connect a vacuum chamber joined on to the horizontal connexion, the roughing line, fine vacuum or air admittance port respectively.

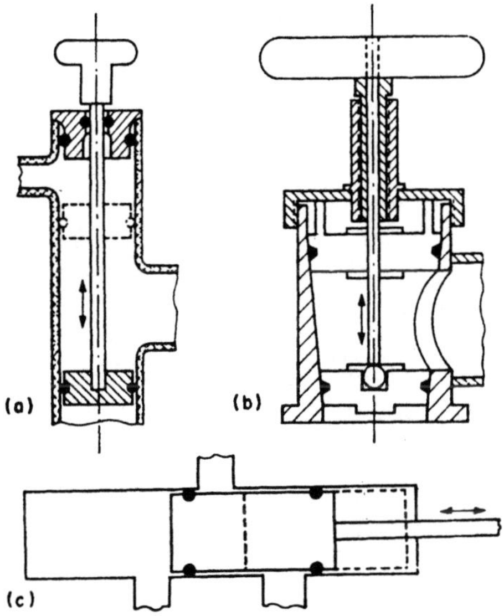

FIG. 6.71 Plunger valves with O-rings: (a) cylindrical plunger (after Stern[1183]); (b) tapered plunger (after Moore[881]); (c) plunger with double O-ring (after Shatford[1126])

The plunger valve may have the closing and sealing O-rings, which are placed on the plunger (Fig. 6.70c), on its cylindrical part and/or on its conical end (Fuller[412], Lach[729]); or some of the O-rings may be seated in the body (Fig. 6.70d) and some of them on the plunger (Stein[1181]).

Figure 6.71 shows some plunger valves using pistons provided with O-rings. Stern[1183] described the simple plunger valve shown on Fig. 6.71a, and Moore[881] described a valve using a slightly tapered piston (Fig. 6.71b) to allow for fine control near the closed position. A two-way valve was described by Shatford[1126] (Fig. 6.71c). Fox[393] described a similar valve, using two pistons mounted on the same shaft permitting the vacuum vessel to be connected to the roughing line, the fine vacuum line or to atmosphere. Lake[729a] described a double piston seal (Fig. 5.21b) which may be used in such a valve.

Barrow[89] described a plunger valve (Fig. 6.72a), which consists of a body (1) and a solid rod (2) which slides within it. Three O-rings (4) are placed inside the body. The channel provided through the rod allows the space between the O-rings and/or the chamber itself to be evacuated, by the connexion (3).

Busen[186a] described a simple valve (Fig. 6.72b), consisting of a glass tube sealed at one end and carrying two O-rings appropriately located with respect to a hole on the side of the tube. The O-rings fit tightly between this tube and the outer tube which forms the body of the valve. The dimensions of the O-rings are chosen so that they roll when the inner tube is moved but form

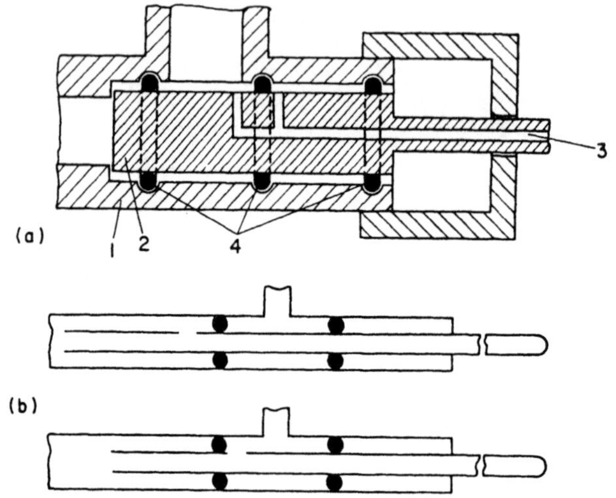

FIG. 6.72 Plunger valves having the connexion through the plunger: (a) with solid plunger. After Barrow[89] *(Courtesy of The Institute of Physics and The Physical Society, London)*; (b) with a tube used as plunger. After Busen[186a] *(Courtesy of Pergamon Press)*

a positive seal at all times. An indication of a good fitting is given when the inner and outer parts of the O-rings are flattened against the glass tubes over a width equal to about half the cross-sectional diameter. In order to open the valve, the rod is pushed until the hole in the inner tube is placed between the O-rings. By providing the outer tube with more outlets and placing more O-rings spaced as required to separate these outlets, a multiple-way valve can be also constructed on the same principle.

Bauer[97] described a similar plunger valve (Fig. 5.30c) using chevron seals instead of O-rings.

Figure 6.73 shows two valve designs where the valve is closed by radially expanding the plunger. The valve in Fig. 6.73a is provided with a piston of heavy-walled elastomer tubing, which is compressed axially until it seals against the wall of the outer tube (Raats[1022]). Figure 6.73b shows a plunger valve construction, where the plunger closes against the body as a result of the difference in thermal expansion between the plunger and the body. If the plunger is made from a material having a small expansion (e.g Invar),

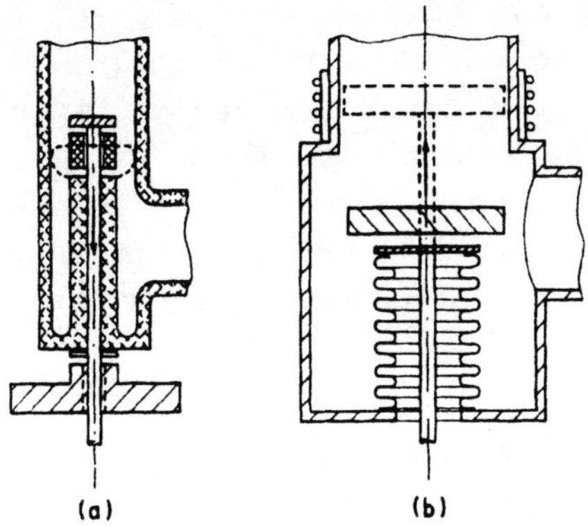

FIG. 6.73 Valves closed by radial displacement of the plunger: (a) with elastomer seal (after Raats[1022]); (b) metal plunger actuated by the thermal expansion difference

and the body is made from a high expansion material (e.g. stainless steel), the valve will be closed when the parts are cold, and will need body heating in order to open it (Ardenne[47]).

Goertz[455] described a valve (Fig. 6.74, 6.75), which is closed by flattening a cone made of a hard spring material until its edge meets the wall. The closure may be repeated in the same position or as close to 0.18 mm apart.

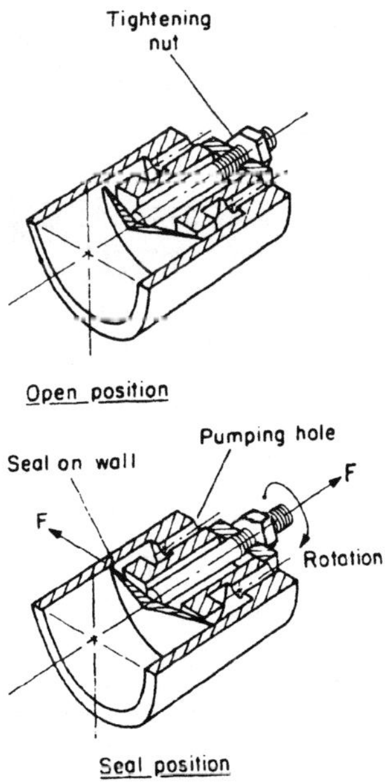

FIG. 6.74 Closing system using an elastic metal washer. Reproduced from Goertz[455] *(Courtesy of Pergamon Press)*

Butterfly valves (Fig. 6.76, Plate 26) consist of a body which receives the valve plate swinging about the stem. An O-ring is placed in a groove cut in the edge of the plate, to provide the seal between the plate and the body when the valve is closed. The centre line of the valve stem is offset from the plane of the plate, by an amount determined by the thickness of the plate and the diameter of the stem. The stem is also displaced parallel to the diameter of the body by the same amount, so that the valve plate is centred in

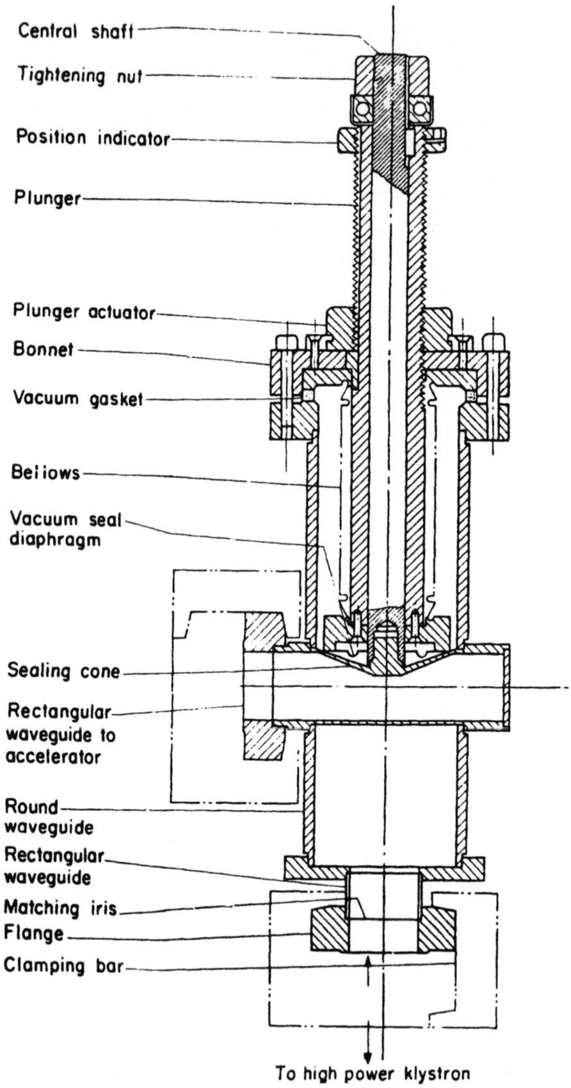

FIG. 6.75 Valve assembly using the closing system from Fig. 6.74. Reproduced from Goertz[455] *(Courtesy of Pergamon Press)*

the body in the open position. By actuating the handle manually or pneumatically, the plate is brought normal to the axis of the body. Closing the valve tends to squeeze the O-ring out of the groove, but by using an O-ring

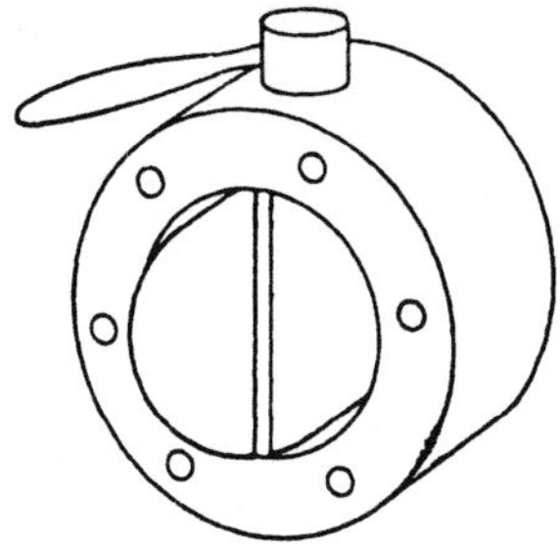

FIG. 6.76 Butterfly valve

which is stretched when on the valve plate, and lubricating the O-ring with vacuum grease, no difficulty is experienced (Holland[585]).

Plate 26 shows the series of butterfly valves supplied by Edwards[328].

PLATE 26. Butterfly valves *(Courtesy of Edwards[328])*

Ball valves (Fig. 6.77) are similar in principle to plug valves (Fig. 6.41) but they use a ball instead of a cylindrical plug. The ball is provided with a large hole, which in the open position of the valve connects the two sides of the valve. As the ball is rotated to 90° the valve is closed by the two O-rings or gaskets sealing against the ball and placed in planes normal to the axis of the valve and its outlets (see also Gilmour[449a]).

Ball valves are available from e.g. National Research Corp.[912], Consolidated Vacuum Corp.[231], Leybold[762] in nominal sizes usually ranging up to

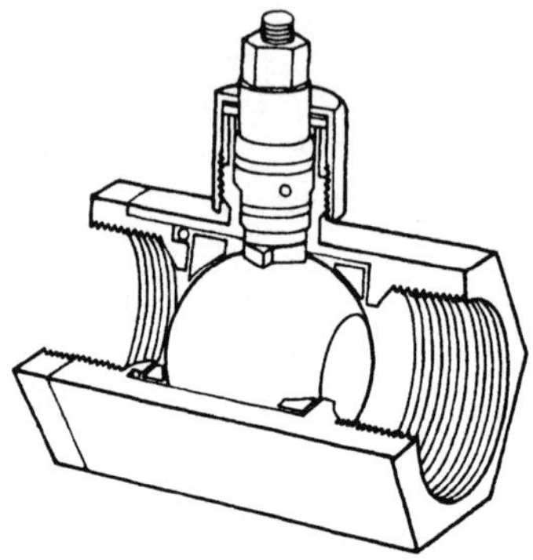

FIG. 6.77 Ball valve (after N.R.C.[912])

2 in. A description of ball valves from various manufacturers is given by Marton[820a].

61.33 Sealing systems of valves. The sealing system of a valve consists (Fig. 6.23) of the parts which seal the inside space of the valve from the surrounding atmosphere, including the seal between the bonnet and body and that between stem and bonnet. The seal between the bonnet and the body may be any static vacuum seal which meets the vacuum requirements of the range where the valve is used (Sections 1.1 and 1.4). The seal between the stem and the bonnet (body) is in fact that which determines the kind of the sealing system of the valve, i.e. if the valve is a *packed* or a *packless* one.

(a) A *packed valve* is defined as a type of valve in which a packing material, placed between the stem and the bonnet, helps to prevent leakage at the point where the stem passes through the bonnet (American Vacuum Society[26]). These valves use gasket seals (Section 51.7) to provide the vacuum tightness of the moving stem.

Packed valves can be used in any vacuum system where the outgassing rate of the elastomer gaskets is tolerable. As motion seals using gaskets may have leak rates less than 1×10^{-2} lusec (Table 6.3), there is no special reason for requiring a packless sealing system in any valve which uses elastomers in its closing system (Section 61.32).

Various packed valves have been described in connexion with their closing systems, see Figs. 6.42–6.46, 6.49–6.50.

(b) In *packless valves* the seal between the stem and bonnet is stationary, and the motion is transmitted using bellows, diaphragms or magnetic transmission.

Bellow-sealed valves have limited motion of the stem since the stroke must correspond to that allowed by the compressibility of the bellows (Section 51.4). If a longer travelling path of the valve plate is required, this must be obtained by using appropriate multiplication systems inside the bellows.

To prevent twisting of the bellows, the stem or plate should have a guided path (see e.g. Fig. 6.78a).

Despite these limitations, bellows-sealed valves are more and more extensively used in high and ultra-high vacuum systems, due to the vacuum-tightness and bakeability of such valves. Bellows-sealed valves have leak rates less than 1×10^{-4} lusec (Table 6.3), i.e. by a factor of at least 100 less than packed valves.

Metal bellows may be used in valves in compression, tension or bending. With valves in the first group the bellows are compressed axially during the closing of the valve (Fig. 6.78a), while in valves in the second group, the bellows are compressed during the opening of the valve (Fig. 6.78b). It is difficult to state if one or the other of these constructions is to be recommended as being more reliable. Logically the choice should depend on the ratio between the time during which the valve is in the closed position, to that when the valve is in the open position. For applications where the valve must be in the closed position for only short periods, the compressed bellows arrangement is advisable; for valves which have to remain closed for long periods, valves where the bellows are compressed during opening, can be recommended.

Due to the mechanical properties of the bellows, valves where the bellows are *bent* are in principle less reliable than those having bellows in compression or tension. Nevertheless such valves were also constructed (Fig. 6.78c) and used (Estermann[359]).

Bellows-sealed valves are shown in Fig. 6.28, 6.47A, 6.48, Plate 24, and Section 61.35f. Other bellows-sealed valves have been described by e.g. Dumond[306], Green[473], Grivet[485], Harvey[517], Lockevitz[778], Reilly[1043], Stohr[1197a], Turnbull[1238].

Valves using glass bellows are described in connexion with Figs. 5.11a and 6.30b, and by e.g. Boll[147], Vaughan[1260].

Diaphragm-sealed valves are defined as those valves in which a thin, flexible disc (Section 51.5), clamped tightly between the body and the bonnet, is interposed between the stem and the plate, thus sealing both the bonnet and the stem, and permitting the transmission of the motion of the stem by its flexure (American Vacuum Society[26], Société Française des Ingénieurs et Techniciens du Vide[1122a]).

Some diaphragm valves use the diaphragm (elastomer) both for sealing and closing the valve. These valves are described in Section 61.32g, and by e.g. Halban[502], Blears[133], Bouring[155], Prugne[1014], Raats[1022].

Valves having the stem sealed with a cylindrical "finger" diaphragm, have been recently manufactured (Röllinger[1074c]).

In other valves, particularly those using metal diaphragms (e.g. Thornes[1224]), the diaphragm serves only as the sealing system of the valve; the closing of the valve is achieved by other means. Warmoltz[1287] described such a valve (Fig. 6.79a) in which the metal diaphragm (1) seals the valve, and its flexi-

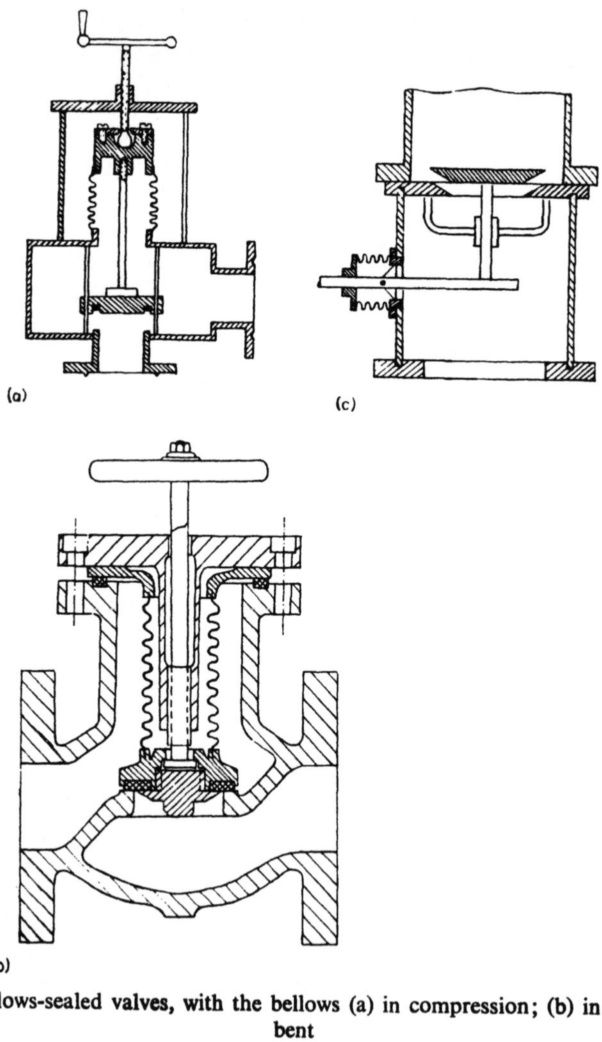

FIG. 6.78 Bellows-sealed valves, with the bellows (a) in compression; (b) in tension; (c) bent

bility permits the copper block (2) to be pressed against the tube (3) to close the valve. Figure 6.79b shows a plate valve (Edwards[328]), where the lever (1) is actuated from the wheel (3) and is sealed by the diaphragm (2).

Magnetically operated valves are discussed in Section 61.34.

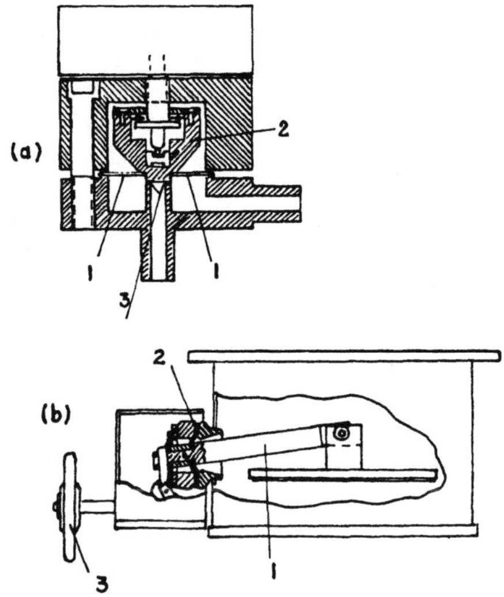

FIG. 6.79 Diaphragm sealed valves; (a) with metal diaphragm. After Warmoltz[1287] *(by permission from Philips Techn. Rev.* **21**, *176, 1959/60)*; (b) with rubber diaphragm (Edwards[328])

61.34 Operating systems of valves. Vacuum valves are operated either manually or by using pneumatic or magnetic actuation. Thermal expansion is also used as an actuating method.

(a) The *mechanically transmitted movement* is specific to manually operated valves, but the pneumatically, and electro-magnetically operated valves (and some of the valves actuated by thermal expansion) also have some mechanical system by which they transmit the motion to the closing system of the valve. Mechanically operated actuation systems of valves are summarized in Fig. 6.80 and Table 6.19.

(b) *Pneumatically operated valves.* Valves may be operated by the pressure of a gas (air) exerted directly onto the closing plate or displacing a piston connected mechanically to the closing system of the valve.

Franks[399] described an isolation valve using a floating piston (1, Fig. 6.81a) sliding with a good fit in the body (2). The body is provided with three connexions: (*A*) to atmosphere, (*B*) to the backing pump, (*V*) to the vacuum system or diffusion pump. When (*A*) opens, the piston is forced upwards and closes the connexion (*V*) from the atmosphere; at the same time air enters the backing pump. On starting the pump (*B*) the space below the piston is partially evacuated, so that the piston sinks downwards to its initial position; but as soon as the pressure is small enough, the piston falls and blocks the connexion (*B*). To avoid this, Servranckx[1122] improved the valve (Fig. 6.81b) by locating nozzles on the base and on the side of the piston. This permits the space below the piston to be evacuated before it falls.

Cameron[192] describes a valve (Fig. 6.82a), in which the plate (1) is moved towards its seat by admitting air through the connexion (4), into the bellows (2), and extending it. The bellows is supported from a spider (3), and the spring (5) helps to contract the bellows.

Meyer[853] used a valve (Fig. 6.82b) consisting of a flat disc (1), which seals the circular edges (2, 3) by means of a rubber gasket. The opening (5) is designed to form the desired area in comparison with the annular opening (4). After the plate (1) is closed on the edges (2, 3), the connexion (5) being eva-

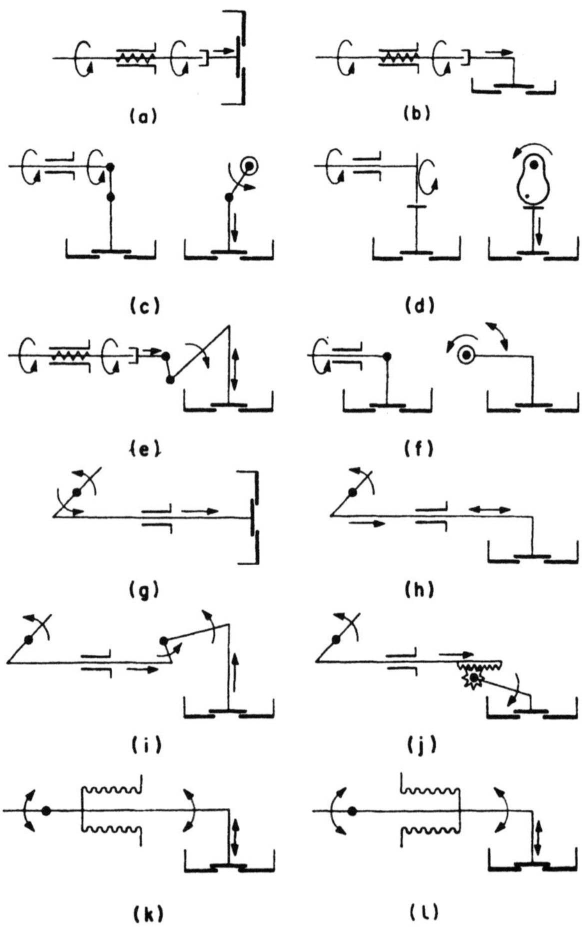

FIG. 6.80 Mechanical operation systems of valves (Table 6.19)

TABLE 6.19. MECHANICAL ACTUATION OF VALVES (Fig. 6.80)

Operation system	Valve (see Table 6.8)	Diagram Fig. 6.80	Examples
Screw and stem	Plate, Cone	a	Figs. 6.38; 6.43*; 6.45; 6.46; 6.78b
Screw and stem	Gate, Plunger	b	Fig. 6.63
Shaft and levers	Plate, Cone, Gate	c	Figs. 6.60a; 6.66a
Shaft and cam	Plate, Cone Gate	d	Fig. 6.51 D'Eustachio[277]
Screw and levers	Plate, Cone	e	Figs. 6.49; 6.50
Shaft** (rotation)	Flap, Butterfly Ball	f	Fig. 6.53; 6.54; 6.56; 6.76
Shaft (sliding)	Plate, Cone, Plunger	g	Figs. 6.26; 6.29; 6.30a; 6.39; 6.47
Lever and shaft	Gate	h	Fig. 6.62
Sliding shaft and levers	Plate, Cone	i	Fig. 6.55
Sliding shaft, rack and pinion	Flap	j	Krohn[720]
Levers	Plate, Cone	k, l	Figs. 6.78c; 6.30b

* Differential screw for fast operation.
** See also Moreau[887], McIntosh[832].

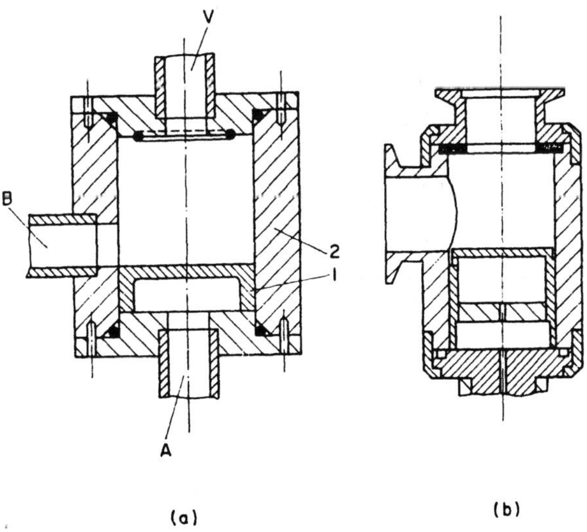

FIG. 6.81 Pneumatically operated valves, having floating pistons: (a) with simple piston. After Franks[399] *(Courtesy of The Institute of Physics and The Physical Society, London)*; (b) piston with impedance. After Servranckx[1122] *(Courtesy of Société Française des Ingénieurs et Techniciens du Vide)*

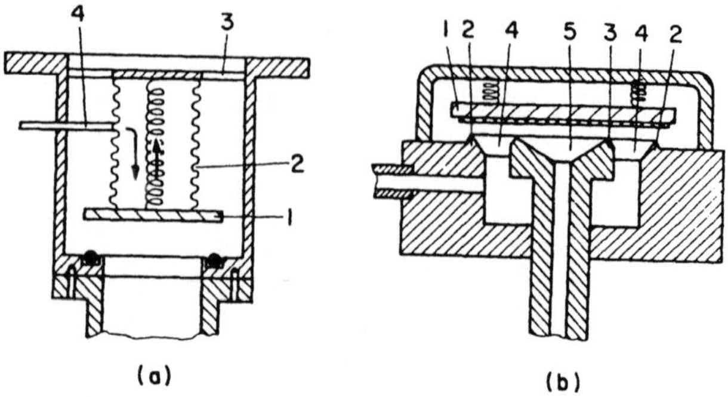

FIG. 6.82 Pneumatically operated valves: (a) bellows-sealed valve (after Cameron[192]); (b) valve operated by the pressure difference. After Meyer[853] *(Courtesy of The American Institute of Physics)*

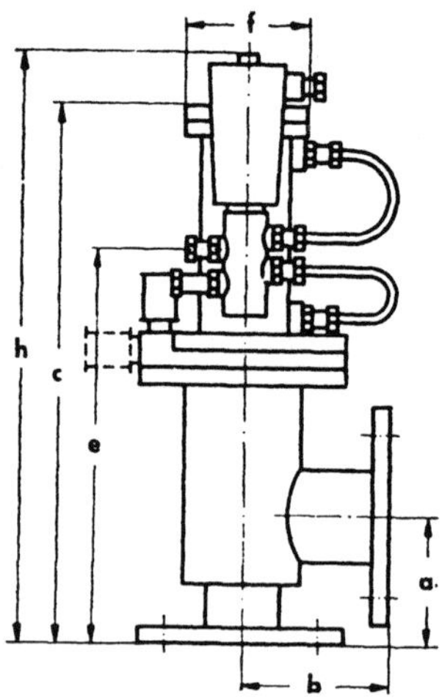

FIG. 6.83 Dimensions of pneumatic valves supplied by Leybold[762] (Table 6.20)

TABLE 6.20. DIMENSIONS (mm) OF THE ELECTRO-PNEUMATIC VALVES* (Fig. 6.83) SUPPLIED BY Leybold[762]

Size	NW 32	NW 50	NW 65	NW 100	NW 150
a, b	77	85	105	130	170
c	312	345	400	510	505
e	240	260	292	352	425
f	75	95	118	165	165
h	405	425	455	515	595

* The compressed air (3–6 at) is supplied through an electromagnetic, fourway control valve. The valve is closed when no current is flowing. The switching time depends on the pre-set throttle valve damping. Normally it is 2 sec.

cuated, the valve remains closed until the pressure in (4) rises to a value which results in the opening of the plate.

Pneumatically or electro-pneumatically operated valves are used especially where quick action of the valve is essential. Round[1089] described a pneumatically operated gate valve operating with a maximum delay of 12 msec.

The electro-pneumatic, right-angle valves supplied by Leybold[762] are listed in Table 6.20, with reference to Fig. 6.83.

Figure 6.84 shows the over-all dimensions of the air or hydraulic fluid operated Varian[1257a] valves.

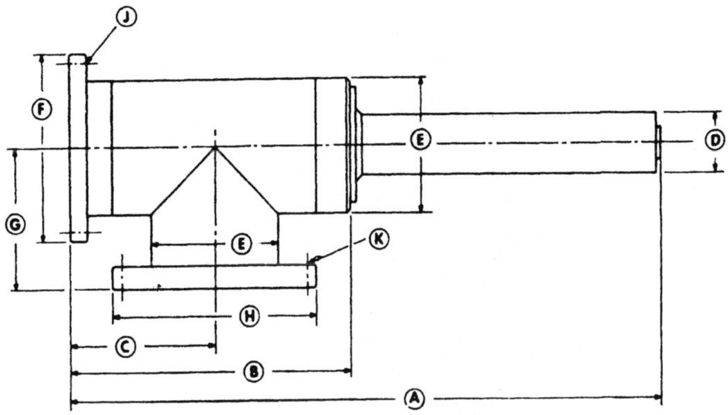

FIG. 6.84 Dimensions of pneumatic valves supplied by Varian[1257a] (Table 6.21)

TABLE 6.21. PNEUMATICALLY OPERATED VALVES* SUPPLIED BY Varian[1257a] (dimensions in inches, see Fig. 6.84)

Size	2 in.	4 in.	6 in.	Size	2 in.	4 in.	6 in.
A	19 1/2	24	32	F	5	7	9 1/4
B	9	12	21 1/2	G* G**	4 4 13/64	5 5–13/64	6 3/8 6–7/16
C** C***	4-1/16 4 1/8	5 1/2 5–11/16	6 3/4 6–13/16	H	5	7	9 1/4
D	1 1/4	2 1/4	3 1/4	J	8×1/4	12×1/4	20×1/4
E	3	5	7	K	on 4 1/4	on 6 1/4	on 8.450
						dia. B.C.	

* Dimensions are for bellows sealed shaft; the valves are available also with double O-ring sealed shaft having a pump-out for guard vacuum.

** With copper shear gasket sealed male flange;

*** Same with female flange.

PLATE 27. Electro-pneumatic elbow valve *(Courtesy of Heraeus[542])*

Plate 27 shows the electro-pneumatic, vacuum valve manufactured by Heraeus[542] with port diameters of 200, 250, 350 and 500 mm.

Bellows-sealed, solenoid-controlled, and air-operated valves with ports of 5/8, 1 and 1½ in. are supplied by Veeco[1262].

(c) *Electromagnetically operated valves* are extensively used in vacuum systems to achieve one or more of the following purposes:

(1) Remote control.

(2) Quick closing or opening motion (Table 6.4) to avoid or allow air inlet in the system in case of power failure or mishandling.

(3) Remote indication of the position (open or closed) of the valve, by means of signal lamps or buzzer (Fig. 6.88).

(4) Pre-set sequence of the opening and closing of various valves in the vacuum plant by means of electric switches, timers or interlocking devices.

(5) Operation of the valve controlled by the main parts of the vacuum plant (e.g. the rotary pump).

(6) Operation of the valves at given pressures, controlled by vacuum gauges joined to the plant.

(7) Operation of the valves at predetermined time intervals by means of cam-operated switches or timers.

The magnetically operated valves are constructed either to be *closed* when energized (air-admittance valves), or to be *open* when energized (isolation valves). The plate of the valve may be actuated, when the valve is not energized, by the pressure difference, the gravity or by springs. The valves based on gravity may be used only in the appropriate position.

Combined isolation and air-admittance valves are also constructed to allow simultaneous closing of the backing line of the diffusion pump and admit air into the rotary pump (Fig. 1.1).

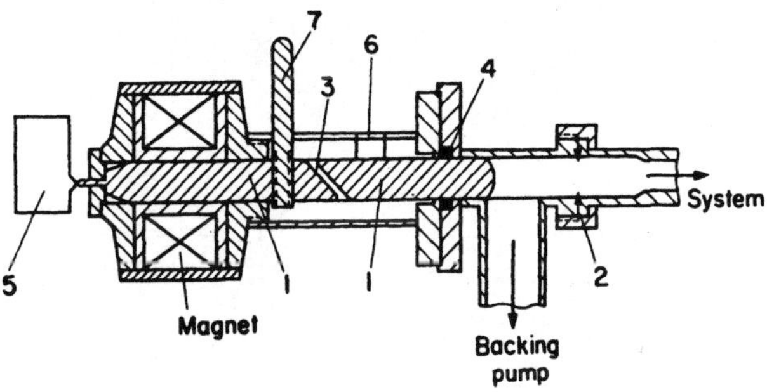

FIG. 6.85 Magnetic isolation and air-admittance valve: (1) plunger; (2, 4) O-rings; (3) oblique hole for air-admittance; (5) micro-switch; (6) opening for air-admittance; (7) handle to withdraw the plunger. After Fisher[379] *(Courtesy of The American Institute of Physics)*

A magnetic valve combining the closing of the system and the admittance of air into the backing line, based on the displacement of a plunger (Fig. 6.85), which closes on an O-ring and admits air through an obliquely drilled hole, has been described by Fisher[379].

Magnetic valves are available in a large range of sizes from the various manufacturers of vacuum equipment (Section 11.3).

Figure 6.86 shows the cross section of the isolation and combined isolation and air admittance magnetic valves supplied by Edwards[328]. Plate 28 illustrates the general aspect of such valves, and Table 6.22 list some characteristics of these valves.

N.G.N. Electrical[926] supplies a series of magnetic valves from 1/2 to 2 in. Magnetically operated valves having nominal bores up to 5 in. are supplied by Genevac[441] (see Fig. 6.87, Table 6.23 and Plate 29).

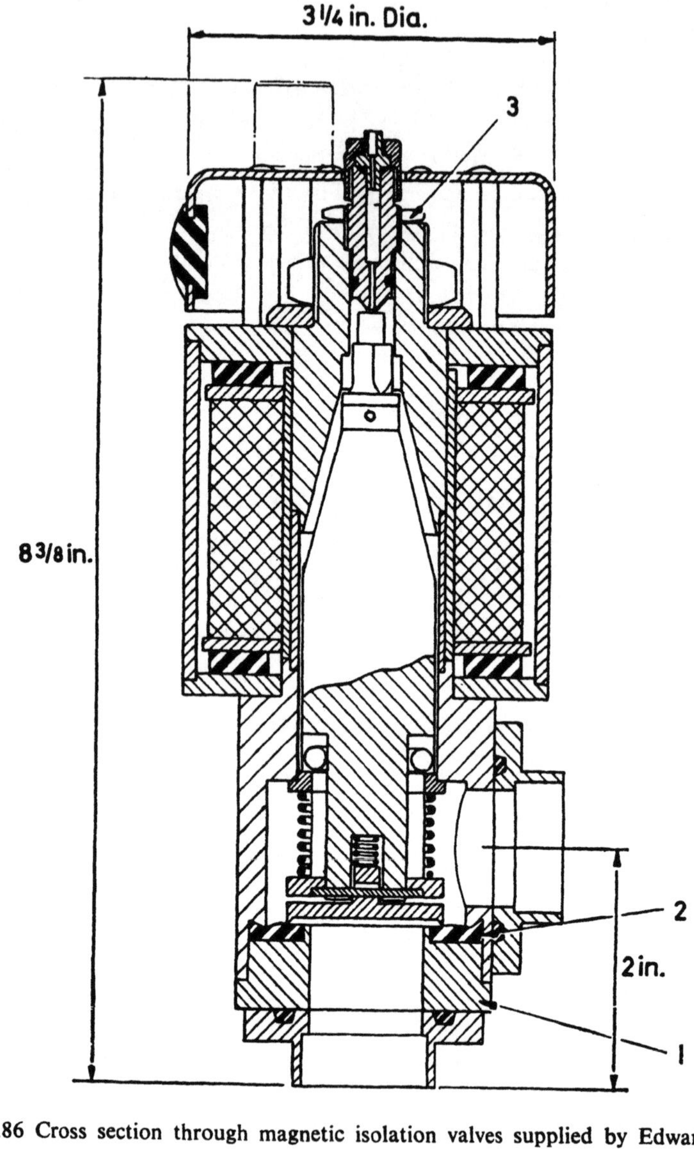

FIG. 6.86 Cross section through magnetic isolation valves supplied by Edwards[328]

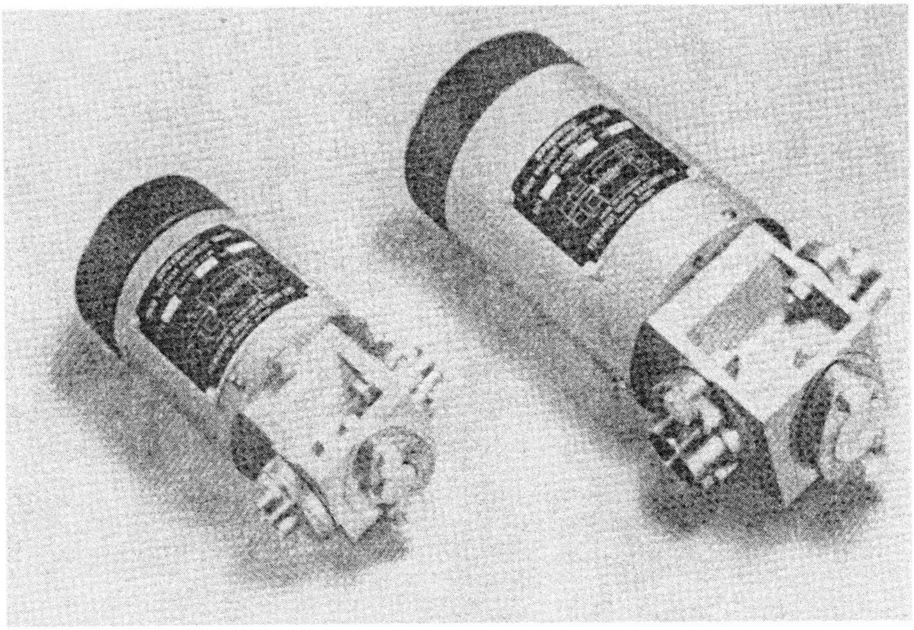

PLATE 28. Magnetic valves *(Courtesy of Edwards[328])*

Magnetic valves are used with separate rectifiers or they have the rectifier built in the body of the valve.

Figure 6.88 shows some circuit diagrams for the use of magnetic valves. To connect the isolation valve (1) to a signal lamp (3) or to a buzzer and an auxiliary relay (4), the scheme shown on Fig. 6.88a may be used. The scheme

TABLE 6.22. MAGNETIC VALVES FOR 200–250 V SUPPLIED BY Edwards[328]

Valve	Height		Diameter		Coil		
	in.	mm	in.	mm	ohms	mA	W
Air admittance 1/16 in.	3 5/16	84	2	51	19 000	8.5–11	1.4–2.3
Air admittance 1/8 in.	3 7/8	98	2 1/8	54	12 500	13–16	2–3.3
Isolation 1/2 in.	7 23/32	196	3 1/4	83	4 300	37–49	6–10
Isolation 1 in.	9 11/16	246	3 15/16	100	2 400	67–87	11–18
Combined isolation and air admittance 1 in.	7 3/4	197	3 1/4	83	4 000	16–20	~4
Isolation and baffle valve	11 16/32	295	4 1/4	108	1 850	87–108	14–22

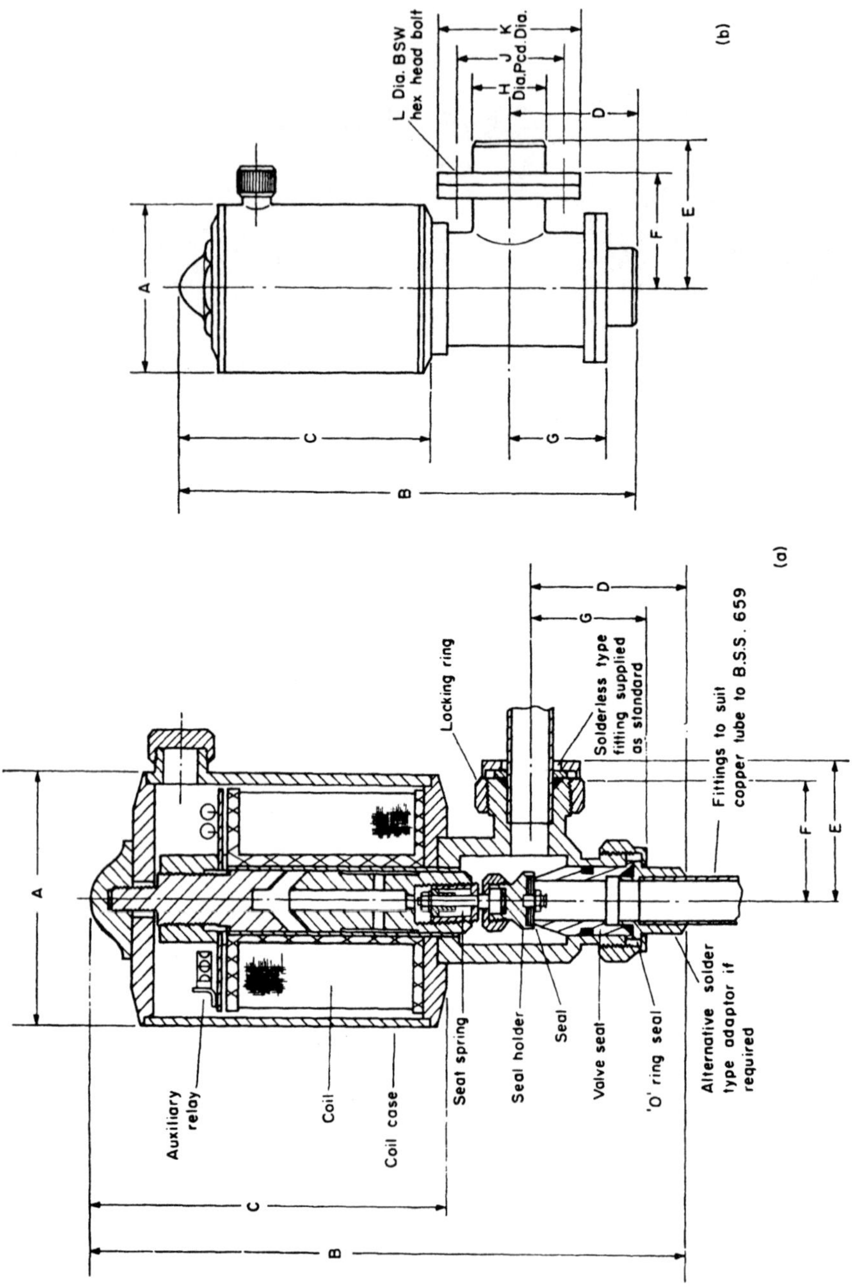

FIG. 6.87 Electromagnetic valves supplied by Genevac[441]; (a) cross section; (b) dimensions (Table 6.23).

TABLE 6.23. ELECTROMAGNETIC VALVES (Fig. 6.87) SUPPLIED BY Genevac[441]
(dimensions in inches)

Nominal bore	*A*	*B*	*C*	*D*	*E*	*F*	*G*	*H*	*J*	*K*	*L*
$1\frac{1}{2}$	$4\frac{1}{2}$	$14\frac{3}{4}$	8	$3\frac{1}{2}$	$3\frac{7}{8}$	$3\frac{1}{8}$	$2\frac{3}{4}$	$2\frac{1}{8}$	$3\frac{3}{8}$	4	$\frac{5}{16}$
2	5	$16\frac{3}{4}$	$9\frac{1}{4}$	$4\frac{1}{8}$	$4\frac{1}{8}$	$3\frac{1}{4}$	$3\frac{1}{4}$	$2\frac{5}{8}$	$3\frac{7}{8}$	$4\frac{1}{2}$	$\frac{5}{16}$
3	$6\frac{1}{4}$	16	$8\frac{3}{4}$	–	–	$5\frac{3}{8}$	$3\frac{5}{8}$	–	$5\frac{1}{2}$	$6\frac{1}{2}$	$\frac{3}{8}$
4	$6\frac{1}{8}$	$17\frac{1}{4}$	$8\frac{3}{4}$	–	–	$5\frac{1}{2}$	$4\frac{1}{2}$	–	7	8	$\frac{3}{8}$
5	$6\frac{1}{4}$	$18\frac{1}{4}$	$8\frac{3}{4}$	–	–	$6\frac{1}{4}$	$4\frac{3}{4}$	–	$8\frac{1}{2}$	$9\frac{1}{2}$	$\frac{3}{8}$

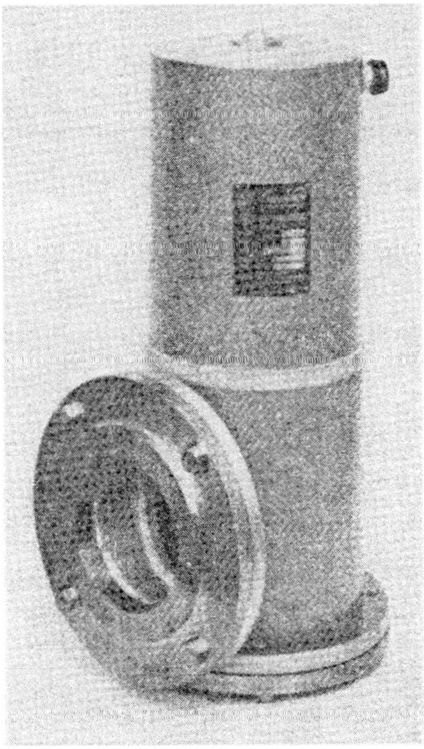

PLATE 29. 3-inch magnetic valve *(Courtesy of Genevac[441])*

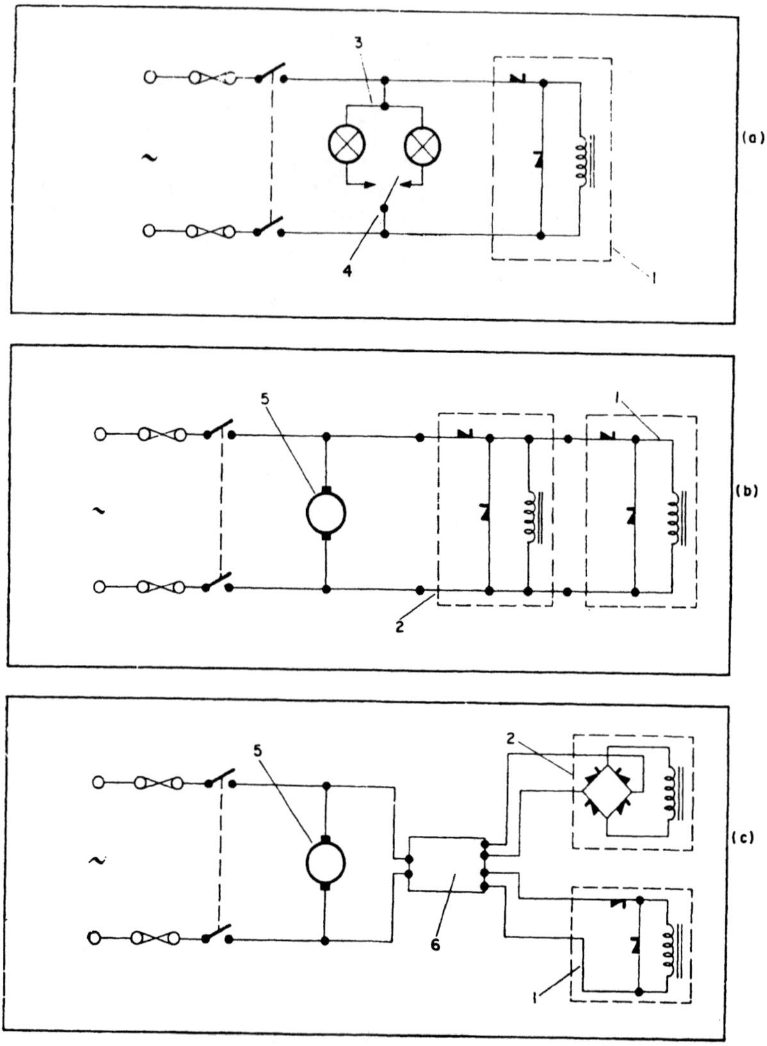

FIG. 6.88 Circuit diagram for magnetic valves

shown on Fig. 6.88b shows the circuit for the connexion of an air-admittance valve (2) and an isolation valve (1) to the motor of the pump (5) for protection against power failure or inadvertent "shut-down". It is often necessary to incorporate a time delay between the opening and closing of the air admittance and isolation valves. In this case (Fig. 6.88c), a delay switch (6) is connected in the circuit.

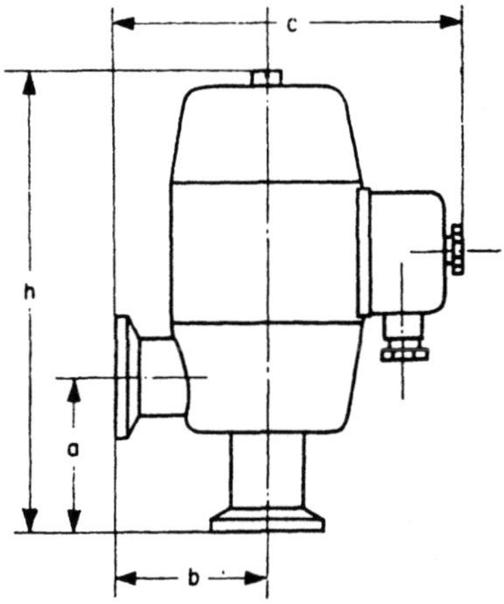

FIG. 6.89 Dimensions of electromagnetic right angle valves supplied by Leybold[762] (Table 6.24)

TABLE 6.24. DIMENSIONS (mm) AND CHARACTERISTICS OF THE ELECTROMAGNETIC VALVES* (Fig. 6.89) SUPPLIED BY Leybold[762]

Size	NW 10	NW 20	NW 32
a, b	30	50	50
c	85	115	125
h	102	152	190
Conductance (at 10^{-1} torr) l./sec	1.6	12	50
Starting current A	1.25	2.1	3.2
Holding current A	0.11	0.25	0.27

* The valves are closed when not energized; the closing time is about 0.05 sec.

Figure 6.89 and Plate 30 show the electromagnetic valves supplied by Leybold[762] and Table 6.24 lists their characteristics.

(d) *Thermal expansion* is rarely used to operate vacuum valves, but the principle is extensively used in variable leaks (Section 61.47).

The actuation of valves by thermal expansion consists in utilizing two materials of different expansion, which by heating or cooling bring the parts closer together or take them further apart.

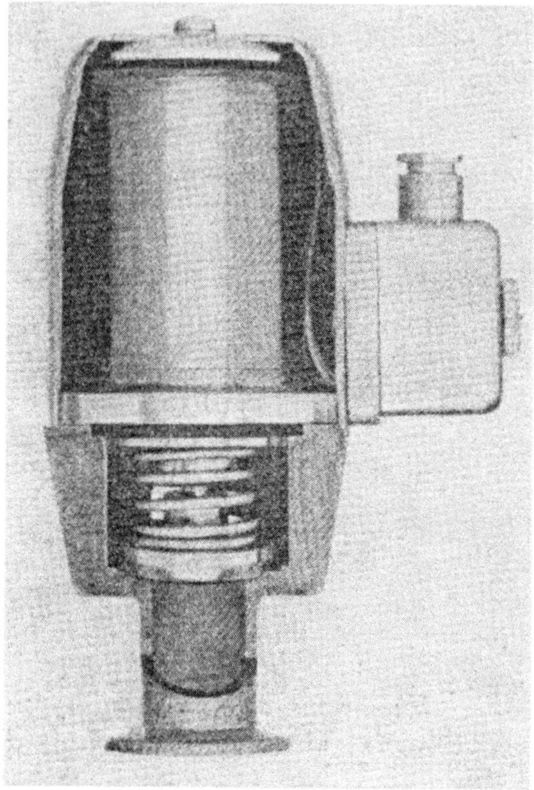

PLATE 30. Sectional view of electromagnetic valve *(Courtesy of Leybold[762])*

Valves operated by thermal expansion are described in Figs. 6.32b, 6.33a, and 6.73b.

61.35 Valves for specific purposes. The various valves should fit the kind of process achieved in the plant and the place where they are connected into the vacuum system.

Valves whose only purpose is separating two spaces at different pressures are known as *isolation valves.* Valves that have also other purposes or that should correspond to special requirements are known as: (a) seal-off valves, (b) throttling valves, (c) air admittance valves, (d) baffle valves, (e) non-return valves, (f) bakeable valves, (g) multiple-way valves.

Isolation valves are fully described in Sections 61.32–61.34 by the various closing, sealing and operation systems included in those valves.

(a) *Seal-off valves* are used to close an evacuated vessel. After closing the valve the sealing system can be taken away, leaving the vessel sealed by the closing system of the valve.

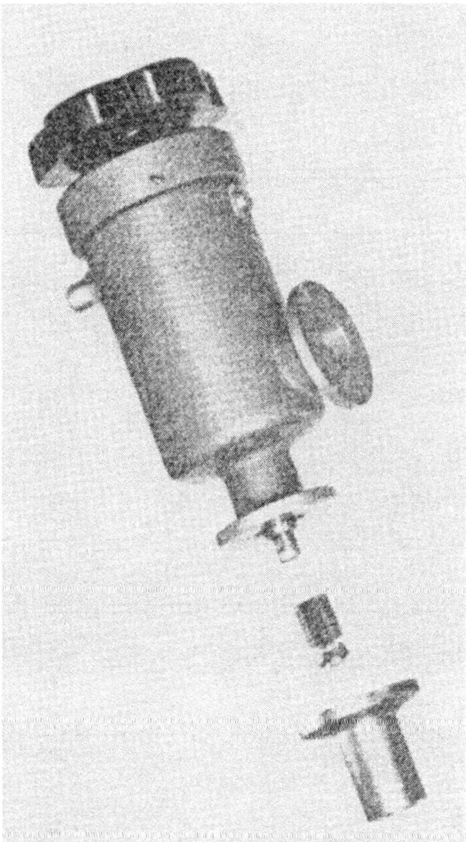

PLATE 31. Seal-off valve *(Courtesy of Leybold[762])*

Plate 31 shows a seal-off valve manufactured by Leybold[762]; its cross section is illustrated in Fig. 6.90 and the dimensions are listed in Table 6.25. A similar seal-off valve is described in connexion with Fig. 2.116 (see also Richards[1053], Anon.[37], Klipping[689]).

b) *Throttling valves* are used to adjust the rate of flow. As most of the throttling valves are constructed for fine adjustement, they are discussed in Section 61.4.

c) *Air-admittance valves* are used to admit air (or other gases) into the evacuated system. If the valve is to be used to admit air, it has only a single sealed connexion, the air flowing in (when the valve is opened) near the screw.

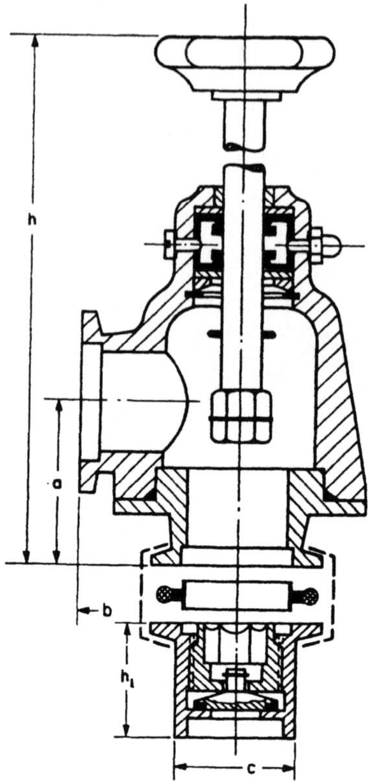

FIG. 6.90 Section view of Leybold[762] seal-off valve (dimensions see Table 6.25)

TABLE 6.25. DIMENSIONS (mm) OF THE SEAL-OFF VALVES* (Fig. 6.90) SUPPLIED BY Leybold[762]

Size	NW 10	NW 32
a, b	30	50
h	104	136
h_1	25	40
c	15	38

* Leak rate less than 1×10^{-2} lusec.

Plate 32 shows air-admittance valves supplied by Edwards[328] for 1/8 in. and 1/4 in. connexions.

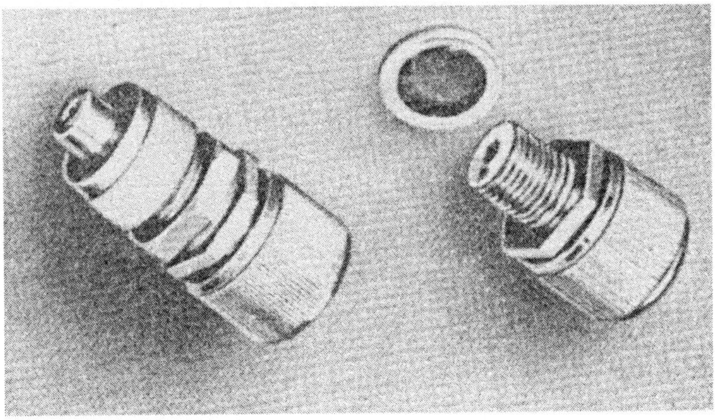

PLATE 32. Air-admittance valves *(Courtesy of Edwards[328])*

PLATE 33. Gas-admittance valves *(Courtesy of Balzers[444])*

When the valve has to admit other gases (argon, helium, hydrogen, etc.) a second connexion to the valve is also required to connect the gas inlet. Plate 33 shows admittance valves manufactured by Balzers[444] with 20 mm diameter nipples. A motor-car tyre valve was used by Jackson[619] as an air-admittance valve.

d) *Baffle valves* are so designed that the disc of the valve remains *in line* with the port, thus it can act as a baffle.* These valves are installed above the

*A *baffle* is a system of cooled plates, placed in a pipe to deviate the flow from the axial direction, or to condense the back-streaming vapour from a pump.

diffusion pump (Fig. 1.1). To increase the efficiency of these valves in stopping the back-streaming vapours, they are often water-cooled.

Figure 6.91 shows baffle valves supplied by Consolidated Electrodynamics Corp.[230] Table 6.26 lists their dimensions.

Baffle valves supplied by Leybold[762] and Edwards[328] are shown in Figs. 6.49–6.50.

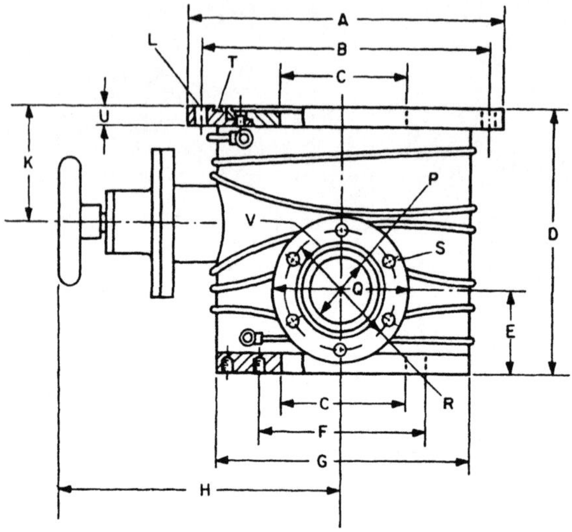

FIG. 6.91 Dimensions of water-cooled baffle valves supplied by C.E.C.[230] (Table 6.26)

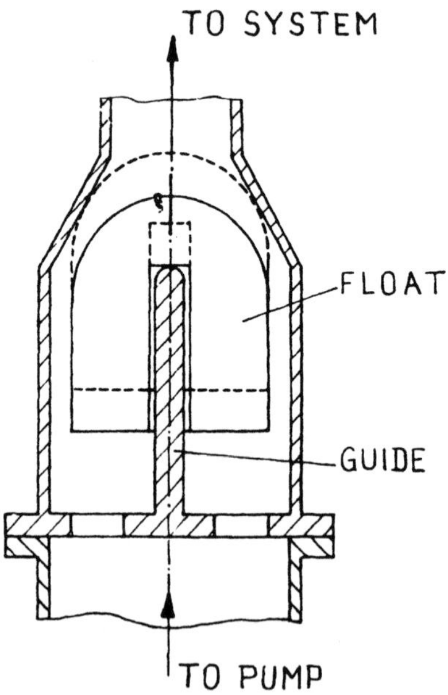

FIG. 6.91A Non-return valve

TABLE 6.26. DIMENSIONS (inches) OF BAFFLE VALVES* (Fig. 6.91) SUPPLIED BY Consolidated Electrodynamics Corp.[230]

Size	4 in.	6 in.	Size	4 in.	6 in.
A	10	13	P	1-11/16	1-11/16
B	9-1/4	12	Q	3-3/4	3-3/4
C	3-7/8	6	R	3-5/16	3-5/16
D	8	9-1/8	S No. Dia.	6 9/32	6 9/32
E	2-1/2	2-1/2			
F	5-1/4	8	T o.d. i.d.	8-7/16 7-11/16	11 10-1/4
G	8	9-7/8			
H	9-1/8	10	U	1/2	5/8
K	3-1/2	4-3/16	V o.d. i.d.	2-15/16 2-3/8	2-15/16 2-3/8
L No. dia.	8 7/16	12 7/16			

* These valves are also available with pneumatically operated actuation.

It is recommended that the water should be so connected to the baffle valves that it flows first through the valve and afterwards through the cooling coils of the diffusion pump (Edwards[328]).

e) *Non-return valves* are used in the inlet pipe of the rotary pumps to prevent the flow of the pump oil into the system, when the pump stops.

The simplest form of a non-return valve (Fig. 6.91A) consists of a spherical or hemispherical float, which is carried by the rising oil, and is trapped by a conical seat held in position by the pressure exerted by the oil.

Non-return valves give protection against the oil for a period of 1–2 hr (Edwards[328]); for longer periods magnetic air-admittance and isolation valves should be used (see e.g. Figs. 6.86, 6.87, 6.89).

Blow-off valves are rarely used in vacuum systems. These valves should open (blow-off) when a given pressure difference is reached between the two sides of the valve. For this purpose cut-offs (Section 61.1) may be used. Kent[663] constructed such a valve by placing a plug into a Wilson seal.

f) *Bakeable (and ultra-high vacuum) valves.* Vacuum systems in which pressures of 10^{-7} torr are to be reached must be baked at at least 200 °C. Systems constructed to reach the range of the ultra-high vacuum must be baked entirely at temperatures between 400–500 °C. Thus the valves used in such vacuum systems must be "bakeable valves".

As a general rule, *bakeable valves* are "all glass" or "all metal" valves, including only materials which can be heated to the baking temperatures; elastomers cannot be used in such valves.

Ultra-high vacuum valves should be bakeable and should meet at least one of the requirements (Adam[3], Amoignon[28], Munday[906]), as follows:

(i) Provide the largest conductance in the open position.
(ii) Have the smallest conductance and leak rate in the closed position.
(iii) Allow the greatest number of opening-closing cycles.
(iv) Require the minimum force for their operation.
(v) Operate fast.

Table 6.27 lists the various kinds of bakeable valves and summarizes the degree in which they meet these various requirements.

The *closing system* of bakeable valves may be based on one of the principles listed in Table 6.27.

Greaseless *ground* joints between highly polished plane, tapered or spherical surfaces are used in glass valves as described by e.g. Adam[3], Yarwood[1340, 1341], Vogl[1271], Decker[270] and shown in Figs. 6.32–6.34. The drawback of such valves is mainly their high leak rate in the closed position. Thus they can be used only in places where a very small pressure difference exists across the valve (e.g. between the ultra-high vacuum pump and the chamber).

The valves closed and opened by *fusing* a glass tube, were described e.g. by Sill[1137] and shown in Fig. 6.31 may have a very limited use (Table 6.27).

The valves closed with *silver chloride* (Table 6.27) have a limited use because of their limited conductance and slow operation.

Molten metal seals have been used to close bakeable valves (Figs. 6.28, 6.29). Their baking temperature is limited by the vapour pressure and/or the melting point of the metals and alloys used (Section 37.6).

The promising principle of the *powdered seals* used in the closing system of bakeable valves, is discussed by Papirov[966b]. The powder seal consists of two parts, the metal filler and the bonding agent. The latter may be a refractory metal (Fe, Ni) and the filler is composed of metals such as gold, silver, tin, indium, copper. The melting point of the filler alloy should be 400–550 °C,

TABLE 6.27. COMPARISON OF THE BAKEABLE VALVES

Valve type	Bake-ability	Open conductance	Leak rate and closed conductance	Number of operations	Force required	Operating speed
Ideal valve	good	large	small	large	small	fast
Ground (polished) joint Figs. 6.32–6.34	good	limited	high	large	small	fast
Fused glass Fig. 6.31	good	small	small	limited	small	slow
Silver chloride Fig. 6.30a	good	limited	small	large	small	slow
Molten metals Figs. 6.28, 6.29	limited	large	medium	large	small	slow
Powder seal Fig. 6.92	good	large	small	large	medium	slow
Knife edge Fig. 6.100	good	limited	small	limited	great	fast
Metal nose on metal seat, Figs. 6.93–6.95	good	medium	small	limited	great	fast
Metal gasket on metal seat Figs. 6.101–6.104	good	medium	small	limited	great	fast

and the filler must wet both the bonding agent and the parts of the valve. The valve (Fig. 6.92) is closed by heating the seal until the parts are wetted; capillary forces prevent the filler from flowing away. The bonding agent-

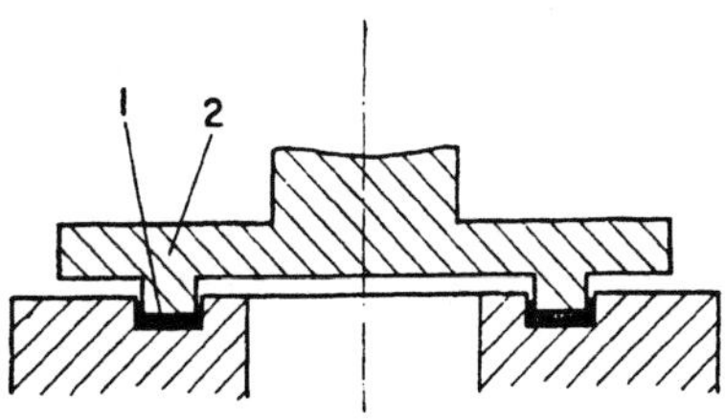

FIG. 6.92 Principle of a closing system with "powdered seal"

filler mixture placed in a groove (1) around the port, forms the joint between the seat (1) and the closing plate (2) during cooling. The valve is opened by heating the mixture up to the melting point of the filler.

TABLE 6.28. ALL METAL ULTRA-HIGH VACUUM VALVES

Valve designation: Closing	Sealing system	Operating	Valve port	Open conductance (l./sec)	Closed conductance (l./sec)	Leak rate (lusec)	Remarks	Reference
Kovar nose on copper	Kovar diaphragm	Differential screw	6 mm dia.	0.3	10^{-11}–10^{-12}	—	see Fig. 6.93	Alpert[17]
Speedivac MCV 1			0.1 in.^{2}	—	10^{-10}–10^{-11}	—	Bakeable up to 450 °C	Edwards[328]
Alpert type valve			—	0.5	—	$<10^{-7}$	Bakeable up to 450 °C; Plate 34a	Balzers[444]
Silver gasket between Monel nose and seat	Nickel diaphragm	Screw	—	1	10^{-14}	—	see Fig. 6.101a	Bills[125]
Monel nose on stainless steel	—	—	—	1	10^{-14}	—	see Fig. 6.94 Bakeable 450 °C	Granville–Phillips[469]
Alpert type valve		—	—	3	$<10^{-10}$	—	see Fig. 6.95 Heated to 200 °C	Mullard, Torrington Place, London
Copper gasket on metal seat	Gasket seals	Lever arm	25 mm	13	10^{-11}	—	see Fig. 6.104 Heated at 150 °C	Heraeus[542]
Aluminium gasket	—	—	—	13	—	10^{-10}	see Fig. 6.101b	Kienel[671]

Gold or silver plate	bellows	—	—	35	3×10^{-14}	$<2\times10^{-8}$	—	Drawin[302]
Monel cone on stainless steel	—	Screw and cam	—	20	—	$<10^{-7}$	Bakeable up to 450 °C	Balzers[444]
Copper gasket on stainless steel seat	—	—	1½ in.	38	—	—	see Fig. 6.102 Bakeable 300 °C	Varian[1257a]
Speedivac MCV 4	—	—	1 in.^{2}	—	10^{-8}–10^{-9}	—	Bakeable 450 °C	Edwards[328]
Cooper gasket on metal seat	—	—	50 mm	55	—	10^{-8}	Up to 150 °C	Heraeus[542]
Copper nose on stainless steel	bellows	mechanical	2 in.	100	10^{-9}	—	Fig. 6.99	Lange[736]
Copper nose on stainless steel	bellows	mechanical	2 in.	—	10^{-10}	—	Bakeable 400 °C	Caldwell[190b]
Metal cone on Monel seat	—	hydraulic	—	140	—	10^{-6}	Bakeable 450 °C Fast acting (1 sec)	Balzers[444] Plate 34.b
Stainless steel spherical nose on silver seat	bellows	—	4 in.	200	10^{-12}	—	—	Mullaney[902a]
Copper poppet on copper seat	bellows	hydraulic	8–3/4 in.	2100	7×10^{-7}	10^{-9}	Fig. 6.96	Parker[969] Dreyer[304]

The *all-metal* valves have the up-to-date constructions used in ultra-high vacuum techniques. The closure of these valves is achieved:

(a) by pressing a *metal nose* or a knife edge onto the metal seat;

(b) by pressing a *metal gasket* between the metal nose and the seat.

The sealing of these valves is achieved by *metal diaphragms* or *bellows.*

The Alpert[17] valve has opened a new series of valve constructions, extending the principle of these valves to progressively larger ports and providing greater conductances of the valve in the open position. Table 6.28 lists the various steps achieved in this series of valves.

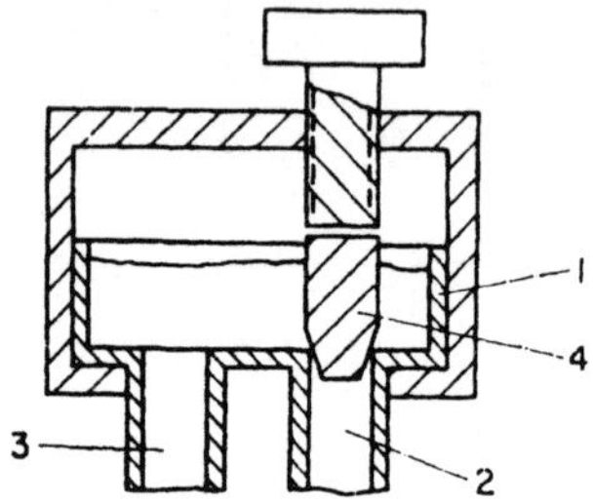

FIG. 6.93 Alpert's ultra-high vacuum valve (principle)

The classic valve due to Alpert[17] (Fig. 6.93) consists of a solid, copper cup (1) 1.75 in. diameter, containing two holes (0.25 in. diameter), connected to Kovar tubes (2, 3). The *closing system* consists of a solid Kovar nose (4) which is forced into one of the holes on the copper cup, thus isolating the connecting tubes from each other. The Kovar nose has a highly polished 45° conical surface (Dutton[318]) which is able to form its own seat on the edge of the hole in the copper cup, when it is closed for the first time. The *sealing system* consists of a flexible Kovar diaphragm to which the closing nose (4) is sealed. The diaphragm is brazed at its periphery to the copper cup, and its flexibility permits a displacement of the nose of about 0.1 in. The *operating system* of the valve is removed during baking, the valve being meanwhile kept in the open position by a bakeout-clamp consisting of a U-shaped stainless steel strap. The actuating mechanism consists of a differential screw giving an axial displacement of the Kovar nose of about 0.01 in. per revolution. The differential screw is turned by a suitable key, enabling forces of 5–10 tons to be exerted by the Kovar nose against the copper seat. The valve gives a very good vacuum seal in the closed position (Table 6.28). In use, due to wear, slight oxidation and collection of minute foreign particles, the conductance of the valve may rise to 10^{-10} (Alpert[17]) or in some cases to 10^{-6} l./sec (Pirani[992]).

Ultra-high vacuum valves of this type which owe the original design to Alpert[17] or which include improvements (e.g. Bills[125], see Fig. 6.101), are supplied by Granville-Phillips[469], Balzers[444], Mullard Ltd.

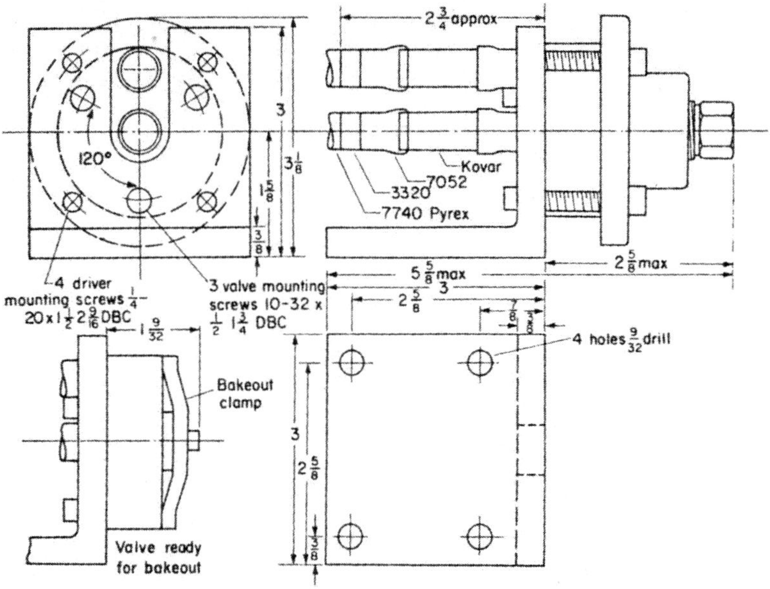

FIG. 6.94 Dimensions of a Granville-Phillips[469] ultra-high vacuum valve

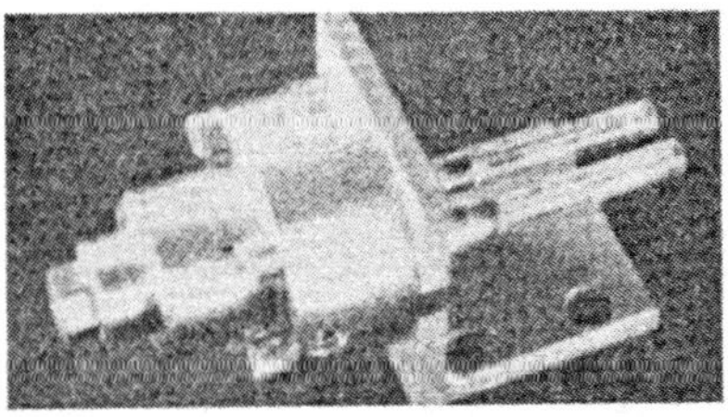

(a)

(b)

PLATE 34. Ultra-high vacuum valves: (a) Alpert type 0.5 l./sec; (b) hydraulically operated 140 l./sec. *(Courtesy of Balzers[444])*

Figure 6.94 shows the dimensions of the valves supplied by Granville-Phillips[469]. Plate 34 shows the valves available from Balzers[444] (Table 6.28).

Brown[174] modified the Alpert valve by changing the Kovar nose to a Monel one, and the Kovar diaphragm to a Monel diaphragm (0.004 in. thick). The diameter of the orifice was increased, and the stainless steel body was nickel plated (0.002 in. thickness) in order to facilitate brazing in a hydrogen furnace. Molybdenum disulphide was used as a lubricant in the valve. Biondi[126] attached the nose assembly to the valve body through a gold wire gasket seal, and used a bellows to seal the valve. Crocker[246] used a polished stainless steel spherical plunger, which is forced via bellows into a soft copper annulus.

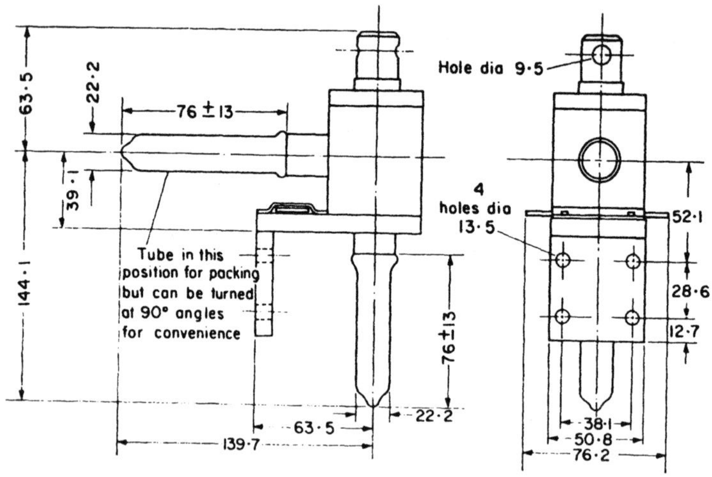

FIG. 6.95 Dimensions of a Mullard ultra-high vacuum valve

Instead of making the two outlets parallel to each other as in Alpert's original design, in some constructions of this valve only one of the outlets is axial, the second one being placed radially as in right-angle valves. Such valves are supplied by Mullard Ltd. (Torrington Place, London, W.C. 1), having the dimensions shown in Fig. 6.95 and the characteristics listed in Table 6.28.

Parker[969] and Dreyer[304] described a large valve (Table 6.28) which is closed by a soft, copper poppet pressed into a stainless steel cutter seat. The radial interference of the two parts, causes the copper of the poppet to flow over the carefully machined contour of the stainless steel seat. The valve body is basically a stainless steel T and is provided with a bellows between the vertical ram shaft and the body (Fig. 6.96). At each closing of the poppet onto the cutting edge (Fig. 6.97), an annular chip is formed (Fig. 6.98) which on an average as quoted by Dreyer[304] is 2 mils per closure from the 2 in. length of

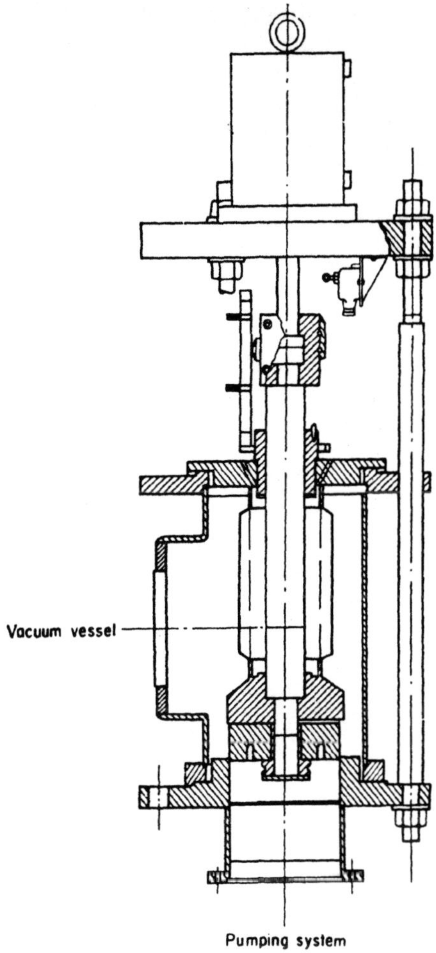

FIG. 6.96 Construction of a large bakeable ultra-high vacuum valve, 8 in. diameter. Reproduced from Parker[969] *(Courtesy of Pergamon Press)*

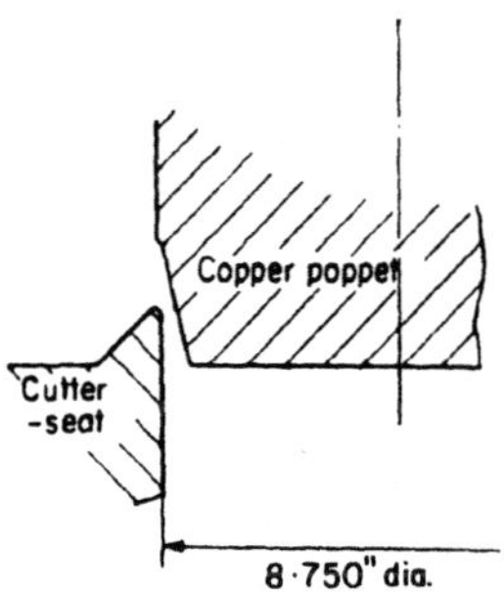

FIG. 6.97 The closing system of the valve from Fig. 6.96. Reproduced from Parker[969]

nose available. The valve requires a force of 25 tons to flow the metal during the closing, and a force of 15 tons (hydraulic system) to hold the seal.

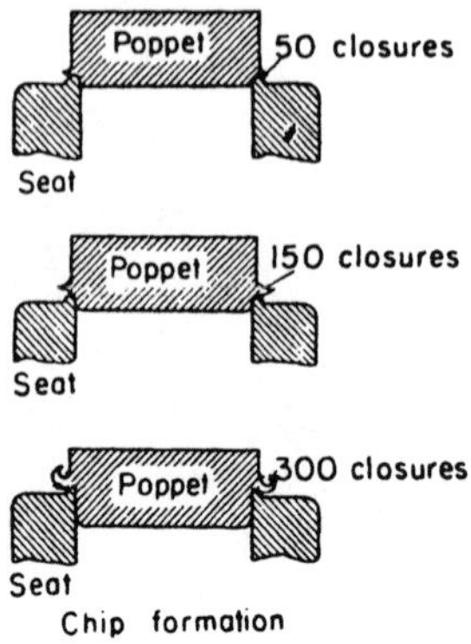

FIG. 6.98 Chip formation in the valve from Fig. 6.96. Reproduced from Parker[969]

An all-metal valve which is bakeable to 500 °C was constructed by Bollinger[148a]. On closing a rounded, hard, steel plate onto a soft copper housing, with each closure of the valve the copper is plastically deformed on a depth of about 3 μ, so that with a copper seat of about 10 mm, approx. 3000 cycles are possible. Leakage rate across the closed valve was less than 10^{-7} lusec, and the force required for repeated closure was claimed to be independent of the number of closing operations.

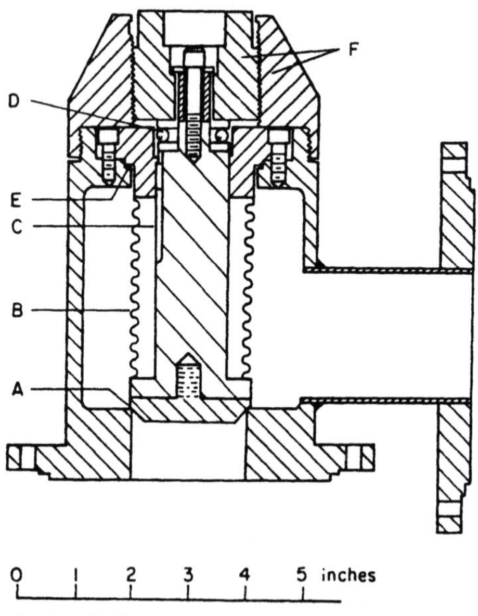

FIG. 6.99 Bellows sealed, ultra-high vacuum valve. After Lange[736] *(Courtesy of The American Institute of Physics)*: (a) Copper nose; (b) bellows; (c) guide; (d) ball bearing; (e) gold wire seal; (f) driving mechanism

Lange[736] has designed a valve (Fig. 6.99) where a copper nose closes against a stainless steel seat. The nose is driven by a stem inside a stainless steel bellows, the bellows being sealed at the upper part by means of a gold gasket seal. In order to reduce the high closing torques required, Lange[739] tested the use of a Teflon nose instead of one of copper, obtained a good seal with torques only 1/5 of those using a copper nose, but found that the baking is limited to a maximum of 350 °C. Caldwell[190b] described some inprovements made on the Lange-type valve in order to increase its open conductance.

The *knife edge* seal (Section 38.56) was also used in the closing system of bakeable valves. Baker[72] describes a valve (Fig. 6.100a) closed by forcing an annealed, flat copper disc against a stainless steel knife edge machined in the valve block. The driving shaft is argon arc welded to one end of a stainless steel bellows, the other end being welded to a guide which is sealed to the valve body by knife edges and a copper gasket. A shim is used to limit the penetration of the knife edges to a depth of 0.007 in. A 60° knife edge is employed, which is first lapped with a brass lap using 15 μ silicone carbide, followed by polishing with a second brass lap using a 3 μ diamond powder to produce a flat perfectly polished surface 0.003–0.010 in. in width. A leak-free seal will only be achieved if no scratches are visible when viewed with a microscope at a magnification of $\times$50. The copper disc and gasket were machined from O.F.H.C copper and annealed in hydrogen prior to assembly. Baker[72] constructed this valve with apertures of 0.6, 2.5 and 7.5 cm having conductances in the open position of 0.3, 15 and 400 l./sec respectively, and having a closed conductances less than 10^{-14} l./sec. Closing torques of 0.5, 2.0 and 7.0 lb.ft were needed. The valves permitted outgassing at about 400 °C, but prolonged outgassing at higher temperatures has resulted in a leak occurring in the body seal.

Eastwell[319] modified this valve (Fig. 6.100b) in order to isolate all moving parts from the vacuum and to simplify the machining.

Faron[370] constructed a bakeable valve (conductance 10^{-18} l./sec in the closed position; bakeable at 200 °C) closed by a nose pressing on the edge of the seat and a knife edge surrounding the nose pressing onto the seat itself.

Eschbach[349] described a bakeable valve of large cross section, closed by pressing the sharp (knife) edge of the valve plate into the seating (groove) covered with indium, tin or silver. The plate is pressed onto the seat by a bellows sealed shaft, which also raises the plate (about 2 mm) when the valve is to be opened. The plate is withdrawn parallel to the port (magnetically). Lichtman[764a] describes a gate valve sealed by bellows and closed on a knife edge. The valve includes two bellows, one sealing the operating system of the gate, and the second providing the displacement of the knife edge (perpendicular to the first one).

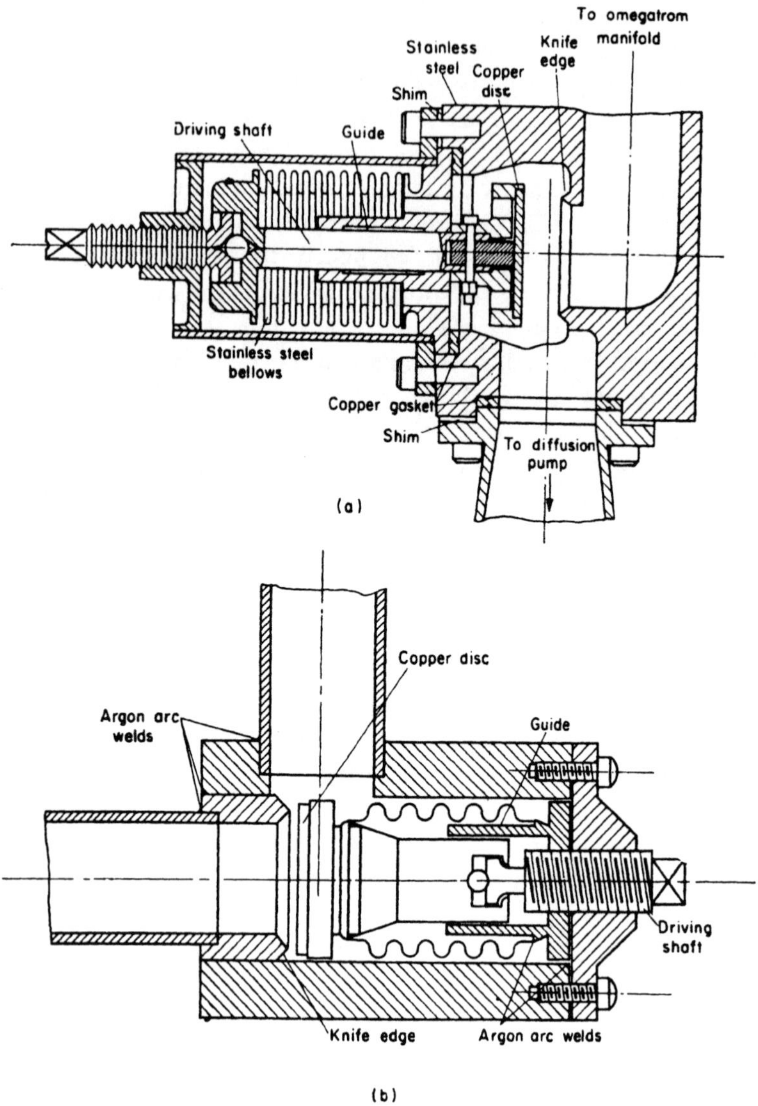

FIG. 6.100 Valves closed by knife edge seal: (a) construction of Baker[72]; (b) construction of Eastwell[319] *(Courtesy of Pergamon Press)*

Mullaney[902a] constructed a bakeable valve (4 in. diameter; conductance 200 l./sec) closed by a spherical stainless steel nose pressed on the edge of a silver seat. When the first closure is made the spherical nose forms the sharp corner of the seat to match its contour. The seal is tight only if the nose is highly polished (8 micro-inch finish) and if the angle between the axis of the plunger holding the spherical nose and the radial line between the sphere centre and the point of contact is 45°. Variations of plus or minus 5° on the angle resulted in irregularities and the need of higher forces to effect the seal. To close the valve (to reach a conductance less than 10^{-12} l./sec) with these imperfections, a force of 5000 lb. was needed instead of the 2280 lb. required when the angle was 45°.

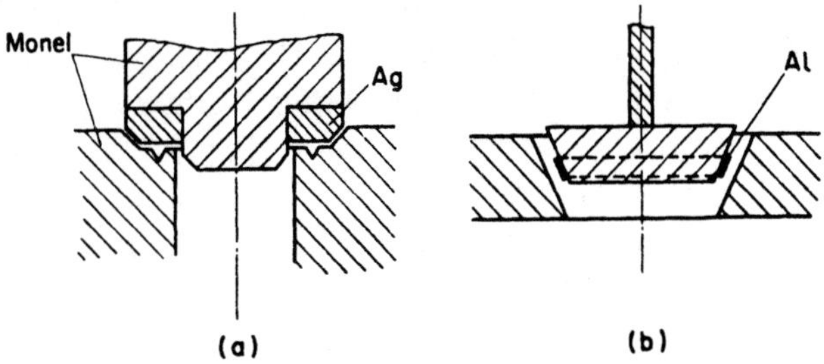

FIG. 6.101 Gasket sealed closing systems for ultra-high vacuum valves: (a) silver ring compressed between hard Monel parts. After Bills[125] *(Courtesy of The American Institute of Physics)*. (b) Aluminium ring in conical joint. After Kienel[671] *(Courtesy of Rudolf A. Lang Verlag, Esch/Taunus, Germany)*

Bakeable (ultra-high vacuum) valves using *metal gaskets* pressed between a moving nose and a seat are described by e.g. Bills[125], Kienel[671], Douglas[204a], Medicus[839a], Ullman[1249].

In order to be able to exert sufficient pressure to seal the microscopic scratches on the surface, Bills[125] confines the softer metal between two harder ones. He uses a vacuum melted pure silver ring (Fig. 6.101a) between the

PLATE 35. Ultra-high vacuum valve *(Courtesy of Varian[1257a])*

harder Monel parts. The silver ring had a width of 0.04 in., a thickness of 0.025 in. and a flat bearing surface along a width of 0.020 in., the edge being bevelled at 45°. The dimensions of the seal are carefully chosen so that the Monel parts do not make contact until the silver is under extremely high pressure, and little, if any, Monel is deformed. The seat is provided with a groove about 0.005 in. wide and 0.010 deep. The small conductance in the closed position (see Table 6.28) is reached using a torque of about 43 lb.ft.

Kienel[671] described an ultra-high vacuum valve, closed by pressing an aluminium ring between the tapered plug and seat (Fig. 6.101b).

Figure 6.102 and Plate 35 show a valve supplied by Varian[1257a] designed to operate down to 10^{-11} torr, closed by impressing the stainless steel seat into a replaceable copper gasket. Gasket life is subject to the temperature history and the cleanliness of the sealing surfaces, but according to Varian[1257a], at least ten bakeout cycles to 300 °C can be imposed before the closing torque

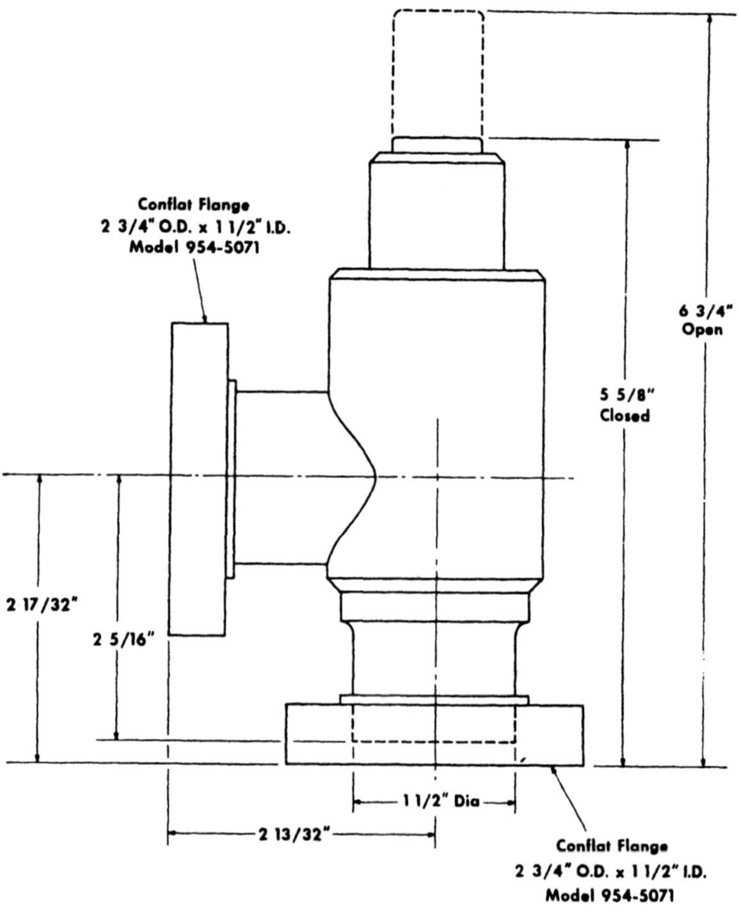

FIG. 6.102 Dimensions of ultra-high vacuum valves supplied by Varian[1257a]

Plate 36. Ultra-high vacuum valve (*Courtesy of Heraeus*[542])

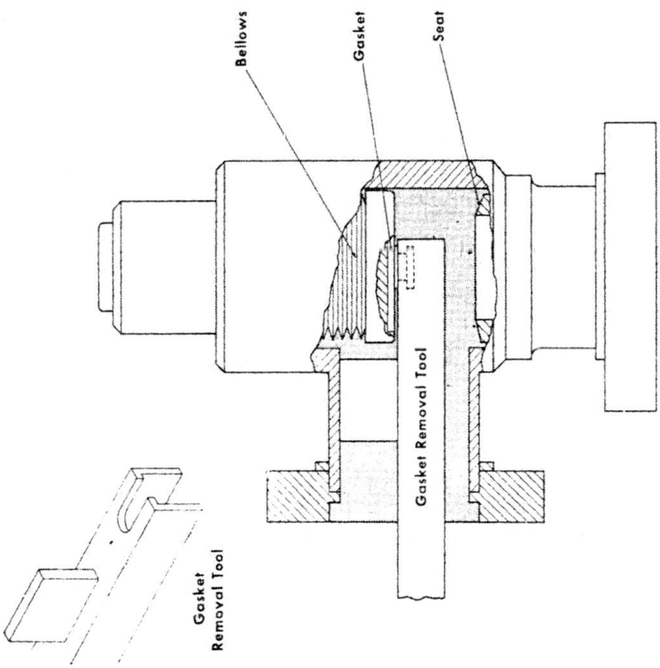

Fig. 6.103 Method of removing the copper gasket from the Varian[1257a] ultra-high vacuum valves

exceeds 100 lb.ft. At this torque level the copper gasket should be replaced. The gasket can be removed through the side port by using a properly shaped tool (Fig. 6.103). To seat a new gasket, an initial torque of 75 lb.ft is required. After the initial seal is made, approximately 30 lb.ft is required to close the valve at room temperature. A torque wrench is necessary for reliable operation of the valve. The valve should be connected to the system by Con-Flat seals (Fig. 3.149).

Figure 6.104 and Plate 36 show the ultra-high vacuum valves supplied by Heraeus[542], with a nominal diameter of 25 mm (dimensions in brackets on Fig. 6.104) and 50 mm. The valves close by compressing a metal gasket between the polished surfaces of the plunger and seat. The valve stem is sealed against atmosphere by two rubber seals filled with oil. Thus the valves can be baked only up to about 150 °C. The gasket may generally be used for several closing cycles, but in some applications it must be replaced after each cycle. In order to change the gasket, the valve insert is pulled out by turning the union nut with handles in the counterclockwise direction, removing the

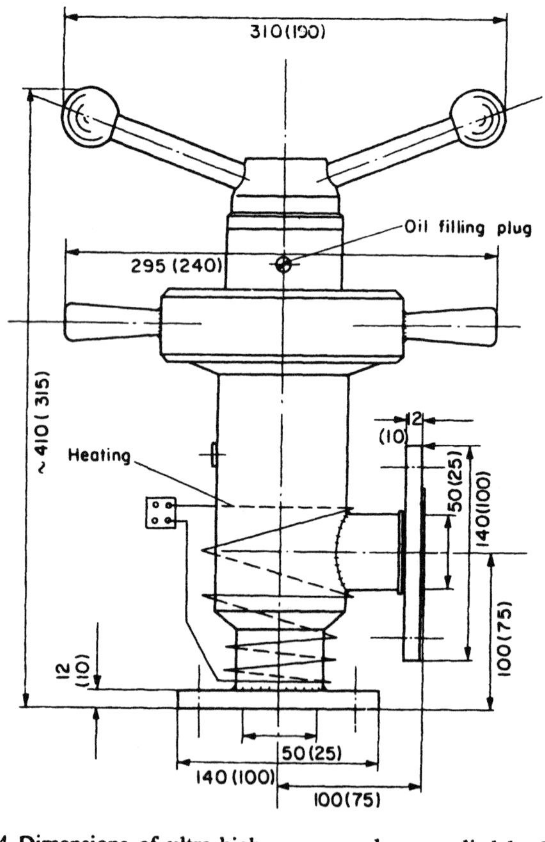

FIG. 6.104 Dimensions of ultra-high vacuum valves supplied by Heraeus[542]

centreing cap on the gasket by pressing down the spring plate and pulling out the pin locating the gasket.

An all-metal, ultra-high vacuum valve which can be operated over the temperature range from 450 °C down to 20 °K is described by Chapin[204a]. This valve is closed by forcing a gold disc against a stainless steel seat, and is sealed by a stainless steel bellows. Medicus[839a] closes his ultra-high vacuum valve by pressing a ring of platinum foil (initially 0.05 mm thick) between a hard seat and a hard, spherically shaped nose (Fig. 6.105). The valve seat (3)

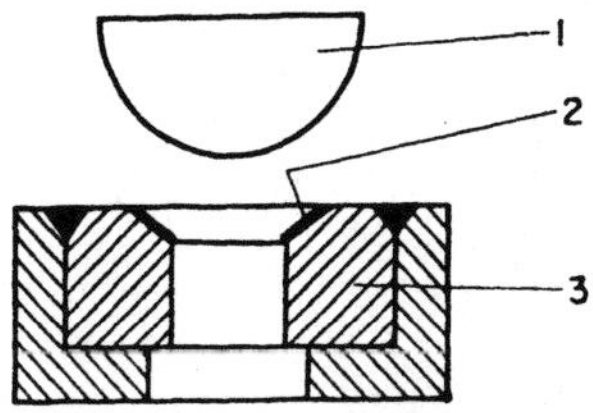

FIG. 6.105 Closing system of an ultra-high vacuum valve: (1) spherical nose; (2) platinum ring; (3) seat. After Medicus[839a] *(Courtesy of The American Institute of Physics)*

is a hard metal (tungsten carbide) ring brazed to the valve body. The platinum ring was diffused to the seat by heating the assembly in hydrogen at 700 °C while pressing the ring (at a force of 20 kg for about 1 hr, for a 9 mm valve) with a sapphire sphere of the same diameter as the nose (1) in the valve. The nose in the valve is made of hard metal (tungsten carbide). The valve allowed 50–100 heating and cooling cycles at 450 °C and simultaneous closing cycles, having in closed position a conductance of about 2×10^{-16} l./sec.

Ullman[1249] described a valve designed to use the "Conoseal"* joint, (Fig. 6.107a) where an originally 35° tapered gasket is axially pressed until it seats in the 10° angle of the seats. Sealing is initiated when the sharp diagonal gasket diameters are forced into intimate contact with the mating seats. It is completed when the gasket yields. In order to limit the freedom of the gasket, this seal was used in the valve (Fig. 6.106) in a modified form (Fig. 6.107b), utilizing a clamp ring which kept the gasket in the seat and restrained vertical movement to a minimum. A 3 in. valve required only 400–700 lb.in. for closing and had a leakage across the seat of less than 3×10^{-5} lusec/cm of gasket.

g) *Multi-way valve blocks* include in a single body various valves for different purposes. The valve blocks are used in connexions where the various valves should be operated in a given sequence, usually by unskilled personnel. The most extensively used valve block includes the baffle valve, the roughing,

* Manufactured by the Aeroquip Corp., Marman Division, Los Angeles 64, California, U.S.A.

the backing and the air-admittance valves of a diffusion-rotary pump unit (Fig. 1.1). Valve blocks are very advantageous to operation; they are locked against mishandling but are more difficult to repair than separate valves.

Multi-way valves are described by e.g. Kinder[672], Garrod[428, 431], Kappen[653]. They are manufactured in various designs being operated with many handles, a single handle or by remote control.

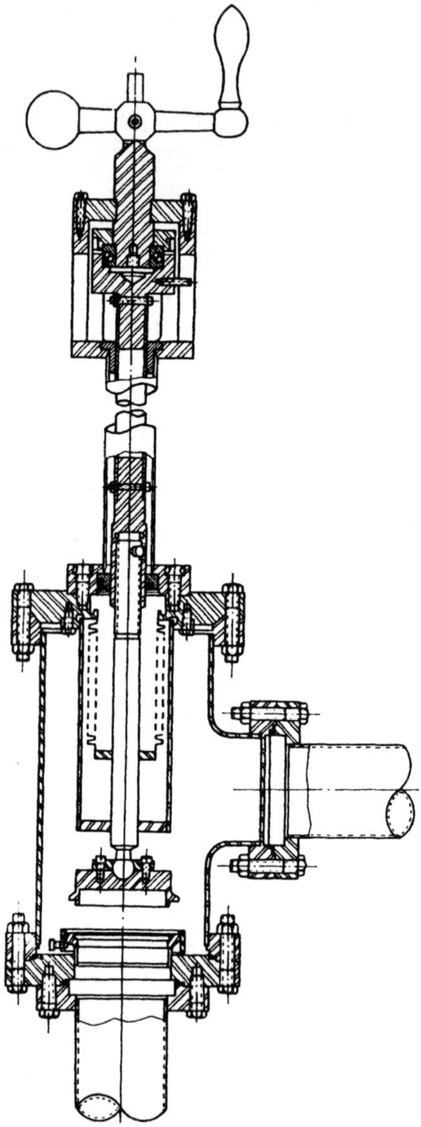

FIG. 6.106 Section through an ultra-high vacuum valve using "Conoseal" joint for its closure. Reproduced from Ullman[1249] *(Courtesy of Pergamon Press)*

Garrod[428] described a very simple multi-way valve consisting of four plate valves built in the same body. Kappen[653] based his multi-way valve on a rotating disc which may connect the required pair of pipes when rotated to various positions (see also Fig. 6.69). Kinder[672] used parallel plungers (Fig. 6.70b) and Fischer[379] used a single plunger (Fig. 6.85).

Figure 6.108 shows schematically the construction and operation of the "valve combination" supplied by Balzers[444]. By placing the main lever as shown in Fig. 6.108a, the main valve is closed, and the by-pass valve connects the vacuum chamber (1) to the roughing pump (2). In position (b) the plate valve is open and the vacuum chamber (1) is connected to the diffusion pump (3), which is now backed by the rotary pump (2). Figures 6.108a and b,

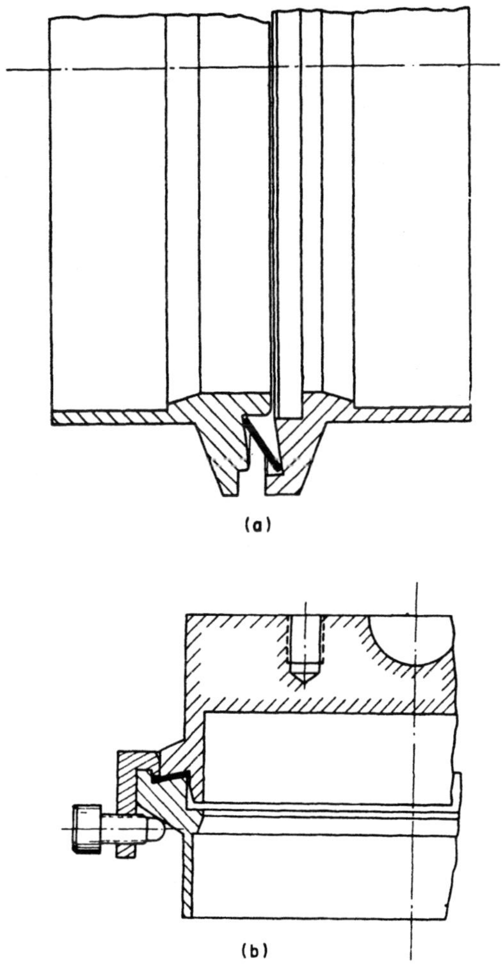

(a)

(b)

FIG. 6.107 Partial section through the 3-in. "Conoseal" used in the valve from Fig. 6.106, showing the added gasket clamp ring. Reproduced from Ullman[1249]

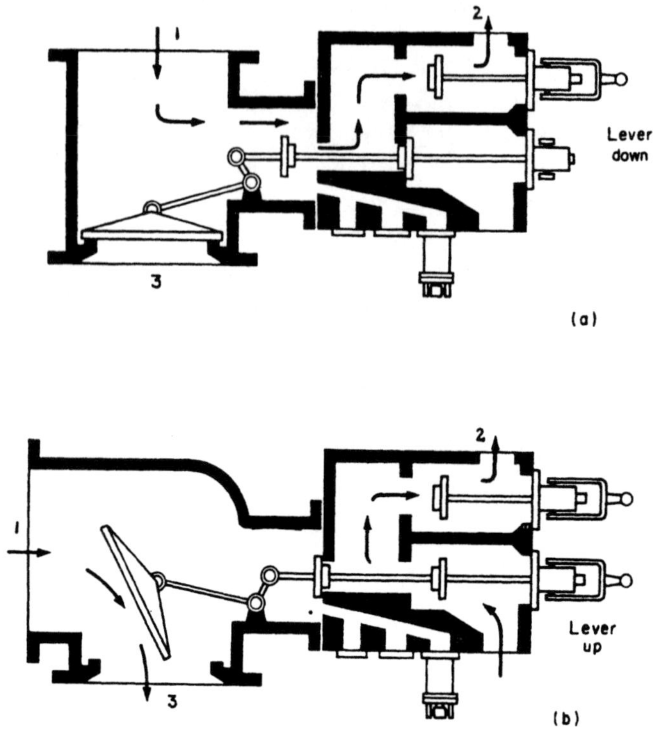

FIG. 6.108 Operation of Balzers[444] valve block: (a) plate valve closed; roughing; (b) plate valve open; backing

PLATE 37. Valve block *(Courtesy of Pfeiffer[987])*

show that such a valve may be connected to the vacuum chamber either in line (a) or at right angles (b). Such valve blocks usually have some other outlets to connect gauges, air inlet valves; the outlets which are not needed may be plugged.

Plate 37 shows a valve block manufactured by Pfeiffer[987]. In this block each valve has its own handle, but their operating system is so designed that by opening one of the valves the others are locked, thus avoiding any possibility of mishandling.

Garrod[431] described a multi-way valve in which three valves are operated by two specially shaped cams rotated on a common handle-driven shaft, thus providing the sequence of operation of the valves. The first 120° rotation

PLATE 38. Shaft and cam operated multi-way valve *(Courtesy of Officine Galileo[953])*

of the shaft isolates the vapour pump from the rotary pump, closes the air inlet and connects the vacuum chamber to the rotary pump (roughing). A further 120° rotation of the shaft, opens the connexion between the rotary and vapour pump (backing). The independent large valve between the vapour pump and vacuum chamber is then opened. Another rotation of 120° of the shaft, admits air into the system.

Plate 38 shows a valve block supplied by Officine Galileo[953] which is operated in the required sequence by cams on a main shaft.

The valve blocks manufactured by Leybold[762] (Fig. 6.109) are operated by a handle which can be turned only in one direction, achieving automatically the correct sequence of closing and opening the various sections of the valve. Table 6.29 lists the positions of the valves at various positions of the lever.

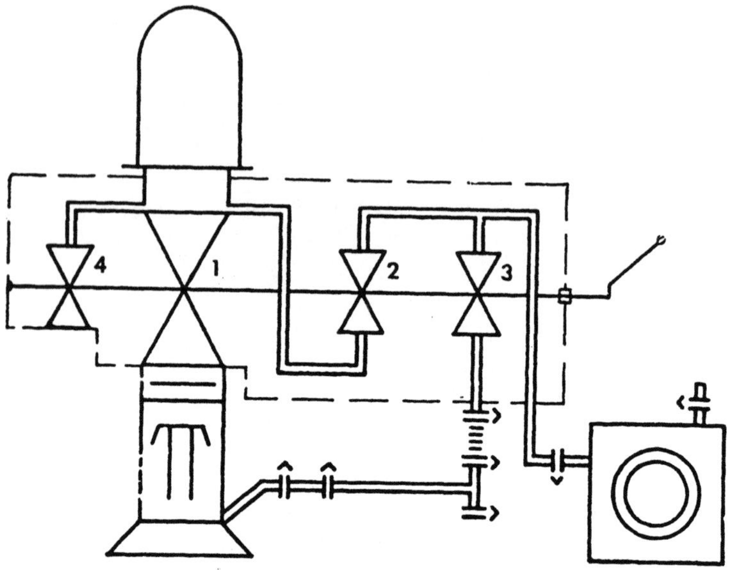

FIG. 6.109 Schematic of a valve block (after Leybold[762])

TABLE 6.29. SWITCHING POSITIONS OF VALVE BLOCKS (Fig. 6.109) SUPPLIED BY Leybold[762]

Position of the lever	Position of the valves			
	1	2	3	4
Roughing	closed	open	closed	closed
Backing	closed	closed	open	closed
High vacuum pumping	open	closed	open	closed
Testing (idle position)	closed	closed	closed	closed
Air admittance	closed	closed	open	open

A three-way valve is available from Cenco[199]. In this valve block the action of a single lever enables the switching to each of the operating positions: closed, rough-down, and diffusion pump.

Valve blocks can be automatically operated by servo-motors, activated by the output of vacuum gauges placed at critical points on the system. Such devices are described by e.g. Jäger[625], Felheimer[372].

61.36 Maintenance of valves. A vacuum valve, carefully chosen to meet the service requirements (Section 61.31), must be properly installed, operated and maintained. To achieve performance according to the characteristics of the valve, it is advisable that the operator knows the principles of the valve and its characteristics (temperature range, conductance, etc.).

To install the valve the operator usually has to deal only with the inlet and outlet connexions, but the maintenance of the valve requires the inspection of the closing, sealing and operating systems of the valve.

A valve should be installed only in the position specified by its design, e.g. the specified connexion must be joined to the low pressure side, gravity closed valves must be installed in vertical position, etc.

Before installing a valve, it must be inspected for foreign materials; the flanges, O-rings and sealing surfaces must be cleaned.

The operating system of the valve must be tested prior to installation. During service, care must be taken not to over-tighten the closing screws; if the valve is constructed for manual operation, the use of keys, wrenches or other such tools will damage the gasket of the closing system or even damage the shaft itself; if the valve is designed to be closed at a given torque (see ultra-high vacuum valves, Section 61.35f) the use of torque wrenches is recommended.

The maintenance problems of valves arise from wear of gaskets, or moving parts, loss of grease, or need of retightening adjusting screws. The operating system must be lubricated, and grease should be added from time to time. Valves used in systems containing mercury vapours, should not be lubricated internally.

The sealing system of the valves (diaphragms, bellows) may crack after a number of cycles. This results usually in a sudden rise in pressure. The valve must be dismantled, and the sealing part replaced by a new one.

The closing system may require a change of the gasket; this applies especially to all-metal valves (see e.g. Fig. 6.103).

The maintenance of valves using ground seals, molten metals, liquid seals, etc. is specific to their construction and results from the operation of these closing systems (Section 61.32).

61.4 Gas-leaks and metering devices

61.41 Principles and classification. "Artificial leaks" are used in order to admit controlled amounts of gases into evacuated systems. These "artificial leaks" are either calibrated for a given throughput or are adjustable in the range of throughputs required. The calibrated or adjustable leaks are used in the filling operation of discharge tubes for the admission of gas samples into mass spectrometers, and in testing the pumping speed of vacuum pumps, etc.

TABLE 6.30. GAS LEAKS

Basic feature of the leak	Characteristics	References
Pinhole	through thin walled glass, quartz or metal	Fig. 6.110 Section 61.42
Orifice	at the end of pipes	Fig. 6.111
Crack	in glass walls; variable by twisting the tube or change of mercury level	Fig. 6.113 Fig. 6.114
Capillary	constant leak	Fig. 6.115
Flattened tube	variable by bending, twisting, squeezing	Fig. 6.117
Porous plug	ceramic, glass frit or metal; variable by changing the mercury level	Figs. 6.118; 6.120
Annular impedance	concentric pipes, cone and seat, washers, O-rings	Figs. 6.93; 6.121; 6.122; 3.126d
Needle valves		Section 61.46
Temperature actuation	longitudinal expansion bimetal radial expansion temperature of the gas	Fig. 6.125a, b Fig. 6.125c Fig. 6.126
Diffusion	for helium, hydrogen, and oxygen	Table 6.32
Pulsed leaks	bubbling, vibrated cap rotating pits vibrated needle	Figs. 6.128a; 6.129b, c Fig. 6.128b Fig. 6.129a

Gas-leaks are in fact impedances placed in the path of the flowing gas to control its throughput. The throughput (Section 12.1) Q (lusec) of such an impedance is determined by

$$Q = C(P_1 - P_2),$$

where C is the conductance (l./sec) of the leak, and P_1, P_2 the pressures (microns) at the inlet and outlet side of the leak respectively. The pressure P_2 at the outlet (vacuum) side of the leak should be taken as equal to that at which the gas is required in the evacuated space. Usually this pressure (P_2) can be neglected when compared with the inlet pressure (P_1). The inlet pressure is usually one atmosphere, but it may be lower or higher.

Thus the throughput can be controlled by the conductance of the leak and/or by the inlet pressure. The conductance of the leak for a given gas is determined (see Appendix A.5) by the shape and dimensions of the opening, the path (length) through the leak, and the temperature of the gas. In viscous flow, the conductance is also a function of the mean pressure $\frac{(P_1+P_2)}{2}$ in the leak, but in molecular flow it is independent of the pressure.

The throughput of a leak is brought to the required value or is varied in the required range by having a small opening, a very long path and/or operating at an appropriate temperature. The various basic features used to construct leaks, are listed in Table 6.30. A review of some leaks was published by Patty[977] and Beynon[120].

An "ideal leak" should meet the following requirements:

1. The leak should meet the general requirements for vacuum valves (Section 61.31) as regards to its freedom from outside leaks and degassing (virtual leaks), mechanical strength and the chemical resistance (as required). Millican[866] described a variable leak (200–2×10^{-2} lusec) which is resistant to uranium hexafluoride.

2. The leak should have the required throughput and/or be variable in the required range. The throughputs obtained with the various gas leaks are listed in Table 6.31.

3. The throughput of the leak should be reproducible as many times as desired.

4. The leak should not be prone to plugging, or if variable, to sticking in any one position in its range. Blockages can sometimes be cleaned by applying a gas pressure in the reverse direction, or as in the case of glass leaks by burning out the blocking material by applying a high frequency discharge through the leak. However any of these treatments may permanently change the throughput. A partially blocked leak, through which the flow conditions are liable to change in a random fashion, should not be used.

5. A variable leak should be capable of having its throughput controlled continously and to a sufficient degree of accuracy.

6. The leak should be non-fractionating, i.e. the composition of a gas mixture passing through the leak should be the same at the inlet and outlet of the leak. Halsted[504] established an approximate expression for the flow of a binary gas mixture, and Kistemaker[680] studied the fractionation effect of

TABLE 6.31. THROUGHPUT RANGE OF GAS LEAKS

Leak	Throughput range (lusec)	Lowest pressure torr)	Remarks	Reference
Orifice through platinum disc	6×10^{-1}–30	–	5–15 μ diameter	Nief[928]
Orifice on the end of tapered capillary	1–9×10^{-4}	–	1.2 μ bore taper 39 μ/cm	Gordon[461]
	1.6	–	24 μ bore taper 30 μ/cm	
Crack on capillary glass tube	2×10^{-3}–1.8	–	varied by twisting the tube (Fig. 6.113)	Hopfield[594]
Slit on glass tube	0–3.8×10^{2}	10^{-8}	varied by mercury level (Fig. 6.114)	Kunzl[722]
Circular cross section capillary	10^{-7}–10^{-2}	–	0.2 to 10 μ dia. 1 cm long see Fig. 6.115	
Flattened copper tube	3×10^{-3}–3	–	varied by rolling the tube see Fig. 6.117b	Nier[930] Veeco[1262]
Glass needle in capillary tube	2×10^{-2}–5	–	see Section 61.43	Hopfield[594]
Porous metal plug	5×10^{-5}–10^{-2}	–	compressed at various loads see Fig. 6.120	Jenkins[631]
Porous material	$> 7\times10^{-1}$	–	varied by the level of the mercury	Rose[1079]
Porous ceramic rod	1×10^{-3}–10	10^{-6}–10^{-3}	see Fig. 6.119	Morrison[890]
Permeation through silicone rubber sheet	$< 1\times10^{-4}$	–	–	Kobayashi[702]

(Table 6.31 Continued)

Leak	Throughput range (lusec)	Lowest pressure (torr)	Remarks	Reference
O-ring seals with variable compression	$> 2\times10^{-3}$	10^{-5}	see Fig. 6.122	Amoignon[29]
Conical, spring washer seal	5×10^{-4}–7×10^{-2}	–	see Fig. 3.126d	Wishart[1327]
Steel ball on seat	7×10^{-1}–7×10^{-3}	–	differential screw, bellows valve	Allensworth[14b]
Needle valves	1×10^{-2}–1.0	–	see Section 61.46	Riggs[1061a]
Pin valve	$> 6\times10^{-5}$	10^{-4}–10^{-6}	bellows sealed	Bouyer[154]
Needle valve	$> 10^{-4}$	–	PTFE seat	Cope[234]
Bakeable all-metal valve	$> 7\times10^{-8}$	–	continuously variable see Fig. 6.93	Alpert[19]
Platinum wire expanding in glass capillary	3×10^{-2}–25	10^{-8}	see Fig. 6.126a	Martin[817]
Tungsten rod expanding in stainless steel body	1×10^{-1}–9.10^{-1}	–	see Fig. 6.126b	Green[474]
Heated capillary	4–40	5×10^{-2}	throughput varied as a result of gas temperature	Smither[1158]

some leaks (used in mass spectrometry) and found that a steady state was reached only after several hours of flow. Boerboom[142] gives graphs establishing the total fractionation and the time required to approach the steady state with leaks of various dimensions.

7. It is desirable to construct the leak on the basis of a previous calculation or at least to predict the throughput for various gases, pressures, etc., of the already constructed leak. The throughput may be initially calculated for simple leak geometries only, as capillaries with circular bore cross section (Guthrie[491], Ochert[951a]), apertures having a particular arrangement (Gordon[461]) or concentric cylinders (Bachman[68]). For other types of leaks (Table 6.30), it is not possible to calculate the throughput and such leaks are constructed only on the basis of experimentally determined values. Nerken[919] gives some formulae established to generalize his results, obtained by measuring the throughput of leaks consisting of flattened tubes. Other formulae (for particular cases) are given by Lawrence[748]. Moller[872] discussed the factors influencing the calibration of standard leaks. Keith[659] gives a method to determine the range of the throughput, based on a maximum and a minimum value corresponding to viscous and molecular flow respectively.

61.42 Pinholes, orifices and cracks used as leaks. Pinholes through glass or metal diaphragms, orifices made in plates or at the end of tapered capillaries, or cracks (slots) made on pipes, were used as gas-leaks by various authors.

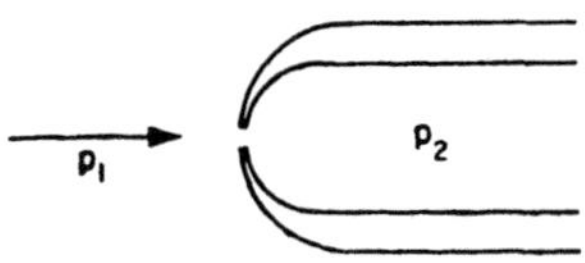

FIG. 6.110 Orifice (leak) through a glass wall

It would be difficult to make an "ideal" opening, since this should be made in an "ideal" diaphragm having an extremely small wall thickness. Nevertheless small openings approaching the ideal (made through diaphragms having a wall thickness which is negligible compared to the diameter of the hole) can be prepared by using various techniques.

Honig[590] prepared openings (of about 25μ diameter) by heating the thin-walled glass tubing at one end and shaping them as shown in Fig. 6.110.

In order to keep the hole from plugging, the leak is oriented in such a way that the gas always flows from the convex to the concave side. According to Beynon[120] very small leaks can be prepared by collapsing the end of a glass tube until a hole of about 20μ remains at the end. The end of the tube is then pressed flat (between an inserted carbon rod and a carbon plate), and ground flat both from the inside and outside to a thickness of about 0.1 mm, followed by fire polishing. A modification of this method for the production of reproducible holes of about 1μ diameter consists of sealing a zinc wire into the end of the glass tube, grinding it as before, and dissolving the zinc with a jet of dilute HCl (Beynon[120]).

Munson[908] produced small holes (10–20μ diameter) in thin walled glass, by punching the heated (300 °C) glass with a high frequency discharge between the sharp edges (0.1 mm) of platinum or gold electrodes. Lossing[748] used this method (with a high frequency Tesla coil) to make small holes in thin quartz walls. Andersen[29c] produced holes of 10–100μ diameter through glass walls by first grinding a conical hole, and when the remaining glass thickness reached 25–50μ, the actual hole was produced by means of a high voltage Tesla discharge.

Eltenton[340, 341] used as a gas leak, a hole pierced in a 20μ thick gold diaphragm, clamped between gold washers against a recessed shoulder in the metal wall of his apparatus. Nief[928] designed a leak which consisted of a platinum disc (0.01 mm thick) placed inside a small bore Pyrex tube. The platinum disc was pierced in the centre with a tungsten needle (5–15μ diameter). This leak was designed to give a throughput of 3–50 cm^3 per hour measured on the intake side (0.6–30 lusec).

To make very small (7–10μ diameter) holes in metals, Marks[807] proposed that a conical tool with a sharp point be pressed into the metal until the point is at a short distance from the opposite side. The material opposite the point of the conical cavity is then removed by grinding until the hole first breaks through.

Stainer[1175a] made holes of about 15μ in 0.1 mm thick tantalum sheets, by using a spark erosion technique (with a tungsten electrode in Kerosene). Baxter[98] described a technique to make very small holes (about 3μ in diameter) by an evaporation technique. An electrolytically pointed and polished needle is placed so as to rest gently, normal to a glass surface coated with a stripping layer (Section 71.36). Vacuum evaporation from a ring source surrounding the needle deposits a uniform film around the point.

Fleischer[383a] described a method of forming fine holes by first exposing the material to high energy particles (fission fragments) and then immersing it in a chemical etching solution (e.g. 20 per cent HF, for mica) which dissolves the damaged material along the paths of the particles. With this technique it is possible to form holes which are as small as 25 Å in diameter and are up to about 10μ long. The hole diameters can be increased at will from the minimum size by continued etching. Fine holes have been formed in this way in glass, mica and other silicate minerals.

Gordon[461] proposed that a slightly tapered glass capillary (Fig. 6.111a) be drawn. Starting from a 0.045 in. bore capillary, the operation will give a fairly sturdy stem with a minimum outside diameter of 1/16–1/32 in. When this is examined under a microscope, a smoothly tapered bore will be seen tapering down to the vanishing point (rate of taper 10–30μ per cm).

Using a calibration curve (Fig. 6.112), the point having the diameter (bore) to give the required throughput is determined and then it is located under a high power microscope. The tube is scored at this point with a glass cutting knife and broken off.

The tapered tube (leak) is enclosed in a glass envelope, and the assembly is used by *placing the small end of the bore in the gas-inlet side*. It is advisable not to touch the leak where it has been broken off.

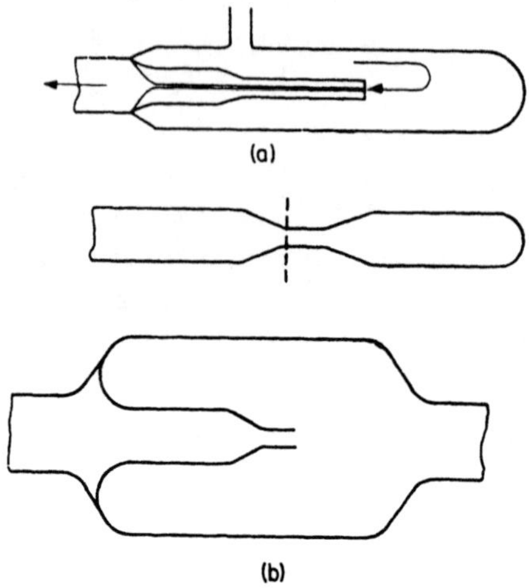

FIG. 6.111 Orifice (leak) on the end of a tapered capillary: (a) glass leak (after Gordon[461]); (b) stages in constructing a glass leak (after Harris[511])

Harris[511] described a method to construct a glass tube which, ends with a short and fine hole (Fig. 6.111b), by starting with a glass tube which is constricted at one end, and evacuated and sealed off, without, however, applying very intense heating. If the resulting pip is tested with a Tesla coil, it frequently proves that it is sealed completely.

Morrison[891] constructed a variable leak consisting of a stainless steel plate provided with four holes of various diameters, which are closed or opened by a smaller stainless steel plate sliding on the large plate. The two plates have been ground flat (lapped) to ensure close contact. The motion was

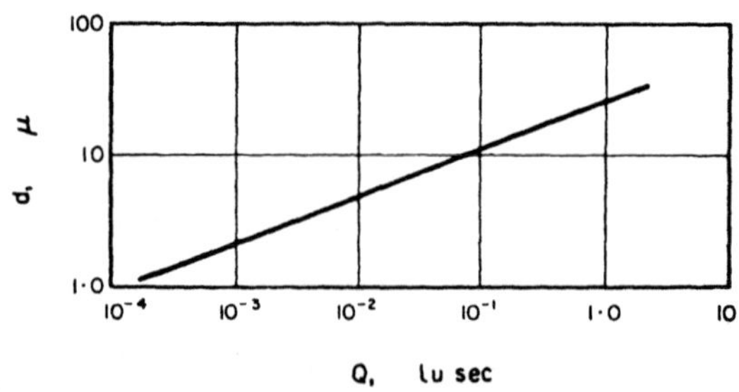

FIG. 6.112 Throughput vs. diameter of the orifice of leaks as in Fig. 6.111a. After Gordon[461] *(Courtesy of The American Institute of Physics)*

transmitted magnetically. The device is bakeable (to about 450 °C) and has a minimum throughput of about 10^{-7} lusec.

Gas leaks based on *cracks* in the pipe were described and used by various authors. Hopfield[594] described such a leak consisting of an inner glass tube (5, Fig. 6.113a) closed at one end (7) and sealed by a ring seal (Section 23.34)

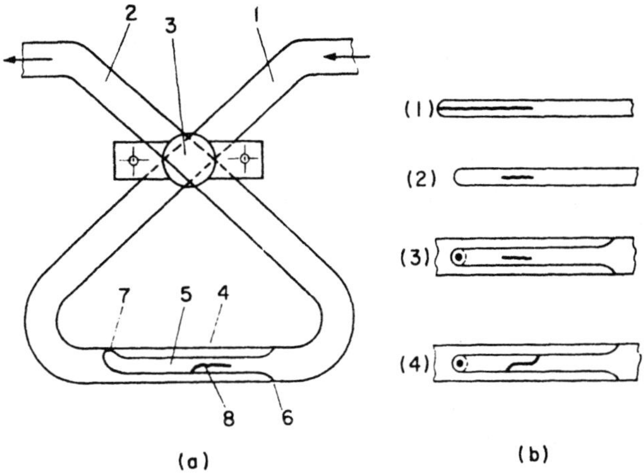

FIG. 6.113 Crack (leak) varied by bending the glass tube: (a) general assembly; (b) stages in constructing the leak. After Hopfield[594] *(Courtesy of The American Institute of Physics)*

at (6) to the outer tube (4). The outer tube receives the gas through the side (1), the gas passes through the crack (8) in the inner tube (5), and flows to the arm (2). The throughput of the crack (8) may be continuously varied by the twist which arises from pressing the arms (1) and (2) together, by a clamp placed at their crossing point (3). The space between the arms at (3) is about 0.5 mm. The construction of the crack (8) is important for the reliability of the leak. Hopfield[534] recommended the following technique:

The end of a tube is heated in a torch and then quenched by dragging it swiftly across a beaker with water. A straight crack (1, Fig. 6.113b) will appear. The end of the tube is sealed-off so that a short axial crack remains (2, Fig. 6.113b), and while hot the tube is quenched in light oil. The oil is then cleaned, the tube (5, Fig. 6.113a) is sealed into the outer tube, and the arms are bent, leaving 1 mm between them at the crossing. While holding the arms squeezed against a metal spacer, the inside tube is sealed to the outside tube on a spot near the closed end (7, Fig. 6.113a). The spot weld is shown on 3, Fig. 6.113b. When the work cools the arms are carefully squeezed towards one another at the crossing point, and a diagonal crack (4, Fig. 6.113b) is obtained. Finally one of the arms is softened above the bend and the spacing at point 3, Fig. 6.113a is reduced to 0.5 mm. Although the throughput of the leak may be varied to very low values (see Table 6.31), it is advisable to leave it open at about the double of the minimum throughput, in order to avoid resealing of the crack.

Kunzl[722] constructed a variable gas leak, modified later by Amoignon[29] (Fig. 6.114). This variable leak consists of a glass tube (1) closed at its lower end, connected at the other end to the vacuum vessel, and having a longitudinal crack (2). The tube (1) is sealed into the tube (3) by means of a tapered ground joint or a rubber gasket. The tube (3) is filled with mercury up to a given level, the level of the mercury being variable by pressing the rubber

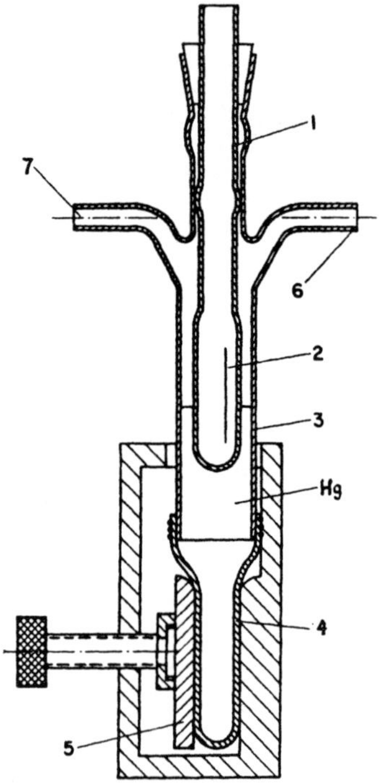

FIG. 6.114 Crack (leak) varied by changing the level of the mercury column. After Amoignon[29] *(Courtesy of Société Française des Ingénieurs et Techniciens du Vide)*

container (4) with the clamp (5). The gas is admitted through (6) and after the space between the two tubes (1 and 3) is flushed by (6 and 7), the mercury, which covered the whole length of the crack (2), is lowered, to leave open the required length of crack (Fig. 6.114).

61.43 Capillaries and flattened tubes used as leaks. The throughput for a given gas through an ideal capillary tube leaving a circular cross section bore (without ellipticity) may be calculated using Knudsen's equation, written by Guthrie[491] and Ochert[951a] in the form:

$$Q = \frac{1}{L}\int_{P_2}^{P_1}\left(aP + b\,\frac{1+c\cdot P}{1+f\cdot P}\right)dP,$$

where

$$a = \frac{\pi d^4}{128\eta}, \quad \text{and for air at 20 °C} \qquad a = 0.182 \cdot d^4,$$

$$b = \frac{1}{6}\left(\frac{2\pi KT}{m}\right)^{1/2} \cdot d^3, \quad \text{and for air at 20 °C} \quad b = 12.1d^3$$

$$c = \left(\frac{m}{KT}\right)^{1/2} \cdot d, \quad \text{and for air at 20 °C} \qquad c = 0.256d,$$

$$f = 1.24\left(\frac{m}{KT}\right)^{1/2}, \quad \text{and for air at 20 °C} \qquad f = 0.316d.$$

d is the diameter of the bore (cm), L the length of the capillary (cm), P the pressure (microns of mercury), Q the throughput (lusec), K Boltzmann's constant (1.38×10^{-16} erg/°K), m the molecular weight of the gas, η the viscosity of the gas, and T the temperature of the gas (°K).

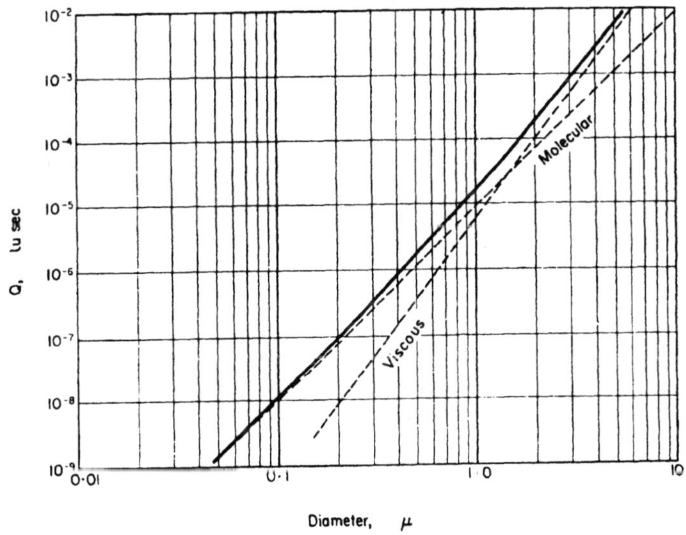

FIG. 6.115 Throughput, for air at 20 °C, vs. the diameter (bore) of the capillary (for length of 1 cm). For other lengths the figures from the diagram should be divided by the length in centimetres

The throughput Q (lusec) resulting from this equation for air at 20 °C is plotted on Fig. 6.115 versus the diameter of the bore (d) together with the curves calculated for pure molecular and pure viscous flow respectively.

Capillary tubes were used as calibration leaks (Graham[468], Harris[511], Laufer[743]) but because of their inconstancy and their tendency to be plugged, these leaks are not recommended. Flattened tubes are usually preferred to the capillaries.

Ochert[951a] recommended that (for throughputs of the order of 10^{-3} lusec) a leak, made by squeezing flat the end of a fine tube of a non-corrosive metal (e.g. aluminium), be used. Since this leak is subject to changes after calibration, it is necessary to check its calibration periodically. The leak is connected to a vacuum union (Section 38.43) on one side and to a suction line at the other end (Fig. 6.116).

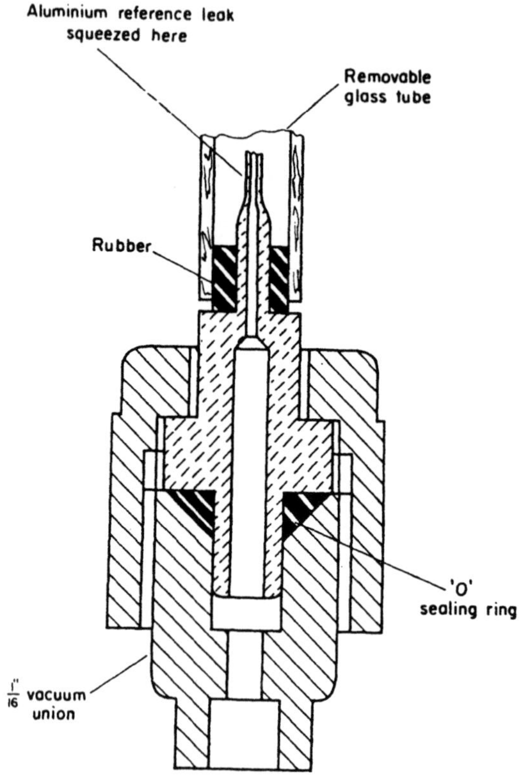

FIG. 6.116 Construction of reference leak with flattened aluminium tube. Reproduced from Ochert[591] *(Courtesy of Pergamon Press)*

Fowler[392] described a leak consisting of a German-silver tube (6 mm diameter, 0.5 mm wall thickness), flattened until the desired throughput was obtained, and bent into a U- shaped spring (Fig. 6.117a). The variation of the throughput was accomplished by opening and closing the spring with a clamp (Fig. 6.117a). The construction of this leak was improved by Nier[930] who built a leak consisting of a copper capillary (12 in. long, 0.125 in. outer diameter and 0.025 in. wall) bent in the form of a hairpin. For a distance of 2.5 in. back from the apex of the hairpin, the tube is flattened. The apex of the hairpin is then soldered to a steel shaft (Fig. 6.117b), and the flattened tubing is wrapped approximately three times about the shaft. The non-flattened ends are then anchored by passing through holes in the block. Nier[930] claimed to have obtained adjustment of the flow over a range of 100/1.

The rotation of the shaft controls the cross section and consequently the throughput (Wright[1333]). Veeco[1262] supplies variable leaks constructed on this principle, having a throughput variable over a range of 1000/1 (see Table 6.31). Obviously, for higher throughputs, the leak can be simply a squeezed

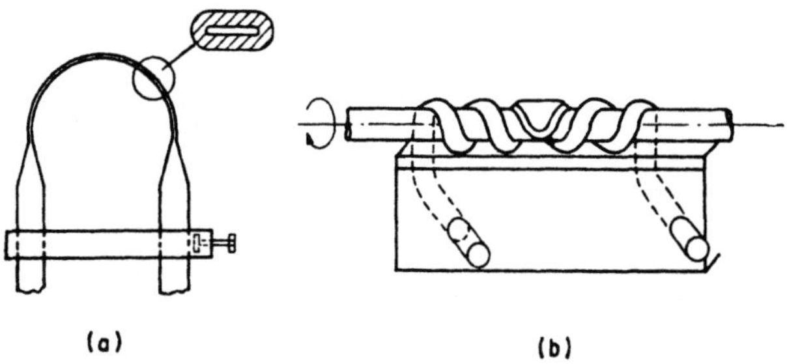

FIG. 6.117 Flattened tubes used as leaks: (a) varied by bending (after Fowler[392]); (b) varied by rolling (after Nier[930]) *(Courtesy of The American Institute of Physics)*

heavy-wall rubber tubing (Wyckoff[1335a]), but such leaks should be opened at short intervals to prevent sticking of the inner surfaces of the rubber tubing. In order to allow more sensitivity near to the cut-off position, a wire (about 0.8 mm) may be inserted axially in the squeezed portion of the rubber (Knauss[695]). A small opening remains along the wire, which may however be closed entirely by applying sufficient pressure with the clamp.

Hopfield[594] described a leak consisting of a capillary tube in which a glass needle is inserted. He obtained a minimum flow of 0.1 cm^3/hr (at STP), which can be increased about 250 fold by moving the needle.

61.44 Porous plugs used as leaks. Gas leaks can be made by allowing the gas to diffuse through *porous materials.* An unglazed filtration grade porcelain tube (Dorn[292], Rose[1079], Smithe[1160], Bazzoni[100], Richards[1052]) or rod (Hagstrum[500], Molnar[873], Morrison[890]) may be used. The throughput can be controlled by changing the level of a mercury column surrounding the porous material (Section 61.12a).

Smithe[1160] and Bazzoni[100] used an 8 cm long tube (0.8 mm i.d.; 1.5 mm o.d) of unglazed porcelain, sealed directly to Nonex glass. The porcelain tube was placed inside a somewhat larger tube of Nonex. Hagstrum[500] used a porcelain rod (1 Fig. 6.118) of about 1.5 mm diameter and 125 mm length, sealed at the lower end to the Pyrex connexion (2) leading to the vacuum chamber. The gas is introduced by the connexion (3) and according to the level of the mercury, more or less gas will diffuse through the rod (1) into the connexion (2).

Molnar[873] determined that the resistance offered to gas flow by the porcelain rod (Fig. 6.118) is proportional to the length of rod immersed in mercury.

Morrison[890] used a gas leak (Fig. 6.119) in which the porous ceramic rod (2) (6 mm or 1.5 mm diameter) is immersed more or less in mercury (1), by lowering or raising the glass cap (3). The glass body (3) contains an iron core (5) beaded in glass wool (6), and is raised magnetically. When (3) is raised

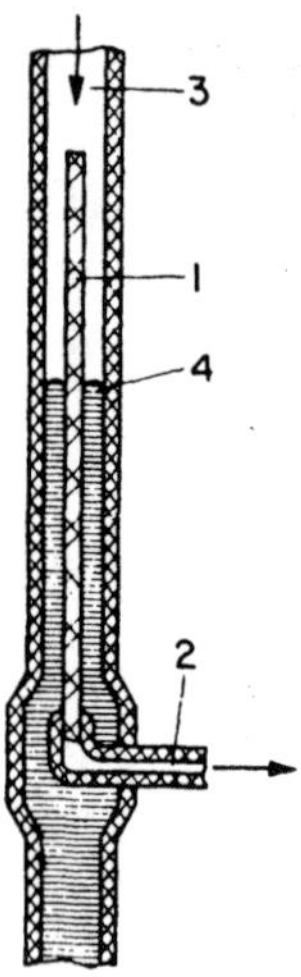

FIG. 6.118 Porcelain rod used as a leak. After Hagstrum[500] *(Courtesy of The American Institute of Physics)*

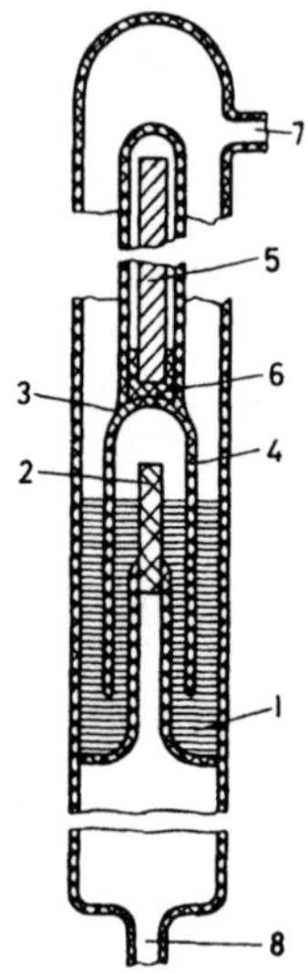

FIG. 6.119 Leak with porcelain rod and magnetically operated plunger. After Morrison[890] *(Courtesy of The American Institute of Physics)*

the level of the mercury drops and the gas entering by (7), flows through (4) and diffusing through the porous rod (2) reaches the connexion (8) to the vacuum chamber. The throughput range is listed in Table 6.31. By raising the mercury level above the top of the porous rod, the leak is completely cut off (D'Amico[258]).

Instead of ceramic porous plugs, sintered glass leaks are also used for pressure control (James[626]), in circulating pumps (Wilson[1323]), in devices for the metering of gas samples (Smith[1156]) or controlled gas transfer (Greenwood[481]).

Jenkins[631] constructed gas leaks by compressing a porous metal in a mild steel block (Fig. 6.120a). The bore of the block (1) was tinned and a small quantity of soft solder powder (2) was then placed in the hole, one end of which was covered by a flat steel plate (3) provided with a small boss filling the hole. The solder powder is compressed with a steel rod (4) having a sliding fit in the hole. When the plate (3) and the rod (4) are removed, a plug of solder powder remains in the hole. Tests were made with various pressed plugs, all having 4 mm diameter and 1 mm thickness, but pressed beginning with various quantities of powder. The leak rates obtained through these

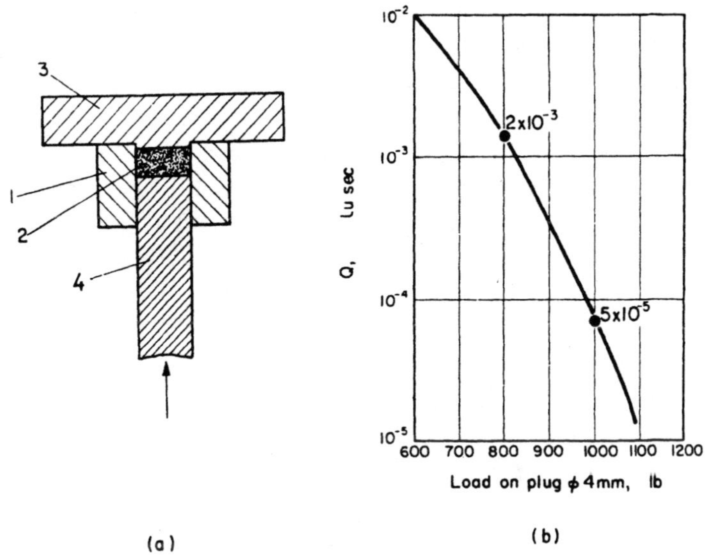

FIG. 6.120 Leak using a porous metal plug: (a) construction; (b) throughput obtained with a 4 mm diameter porous plug pressed with various loads to a final thickness of 1 mm. After Jenkins[631] *(Courtesy of The Institute of Physics and The Physical Society, London)*

plugs are plotted on Fig. 6.120b as a function of the load required to reach the same thickness (1 mm). Larger leaks could be made by using thinner plugs or larger diameter plugs pressed with a smaller load.

Sintered silicon carbide ("Metrosil" of Metropolitan Vickers Ltd.) was found to be suitable for the construction of constant leaks. Some metals (Cu, Ni) can be made porous (Adam[2]) and used as leaks.

61.45 Annular impedances used as leaks. The annular spaces between a pipe and a rod spaced inside, between two mating conical or spherical surfaces, etc., were also used as fixed or variable leaks.

Amariglio[25] described a leak consisting of a ground glass tube (e.g. 10 mm i.d.) fitted with a glass rod of very close clearance (e.g. 0.005 mm). The annular space between the tube and the rod serves as the leak, the throughput being controlled by the length of rod pushed into the tube. The throughput is linear with this length (l); for the values given as example (10 mm i.d. with 0.005 mm clearance) the throughput was $V=50/l$ mm^3/sec N.T.P.

Shields[1132] described a flow regulating valve which controls minute flows of gases by allowing the gas to diffuse between two mating surfaces. Harrison[512] used conical or hemispherical ground glass joints actuated by the differential expansion of the coaxial glass tubes supporting the ground members.

Nier[931] used a leak which was adjusted by closing and opening an annular space between the parts and actuated by turning a handle.

In the leak designed by Mueller[901] (Fig. 6.121) the throughput is controlled by varying the pressure applied to a series of alternate rubber and metal washers, placed in a cylindrical body into which fine grooves have been machined. The plunger exerting the required pressure can be moved by rotating a threaded sleeve, through a differential gear. Under the compression thus obtained, the rubber washers expand radially into the grooves, and change the degree of leakage through the valve.

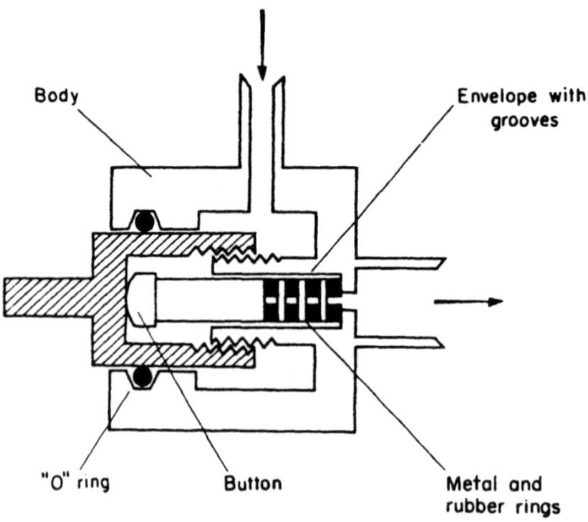

FIG. 6.121 Variable leak consisting of alternate rubber and metal washers. Reproduced from Mueller[901] *(Courtesy of Pergamon Press)*

Amoignon[29] placed the grooved component (1, Fig. 6.122) in the centre of the device in order to allow easy changing of this part. A cylinder of aluminium alloy (dural) was used, with a diameter of 17.2 mm, and having longitudinal scratches (about 0.1 mm deep). This cylinder is surrounded by three stainless steel rings (2, 3, 4), with edges tapered to about 10° (see Fig. 6.122). The rubber O-rings (5) placed between the rings (2–3 and 3–4) are pushed towards the surface of the cylinder (1) by tightening the screwed part (8) actuated through the shaft (9) sealed by a Wilson seal (10). The rotation of (8) is transmitted to the ring (2) as a translation (without rotating). Depending on the pressure exerted in this way on the O-rings, the throughput of gas entering by (7) and flowing through (6) can be accurately controlled for throughputs greater than 10 mm^3/hr N.T.P.

Amoignon[29] mentions that the O-rings used should have an inside diameter smaller than the diameter of the cylinder (1) in order to be seated permanently on the cylinder. The shape and depth of the scratches on the cylinder is critical.

Wishart[1327] used a conical spring washer seal (Fig. 3.126d) as a controlled leak, and obtained good reproducibility in the range between 7×10^{-2}–-5×10^{-4} lusec of helium. At throughputs below this (7×10^{-7} atm. cm^3/s of He), a non-linearity and hysteresis appear.

Allensworth[14b] replaced the original Kel–F seat of a (Hoke No. 482) packless bellows type valve, with a 0.250 in. steel ball, and by controlling

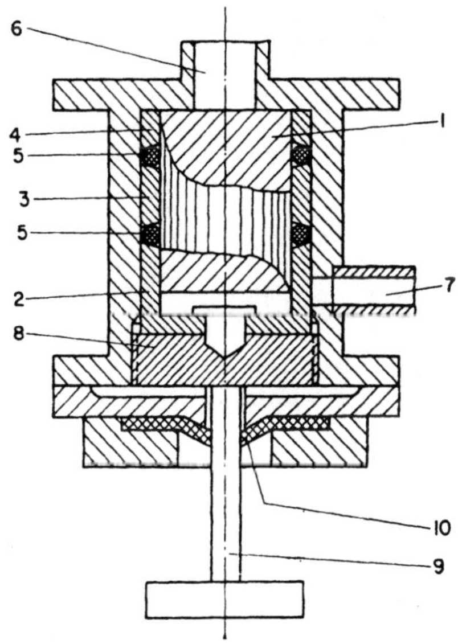

FIG. 6.122 Variable leak consisting of O-rings compressed between metal rings. After Amoignon[29] *(Courtesy of Société Française des Ingénieurs et Techniciens du Vide)*

the motion of the ball with a differential, screw-driven actuator, obtained throughputs in the range of 7×10^{-1}–7×10^{-3} lusec.

The Alpert-type ultra-high vacuum valve (Fig. 6.93) can be successfully used as a variable leak, especially when gas should be introduced in spaces evacuated to less than 10^{-6} torr, and the pressure should rise not more than up to 10^{-5} torr. These valves can be used down to 10^{-9} torr, at very small throughputs (Table 6.31). Bouyer[154] describes a bellows-sealed, all-metal valve, closing on a blunt, non-rotating pin (Table 6.31). Granville-Phillips[469] supplies a bakeable leak with the conductance continuously variable from 10^{-10} cm³/sec to 10 cm³/sec.

61.46 Needle valves. A needle valve is defined (American Vacuum Society[26], Société Française des Ingénieurs et Techniciens du Vide[1122a], D.I.N.[281a]) as a throttling valve (Section 61.35b) in which a needle can be moved against a long, needle-shaped seat. The closing (throttling) system of these valves is based on the very minute changes in the width of the conical space left between the needle and the seat. The needle (1–5 mm diameter) is tapered at 1/30 to 1/50 in order to obtain the minimum change in throughput for a given axial displacement. For 0.1 mm axial displacement of a needle having a taper of 1/50, the width of the conical space (impedance) changes by about 1 μ.

A method for calculating the throughput of needle valves, from the geometry of their closing system, was described by Soa[1162c].

Besides the construction of needle valves having a tapered needle (Edwards[326], Stallmann[1172]) closing on a tapered seat (Fig. 6.123a), other constructions use different needle-seat arrangements.

A tapered needle ending in a cylindrical portion (Fig. 6.123b) sealing on a seat of the same shape, adds the advantages of the annular impedances

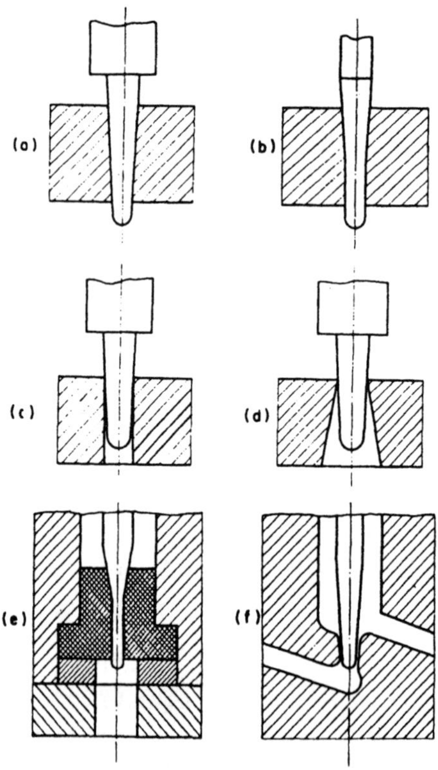

FIG. 6.123 Closing systems of needle valves

(Section 61.45) to those of the needle valve itself, thus allowing a fine variation of the throughput near the cut-off position of the valve.

When a tapered needle is used in conjunction with a cylindrical seat (Fig. 6.123c) or a conical seat (Azan[60b]) (Fig. 6.123d) then the needle forms the seat, in an exact replica of its own shape.

The seat of the needle valves is usually metal (softer than the needle), but also constructions using elastomer seats (Fig. 6.123e) are possible. Needle valves using PTFE (Teflon) seats are described by Cope[234], Rueger[1091], Duncan[308].

The gas flow in the needle valves is usally axial, but for better cut-off conditions, needle valves having the direction of flow at 45° (Fig. 6.123f) are recommended by some authors (Riggs[1061a], Topanelian[1233]).

For better and more reproducible operation of the needle valves, it is recommended (Bogg[145]) that the needle be not free to rotate, but be free to align itself with the seat.

The inlet of the needle valve is protected in some constructions (e.g.

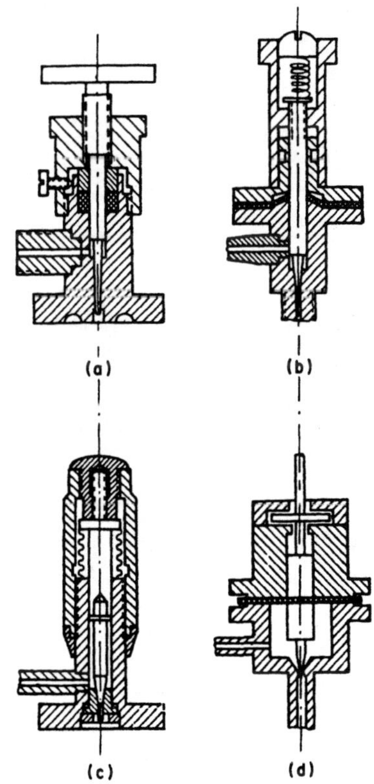

FIG. 6.124 Needle valves: (a) with cylindrical elastomer shaft seal; (b) with Wilson seal (Edwards[328]); (c) with bellows and differential screw; (d) with diaphragm seal (Alexander[13])

Leybold[762]) by a porous filter, in order to avoid plugging or irreproducible operation of the valve.

The needle valves are constructed with packed or packless sealing systems (Section 61.33). The valves using elastomer compression seals (Fig. 6.124a) or Wilson seals (Holton[589], Duncan[303], and Fig. 6.124b) are to be used only in applications where the required purity is not excessive. For cleaner gas transfer, diaphragm-sealed valves (see Alexander[13], Babelay[62], and Fig. 6.124d) or preferably bellows-sealed valves (Fig. 6.124c) are to be used (Bouyer[154], Krieg[718], Edwards[326], Wagner[1280], Topanelian[1233]).

The operating system of needle valves is usually based on a screw. Where a very accurately controlled displacement is required a differential screw system (Fig. 6.124c) may be used (Bouyer[154], Kersten[667]). A spring ratchet device may be included in the actuation system of the needle valve (Edwards[326]) to prevent the use of excessive force in seating the needle, which might damage both the needle and the seat. A magnetically operated needle valve is described by Cope[234].

Holton[589] mentioned a needle valve where the throughput may be set at a predetermined value by depressing a lever. For sensitive control of the throughput it is advisable (Johnson[637], Morrison[892]) to connect two needle valves in series by means of a small volume. In this way it is possible to overcome the troubles given by the hysteresis,* which arise when a single needle valve is used to control small throughputs (3×10^{-4} lusec). The operation procedure is to set one valve at a throughput somewhat higher than the required, and to vary the other valve to obtain the required flow. Using this method John-

PLATE 39. Needle valves *(Courtesy of Edwards[328])*

* Hysteresis curves of needle valves were drawn by Babelay[62].

Harrison[512] used conical or hemispherical ground glass joints, having their parts individually supported by coaxial glass tubes. The leak was actuated by heating the outer tube. Throughputs of the order of 0.5–1 lusec were achieved.

The leak constructed by Flinta[384] had a movable arm (1 Fig. 6.125b) carrying a hard rubber pad (2) which bears on the mouth of the gas supply tube (3). The pad is held tight against the tube by means of the stainless steel wire (4) which is stretched by the adjusting screw (5). The lower end of the

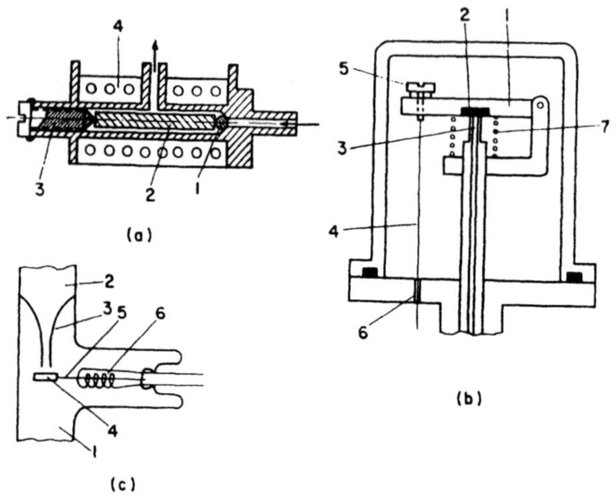

FIG. 6.125 Temperature actuated leak valves: (a) expansion of the stem (after White[1305a]); (b) expansion of a wire (after Flinta[384]); (c) expansion of a bimetal (after Becker[102a])

wire is attached to a Kovar-glass lead-through (6). The arm (1) is pressed upwards by the spring (7). The wire (4) is heated by current (10 V), and as the wire expands, the spring raises the arm so that the gas inlet opens gradually. In this way a continuous variation of the throughput (up to several hundred cm^3 N.T.P./hr) was possible.

Becker[102a] actuated the leak valve shown on Fig. 6.125c by the deflection of the bimetal strip (5), heated by the coil (6). This deflection moves the rubber pad (4) from its seat (3), allowing a controlled gas flow from (1) to (2).

Figure 6.126 shows some leak valves; their operation is based on the increase or decrease of a small gap between materials having different expansion coefficients.

Martin[817] inserted a platinum wire (1, Fig. 6.126a) into a glass capillary (2) and collapsed the glass onto the wire by heating the assembly. During cooling the wire shrank away from the inner surface of the glass tube to form a very small passage for the gas. The capillary is mounted in an appropriate tubulation and platinum wire taken through the seals (3). With a platinum wire of

0.032 in. diameter in contact with the capillary on a length of 1.5 cm a maximum throughput of 25 lusec was obtained. With a platinum wire of 0.010 in. embedded in the capillary on a length of 6 cm, the throughput was in the range of 3×10^{-2} lusec. The throughput may be varied by heating the platinum wire, but the reproducibility is sometimes poor.

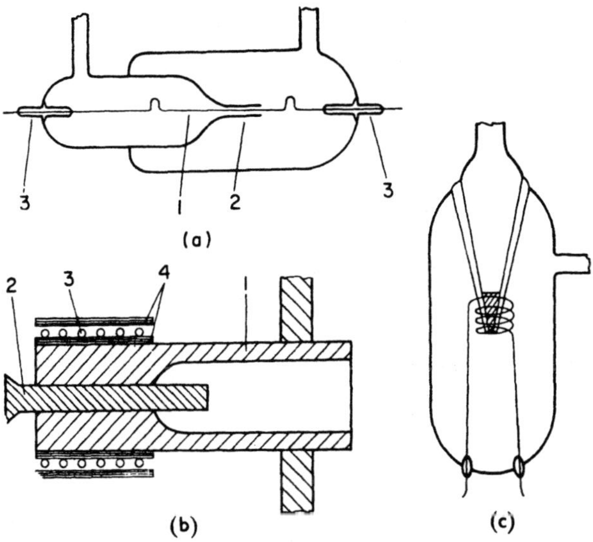

FIG. 6.126 Leak valves actuated by differential expansion: (a) platinum wire in glass capillary (Martin[817]); (b) tungsten rod in stainless steel body (Green[474]); (c) metal plug in tapered glass tube (Nester[921])

Müller[904] sealed a 0.5 mm diameter platinum wire in Jena 20 glass, so as to produce a leak at room temperature, and succeeded in controlling the leak rate to a ratio of 4 to 1 by heating the platinum.

Green[474] constructed a variable leak (Fig. 6.126b) consisting of a stainless steel tube (1) into which is shrunk a tungsten rod (2) so that at room temperature it is completely sealed into the stainless steel. A NiCr heater (3) insulated from the two sides (4) provides the possibility of heating up the assembly. After heating, a gap is created between (1) and (2), permitting a small throughput of gas (Table 6.31).

Nester[921] used a leak (Fig. 6.126c) consisting of a tapered metal (gold) plug cast inside a tapered glass tube, the tip of which is surrounded by a resistance heater. The leak is kept closed by heating the assembly (300 °C) so that the metal plug expands tight against the walls of the (Nonex) glass tube. To permit gas flow the temperature is decreased so that the plug contracts away from the walls of the glass tube.

Bayley[99] and Smither[1158] controlled the gas flow through capillary leaks by regulating the temperature of the gas as it passes through the capillary. In practice a Pyrex capillary (50–100 μ bore, and 50–80 mm length) wound with NiCr wire may be used. As the temperature of the gas is increased, with the pressure hold constant, the viscosity of the gas will increase and the density will decrease. These two effects combine to restrict the throughput when the capillary is heated. This type of control is most useful in the case of viscous flow (see Appendix A 5) where the throughput is proportional to $T^{-1.6}$ to $T^{-1.8}$ (T is absolute temperature) depending upon the kind of gas. It was possible to vary such a leak continuously between 4 and 40 lusec.

61.48 Diffusion leaks. The diffusion (or permeation) leaks are used when *specific gases* are to be introduced in highly evacuated systems. The rates of diffusion of various gases through various materials are summarized in Table 6.32, where D is expressed in cm^3 of gas (N.T.P.) which flows per second per cm^2 of wall surface, per mm of wall thickness at a pressure difference between the two sides of the wall of 1 torr. Gas diffusion leaks may be constructed to operate over a wide range of throughputs. The characteristics of some of these leaks, such as those of glass and nickel, change very little with time. Other materials undergo irreversible changes in use, but the leak rate is in all cases very dependent on temperature (see Figs. 2.5–2.11).

A small standard leak ($<1\times10^{-4}$ lusec) utilizing the permeation of a probe gas through a silicone rubber sheet, was used by Kobayashi[702].

Jenks[633] described a diffusion leak for helium (Fig. 6.127a) constructed from a silica (quartz) glass bulb (1) about 1 cm in diameter and 0.1 mm wall thickness, sealed with a graded seal (Section 23.35) to the Pyrex, helium container (3), and enclosed in a Pyrex vessel. This device was able to deliver 5×10^{-8} cm^3/sec of helium.

Young[1350, 1351] used a quartz tube 1/4 in. diameter, 6 in. long and having a wall thickness of 0.010 in., surrounded by a NiCr heater wire, which was capable of heating the tube to 700–800 °C. The quartz tube was enclosed in an outer jacket through which tank helium passed. The pure helium diffusing through the quartz was collected at the outlet of the quartz tube. The maximum throughput for this helium leak was 0.1 lusec, when the temperature of the quartz was about 750 °C.

If the tank helium contains hydrogen some care must be exercised since the hydrogen permeation rate through the quartz is about one tenth that of helium. The impurities in the helium, which diffused through the leak were less than 1 ppm except for hydrogen which was higher (10 ppm).

Works[1330] described some experiments on the constancy of quartz helium leaks.

Sundheim[1209b] constructed an "absolute leak" consisting of a radioactive source emitting alpha particles. As these particles capture electrons, they

TABLE 6.32. RATE OF DIFFUSION OF GASES

Gas	Wall	Temp. (°C)	D	Reference
Helium	Quartz	927	3.2×10^{-6}	Jossem[647]
	Glass	700	2.1×10^{-8}	Norton[944, 947]
	Perbunan	25	3.8×10^{-9}	Norton[944, 947]
	Neoprene	25	4.5×10^{-9}	Norton[944, 947]
	Natural rubber	25	2.3×10^{-8}	Norton[944, 947]
Neon	Quartz	927	1.2×10^{-7}	Jossem[647]
	Glass	700	4.2×10^{-10}	Norton[944], Amerongen[27]
Xenon	Perbunan	25	8.0×10^{-10}	Norton[944, 947]
	Neoprene	25	1.0×10^{-8}	
	Natural rubber	25	4.3×10^{-8}	
Argon	Glass	700	$<1.0\times10^{-15}$	Norton[944]
Hydrogen	Nickel	927	1.4×10^{-3}	Jossem[647], Gorman[464]
	Platinum	927	1.2×10^{-4}	
	Molybdenum	927	6.3×10^{-5}	
	Copper	927	7.4×10^{-5}	
	Aluminium	927	2.7×10^{-4}	
	Iron	927	8.5×10^{-4}	
	Palladium	927	1.9×10^{-1}	Jossem[647], Davis[265]
	Glass	700	2.1×10^{-9}	Norton[944]
Oxygen	Silver	927	9.7×10^{-5}	Jossem[647]
	Glass	700	$<1.0\times10^{-15}$	Norton[944]
Nitrogen	Molybdenum	927	1.6×10^{-8}	Jossem[647]
	Iron	927	7.0×10^{-6}	
	Glass	400	6.0×10^{-14}	
Carbon monoxide	Iron	927	1.7×10^{-5}	Jossem[647]

become helium atoms. From 1 millicurie of polonium, 1.2×10^{-12} cm^3 (STP)/sec of helium is produced.

Turner[1241] used a small Pyrex glass chemistry flask (1, Fig. 6.127b), coated with about 3 mm thickness of wax (2) and then embedded the flask in epoxy resin (3) in such a way that the volume occupied by the wax layer connects with an external vacuum fitting (4). After the epoxy coating is cured, the wax is melted out. In use, the flask is filled with helium and closed with a stopper (5). The helium diffuses slowly through the glass out to the fitting (4).

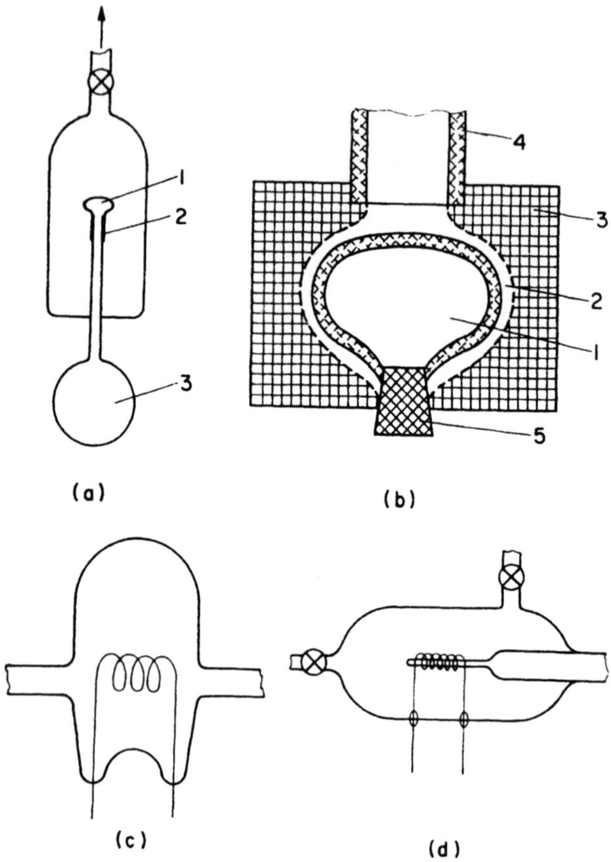

FIG. 6.127 Diffusion leaks: (a) with graded glass seal (Jenks[633]); (b) with resin seal (Turner[1241]); (c) with nickel spiral (Landecker[731]); (d) with palladium or platinum tube (Jossem[647])

For the diffusion of hydrogen, nickel or palladium tubes are used, and silver is the best material for the diffusion of oxygen (see Table 6.32).

Harrison[513] described a leak consisting of a nickel tube (0.03 in. bore, 0.005 in. wall thickness, and 60 in. total length) wound on a helically grooved silica tube (3/4 in. o.d. and 3 in. long). The assembly is supported in a glass envelope, with the ends of the nickel brought out through two Kovar tubes terminated in suitable unions. An electric current heats the nickel directly. The hydrogen diffuses *from the inside* of the nickel tube *to the outside*. It is not advisable to allow the hydrogen to diffuse from the outside into the nickel tube, since in this case a much larger power input is required as a result of heat conduction to the envelope. Harrison[513] claimed that the nickel tube was unaffected by prolonged heating up to 600 °C, and that the diffusion rate was constant if a small continuous flow of hydrogen was maintained through the tube, to avoid the accumulation of impurities. Without this steady flushing, the rate of diffusion falls by some 50 per cent in one hour. Weisbeck[1292a] used a nickel tube (1.5 m long, 1.0 mm o.d. 0.1 mm wall thickness) wound in a spiral (of about 30 mm diameter), placed in a glass container, and heated by direct electric heating. The hydrogen diffuses from the inside of the tube to the outside (glass container).

Landecker[731] used a similar arrangement (Fig. 6.127c) for the diffusion of deuterium. The most efficient diffusion leaks for hydrogen are those using palladium (Juenker[648], Jossem[647]). Figure 6.127d shows the diffusion leak described by Jossem[647]. Palladium presents the difficulty that it hardens as a result of repeated heating and cooling in hydrogen. To overcome this difficulty, in some palladium leaks the adopted procedure is to evacuate the leak, and to admit the hydrogen only after it reaches the service temperature.

Instead of pure palladium (Connor[228a]), palladium alloys were used for diffusion leaks. It has been found that an alloy of 23 per cent silver and 77 per cent palladium gives a maximum rate of diffusion of hydrogen. These leaks* are operated at 350–400 °C.

Silver tubes are used for the diffusion of oxygen. Taylor[1216] used a silver tube of 3 mm diameter and 0.12 mm wall thickness, and Whetten[1305] described a diffusion leak for oxygen with a silver tube 6 in. long, 0.125 in. i.d. and 0.010 in. wall thickness, which was operated at 600–700 °C.

61.49 Pulsed leaks. In order to control the throughput of a leak, a *pulsed* system can be used, in which the gas is admitted in pulses to the vacuum chamber. Pulsed leaks may use the bubbling of the gas through liquids, the repeated opening of an orifice, the rotation of a shaft with pits, or the vibration of a needle, nose or disc.

Small, measured quantities of gas may be admitted into a vacuum system by bubbling the gas through a liquid column. The liquid used is generally mercury (see e.g. Mignolet[857], Glocker[453], Littmann[774a], Osborne[957], Wyllie[1337]), but other liquids may also be used (Avera[57]).

Bachman[66] described a pulsed leak (Fig. 6.128a) consisting of a cam (1), which moves the spring on which the rubber cap (2) is carried, thus opening and closing the tube (3). For normal cam speeds the throughput is not considerably influenced by the rotational speed. For a given tube size (3), the time

* Available from Johnson Matthey & Co. Ltd., Hatton Garden, London, E.C.l.

ratio for the open and closed positions determines the throughput and this is set by the shape and position of the cam. An alternative construction of a pulsed leak (Fig. 6.128b) consists of a metal shaft (1), having pits (2) drilled or bored into its cylindrical surface. The tube leading to the vacuum system is inserted into the rubber collar (3) which is shaped to fit the shaft. The collar is held against the shaft with enough pressure to give a vacuum-tight seal between the pits, and the seal is lubricated with vacuum oil or grease. As the shaft is rotated, each pit delivers its volume of gas to the evacuated tube and moves on to be recharged. The throughput is determined by the volume of the pits and the pulse frequency.

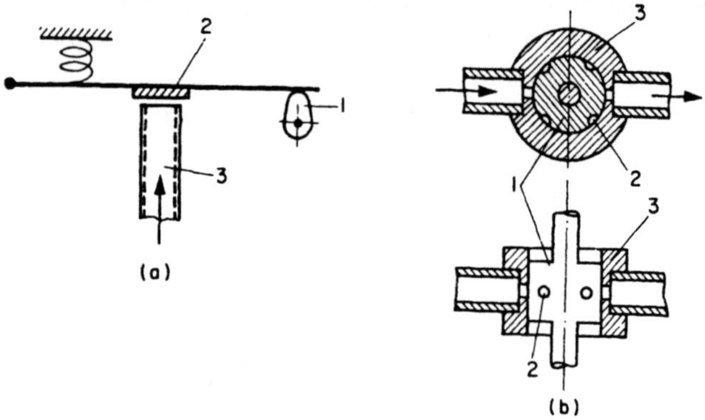

FIG. 6.128 Pulsed leaks: (a) with cam operated cap; (b) using a shaft with pits. After Bachman[66] *(Courtesy of The American Institute of Physics)*

Stinnett[1191] used a needle valve (Fig. 6.129a) and vibrated the needle axially by an alternating current operated magnet coil. This alternatively opens and closes the valve at a frequency double that of the current supplied. By limiting the length of the stroke of the needle with the adjusting screw, the throughput can be controlled accurately.

In order to achieve very short pulses (1 cm^3 of gas per 100 μsec), Marshall[810] used a valve (Fig. 6.129b) which opens as a result of a longitudinal sound wave created in an anvil at one end of a steel rod and transmitted to a moving valve member at the other end.

Gorovitz[465] was able to introduce gas quantities as small as 0.009 cm^3 (STP) using the system shown in Fig. 6.129c. This device is operated by the discharge of a capacitor (10 kV, 3 μf) through the coil, which creates a magnetic field. This field induces eddy currents in the metallic disc (4), and a mutual repulsion results, forcing the disc away from its O-ring seat and admitting a quantity of gas.

A very fast acting (70 μsec) leak valve is described by Lowder[786a].

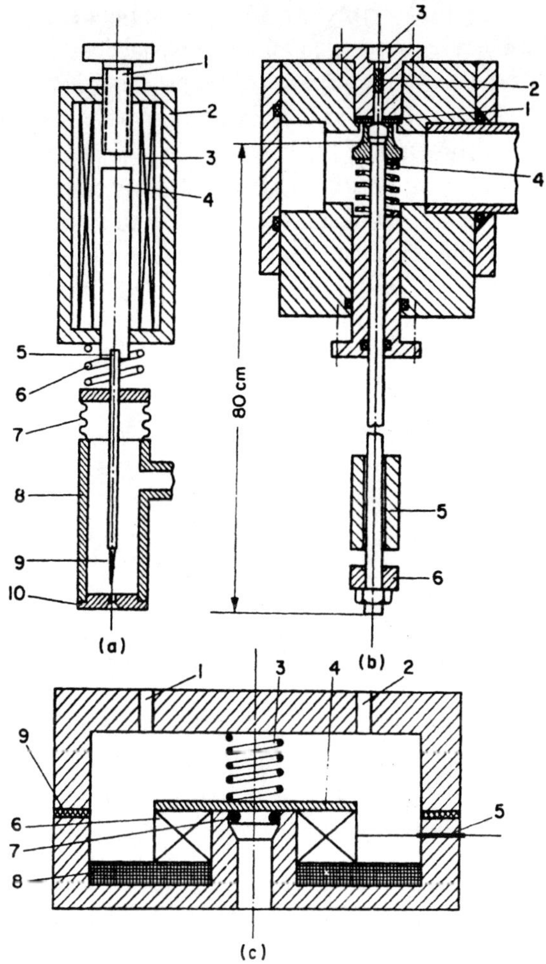

FIG. 6.129 Pulsed leaks: (a) with needle vibrated by a magnetic coil; (1) adjusting screw; (2) steel frame; (3) magnet coil; (4) steel core; (5) loose joint; (6) bronze spring; (7) bellows; (8) brass tube; (9) hardened steel needle; (10) brass needle seat (after Stinnett[1191]); (b) valve actuated by a sound wave; (1) Teflon gasket; (2) socket drive set screw; (3) gas inlet; (4) spring; (5) hammer; (6) anvil block (after Marshall[810]); (c) valve with a disc vibrated by magnetic coil; (1) gas inlet; (2) pump outlet; (3) spring; (4) valve disc; (5) lead-in wire; (6) valve acting coil; (7) Teflon seat; (8) potting compound; (9) rubber gasket (after Gorovitz[463]) *(Courtesy of The American Institute of Physics)*

61.5 Techniques for opening sealed gas container

In order to fill a container with a given pure gas (or vapour) at a particular pressure, or to measure the quantity, pressure, or purity of the gas contained in a vacuum device (e.g. electric lamp, electron tube), it is required to transfer the pure gases or vapours from the sealed container into an evacuated vessel. The transfer of these gases (or vapours) maintaining their original purity requires the opening under vacuum of the sealed container.

The techniques used for this purpose consist in enclosing the whole container (flask, ampoule) or the side to be opened, in a space which is evacuated before the container is opened. If the container has special provisions to be opened under vacuum (break-off seals) only the outlet of this seal is evacuated, but if the container has no such opening devices, it must be enclosed entirely in a vacuum chamber and opened by breaking it under vacuum. The opening (breaking) may be done by striking (Section 61.51), bending (Section 61.52), punching (Section 61.53) or squeezing (Section 61.54).

61.51 Opening by striking. Rare or very pure gases are supplied in glass containers, sealed with "break-off" seals (2, Fig. 6.130a). The technique in opening such a container is to introduce a small* cylindrical iron part (3) (or

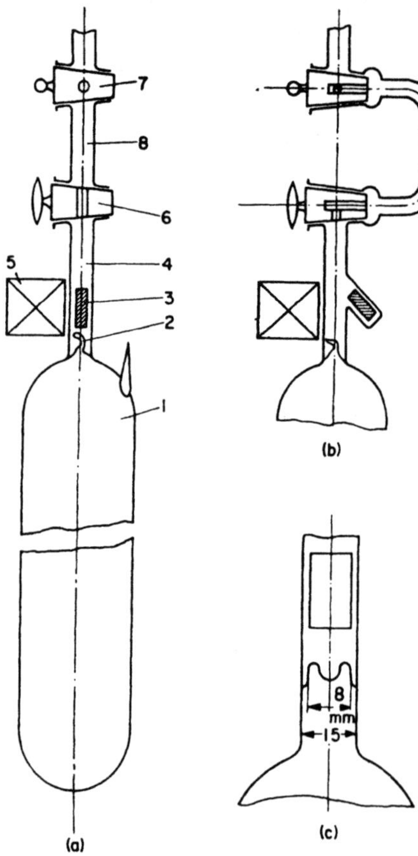

FIG. 6.130 Devices for the opening of break-off seals by axially moved iron core: (a) with iron core in connecting tube; (b) iron core in side arm; (c) modified break-off seal (Wertz[1296])

* About 8 mm diameter and 15 mm length.

preferably an iron core, enclosedin glass) into the connecting tube (4), and to seal the tube (4) to a glass pipe, which is provided with two stopcocks (6, 7). After evacuating the spaces (8) and (4), both stopcocks are closed. The iron core (3) is lifted by an external magnet (5) and left to fall (from about 2 cm) on the break-off seal. This breaks the thin end of the seal, and the gas expands into the space (4).

In order to prevent untimely breaking of the seal, it is advisable to place the iron core in a side arm (Fig. 6.130b), and to bring it in the main tube only after this space has been evacuated, and the stopcocks are closed. For the double stopcock system required, it is recommended that high vacuum stopcocks with closed shell (Section 61.21) be used.

Wertz[1296] criticized the *septum* of the usual break-off seal (Fig. 6.130a, b), objecting that after breaking such a seal, the remaining capillary opening throttles the gas flow. He also maintained that the shape of the usual break-off seal prevents effective cleaning of the surface before opening. In order to avoid these drawbacks Hertz[1296] proposed that the break-off seal be constructed as shown in Fig. 6.130c.

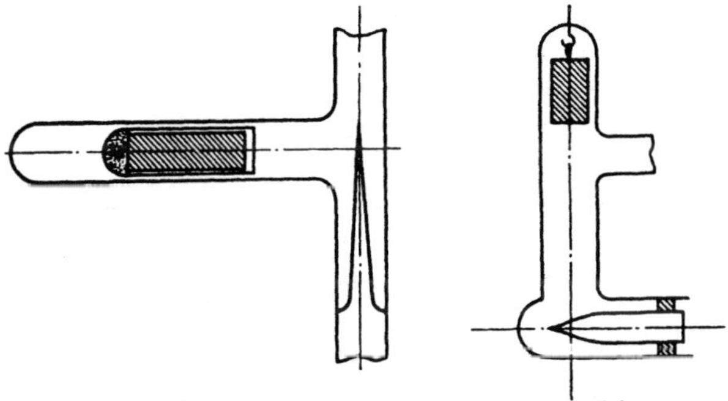

FIG. 6.131 Devices for the opening of containers by radial stroke: (a) break-off seal; (b) enclosed tube or ampoule

Instead of opening the container by an axial motion of the core, in some cases it is more advantageous to use iron cores radially striking the break-off seal (Fig. 6.131). In this construction, a pointed break-off seal (Fig. 6.131a) is required, having the point in front of a side tube in which the iron core moves. Morrison[191a] described a device (Fig. 6.131b) in which the iron breaker is magnetically lifted from its supporting hook, lowered to an appropriate height above the tube to be broken and released. It falls freely and breaks the tube (see also Fig. 6.136).

Gas containers (Fig. 6.130) may be also opened without using magnets by disconnecting the container with the two stopcocks from the vacuum system and moving the iron core (or just a glass rod in this case) by tilting or shaking the assembly. In a technique used to fill metal thyratrons (Espe[354]) with metered quantities of gas (Kr–He mixture) a small glass container, with the required quantity of gas, is placed in a side container of the thyratron. After the thyratron has been evacuated (together with the side container) and sealed-off, the glass container is broken by shaking the assembly. The gas expands in the thyratron and then the side container is separated (Section 26.31) from the thyratron.

61.52 Opening by bending. The break-off part of a gas container can be opened in vacuum by bending it until it breaks. The break-off should be constructed as a long (pointed) tube with thin walls, or if the wall is thick, it should be scratched at the place where it has to be cracked.

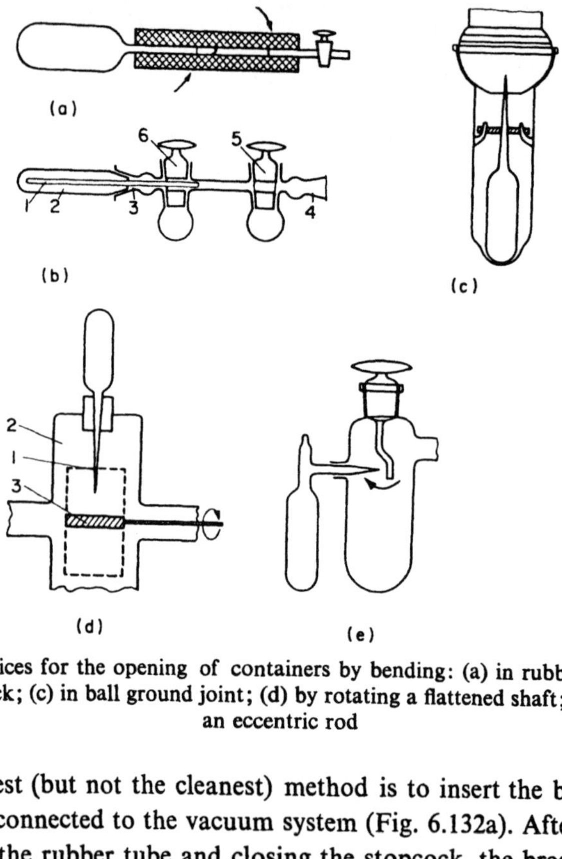

FIG. 6.132 Devices for the opening of containers by bending: (a) in rubber tube; (b) in tapered stopcock; (c) in ball ground joint; (d) by rotating a flattened shaft; (e) by rotating an eccentric rod

The simplest (but not the cleanest) method is to insert the break-off in a rubber tube connected to the vacuum system (Fig. 6.132a). After evacuating the space in the rubber tube and closing the stopcock, the break-off is broken by bending it together with the rubber tube.

A sample tube (1, Fig. 6.132b) of the appropriate dimensions can be broken by inserting it into a storage vessel (2), connected by a conical joint (3) to a double stopcock system. After evacuating the system connected by (4) to a vacuum pump, and closing stopcock (5), the pointed break-off, which protrudes into the hole of the stopcock plug (6), is broken by rotating the plug (6) (von Ubisch[1246]). For the same purpose, instead of a stopcock, a spherical ground joint (Fig. 6.132c), may also be used (Snow[1161]).

Highhouse[560] described a device where the scratched ampoule (1, Fig. 6.132d), introduced in a vacuum lock (2), is broken by rotating the flat piece (3). An alternative technique is to use an eccentric rod, fixed on a tapered ground joint (Fig. 6.132e).

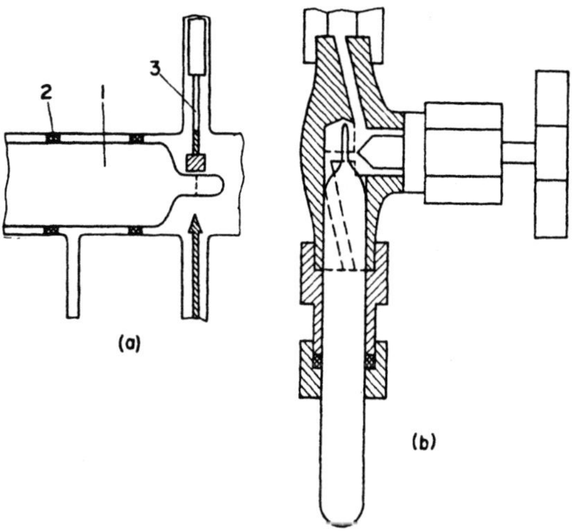

FIG. 6.133 Devices for the opening of containers: (a) by using a pushing nose (Charlton[206]); (b) by using a converted needle valve (Mahler[794a])

Charlton[206] opened the gas container (1) under vacuum (electron tube) by sealing it with a double O-ring seal (2) (Fig. 6.133a) and breaking the exhaust tube with the arm (3) pushed towards it.

Mahler[794a] converted a needle valve to an opener (Fig. 6.133b) by removing the valve seat (removed portion indicated by dotted lines).

61.53 Opening by punching. Schacher[1101] described a technique for opening small capsules under vacuum by punching them (see Fig. 5.7b). Charlton[206] opened sealed electron devices in vacuum using the device shown on Fig. 6.134a, in which the electron tube (5) receives a very powerful, sharp stroke from the sharp nose of the rod (2), which moves upwards when the solenoid (4) is energized. The upper rod (1), which is attached to a shaped anvil, has the solenoid (3) positioned so that when it is energized (simultaneously with solenoid 4), the rod (1) experiences a downward pull and therefore acts as a shock absorbant for the impact of the sharp nose.

An alternative technique consists of enclosing the material (3, Fig. 6.134b) in an ampoule, together with an iron core (2), packed in a soft material (1), and to move this ampoule (by magnet 5) towards a pointed part (6). By this technique the thinner wall (4) of the ampoule is punched.

Thorness[1223a] described a system of opening (remote handling) a capsule by cutting it with a mechanism similar to a pipe cutter.

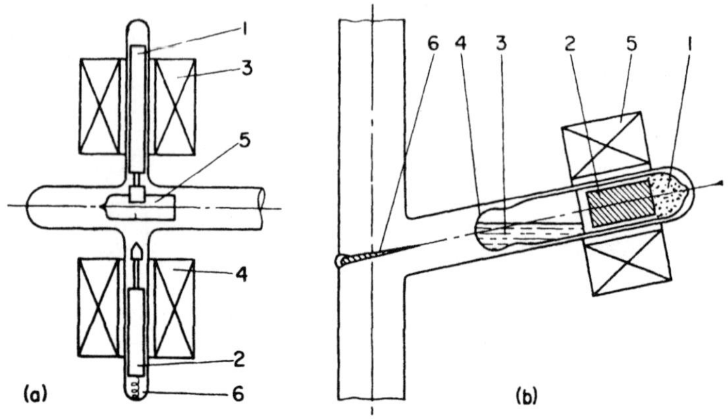

FIG. 6.134 Devices for the opening of containers by punching: (a) by moving the punch; (1, 2) mild steel rods; (3, 4) solenoids; (5) container to be opened; (6) spring (after Charlton[206]); (b) by moving the ampoule

61.54 Opening by squeezing. A simple method of opening a gas filled ampoule under vacuum is to introduce it into a rubber tubing (2, Fig. 6.135), to plug (3) the tubing, and to evacuate it through the stopcock (4). After closing the stopcock (4) the ampoule is squeezed through the rubber by the clamp (5).

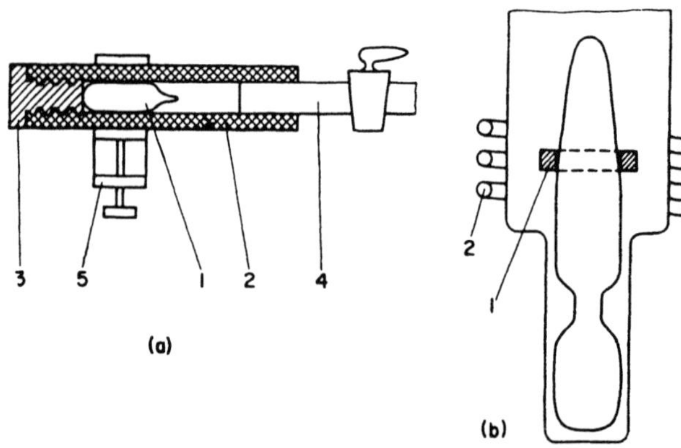

FIG. 6.135 Devices for the opening of containers by squeezing: (a) in rubber tube; (b) with r. f. heated collar (after Blake[129b])

Norton[945] described a technique, in which the glass ampoule is placed into a thin-walled metal (copper) tube closed at its end. The other end of the tube is connected to a vacuum system, the metal tube is evacuated, and the ampoule is broken by squeezing the metal tube (in a vice).

Blake[129b] opened the gas containing ampoules by fitting a metal (Nimoniac alloy) collar (1, Fig. 6.135b, with 35 mm o.d., 24 mm i.d. and 6 mm thick) over the conically shaped, upper part of the ampoule. The collar is heated by a radio frequency coil (2) to 800 °C so that it slips tightly around the tapered end of the ampoule. On cooling, the collar contracts and fractures the glass. The system has the advantage that it can be handled from a distance.

6.2 TRANSFER OF LIQUIDS THROUGH SEALS

The transfer of liquids into evacuated spaces is a rarely used operation in vacuum technique, nevertheless there are some cases where such a technique is required.

62.1 Continuous transfer

Liquids can be transferred continuously into evacuated vessels, using stopcocks or valves placed at the bottom of a liquid column connected to the vessel. The liquid should fill up all the space above the stopcock or valve, while the opposite side of the stopcock or valve is evacuated through the vessel. When the stopcock or valve is opened, the liquid flows into the vacuum vessel due to the pressure difference existing between the two ends of the column. This kind of transfer does not allow a high purity to be obtained due to the fact that a considerable amount of gas is also transferred together with the liquid into the vacuum vessel. The transfer should occur slowly in order to allow control of the flow and to permanently keep some liquid over the stopcock or valve. In order to control the flow Perrine[962a] incorporated a needle valve in the plug of the stopcock. The stainless steel needle was placed axially in the Teflon plug of the stopcock, and permitted to change the throughput in the bore of the plug. Biddulph[123] introduced highly reactive liquids into the evacuated space, using an all-metal valve, which closed with a ball bearing pressed against the seat, and was sealed with a copper diaphragm.

62.2 Metered transfer

Metered quantities of liquids may be transferred into an evacuated chamber by:

(1) Breaking (opening) in vacuum an ampoule containing the required quantity of liquid.

(2) Using a feeder.

62.21. The technique which should be used in order to transfer a liquid by *breaking the ampoule* under vacuum depends on the shape of the ampoule or container and on the requirements concerning the purity. Figure 6.136 shows some techniques which may be used for this purpose. Ellis[337a] described a technique for the vacuum-filling of glass capillaries.

If the ampoule is a long capillary tube (Fig. 6.136a), it is placed so as to project into the hole of the plug of a tapered ground joint. After sealing the system, and evacuating it, the tip of the capillary sample tube is broken by rotating the plug. The system shown in Fig. 6.136a is sealed with mercury (see also Section 37.2); if the purity conditions of the process permit, greased ground joints or gasket joints can also be used. A ball-ground joint for this purpose is shown by Fig. 6.132c.

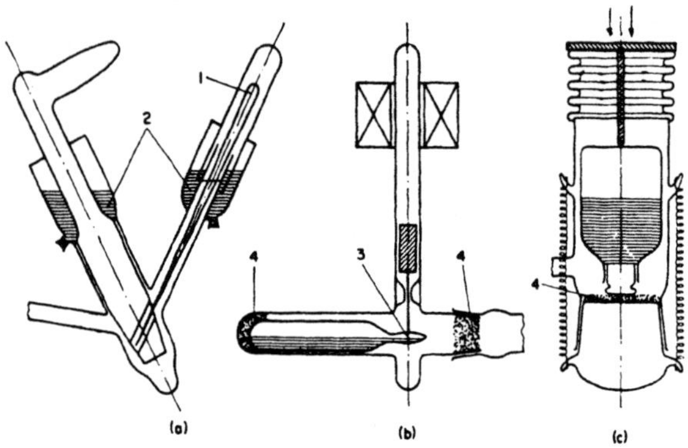

FIG. 6.136 Transfer of liquids in vacuum by breaking the containers: (a) sample tube broken in a stopcock; (1) sample tube; (2) mercury; (b) ampoule, broken by a falling weight; (3) scratch; (4) glass wool; (c) bottle broken by pointed breaker after Beynon[120]

For heavier ampoules or containers, specially constructed breakers must be used (Beynon[120]). The breaker shown in Fig. 6.136b is magnetically operated. The weight is raised by an external magnet, and in falling its needle breaks the ampoule. In order to facilitate the breaking of the ampoule, it is advisable to scratch it before inserting it into the system.

Figure 6.136c shows a system for breaking a glass bottle under vacuum. The supporting structure of such a system should be strong enough to withstand the shock produced when the breaker is hit with a hammer.

The ampoule may be broken also by high frequency heating (Fig. 6.135b). In this case it should be provided with an external "short circuit ring", which enables the ampoule to crack using the energy of a high frequency outside coil.

62.22. The transfer of metered quantities of liquids into evacuated spaces may be repeated, or made continuously, if *feeding* systems are used.

A simple device (Fig. 6.137) for the accurate metering of small quantities of mercury was described by Risz[1066]. Here the piston (1) is displaced by the micrometric screw (2), which pushes the mercury in the capillary pipe (3), forming drops, which can be driven to the place where they are required. Risz[1066] places these drops in the small cavities (1, Fig. 6.138a), provided on a long plug (2), protruding in the vacuum line from a conical ground joint. After the containers (4) to receive the mercury (lamps) were evacuated, the plug (2) is rotated with 180°, and by this the metered quantities of mercury are poured through the connecting pipes (5) into the containers (4).

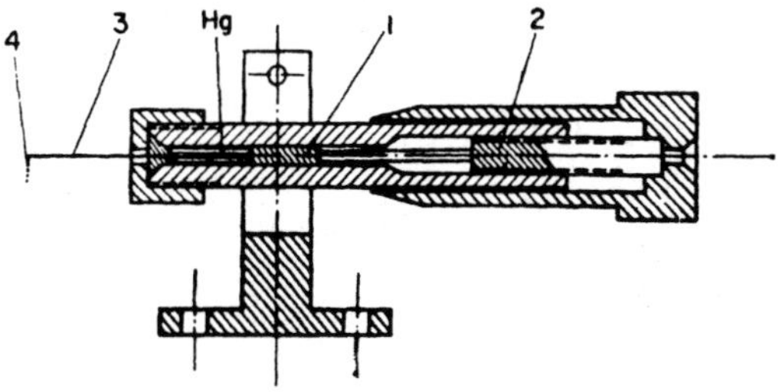

FIG. 6.137 Metering device for small quantities of mercury: (1) capillary; (2) plunger; (3) hypodermic needle; (4) mercury drops (after Risz[1066])

Berger[114] described a similar transfer system (Fig. 6.138b), but which was made to work more continuously. Here the shaft (1) sealed with a packing (2) (Section 51.7) has a small cavity filled up by the liquid (3). When the shaft is rotated the cavity fills up with liquid, and pours it into the delivery pipe (4). After a complete rotation the cavity is up again, the liquid flows into it and the device can repeat the operation.

Zovac[1356] described a transfer device (Fig. 6.138c) consisting of a shaft (1), having a pit (2). By the sliding of the shaft, the liquid contained in the pit is transferred through the seal into the evacuated space, and, hereby rotating the shaft, the liquid is poured. The device also introduces, together with the liquid, small quantities of air, if the outside of the shaft is at atmospheric pressure. To avoid this a vacuum lock (Fig. 6.139) should be added to the seal.

The author constructed and used, for the feeding of metered quantities of very pure mercury, the device shown in Fig. 6.138d. This device consists of a container (1) with a tapered bottom, sealed to a bonnet (2), holding a guiding rod (3). This chamber contains an annular container (4) filled with mercury. A pipe (5) can slide axially between the container (4) and the rod (3). This pipe has an oblique hole (6), which in the raised position of the pipe (5) meets the hole provided near the bottom of the container (4). Since the whole chamber is evacuated with the pipe (5) in its lowered position, when this pipe is raised with the aid of the magnetic motion system (7) (Section 5.2), the mercury fills up the hole (6). When in the next sequence the pipe (5) is lowered, the hole (6) comes below the end of the rod (3), and the metered quantity of

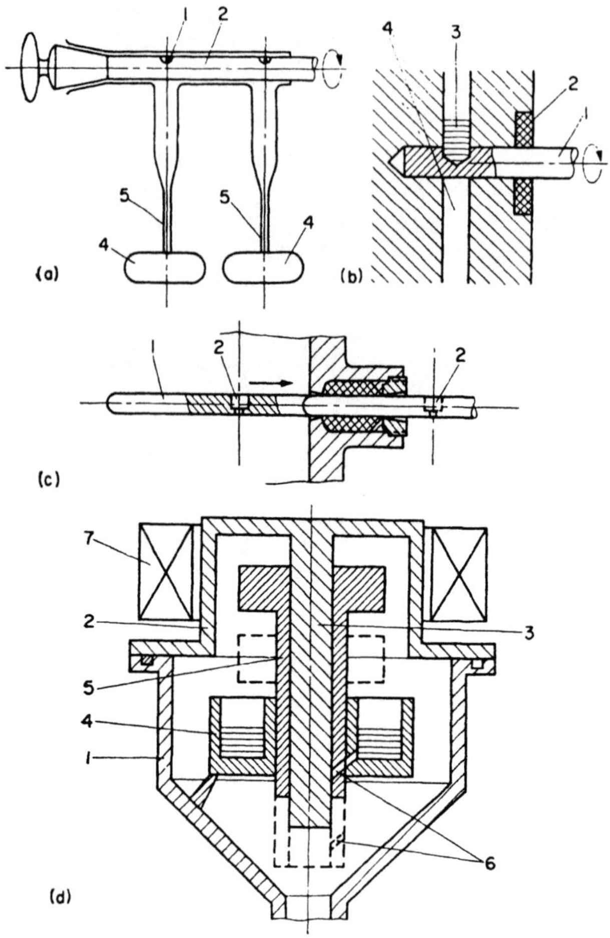

FIG. 6.138 Metering devices with pits: (a) rotation of a glass plug with pits; (b) rotation of a shaft (1) with pits (2); (c) sliding shaft with a pit; (d) electromagnetically operated mercury metering device

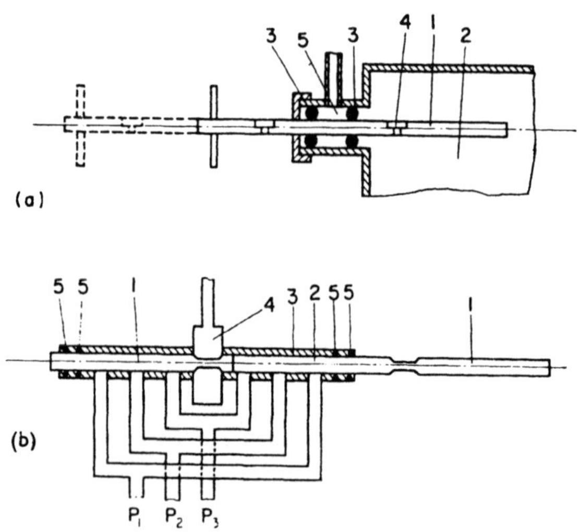

FIG. 6.139 Vacuum locks, using a sliding rod: (a) with simple pumping. After Delsasso[273] *(Courtesy of The American Institute of Physics)*; (b) with cascade pumping (after Stevens[1188])

mercury drops into the vacuum vessel below. The device allows both continuous degassing of the mercury, since the pool (4) is under vacuum, and its continuous transfer in metered quantities (depending on the sizes of hole 6). The device was connected to a control system, and its cycles synchronized with an exhaust machine for mercury discharge lamps.

6.3 TRANSFER OF SOLIDS THROUGH SEALS

In some vacuum applications, it is required to transfer solid parts as: specimens (Garrod[433], Leisegang[756]); photo plates (Brunnee[178], Thomson[1223], Ardenne[47], Zovac[1356]); cathodes (Kofoid[704a]); or windows (Saxon[1099]) into the evacuated chamber; or to withdraw them to the atmosphere without breaking the vacuum in the chamber. To achieve the transfer of such parts, *vacuum locks* are used (Section 63.1). When materials must be transferred from vacuum to vacuum, they are closed in evacuated ampoules, and these are transferred. Sodium or potassium, which cannot be handled in air, may be introduced into evacuated glass vessels by *electrolytic* transfer through the wall of the vessel (Section 63.2).

63.1 Vacuum locks

A vacuum lock consists of a space (chamber), adjacent to the vacuum system, and which can be opened either to the atmosphere or to the vacuum system. During the transfer of an object into the vacuum chamber, the vacuum lock is operated in steps as follows:

(1) The lock is opened to the atmosphere (being sealed towards the vacuum chamber) and the object is introduced into the lock.

(2) The port to the atmosphere is closed and the vacuum lock is evacuated (through its own pumping line).

(3) The port to the vacuum chamber is opened and the object is transferred from the lock into the vacuum chamber.

(4) The port to the vacuum chamber is closed. To withdraw an object from the vacuum chamber the lock is evacuated, the object is transferred to the lock, and after closing the port to the chamber, the object is withdrawn to the atmosphere.

Vacuum locks may be based on:

(1) sliding rods (Section 63.11);

(2) rotating plugs (Section 63.12);

(3) chamber with double port (Section 63.13).

A review of vacuum locks is given by Brunnee[178].

63.11 Vacuum locks using sliding rods. These vacuum locks consist of a rod (1, Fig. 6.139a) which may be pushed into the vacuum chamber (2) through a seal (3) (Section 51.7). The rod (1) carries the objects to be transferred in a pit (or pits) (4) of appropriate shape. The construction is similar to that of the device on Fig. 6.138c, but it includes also an intermediate chamber which may be evacuated.

The seal of the rod may be effected with elastomer gaskets (Delsasso[273], Leisegang[756]), O-rings (Garrod[433]), Wilson seals (Garrod[433], Schulz[1111]).

Stevens[1188] and Echo[319a] used a vacuum lock (Fig. 6.139b), consisting of a long rod made of three sections; the two end sections (1) are carriages and the middle section (2) is a blank rod. The sample is mounted inside a carriage, and the rod is moved through the channel (3). Three pumping lines (P1, P2, P3) are connected to both sides of the channel, assuring the pressure gradient (guard vacuum) from the atmosphere to the vacuum chamber (4). The clearance* between rod (2) and channel (3) was 0.05–0.1 mm. At the two ends of the rod, double O-ring seals (5) were provided, the two O-rings of a seal being spaced so that at all times at least one O-ring provides a seal.

The same system of *cascade pumping* was proposed for a vacuum lock having a continuously travelling tungsten strip instead of a rod (Brunnee[178]).

Colombani[223a] described a vacuum lock which was sealed by elastomer gaskets, and was evacuated in cascade by a rotary and a diffusion pump.

63.12 Vacuum locks using rotating plugs. In these arrangements, the plug receives the object from one side, and by its rotation allows the transfer of the object on the other side. The *path* of the object during the transfer may be *radial, axial* or *tangential* (peripheric) with respect to the plug.

In the technique shown on Fig. 6.140a, the object is pushed into the plug (4) through the opening (1). By the clockwise rotation of the plug, its opening reaches the outlet (2) through which the space in the pit is evacuated. By further rotation of the plug, the object is transferred into the chamber (3).

In an alternative construction (Fig. 6.140b) the plug (1) may be withdrawn lrom the shell, since a second rotating shell (3) is included in the vacuum fock. After inserting the object into the pit (2), the plug is placed again in the shell. By the rotation of plug (1) and shell (3), the pit is brought to (4) and evacuated. By further rotation of plug and shell the object is transferred into the vacuum chamber (5).

Figure 6.140c shows an arrangement for the transfer of photo-plates into an evacuated chamber. The plug (1) is first closed and the plate (2) is placed in its chamber. This is evacuated (outlet is not shown on the Figure), the plug

* Boerboom[142a] constructed the same type of vacuum lock, with a clearance of only 0.002 mm; this was possible by hard-chromium plating the rod, polishing the sliding surface to a finish of 2 micro-inches, and lubricating them with molybdenum-disulphide.

(1) is rotated so that the slit is in the horizontal position, and the plate is transferred into the vacuum chamber using the mechanical device (3).

For small samples, stopcocks or plug valves may be used as vacuum locks (Roberts[1068], Estle[360]). The stopcock or valve is first closed, the object is

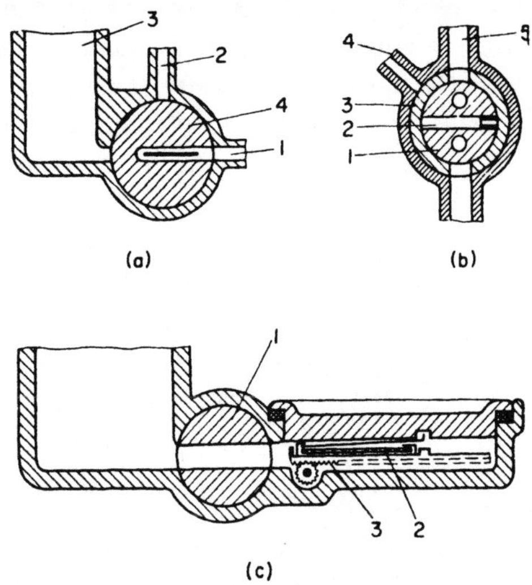

FIG. 6.140 Vacuum locks, using plugs. Object travels radially. (a) Plug with pit; (b) plug with pit and double shell; (c) straight-through plug and evacuation chamber

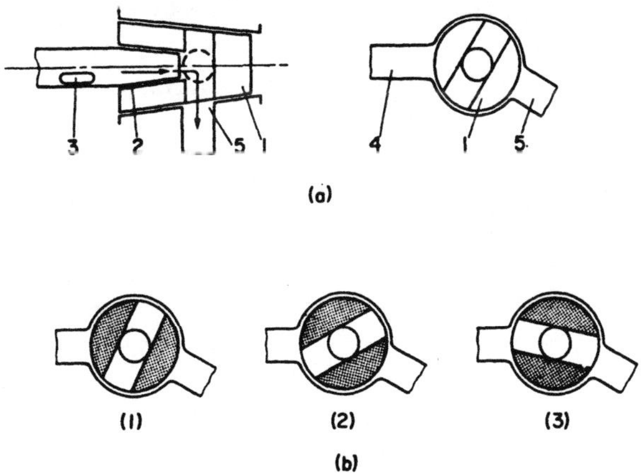

FIG. 6.141 Vacuum lock with plug. Object introduced axially and transferred radially (after Feichtinger[371a])

introduced in a space before the valve which is subsequently evacuated. The stopcock is rotated in open position, and through its bore the sample is transferred into the vacuum chamber.

A vacuum lock arrangement (Fleichtinger[371a]), where the object enters *axially* into the plug and leaves it radially, is shown in Fig. 6.141. The plug

(1, Fig. 6.141a) has a T-bore, and on its axial side the bore is finished with a ground conical joint (2). The shell of the vacuum lock has a connexion (4) to the pumping system and a connexion (5) to the vacuum vessel. With the plug in position (1, Fig. 6.141b), the sample (3) is brought to the system in a container fitting the conical joint (2). By rotating the plug in position (2, Fig. 6.141b), the space containing the sample is evacuated, without breaking the vacuum in the vessel (connected by 5). By further rotation in position (3, Fig. 6.141b), the vacuum vessel and sample container are connected to the pump. In this position the sample is transferred first axially, then radially through the plug into the vacuum vessel (5).

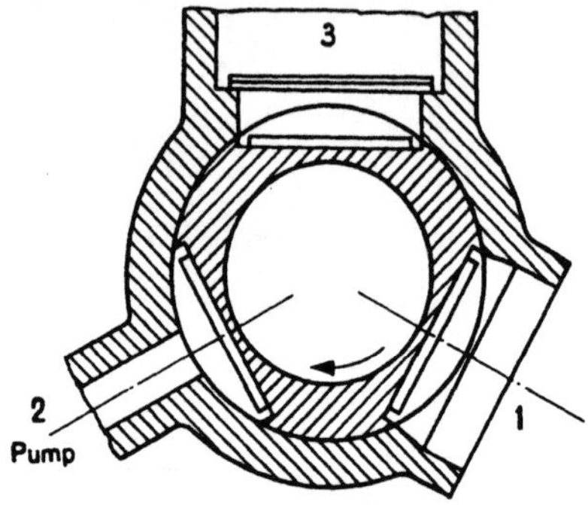

FIG. 6.142 Vacuum lock with plug. Object tangential to the plug

A vacuum lock which transfers the object *tangentially* to the plug is shown in Fig. 6.142. The object (photo-plate) is placed in position 1, transferred by rotation of the plug to the evacuation position (2) and from here into the vacuum space (3).

61.13 Vacuum locks using chambers with double ports. The vacuum lock consists of a space connected on one side to the vacuum chamber through a port, and on the other side through another port to the atmosphere. The lock has also a connexion to a pumping line.

The connecting port is usually a gate valve (Fultz[414]) (Section 61.32h) or a plate valve, but other arrangements have also been used in the construction of the various vacuum locks. Duncan[307] used a cone (nose) closing system in his vacuum lock. In the vacuum lock shown in Fig. 6.143, the object is introduced in the lock by the port (1), and the lock is evacuated through the connexion (2). The bellows sealed (see Sec. 61.33b) plate valve (3) is opened, the object is transferred over the plate of the valve, and by closing the valve, it is pushed into the evacuated chamber (4).

The principle of a double port chamber used as a vacuum lock in electron microscopes is shown in Fig. 6.144. Here the probe (1) is introduced in the

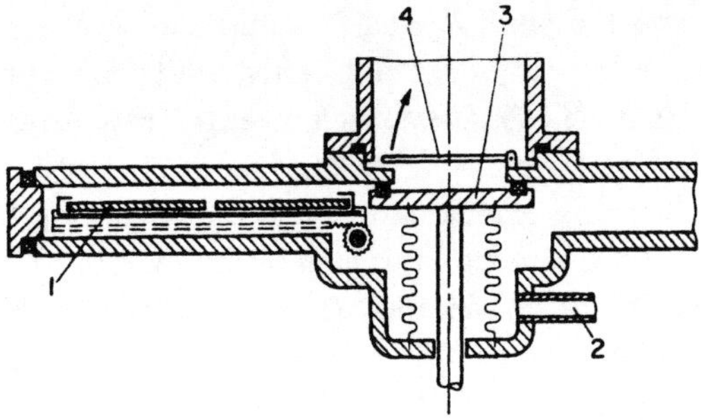

FIG. 6.143 Vacuum lock with bellows-sealed plate valve

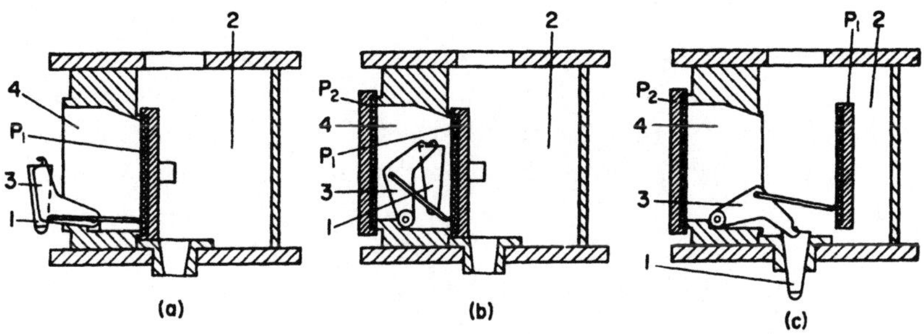

FIG. 6.144 Vacuum lock with double port: (a) the object in its holder; (b) the object in the closed vacuum lock; (c) the object transferred into the vacuum chamber

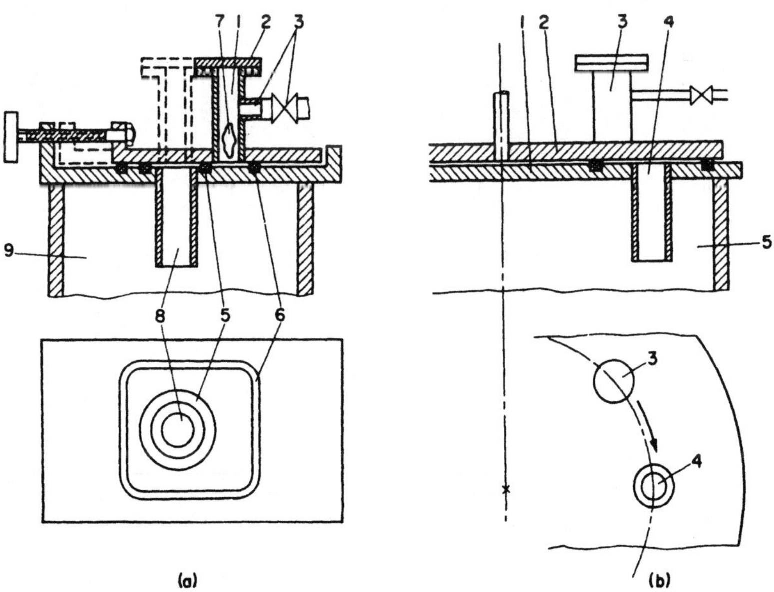

FIG. 6.145 Vacuum lock with moving chamber: (a) translation; (b) rotation

chamber (4) between the ports P_1 and P_2, and then, by opening the port P_1, into the chamber (2). Instead of opening and closing double ports, the airlock may itself move. Figure 6.145 shows such arrangements. In the arrangement shown on Fig. 6.145a (Porteous[1000a]), the airlock (1) receives the object (7) by the end seal (2) and is evacuated through the connexion (3). The airlock may slide on the lapped surface (4), being sealed by the O-rings (5) and (6). When the airlock (1) arrives over the port (8), the object falls into the evacuated chamber (9).

A similar arrangement (Fig. 6.145b) may be constructed using two lapped plates (1) and (2) sliding on each other. In this way the vacuum lock (3) receives the object and after being evacuated moves over the opening (4), where the object falls in the evacuated space (5). The seal of such sliding plates may be assured as in rotating gate valves (Section 61.32b) or with an oil-lubricated, lapped seal (Sections 36.1 and 37.4). Such a vacuum lock is quoted by Brunnee[178], who also described a lock permitting the introduction of an object and the simultaneous withdrawal of another object from the evacuated space.

63.2 Electrolytic transfer through glass walls

Sodium or potassium may be introduced into an evacuated glass bulb, by electrolysis, through the walls of the bulb (Strong[1207], Roth[1083]).

The electrolysis of sodium is done (Fig. 6.146) by immersing the bulb (1) (made of a sodium glass) into a bath (2) made of iron (steel) in which a

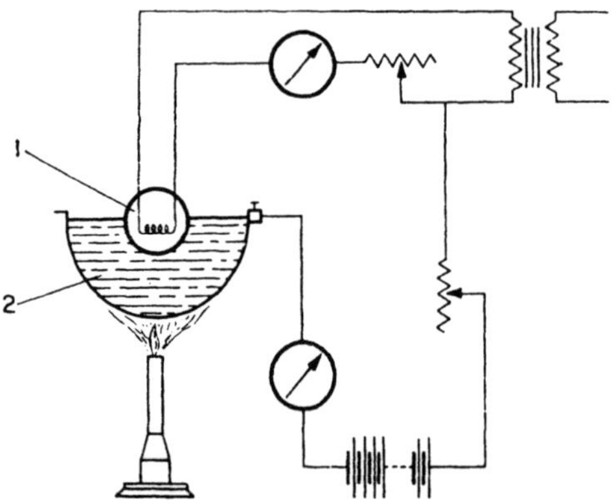

FIG. 6.146 Device for the electrolytic transfer of sodium into evacuated glass chambers

mixture of sodium nitrite ($NaNO_2$) and sodium nitrate ($NaNO_3$) mixed 1 to 1 is heated to about 300 °C (molten). For potassium glass, the same potassium salts are used. The evacuated bulb is filled with an inert gas (e.g. argon at 1–3 torr). The voltage is applied between the iron bath (anode) and an electrode of the bulb (see Fig. 6.146), and a glow discharge is obtained in the bulb. On the inside surface of the glass wall sodium (potassium) is deposited. The process should use current densities of max. 0.5 mA/cm^2 of glass area immersed in the molten mixture of the bath.

If there is a cold metal, in the electrode inside the bulb, voltages of 500–1000 V are necessary for the process, but if an incandescent tungsten filament is used (Fig. 6.146) 200–300 V is usually enough.

CHAPTER 7

SEALS USED IN TRANSMITTING RADIATION

7.1 WINDOWS

71.1 Selection according to the radiation

Radiation very often has to be transmitted from the surrounding space into the vacuum vessel, or from the vacuum vessel outside. The usual cases are: the lighting of the inside of a vacuum vessel from a source placed outside it or the transmission of light, ultraviolet, infrared or other radiation from the vacuum vessel where it is produced, to the surrounding space.

In order to meet this task, the whole vacuum chamber or a part of it (window) should be made of a material with good transmission for the given radiation. The requirements for such chambers or windows are:

(1) They should be able to withstand the pressure difference (Section 21.1).

(2) They should seal the evacuated space, without allowing real or virtual leaks (Section 1.3).

(3) They should transmit the radiation at the intensity and distribution as required.

(4) They should not transmit other (undesired) radiations.

(5) They should have good resistance to radiation or should be protected from damaging radiation, or other causes leading to changes in their transmission characteristics (Vaughan[1261], Feathes[371]).

These chambers and windows can be grouped according to the kind of the radiation which they should transmit: light (Section 71.11), ultraviolet (Section 71.12), infrared (Section 71.13), alpha particles (Section 71.14), beta particles or electrons (Section 71.15), gamma or X rays (Section 71.16), neutrons (Section 71.17).

71.11 Windows for light. Glass is the material usually used in most of the applications where light should be transmitted. In vacuum applications either the chamber itself is made of glass or a glass window is applied to the metal chamber.

Regular non-coloured glass transmits light with a loss of 6–20 per cent when the thickness is 1–2 mm, 10–30 per cent for 2–4 mm thickness and 20–40 per cent for 4–8 mm thickness. If optical glasses are used, the same amount of losses (reflection and absorption) are from much thicker glasses (about 20 times). The transmission through glasses decreases if a weathering process (Sections 21.3 and 23.21) occurs. Weathered glasses may have losses up to 90 per cent (Stockdale[1193]).

For some vacuum devices (e.g. lamps) coloured glasses (Fritz[407], Knapp[693]) or opal glasses (Weigel[1290], Hampton[507]) are used.

For light transmission, besides glass, some other materials, e.g. polyvinyl chlorides or acrylic plastics are also used (Section 21.4).

71.12 Windows for ultraviolet. As transmitting materials for the near ultraviolet (3000–4000 A°), mica, special glasses and acrylic plastics (Table 2.14) can be used. For the shorter ultraviolet (down to $\lambda = 2000$ Å) only quartz, sapphire or some salts (ClNa, ClK, FCa_2, LiF) are useful as transmitting materials.

The glasses for ultraviolet transmission (e.g. Corning: 9700, 9741; Osram: Uviol glass; Chance: Vita glass) are and should be free of Fe, Cr, Pb, Ti, Sb (Stanworth[1176]), or have special compositions such as phosphate glasses (e.g. Corning: Corex glass).

Most glasses used in high intensity, ultraviolet radiation, develop a tarnishing after some time, known as "solarization" (Weyl[1300], Nordberg[938], White[1307]). The transmission decreases due to solarization. This can be increased again to some extent by submitting the glass to a heat treatment. Plastics show a change in colour after longer exposure to ultraviolet (Blackmon[129a]). The ideal material for ultraviolet transmission is quartz (Sections 71.2; 71.31).

71.13 Windows for infrared. Window materials for infrared should be chosen to give satisfactory transmission in the desired range of wavelengths. Table 7.1 lists the various materials indicating the maximum wave length (microns) at which a 5 mm thick plate transmits about 50 per cent of the incident radiation (Gray[471a], Mönch[879], Florence[385]).

The transmission of infrared through glasses depends on their content of iron oxides. Glasses free of FeO have a good transmission up to wavelengths of about 2 μ. Borosilicate glasses (Section 21.3) can be considered as a transmitter for infrared up to wavelengths of 3.5 μ (Douglas[297]).

For infrared of greater wavelength, quartz, mica, sapphire or salt windows (Table 7.1) should be used.

TABLE 7.1. TRANSMISSION FOR INFRARED

Material	Wave length (μ) at which a 5 mm thick plate has 50% transmission	Remarks
Glass	2.2–2.5	
$CaCO_3$	2.5	
Quartz	3.8–4.4	
Mica	5.3	
Sapphire (Al_2O_3)	6.5	
LiF	7.0	
MgO	7.0	
CaF_2	10.0	
NaF	10.0	
NaCl	20.0	hygroscopic
KCl	25.0	hygroscopic
AgCl	28.0	
KBr	32.0	hygroscopic
KI	37.0	hygroscopic
CsBr	50.0	
CsI	60.0	

71.14 Windows for alpha particles. Alpha particles have a range* of 1–8 cm in air according to their energy (e.g. 1.1 cm at 2 MeV; 3.6 cm at 5 MeV and 7.4 cm at 8 MeV). The range of alpha particles in various solid materials is of the order of microns up to a few tens of microns (Korff[711a]). For example, alpha particles of 5.4 MeV have a range of 25 μ in cellophane, 22 μ in aluminium, 17 μ in mica, 9.3 μ in tantalum and 8.4 μ in gold. Thus foils of micron thicknesses are required for the transmission of alpha particles.

Mica can be split to foils of a thickness less than 4 μ (Section 71.34), which are able to transmit alphas at least at 80 per cent of their original energy.

Aluminium is obtainable as thin as 2.5 μ, however it is not free from pinholes, and cannot withstand the oxidizing atmosphere produced in air by the alpha radiation. Thus for this purpose, aluminium oxide is preferred (Section 71.35c).

Tantalum is available at about 6 μ thickness, which allows about 30 per cent transmission for alphas.

* The distance that a particle will penetrate in a given material before its kinetic energy is reduced to a value below which it can no longer produce ionization.

Gold foils of about 1 μ can be produced but they are always porous.

Thin glass windows, having a stopping power equal to a few millimetres of air, were described by Rosenblum[1080]. For techniques in making such thin glass windows, see Section 71.31b.

71.15 Windows for beta particles. The range of beta particles (electrons) in materials is a function of their energy and of the product $\varrho \cdot t$ (specific gravity$\times$thickness) of the material, and is given in terms of:

$$\frac{\text{mg}}{\text{cm}^3}\times\text{cm} = \frac{\text{mg}}{\text{cm}^2}.$$

For example, beta particles with a maximum energy of 0.3 MeV have a range of 77 mg/cm², those with 1 MeV–400 mg/cm², and with 3 MeV–1500 mg/cm². As a comparison value, 1 m of air is equivalent to about 120 mg/cm². Table 7.2 lists the thicknesses (μ) of various materials which would allow 80 or 50 per cent transmission of the electron beams at various energies.

TABLE 7.2. TRANSMISSION FOR ELECTRON BEAMS

Energy of electrons keV	80% Transmission				50% Transmission			
	10	20	40	80	10	20	40	80
$\varrho \cdot t$ (mg/cm²)	0.02	0.1	0.28	0.96	0.08	0.26	0.85	2.6
Plastics $\varrho = 1.45$	0.12	0.6	1.9	6.6	0.55	1.8	5.8	17.8
SiO_2 $\varrho = 2.3$	0.08	0.43	1.2	4.2	0.35	1.1	3.7	11.3
Al $\varrho = 2.7$	0.07	0.37	1.0	3.6	0.3	0.96	3.1	9.6
Al_2O_3 $\varrho = 3.8$	0.05	0.26	0.74	2.5	0.2	0.68	2.2	6.8

Metal foils can be used as windows for electrons, but they must be very thin (about 10 μ). They should be used in multiple layers in order to avoid the effect of pinholes, Hoffmann[574] used Al_2O_3 foils of about 0.25 μ thick to transmit 98 per cent of 16 KeV electrons. Young[1346] found that the practical range-energy relation of Al_2O_3 for electrons is given by

$$R_{(\text{mg/cm}^2)} = 0.0115 \cdot E^{1.35}_{(\text{Kev})}.$$

Wu[1335] recommended the use of mica for beta-ray windows.

Nylon films (0.02–0.1 mg/cm²) were used by Sturcken[1209a] as transmitting windows for electrons. Colloidon and Formvar films were also used for this purpose. Turner[1241] recommended the use of Mylar (Polyester films).

Thin, glass windows of 0.1–1 μ constructed as described in Section 71.31 can also be used as windows for electrons (Lloyd[777], Seren[1119], Böhm[146],

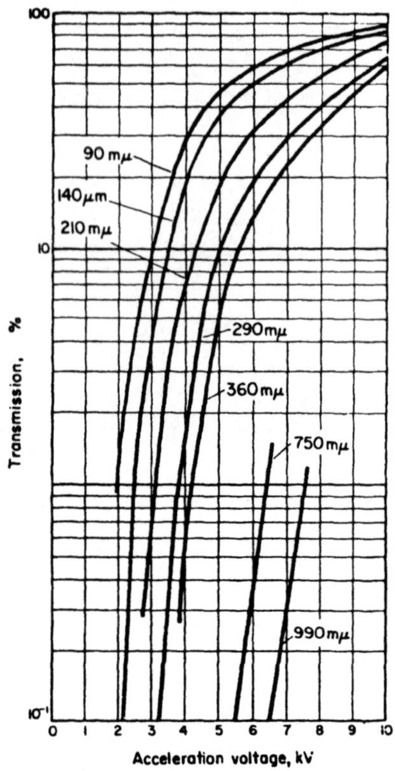

FIG. 7.1 Transmission of glass sheets for electrons. Plotted after data from Batthow[94]

Bathow[94], Rosenblum[1080]). The transmission of glass sheets for electrons is shown on Fig. 7.1.

71.16 Windows for gamma and X rays. Thin foils of *light materials* such as light metals, mica, plastics and special glasses are used for the transmission of gamma and X rays.

Provan[1013] used aluminium foils of 0.05 mm thickness, but beryllium windows are more suitable and extensively used, especially for X rays (Heberle[526a], Brackney[158], Papacosta[966], Mueller[900]). Wu[1335] recommended the use of mica windows, for soft X rays; Hodkinson[571] used Polythene windows.

Regular glasses present a colour change ("solarization") after exposure to X rays, becoming yellow-brown (Kersten[668], Zunick[1358], Nürnberger[949], Smelt[1150]). For X rays and particularly soft X rays, the windows should be made of glasses containing only light metal oxides. These glasses contain mainly boron, lithium, or beryllium; amongst these the Lindemann[796] window containing B_2O_3, $Li_2\ B_4O_7$ and BeO is the best known. The Lindeman glass can be sealed with soft glasses (Pt glasses, see Section 24.31) as a circular window, but must be protected from the atmosphere by a layer of lacquer painted on its surface. Lindemann windows are usually constructed in concave or convex shape, having thicknesses of 0.1–0.25 mm (Bleeksma[136], Friedman[405]).

For the absorption of gamma and X rays, windows of glasses containing heavy metal oxides (Pb, Ba, Bi, Th, U, Ta, W) are to be used (see Gott[466], Sun[1210], Rothermel[1087]).

71.17 Windows for neutrons. Regular glasses containing SiO_2, Al_2O_3, PbO, MgO, etc., have a fair transmission for neutrons (Sun[1211]), i.e. a 200–400 times higher transmission than cadmium (taken as the reference absorbent material for high energy neutrons).

Glasses with neutron absorbing properties must contain CdO, In_2O_3, Li_2O, Eu_2O_3, Gd_2O_3 (Sun[1211], Melnick[894]).

71.2 Chambers made of transmitting materials

Electric lamps are the main example of sealed vacuum devices, where the chamber wall itself is used as the transmitting material for the radiations produced in the lamp (see also Section 21.3). In regular incandescent lamps or in the discharge lamps, the radiations to be received outside the lamp are in the visible range. Thus, glass envelopes are used (bulbs, tubes). These bulbs or tubes are sealed (Section 21.1) to the stems (Section 42.2).

When ultraviolet radiations are to be transmitted to the surroundings (e.g. high pressure mercury vapour lamps) quartz tubes must be used (Section 71.12) and sealed at the current lead-ins (Sections 24.42 and 42.4).

In other applications the demountable vacuum chamber is used to transmit radiations; e.g. in the case of glass bell-jars used in vacuum evaporation or other vacuum techniques. The bell jars may be sealed to the base plates by wax (Section 31.2), liquid seals (Section 37.2) or preferably by gasket seals (Section 3.8).

71.3 Attached windows

In most applications the vacuum chamber does not transmit the radiations and the transmitting surface is limited only to the area of the attached windows, sealed at the required place to the chamber.

The attached windows are usually made of glass or quartz (Section 71.31), but for other than visible radiations, special windows of ceramic (Section 71.32), various salts (Section 71.33), or mica (Section 71.34) are used. Very thin plastic, metal or metal oxide windows are constructed using "self supporting films" (Section 71.35).

71.31 Glass and quartz windows. Glass windows are used either permanently sealed to the end of glass, ceramic or metal tubes, or connected with demountable seals to the vacuum vessels or pipes. The methods used for the sealing of demountable glass windows are not specific for these windows (Section 7.2).

Permanently sealed glass windows are of two main types: thick and thin windows.

(a) *Thick glass windows* are usually plain ones, thick enough to withstand the atmospheric pressure, when the chamber is evacuated (for dimensions see Table 2.2).

The windows are sealed to metal frames using matched (Section 24.3) or unmatched (Section 24.4) seals, or to ceramic bodies (Section 2.5). Matched seals are used mainly with hard glass windows, sealed usually to the end of Kovar tubes, which are brazed to strong flanges (Tennent[1218], Conor[229]). Hard or soft glass windows can be sealed to copper tubes using the Housekeeper feather edge technique (Section 24.41). Glass windows sealed by matched or unmatched seals can be baked usually up to 400 °C. Thus they can be used in high and ultra-high vacuum plants.

Doyle[301a] described two methods for sealing alumino-silicate glass windows to molybdenum cups. He used the chromallizing (impregnating the molybdenum with chromium) method or the direct sealing of the glass to the molybdenum in an argon atmosphere.

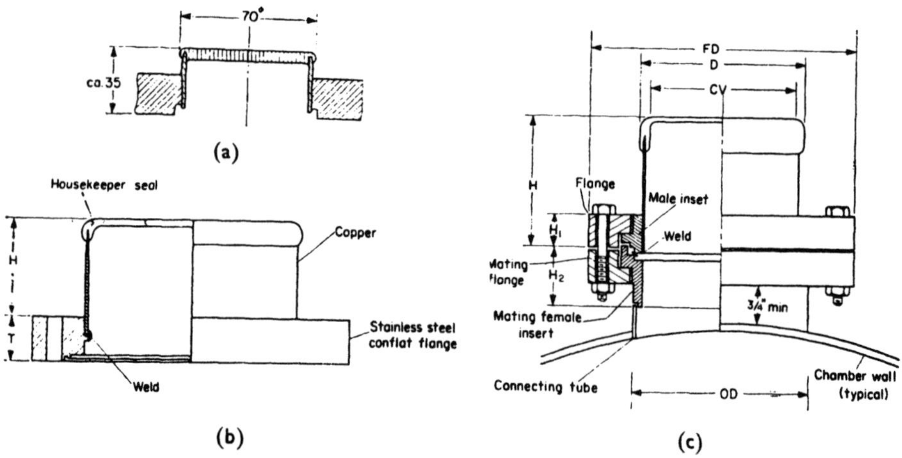

FIG. 7.2 Permanently sealed glass windows: (a) supplied by Leybold[762]; (b) supplied by Varian[1257a]; (c) supplied by Ultek[1251a].

Permanently sealed windows mounted on flanges are available ready-made. Figure 7.2 shows such windows, from Leybold[762] (Fig. 7.2a), Varian[1257a] (Fig. 7.2b) and Ultek[1251a] (Fig. 7.2c). The dimensions of these windows are obtained from Fig. 7.2 and Tables 7.3 and 7.4.

PLATE 41. Bakeable (ultra-high) vacuum inspection window *(Courtesy of Leybold*[762]*)*

A thick glass (Pyrex) window sealed permanently to OFHC copper* which is welded to a stainless steel flange is shown in Plate 4.1. The flange can be sealed to the vacuum vessel, using metal (copper) gaskets (Section 38.56).

TABLE 7.3. HIGH VACUUM VIEWING PORTS (Varian[1257a]) (see Fig. 7.2b)

Port diameter		Port dimensions				Remarks
		H max		T		
inch	mm	inch	mm	inch	mm	
1 1/2	38	2	51	1/2	12.7	To be used with Conflat seal (Section 38.56b)
4	102	2	51	25/32	19.8	

The glass face plates sealed onto the front of television picture tubes should be considered as permanently sealed windows. Bender[107] described the sealing technique of the glass face plate of television tubes to the hollow truncated cone of (17 per cent Cr.) steel.

(b) *Thin glass windows* sealed to the end of glass pipes can be prepared in thicknesses in the range of 10–0.1 μ.

* When baking at high temperatures for long periods of time it is necessary to protect the copper from oxidation by surrounding the outside of the port with a neutral or slightly reducing gas.

TABLE 7.4. VIEW PORTS FOR HIGH VACUUM SYSTEMS (Ultek[1251a]) (see Fig. 7.2c) (dimensions in inches)

Nominal size	o.d.	CV	D	FD	H	H_1	H_2	Number of bolts
1	1.50	13/16	1 1/8	2 7/8	1 1/8	1/2	7/8	6
1 1/2	2.00	1 1/4	1 5/8	3 1/4	1 5/8	1/2	7/8	6
2	3.00	1 3/4	2 3/16	4 1/2	2	1/2	7/8	8
3	4.00	2 3/4	3 3/16	5 1/2	2 1/4	1/2	7/8	8
4	4.50	3 3/4	4 3/16	6	2 1/4	5/8	1	8
4	6.00	3 3/4	4 3/16	7 3/4	2 1/4	3/4	1 1/4	12

Seren[1119] was able to blow, on the end of Pyrex glass pipes (12 mm diameter, 3 mm wall thickness), 1–0.8 mm thick hemispherical windows which were able to withstand the pressure differences occurring on the walls of vacuum vessels.

Hemmendinger[536a] used a special technique in order to obtain very thin glass windows (down to 0.1 μ). In this technique, the sealed-off ends of a Y-shaped glass pipe are brought together (1, Fig. 7.3a), and the heated junction is blown to produce a bubble with a glass diaphragm (2, Fig. 7.3a) across an equatorial plane. One side of the bubble is then heated and opened (3, Fig. 7.3a), and finally the foil is fused (4, Fig. 7.3a) to the end of a glass tube.

An alternative technique to seal very thin glass windows onto the end of a capillary is described by Rosenblum[1080]. A thin glass bubble is blown onto the end of a glass tube (1, Fig. 7.3b). A second glass tube is then drawn so that its end is 1–2 mm inside diameter and it has thickened walls. While the thickened end is still hot, it is brought close to the bubble (2, Fig. 7.3b), and with a light suction the thin glass wall of the bubble is drawn onto the heated end of the capillary, and fused to it (3, Fig. 7.3b). In this arrangement, the fragile window is protected by the robust wall of the capillary. Lloyd[777] constructed recessed thin glass windows about 1 μ thick and 18 mm diameter. He recommended the use of soda glass (soft glass) for this purpose.

Bathow[94] described a procedure for constructing thin glass windows. This consisted of first blowing a thin bulb (1, Fig. 7.3c) on the end of a pipe (e.g. soft glass, 15 mm diameter, 0.5–1.0 mm wall thickness). On being heated this thin bubble becomes more or less flat in shape (2, Fig. 7.3c). If it is sucked it takes on a recessed form (3, Fig. 7.3c). If needed, in this phase a second pipe can be sealed to the joint (4, Fig. 7.3c) in order to protect the window or to connect it onto the opposite side. With this technique, windows as thin as 0.1 μ can be obtained, but windows as this must be maintained with under-pressure on their convex side, since even equal pressure on both sides will break the foil.

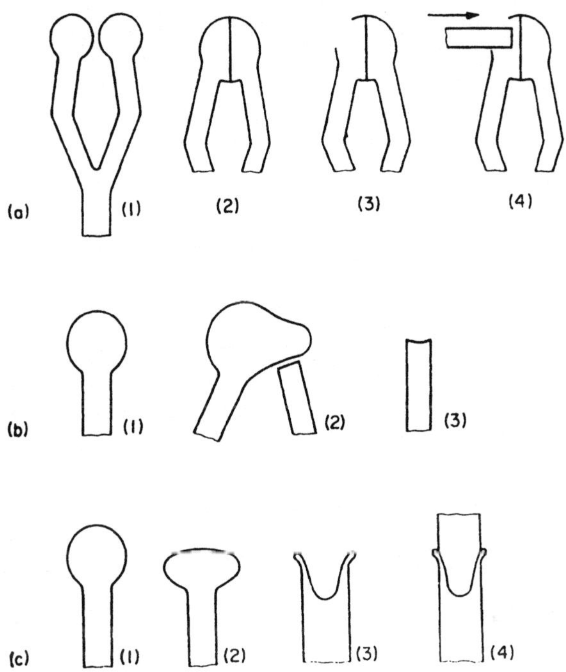

FIG. 7.3 Techniques for constructing thin glass windows: (a) straight glass diaphragm, transferred onto the end of a pipe. After Hemmendinger[536a] *(Courtesy of The American Institute of Physics)*; (b) thin-walled bubble transferred onto the end of a pipe. After Rosenblum[1080] *(Courtesy of The Institute of Physics and The Physical Society, London)*; (c) diaphragm, sucked into the pipe. After Bathow[54] *(Courtesy of Rudolf A. Lang Verlag, Esch/Taunus, Germany)*

(c) *Quartz windows* can be permanently sealed to glass chambers, using graded seals (Section 23.35). An older technique had a quartz window (Fig. 7.4a) on the end of a pipe, made of successive segments with graded expansions from the quartz to the glass of the container. This graded pipe is then sealed to the glass chamber; but the resulting construction always possessed the unpleasant shape of the protruding pipe.

An improved technique, described by Görlich[463], permits sealing of the quartz window at the same level as the surface of the glass vessel. According to this technique, a quartz pipe is sealed to the glass vessel, using a conical graded seal (Fig. 7.4b1). Subsequently the quartz tube is sealed-off, and the graded seal is blown to a spherical shape, which has the quartz window in the middle (Fig. 7.4b2).

Quartz windows can be sealed into metal tubes using compression seals (Fig. 2.87a).

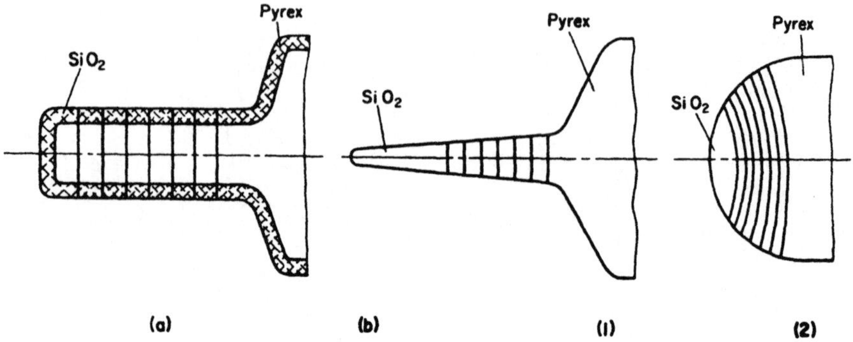

FIG. 7.4 Techniques for sealing quartz windows onto glass chambers: (a) quartz window sealed onto the end of a graded glass pipe; (b) quartz window sealed onto the spherical surface of a glass container. After Görlich[463] *(Courtesy of Rudolf A. Lang Verlag, Esch/Taunus, Germany)*

71.32 Ceramic windows. Ceramic-to-glass and ceramic-to-metal sealing techniques (Section 25.2) are used to construct ceramic (sapphire) permanently sealed windows.

Sapphire windows have been sealed on to glass or metal pipes, using fusion techniques (Chasmar[207], Rawson[1031], Meechan[840]).

Chasmar[207] put a sapphire disc (window) horizontally on to a pyrophillite block in an oven, and, over the sapphire placed a glass ring (about 2 mm thick) with an equal diameter. The glass surface was highly polished.

By raising the temperature up to 900 °C, the glass ring collapsed and formed a wall of molten glass around the sapphire. The temperature was decreased to about 780 °C, and a tubular, graded seal (Section 23.35) was lowered over the sapphire; this resulted* in a sapphire window sealed to a glass pipe with a Pyrex bottom.

Rawson[1031] placed the sapphire disc (1, Fig. 7.5a) on the end of a carbon rod (2) positioned centrally in a high frequency induction coil (3). A glass tube (4) made up of a graded seal (ending in a fire-polished B.T.H. C79 glass, see Table 2.10) was clamped so as to leave a 2–3 mm gap from the sapphire. After the carbon was heated to about 800 °C, the graded seal was lowered** onto the sapphire, gently pressed down on to the disc and rotated slowly

* A good annealing (Section 23.13) is required.

** There is a danger of the sapphire cracking if its temperature is not uniform when the graded seal is lowered onto it.

about the vertical axis. The seal was lifted from the carbon rod, and annealed by heating it at 560 °C in an electric furnace for 10 min, and then cooling it at a rate of about 3 °C/min.

Sapphire windows sealed on glass tubes are available from Fredericks Co., Bethayres, Pa., U.S.A.

Meechan[840] sealed a sapphire rod in a metal (copper) block by vacuum casting it in a copper melt. As the "window" was constructed for use in low temperature work (below 80 °K), a ring section of the copper was machined (etched) out (Fig. 7.5b), and the two metal ends of the unit were sealed to the two walls of a double-walled, cryogenic chamber (Dewar).

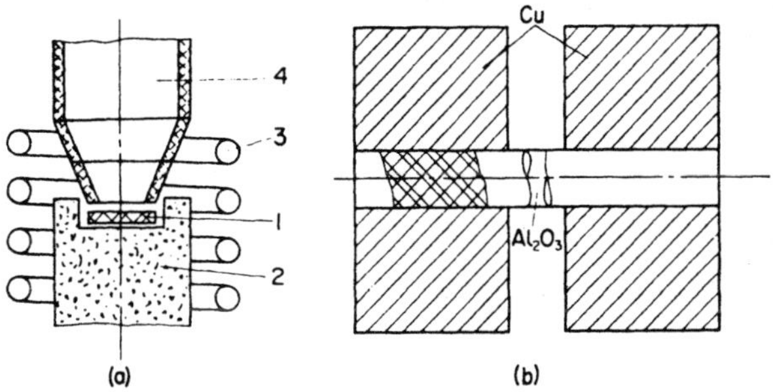

FIG. 7.5 Sapphire window sealed (a) to glass (Rawson[1031]); (b) to metal (Meechan[840])

71.33 Salt windows. Sodium chloride, lithium fluoride, potassium bromide, calcium fluoride, silver chloride and other salts are used as windows (see Table 7.1). These windows have the disadvantage of being generally soft and some of them are hygroscopic (Hyman[613]).

Salt windows can be sealed into the vacuum chamber only by using epoxy resin (Section 3.3) or silver chloride (Section 3.4) seals. Table 7.6 lists various examples of such seals.

71.34 Mica windows. Mica can be split into extremely thin and uniform layers, which are still strong enough to withstand atmospheric pressure over a considerable area. Mica of 5 mg/cm² (about 20 μ) withstands atmospheric pressure up to a diameter of 28 mm (Wu[1335]); for a window of 1.1 cm diameter a thin, mica foil of 0.97 mg/cm² (about 3 μ) was strong enough to withstand the same pressure difference (Arnold[51a]).

In order to split the mica sheet, it must be initially cut to a square or rectangular shape with clean edges. A corner is then frayed out by rubbing it, and a dissecting needle is introduced so as to divide the thickness of the sheet. A drop of water is introduced into the cavity and the mica is then split all around the edges with the needle travelling, point first, at an angle of about 30° along the edge of the sheet. This starts the edge splitting inside, from where it progresses outwards, in such a way as to ensure that the needle remains in the same cleavage plane. The two sheets are held apart at one corner and a drop or two of water is added, whereupon the sheets may be separated easily (Strong[1205a]).

The difficulty in washing the mica sheets is that they cannot be seen in water. Aharoni[10] proposed illuminating the bath with polarized light; the mica appears much brighter than the glass or water (which are dark). The polarized light also helps to distinguish differences in thickness.

Mica windows can be cut to the desired shape by punching. The punch should be very sharp in order to obtain clean and straight edges. Instead of cutting Meunier[852a] recommended shaping by burning. The procedure consists of inserting the mica foil between the butted ends of two steel pipes, cut perpendicular to the axis and having polished edges. The annulus of mica, protruding outside the pipes, is burned out by a strong flame.

For degassing properties and permeability to gases see Sections 21.13 and 21.12 (and Wagener[127]) respectively.

For sealing procedures used with mica windows see Table 7.6.

71.35 Self-supporting windows. The self-supporting* films are thin foils produced by depositing a thin layer of the material on a substrate, and subsequently removing and transferring the resulting film on a holding frame. Thus, the film is finally held by the frame around the periphery, the middle of the film forming a self-supporting window.

Such thin windows are made up from plastics, metals or oxides.

As the self-supporting windows are very thin, the main problem is obviously the mechanical strength to withstand pressure differences. Formulae for the calculation of the minimum thickness, maximum pressure or stress are given by Seehof[1117a], Ardenne[47], Hauser[520].

The diameter (D) to thickness (t) ratio for self-supporting windows, is given by

$$\frac{D}{t} \leqslant 200 \frac{\sigma}{P}$$

where σ is the tensile strength (kg/mm^2) and P the pressure difference (atm). Thus the thickness of windows used for 1 atm pressure difference must be greater than the values listed in Table 7.5.

(a) Self-supporting plastic films. Two main techniques are used for the preparation of thin plastic films:

The first technique consists in pouring the solution of the plastic onto the surface of a liquid (water) in drops. After the drops spread over the liquid and the film solidifies, it is lifted from the liquid.

The second technique is to paint (by dipping, brushing or spraying) the solution of the plastic onto a plane surface (e.g. glass), and after drying, to

*Sometimes called "unsupported" or "unbacked" films.

TABLE 7.5. MINIMUM THICKNESS OF SELF-SUPPORTING WINDOWS, FOR $P = 1$ ATM

Material	(kg/mm²)	D/t max	t min. (μ) for $D = 10$ mm	t min. (μ) for $D = 1$ mm
Al	9	1800	5.5	0.55
Al_2O_3	26	5200	1.9	0.2
Mica	40	8000	1.25	0.13
Au	10	2000	5.0	0.5
Be	20	4000	2.5	0.25
Colloidon	4.5	900	11.0	1.1
Mylar	16	3200	3.1	0.3

float the plastic film by the immersion of the film (with its support) into a bath (see Fig. 7.7).

The *drop technique* may be used to prepare films of colloidon or Nylon. Ference[373] prepared thin* (10–0.07μ) colloidon films, by allowing a drop or more of colloidon dissolved in amylacetate (1 : 4) to spread over a clean surface of water. The films were lifted from the water by means of a circular holder, and immerdiately dried on a dust-free oven.

Nylon can be dissolved in isobutyl alcohol. Brown[172a] made thin* nylon films (about 0.01 mg/cm²) by using a drop of dissolved Nylon, spread on the surface of a water bath.

Polyvinyl acetate, polystyrene and polyethylene films may be prepared using the flotation technique (Barreau[82], Vuccino[1275], Amsel[29a]).

Polyvinyl acetate (Formvar) is first dissolved (0.1–0.5 per cent) in chloroform or dioxane. The solution is poured in drops on a glass slide, and is spread over the surface to obtain a uniform layer of liquid. The solvent evaporates quickly (2–3 min). By immersing the slide obliquely into a water bath the Formvar film will float onto the surface of the water.

In order to prepare polystyrene films Amsel[29a] recommended the use of a solution of this material in chloroform. The concentration to be used depends on the required thickness of the film (Fig. 7.6). To prepare the film, first a clean glass plate is immersed in detergent (Teepol) solution, then wiped to obtain a very thin film of detergent on the glass. By dipping the so

* For the measurement of the thickness by the interference colour see e.g. Methfessel[850], Mönch[879].

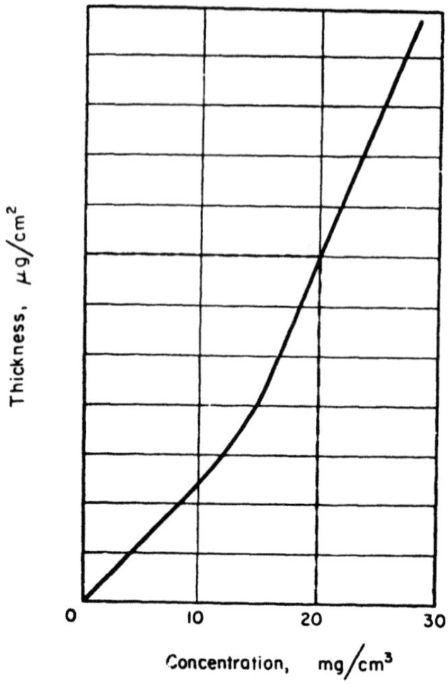

FIG. 7.6 Thickness of polystyrene films as a function of the concentration of polystyrene–chloroform solutions (after Amsel[29a])

prepared plate in the solution of polystyrene, a thin film of this material is obtained on the plate. After drying, the edges of the glass plate are scraped clean of polystyrene and the plate is inserted obliquely in water. By this treatment the detergent dissolves gradually and water enters under the film, lifting it by surface tension, until it floats (Fig. 7.7).

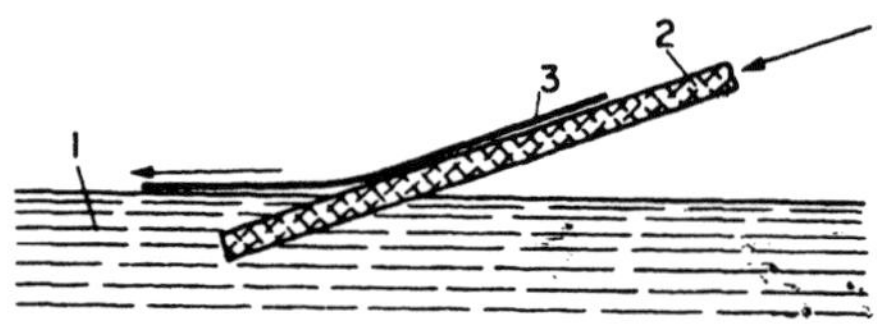

FIG. 7.7 The floating technique of plastic films by oblique immersion in water: (1) water; (2) glass plate; (3) plastic film

Two brass frames are prepared (Watts[1288a] used aluminium holders) having flat and smooth surfaces, good circular symmetry, and the ability to be superimposed and screwed together. One of the frames is immersed in the bath moved below the film, and the film is lifted by withdrawing the frame obliquely from the water. After gently depressing the film by blowing, in order to produce a slightly spherical surface, the second frame is fitted on.

Watts[1288a] described a similar technique; he sprayed Parlodion solved in pyroxylin, or brushed a Parlodion (0.5 g) solution in amylacetate (10 g), and ethyl alcohol (30 g) on a glass plate, and after drying floated it in water (see Fig. 7.7). The films prepared in this way were very thin (100–400 μg/cm^2).

Reid[1042] obtained thin (600 μg/cm^2) Polythene films by pouring on a 3×3 in. glass a warm (80 °C) solution of Polythene (30 mg) in toluene (5 cm^3) and (after drying) soaking the Polythene film off in warm water.

(b) Self-supporting metal (carbon) foils. Thin metal foils can be prepared by depositing the metal layer on a supporting material and then separating it by flotation on water (Fig. 7.7) or by dissolving away the support.

For the deposition of a thin metal layer, generally vacuum evaporation or sputtering is used (Holland [581], Methfessel[850]).

Hast[517a] prepared self-supporting beryllium foils down to 20Å, and aluminium foils down to 12 Å thickness, by depositing the metal in the vapour phase onto a cellulose nitrate film, and dissolving away the supporting film. Hast[517b] also described a technique consisting of covering a glass slide with a layer (less than 0.5 mm) of glycerol, and evaporating aluminium or beryllium onto this layer. The glass slide is next put in a container into which distilled water is introduced. The glycerol is dissolved and the metal foil floats on the surface of the water.

Hast[517b] mentioned that by evaporating the metal onto the surface of vacuum oil, or mercury, he was unable to obtain a continuous foil.

The deposition of thicker (1.–0.1μ thick, 26 mm diameter) beryllium foils by evaporation technique is mentioned by Bradner[158a].

Amsel[29a] mentions aluminium foils of 0.04μ prepared by evaporation in vacuum onto polystyrene films (Fig. 7.6). After evaporation the polystyrene film was removed by dissolving it in dioxane. Since the foil obtained by this technique is very fragile, the dissolving process must be done by suspending the coated film (in its frame) into a U-bath (Fig. 7.8), thus allowing the liquid to rise slowly and cover the film. Amsel[29a] quotes self-supporting aluminium foils as thin as 120 Å on an area 8 mm in diameter. To obtain enough strength for handling such thin foils, the annulus around the window was coated more heavily with aluminium than the central part.

A technique to produce self-supporting copper foils is described by Brunner[178b]. The copper was evaporated on a detergent coated* glass substrate. The copper film (about 1.8 mg/cm^2) was removed from the substrate by immersion in warm water, where the detergent dissolves, allowing the copper film to float freely.

Bashkin[93b] prepared self-supporting nickel foils down to 250 Å thickness.

* The glass slide was first cleaned by vigorous scrubbing with a paste of Shamva and water, and then rinsed in distilled water. Detergent was spread over one face of the slide, using another slide to distribute the detergent evenly. The film was allowed to dry for a moment and then burnished to a thin, uniform layer with a gauze pad.

In order to prepare cobalt, nickel, manganese, vanadium, chromium, titanium, beryllium or iron foils, Brunner[178a] recommended that the evaporation of these metals be made onto substrates previously carefully coated with water soluble salts, having melting points above 800 °C (e.g. sodium chloride). Brunner used as a substrate a nickel disc of 3 in. diameter, 3/16 in. thick, polished to a surface finish of 8 μin., and coated by evaporation with a

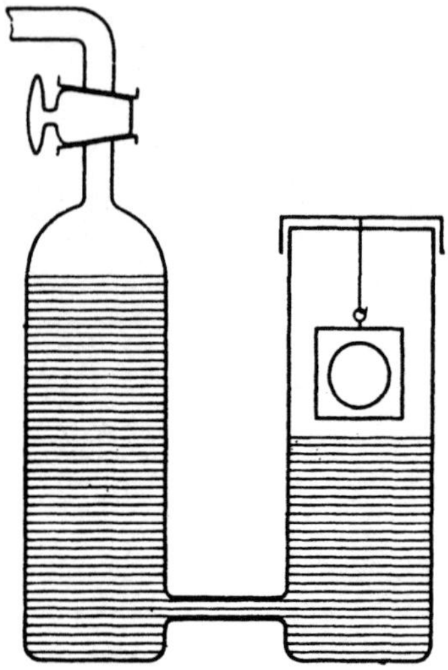

FIG. 7.8 U-bath for the chemical treatment of very thin foils

layer of 0.1–0.15μ sodium chloride. During the subsequent evaporation of the metal layer onto this substrate, the nickel disc was heated to 650 °C in order to provide the required ductility of the film and to anneal it. Substrate and film were then separated by floating the film (Fig. 7.7). Brunner recommended placing the disc obliquely (film-side up) in a shallow pan, and gradually admitting warm water. As the water level reaches the film, the salt dissolves, allowing the film to float on the surface of the water. If at any time the foil resists parting, the water flow should be stopped for a short time, to prevent undue strain on the foil.

Massey[823a] prepared self-supporting carbon foils (50–100 Å thick) by evaporating the carbon onto detergent coated glass slides or onto nickel foils. The carbon film is separated from the detergent-coated glass slide by floating it on water (Fig. 7.7). The carbon film deposited onto a nickel foil (about 6 μ thick is separated by etching the nickel with a sulpho-nitric acid solution (1 pbw H_2SO_4, 2 pbw HNO_3, 2 pbw H_2O). After dissolution of the nickel, the carbon film should be rinsed with dilute nitric acid and then with distilled water. During the entire etching and rinsing operations, only one drop of liquid should be on the film at any time.

Muggleton[901a] used carbon foils to evaporate tungsten oxide on them.

(c) Self-supporting oxide films. Oxide films may be prepared by evaporation techniques or by anodic oxidation.

The preparation of silicon oxide foils (8 μg/cm^2 on 8 mm diameter) by evaporation onto Zapon film backing is described by Sawyer[1098a].

Various techniques for the preparation of aluminium oxide, self-supporting films are described by Strohmaier[1205], Hennig[537], Hoffmann[574], Harris[511a], Young[1349a], Amsel[29a].

The self-supporting oxide film is prepared by anodic oxidation of the metal foil and by etching-off the remaining metal. If the anodic oxidation is done on both sides of the metal foil, it should be etched from one side, removing first the oxide layer and then the metal layer, so that only the second oxide layer remains.

With this method, oxide foils (windows) of aluminium, tantalum, silicon, niobium (and probably nickel, telurium, zirconium, beryllium) oxides can be prepared. The method is unsuitable for iron or copper oxides.

The stages of a typical process (for Al_2O_3) are.

(1) The shaping of the metal foil in a frame:

(2) The transfer of the metal foil from the frame into the anodizing bath.

(3) The anodizing process.

(4) The removal of the anodized foil from the bath and its subsequent washing and drying.

(5) The mounting of the foil on a second frame and the etching of the oxide and metal layers.

Shaping of the foil. Since the oxide is not elastic, a sag should be formed on the metal foil before oxidizing it. This is done by mounting the metal foil in a frame and producing a hemispherical surface by suction. For an aluminium foil of 15 mm in diameter and 20 μ thickness, the sag is best made by subjecting the foil to a pressure difference of about half an atmosphere.

Using, commercial, household aluminium foil, the mat side of the foil should be made concave, since the oxide window will remain on the bright side.

In order to allow the foil to be mounted later in a second frame, positioning dowel holes should be made in the foil, *after forming the sag*, and prior to oxidation.

The electrodes of the anodizing bath. The anode is the aluminium foil to be oxidized. Household aluminium foil (99.7–99.8 per cent Al) is preferred to absolutely pure aluminium, because of its ability to be dissolved easily (Harris[511a], Amsel[29a]). Foils of 10–20 μ are used (Strohmaier[1205]). The foil is hung vertically in the electrolyte so that both sides and the edges be anodized simultaneously (Harris[511a]).

Stohmaier[1205] placed the aluminium foil so that it would float on the surface of the bath, in order to anodize only one face* of the foil. From the prepared foil the subsequent etching of the aluminium (see etching) has to be performed. During the immersion of the aluminium foil into the electrolyte solution, care should be taken that no other metal (holders) touches the solution.

The cathode may be of platinum or of the same material as the anode (e.g. Al). Strohmaier[1205] also reported the use of carbon as a cathode material.

The anodizing bath. The electrolyte may be a citric, tartaric or boric acid solution.

Strohmaier[1205], Harris[511a], and Amsel[29a] used a 3 per cent ammonium citrate solution prepared with *very pure, fresh, chloride free* deionized water (resistivity of about 100 kohm/cm). In order to avoid the formation of a porous oxide layer, the use of large volumes (more than 250 cm^3) of electrolyte and/or the presence of an excess of NH_3 is recommended (pH of the bath 5.5).

For high voltages (see anodizing process) the pH may have to be raised to 9 by adding KCN instead of NH_3.

The anodizing process. The direct current required for the anodizing process should be supplied from a large source (battery), since on applying the voltage to the bath, it momentarily takes a very high current. As the voltage rises the current falls rapidly (1–2 min) to about 0.1 mA/cm^2. Amsel[29a] recommended that the initial current density be limited to about 100 mA/cm^2 by a resistance.

The thickness of the oxide film formed is practically proportional† to the applied voltage. The ratio of the thickness (Å) to the voltage is : 13.7 Å/V for Al_2O_3; 16.0 Å/V for Ta_2O_5 and 4 Å/V for SiO_2.

The strength of Al_2O_3 films is calculated by Hauser[520] in the formula

$$\frac{p_{\max} \cdot d}{\log_{10}\left(\dfrac{V}{24}\right)} = 400,$$

where $p_{\max}$ is the breaking pressure (torr), V is the anodizing voltage ($50 \leqslant V \leqslant 500$), and d is the diameter of the window (mm).

With regular anodizing techniques Al_2O_3 foils down to 600 Å thickness (44V) can be made. To obtain thinner films (down to 250 Å) the central part of the aluminium foil is first masked (on the bright side) and an annulus around this part is anodized to about three times the thickness (750 Å). Sub-

* Prior to this the foil was immersed and anodized successively on the four sides of the square.

† Bahn[70] found that for Al_2O_3 the thickness/voltage ratio is 18.0 Å/V at 25 V; 14.5 Å/V at 100 V and 12.0 Å/V at 200 V.

sequently the central masking is removed and the whole foil is anodized to 250 Å. After etching, a thin (250 Å) window is obtained in the centre, held by a heavier Al_2O_3 annulus.

With the 3 per cent ammonium citrate electrolyte the voltage is limited to 150–200 V (Hennig[537]); at higher voltages breakdown occurs. To form thicker films, less concentrated ammonium citrate solutions or boric acid–borax (Young[1349a]) solutions should be used.

The holding of the foil. After its removal from the anodizing bath, the foil should be washed by immersion in distilled water and then in reagent—pure acetone (Harris[511a]). The dried, anodized foil is sandwiched between two 3 mm thick glass plates (Harris[511a]) having central holes (50–75 mm diameter), or is placed in brass frames (Amsel[29a]) screwed together. The frames have positioning pins for the foil, placed identically with the pins of the shaping frame (see shaping).

The etching. According to the technique described by Amsel[29a], a rubber O-ring (3, Fig. 7.9) is laid on the concave (mat) side of the foil (2) in such a way that a one millimetre wide annulus is left clear between the O-ring and

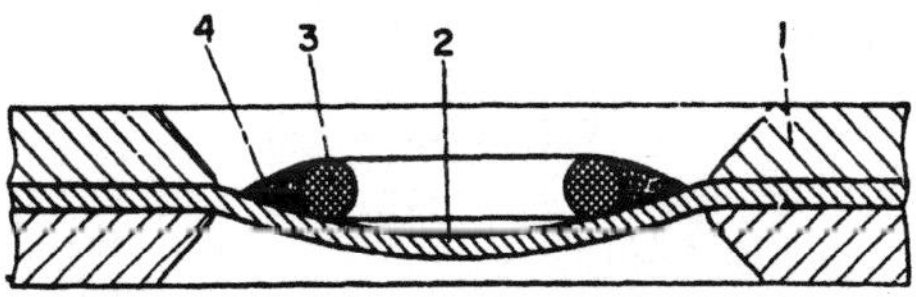

FIG. 7.9 Etching process: (1) frame; (2) anodized foil; (3) O-ring; (4) varnish

the hole in the frame (1). The O-ring is sealed to the foil by applying, around the outer surface, an Apiezon wax or Picein (Sections 3.1 and 3.2) dissolved in trichlor-ethylene (see 4, Fig. 7.9). The etching process is performed after a drying period of about 30 min.

The *oxide film* on the concave side is dissolved using sodium hydroxide solution (8N). To avoid stresses it is required to dissolve the oxide film, in a perfectly continuous and symmetric contour.

The time (minutes) required to dissolve the oxide layer is approximately equal to 1/10 of the voltage (volts) applied in the anodizing process, e.g. a film formed at 120 V takes about 12 min to be dissolved (Amsel[29a]).

After the etching of the oxide layer is completed, the O-ring is removed by dissolving the wax (e.g. in trichlor-ethylene, by washing for 15 min).

The aluminium layer, exposed as a result of the etching of the oxide layer is dissolved using a 8N. hydrochloric acid (1.2 g/ml) solution containing 0.5 g/litre $CuCl_2$ (Harris[511a]). The dissolving process is best done in a U-bath (Fig. 7.8), permitting the slow motion of the liquid level. The etching of the

TABLE 7.6. METHODS OF SEALING WINDOWS

Seal	Material of the window					
	Metal foils	Glass or quartz	Ceramics (sapphire)	Plastics	Salts	Mica
Welded (Section 2.2)	Shon[1134] Fig. 7.10	–	–	–	–	–
Fusion (Sections 2.3 and 2.5)	–	Tennent[1218] Connor[229] Figs. 7.2–7.4	Chasmar[207] Rawson[1031] Anderson[32] Fig. 7.5	–	–	–
Brazed (Section 22.2)	Papacosta[966] Brackney[158] Griffoul[484] Heberle[526a]	Nicolaev[926a]	Chen[211] Grove[489] Fig. 7.11 Section 2.5	–	–	–
Soldered (Section 35.1)	Groom[485a]	Marsden[808] Eckstein[320] Fig. 2.102A	Marsden[808] Fig. 7.12 Pakswer[963] Fig. 3.40a	–	–	–
Waxed (Section 3.1)	–	Dushman[314] Fig. 3.4A Fig. 7.13	–	Wars-hauer[1288] Corelli[237]	–	Copp[235]
Solder glass (Section 35.2)	–	Fiske[380] Fig. 7.14	Anderson[32] Spindler[1170]	–	–	see Table 7.7
Epoxy (Section 3.3)	Provan[1013]	Sayers[1100] Roberts[1071] Quarring-ton[1020] Figs. 7.16; 7.17 Benson[111] Fig. 3.10	Roberts[1070] Martz[822]	–	Rauch[1208a] Roberts[1070] Schwartz[1112] Fig. 7.16	–
Silver chloride (Section 3.4)	–	Weber[1189] Benson[111] Fig. 3.10 Fig. 7.17	–	–	Section 72.6	–
Compression (Section 72.7)	–	Armand[49c] Section 24.23	Section 25.24	–	–	–

(Table 7.6 Continued)

Seal	Material of the window					
	Metal foils	Glass or quartz	Ceramics (sapphire)	Plastics	Salts	Mica
Elastomer gaskets (Section 72.8)	Mueller[900]	Section 38.4	—	Sturcken[1209a] Hodkinson[571] Figdor[374] Fig. 7.22	—	—
Metal gaskets (Section 38.5)	Perry[985] Craig[293a] Fig. 7.23a	Scott[1114] Bridge[167a] Willis[1321] Ehlers[332] Smiley[1150a] Figs. 7.23b; 7.26; 7.27	Brice[167] Fig. 7.25	—	—	Strnad[1204] Lindsay[770] Sterzer[1184] Fig. 7.24

aluminium takes about two minutes, the end of the process being indicated by the transparency of the remaining Al_2O_3 film. The acid is then washed off the foil with distilled water in a second U-bath.

7.2 WINDOW SEALING TECHNIQUES

Windows may be sealed to the chambers, using any general sealing technique, but for each particular window some specific sealing techniques are preferred. Table 7.6 is a guide listing the sealing techniques preferred for the various kinds of windows.

72.1 Welded and fusion seals

Welded seals can be used only for metals, but in the case of windows the use of these seals is limited by the difficulties brought by the small thickness of the window foils, and the great difference between the thickness of the window and that of the holder (port).

Nevertheless some metal windows are sealed by welding, using the edge-welding technique (Figs. 2.35, 2.36). Shon[1134] described a technique used to weld (heliarc, see Section 22.11b) an aluminium window (0.05 mm thick) onto a stainless steel pipe (18 mm inside diameter, and 3 mm wall thickness). This was done by machining the ends of the pipes till they were smooth (1, 2, Fig. 7.10), and then turning a groove with a feather edge close to the ends of each of the pipes. These ends were butted with the aluminium foil (3) between them, and then heliarc welded.

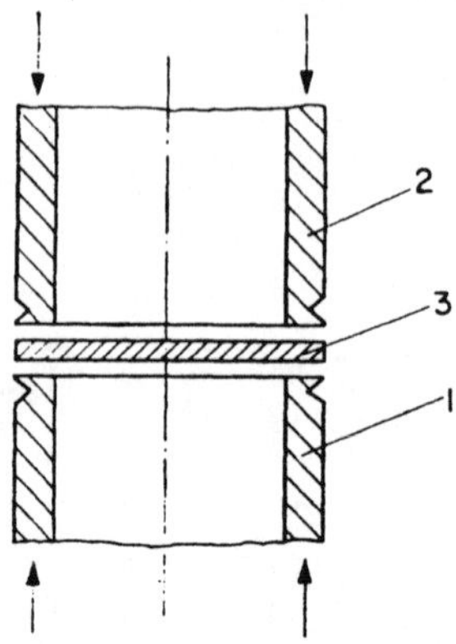

FIG. 7.10 A method for the welding of thin windows to heavy frames: (1, 2) stainless steel; (3) aluminium foil. After Shon[1134] *(Courtesy of The American Institute of Physics)*

Fusion seals are used to connect glass windows to glass chambers or to metal frames, or ceramic windows to glass vessels. The techniques used for such seals are described in Sections 71.31 and 71.32.

72.2 Brazed and soldered windows

The general features of vacuum sealing by brazing are described in Section 22.2; soldering is treated in Section 35.1. These basic features are to be completed here with some examples of window seals using brazing and soldering techniques.

Brazed joints may be used to seal metal, quartz, or ceramic windows.

Beryllium windows are sealed by brazing, and such seals are described by various authors. Brackney[158] used a silver–copper eutectic in order to seal the beryllium window onto a Kovar ring sealed to the X-ray tubes. The brazing was done in a protective hydrogen atmosphere or under vacuum. Heberle[526a] brazed a beryllium window (1 mm thick) to a Monnel frame by using a gold–beryllium alloy. Papacosta[966a] described a technique for sealing thin (0.1 mm) beryllium windows by brazing them onto a thin copper frame, which is then sealed to ceramic parts. The beryllium was first copper plated by evaporation, and then brazed with an indium–copper–silver alloy onto the copper frame.

Quartz and ceramic windows can be brazed to metal ports, if they are first metallized on their sealing surfaces. Such quartz window seals are reported by Eimac[334], Nicolaev[926a].

Chen[211], Benichou[109] described similar ceramic window (1 mm thick) seals. Grove[489] described a sapphire window (Fig. 7.11) sealed to a metal sleeve.*

* Such window assemblies are available from Ceramaseal Co., P.O.B. 25, New Lebanon Center, New York.

The sapphire is first metallized with titanium hydride (Section 25.23a) and then dry hydrogen brazed to the metal sleeve, which is welded to the flange (see Fig.7.11).

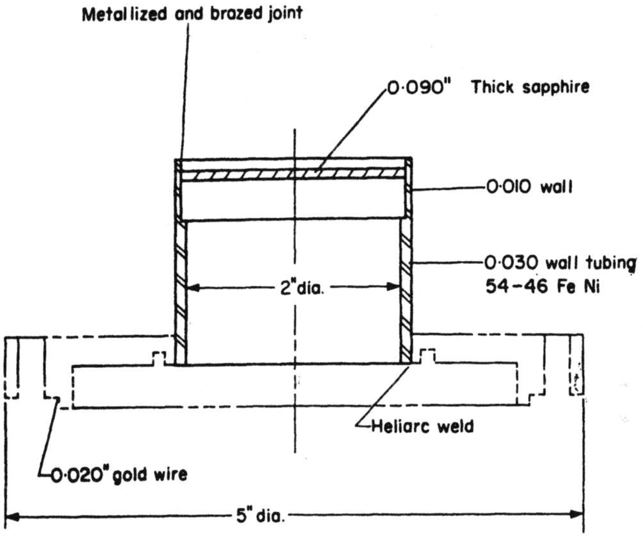

FIG. 7.11 Sapphire window sealed by titanium hydride process. Reproduced from Grove[489] *(Courtesy of Pergamon Press)*

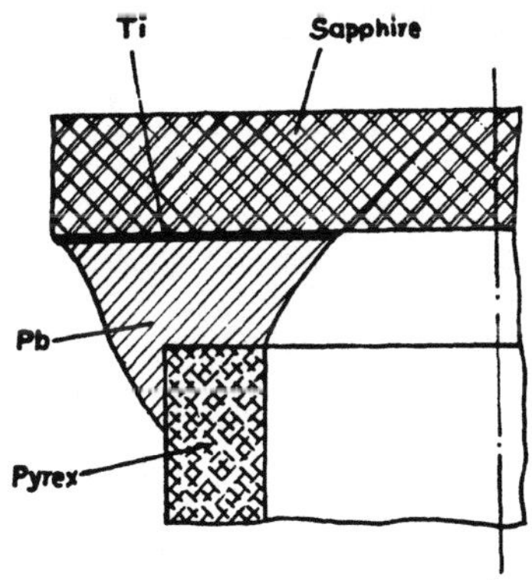

FIG 7.12 Sapphire window sealed to Pyrex using lead. After Marsden[808] *(Courtesy of Rudolf A. Lang Verlag, Esch/Taunus, Germany)*

Groom[485a] made vacuum-tight seals of aluminium foils by using *soft soldering*. The aluminium foil was first coated with zinc by immersion in sodium-zincate, and then it was copper plated. The aluminium foil so prepared was soft soldered onto a copper tube.

Marsden[808] recommended that metallized sapphire windows (Section 25.23a) be sealed by using lead (Fig. 7.12).

72.3 *Waxed windows*

Waxed seals (Section 3.1) can be used to seal glass and quartz windows (Fig. 3.4A), which are not subjected to heating (baking).

Waxes or paints have also been used to seal plastic windows or even mica, in temporary systems. Polyethylene or other plastic films can be sealed to the supporting window port by coating them with a thin layer of high-vacuum grease, and clamping them between greased retainer plates (Warshauer [1288]). Corelli[237] described a procedure to make a vacuum-tight seal from a Mylar foil onto a brass support. The brass (cleaned with alcohol and ether) is coated (painted) with a cement (Du Pont Polyester No. 4695), and allowed to dry (8–10 min). The Mylar is applied onto the brass, and an intimate contact is assured by smoothing with a stick or by placing a heavy weight on top. A curing of 5–8 min at 160 °C, taking care to raise and drop the temperature slowly, completes the seal.

Copp[235] sealed a mica window (0.02 mm thick) onto a metal flange using De Khotinsky cement (Sections 3.1, 3.2).

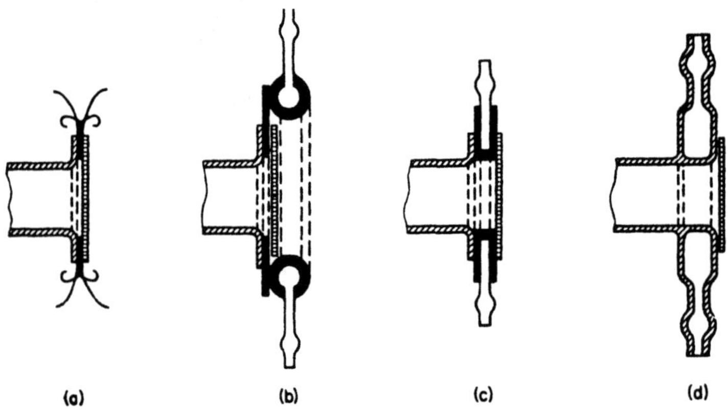

FIG. 7.13 Cooling devices for waxed windows: (a) copper ring with fins; (b) copper ring with cooling pipe; (c) metal cooling jacket; (d) glass cooling jacket

In order to cool waxed windows, it is often sufficient to blow cold air or to surround them with a wet cloth. When stronger cooling is needed, the window is provided with an air-cooled copper ring having fins (Fig. 7.13a) or with a lead pipe (Fig. 7.13b). A more efficient cooling is obtained by providing metal (Fig. 7.13c), or glass (Fig. 7.13d), water-cooled jackets.

72.4 *Solder glass sealed windows*

Solder glasses (Section 35.2) are extensively used for the sealing of windows, especially mica windows.

Fiske[380] sealed glass windows onto metal (FeNiCo) frames by spraying the surfaces to be sealed with a suspension in alcohol of the powdered solder glass (Corning 705A0, 60 mesh). The seal was completed by pressing the glass on the metal frame between two graphite blocks, at 950 °C (for about 15 min). Fiske[380] tested various geometries for the window-frame contact surface (Fig. 7.14a) and found that the step-shaped edge (Fig. 7.14.a4) gives the best adhesion. Figure 7.14b shows the construction of a window using step-shaped edges.

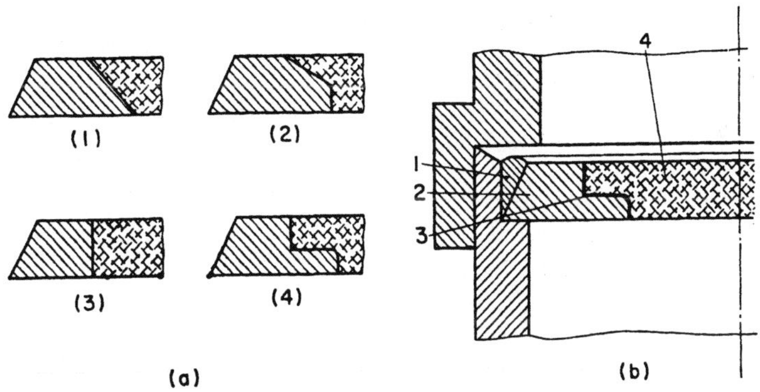

FIG. 7.14 Glass window, sealed with solder glass: (a) geometries of the edge of the window (Nr. 4 is recommended); (b) detail of window seal; (1) solder; (2) FeNiCo frame; (3) glaze; (4) glass window. After Fiske[380] *(Courtesy of The American Institute of Physics)*

The solder glass seal was extensively used for mica windows; various authors constructed such seals differing in the dimensions of the mica window, the material and shape of the port and the geometry of the seal. Table 7.7 and Fig. 7.15 present a summary of such types of seals, showing comparatively their basic features.

In order to show the techniques used, it is considered useful to present some of these mica window seals in more detail. Donal[284] (see also Table 7.7) prepared the surface of the (glass) port by grinding it flat. The mica was cut to overlap the glass slightly. The solder glass suspension (in water) was painted on the ground glass surface; the mica was pressed into position on the glass paste and some extra glass suspension painted over the edges of the mica, to prevent subsequent splitting. The parts (with the mica placed horizontally) are placed in an oven and the assembly is heated to about 600 °C (5–10 min) and finally cooled slowly (30–60 min) to room temperature. It is not recommended (Donal[284]) that the heating be done with a torch instead of an oven.

TABLE 7.7. SOLDER GLASS-SEALED MICA WINDOWS

Mica window		Window port	Solder glass (see Table 3.6)	Shape of the seal (Fig. 7.15)	Sealing technique	Remarks	Reference
Thickness (μ)	Diameter (mm)						
4–12	30	soft glass	M-130, suspension in xylol or alcohol-water solution	g	3 min at 630–650 °C; cooled slowly (20 min)	The inner edge of the port was ground and graphite coated, to avoid flow of solder	Meunier[852a]
4–12	–	CrFe (25/75)	D-6, 100 mesh in distilled water	h	30 min at 550–600 °C in oxidizing atmosphere; 30 min annealing at 400 °C.	–	Anton[42]
20	28	soft glass, ground surface	Lead-borate-silicate ($\alpha = 98.10^{-7}$); Softening Point 450 °C) in water	a	Heated to 570 °C in 60 min, slow cooling	–	Wu[1335]
10–40	30	soft glass, FeNi (50/50) or CrFe (30/70)	M-130 with 5% Pyrex powder	d	Solder glass applied in two layers. Heated to 600 °C in 60 min; maintained 15 min at 600 °C, then cooled to 20 °C	Bakeable up to 300 °C	Labeyrie[726]
25–75	–	Titanium flange; Mica covered with Ti washer	7572 Pyroceram (Corning)	f	60 min at 450 °C, in air	Groove to avoid flow of the solder glass. Bakeable to 500 °C	Anderson[30]

(Table 7.7 Continued)

Mica window		Window port	Solder glass (see Table 3.6)	Shape of the seal (Fig. 7.15)	Sealing technique	Remarks	Reference
Thickness (μ)	Diameter (mm)						
–	40	soft glass or copper plated FeNi (50/50)	Lead-borate-silicate	c	Some minutes at 500 °C, cooling 5°C/min	–	Dale[253]
125–500	50	soft glass	Lead-borate-silicate $\alpha = 98.10^{-7}$; Softening Point 450 °C)	b	5–10 min at 600°; cooling 30–60 min	–	Donal[284]
130	8	FeNiCr (52/42/6) reduced in dry H_2 15 min at 1100 °C, then oxidized in wet H_2 15 min at 1100 °C	D-6; Corning 7570	e	15 min at 600 °C	–	Collins[222a]

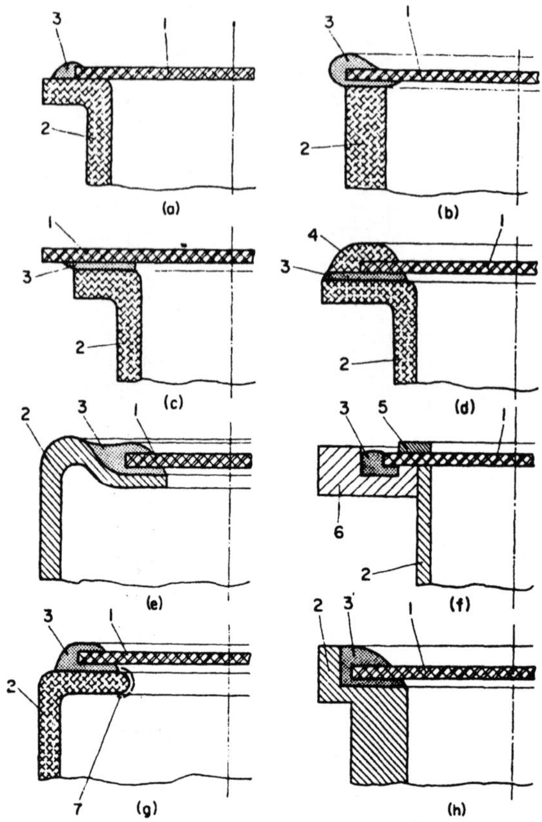

FIG. 7.15 Mica window sealed with solder glass (Table 7.7): (1) mica window; (2) port; (3) glass solder; (4) glass solder; (5) sealing ring; (6) flange; (7) graphite coating

Wu[1335] cut the mica to a diameter slightly less than the flange (Fig. 7.15a), applied glass solder paste on the flange (brushing), placed the mica on the top of the paste, and applied some more paste around the edge of the mica. The heating was done as mentioned in Table 7.7.

72.5 *Epoxy resin sealed windows*

Epoxy resins (Section 3.3) may be used to seal metal, glass, quartz, ceramic or salt windows (Table 7.6).

Provan[1013] sealed with Araldite 1, aluminium (with 5 per cent silicon) foil of 0.05 mm, to an X-ray camera slit, and obtained 10^{-5} torr at temperatures of the window up to 200 °C.

Roberts[1070] sealed windows of glass, sapphire, silver chloride (coolable down to 20 °K) and silica, calcium fluoride windows (cooled down to 77 °K) of diameter 20–25 mm (about 0.5–1.5 mm thickness), using an epoxy resin.

This window seal is based on a thin (0.08 mm) copper diaphragm (1 Fig. 7.16a) which has a circular aperture with 3 mm smaller diameter than the window. The window (3) is placed in position over the shaped (Fig. 7.16a), annealed, and thoroughly cleaned copper diaphragm. The angle (4) is filled with Araldite and cured (Section 3.3). The diaphragm fits into the copper tube (2), to which it is sealed (5) with Wood's metal (Section 35.11). To avoid the soldering operation after the window was mounted on the copper diaphragm, in an alternative construction Roberts[1071] sealed the window using a

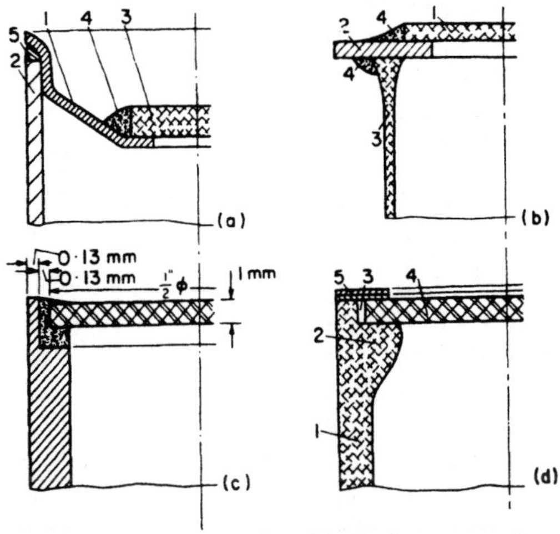

FIG. 7.16 Windows sealed with epoxy resins: (a) elastic construction for differential expansion (Roberts[1077]); (b) quartz window on glass tube (Sayers[1100]); (c) salt window on brass holder (Rauch[1028a]); (d) salt window seal (Schwartz[1112])

Polythene gasket (0.1 mm thick, 10 mm wide), sealed between the diaphragm and a small flange. This construction was used at liquid hydrogen and liquid helium temperatures.

Sayers[1100] succeeded in sealing with Araldite (4), a quartz window (1) to stainless steel and this to kodial glass (3) (Fig. 7.16b). For this seal, the parts were ground and polished, baked in air at 300 °C and then assembled in jigs. After heating the assembly to about 120 °C, Araldite No. 1 was applied to the outside of the joint which was cured at 200 °C for 20 min. The rate of heating between 100–200 °C should be maximum 4 deg·C/min in order to avoid bubles in the joint.

Martz[822] sealed sapphire windows onto glass, in the construction of a Dewar vessel for liquid nitrogen using Epon 828 (10 pbw), with a filler of lithium aluminium silicate (4 pbw), and with Diethylenetriamine (1 pbw) as a hardener. This filler has a negative thermal expansion, and the resultant expansion is well below that of the resin alone. This permits good seals even under adverse conditions.

The same epoxy resin (Epon 828) was used by Rauch[1028a] to seal calcium fluoride windows on brass holders (Fig. 7.16c).

Schwartz[1112] sealed alkali halide windows by using a foil of yieldable metal (e.g. silver, aluminium 25 μ thick) cemented with Araldite to the outside of the window and to the window port. Figure 7.16d. shows such a seal on a circular window sealed to the end of a pipe. The glass tube (1) is thickened at the end (2), and the window (4) is received by a recess (3), ground to a close radial fit and with a depth so that the outer face of the window is exactly level with the end of the glass tube. The sealing foil (5) is a flat ring sealed with Araldite to the edge of the window, and to the end face of the glass tube.

Schwartz[1112] also described the seal of a rectangular window fixed into the side of a glass tube using the same technique.

72.6 Silver chloride sealed windows

Silver chloride may be used to seal quartz and especially salt windows. The general techniques used for silver chloride seals are described in Section 3.4.

Quartz windows may be sealed to Pyrex tubes using one of the techniques shown on Fig. 7.17. In one technique (Weber[1289]), the Pyrex tube (1, Fig. 7.17a) is flared outwards at the end, and a short length of Pyrex (2) is sealed

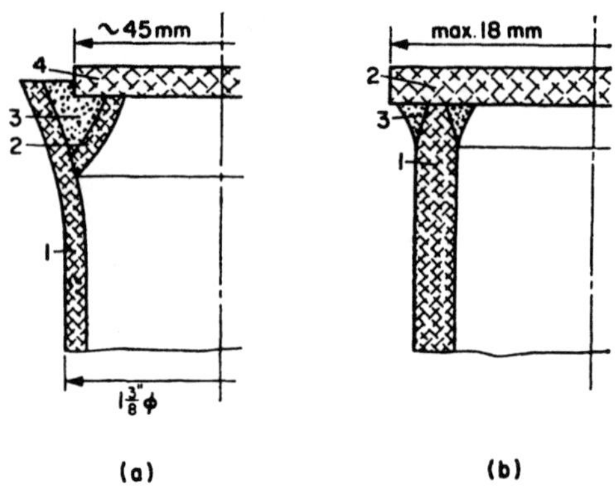

FIG. 7.17 Quartz windows sealed with silver chloride: (a) silver chloride in channel Weber[1289]); (b) silver chloride on tapered end (Benson[111])

inside to form a channel (3) about 3 mm deep. The edge of the tube (2) is ground so that it is below that of tube (1). The quartz window (4) is placed onto the ground surface of the tube (2) and sealed with silver chloride which is placed in the channel (3) (Section 3.4).

In another technique (Benson[111]), the Pyrex tubing (1, Fig. 7.17b) is first ground to a flat edge and then the edge is bevelled both on the inside and outside, to give a taper of about 75° to the length of the tube, until the flat end is reduced approximately to half of its initial width. The tube is fire-polished, and the ground surface of the tube plus the annular portion of the quartz window are platinized. The platinized end of the tube is heated and dipped into a bath of molten silver chloride. This operation is repeated until a thickness of about 1 mm of silver chloride is deposited on to the lip. The tube is clamped above the window, and the window heated until the silver chloride melts and wets the quartz window.

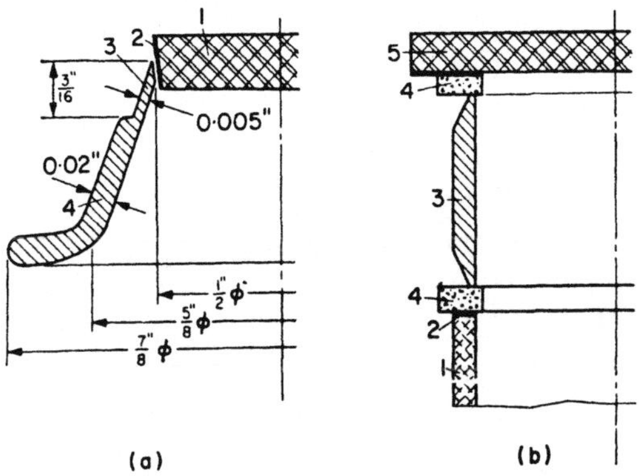

FIG. 7.18 Salt windows sealed with silver chloride: (a) LiF window, side seal (Lord[783]); (b) salt window, edge seal (Wildy[1315])

Salt windows were sealed with silver chloride, using side seals (Fig. 7.18a), channel seals, step seals (Fig. 3.10) or edge seals (Fig. 7.18b).

Lord[783] sealed a lithium fluoride window (1, Fig. 7.18a), by coating it first with platinum (2), and then by joining it with silver chloride (3) to an elastic silver tube (4). The silver tube acts as a compensator for the differential expansion, and so the seal can be used at temperatures down to liquid helium. The thin edge of the silver tube was degreased with acetone, and etched in concentrated nitric acid for a few seconds. The silver chloride wets the etched silver. The edge of the lithium fluoride was tapered (about 10°), cleaned, coated with platinum, the window pressed in the silver thimble, and heated at 500 °C. A silver chloride rod was then applied as a solder to the crack between the lithium fluoride and the silver.

Simard[1138] used the silver chloride in form of gaskets (44 mm i.d., 60 mm o.d., 0.3 mm thick) placed between the sodium chloride window and the flared walls of a Pyrex tube. This window must be kept above 150 °C, otherwise stresses resulting from expansion differences will cause cleavage of the salt window.

Palmer[964] sealed calcium fluoride windows to glass, by placing the silver chloride in a channel around the window (Fig. 3.10). The temperature is raised to 450 °C, gradually increasing, while the window is watched. As the silver chloride melts, a thin layer runs in suddenly between the window and its seat; the heating should be stopped immediately. The joint is vacuum tight even at bakeout to 430 °C.

Greenblatt[479] sealed a calcium fluoride window (18 mm diameter 1 mm, thick) to a glass pipe, by providing a step (Fig. 3.10) on the end of the pipe, by painting the seat with platinum,* and sealing with silver chloride.

Vogl[1271a] described a barium fluoride window (with transmission range from 1800 Å to 12.5 μ), sealed into a suitably formed, fine silver spinning with silver chloride. In order to make a reliable seal, the window blank was first washed with acetone and then inserted in the previously degreased silver frame, its bottom edge sitting on the shoulder of the frame. A strip is then cut from a sheet of 0.020 in. thick, clear silver chloride of such a width that when inserted edgewise its top is approximately 0.020 in. below the top of the barium fluoride. Caution should be exercised to avoid contact between the silver chloride and the polished surfaces of the barium fluoride since such contact tends to enhance the undesirable effect of coating of the barium fluoride. The assembly is then placed in a controlled atmosphere furnace (e.g. argon) and the temperature increased. For a 3 1/4 in. diameter by 1/2 in. thick window, a heating and cooling rate between 1–1.5 deg. C/min is satisfactory, but thinner windows could be heated more rapidly. Heat is applied until the frame reaches a temperature of 460–465 °C, and the assembly is held at this temperature for about 15 min. The temperature is then lowered at the same rate as it was increased. In the sealing process, a thin film of silver chloride forms over both faces of the window. Exposure of the assembly to light from a fluorescent lamp for about half an hour will solarize the film and render it easily visible. The film is removed by hand-polishing (with rouge paste on a microcloth pad).

Frank[398] sealed rocksalt windows (50 mm dia. 7 mm thick) and Wildy[1315] lithium fluoride windows, using an edge seal with silver chloride (Fig. 7.18b). The Pyrex tube (1) was ground flat and the ground surface was coated (2) with liquid bright platinum* and fired at 570 °C. A thin-walled, seamless silver tube (3) (10–12 mm long, and 0.35–0.50 mm wall thickness) was chamferred at its edges (to a wall thickness of 0.12 mm). The silver tube was degreased in 5–30 per cent nitric acid, washed in distilled water and dried with acetone or ether. Two rings (4) were cut from a 0.5 mm thick, silver chloride sheet, so that their inside diameters were slightly less than the bore of the Pyrex tube, and their outside diameters were slightly greater than the external diameter of the Pyrex tube. The cut silver chloride rings were carefully scraped with a razor blade, to remove any black traces of silver sulphide,

*Liquid Bright Pt #5 (Hannovia Chem. and Mfg. Co., Newark, N.J., U.S.A.).

thoroughly degreased in ether and pressed flat between filter papers. The whole assembly (Fig. 7.18b) was mounted vertically (without clamping), heated gradually to 500 °C, and allowed to cool slowly (5 deg. C/min) to room temperature. This window seal is bakeable to 420 °C, and is demountable, since heating to 500 °C allows the window (5) to be removed, and another platinized window to be replaced without the addition of a further silver chloride ring.

If this seal is to be made on metal instead of glass tubes, the silver tube is brazed at one end to the metal pipe, and has at the other end the silver chloride ring. In this case it is recommended to paint a thin ring of Aquadag on the inside of the silver ring (about 3 mm from the end) in order to prevent the flow of silver chloride to the metal pipe, which can be strongly corroded.

The sealing of *silver chloride* windows can be done by coating the end of the tube with silver chloride. This coated end is reheated until the coating is molten and then pressed immediately against the silver chloride window (Fugassi[411]). Silver chloride windows are plastic enough (Goodman[460], Fuoss[415]) to give a good seal when compressed by lock rings. Fuoss[415] made the seal by rubbing a minute amount of Apiezon grease M on the seat, and by using Cellophane washers between the lock ring and the window.

Small mica windows (up to 10 mm diameter) can also be sealed using silver chloride. The mica should be scratched since the silver chloride wets only rough mica surfaces (Espe[354]).

72.7 *Windows joined with compression seals*

Glass, quartz or ceramic windows can be sealed using the compression-seal technique (Sections 24.43, 25.24).

The quartz window, shaped as shown on Fig. 7.19, is sealed in a titanium ring, and brazed with silver–copper eutectic (melting at 779 °C). At 850 °C

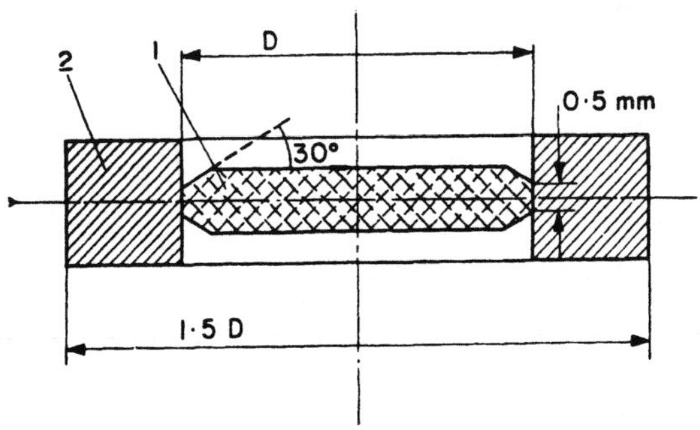

FIG. 7.19 The shape of a quartz window (1) sealed by a compression seal in a titanium ring (2). After Armand[49c] *(Courtesy of Société Française des Ingénieurs et Techniciens du Vide)*

this alloy dissolves enough titanium to be able to wet the quartz. The titanium ring is then sealed to the vacuum chamber by gaskets or brazing. As the expansion of the titanium is higher than that of the quartz, the inside diameter of the titanium ring should be machined smaller than the outside diameter of the quartz window, and the quartz window fitted inside at high temperatures. For the arrangement in Fig. 7.19, Armand[49c] finds that the optimal (negative) clearance on the diameter is:

0.12 mm for quartz of 30 mm diameter (baked at max 500 °C)

0.24 mm for quartz of 45 mm diameter (baked at max 400 °C)

0.61 mm for quartz of 90 mm diameter (baked at max 300 °C).

72.8 Windows sealed with elastomer gaskets

When windows are to be sealed using elastomer gaskets, the O-ring is the most convenient form of gasket. The two most common forms of O-ring seals used on windows are the *groove seal* (Section 38.41) and the *conical seal* (Section 38.43). Figure 7.20 shows the arrangements for such seals used on win-

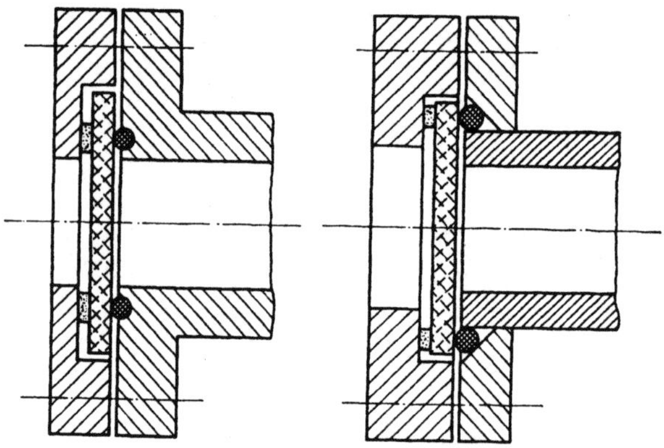

(a) (b)

FIG. 7.20 Windows sealed by O-rings: (a) O-ring in groove; (b) conical O-ring seal

dows. Instead of O-rings with circular cross sections square or trapezium cross section gaskets may also be utilized.

In order to construct reliable window seals, two basic conditions should be met:

(1) The window should have *no bending* strain, i.e. the force should be applied on both sides of the window on the same diameter.
(2) The window should *not be pressed against any hard material.*

Elastomer gasket seals may be used to seal glass, quartz, metal or plastic windows. The technique used to construct static gasket seals on plane glass or quartz windows is shown in Fig. 7.20, and is the same as the general technique described in Section 3.8. If the window has to be opened frequently it can be constructed (Tomboulian[1231]) as the sliding part of a gate valve (Section 61.32h) or sealed on an access door (Simons[1140]).

Mueller[900] sealed a beryllium (0.25 mm thick) window onto a cylindrical opening (Fig. 7.21) on a 4 in. diameter vessel. A shoulder was cut in the wall (1) of the vessel, and a small groove was then machined in this shoulder around the window opening. A continuous O-ring (2) was then made (Sections 38.41e and 38.48) to fit the groove around the window. A stainless steel frame (3) clamped the beryllium window (4). Since beryllium is hard and brittle it is necessary to first insert the centre screws and then to tighten the frame towards the ends.

The vacuum-tightness of the seals on metal windows can be tested very easily with helium, since helium does not permeate metals (Section 21.12). This ascertains that any helium detected on the other side of the seal has passed only through the seal (Klumb[692]).

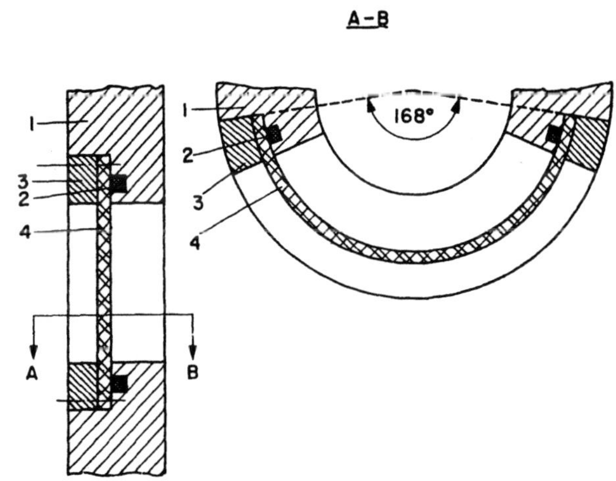

FIG. 7.21 Cylindrical beryllium window, sealed with O-rings. After Mueller[900] *(Courtesy of The American Institute of Physics)*

Nylon, Polythene or other plastic windows may be sealed using elastomer O-rings (Sturcken[1209a]). Hodkinson[571] described a technique used to seal a Polythene window on the side of a glass tube (Fig. 7.22a). A hole (1) of the required size is made in the glass tube by grinding, and the sharp edges of the opening are removed. An O-ring (2) of such a size is selected so that it will just rest on the circumference of the hole (cross section about 3.5 mm).

The O-ring is glued (with a rubber adhesive) onto the glass pipe around the hole; it will take the cylindrical shape of the contour of the glass tube. A Polythene tape (3) (0.25 mm thickness) is wrapped around the tube over the O-ring, and the ends are sealed together (with a soldering iron) on the other side of the tube. Hodkinson[571] obtained 5×10^{-6} torr in a tube having such a window.

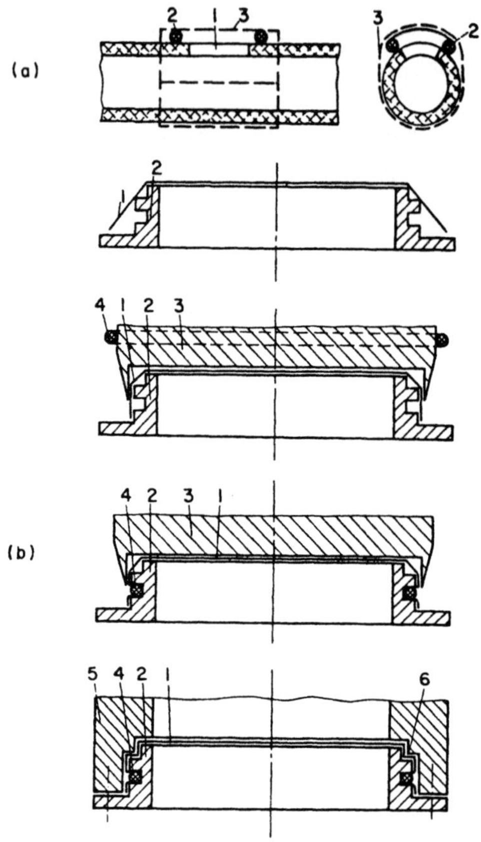

FIG. 7.22 Thin, plastic windows, sealed by O-rings: (a) side window of wrapped plastic tape (after Hodkinson[571]); (b) stretching of a plastic foil. After Figdor[374] *(Courtesy of The Institute of Physics and The Physical Society, London)*

Figdor[374] described a method of sealing thin plastic windows, by very tightly stretched membranes free of wrinkles. The window (1, Fig. 7.22b) is fixed onto a support (2), by laying it onto this part, and placing over it a tapered cylinder (3), which enables an O-ring (4) to slide into the groove of the part (2). The O-ring (4) holds the membrane strongly in position, and by pressing the part (5) over part (2) and screwing it down, the slightly rounded edge (6) stretches the membrane.

Elastomer gasket seals are also used to seal glove box windows of Perspex, PVC or glass. These seals are described in detail by Walton[1285], Barton[91], White[1306a], Kelman[660].

72.9 *Windows sealed with metal gaskets*

In applications where the window must resist high or low temperatures, the demountable seals should be made with metal gaskets.

For low temperature duty, metal gaskets made of lead, tin, indium or their alloys are used. For high temperatures, gold, copper and aluminium gaskets are used. The detailed shapes, dimensions and mounting techniques concerning the various metal gaskets are given in Sections 38.4–38.6.

Craig[243a] sealed an aluminium alloy (Dural 6061) window (6 Fig. 7.23a) to a cryostat (2), by using an In–Pb (50/50) O-ring (about 1.5 mm cross-sectional diameter). A brass collar (1) was soldered to the cryostat wall (2). The In–Pb O-ring was placed around the collar (1); the aluminium retainer (3) with the window (6) was then slipped over the collar (1) and the O-ring, and compressed by the clamp ring (4). Between collar (1) and retainer (3) a radial clearance of 0.05 mm is left. The window was provided with a lip (5) in order to transfer the stresses to the thin neck (7) which deforms elastically.

Perry[985] sealed beryllium windows by placing a soft metal gasket between the frame and the supporting flange.

Scott[1114] described the seal of a glass window made with double lead gaskets, and a guard vacuum (Section 38.23). The lead gaskets were mounted (Fig. 7.23b) on rings pressed by an inflatable stainless steel tubing, filled with helium.

Indium gaskets were used by Smiley[1150a] and Willis[1321] to seal quartz windows. Smiley[1150a] used an O-ring made of 2.5 mm indium wire, placed in a groove of semicircular cross section (Section 38.41d). The quartz window was sealed inside with this indium seal, and on the opposite (outside) side of the window with a Viton O-ring.

Willis[1321] used an indium gasket ring of rectangular cross section (1.9 mm wide and 1.4 mm thick) and compressed the window with spring washers (Section 38.58).

Gold washers have been used by Strnad[1204] and Lindsay[770] for sealing mica windows (Fig. 7.24). Strnad[1204] placed the mica window (1. Fig. 7.24a) between a Kovar part (2) and a gold washer (3) (0.05 mm thick), pressed on the other side by a second Kovar part (4). The two Kovar parts are joined in hot state to the outer stainless steel part (5). The gold washer is squeezed into intimate contact with the mica on one side and the rounded Kovar edge

(4) on the other side, due to the compression given by the differential contraction of the Kovar and the stainless steel.

Lindsay[770] sealed the mica window (1), between a molybdenum or Kovar washer (2 Fig. 7.24b), and a gold foil (3). The 0.025 mm gold foil (3) is pressed by a chamfered copper ring (4) between the flanges (5, 6).

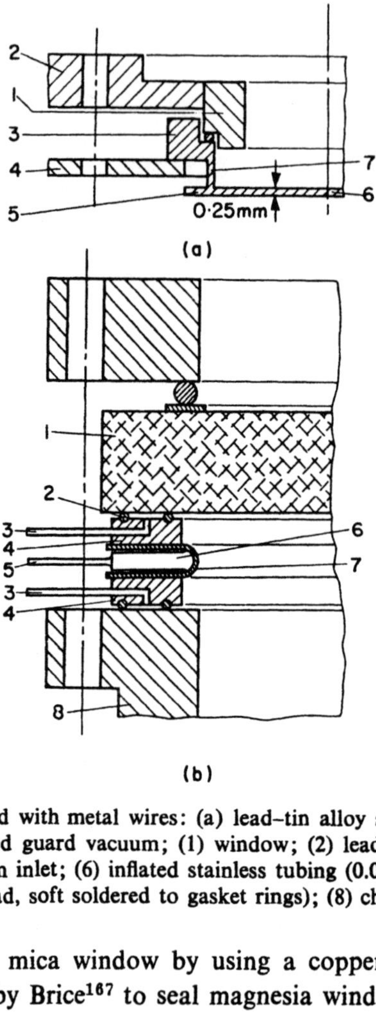

FIG. 7.23 Windows, sealed with metal wires: (a) lead–tin alloy seal (Craig[243a]); (b) seal with double lead wire and guard vacuum; (1) window; (2) lead-gaskets; (3) pumpout; (4) gasket ring; (5) helium inlet; (6) inflated stainless tubing (0.010 in. wall); (7) sealing diaphragm (0.015 in. lead, soft soldered to gasket rings); (8) chamber wall (Scott[1114])

Sterzer[1184] sealed a mica window by using a copper gasket. Copper or aluminium was used by Brice[167] to seal magnesia windows (Fig. 7.25). The magnesia window (12 mm diameter, 6 mm thick) was ground and polished to a 1° taper. A heavy bar of aluminium (or copper) was then bored out with an identical inside taper and the window pressed firmly into it. Then the aluminium was turned down on the outside, to leave a thickness of 0.5 mm. This was pressed into a steel tube with the same inside taper.

Coined copper gaskets have been used by Bridge[167a] and Ehlers[332] for sealing glass or quartz windows (Fig. 7.26). Bridge[167a] used a copper gasket of 4 × 10 mm (see Fig. 7.26a), and made the seal by screwing the flanges with a 300 kg/cm length of gasket, on a diameter of 34 cm. Ehlers[332] sealed a window (1) on his ultra-high vacuum, double-walled chamber, by compressing a specially shaped (Fig. 7.26b) copper gasket (2) to the window. The copper

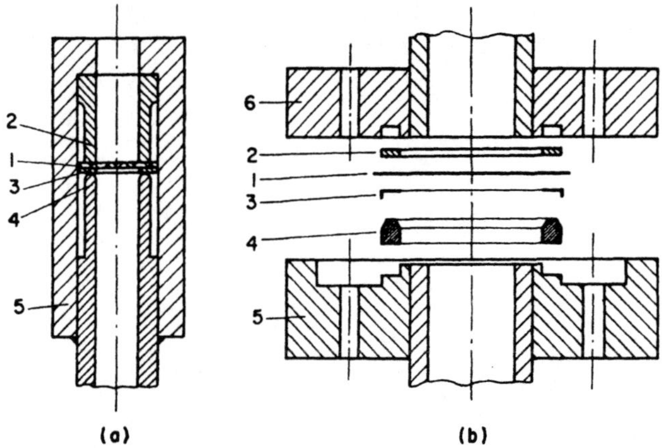

FIG. 7.24 Window seals with gold foil: (a) mica window, sealed with gold washer pressed by Kovar nose; (1) mica window; (2) Kovar tube; (3) gold washer; (4) Kovar tube; (5) stainless steel tube (after Strnad[1204]); (b) mica window, sealed with gold washer pressed by profiled copper gasket; (1) mica window (0.001 in.); (2) molybdenum washer (0.015 in.); (3) gold foil (0.001 in.); (4) copper gasket (0.015 × 0.015 × 0.004 in.) (50 flange. After Lindsay[770]) *(Courtesy of The American Institute of Physics)*

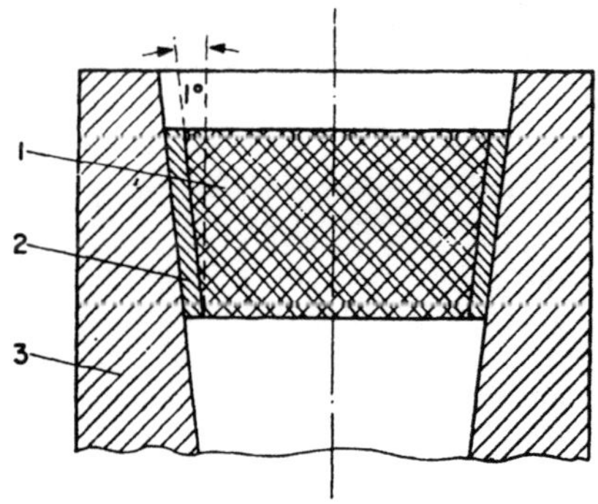

FIG. 7.25 Conical seal on MgO window: (1) MgO window; (2) aluminium or copper; (3) steel. After Brice[167] *(Courtesy of The American Institute of Physics)*

gasket is the seal between the guard vacuum and the ultra-high vacuum (Fig. 7.26b). Tobin[1225a] used coined copper gaskets and Conflat seals (Section 38.56c).

Higatsberger[559] sealed a glass window (1, Fig. 7.27) (3.5 in. diameter, 7/16 in. thick) by depositing onto it a layer (0.1 μ) of chronium (2), on which another layer of Cr–Cu (3) of about the same thickness is formed by simultaneous evaporation of chromium and copper. Finally the chromium evaporation is stopped, and only the copper evaporation is continued, in order to obtain copper on the surface. Onto this copper layer, a 0.1–0.5 mm copper is electroplated (4), onto which again an evaporated chromium (0.1 μ) layer (5) is deposited. Onto this layer, an aluminium gasket (6) with a knife edge seal was used. The aluminium gasket sealed well to the formed layer, but aluminium does not seal well directly on glass.

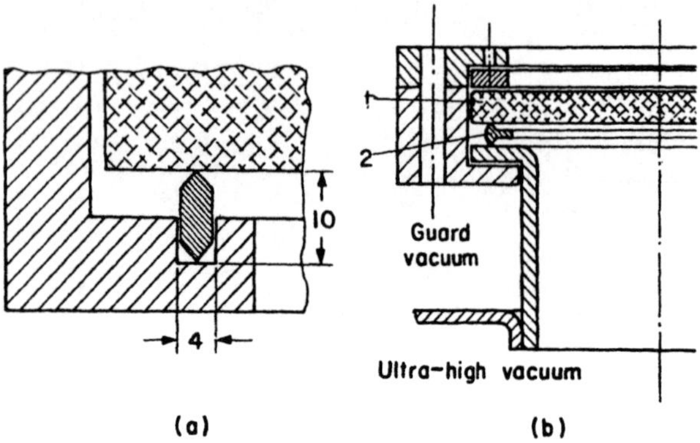

FIG. 7.26 Bakeable windows, sealed with profiled metal gaskets: (a) in groove. Reproduced from Bridge[167a]. (b) Double knife edge gasket. Reproduced from Ehlers[331] *(Courtesy of Pergamon Press)*

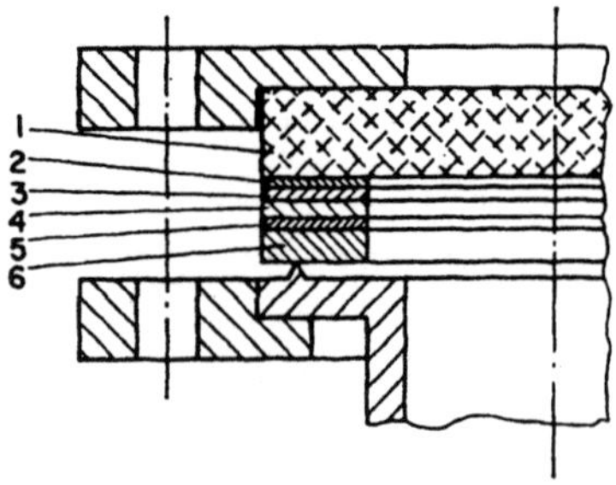

FIG. 7.27 Window sealed with knife edge seal and sealing metal coatings. After Higatsberger[559] *(Courtesy of The American Institute of Physics)*

APPENDICES

Appendix A.1. Conversion Factors (n) for Pressure Units* (1.X = n.Y)

X \ Y	dyne/cm²	microns of Hg	N/m²	mm of water or kg/m²	millibar	cm of water	torr	in. of Hg	lb/in² (psi)	at (kg/cm²)	Bar	atm
dyne/cm² or microbar	1	7.5×10^{-1}	1×10^{-1}	1.01×10^{-2}	1×10^{-3}	1.01×10^{-3}	7.5×10^{-4}	2.95×10^{-5}	1.45×10^{-5}	1.01×10^{-6}	1×10^{-6}	9.8×10^{-7}
microns of Hg	1.33	1	$1\times33\times10^{-1}$	1.35×10^{-2}	1.33×10^{-3}	1.35×10^{-3}	1×10^{-3}	3.93×10^{-5}	1.93×10^{-5}	1.35×10^{-6}	1.33×10^{-6}	1.31×10^{-6}
N/m² (Newton per m²)	10	7.5	1	1.01×10^{-1}	1×10^{-2}	1.01×10^{-2}	7.5×10^{-3}	2.95×10^{-4}	1.45×10^{-4}	1.01×10^{-5}	1×10^{-5}	9.8×10^{-5}
mm of water or kg/m²	98	73	9.8	1	9.8×10^{-2}	1×10^{-1}	7.3×10^{-2}	2.89×10^{-3}	1.42×10^{-3}	1.10^{-4}	9.8×10^{-5}	9.6×10^{-5}
millibar (mb)	1×10^{3}	750	100	10.1	1	1.01	7.5×10^{-1}	2.95×10^{-2}	1.45×10^{-2}	1.01×10^{-3}	1×10^{-3}	9.8×10^{-4}
cm of water or Ger (Guericke)	980	730	98	10	9.8×10^{-1}	1	7.3×10^{-1}	2.89×10^{-2}	1.42×10^{-2}	1×10^{-3}	9.8×10^{-4}	9.6×10^{-4}
torr	1.33×10^{3}	1×10^{3}	133.3	13.59	1.33	1.35	1	3.93×10^{-2}	1.93×10^{-2}	1.35×10^{-3}	1.33×10^{-3}	1.31×10^{-3}
in. of Hg	3.3×10^{4}	2.54×10^{4}	3386	340	33	34	25.4	1	4.9×10^{-1}	3.4×10^{-2}	3.3×10^{-2}	3.3×10^{-2}
lb/in² (p.s.i.)	6.8×10^{4}	5.17×10^{4}	6894.7	700	68	70	51.7	2.03	1	7×10^{-2}	6.8×10^{-2}	6.8×10^{-2}
Techn. atmosphere (at) kg/cm²	9.8×10^{5}	7.3×10^{5}	9.81×10^{4}	1×10^{4}	980	1×10^{3}	735	28.9	14.2	1	9.8×10^{-1}	9.6×10^{-1}
Bar	1×10^{6}	7.5×10^{5}	1×10^{5}	1.01×10^{4}	1×10^{3}	1.01×10^{3}	750	29.5	14.5	1.01	1	9.8×10^{-1}
Physical atmosphere (atm)	1.01×10^{6}	7.6×10^{5}	1.01×10^{5}	1.03×10^{4}	1.01×10^{3}	1.03×10^{3}	760	29.92	14.7	1.03	1.01	1

* 1 Gd (Gaede) = 10^{-6} Newton/m².
1 Pascal = 1 Newton/m² = 10 microbar.
1 pz (pieze) = 1000 Newton/m².

Appendix A.2.
Conversion Factors (n) for Conductance and Pumping Speed Units (1.X = n.Y)

X \ Y	cm^3/sec	l./min	m^3/hr	ft^3/min	l./sec
cm^3/sec	1	6×10^{-2}	3.6×10^{-3}	2.1×10^{-3}	1×10^{-3}
l./min	16.67	1	6×10^{-2}	3.53×10^{-2}	1.67×10^{-2}
m^3/hr	277.8	16.67	1	5.89×10^{-1}	2.78×10^{-1}
ft^3/min	471.95	28.32	1.699	1	4.7×10^{-1}
l./sec	1000	60	3.6	2.12	1

Appendix A.3. Conversion Factors (n) for Throughput or Leak Rate Units (1.X = n.Y)

X \ Y	cm^3 (STP)/ /yr	atm.cm^3/hr	μ.ft^3 per min.	ft^3(STP)/yr	lusec	cm^3(STP) /min	g/sec (air)	cm^3. atm/sec	torr.l./sec	ft^3(STP)/hr	kg/hr (air)
cm^3 (STP)/yr	1	1.12×10^{-4}	5.05×10^{-5}	3.5×10^{-5}	2.4×10^{-5}	1.88×10^{-6}	3.74×10^{-8}	3.1×10^{-8}	2.4×10^{-8}	4.05×10^{-9}	1.47×10^{-10}
atm.cm^3/hr	8.9×10^{3}	1	4.47×10^{-1}	3.1×10^{-1}	2.11×10^{-1}	1.62×10^{-2}	3.2×10^{-4}	2.78×10^{-4}	2.11×10^{-4}	3.57×10^{-5}	1.29×10^{-6}
micron.ft^3/min	1.98×10^{4}	2.236	1	6.9×10^{-1}	4.72×10^{-1}	3.75×10^{-2}	7.4×10^{-4}	6.2×10^{-4}	4.72×10^{-4}	7.95×10^{-5}	2.88×10^{-6}
ft^3 (STP)/yr	2.86×10^{4}	3.24	1.45	1	6.82×10^{-1}	5.8×10^{-2}	1.07×10^{-3}	8.9×10^{-4}	6.82×10^{-4}	1.15×10^{-4}	4.18×10^{-6}
lusec (l.μ/sec)	4.2×10^{4}	4.74	2.12	1.47	1	7.9×10^{-2}	1.56×10^{-3}	1.32×10^{-3}	1×10^{-3}	1.69×10^{-4}	6.12×10^{-6}
cm^3 (STP)/min or std.cc/min	5.36×10^{5}	60.4	27	18.6	12.7	1	1.98×10^{-2}	1.68×10^{-2}	1.27×10^{-2}	2.14×10^{-3}	7.75×10^{-5}
grams of air per sec	2.7×10^{7}	3040	1360	940	640	50.4	1	8.4×10^{-1}	6.4×10^{-1}	1.08×10^{-1}	3.93×10^{-3}
std.cc/sec, or cm^3 (STP)/sec, or cm^3.atm/sec	3.22×10^{7}	3600	1620	1120	760	58.2	1.16	1	7.6×10^{-1}	1.29×10^{-1}	4.64×10^{-3}
torr. l./sec	4.2×10^{7}	4738	2120	1470	1000	79	1.56	1.32	1	1.69×10^{-1}	6.12×10^{-3}
ft^3 (STP)/hr	2.5×10^{8}	2.84×10^{4}	1.27×10^{4}	8750	5970	471	9.32	7.9	5.97	1	3.64×10^{-2}
kg of air/hr	6.9×10^{9}	7.73×10^{5}	3.46×10^{5}	2.39×10^{5}	1.63×10^{5}	1.26×10^{4}	255	214	163	27.5	1

APPENDIX A.4. CONVERSION FACTORS (N) FOR PERMEABILITY COEFFICIENTS (1.X = n.Y)

X \ Y	$\frac{cm^3(STP)mm}{cm^2/sec/torr}$	$\frac{cm^3(STP)mm}{cm^2/sec/atm}$	$\frac{l.(torr.\ 0\ °C)mm}{cm^2/hr/torr}$	$\frac{lusec\ mm}{cm^2/atm}$	$\frac{l.(torr.0\ °C)mm}{cm^2/hr/atm}$	$\frac{l.(\mu.0\ °C)mm}{cm^2/min/atm}$	$\frac{cm^3(torr.\ 0\ °C)mm}{dm^2/hr/torr}$	$\frac{cm^3(torr.\ 0\ °C)mm}{dm^2/hr/atm}$
$\frac{cm^3(STP)mm}{cm^2/sec/torr}$	1	760	2.74×10^{3}	5.7×10^{5}	2.08×10^{6}	3.4×10^{7}	2.74×10^{8}	2.08×10^{11}
$\frac{cm^3(STP)mm}{cm^2/sec/atm}$	1.32×10^{-3}	1	3.6	760	2.74×10^{3}	4.5×10^{4}	3.6×10^{5}	2.74×10^{8}
$\frac{l.(torr.0\ °C)mm}{cm^2/hr/torr}$	3.67×10^{-4}	2.78×10^{-1}	1	210	760	1.27×10^{4}	1×10^{5}	7.6×10^{7}
$\frac{lusec\ mm}{cm^2/atm}$	1.74×10^{-6}	1.34×10^{-3}	4.76×10^{-3}	1	3.64	61	476	3.64×10^{5}
$\frac{l.(torr.0\ °C)mm}{cm^2/hr/atm}$	4.8×10^{-7}	3.56×10^{-4}	1.31×10^{-3}	2.74×10^{-1}	1	16.7	131	1×10^{5}
$\frac{l.(\mu.0\ °C)mm}{cm^2/min/atm}$	2.9×10^{-8}	2.2×10^{-5}	7.9×10^{-5}	1.68×10^{-2}	6×10^{-2}	1	7.9	6×10^{3}
$\frac{cm^3(torr.0\ °C)mm}{dm^2/hr/torr}$	3.67×10^{-9}	2.78×10^{-6}	1×10^{-5}	2.1×10^{-3}	7.6×10^{-3}	1.21×10^{-1}	1	760
$\frac{cm^3(torr.0\ °C)mm}{dm^2/hr/atm}$	4.8×10^{-12}	3.65×10^{-9}	1.31×10^{-8}	2.74×10^{-6}	1×10^{-5}	1.67×10^{-4}	1.31×10^{-3}	1

APPENDIX A.5. CONDUCTANCE FORMULAE*

	Molecular flow	Intermediate flow	Viscous flow
Criterion (10% error)	$\bar{p}\cdot d < 1.5\times 10^{-2}$	$1.5\times 10^{-2} < \bar{p}\cdot d. < 0.5$	$p\cdot d > 0.5$
Long pipe with circular cross section	$C=k_m\cdot d^3/l$	$C = k_m\cdot J\cdot d^3/l$	$C = k_v\cdot \bar{p}\cdot d^4/l$
Orifice	$C = k_0\ A$		
Duct with rectangular cross section	$C = k_r \dfrac{a^2\cdot b^2}{(a+b)l} K$		$C = k_p\cdot Y\cdot \bar{p}\dfrac{a^2\cdot b^2}{l}$
Short pipe ($l < 20.d$)	$C = k_m\cdot e\cdot d^3/l$		

Long pipe with circular cross section (intermediate flow):

$\bar{p}\cdot d.$	0.02	0.06	0.1	0.4
J	1.1	1.7	2.3	6.9

Duct with rectangular cross section:

b/a	1	0.7	0.5	0.2	0.1
K	1.1	1 12	1.15	1.3	1.4
Y	1.0	0 95	0.82	0.4	0.23

Constants:

	Ne	A	Air	N_2	He	vap. H_2O	H_2
k_v	106	150	182	190	169	345	382
k_m	14.4	10.3	12.1	12.2	32.2	15.2	45.6
k_0	13.8	9.9	11.6	11.7	31.0	14.6	43.7
k_r	36.6	26.3	30.9	31.1	82.5	39.0	117.0
k_p	152	214	260	270	241	494	545

Short pipe:

l/d	0.05	0.1	0.4	0.8	1.0	2	10	50	100
e	0.036	0.07	0.21	0.30	0.38	0.54	0.84	0.96	1

* $\bar{p}$ = mean pressure (torr) = $\dfrac{p_1+p_2}{2}$

A = area of the orifice (cm²)

d = diameter (cm); l = length (cm)

a, b = sides of the rectangle (cm)

C = conductance (l./sec).

For more detailed formulae and graphs see e.g. Delafosse[271], Dushman[314, 315], Roth[1086], Rothstein[1088], Santeler[1097], Müller[905], Heinze[533], Guthrie[491].

APPENDIX B.1a. MECHANICAL PROPERTIES OF METALS AND ALLOYS

Material	Characteristics	Specific gravity	Tensile strength (kg/mm^2)		Elongation (%)	Modulus of elasticity (kg/mm^2)		Modulus of torsion (kg/mm^2)	Hardness*
			20 °C	500 °C		20 °C	500 °C		
Aluminium	Pure (99.6%); annealed	2.7	9	0.7	30–45	7 400	5 000	2 700	M-2; B-16
Beryllium	Pure (99.5%)	1.85	21	—	6	19 000	—	—	M-7.8; B-60/120
Iron	Pure	7.87	25	8	50	21 200	12 000	8 000	M-4.5; B-45/80; R-40B
Copper	—	8.93	30	3	30	12 000	4 000	4 000	M-2/4; B-50; R-40E
Nickel	—	8.9	60	34	40	20 500	15 000	7 000	M-3.8; B-200; V-210
Gold	—	19.3	15	—	25	8 000	—	—	M-2.5; B-20/30; V-20
Silver	—	10.4	14	—	68	8 000	—	2 800	M-2.6; B-20/40; V-25
Platinum	—	21.5	24	18	35	16 500	—	6 600	M-4.5; B-55; V50
Palladium	—	12.0	36	—	4	11 500	—	5 000	M-4.8; B-49; V-38
Titanium	—	4.5	66	—	10	12 000	—	—	B-185/250; R-91
Zirconium	—	6.5	95	—	2	9 000	—	—	M-6.5; B-80; R-B87
Molybdenum	—	10.2	95	58	15	32 000	—	16 000	M-5.5; B-147; V-160; R-26/80
Tantalum	—	16.6	93	—	15	19 000	—	6 300	M-6.3; B-45/350; V-70
Tungsten	—	19.3	115	55	3	36 000	—	22 000	M-5/8; B-350; R-37C; V-400

Appendix B.1a. Mechanical Properties of Metals and Alloys

Material	Characteristics	Specific gravity	Tensile strength (kg/mm^2)		Elongation (%)	Modulus of elasticity (kg/mm^2)		Modulus of torsion (kg/mm^2)	Hardness*
			20 °C	500 °C		20 °C	500 °C		
Platinit	Fe/Ni/C (51.85/48/0.15)	8.3	65	–	20	15 000	–	6 500	B-100; R-B70
Invar	Fe/Ni/C (63.6/36/0.4)	8.1	55	–	30	15 000	–	–	B-110/160
Fernico (Kovar)	Fe/Ni/Co (55/28/17)	8.2	62	–	35	16 000	–	–	B-150; R-B80
Monel	Ni/Cu/Fe/Mn/Si (64.3/30/2.4/2/0.5)	8.8	70	50	20	20 000	–	–	B-120/180; R-B60
Constantan	Cu/Ni (55/45)	8.9	45	35	40	14 800	–	–	B-80
Iron–Chromium	Fe/Cr (70/30)	7.5	60	–	20	20 000	–	–	B-140/200; V-140; R-B90
Stainless steel	Fe/Cr/Ni/Si/Mn/C (73/18/8)	7.9	65	36	50	19 500	–	–	B-160; R-B72
Nichrome	Ni/Cr (80/20)	8.3	60	45	30	21 000	–	–	B-90; R-B90
Inconel	Ni/Cr/Fe/Mn/Cu/Si (78/13/7/1)	8.6	80	–	30	22 000	–	–	B-120; R-B65
Tombak	Cu/Zn (72/28)	8.6	55	–	10	11 000	–	–	B-120; R-B93
Brass	Cu/Zn (63/37)	8.4	32	–	5	9 000	–	–	B-50
Bronze	Cu/Sn (92/8)	8.8	40	–	60	10 000	–	–	B-80; R-B84

* M-Mohs; B-Brinell; R-Rockwell; V-Vickers.

APPENDIX B.1b. MECHANICAL PROPERTIES OF GLASS, CERAMICS AND ELASTOMERS*

Material characteristics		Specific gravity	Tensile strength (kg/mm²)	Elongation (%)	Compression strength (kg/mm²)	Bending strength (kg/mm²)	Torsion strength (kg/mm²)	Modulus of elasticity (kg/mm²)	Hardness*	Impact strength (kg/mm²)
Soft glasses	Soda-lime	2.4—2.6	5—13	—	60—90	5	8—9	5 000— 8 000	M-4/5	—
	Lead silicate	3.1—3.9	5—8	—	60—90	4—5	7—8	5 000— 6 000	M-4/4.5	—
Hard glasses	Borosilicate	2.2—2.4	8—15	—	60—100	6—8	8—9	6 500— 8 500	M-7/7.5	—
	Alumino-borosilicate	2.2—2.7	8—15	—	75—100	6—8	8—9	6 500— 9 000	M-7/8	—
Quartz	—	2.0—2.2	7—12	—	100—200	7	3	6 200— 7 200	M-7	—
High voltage porcelain	—	2.3—2.5	2—4	—	40—50	7—8	—	5 500— 8000	M-7	1.5—2.0
Steatite	—	2.6—2.7	5—6	—	60—80	12—16	—	8 000—11 000	M-7.5	3.0—3.5
Forsterite	—	2.7—2.9	7—8	—	70—90	11—16	—	11 000—13 000	M-7.5	3.5—4.0
Alumina	85–99% Al_2O_3	3.4—3.8	12—18	—	100—200	25—40	—	25 000—38 000	M-8/9	5.0—6.0
Zircon (ceramic)	—	3.1—3.4	5—10	—	60—120	11—18	—	17 000	M-8	5.0—6.0
Mica	—	2.1—3.2	40—75	—	10 (normal)	—	—	16 000—21 000	M-2.5/3	—
Natural rubber	—	0.9—2.0	0.8—2.5	350—1000 max. allowed 100	0.5	—	—	0.1—0.7	S-30/90	—
Neoprene	—	1.1—1.24	1—1.5	200—800	see Fig. 3.52	—	—	—	S-40/60	—
Silicone rubber	—	1.2—2.1	0.4—0.7	100—800	see Fig. 3.52	—	—	—	S-40/80	—
PVC	—	1.2—1.6	5—7	10—50	7—9	—	—	150—350	—	—
Teflon	—	2.1—2.3	1.4—3.8	200—400	4—6 (see Fig. 3.53)	1.3	—	43—52	S-50/65	—
Perspex	—	1.15—1.18	4—8	1—10	6—12	—	—	300—350	—	—
Polystyrene	—	1.05	4.7	1—4	8—12	—	—	100—300	—	—
Saran	—	1.65—1.72	3—4	6—9	4—5	—	—	350—450	—	—

* Hardness: M-Mohs; S-Shore (Durometer).

Appendix B.2a
Thermal Properties of Metals and Alloys

Material	Charac-teristics	Melting Point (°C)	Specific heat (cal/g/deg.C)		Thermal con-ductivity (cal/cm/sec/deg. C)		Coefficient of thermal expan-sion (linear) ($\alpha \times 10^{-7}$)	
			20 °C	500 °C	20 °C	500 °C	20 °C	500 °C
Aluminium		659	0.21	0.26	0.53	0.40	231	300
Beryllium	see	1284	0.40	0.60	0.43	—	140	180
Iron	App.	1535	0.11	0.16	0.17	0.11	115	170
Copper	B.1a.	1084	0.09	0.11	0.94	0.85	164	200
Nickel		1453	0.10	0.13	0.20	0.14	125	172
Gold		1063	0.031	—	0.71	—	142	—
Silver		960	0.056	0.058	1.0	0.9	187	—
Platinum		1773	0.033	0.035	0.17	0.19	89	100
Palladium		1554	0.054	0.065	0.17	—	106	126
Titanium		1690	0.126	—	0.045	—	88	—
Zirconium		1857	0.07	0.08	0.038	—	63	83
Molybdenum		2630	0.062	0.068	0.38	0.34	51	62
Tantalum		2996	0.034	0.036	0.13	0.14	66	68
Tungsten		3395	0.034	0.037	0.39	0.31	44	47
Platinit		1470	0.12	—	0.038	—	89	140
Invar		1495	0.122	0.130	0.026	—	8	175
Fernico (Kovar)	see App.	1450	—	—	0.05	—	45	50
Monel	B.1a.	1325	0.12	—	0.06	—	140	—
Constantan		1260	0.098	—	0.05	—	150	176
Iron-Chromium	see	1480	—	—	0.05	—	95	117
Stainless steel	App. B.	1400	0.118	—	0.035	—	158	210
Nichrome	1a.	1450	0.106	—	0.032	0.050	136	175
Inconel		1417	—	—	0.04	—	36	—
Tombak	see	950	0.09	—	0.27	—	186	—
Brass	App. B.	905	0.09	—	0.20	0.34	184	—
Bronze	1a.	1020/880	0.09	—	0.15	—	182	—

APPENDIX B.2b
THERMAL PROPERTIES OF GLASSES, CERAMICS AND ELASTOMERS

Material	Characteristics	Softening Point (°C)	Specific heat (cal/g/deg.C)	Thermal conductivity (cal/cm/sec/deg.C)	Coefficient of thermal expansion (linear) ($\alpha \times 10^{-7}$)	Working temperature range (°C)
Soft glasses	Soda–lime	see App. B.3	0.08–0.23 (20 °C)	0.0024	68–110	see App. B.3
	Lead-silicate		0.08–0.23 (20 °C)	0.0018	80–90	
Hard glasses	Borosilicate		0.08–0.23 (20 °C)	0.0030	35–60	
	Alumino-borosilicate		0.08–0.23 (20 °C)	0.0024	30–60	
Quartz			0.20 (100 °C); 0.29 (1000 °C)	0.0029 (0 °C); 0.0064 (950 °C)	5.8–5.9	
High voltage porcelain	see Table 2.40	1300–1500	0.19–0.21 (20–100 °C)	0.0028–0.0040	40–50	up to 1100
Steatite		1300–1400	0.19–0.22 (20–100 °C)	0.0060–0.0070	80–100	up to 1100
Forsterite		1400–1500	0.19–0.22 (20–100 °C)	0.0060–0.0080	90–120	up to 1100
Alumina		1400–1900	0.20–0.23 (20–100 °C)	0.030 –0.070	70–90	up to 1600
Zircon (ceramic)		1400–1500	0.12 (20–100 °C)	0.007 –0.0120	50–60	up to 1300
Mica		—	0.206	$0-14 \times 10^{-4}$	85–100	500–600
Natural rubber		see Working temp. range	0.4–0.5	$3-5 \times 10^{-4}$	1400–1900 (20–120 °C)	−65 to 75 (instant); −30 to 60 (permanent);
Neoprene			0.4	5×10^{-4}	1800	−40 to 100
Silicone rubber			0.3 (90 °C)	$5-17 \times 10^{-4}$	2500	−50 to 180 (permanent) −85 to 250 (instant)
PVC	see Table 2.14	see Working temp. range	—	$4-5 \times 10^{-4}$	600–800	up to 80
Teflon			0.25	6×10^{-4}	800–2000	−50 to 200 (permanent); −150 to 250 (instant)
Perspex			—	4×10^{-4}	700–1000	up to 60
Polystyrene			—	$2-4 \times 10^{-4}$	600–800	up to 90
Saran			—	3×10^{-4}	1600–1700	up to 80

APPENDIX B 3.
THERMAL POINTS OF GLASSES**

Designation of the glass		Manufacturer***	Coefficient of thermal expansion ($\alpha.10^7$).	Transformation* Point ($10^{13.3}$ poises) (°C)	Strain Point ($10^{14.5}$ poises) (°C)	Annealing Point (10^{13} poises) (°C)	Deformation Point(10^{11}–$10^{11.5}$ poises) (°C)	Softening Point ($10^{7.65}$ poises) (°C)	Working Point (10^4 poises) (°C)
7740	Pyrex	C	32 (0–300 °C)	—	515	555	—	820	1220
C 38	—	BTH	32 (50–400 °C)	450	—	505	—	—	—
8330	Duran	J	32 (20–300 °C)	520	510	568	—	815	1245
C 9	Tungsten	BTH	36 (50–400 °C)	—	480	525	575 (10^{12} poises)	775	—
Bluesil	Tungsten	P	37 (0–300 °C)	—	—	570	—	—	—
7720	Nonex	C	36 (0–300 °C)	—	484	518	—	755	1110
712b	Tungsten	O	42 (25–75 °C)	565	—	—	635 (10^{11} poises)	738 (10^9 poises)	—
W 1	Tungsten	GEC	37 (20–350 °C)	—	510	570	—	760	—
GSD	Tungsten	Ch	38 (10–100 °C)	—	—	570	605 (10^{12} poises)	780	—
7050	Clear sealing	C	46 (0–300 °C)	—	461	496	—	703	1025
GSB	Kovar	Ch	49 (10–100 °C)	—	390	450	500 (10^{12} poises)	710	—
7040	Kovar	C	47 (0–300 °C)	—	450	484	—	702	1080
Dial 43		P	42 (0–300 °C)	—	—	560	—	—	—
C 11	Molybdenum	BTH	45 (50–400 °C)	530	500	575	—	795	—
637h	Molybdenum	O	48 (25– 75 °C)	550	—	—	628 (10^{11} poises)	702 (10^9 poises)	—
3072	(G 20)	J	46 (20–100 °C)	558	—	—	608 (10^{12} poises)	779 (10^8 poises)	—
Kodial		P	49 (0–300 °C)	—	—	535	—	—	—
C 40	Kovar	BTH	48 (50–400 °C)	480	455	505	535 (10^{12} poises)	710	—
K 650		K	51 (0–300 °C)	—	472	502	—	705	1015
7060	Kovar	C	50 (0–300 °C)	—	463	495	—	690	—

HH	Molybdenum	GEC	47 (20–450 °C)	—	500	590	—	780	—
362a	Tungsten	O	39 (25– 75 °C)	522	—	—	548 (10^{11} poises)	664 (10^{9} poises)	—
1447	Molybdenum	J	51 (20–300 °C)	520	483	529	—	725	1080
GS 4	Molybdenum	Ch	44 (20–350 °C)	—	(590)	620	—	—	—
0 120	Soda lead	C	89 (0–300 °C)	—	400	433	—	630	975
C 12	Dumet	BTH	91 (50–400 °C)	—	380	435	465 (10^{12} poises)	630	960
L 1	FeCr, Dumet	GEC	91 (20–320 °C)	—	340	430	—	610	—
0010	Soda lead	C	91 (0–300 °C)	—	397	428	—	626	970
GWB	Dumet	Ch	86 (10–100 °C)	—	—	400	435 (10^{12} poises)	610	—
C 31	FeCr	BTH	97 (20–350 °C)	—	—	442	—	—	—
0240	FeCr	C	96 (0–300 °C)	—	397	425	—	607	—
M 6	Neutral	GEC	73 (20–350 °C)	—	450	580	—	600	—
16 III	Normal	J	83 (20–200 °C)	550	—	—	589 (10^{12} poises)	693 (10^{8} poises)	—
0080	Soda lime	C	92 (0–300 °C)	—	478	510	—	696	1000
GWA	FeCr, Platinum	Ch	91 (20–350 °C)	—	—	520	545 (10^{12} poises)	640–710	—
548d	Apparate-glass	O	88 (25– 75 °C)	530	—	—	578 (10^{11} poises)	646 (10^{9} poises)	—
X 8	FeCr	GEC	96 (20–350 °C)	—	400	520	—	690	—
C 19	FeCr	BTH	95 (50–400 °C)	550	—	530	550 (10^{12} poises)	710	1025
C 22	Soda-lime	BTH	104 (50–400 °C)	530	—	505	525 (10^{12} poises)	680	—
105	Magnezia	O	89 (25– 75 °C)	508	—	—	555 (10^{11} poises)	627 (10^{9} poises)	—

* Measured as shown in Fig. 2.46.

** Section 23.11.

*** BTH — The British Thomson-Houston Co. Ltd., Rugby, England; Ch — Chance Brothers Ltd., Birmingham; C — Corning Glass Works, Corning, N.Y.; GEC — Osram-G.E.C. Glass Works, East Lane, Wembley, Middlesex; O — Osram, Berlin W.; J — Jenauer Glaswerk Schott u. Gen., Mainz (W. Germany); P— Plowden-Thomson, Stourbridge, England; K — Kimble, Owens Illinois Glass Co.

Appendix B.4a
Outgassing of Metals

Material	Total gas content (cm^3STP/cm^3 material)	Outgassing (lusec/ /cm^2 at 20 °C, after 2 hr pumping)	Rate of evaporation at 10^{-5} torr (g/cm^2/ /sec)	Temperature (°C) at which the vapour pressure is 10^{-8} torr	10^{-5} torr	10^{-2} torr	1 torr
Aluminium	0.1–0.5	7×10^{-6}	9×10^{-8}	677	882	1207	1547
Beryllium	–	–	5×10^{-8}	699	902	1212	1567
Iron	0.1–1.0	2×10^{-6}	1×10^{-7}	877	1107	1467	1847
Copper	0.04–0.7	1×10^{-5}	1×10^{-7}	732	1042	1272	1622
Nickel	0.4–40	5×10^{-6}	1×10^{-7}	912	1142	1497	1877
Steel (1% C)	150–200	–	–	about as iron			
Gold	–	–	2×10^{-7}	772	987	1332	1718
Silver	–	3×10^{-6}	1×10^{-7}	579	757	1032	1338
Platinum	–	–	2×10^{-7}	1287	1602	2077	2587
Palladium	800(H_2; 100 °C)	–	1.6×10^{-7}	907	1157	1546	1967
Titanium	1000(H_2; 500 °C)	–	1×10^{-7}	1056	1328	1728	2177
Zirconium	1200(H_2; 560 °C)	–	1×10^{-7}	1472	1838	2397	2977
Molybdenum	0.01–0.1	6×10^{-6}	1×10^{-7}	1582	1987	2627	3300
Tantalum	180(H_2; 500 °C)	5×10^{-6}	1.5×10^{-7}	1957	2397	3067	3737
Tungsten	0.004–0.008	1×10^{-6}	1.5×10^{-7}	2067	2547	3297	–

Appendix B.4b
Outgassing of Glasses, Ceramics and Elastomers

Material		Outgassing rate* (lusec/cm^2, at 20 °C after 3 hr pumping)	Total gas evolution* (cm^3 STP/dm^2)	Water absorption (%)
Soft glasses;	Soda–lime	–	20 (max.at 250 °C)	–
	Lead-silicate	–	max.at 160 °C	–
Hard glasses	Borosilicate	–	6 (max.at 320 °C)	–
	Aluminosilicate	–	max.at 350 °C	–
Steatite		4×10^{-5}	–	–
Natural rubber**		2.10^{-5}†	–	–
Neoprene**		5×10^{-5}†	–	–
Silicone rubber**		4×10^{-6}†; 1×10^{-4}	–	1 (after 7 days)
PVC		4×10^{-4}	–	0.1–0.5
Teflon		2×10^{-5}†; 1×10^{-4}	–	less than 0.01
Perspex		1×10^{-3}	–	less than 0.5
Viton A		2×10^{-5}†; 2×10^{-4}	–	–

* See also Figs. 2.13 and 2.16.
** See also Table 2.7.
† After Farkass[367].

Appendix B.5
Vapour Pressure of Water (Ice) and Mercury (torr)

Temperature (°C)	Water	Mercury	Temperature (°C)	Water	Mercury
−183	1.4×10^{-22}	3.48×10^{-32}	30	31.82	2.8×10^{-3}
−150	7.4×10^{-15}	–	40	55.32	6.1×10^{-3}
−140	2.9×10^{-10}	–	50	92.51	1.27×10^{-2}
−130	6.98×10^{-9}	–	60	149.3	2.52×10^{-2}
−120	1.13×10^{-7}	–	70	233.7	4.82×10^{-2}
−110	1.25×10^{-6}	–	80	355.1	8.88×10^{-2}
−100	1.1×10^{-5}	2.39×10^{-11}	90	525.7	1.58×10^{-1}
−90	7.45×10^{-5}	–	100	760	2.72×10^{-1}
−80	4.1×10^{-4}	2.38×10^{-9}	150	3570.4	2.80
−70	1.98×10^{-3}	1.68×10^{-8}	200	11 650	17.28
−60	8.1×10^{-3}	9.89×10^{-8}	250	29 817	74.37
−50	2.9×10^{-2}	4.94×10^{-7}	300	64 432	246.8
−40	9.7×10^{-2}	2.51×10^{-6} (−38.9°C)	400	–	1574.0
−30	2.9×10^{-1}	4.78×10^{-6}	500	–	7691
−20	7.8×10^{-1}	1.81×10^{-5}	600	–	22.8 atm
−10	1.95	6.06×10^{-5}	700	–	52.5 atm
0	4.58	1.85×10^{-4}	800	–	103.3 atm
10	9.2	4.9×10^{-4}	900	–	180.9 atm
20	17.54	1.2×10^{-3}	1000	–	290.5

APPENDIX B.6a

ELECTRICAL PROPERTIES OF METALS AND ALLOYS

Material	Characteristics	Electrical resistivity (Ohm.cm × 10⁶) at 20 °C	Electrical resistivity (Ohm.cm × 10⁶) at 500 °C	Curie Point (°C)
Aluminium	see App.B.1a.	2.80	8.40	–
Beryllium		6.6	–	–
Iron		9.0	45	768
Copper		1.63	5.3	–
Nickel		6.18	36	350–360
Gold		2.3	6.8	–
Silver		1.57	3.0	–
Platinum		9.8	28	–
Palladium		11	20	–
Titanium		32	95	–
Zirconium		41	115	–
Molybdenum		5.1	18	–
Tantalum		12.4	37	–
Tungsten		5.0	18	–
Platinit	see App. B.1a.	44	110	460
Invar		75	109	250
Fernico (Kovar)		49	–	435
Monel		48	–	43–60
Constantan		49	–	–200
Iron–chromium	see App. B.1a.	55	69	–
Stainless steel		73	–	–
Nichrome		110	115	–
Inconel		98	–	–40
Tombak	see App. B.1a.	7	–	–
Brass		6.5	–	–
Bronze		13	–	–

APPENDIX B.6b. ELECTRICAL PROPERTIES* OF GLASSES, CERAMICS, AND ELASTOMERS

Material	Electrical resistivity at 20 °C (Ωcm)	Temperature for 10^8 Ωcm (°C)	Electrical breakdown (kV/mm)	Dielectric constant (ε)	Dielectric strength (tg.δ·10^4)
Soft glasses Soda–lime	$10^{13}-10^{15}$	150–250	3.0–4.0	7.9–8.5	50–100 (20 °C, 10^6 cycles)
Lead silicate	10^{17}	300–425	2.0–3.0	6.7–10	5–10 (20 °C, 10^6 cycles)
Hard glasses Borosilicate	$10^{14}-10^{18}$	125–360	3.8–4.2	4.5–8	20–40 (20 °C, 10^6 cycles)
Alumino-borosilicate	10^{18}	185–580	3.6–4.0	6–7	60–75 (20 °C, 10^6 cycles)
Quartz	$10^{17}-10^{18}$	600	2.5–4.0	3.5–3.7	1.5 (20 °C, 10^6 cycles)
High voltage porcelain	$10^{11}-10^{12}$	500 (10^6 Ωcm)	8–10	5.8–6.0 (10^6 cycles)	30–40 (20 °C, 10^6 cycles)
Steatite	$10^{11}-10^{13}$	750 (10^6 Ωcm)	9–10	5.8–6.2 (10^6 cycles)	13–21 (20 °C, 10^6 cycles)
Forsterite	$10^{12}-10^{13}$	100 (10^6 Ωcm)	9–10	6.0–6.2 (10^6 cycles)	4–5 (20 °C, 10^6 cycles)
Alumina	$10^{12}-10^{13}$	900 (10^6 Ωcm)	10–12	8.0–8.3 (10^6 cycles)	7–8 (20 °C, 10^6 cycles)
Zircon (ceramic)	$10^{12}-10^{13}$	870	9–10	7.2–8.8 (10^6 cycles)	7–10 (20 °C, 10^6 cycles)
Mica	$10^{15}-10^{17}$	500–800	20–70	6–8 (10^6 cycles)	1.6–2 (20 °C; 10^7 cycles)
Natural rubber	$10^{13}-10^{16}$	–	16–24	2–4	90–200 (1000 cycles)
Neoprene	10^{12}	–	23	7–7.5	250–300
Silicone rubber	$10^{13}-10^{14}$	–	26–30	3–10	12–200 (10^2-10^8 cycles)
PVC	$10^{14}-10^{16}$	–	20	3.1–3.4	150–200 (10^2-10^6 cycles)
Teflon	$10^{15}-10^{19}$	–	16–20	2.0–2.1	3–5
Perspex	$10^{14}-10^{18}$	–	20	2.7–4.0	200–500
Polystyrene	$10^{17}-10^{20}$	–	20–30	2.5–2.7	1–5
Saran	$10^{14}-10^{16}$	–	10–15	4–6	300–800 (100 cycles)

* For the electrical constants at various temperatures see Section 41.1.

REFERENCES

1. ADAM H., The theoretical foundation of glass compression seals and their practical consequences, *Trans. Soc. Glass Technol.*, **38**, 285, (1954).
2. ADAM H., Neue Werkstoffe — Metalle vorgegebener Porositat, *Vakuum-Technik*, **4** (3), 65 (1955).

2a. ADAM H., Vakuumdichte Molibden-Glas Verschmeltzungen, *Vakuum-Technik*, **8**, 59, (1959).

3. ADAM H., *Ultrahochvakuumtechnik E.* Leybolds. Köln, 1960.
4. ADAM H., Fortschritte an vakuumdichten Druckglaseinschmelzungen, *Vakuum-Technik*, **10**, 65, (1961). See also: ADAM H., *Feinmechanik u. Präzision*, **56**, 1, (1952) and ADAM H., ESPE W., and SCHWARTZ-BERGKAMPF E., *Glas-und Hochvakuumtechnik*, **1**, 123. (1952).
5. ADAM H., KAUFMAN S., and LILEY B. S., Indium seals for dismountable vacuum system, *J. Sci. Instr.*, **34**, 123. (1957).
6. ADAMS G. D. and SHERWIN W. C., On the vacuum properties of silastic, *Rev. Sci. Instr.*, **20**, 957, (1949).
7. ADAMS L. H., The annealing of glass as a physical problem, *J. Franklin Inst.*, **216**, (7), 39 (1933).
8. ADAMS L. H., and WILLIAMSON E. D., The annealing of glass, *J. Franklin Inst.*, **190**, 597 (1920).
9. ADDIS R. R., PENSAK L., and SCOTT N. I., Evaluation of a new fluoroelastomer as a gasketing material for high vacuum systems, *Trans. AVS. Nat Vac. Symp.* **1960**, p. 39, Pergamon Press, Oxford, 1961.

9a. Advanced Prod. Co. (59 Broadway, North Haven, Conn., U.S.A.), *Metal O Ring Design Manual*, 980—2.62 M.

10. AHARONI A., and FREI E. H., The cleaving of mica in water, *J. Sci. Instr.*, **34**, 513 (1957).
11. AHN S. H., and STRAIT E. N., Vacuum seals at liquid-nitrogen temperature, *Rev. Sci. Instr.*, **30**, 200 (1959).
12. AITKEN P. B., Vacuum sealing of irradiation containers by cold welding, *Nucleonics*, Nov., 89 (1952).
13. ALEXANDER P., Instrument for the measurement of pumping speed of high-speed vacuum pumps, *J. Sci. Instr.*, **21**, 216 (1944).
14. ALLEN P. W., and *al.*, A rubber gasket cutter, *J. Sci. Instr.*, **28**, 386 (1951).

14a. Alley Co. (Newlands Road, Cathcart, Glasgow S. 4., Scotland), *Vacuum Unions*.

14b. ALLENSWORTH D. I., Modification of a Hoke No. 482 valve for use as a variable leak, *Rev. Sci. Instr.*, **34**, 448 (1963).

15. ALMA G., and PRAKKE F., New series of small radio valves, *Philips Techn. Rev.*, **8**, 289 (1946).
16. ALMOND I., Simple gas inlet for use with vacuum systems, *J. Sci. Instr.*, **35**, 70 (1958).
17. ALPERT D., Vacuum valve for the handling of very pure gases, *Rev. Sci. Instr.*, **22**, 536 (1951).
18. ALPERT D., New developments in the production and measurement of ultra-high vacuum, *J. Appl. Phys.*, **24**, 860 (1953).
19. ALPERT D., Ultra-high vacuum technology, *Trans AVS Nat. Vac. Symp*, 1954 p. 69, Pergamon Press, Oxford, 1955.

20. ALPERT D., Recent advances in ultra-high vacuum technology, **1960**, *Adv. Vac. Sci. Technol.*, p. 34, Pergamon Press, Oxford, 1961.
21. ALPERT D., CARMICHAEL J. H., and *al.*, *Ultra-High Vacuum Techniques*, Westinghouse Progr. Rep. AT-11-1; Res. Rep. 100 FF-1054-R1 (1956).
22. ALPERT D., CARMICHAEL J. H., LANGE W. J. and *al*, *Ultra-High Vacuum Techniques*, Westinghouse Electr. Corp. Rep. AECU 3888 (1956), 3889 (1957), and 3890 (1957).
23. ALTAISKII YU. M., EZHOV N. M. and SIDYAKIN V. G., A metal stopcock for vacuum systems, *Pribory i Tekh. Eksper.*, (USSR), **6**, 119 (1958)., *Instr. exper. Tech.*, (USA), **6**, 830 (1958).
24. ALVAREZ L. W., BRADNER H., and *al.*, Berkeley proton linear accelerator, *Rev. Sci. Instr.*, **26**, 111 (1955).
25. AMARIGLIO I., and BENARIE M. M., A corrosion proof vacuum controller for pressures under 1 mm of mercury, *J. Sci. Instr.* **35**, 385 (1958).
26. American Vacuum Soc., *Glossary of Terms Used in Vacuum Technology*, Pergamon Press, London, 1958.
27. AMERONGEN G. I. VAN, The permeability of different rubbers to gases and its relation to diffusivity and solubility, *J. Appl. Phys.*, **17**, 972 (1946).
28. AMOIGNON J., Exploitation industrielle de l'ultra-vide, *Ind. Atomiques*, **5** (9/10), 107 (1961).
29. AMOIGNON J., DELDIER J., and GELLER R., Robinets a microfluites ajustables, *Le Vide*, **12**, 176 (1957).
29a. AMSEL G., *Techniques for Producing unsupported Thin Metal Films*, I.A.E.C. Rep. I.A., 633 (1961).
29b. ANASTASIO T. A. and POKORSKI R. G., High-vacuum motor, *Rev. Sci. Instr.*, **34**, 1133 (1963).
29c. ANDERSEN D. G., BERGER R. A., and SAUTER G. F., Method of producing extremely small holes in glass walls, *Rev. Sci. Instr.*, **35**, 412 (1964).
30. ANDERSON J. M., Thin vacuum-tight mica windows suitable for baking at 500 °C, *Rev. Instr.*, **31**, 898 (1960).
31. ANDERSON N. E., Arcwelding with Ar or He, *Welding Journal*, **28**, 222 (1949).
32. ANDERSON R. A. and STEPP E. E., Glass-to-sapphire end window seals, *Rev. Sci. Instr.*, **33**, 119 (1962).
32a. ANDERSON J. S. and KAMPHAUSEN H. A., Optisch nahezu flache Diffusionsverschlusse aus Glas, *Vakuum-Techn.*, **12**, 152 (1963).
33. ANDRÉ M. P. Propriétés de l'Araldit, *Le Vide*, **7**, 1200 (1952).
34. ANGERER E. V., and EBERT H., *Technische Kunstgriffe bei physikalischen Untersuchungen*, F. Vieweg & Sohn, Braunschweig, 1959.
34a. ANGUS G. & Co., (Coast Road, Wallsend-on-Tyne, England), *CACO, O-Rings and Hydraulic Packings*, Cat. 244/4/62.
34b. ANGUS G. & Co. (Coast Road, Wallsend-on-Tyne, England), *Fluocarbon Plastics*, Cat. 416/10/61.
35. Anon., Powdered glass for better seals to metal, *Instruments*, **19**, 89 (1946).
36. Anon., Glas als Werkstoff, *Die Technik*, **8**, 675 (1953)
37. Anon., Seel-off valve for vacuum systems, *J. Franklin Inst.* **258**, 401 (1954).
38. Anon., Bonding of Teflon, *Mater. & Methods*, **43** (4), 119 (1956).
39. Anon., Soldering difficult materials, *Mater. & Methods*, **44** (1), 114 (1956).
40. Anon., Ground glass joints in the laboratory, *Lab. Practice*, **6**, 465 (1957).
41. Anon., New high temperature epoxy resin, *Mater. in Design Engng.*, **51** (1), 9 (1960).
42. ANTON N., Fused vacuum-tight, metal-to-ceramic, ceramic-to-glass, metal-to-glass and metal-to-mica sealing by powdered glass technique, *Ceramic Age*, **63** (6), 15 (1954).
43. APKARIAN H., *Gas Leakage Tests of Shaft Seals*, Gen. Electric Co. Rep. AECU-555 (1949).
44. APKARIAN H., *Gas Leakage Test on Shaft Seals*, Rep. AECU-555 (1954).
45. ARCHARD G. D., A method of preventing mercury from sticking to glass, *J. Sci. Instr.* **29**, 412 (1952).

46. ARCHER R. M., Notes on the design of vacuum joints in metal apparatus, *J. Sci. Instr.*, **13,** 161 (1936).
47. ARDENNE M. v., *Tabellen der Elektronenphysik, Ionenphysik und Übermikroskopie,* Dtche. Verl. Wissenschaften Berlin, 1956.
48. ARKINS W. F. and JENKINS G. C. H., The design of a large vacuum valve for application to an electric furnace, *Machinery,* **9,** 391 (1950).
49. ARMAND G., LAPUJOULADE J., and PAIGNE J., Correlation between leak rate and some phenomena observed in metal-to-metal contact, Trans AVS Nat. *Vac. Symp.* **1961,** p. 1091, Pergamon Press, Oxford, 1962.
49a. ARMAND G. and PAIGNE J., Joints étuvables de grand diamètre, *Le Vide* **18,** 35 (1963).
49b. ARMSTRONG D. E., and BLAIS N., Large rotary vacuum seal, *Rev. Sci. Instr.*, **34,** 440 (1963).
49c. ARMAND G., and LAPUJOULADE J., Liaisons quartz–metal étuvables, *Le Vide,* **18,** 318 (1963).
49d. ARMAND G., LAPUJOULADE J., and PAIGNE J., A theoretical and experimental relationship between the leakage of gases through the interface of two metals in contact and their superficial micro-geometry, *Vacuum,* **14,** 53 (1964).
50. ARNOLD W., *Handbuch fur das Kupferschmiedgewerke, Rohrleitungs- und Apparatebau,* Verl. S. Schultz, Hanover, (1960).
51. ASOO S., and MURAMATSU K., A large unwelded vacuum gate valve, *J. Sci. Instr.*, **30,** 209. (1953).
51a. ARNOLD W., Thin mica windows, *Rev. Sci. Instr.*, **10,** 197 (1939).
52. Associated Electrical Ind. Ltd., (Trafford Park, Manchester 17), *Descriptive List,* 2009 (1960).
53. ASTM., *Standard Definitions of Terms Relating to Glass and Glass Products,* Standard C 162-52/56.
53a. Atlas-Werke (Bremen), *Hochvakuumtechnische Gerate,* Cat. X-1135 (1961).
54. ATTA C. M. VAN, *The design of high vacuum systems and the application of Kinney high-vacuum pump.* Kinney Mfg. Div., N.Y. Air Brake Co., Boston, 1955.
55. AUSTIN J. B., The Thermal expansion of some refractory oxides, *J. Am. Ceram. Soc.* **14,** 795 (1931).
56. AUWÄRTER M., *Ergebnisse der Hochvakuumtechnik und der Physik dünner Schichten,* Wissenschaftliche Verlag, Stuttgart, (1957).
57. AVERA C. B. Simple flow regulator for extremely low gas flow, *Rev. Sci. Instr.*, **32,** 985 (1961).
58. AVERY A. W. Sealing against pressure and vacuum, *Brit. Chem. Eng.*, **5,** 658 (1960).
59. AXELROD N. N., Ultra-high vacuum valve, *Rev. Sci. Instr.*, **30,** 944 (1959).
60. AXILROD B. M., and LAMB J. J. Tensile properties of rolled silver chloride, *J. Appl. Phys.*, **19,** 213 (1948).
60a. AYER F., Suppressing mercury vapour in vacuum systems, *Rev. Sci. Instr.*, **21,** 496 (1950).
60b. AZAN and ORTEL, Robinets à fuite, *Le Vide,* **5,** 914 (1950).
61. BABAT G., and LOSINSKY M., Concentrator of eddy-current for zonal heating of steel parts, *J. Appl. Phys.*, **11,** 816 (1940).
62. BABELAY E. F., and SMITH L. A., A needle valve type of variable gas leak for mass spectrometers, *Rev. Sci. Instr.*, **24,** 508 (1953).
63. BACK D. R., CHILDS W. J. *et al.*, Cyclotron instrumentation for nuclear spectroscopy at medium resolution in energy, *Rev. Sci. Instr.*, **27,** 516 (1956).
64. BACK B. B., DAWSON J. V., and SMITH L. W. L., Ball valves, *J. Sci. Instr.*, **31,** 343 (1954).
65. BACHMAN C. H., *Techniques in Experimental Electronics,* John Wiley, New York, 1948.
66. BACHMAN C. H., New Principle in controlled vacuum leaks, *Rev. Sci. Instr.*, **20,** 219 (1949).
67. BACHMAN C. H., and BENSIN I. B., Tyre valve in vacuum application, *Rev. Sci. Instr.*, **13,** 534 (1942).

68. BACHMAN C. H. and SILBERG P. A., Gas flow in capillaries of non-circular cross section, *Trans. AVS Nat. Vac. Symp.* **1954**, p. 63, Pergamon Press, Oxford, 1955.

69. BACHMAN J. S., Use of Araldite in vacuum, *J. Sci. Instr.*, **31**, 136 (1954).

69a. BACHMAN K. C., Greaseless stopcock for use at subatmospheric pressure, *Anal. Chem.*, **23**, 1724 (1951).

69b. BACQUET E. H., Analyse critique des pompes à vide dites préliminaires, *Le Vide*, **3**, 409 (1948).

70. BAHN R., and BÖTGER O., Die Brechzahl und die Dicke dünner anodische erzengter Aluminiumonydschichten, *Zschr. fur Physik*, **135**, 376 (1953).

71. BAIRD & TATLOCK (London), *General Apparatus, Instruments Medical and Industrial Sciences*, Catalogue (1958).

72. BAKER D., An all-metal, high-vacuum valve, *Vacuum*, **12**, 99 (1962).

73. BALAIN K. S., High and low temperature vacuum seals, *Trans AVS Nat Vac. Symp* **1961**, p. 1300, Pergamon Press, Oxford, 1962.

74. BALAIN K. S.. and BERGERON C. J., High pressure and vacuum seals at liquid helium temperatures, *Rev. Sci. Instr.*, **30**, 1058 (1959).

75. BALDOCK R., and SITES J., A versatile aluminium–metal 60° mass spectrometer, *Phys. Rev.*, **81**, 294 (1951).

76. BALLARD W. E., *Metal Spraying and Sprayed Metal*, London, 1948.

77. BALLIF I. L., *A high-vacuum application of inflatable seals for the processing refabrication experiment*, Atomes International, NAA-SR-2544 (Jan. 1959).

78. BALMFORTH W., and HOLLAND A. I., Photoelectric constant of glasses, *J. Soc. Glass Technol.*, **29**, 111 (1945).

79. BANKI L., and HOLLO I., *Vákuumtechnikai gépek* (Machinery for Vacuumtechnique), Nehézipari K. Budapest, 1954.

80. BANNOCK R. R., Molecular sieve pumping, *Vacuum*, **12**, 101 (1962).

81. BARR W. E., and ANHORN V. J., *Scientific and Industrial Glass Blowing and Laboratory Techniques*, Instrument Publ. Co., Pittsburg, 1949.

82. BARREAU P., LÉGER P., *et al.*, Préparation de films minces de formvar, utilisés comme supports de couches métalliques, en physique nucléaire, *J. Phys. et Radium*, **15**, 4A (1954).

83. BARRER, R. M., Stationary and non-stationary states of flow of hydrogen in palladium and iron, *Trans. Faraday Soc.*, **36**, 1235 (1940).

84. BARRER R. M., *Diffusion in and through solids*, Cambridge Univ, Press, Cambridge, 1951.

85. BARRER R. M., and GROVE D. M., Flow of gases and vapors in a porous medium and its bearing on adsorption problems, *Trans. Faraday. Soc.* **47**, 826 (1951); and **47**, 837 (1951).

86. BARETT A. S. D., and POWER B. D., *The Production of Vacuum*, in Cremer–Davies, *Chemical Engineering Practice*, Vol. 5, p. 299, Butterworth, London, 1958.

87. BARETT A. S. D., and HARPER M. E., Selection of pumping systems for metallurgical processes, *J. Inst. of Metals*, **87**, 227 (1959).

88. BARETT C. S.: *Structure of Metals*, McGraw-Hill, New York, 1943.

88a. BARRINGTON A. E., *High Vacuum Engineering*, Prentice-Hall, Englewood Cliffs, New Jersey, 1963.

89. BARROW C. A simple air lock for ionization gauge, *J. Sci. Instr.*, **33**, 83 (1956).

90. BARTELS H., Spannungsmessung in Glas, *Feinmech. u. Prazision*, **47**, 219 (1939).

91. BARTON C. I., *A Review of Glove Box Construction and Experimentation*, ORNL-3070 (1961).

92. BARTON D. M., Vacuum O-ring reals, *Vacuum*, **3**, 51 (1953).

93. BARUCH P., The use of the electron accelerator in solid state physics. *Nucl. Instr. & Meth.*, **11**, 196 (1961).

93a BAS E. B. and CREMOSNIK G., Schweissen im Hochvakuum mit Elektronenstrahlen, *Vakuum Techn.*, **8**, 181 (1959).

93b. BASHKIN S., and GOLDHABER G., A note on thin nickel films, *Rev. Sci. Instr.*, **22**, 112 (1951).

94. BATHOW G., Glasfolien hoher Druckbestandigkeit mit einer Dicke unter 100 mμ., *Vakuum-Techn.*, **5**, 156 (1956).
95. BATZER TH. H. *A Large Bakeable Vacuum Valve*, Lavrance Rad. Lab., UCRL-5649 (Aug. 1959).
96. BATZER TH. H., A large bakeable vacuum valve, *Trans AVS Nat Vac Symp* **1959**, p. 265, Pergamon Press, Oxford, 1960.
97. BAUER S. H., Compact multiple way metal stopcock, *Rev. Sci. Instr.*, **19**, 52 (1948).
98. BAXTER A. S., Production of small apertures by evaporation techniques, *Vacuum*, **1**, 115 (1951).
99. BAYLEY D. S., Variable capillary tube leak, *Rev. Sci. Instr.*, **13**, 299 (1942).
100. BAZZONI C. B., A modified Smythe vacuum leak, *Rev. Sci. Instr.*, **8**, 171 (1937).
101. BEACH J. G., and FAUST C. L., Electroplating of Be, *J. Electrochem. Soc.*, **100**, 276 (1953).
101a. BEAUDOIN ETS., (1—3, rue Rataud, Paris 5, France), *Robinets à vide*, Catalogue.
101b. BEAVIS I. C., and STEWART D. I., Two ultra high vacuum radio-frequency feed-throughs, *Rev. Sci. Instr.*, **34**, 449 (1963).
102. BECKER J. A., Ultra-high Vacua, *Bell Lab. Rec.*, January 1955, 1.
102a. BECKER K., and PIRANI M., Gaseinlassventil mit Bimetallstreifen, *Techn.-wiss. Abh. Osram-Ges.*, **3**, 61, 1934.
102b. BECKMANN W., Gas description of some rubber-type materials, *Vacuum*, **13**, 349 (1963).
103. BEGGS J. E., Sealing metal and ceramic parts by forming reactive alloys, *Trans IRE*, **CP4**, (March) 28 (1957).
103a. BEHRNDT K. H., A demountable ultra-high vacuum glass system and its components, *Trans AVS Not Vac. Symp.* **1959**., p. 255, Pergamon Press, Oxford, 1960.
104. BELSER R. B., A technique of soldering to thin metal films, *Rev. Sci. Instr.*, **25**, 180 (1954).
105. BELSER R. B., Frictional adhession of metal-to-glass, quartz and ceramic surfaces, *Rev. Sci. Instr.*, **25**, 862 (1954).
105a. BELSER R. B., *Indium — A versatile Solder to Metals and non-Metals*, in *Indalloy Intermediate Solders*, Indium Corp. of America, 1957.
106. BENBENECK J. E., and HONIG R. E., Method for sealing stainless steel to glass, *Rev. Sci. Instr.*, **31**, 460 (1960).
107. BENDER B. G., How to seal glass to metal in mass output, *Ceram. Ind.*, **57** (6), 76 (1951).
108. BENDER H., High temperature metal–ceramic seals, *Ceram. Age*, **63** (4), 15 (1954).
109. BENICHOU R., High melting-point seals, *Le Vide*, **12**, 141 (1957).
110. BENNETT W. P., One-kilowatt tetrode for UHF transmitters, *Proc. IRE*, **41**, 13 (1953).
111. BENSON S. W., Sealing quartz windows on Pyrex tubes, *Rev. Sci. Instr.*, **13**, 267 (1942).
112. BENZ W. G., and SOHN J. S., Aircomatic welding of Cr-Ni Steel, *Welding J.*, **30**, 911 (1951).
113. BERGER E., Bestimmung der Transformations-Temp. von Glas durch Ausdehnungs- und Widerstandsmessung, *Glastechn. Ber.*, **12**, 172 (1934).
114. BERGER K., OETJEN G. W., and Holland-Merten E. L., *Druck und Vakuumtechnick*, in *Ullmanns Encyklopadie der Technischen Chemie*, Urban & Schwarzenberg, München, 1951.
115. BERNHARD F., and BUMM H., Das Löten in Vakuum, *Vakuum Technik*, **7**, 153 (1958).
116. BERTELE H. v., Cooled Hg cathode, *Direct Current*, **1**, 43 (1952).
117. BESKE H. E., Ein ausheizbares Hochvakuumventil geringer Bauhöhe, *Vakuum-Technik*, **10**, 212 (1961).
118. BESKE H. E., Ein doppelfokusierender Massenpektrograph nach Mattinek–Herzog Zur Spurenanalyse von Feststoffen, *Zeitschr. Angew. Phys.*, **14**, 31 (1962).

119. Ben K. E., Automatic switch for protecting water cooled X-ray tubes, *Rev. Sci. Instr.*, **18**, 926 (1947).
120. Beynon I. H., *Mass Spectrometry and its Applications to Organic Chemistry*, Elsevier Publ. Co., Amsterdam, 1960.
121. Bezold D. v., and Schuler C., Ersatz fur hochbelastbare elektrische Einführung in Glass Hochvakuum Apparaturen, *Vakuum-Technik*, **5**, 191 (1956).
122. Bicknell C. B., An improved helium-only sensitive method for calibrating silica membrane helium leaks, *Trans AVS Nat Vac. Symp. 1959, p. 97*, Pergamon Press, Oxford, 1960.
123. Biddulph R. H., and Plesch P. H., A corrosion resistant, high-vacuum valve for liquids, *Chem. & Ind.*, **30**, (6), 569 (1956).
124. Billett E. A. and Bishop J., Greaseless vacuum seal for rotating shafts, *J. Sci. Instr.*, **35**, 70 (1958).
125. Bills D. G., and Allen F. C., Ultra-high vacuum valve, *Rev. Sci. Instr.*, **26**, 654 (1955).
126. Biondi M. A., Fox R. E., *et al.*, *Ultra-high Vacuum Techniques*, An. Rep. NYO 2904.
127. Biram J. G. S., Some aspects of handling mercury, *Vacuum*, **5**, 77 (1955).
128. Biram J. G. S., and Jacques T. A. J., Rubber in vacuum system, V*acuum*, **4**, 341 (1954).
129. Bishop J., Vacuum technique in mass spectrometry, *Trans. AVS Nat. Vac. Symp. 1958*, p. 484, Pergamon Press, Oxford, 1959.
129a. Blackmon P. H., Clauss F. J., *et al.*, Materials evaluation under high vacuum and other satellite environmental conditions, *Trans. AVS Nat. Vac. Symp. 1961*, p. 1244, Pergamon Press, Oxford, 1962.
129b. Blake A. R., and Redford R. A., The technique for the safe handling and controlled opening of sealed glass bulbs containing gas at a pressure of 280 lb/sq. in., *J. Sci. Instr.*, **40**, 200 (1963).
130. Blanaru L., High vacuum valve, *J. Sci. Instr.*, **35**, 184 (1958).
131. Blanc B., Elements standards, *Le Vide*, **13**, 135 (1958).
131a. Blanc B. L., Robinets de vide primaire et secondaire, commande automatique-Robinets á fuite, *Le Vide*, **19** (110), 27 (1964).
132. Bleakney W., Cummings C. S., A simple and rugged mass spectrograph for routine analysis, *Phys. Rev.*, **57**, 1072 (1940).
133. Blears J., *Improvements r.t. Manually Operated Diaphragm-Type Valves*, Brit. Pat. 759, 492 (1957).
134. Blears J., and Leck J. H., General principles of leak detection, *J. Sci. Instr. (Suppl. 1)*, **28**, 21 (1951).
134a. Blears J., and Leck J. H., Differential methods of leak detection, *Brit. J. Appl. Phys.* **2**, 227 (1951).
135. Bleecher H., Valve for Glass blowing on contaminated apparatus, *Rev. Sci. Instr.*, **31**, 997 (1960).
136. Bleeksma J., G. M. Zahlrohre mit Lindemann — Glass-Fenstern, *Philips Techn. Rdsch.*, **10**, 1 (1948).
137. Bleil D. F., and Hause C. D., Insulated vacuum leads, *Rev. Sci. Instr.*, **12**, 369 (1941).
138. Bleuze J., and Dussaussory P., Les techniques de scellement par poli optique, *Le Vide*, **7**, 1182 (1952).
139. Block H., A small mercury cut-off withstanding large pressure differences, *Nature*, **178**, (4545), 1307 (1956).
140. Blodgett K. B., Superficial conductivity of lead glasses after reduction of H2, *J. Am. Ceram. Soc.* **31**, 89 (1948) and **34**, 14 (1951).
141. Bode H., Corrosion-resistant Roots pumps, *Trans. AVS Nat. Vac. Symp. 1959*, p. 268, Pergamon Press, Oxford, 1960.
142. Boerboom A. J. H., and Tasman H. A., On the fractionation of gaseous samples in a mass spectrometer viscous leak, *Physica*, **24**, 683 (1958).

142a. Boerboom A. J. H., de Jongh A. P., and Kistemaker J., Two versions of a sliding bar vacuum lock, *Rev. Sci. Instr.* **35**, 301 (1964).
143. Boersch H., Hamisch H., and Wiesner S., Über elektrische Mikroüberschläge in Vacuum, *Zschr. Angew. Physik*, **13**, 450 (1961).
144. Boettcher A., Gase auf Glasoberflachen und ihre Entfernung, *Glastechn. Ber.*, **25**, 347 (1952).
145. Bogg W., An improved form of high vacuum leak, *J. Sci. Instr.*, **14**, 412 (1937).
146. Böhm H., Geiger–Müller–Zahlrohre, *Siemens Ztschr.*, **29**, 324 (1955).
147. Boll H. J., One-way, all-glass valve for high-vacuum systems, *Rev. Sci. Instr.*, **32**, 1415 (1961).
148. Bollenrath F., Optisches Differentialdilatometer, *Metallkunde*, 3/1934.
148a. Bollinger H., Ein Ausheizbares Ganzmetallventil, *Exp. Techn. der Physik*, **11**, 369 (1963).
149. Bolton M. I., *Solder for ceramics*, U.S. Pat. 2,636, 820 (1953).
150. Bondelid R. O., Modification to the vane valve, *Rev. Sci. Instr.*, **24**, 546 (1953).
151. Bondley R. J., Metal–ceramic brazed seals, *Electronics*, **20** (7), 97 (1947).
152. Bottomley G. A., A valve for the grease-free manipulation of mercury, *J. Sci. Instr.* **34**, 369 (1957).
153. Boulloud J. P., and Schweitzer J., Etude expérimentelle de joints métalliques pour ultra-vide, *Le Vide*, **14**, 241 (1959).
154. Bouyer P., Cassignol C., and Lazeyras P., Un robinet à fuites de construction entierement métallique, *Le Vide*, **15**, 297 (1960).
155. Bowring R. W., and Davies R. H., *A Pressure-Vacuum Diaphragm Valve*, AERE Rep. GP/R 1481 (1954).
156. Boyd G. A., Sliding vacuum gate for the cyclotron, *Rev. Sci. Instr.*, **18**, 250 (1947).
147. Boyd R. L. F. Little atomarc apparatus with corona discharge, *Electronic Engng.* **26** (312), 78 (1954).
158. Brackney H. and Atlee Z. J., Beryllium windows for permanently evacuated X-ray tubes, *Rev. Sci. Instr.*, **14**, 59 (1943).
158a. Bradner H., The production of thin Be foils, *Rev. Sci. Instr.*, **19**, 662 (1948).
159. Brady G. S., *Materials Handbook*, McGraw-Hill, New York, 1956.
160. Brand F., Jacobs H., and Locascio C., Experiments in the design of ceramic electron tubes, *Ceramic Age*, **63**, (5), 18 (1954).
161. Brannen E., and Ferguson H. I. S., Simple universal vacuum joint, *Rev. Sci. Instr.*, **25**, 836 (1954).
162. Branson L. K., and Sutherland R. J., Vacuum application of industrial glass pipe, *Rev. Sci. Instr.*, **31**, 665 (1960).
163. Breadner R. L., and Simms C. H., The processing of glass in the lamp and radio-valve industries, *G.E.C. Journ.*, **12**, (4), (1963).
164. Breadner R. L., and Simms C. H., The planning of a glassworking department, *J. Sci. Instr.*, **21**, 169 (1944).
165. Brebner J. L., and Mooser E., Cryostat for optical measurements between 300 °K and 4.2 °K, *J. Sci. Instr.*, **39**, 69 (1962).
165a. Brennan D., and Haynes F. H., Gas handling apparatus for $10^{-2}\mu$ torr based on greased stopcocks, *J. Sci. Instr.*, **39**, 534 (1962).
166. Brewster, Sir D., Results of some recent experiments on the properties impressed upon light by the action of glass raised to different temperatures and cooled under different circumstances, *Phil. Trans. Royal Soc. London, 436 (1814).*
167. Brice R. T., A method for scaling β-magnesia windows into steel spectroscopic absorption furnace and a needle valve for controlling the evacuation of such furnaces, *Rev. Sci. Instr.*, **8**, 209 (1937).
167a. Bridge H., Budde R., *et al.*, Some vacuum problems at low temperatures, *Adv. Vac. Sci. Technol.*, p. 481, Pergamon Press, Oxford, 1960.
168. Briggs W. E., Jones A. C., and Roberts J. A., Leak detection techniques, *Trans. AVS Nat. Vac. Symp. 1958*, p. 129, Pergamon Press, Oxford, 1959.

169. British Standard — 1806/1951, *Dimensions of Toroidal Sealing Rings*, Brit. Stand. Institution London, 1951.
169a. British Standard — 2951/1958, *Glossary of Terms used in High Vacuum Technology*, Brit. Stand. Institution, London, 1958.
170. BROMBACHER W. G., *Bibliography and index on vacuum and low pressure measurement*, Nat. Bur. Stand Monogr. No. 35 (1961).
170a. BRON W. E., MANNHEIMER R., and TAYLOR G., High voltage vacuum lead through, *Rev. Sci. Instr.*, **33**, 1287 (1962).
170b. BROOK F., European aspects of tube techniques, *1962 Advances in Electron Tube Techniques*, p. 1, Pergamon Press, Oxford, 1963.
171. BROSE H. L., and KEYSON J. E., Vacuum levers, *J. Sci. Instr.*, **7**, 19 (1930).
172. BROWN F. C., Basic technique in design and construction of the vacuum plant, *Trans. AVS Nat. Vac. Symp. 1958*, p. 93, Pergamon Press, Oxford, 1959.
172a. BROWN J. *et al.*, On making thin nylon films, *Rev. Sci. Instr.*, **19**, 818 (1948).
173. BROWN R. A. and SKAHAN D. I., Use of Toggle valves and Teflon in hydrocarbon analysis, *Anlyt. Chem.*, **26**, 788 (1954).
174. BROWN S. C., and COYLE J. E., All-metal vacuum valve, *Rev. Sci. Instr.*, **23**, 570 (1952).
174a. BROWNELL L. E., and YOUNG E. H., *Process Equipment Design*, John Wiley, New York, 1959.
175. BRUCKNER F., (Brown, Bovery & Co.), *Glass for vacuum-tight metal connections*, Germ. Pat. 1,009,770 (1957).
176. BRUESCHKE E. E., Ultra-high vacuum seal for space simulation system, *Rev. Sci. Instr.*, **32**, 732 (1961).
177. BRUESCHKE E. E., A leak-free method of transmitting motion into ultra-high vacuum *Vacuum*, **11**, 255 (1961).
178. BRUNEE C., Eine neue Vakuumschleuse fur Massenspektrometer, *Ztschr. Instrkde.*, **68** (5), 97 (1960).
178a. BRUNNER W. F., and PATTON H. G., Vacuum deposition of ductile metallic films, *Trans. AVS Nat. Vac. Symp.* **1961**, p. 895, Pergamon Press, Oxford, 1962.
178b. BRUNNER W. F., and PATTON H. G., *Preparation of Thin Self-supporting Films*, UC RL Rep. 6225 (1961).
179. BRYMNER R., and STECKELMACHER W., Demountable vacuum seal for operation at temperatures from —180 °C to 800 °C, *J. Sci. Instr.*, **36**, 278 (1959).
180. BUCK H. M., A modified flap-type vacuum valve, *J. Sci. Instr.*, **33**, 361 (1956).
180a. BUCH S., *Einfuhrung in die allgemeine Vakuumtechnik*, Wissenschaftl. Verl., Stuttgart, 1962.
181. BUCHMAN K. C., Greaseless stopcock for use at sub-atmospheric pressures, *Analyt. Chem.*, **23**, 1724 (1951).
181a BUNSHAH R. F., *Vacuum Metallurgy*, Reinhold, New York, 1958.
182. BUREAU A. J., LASLETT L. J., and KELLER J. M., Pumping speed of a circular aperture in a diaphragm across a circular tube, *Rev. Sci. Instr.*, **23**, 683 (1952).
182a. BURGERJON J. J. *et al.*, The Pretoria cyclotron, *Nucl. Instr.*, **3**, 323 (1958).
182b. BURKARD P. N., Cleaning of Metals, in Bolz R. W., *Metals Engineering Processes*, McGraw Hill, New York, 1958.
183. BURNETT R. E., and NORDLANDER B. W., (General Electric), *Sealing Material for Porous Surfaces*, U.S. Pat. 2, 628, 178 (1953).
184. BURNSIDE D. G., Ceramic–metal seals of the tungsten–iron type, *R.C.A. Rev.*, **15** (3), 46 (1954).
185. BURROWS G., (Metropolitan Vickers), *Improvements r.t. the Sealing of Joints in High Vacuum and the like Systems*, Brit. Pat. 701,614.
186. BURROWS G., and STARLING P. P., (Metropolitan Vickers), *Improvements relating to Sealing Glands*, Brit. Pat. 703,696.
186a. BUSEN K. N., Simple greaseless high-vacuum valve using a rolling O-ring seal, *Vacuum*, **13**, 475 (1963).
187. BUSH S. H., Electrolytic polishing of metals, *Mater. & Meth.*, **39** (6), 137 (1954).

188. BUTLER C. P., and CARPENTER F. E., A motor for use in vacuum systems, *Rev. Sci. Instr.*, **21**, 103 (1950).
189. BUTLER I. W., High efficiency, long-lived vacuum cold trap, *Nucl. Instr. & Meth.* **8**, 211 (1960).
190. CADENHEAD A. F. G., Indium — its properties and uses, *Can. Metals*, **17**, (7), 16 (1954).
190a. CALDWELL W. C., Vacuum evaporation of metals, *J. Appl. Phys.*, **12**, 799 (1941).
190b. CALDWELL B. A., and KLEMPERER D. F., Large ultra-high vacuum valve, *Rev. Sci. Instr.*, **33**, 1458 (1962).
191. CAMACK W. G., and ELLIS C. A., A high vacuum orbital simulator, *Trans. AVS Nat. Vac. Symp.* **1961**, p. 1260, Pergamon Press, Oxford, 1962.
192. CAMERON A. E., Compact high-vacuum valve, *Rev. Sci. Instr.*, **25**, 1027 (1954).
193. CAMPBELL J. B., A glossary of physical properties and tests, *Mater. in Design Engng.*, **49** (6), 91 (1959).
194. CARMICHAEL J. H., LANGE W. J., and TRENDELENBURG E. A. T., *Ultra-high Vacuum Techniques*, Rep. AECU 3888 (1956).
195. CARPENTER R., A new version of the knife edge vacuum seal, *J. Sci. Instr.*, **39**, 533 (1962).
195a. CARSON R. S., HENDRICKS C. D., Vacuum seals, *Rev. Sci. Instr.*, **34**, 118 (1963).
196. CASSEL A. D., HOCKLY D. A., Adaptation of a gas and water union to high vacuum circuits, *J. Sci. Instr.*, **33**, 279 (1956).
197. CASWELL H. L. An oil-free, ultra-high vacuum system for deposition of thin films, *Trans. AVS Nat. Vac. Symp. 1959*, p. 66, Pergamon Press, Oxford, 1960.
197a. CASWELL H. L., *Ultra-High Vacuum Evaporators and Residual Gas Analysis* in Hass G., *Physics of Thin Films*, Academic Press, New York, 1963.
197b. CATALANO E., and WRENN E. W., Technique in the use of two low melting point solders, *Rev. Sci. Instr.*, **34**, 1260 (1963).
198. CATTON N. L., *The Neoprenes*, Wilmington, Delaware, 1953.
199. CENCO S. A. (Fribourg, Switzerland), *Instruments for the Laboratory Sciences*, Catalog J-300.
200. Cerro de Pasco Corp. (300 Park Ave., N.Y., 22), *Cerro Alloys; Glass-to-Glass and Glass-to-Metal Joints;* Cat. E9—10—59—5M; E14—10—59—5M; I2—9—61—5M BK.
201. CHAILLON A., and OLETTE M., Joints plastiques pour vide poussé (Wilson), *Le Vide*, **12**, 383 (1957).
202. CHAMPEIX R., *Physique et Technique des Tubes Electroniques*, Dunod, Paris, 1958.
203. CHAMPEIX L., and DUFLO J., Joint coulissant pour appareil sous vide, *Le Vide*, **13**, 191 (1958).
204. CHANDRA M. N., Ein neues Verfahren für eine vakuumdichte Glass-Aluminium Verbindung, *Vakuum-Technik*, **11**, 123 (1962).
204a. CHAPIN D. S., Packless all-metal, ultra-high vacuum valve for temperatures from —196 °C or below, to 450 °C, *Rev. Sci. Instr.*, **34**, 47 (1963).
205. CHARLTON M. G., DEWSON D., and WHITCHURCH P. J., An automatic apparatus for the testing of getter adsorption, *Vacuum*, **10**, 189 (1960).
206. CHARLTON M. G., and PERKIN G. M. C., Method for breaking open-sealed electron devices *in vacuo*, *J. Sci. Instr.*, **36**, 49, (1959).
207. CHASMAR R. P., CRASTON J. L., *et al.*, A method of sealing sapphire to glass and its application to infra-red photo-cells, *J. Sci. Instr.*, **28**, 206 (1951).
208. CHATTFIELD C. H., Silver brazing of refractory metals, *Welding J.*, **33**, 865 (1954).
209. CHAUDHRI R. M., A mercury safety valve for electrolysis of heavy water, *Proc. Phys. Soc.*, **68B**, 1157 (1955).
210. CHELIUS J., Processing of molybdenum, *Mater. & Meth.*, **32**, 45 (1950).
211. CHEN T. S., Output windows for turnable magnetrons, *Electronics*, **27** (5), 170 (1954).
212. CHEVENARD P., Dilatometre avec registrateur mecanique, *J. Phys. Radium*, **7**, 240 (1926).

212a. CHICK A. J. *et al.*, *Ceramic-Metal Seal*, U.S. Pat. 2,708,787 (1955).
213. CHILCOT A. L., Leakage of water through tungsten seals, *Electronic Eng.*, **17**, 693 (1945).
214. CHOPINET-DUMON, Sur la soudure par argon-arc dans la technique des tubes electroniques, *Le Vide*, **15**, 63 (1960).
215. Ciba Co. (Basel, Switzerland), *Araldite. (Catalog)*
216. CLARK E. J., Vacuum diffusion joining of titanium, *Welding J.*, **38**, 251 (1959).
217. CLAUSER H. R., Soft and hard solders, *Mater. & Meth.* **34** (Mars), 105 (1952).
218. CLOUD W. R., and PHILIP S. F., Vacuum tests of rubber, lead and Teflon gaskets and vinyl acetate joints, *Rev. Sci. Instr.*, **21**, 731 (1950).
219. COBINE J. D., Arcatom with supply of usw., *Mater. & Meth.*, **31**, 47 (1950).
219a. COENRAADS C. N. and LAVELLE I. E., High speed drives in high vacuum, *Rev. Sci. Instr.*, **33**, 879 (1962).
220. COHEN J., Resistance heated high vacuum furnace for temperatures up to 1400 °C, *Rev. Sci. Instr.*, **31**, 267 (1960).
221. COHN CH. E., Comment on "Vacuum application of industrial glass pipe", *Rev. Sci. Instr.*, **31**, 1165 (1960).
221a. COLGATE S. O., and WHITEHEAD E. C., Sealing Pyrex glass to stainless steel and inconel, *Rev. Sci. Instr.*, **33**, 1122 (1962).
222. COLLINS A., and TURNBULL J. C., Degassing and permeation of gases in tube materials, *Vacuum*, **11**, 115 (1961).
222a. COLLINS G. B., *Microwave Magnetrons*, McGraw-Hill, New York, 1948.
223. COLNOT P., and GALLET G., *Le Verre et la Céramique dans la Technique du Vide*, Gauthier-Villars, Paris, 1962.
223a. COLOMBANI A., and RANC G., Sas automatique, *C.R. Acad. Sci. (France)*, **257** (No. 10), 1682 (1963).
224. Compagnie Général de Radiologie (Issy-les-Moulineaux, 48 Bd. Gallieni, France), *Pour tous vos Problèmes de Vide*, Cat.
225. Compagnie Générale de Telegraphie sans Fil, *Vacuum-tight Joints*, Brit. Pat. 705,003; 732,853.
226. Compagnie Générale de Telegraphie sans Fil, (Corbeville-France), *Techniques du Vide*, Cat. 6203/C7 (1962).
227. COMSA G., LUNGU S., and SIMIONESCU C., Metal-to-metal and ceramic-to-metal bakeable vacuum seals, *Trans. AVS Nat. Vac. Symp.* **1961**, p. 1319, Pergamon Press Oxford, 1962.
227a. COMSA G., and SIMIONESCU C., Ceramic–metal sealing techniques for a bakeable, multiple electrical lead-through in metal, ultra-high vacuum systems, *J. Sci. Instr.*, **41**, 352 (1964).
228. CONABOY Y., Precision ends for fused silica tubes, *J. Sci. Instr.*, **36**, 148 (1959).
228a. CONNOR H. Palladium alloy diffusion cells, *Platinum Metals*, **6**, 130 (1962).
229. CONNOR R. J., BURITZ R. S., and ZWECK T. VON, Ultra-high vacuum system for a large electromagnetic isotope separator, *Trans. AVS Nat. Vac. Symp. 1961*, p. 1151, Pergamon Press, Oxford, 1962.
230. Consolidated Electrodynamics Corp. (Rochester, New York) *High Vacuum Valves, Baffles, Traps*, Bulletin 10-1 (1959).
231. Consolidated Vacuum Corp., (1775 Mt. Read Brd., Rochester, New York), *Bulletin* **6** (1960).
232. COOK L., *Vacuum Properties of Plastic Materials*, Rep. MDDC–588 (1946).
233. COOK D. B., and DANBY C. J., A simple diaphragm micromanometer, *J. Sci. Instr.*, **30**, 238 (1953).
234. COPE J. O., Magnetically operated needle valve, *Rev. Sci. Instr.*, **29**, 232 (1958).
235. COPP D. H., and GREENBERG D. M., A mica window Geiger counter tube measuring soft radiations, *Rev. Sci. Instr.*, **14**, 205 (1943).
236. CORAK W. S. and WEXLER A., Metal-to-glass seals for sub λ point helium apparatus, *Rev. Sci. Instr.*, **24**, 994 (1953).

237. CORELLI J. C., LIVINGSTON M., and SEIDLITZ L., Construction of thin Mylar gas target holders, *Rev. Sci. Instr.*, **28**, 471 (1957).

238. CORNELL L. W., Properties of PTFE, *Mech. Engng.*, **75**, 883 (1953).

239. Corning Glass Works (Corning, N.Y., U.S.A.) *Properties of Selected Commercial Glasses*, Cat. B-83 (1959).

239a. CORNWALL E. P., Welded joints for hard-vacuum systems, *Machine Design*, **35** (19), 135 (1963).

240. COUNTRYMAN M. A., Alloying and wetting of liquid metals, *J. Appl. Phys.*, **8**, 432 (1937).

240a. COURANT H. F., Elektrisches Widerstands-Glassschweissen, *Elektrot. Ztg.* (ETZ), **B9**, 271 (1957).

241. COWIE D. B., High vacuum seal, *Rev. Sci. Instr.*, **15**, 46 (1944).

242. COWIE D. B., and GREEN G. K., A high vacuum valve, *Rev. Sci. Instr.*, **12**, 556 (1941).

243. COXE C. D., *Filler Rods for Joining Nonmetallic Materials* U.S. Pat. 2,739,375 (1956).

243a. CRAIG P. P., STEYERT W. A., and TAYLOR R. D., High vacuum brass-to-aluminium seal for use at cryogenic temperatures, *Rev. Sci. Instr.*, **33**, 869 (1962).

244. CRAWLEY D. J., High vacuum techniques for nuclear industry, *Atomic World*, **9**, 423 (1958).

245. CRAWLEY R. H. A., A universal reagent for cleaning glassware and silica., *Chem. Ind.*, **45**, 1205 (1953).

246. CROCKER A., All-metal high conductance vacuum tap, *J. Sci. Instr.*, **36**, 447 (1959).

247. CROES M. DE, PARKER W., and SEVIER K., Thin film thickness distribution by alpha absorbation, *Nucl. Instr. & Meth.*, **7**, 160 (1960).

248. CROFT A. J., A diffusion pump protection Switch, *J. Sci. Instr.*, **25**, 386 (1948).

249. CROSS S. H., System design and the choice of materials for the NIMROD Vacuum system, *Vacuum*, **10**, 86 (1960).

250. CROSS S. H., and STECKELMACHER W., Leak detection by vacuum techniques, *Research*, April, 124 (1956).

251. CROZE M. W. and HENDRICK P. L., *Method of Coating PTFE*, U.S. Pat. 2,689,805 (1956); Brit. Pat. 756,814 (1956).

252. CURTIS F. W., *High Frequency Induction Heating*, McGraw-Hill, New York, 1950.

253. DALE A. E., Sealing to ductile metal with "Solder" glass powder, *J. Soc. Glass Technol.*, **33**, 167 (1949).

254. DALE A. E. and STANWORTH J. E., A note on very soft glasses and some of their electrical applications, *J. Soc. Glass Technol.*, **32**, 147T (1948).

255. DALE A. E. and STANWORTH J. E. The development of some very soft glasses, *J. Soc. Glass Technol.*, **33**, 167 (1949).

256. DALTON R. H., *Welding of Soft. Glass Articles*, U.S. Pat. 2,642,633 (1953).

257. DALTON R. H., Solder glass sealing, *J. Am. Ceram. Soc.*, **39**, 109 (1956); see also *J. Soc. Glass Technol.*, **35**, 4 (1951).

258. D'AMICO C., and HAGSTRUM H. D., Improvement in the use of the porcelain rod gas leak, *Rev. Sci. Instr.*, **28**, 60 (1957).

259. DANZIN A., and DESTOIS E., Optically polished surfaces for vacuum tube seals, *Ann. Radioelectr.*, **3**, 281 (1948).

260. DARTNELL R. C., FAIRBANKS H.V., and TOEHLER W. A., Investigation of the adherence of glass to metals and alloys, *J. Am. Ceram. Soc.*, **34**, 357 (1951).

261. DAVIES A. C. *Welding*, Cambridge Univ. Press, Cambridge, 1956.

261a. DAVIES J. W., Glass shaping machines, *Inst. Mech. Eng. Proc.*, **158**, 352 (1948).

262. DAVIES A. J., A note on the use of polytetrafluoroethylene in vacuum seals, *J. Sci. Instr.*, **35**, 378 (1958).

263. DAVIS E. J., Metal through glass seal, *J. Sci. Instr.*, **35**, 308 (1958).

264. DAVIS R. E., and MCCREA I. M., Liquid sample inlet system for gas chromatographs, *Analyt. Chem.*, **29**, 1114 (1957).

265. DAVIS W. D., *Diffusion of Gases through Metals*, AEC Rep. KAPL 1227 (1954); KAPL 1375 (1955).
266. DAVY I. R., *Industrial High Vacuum*, I. Bitman & Sons, London, 1951.
267. DAWTON R. H. V. M., *High Vacuum Shaft Seals, Flanged Joints and the Glassing and Permeability of Rubber-like Materials*, AERE-GR-393 (1949); GP/R-1875 (1956); *Brit. J. Appl. Phys.*, **8**, 414 (1957).
268. DAYTON B. B., Outgassing rate of contaminated metal surfaces, *Trans. AVS Nat. Vac Symp.* **1961**, p. 42, Pergamon Press, Oxford (1962).
269. DECECCO N. A. and PARKS J. M., The brazing of Titanium, *Welding J.*, **32**, 1071 (1953).
270. DECKER R. W., All-glass valves for use in obtaining ultra-high vacua, *J. Appl. Phys.*, **25**, 1441 (1954).
271. DELAFOSSE J., and MONGODIN G., Les calculs de la technique du vide, *Le Vide*, **92**, (1961).
272. DELLA PORTA P., Apparatus and techniques for measurements of the adsorption of gases by evaporated getters, *Vacuum*, **10**, 182 (1960).
273. DELSASSO L. A., and CREUTZ E. C., Simple quick acting vacuum lock, *Rev. Sci. Instr.*, **12**, 450 (1941).
274. DENTON E. P. and RAWSON H., Solder glass for hard glass, *J. Soc. Glass Technol.*, **40**, 252 (1956).
275. DERUYTTER A., A 4π-fission detector, *Nucl. Instr. & Meth.*, **7**, 145 (1960).
275a. DESCARSIN M., Fusion et façonnage du verre par courants de très haute fréquence *Bull. Inst. du Verre*, **1**, 17 (1946).
276. DETTER W., Electrolytical and chemical polishing, *Metall*, **7**, 325 (1953).
276a. D'EUSTACHIO D., A useful seal for dynamic vacuum systems, *Rev. Sci. Instr.*, **16**, 377 (1945).
277. D'EUSTACHIO D., A cam-operated vacuum valve, *Rev. Sci. Instr.*, **18**, 369 (1947).
278. Deutsche Vakuumapparate, (Sangerhausen, Germany) *Fine and High Vacuum*, Catalogue No. 5.
279. DICKINSON J. M., High temperature resistance furnace for operation at 3000 °C, *Trans. AVS Nat. Vac. Symp.* **1958**, p. 194, Pergamon Press, Oxford, 1959.
280. DIELS K., and JAECKEL R., *Leybold Vakuum-Taschenbuch*, Springer, Berlin 1962.
280a. DILLON J. A., The clean surface approach to adsorption studies, *Trans. AVS Nat. Vac. Symp.* **1961**, P. 113, Pergamon Press, Oxford, 1962.
281. DIMITRIEVSKAYA T. I., KRAVTSEV V. V., and TSVETAEVA, Application of end-window counters to the measurement of low-activity betaradiation, *Instr. & Exp. Techn. USSR.*, **2**, 221 (1960).
281a. Deutsche Normen, DIN 28400 (1963), *Vakuumtechnik Fachausdrucke und Begriffe fur Vakuum-Systeme und-Messgerate, Lecksuche, Vakuum-Absperr und Drossel-organe und Dichtungen.*
282. DITSBURN R. W., An improved cut-off for high vacuum work, *J. Sci. Instr.*, **8**, 267 (1931).
283. DOBKE G., The vakuum technique of the mercury vapour iron rectifier — Part I, *Glass & Hochvakuum Techn.*, **2**, 249 (1953).
284. DONAL J. S., Sealing mica to glass or metal to form a vacuum-tight joint, *Rev. Sci. Instr.*, **13**, 266 (1942).
285. DONALDSON E. E., Particulate contamination from glass, *Vacuum*, **12**, 11 (1962).
286. DONG W., and BROMLEY L. A., Vacuum flow of gases through channels with circular, annular and rectangular cross-sections, *Trans. AVS Nat Vac. Symp.* **1961**, p. 1116, Pergamon Press, Oxford, 1962.
287. DOOLITTLE H. D., ETTRE K., *et al.*, *Ceramic-Metallizing Tape for Reliable Metal – Ceramic Seals*, IRE. Trans. of Electron Devices ED-8, 390 (1961).
288. DORÉ R., Quelques aspects industriels des traitements de surface en technique du vide, *Le Vide*, **17**, 208 (1962).
289. DORGELO E. G., Sintered Glass, *Philips, Techn. Rev.*, **8**, 2 (Jan. 1946).

290. DORGELO E. G., Technologie von Magnetrons und Klystrons, *Vakuum-Techn.*, **5**, 180 (1956).
291. DORGELO E. G., and STEVELS J. M., Pulverglas-Einschmelzungen, *Philips Techn. Rdsch.*, **10**, 273 (1949).
292. DORN J. E., and GLOCKNER G., An adjustable leak valve for gas X-ray tubes, *Rev. Sci. Instr.*, **7**, 319 (1936).
293. DORRAT M., Arcatom, *Welding J.*, **31**, 306 (1943).
294. DORSTEN A. C. *et al.*, Electron microscope, *Philips Tech. Rev.*, **12**, 38 (1950).
295. DOTY W. R., O-ring seals for glass apparatus, *Rev. Sci. Instr.*, **30**, 1053 (1959).
296. DOTY W. R., and RYASON P. R., Greaseless vacuum valve useful in kinetic studies, *Rev. Sci. Instr.*, **32**, 89 (1961).
297. DOUGLAS R. W., The use of glass in high-vacuum apparatus, *J. Sci. Instr.*, **22**, 81 (1945).
298. DOUGLAS R. W., Stresses in glass–metal seals, *J. Soc. Glass Technol.*, **29**, 92 (1945).
299. Dow Corning Corp. (Midland, Mich., U.S.A.), *Silicone Laboratory Lubricants*, Cat., 1957.
300. Dow Corning Corp. (Midland, Mich., U.S.A.), *What's Silicone?*, Cat.
301. DOWD J. D., Soldering of Aluminium, *Welding J.*, **33**, 113s. (1954).
301a. DOYLE E. M., High temperature-high power microwave windows, **1962** *Advances in Electron Tube Techniques*, p. 11, Pergamon Press, Oxford, 1963.
302. DRAWIN H. W., Ausheizbares Ultra-Hochvakuum-Ventil mit grossen Öffnungs-querrschnitt, *Vakuum-Technik*, **9**, 199 (1960).
303. DRAWIN J. W., and BRUNNÉE C., Messung niedriger Partialdrucke mit dem Massem-spektrometer, *Vacuum-Technik*, **9**, 66 (1960).
304. DNEYER K., and MARK I. T., The ultra-high vacuum system for the C-stellarator, *R.C.A., Rev.* **21**, 508 (1960).
305. DROWART J., and GOLDFINGER P. A., A new demountable ultra-high vacuum joint, *J. Sci. Instr.*, **34**, 248 (1957).
305a. DUFOUR CH., Technique de l'évaporation des couches minces multiples, *Le Vide*, **3**, 480 (1948).
306. DUMOND J. W. M., Two applications of the Sylphon bellows in high-vacuum plumbing, *Rev. Sci. Instr.*, **6**, 285 (1935).
307. DUNCAN J. F., and WARREN D. T., Rotary vacuum reaction chambers, *J. Sci. Instr.*, **30**, 462 (1953).
308. DUNCAN J. F., and WARREN D. T., High-vacuum application of Polythene, *Brit. J. Appl. Phys.*, **5**, 66 (1954).
309. DUNKEL M., and LEYBOLDS NACHF., *Vacuum-tight Joint between High-vacuum Pipes*, Brit. Pat. 783,397 (1957).
310. DUNOYER L., Commentaires sur les formules pour l'écoullement à basse pression des gaz dans les conduites, *Le Vide*, **5**, 881 (1950).
311. DUNOYER L., *Le Vide et ses Applications*, Presse Univ. de France, Paris, 1951.
312. Dupont de Nemours & Co. Inc. (Wilmington 98, Del., U.S.A.) *Teflon, Mylar*, Cat. 1961.
313. DURAN I. C., MARTINUS A. M. *et al.*, *Seals for Electric Discharge Tubes*, U.S. Pat. 2,873,510 (1959).
314. DUSHMAN S., *Scientific Foundation of Vacuum Technique*, John Wiley, New York, 1949.
315. DUSHMAN S. (LAFERTY I. M., Ed.) *Scientific Foundation of Vacuum Technique* (2nd ed.), John Wiley, New York, 1962.
316. DUSSING W., Vakuumdichte Glass-Metallverschmelzung fur die Elektrotechnik, *Glastechn. Ber.*, **26**, 232 (1953).
317. DUSSING W., Glass–Kovar Einschmelzungen fur Hochvakuumrohren, *Telefunken Zeitung*, **26**, 111 (1953).
318. DUTTON D. J., Production and measurement of ultra-high vacua, *Research (London)*, **12**, 103 (1959.)

319. EASTWELL R. A., Comments on "An all-metal, high-vacuum valve", *Vacuum*, **12**, 226 (1962).

319a. ECHO M. W., and MORGAN T. D., A surface ionization mass spectrometer for production control, *Analyt. Chem.*, **29**, 1593 (1957).

320. ECKSTEIN N. A., FITZGERALD I. W. *et al.*, Method for making glass–metal seals, *Rev. Sci. Instr.*, **21**, 398 (1950).

321. ECONOMOS G., and KINGERY W. D., Metal–ceramic interactions, *J. Am. Ceram. Soc.*, **36**, 403 (1953).

322. EDENS A. H., A method for sealing the window and cone of television picture-tubes, *Philips Techn. Rev.*, **19**, 318 (1957/58).

323. EDGE J., and OLDFIELD L. F., The Pretreatment of Glass Surfaces for the Application of Silicones. *A Class of High Alkali Oxide Content.*, G.E.C. Communic. 89 (1958).

324. EDWARDS P. L., Rolling O-ring seal, *Rev. Sci. Instr.*, **31**, 1356 (1960).

325. EDWARDS W. D., An insulated vacuum lead-in using an O-ring, *J. Sci. Instr.*, **35**, 111 (1958).

326. EDWARDS W. L. and MAXWELL L. R., An adjustable leak valve, *Rev. Sci. Instr.*, **9**, 201 (1938).

327. Edwards High Vacuum Ltd. (Manor Royal, Crawley, England), *Electrodes, Flexible Vacuum Couplings* and *Rotary Shaft Sels*, Catalogue C. 128/2.

328. Edwards High Vacuum Ltd. (Manor Royal, Crawley, Sussex, England), *High Vacuum Control Valves and Devices*, Catalogue C. 130/2.

329. Edwards High Vacuum Ltd. (Manor Royal, Crawley, Sussex, England), *High Vacuum Unions and Joints*, Catalogue C. 124/6; *Elastomer O-Rings, Sheet and Tubing* Catalogue 107/8.

330. Edwards High Vacuum Ltd., (Manor Royal, Crawley, Sussex, England), *Vacnique*, Vol. 1, No. 1, Jan. 1961.

331. EHLERS H., and MOLL J., Results with ultra-high vacuum metal systems including windows, evaporators and lead-ins, *Trans AVS Nat. Vac. Symp.* **1959**, p. 261, Pergamon Press, Oxford, 1960.

332. EHLERS H., and MOLL J., Eine Metallaparatur für Versuche im Ultrahochvakuum, *Ztschr. Angew. Phys.* **12**, 324 (1960).

333. EHRENBERG W., Insulating joints in Vacuum tubes, *J. Sci. Instr.*, **16**, 162 (1939).

334. Eimac (Eithel McKullough Inc.), La technique du scellement quartz–metal perfectionne, *Le Vide*, **15**, 286 (1960).

334a. EINSPRUCH N. G., and WISSEMAN W. R., High-vacuum, insulated electrical feedthrough for use in cryostats, *Rev. Sci. Instr.*, **34**, 929 (1963).

335. Eisler Engng. Co., (750 South 13th Str., Newark, New Jersey, U.S.A.) *Catalogue No. 59-FL* (1959).

336. ELDRED B. E., *Copper-Sheeted Wire for Glass Seals*, U.S. Pat. 1,140,136 (1915).

337. ELLIOT A. E., and DICKSON J. H., *Laboratory Instruments*, Chapman & Hall, London, 1959.

337a. ELLIS D. R., Vacuum filling of "one entrance" apparatus, *J. Sci. Instr.*, **40**, 604 (1963).

338. ELONKA S., Gaskets, *Power*, p. 105, March 1954.

339. ELSWORTH A, HOLLAND L. and LAURENSON L. Further experiences with Al wire seals for bakeable vacuum system, *J. Sci. Instr.*, **37**, 449 (1960).

340. ELTENTON G. C., The study of reaction intermediates by means of a mass spectrometer, *J. Chem Phys.*, **15**, 455 (1947).

341. ELTENTON G. C., The mass spectrometric detection of free radicals, *J. Phys. Colloid. Chem.*, **52**, 463 (1948).

342. ENGEL F., Zur optimalen Anpassung von Verschmelzpartnern im Elektronenröhrenbau, *Glastechn. Ber.*, **29**, 209 (1956).

343. ENGEL F., Messungen von Spannungen in Glass-Metal Verschmelzungen, *Glastechn. Ber.* **31**, 133 (1958).

344. ENNOS A. E., The sources of electron-induced contamination in kinetic vacuum systems, *Brit. J. Appl. Phys.*, **5**, 27 (1954).
345. EPSTEIN G., *Adhesive Bonding of Metals*, Reinhold Publ. Co., New York, 1954.
346. ESCHBACH H. L., *Praktikum der Hochvakuumtechnik*, Akad. Verlag, Leipzig, 1962.
347. ESCHBACH H. L., and JAECKEL R., Wände für Ultrahochvakuumbehalter aus Schichtstoffen, *Z. Naturforsch.*, **15a**, 267 (1960).
348. ESCHBACH H. L., JAECKEL R., and MÜLLER D., Molekulare Strömung bei sehr niedrigen Drucken, *Trans. AVS Nat. Vac. Symp.* **1961**, p. 1110, Pergamon Press, Oxford, 1962.
349. ESCHBACH H. L., and PEPERLE W., Zerlegbares Ultrahochvakuum-Ventil mit grossem Öffnungsquerschnitt, *Vakuum-Technik*, **10**, 210 (1961).
350. ESPE W., *Werkstoffe der Elektrotechnik in Tabellen und Diagramen*, Akad. Verl., Berlin, 1954.
351. ESPE W., Ferritsches Chromeisen für Vakuumdichte Glass–Metal–Verschmelzungen, *Vakuum-Technik*, **4**, 82 (1955).
352. ESPE W., Schweiss-und Lötverbindungen metallischer Baustoffe in der Hockvakuum technik, *Vakuum-Technik*, **4**, 51 (1955).
353. ESPE W., Kupfer-Glass-Verschmelzungen, *Nachrichtentechnik*, **6**, 401 (1956).
354. ESPE W., *Werkstoffkunde der Hochvakuumtechnik, Vol. I—III*, Deutscher Verlag der Wissenschaft, Berlin, 1960—61.
354a. ESPE W., Uber Waschprozesse in der Hochvakuumtechnik und in Feingeratebau, *Vakuum-Technik*, **12**, 145 (1963).
355. ESPE W., LEM T., and VAGNER St., (Tesla Elektronic, Praha), *Entladungsröhre besonders als Überspannungsableiter*, CSR. Pat. 85,533 (1951).
356. ESPE W., and KNOLL M., *Werkstoffkunde der Hochvakuumtechnik*, Springer, Berlin, 1936.
357. ESPE W., and SLOSIAR J., Über Vakuumdichtes Sinterglas, *Vakuum-Technik*, **8**, 209 (1959); and **9**, 7 (1960).
358. ESSIG S. F. A mercury-sealed gas valve, *Rev. Sci. Instr.*, **3**, 762 (1932).
359. ESTERMANN J., and FONER S. N., A vacuum valve, *Rev. Sci. Instr.*, **18**, 64 (1947).
360. ESTLE T. L., and WALTERS G. K., Air lock for changing samples in low temperature electron spin resonance experiments, *Rev. Sci. Instr.*, **32**, 1058 (1961).
361. EVANS E. C., and BABELAY E. F., Removable bowl cold trap, *Rev. Sci. Instr.*, **23**, 249 (1952).
362. EVANS E. C., and LANGDON A., *Modified Hoke Connector*, Carbide and Carbon Chem. Co., Rep. K-1031 (1953).
363. EVANS R. M., High temperature brazing *Batelle Techn. Rev.*, **5**, 8 (Aug. 1956).
363a. EVERETT M. H., and GILETTE H. G., Squeezed-type molded packings, in *The Seals Book*, Machine Design, Jan. 1961.
364. FABIAN R. J., Materials for gaskets packings and seals, *Mater. in Design Engng.*, **50** (7), 111 (1959).
364a. FALCE L. R., A practical method of ceramic-to-metal seal testing, **1962** *Advances in Electron Tube Techniques*, p. 51, Pergamon Press, Oxford, 1963.
365. FALOON J. E., Glass bonded mica, *Mater. in Design Engng.*, **51** (2), 96 (1960).
366. Fansteel Prod., Welding Ta under protection liquid, *Electr. World*, **104**, 1184 (1934).
367. FARKASS J. and BARRY E. J., Improved elastomer seal design for large metal, ultra-high vacuum systems (to 10^{-10} torr), *Trans. AVS Nat. Vac. Symp.* **1960**, p. 35, Pergamon, Press, Oxford, 1961.
368. FARKASS J., and BARRY E. J., The origins and composition of the limiting gas load in ultra-high vacuum systems, *Trans. AVS Nat. Vac. Symp.* **1961**, p. 66, Pergamon Press, Oxford, 1962.
369. FARKASS J., and VANDERSCHMIDT G. F., Ultra-high vacuum in large metal systems, *Trans. AVS Nat. Vac. Symp.* **1959**, p. 46, Pergamon Press, Oxford, 1960.
370. FARON M. J., BOTTAZZI H. J., and TEICHNER S. J., Un nouveau dispositif pour la réalisation des vides très poussés, *J. Chim. Phys.*, **57**, 600 (1960); and **57**, 793 (1960).

371. FEATES F. S., Magnetically operated shutters for use in vacuum systems, *J. Sci. Instr.*, **38,** 509 (1961).
371a. FEICHTINGER H., Weiterentwicklung bei der Vakuum-Heissextraktionsanalyse, *Berg- u. Hüttenmannische Monatshefte*, **100** (7/8), 230 (1955).
371b. FELDMAN W. D., *et al.*, *Ultra-High Vacuum Techniques*, Res Rep. (Westinghouse Res. Lab.), TID-18283 (1963).
372. FELHEIMER C., and LITWAK A. A., Automatic valving system for vacuum apparatus *Trans. AVS Nat. Vac. Symp.* **1954**, p. 35, Pergamon Press, Oxford, 1955.
372a. FEJNA J. E., *Spai steklo & metallom* (Glass–metal seals), Soviet Radio, Moscow, 1951.
373. FERENCE M., and STEPHENSON R. J., An apparatus for determining the transmission of slow electrons through thin colloidon films, *Rev. Sci. Instr.*, **9,** 246 (1938).
374. FIGDOR H., Simple procedure for strechting circular membranes, *J. Sci. Instr.*, **39,** 241 (1962).
375. FINCKE G. C., and TAYLOR G. W., A method of improving the glass-to-metal seal resistance to high humidity deterioration in electron tubes, *Proc. of 4th National. Conf. on Tube Technique*, p. 34, Univ. Press, New York, 1959.
376. FINE J., A sliding ground-glass joint, *Rev. Sci. Instr.*, **7,** 192 (1936).
377. FINK C. G., *Copper-Sheeted Sealing Wire*, U.S. Pat. 1,498,908 (1924).
378. FISCHHOFF E. *et al.*, Vide et ultra-vide dans une experience d'injection d'ions et miroirs magnetiques, *Le Vide*, **17,** 195 (1962).
379. FISCHER P., Magnetically operated vacuum valve, *Rev. Sci. Instr.*, **27,** 271 (1956).
380. FISKE M. D., Resonant windows for vacuum seals in rectangular wave guides, *Rev. Sci. Instr.*, **17,** 478 (1946).
381. FLANSCHEN S. S., PEARSON A. D., and NORTHOVER W. R., Low-melting inorganic glasses with high melt fluidities below 400 °C, *J. Am. Ceram. Soc.*, **42,** 450 (1959).
382. FLECKEN F. A., and NOLLER H. G., Gasabgabe Verschiedener Metallischer Hochvakuum-Baustoffe, *Trans. AVS Nat. Vac. Symp.* **1961**, p. 58, Pergamon Press, Oxford, 1962.
383. FLEISCHMANN W. L., and HERD A. J., *Leak Testing of Type 347 Stainless Steel Welds*, Knolls Atomic Power Labs. Rep. KAPL-M-AJH-Z (1955).
383a. FLEISCHER R. L. *et al.*, Method of forming fine holes of near atomic dimensions, *Rev. Sci. Instr.* **34,** 510 (1963).
384. FLINTA I., A gas leak, *J. Sci. Instr.*, **31,** 388 (1954).
385. FLORENCE J. M., Infrared transmission of glasses, *J. Res. Nat. Bur. Stand.*, **45,** 121 (1950).
386. FLORESCU N. A., Sur la conductance des appareils à écoulement moléculaire des gas, *Le Vide*, **15,** 197 (1960).
386a. FLORESCU N. A., First types of demountable vacuum joints, *Vacuum*, **12,** 316 (1962).
387. FLÜGGE S., *Handbuch der Physik* (Vol. 20), Springer, Berlin, 1957.
388. FOLEY R. T., ALVORD R. D., and EASLEY J. K., The electrodeposition of Ag on FeNi alloy glass-to-metal arrangement, *J. Electrochem. Soc.*, **99,** 349 (1952).
389. FOOTE T., and HARRINGTON D. B., Demountable metal vacuum seals, *Rev. Sci Instr.*, **28,** 585 (1957).
390. FORMAN R., Vacuum valve to provide small controlled vacuum leak rates, *Rev. Sci. Instr.*, **24,** 326 (1953).
391. FORTEY J. A. projection type of strain-viewer, *J. Soc. Glass Technol.*, **29,** 124 (1945).
392. FOWLER R. D., A variable capillary gas leak, *Rev. Sci. Instr.*, **6,** 26 (1936).
393. FOX G. L., Large diameter reciprocating-action vacuum valve, *Rev. Sci. Instr.*, **25,** 616 (1954); Rep MLM 850 (1953).
394. FOX G. L., *A Glass-Metal Vacuum System*, AEC, Res. & Develop. Rep. MLM 1015 (Oct. 1954).
395. FRAENKEL V. H., Pyrex-glass seals, *Rev. Sci. Instr.*, **7,** 395 (1936).
396. FRANCK J., *Optical Windows and Seals for Hydrogen Bubble Chamber*, Rep. UCID-71 (1956).

397. FRANK R. C., Some observations regarding the present status of measurement of the diffusion coefficients of hydrogen in iron and mils stell, *J. Appl. Phys.*, **29**, 1262 (1958).
398. FRANK R., and STOW R. L., Securing rock-salt windows to sealed-off vacuum tubes, *Rev. Sci. Instr.*, **25**, 514 (1954).
398a. FRANKE H. G., and KRAFT O., An easily exchangeable lead-in electrode used in high vacuum technique, *J. Sci. Instr.*, **40**, 502 (1963).
399. FRANKS A., An automatic vacuum isolation valve, *J. Sci. Instr.*, **34**, 122 (1957).
400. FRASER D. B., Special indium seal for cryogenic use, *Rev. Sci. Instr.*, **33**, 762 (1962).
401. FREMLIN J. H., A flexible vacuum joint, *J. Sci. Instr.*, **29**, 267 (1952).
402. FREMLIN J. H., A quick-release seal for vacuum desiccators, *J. Sci. Instr.*, **29**, 236 (1952).
403. FRENCH E. P., and NEWTON A., A high-vacuum rotary seal, *Nucleonics*, **10** (3), 66 (1952).
404. FRIAUF R. J., Vacuum-tight sample holder for high temperature conductivity measurements, *Rev. Sci. Instr.*, **31**, 1161 (1960).
405. FRIEDMAN H., G. M. tube with Lindemann-window, *Electronics*, April, 132 (1945).
406. FRIEDRICHS F., Das *Glas im chemischen Laboratorium*, Springer, Berlin, 1960.
407. FRITZ-SCHMIDT M., Strahlendurchlässigkeit von Glas für UR, UV, und sichtbares Licht, *Ztschr. Techn. Phys.*, **11**, 289 (1930).
408. FROST J. A., Glass blowing, *Lab. Pract.*, **5**, 134 (1956).
409. FROST J. A., Valve for the grease free manipulation of mercury, *J. Sci. Instr.*, **35**, 268 (1958).
410. FRY H., *Leak Detector using Nekal-coating*, BIOS Rep. 1183 (1947).
410a. FUCHS K. Einfluss der Viskosität des Glasses auf seine Verarbeitungsgeschwindigkeit, *Glass-Email-Keramo-Tehn.*, **9**, 366 (1958).
411. FUGASSI P., and MCKINNEY D. S., The preparation of silver chloride films, *Rev. Sci. Instr.*, **13**, 335 (1942).
412. FULLER J., A simple gas valve, *J. Sci. Instr.*, **33**, 160 (1956).
413. FULRATH R. M., MITOFF S. P., *et al.*, Fundamentals of *Glass-to-Metal Bonding*, Rep. AECU-3063 (1955); *J. Am. Ceram. Soc.*, **40**, 269 (1957).
414. FULTZ C. R., *A Metal Vacuum Lock for a Mass Spectrometer*, Instrumentation Rep. Y-990 (1953)
415. FUOSS R. M., Silver chloride windows for infrared cells, *Rev. Sci. Instr.*, **16**, 154 (1945).
416. FUSCHILLO N., Sliding and rotating mechanical and electrical feed-through seals into vacuum, *Am. J. Phys.* **26**, 400 (1958).
417. FUSCHILLO N., and GILVRY J. A simple demountable vacuum seal for electrical connections, *J. Sci. Instr.*, **33**, 402 (1956).
418. GALE A., Vacuum techniques in electrostatic particle accelerator, *Trans. AVS Nat. Vac. Symp. 1954*, p. 101, Pergamon Press, Oxford, 1955.
419. GALE A. J., Cold sealed getter-ion pumped super-voltage X-ray tubes, *Trans. AVS Nat. Vac. Symp. 1956*, *p.* 12, Pergamon Press, Oxford, 1957.
420. GALE R. F., and MACHIN C. F., The use of silicone rubber gaskets for high vacuum work, *J. Sci. Instr.*, **30**, 97 (1953).
421. GALL J. F., and MILLER H. C., *Etching of Tantalum*, U.S. Pat. 2,466,095 (1949).
422. GALLUP I., and DINGWALL A. G. F., Properties of low-temperature solder glasses, *Ceram. Bull.* **36**, 47 (1957).
423. GANDY H. W., High vacuum apparatus for studying the luminescence of solids, *Rev. Sci. Instr.*, **32**, 1019 (1961).
424. GARBE H. W., HF Brazes high vacuum joints, *Machinist*, **16**, 799 (1953).
425. GARBE S., and CHRISTIANS K., Zur Gasabgabe von Gläsern, *Vakuum-Technik*, **11**, 9 (1962).
426. Garlock Inc., (Palmira, New York, U.S.A.). *Teflon, Nylon, Delrin, Kel-F*, Cat. 1961.

427. Garner L. P., Machined metal stuffing box seal adapted to high vacuum technique, *Rev. Sci. Instr.*, **8**, 329 (1937).
428. Garrod R. J., A multiple high vacuum valve, *J. Sci. Instr.*, **23**, 191 (1946).
429. Garrod R. J., A low-impedance high vacuum valve, *J. Sci. Instr.*, **27**, 205 (1950).
430. Garrod R. J., A compact sliding vacuum seal, *J. Sci. Instr.*, **28**, 187 (1951).
431. Garrod R. J., and Coyle R. A., A multiple vacuum valve unit, *J. Sci. Instr.*, **27**, 228 (1950).
432. Garrod R. J., and Gross K. A., A universal coupling for use in vacuum systems, *J. Sci. Instr.*, **26**, 279 (1949).
433. Garrod R. J., and Hart R. K., Transmission specimen holder electron diffraction, *J. Sci. Instr.*, **36**, 44 (1959).
434. Garrod R. J., and Nankivell I. F., Leak testing of large welded vacuum vessels, *Vacuum*, **11**, 139 (1961).
435. Garton W. R. S., A demountable flesh-discharge tube for production of the Lyman continuum, *J. Sci. Instr.*, **30**, 119 (1953).
436. Gaunt A. J., and Redford R. A., Cap-and-cup mercury vacuum seal, *J. Sci. Instr.*, **36**, 377 (1959).
437. Geller R., Dégazage des matériaux sous vide, *Le Vide*, **13**, 71 (1958).
438. Geller R., and Mongodin G., *Pompage au travers d'une Canalisation en Régime de Poiseuille*, CEA-595 (1956); CEA-1476 (1960) and *Le Vide*, **11** (63), 106 (1956).
439. General Mills (Kankakee, Ill. U.S.A.), *Structural Adhesives add Sealants*, Techn. Bull. 11-F-2 (1960).
440. General Radiological Ltd., (Honeypot Lane, Stanmore, England), *Ionization Chamber Type NE* **006** — Catalogue.
441. Genevac Ltd. (Radcliffe, Manchester, England), *Electro-Magnetic Valves*, (Bulletin 109).
442. Genevac Ltd., (Radcliffe, Manchester, England), *Hand-operated Valves and Universal Coupling*, (Catalogue).
443. George W. R., and Winkle R. V., Seal between silica tubing and glass apparatus, *J. Sci. Instr.*, **38**, 375 (1961).
444. Geraetebau-Anstalt Balzers (Balzer-Lichtenstein) *Vakuumtechnik*, Catalogue (P. 10-06d).
444a. Gerber J. F., High vacuum dual-motion feedthrough, *Rev. Sci. Instr.*, **34**, 1111 (1963).
445. Gerrow G. P., High vacuum equipment, *Iron Age*, **173**, 142 (18. 2. 1954).
446. Giaimo E. C. Jr., Ring type Teflon gasket, *Rev. Sci. Instr.*, **26**, 520 (1955).
447. Giauque W. F., Geballe T. H., *et al.*, Some properties of plastics and the use of plastics in low temperatures, *Rev. Sci. Instr.*, **23**, 169 (1952).
448. Gibbons W. F., How to make ceramic seals, *Ceram. Ind.*, **72**, (5), 126, 154 (1959); and **72** (6), 88 (1959).
449. Gibson G. J., Argonarc and heliarc welding, *Welding J.*, **26**, 282 s. (1947)., and **27**, 496s. (1948).
449a. Gilmour A. S., and Hallock D. D., A demountable beam analyser for studying magnetically confined electron beams, *1962 Advances in Electron Tube Techniques*, p. 153, Pergamon Press, Oxford, 1963.
450. Gilvey J. P., Modified plumbing unions for vacuum connections, *Rev. Sci. Instr.*, **24**, 984 (1953).
451. Glaister R. M., A magnetically operated vacuum valve, *J. Sci. Instr.*, **33**, 34 (1956).
452. Gleason W. A., Five ways to seal glass to metal, *Mater. in Design. Engng.*, **51** (4), 120 (1960).
453. Glockler G., and Horst H. Ven., A micro-gas burette, *Rev. Sci. Instr.*, **23**, 563 (1952).
454. Godron G. Y., Wettability of glass/metal and of metal/ceramics, *Silicates Ind.*, **24**, 539 (1959).
454a. Goebel R., Heinze E. *et al.*, Elektronenstrahl Schweissung, *Kernenergie*, **5**, 104 (1962).

455. GOERTZ D. J., Ultra-high vacuum components for the proposed Stanford two-mile linear electron accelerator, *Trans. AVS Nat. Vac. Symp. 1960*, p. 16, Pergamon Press, Oxford, 1961.

456. GOLDSTEIN N. R., Unusual sealing techniques used at the new tyratron, *Ceram. Age*, **65** (5), 16 (1955).

456a. GOMER R., Preparation and some properties of conducting transparent glass, *Rev. Sci. Instr.*, **24**, 993 (1953).

457. GOMER R., Electrical contact for high vacuum system, *Rev. Sci. Instr.*, **27**, 544 (1956).

457a. GONDET H., Emploi du caoutchouc de silicone (Silastic) pour la confection de joints étanches au vide, pouvant etre utilisés à des températures atteignant 200 °C, *Le Vide*, **4**, 676 (1949).

458. GONSIOR B., Eine Thermodiffusionsanlage für Tritium zur Verarbeitung geringer Tritiumkonzentrazionen, *Ztschr. f. Angew. Phys.*, **13**, 545 (1961).

459. GOODMAN I. S., Soldering in the tube construction, *Trans. Electrochem. Soc.*, **87**, 261 (1946).

460. GOODMAN I. E., and SUTHERLAND J. G., Elasto-plastic stress-optical effect in silver chloride single crystals, *J. Appl. Phys.*, **24**, 577 (1953).

461. GORDON S. A., Construction of small fixed leaks of predictable throughput, *Rev. Sci. Instr.*, **29**, 501 (1958).

462. GORE G. W., A simple glass sliding joint for vacuum work, *J. Sci. Instr.*, **34**, 459 (1957).

463. GÖRLICH P., and KROHS A., Über ein neues technologisches Verfahren für Aufbringung von Quarzfenstern auf Entladungsröhren, *Vakuum-Technik*, **5**, 12 (1956).

463a. GORMAN J. M., An accurate pressure regulating device, *Anal. Chem.*, **19**, 506 (1947).

464. GORMAN J. K., and NARDELLA W. R., Hydrogen permeation through metals, *Vacuum*, **12**, 19 (1962).

465. GOROWITZ B., MOSES K., and GLOERSEN P., Magnetically driven fast-acting valve for gas injection into high vacua, *Rev. Sci. Instr.*, **31**, 146 (1960).

466. GOTT O., Gläser mit besonderen Spektraleigenschaften, *Glastechn. Ber.*, **24**, 36 (1951).

467. GOULD C. L., *Vacuum System for a 30 BEV Accelerator*, Brookhaven Nat. Lab. BNL-2930 (1956).

468. GRAHAM R. L., HARKNESS, *et al.*, An a.c. operated mass spectrometer for isotope abundance measurements, *J. Sci. Instr.*, **24**, 119 (1947).

469. Granville-Phillips Co. (Box 1290 Boulder, Colorado, U.S.A.), *Type C Ultra-high Vacuum Valves* (Cat.)

470. GRAVES B. B., Platinum-to-Pyrex tubular Housekeeper seals, *Rev. Sci. Instr.*, **31**, 349 (1960).

471. GRAY R. V., Window shield for vacuum melting operations, *Rev. Sci. Instr.*, **23**, 309 (1952).

471a. GRAY D. E., (Editor), *American Institute of Physics Handbook*, McGraw-Hill, New York, 1957.

472. GRAZIOTTI R., and COMBOURIEU J., Robinet à voie droite de grande ouverture, *Le Vide*, **12**, 220 (1957).

473. GREEN G. W., An electro-pneumatic vacuum valve, *J. Sci. Instr.*, **26**, 206 (1949).

474. GREEN G. W., An electrically operated all metal capillary vacuum valve for adjustable gas flow, *J. Sci. Instr.*, **30**, 171 (1953).

475. GREEN G W., Simplified construction of a straight-through, metal, O-ring sealed, vacuum stopcock, *J. Sci. Instr.*, **31**, 473 (1954).

476. GREEN G. W., A simple metal-to-metal vacuum valve, *J. Sci. Instr.*, **32**, 192 (1955).

477. GREEN M., and MAXWELL K. H., A cut-off for the vacuum manipulation of chlorine, *J. Sci. Instr.*, **37**, 303 (1960).

478. GREEN L. A., and MILES H. T., A dismountable vacuum joint suitable for bakeable glass vacuum system, *J. Sci. Instr.*, **36**, 324 (1959).

479. GREENBLATT M. H., Sealing a calcium fluoride window to glass, *Rev. Sci. Instr.*, **29,** 738 (1958).

479a. GREENHOUSE H. M., and VERGARA W. C., Versatile high vacuum rotary feedthrough, *Rev. Sci. Instr.*, **35,** 129 (1964).

480. GREENE S. A., Glass check valve, *Rev. Sci. Instr.*, **26,** 731 (1955).

480a. GREENLAND K. M., Multi-electrode ring for a vacuum plant, *J. Sci. Instr.*, **28,** 324 (1951).

481. GREENWOOD N. N., A mercury cut-off for transferring gas at atmospheric pressure into a high vacuum, *J. Sci. Instr.*, **33,** 318 (1956).

482. GREINER A., *Quartz-to-Metal Seal,* U.S. Pat. 2,504,522 (1950).

483. Griffin & George Ltd. (Ealing Road, Alperton, Wembley, Middlesex, England), *Chemical Laboratory Apparatus*, Catalogue 56 S.

484. GRIFFOUL M. R., La soudure de Kovar-Be avec anneau de Monel interposé, *Le Vide*, **6** (33), 995 (1951).

485. GRIVET P., and BLATTMANN H., Robiet à vide pour canalisation à grand débit, *Le Vide*, **2,** 47 (1946).

485a. GROOM E. J., Aluminium-foil soldering on copper pipe for G. M. tubes, *Light Met.* **9,** 464 (1946).

485b. GROSS M. J., *New Tube Techniques of Telefunken Berlin,* FIAT Rep. No. 560 (1945).

486. GROSS R. C., Design evaluation considerations for vacuum brazing, *Trans. AVS Nat. Vac. Symp. 1958*, p. 175, Pergamon Press, Oxford, 1959.

487. GROSZKOWSKI J., Gas desorption on rubbing the surface in vacua, *Trans. AVS Nat Vac. Symp. 1961*, p. 186, Pergamon Press, Oxford, 1962.

488. Groupe des Techniques nucléaires — Saclay, Mode de serrage de joints métalliques étanches au vide élevé et à la pression pour les températures variant de 4 °K à 800 °K, *Le Vide*, **15,** 23 (1960).

489. GROVE D. J., The application of ultra high-vacuum techniques to controlled therminuclear devices, *Trans. AVS Nat. Vac. Symp. 1958*, p. 10, Pergamon Press, Oxford, 1959.

489a. GUILDNER L. A., and STIMSON H. F., Vacuum-tight cylinder joints and ball and socket joints. *Rev. Sci. Instr.*, **34,** 658 (1963).

490. GUTHRIE A., Vacuum Technique, in *Handbook of Physics* (pp. 5–78), McGraw-Hill, New York, 1958.

490a. GUTHRIE A., *Vacuum Technology*, John Wiley & Sons, New York, 1963.

491. GUTHRIE A., and WACKERLING R. K., *Vacuum Equipment and Techniques*, McGraw-Hill, New York, 1949.

492. GUYER E. M., Electronic welding of glass, *Electronics*, **18,** 92 (1945).

493. GUYER E. M., Electric welding of glass, *Electr. Eng.*, **67,** 446 (1948).

494. HAALAND C. M., Metal bakeout valve for ultra-high vacuum, *Rev. Sci. Instr.*, **30,** 947 (1959).

495. HAAS N. DE, Metal–glass vacuum seal for use at low temperature, *Rev. Sci. Instr.*, **30,** 594 (1959).

496. HAASE M., Filterpolarisatoren und ihre Anwendungen, *Glastech. Ber.*, **15,** 295 (1937).

497. HAASE A. P., and HORTON C. E., Sealing techniques for miniature tubes, *Ceram. Age*, **66,** 16 (Nov. 1955).

498. HACK W., Which brazing alloys for vacuum systems, *Mater. in Design. Engng.*, **52,** (1), 116 (1960).

499. HADDEN R. J. B., and KIRK A., Leakage within glass-fibre reinforced epoxy-resin laminates, *Vacuum*, **11,** 147 (1961).

500. HAGSTRUM H. D., and WEINHART H. W., New porcelain rod leak, *Rev. Sci. Instr.*, **21,** 394 (1950).

501. HAGY H. E., and RITLAND H. N., Viscous flow in glass-to-metal seals, *J. Am. Ceram. Soc.*, **40,** 58 (1957).

502. HALBAN H., and WILSON R., A simple pressure and vacuum tap, *J. Sci. Instr.*, **26,** 414 (1949).

503. HALL C., Demountable all-metal, high vacuum coupling, *Rev. Sci. Instr.*, **33,** 131 (1962).
504. HALSTED R. E., and NIER A. O., Gas flow through the mass spectrometer viscous leak, *Rev. Sci. Instr.*, **21,** 1019 (1950).
505. HAMMOND V. J., A simple magnetically controlled mercury cut-off, *J. Sci. Instr.*, **31,** 258 (1954).
506. HAMPEL C. A., (Editor), *Rare Metals Handbook*, Reinhold, New York, 1954.
507. HAMPTON W. M., Opalglass, *Illum. Engng.*, **26,** 48 (1933).
508. HANCHER R. L., Electropolishing, *Metallurgia*, **49,** 47 (1954).
508a. HANDEN J., and HRUDA R. M., Glass-to-metal seals using induction heating, *Electr. Manuf.*, **55,** 129 (1955).
508b. HANDY & HARMAN (850 Third Ave., N. Y. 22, U.S.A.), *Brazing Alloys*, Bull. T1, T2, T5, 25.
509. HANDY & HARMAN, Nomograph for selecting size of wire for brazing, *Mater. & Meth.*, **43** (1), 129 (1956).
509a. HANSON W. B., Ring magnet vacuum chamber of the Argonne National Lab., Zero gradient synchrotron, *Trans. AVS Nat. Vac. Symp. Trans.* **1961,** p. 1140, Pergamon Press, Oxford, 1962.
510. HARE M. D., KELLER R. F., and MENESES H. A., Electroformed ceramic-to-metal seal for vacuum tubes, *Proc. 4th Nation. Conf. on Tube Technique*, p. 25, Univ. Press, N. Y., 1959.
510a. HARINGX J. A., Conical disc springs, *Philips Techn. Rev.*, **10** (Aug), 61 (1948).
511. HARRIS E. J., A gas inlet for mass spectrometry, *J. Sci. Instr.*, **25,** 417 (1948).
511a. HARRIS L., Preparation and infrared properties of aluminium oxide films, *J. Opt. Soc. Am.*, **45,** 27 (1955).
512. HARRISON E. R., Glass leak and control valves, *J. Sci. Instr.*, **30,** 170 (1953).
513. HARRISON E. R., and HOBIS L. C. W., Hicked diffusion leak for hydrogen, *Rev. Sci. Instr.*, **26,** 305 (1955).
514. HARROWER G. A., BEST F. S., and MACHALETT A. A., Shielded electrical connection into vacuum, *Rev. Sci. Instr.*, **26,** 404 (1955).
515. HART D. M., A demountable glass vacuum system using electron bombardment in the vacuum deposition of ferromagnetic films, *Trans. AVS Nat. Vac. Symp. 1958*, p. 230, Pergamon Press, Oxford, 1959.
516. HARTMAN P. L. Simple vacuum valve, *Rev. Sci. Instr.*, **27,** 871 (1956).
517. HARVEY R. B., and MCCULLUM, Vacuum evaporation apparatus, a new vacuum valve and an ion gauge protective circuit. *Abstr. Vacuum*, **4,** 245 (1954) from *Suffield Tech. Pap.* No. 34 (1954).
517a. HAST M., Preparation of thin specimen films, *Nature*, **159** (4037), 370 (1947).
517b. HAST N., Production of extremely thin metal films by evaporation onto liquid surfaces, *Nature*, **162,** (4127), 892 (1948).
518. HAUFFE K., *Reaktionen in und an festen Stoffen*, Springer, Berlin, 1955.
519. HAUL R. A. W., and SWART E. R., A high vacuum magnetic mercury cut-off, *J. Sci. Instr.*, **33,** 243 (1956).
520. HAUSER U., and KERLER W., Easily mounted aluminium oxide foils for windows and backings, *Rev. Sci. Instr.*, **29,** 380 (1958).
521. HAYNES J. R., Technique for obtaining increased range and mobility of free electrons in silver chloride, *Rev. Sci. Instr.*, **19,** 15 (1948).
522. Haynes Stellite Co. (Kokomo, Ind., U.S.A.), *Metal Ceramics*, (Publ. July 1959).
523. HAYWARD A. G., Simple rotary vacuum seal, *J. Sci. Instr.*, **29,** 410 (1952).
524. HAYWARD A. G., Simple vacuum seals, *Vacuum*, **2,** 262 (1952).
525. HEARST I. R., AHN S. H., STRAIT E. N., Vacuum seals at liquid nitrogen temp., *Rev. Sci. Instr.*, **30,** 200 (1959).
526. HEATHCOTE V. A. and READ W. E., A demountable seal for high vacuum work *J. Sci. Instr.*, **34,** 247 (1957).
526a. HEBERLE J., Beryllium window for gamma rays in a liquid helium cryostat, *Rev. Sci. Instr.*, **33,** 1476 (1962).

527. HEERDEN P. J. VAN, Metal gaskets for demountable vacuum systems, *Rev. Sci. Instr.*, **26**, 1131 (1955).

528. HEERDEN P. J. VAN, Metal gaskets for demountable vacuum systems, *Rev. Sci. Instr.*, **27**, 410 (1956).

529. HEES G. W., EATON W., and LECH J., The knife-edge vacuum seal, *Vacuum*, **4**, 438 (1954).

530. HEES G. W., EATON W., and LECH J., The knife-edge vacuum seal, *Trans. AVS Nat. Vac. Symp. 1955*, p. 75, Pergamon Press, Oxford, 1956.

531. HEES G. W., and EARLEY K. D., The use of aluminonsilicate glasses in experimental vacuum devices, *Vacuum*, **11**, 64 (1961).

532. HEINRICH J. T., Simple vacuum coupling, *Rev. Sci. Instr.*, **29**, 1053 (1958).

533. HEINZE W., *Einführung in die Vakuumtechnik*, Vol. I., Verlag Technik, Berlin, 1955.

534. HELDMAN J. D., *Techniques of Glass Manipulation in Scientific Research*, Prentice-Hall, New York, 1947.

535. HELLER R. B., Cutting flat vacuum gaskets on a lathe, *Nucleonics*, **6** (2), 81 (1950).

536. HEMMENDINGER A., Electrodes in metal vacuum systems, *Rev. Sci. Instr.*, **10**, 389 (1939).

536a. HEMMENDINGER A., and ROENSCH A. P., Thin glass windows *Rev. Sci. Instr.*, **26**, 562 (1955).

537. HENNIG M., Über die Dicke und die Dispersion der Brechzahl dünner Aluminium-oxydschichten, *Ztschn. f. Physik*, **144**, 296 (1956).

538. HENRY H. G., Cold welding of Teflon, *Mater. and Meth.*, **38**, 114 (1953).

539. HENRY R. P., La normalisation française des tubes et raccords pour vide poussé, *Adv. Vac. Sci. Technol.*, p. 349, Pergamon Press, Oxford, 1960.

540. HENRY R. P., and BLAIVE I. C., Les joints métalliques pour le vide élevé et leurs utilisations industrielles, *Adv. Vac. Sci. Technol.*, p. 345, Pergamon Press, Oxford, 1960.

541. HENRY R. P., and BLAIVE I. C., Joints métalliques pour installations de vide industriel, *Le Vide*, **13**, 172 (1958).

542. HERAEUS W. C. GmbH. (Hanau, Germany), *High Vacuum Valves and Components*, Cat.

543. HERITAGE R. J., and BALMER I. R., The metallization of glass and ceramics, *Metallurgia*, **47**, 171 (1953).

544. HERNE R. I., A vacuum "Lead-in", *J. Sci. Instr.*, **23**, 244 (1946).

545. HERRMANN H., Messung von Glasspannungen bei höheren Temp., mittels Polarisationmikroskopes, *Ztschr. Angew. Phys.*, **7**, 174 (1955).

546. HERRMANN H., Beobachtungen über den Zusammenhang Zwischen Wärmeausdehnung und elastischen Spannungen bei Glass-Metall-Verbindungen, *Vakuum-Technik*, **4**, 115 (1955).

547. HERRMANN H., and THOMAS H., Fe-Ni-Co Legierungen für Glass-Metall Einschmelzungen, *Ztschr. Metallkunde*, **48**, 582 (1957).

548. HERZOG L. F., ESKEW T. J. *et al.*, Residual gases in closed high-vacuum systems, determined by spectrometry, *Trans. AVS Nat. Vac. Symp. 1961*, p. 581, Pergamon Press, Oxford, 1962.

549. HESS E. F., High strength adhesives, *Mater. in Design Engn.*, **50** (1), 104 (1959).

550. HESS W. T., LANDER H. J., *et al.*, Fusion welding of beryllium in a vacuum atmosphere, *Trans. AVS Nat. Vac. Symp. 1961*, p. 679, Pergamon Press, Oxford, 1962.

551. HETENYI M., *Handbook of Experimental Stress Analysis*, John Wiley, New York, 1950.

552. HEYES J., Elektrolytisches Polieren, *Werkstofftechn. u. Maschinenbau*, **43**, 122 (1953).

553. HEYLEN A. E. D., Bakeable ionization chamber, *J. Sci. Instr.*, **39**, 122 (1962).

554. HEYWOOD W. A., *Report on Static Seal Test*, Knalls At. Power Lab. KAPL-501 (1951).

555. HEYWOOD W. A., *Static Seal Studies*, Rep. KAPL-974 (1953).

556. HEYWOOD W. A., *Control Rod Gas Seals*, Rep. KAPL-1161 (1955).

557. HICKMAN K. C. D., High vacuum with the polyphenol-ethers—a self-contained

technology, *Trans. AVS Nat. Vac. Symp. 1961*, p. 307, Pergamon Press, Oxford, 1962.

557a. Hickman W. M., Design for a metal mass spectrometer tube, *Rev. Sci. Instr.*, **20,** 472 (1949).

558. Higatsberger M. I., High-vacuum gaskets capable of being degassed, *Acta. Phys. Austriaca,* **10,** 181 (1956).

559. Higatsberger M. I., and Erbe W. W., Improved metal-to-metal vacuum seals, *Rev. Sci. Instr.*, **27,** 110 (1956).

560. Highhouse F., and White J., A convenient gas sample tube holder and break-off applicable to mass spectrometers, *Rev. Sci. Instr.*, **21,** 101 (1950).

560a. Hill D. E., and King G. D., A method of sealing evacuated, wide-bore tubing, *J. Sci. Instr.*, **41,** 187 (1964).

561. Hill M. L., Automatic apparatus for measuring gas evolution rates, *Trans. AVS Nat. Vac. Symp. 1961,* p. 813, Pergamon Press, Oxford, 1962.

562. Hill M. L., Holman R., and Kulinski E., Equipment for hot rolling strip in vacuum, *Trans AVS Nat. Vac. Symp. 1959*, p. 124, Pergamon Press, Oxford, 1960.

563. Hillier M., Bell R. L., Aluminosilicates and the high temperature processing of microwave vacuum tubes, *Brit. J. Appl. Phys.*, **9,** 94 (1958).

564. Hinterberger H., Erfahrungen mit Metallfolien als Hochvakuumdichtungen, *Ztschr. Naturforsch.*, **6** A, 459 (1951).

564a. Hippel R. von, *Dielectric Materials and Applications*, John Wiley, New York, 1954.

565. Hirayama C., The sag point of glasses, *J. Am. Ceram. Soc.*, **45** (3), 113 (1962).

566. Hoare F. E., Jackson L. C., and Kurti N., (Edit.), *Experimental Cryophysic,* Butterworth, London, 1961.

567. Hobson J. P., Physical adsorption in ultra-high vacuum systems, *Trans. AVS Nat. Vac. Symp. 1961*, p. 146, Pergamon Press, Oxford, 1962.

568. Hoch H., Ausheizbare Verbindungen an Hochvakuum-Apparaturen, *Vacuum-Technik*, **10,** 235 (1961).

569. Hoch M., New installation for radiofrequency leads in high vacuum work, *Rev. Sci Instr.*, **23,** 651 (1952).

570. Hodder B. H., Vacuum or low pressure seal utilizing modified standard refrigeration type flare tube fittings, *J. Sci. Instr.*, **35,** 182 (1958).

571. Hodkinson I., Polythene windows for vacuum systems, *Vacuum*, **3,** 414 (1953).

572. Hoechst A. G. (Frankfurt/M, Germany), *Rare Gases*, Catalogue.

573. Hoeprich P. D., and Whitesides J. N., Protection of electrodes by local potting, *Rev. Sci. Instr.*, **32,** 458 (1961).

574. Hoffmann O., Durchlässigkeit von 1000—3000 Å dicken Aluminium-oxydfolien für Elektronen unterhalb 16 KeV., *Ztschr. Phys.*, **143,** 147 (1955).

574a. Hofmann W., and Ruge J., Kaltpressschweissung von Metallen, *Ztschr. Metallkunde*, **43,** 133 (1952).

575. Högberg L., and Vogel K., Experiments with electrodeless generation and acceleration of plasma rings, *Nucl. Instr. & Meth.*, **10,** 95 (1961).

576. Hogg B. G., and Duckworth H. E., Vacuum properties of synthetic dielectrics, *Rev. Sci. Instr.*, **19,** 331 (1948).

577. Hogg I. H., Welding techniques in atomic energy plants, *Atomics*, **8,** 77, (1957).

578. Holden I., Holland L., and Laurenson L., Bakeable vacuum seals using aluminium wire gaskets, *J. Sci. Instr.*, **36,** 281 (1959).

579. Holland L., The cleaning of glass in a glow discharge, *Brit. J. Appl. Phys.*, **9,** 410 (1958).

580. Holland L., Theory and design of getter-ion pumps, *J. Sci. Instr.*, **36,** 105 (1959).

581. Holland L., *Vacuum Deposition of Thin Films*, Chapman & Hall, London, 1960.

582. Holland L., Problems in the construction and operation of bakeable and non-bakeable vacuum systems, *Trans AVS Nat. Vac. Symp. 1960*, p. 172, Pergamon Press, Oxford, 1961.

583. Holland L. Aluminium bakeable vacuum seal, *J. Sci. Instr.*, **38,** 339 (1961).

584. Holland L., and Bateman S. K., A low pressure evaporation plant for rapid

operation, *Trans. AVS Nat. Vac. Symp. 1961*, p. 1201, Pergamon Press, Oxford, 1962.

584a. HOLLAND L., *The Properties of Glass Surfaces*, Chapman & Hall, London, 1964.

585. HOLLAND R. E., and MOORING F. P., High vacuum butterfly valve, *Rev. Sci. Instr.*, **26**, 989 (1955).

586. HOLLANDER M. B., Welding metals by friction, *Mater. & Engng. Design*, **55** (2), 79 (1962).

587. HOLLAND-MERTEN E. L., *Handbuch der Vakuumtechnik* W. Knapp, Halle, 1953.

588. HOLMES R. W., Cementing metal screens to cathode ray tubes, *Electronics*, **27**, 300 (Febr. 1954).

589. HOLTON P., A gas control valve for flow proportional counters with facilities for flushing, *J. Sci. Instr.*, **29**, 300 (1952).

590. HONIG R. E., Gas flow in the mass spectrometer, *J. Appl. Phys.*, **16**, 646 (1945).

591. HONIG R. E., A greaseless gas flow valve, *Rev. Sci. Instr.*, **21**, 1024 (1950).

592. HONIGMANN L., Spannungsmessungen in Glas-Metal-Verschmelzungen, *Glastechn. Ber.*, **12**, 372 (1934).

593. HOPFIELD J. J., Capillary valves for gases, *J. Opt. Soc. Am.*, **12**, 391 (1926).

594. HOPFIELD J. J., Glass-variable microleaks for gases, *Rev. Sci. Instr.*, **21**, 671 (1950).

594a. HORIKOSHI G., and TANAKA J., A thin vacuum gate valve for large aperture, *Japan. J. Appl. Phys.*, **2** (2), 135 (1963).

595. HORN O., and STARCK W., Fluorhaltige Kunststoffe, *Angew. Chem.*, **64**, 533 (1952).

596. HORNING D. O., *Vacuum Vessel Construction Technique and Sealing Methods*, Instr. Eng. Res. Univ. Calif. Tech. Rep., HE 150–84 (1951).

597. HORSELING J., A high vacuum tap with short outgassing time, *Philips Techn. Rev.*, **17**, 184 (1955).

598. HORWITZ N. H., and BOHN H. V., Metal-to-glass vacuum seal for low temperatures, *Rev. Sci. Instr.*, **32**, 857 (1961).

599. HOUSEKEEPER W. G., The art of sealing base metals through glass, *J. Am. Inst. Electr. Engrs.*, **42**, 954 (1923).

600. HOUTEN G. R. VAN, A survey of the bonding of cermets to metals, *Welding J.*, **37**, 558 s. (1958).

601. HOUTEN G. R. VAN, Transition layer is key to good ceramic-to-metal bonds, *Mater. in Design Engn.*, **48**, (7), 112 (1958).

602. HOUTEN G. R. VAN, A survey of ceramic-to-metal bonding, *Bull. Am. Ceram. Soc.*, **38**, 301 (1959).

602a. HOWARD F. L., Leak-resistant rotation seal for vacuum applications, *J. Res. Natl. Bur. Stand.*, **67C** (4), 335 (1963).

603. HOWE P. G., Greaseless vacuum rotor seal, *Rev. Sci. Instr.*, **26**, 625 (1955).

604. HRZEK B., and ESPE W., *Vacuum-tight Glass-to-Metal Bonding*, Austr. Pat. 179, 035 (1954).

605. HUBER H., La technologie des tubes électroniques dans l'industrie atomique, *Ind. Atomiques*, **5** (9/10), 85 (1961).

606. HUFF H. A., and KUGLER A. N., Argonarc Welding methods for Al, Cu.,Mg, Ag and stainless steel, *Welding J.*, **28**, 128 (1949).

607. HUGHES A. L., and PAO C. S., The use of Wood's metal in vacuum systems. *Rev. Sci. Instr.*, **14**, 254 (1943).

608. HÜHN C. F., Bildröhrenfertigung, *Telefunken Ztg.*, **28**, 215 (1955).

609. HULL A. W., *et al.*, Glass-to-metal seals, *J. Appl. Phys.*, **12**, 698 (1941).

609a. HULL A. W., Stresses in cylindrical glass–metal seals with glass inside, *J. Appl. Phys.* **17**, 685 (1946).

610. HULL A. W., BURGER E. E., Glass-to-metal seals, *Physics*, **5**, 384 (1934).

611. HULL A. W., and BURGER E. E., Stress measurement with quartz key and polarization filter, *Rev. Sci. Instr.*, **7**, 98 (1936).

612. HUME G. W., Aluminium–ceramic bond as hermetic seal, *Mater. & Meth.*, **41** (4), 110 (1955).

612a. HUNTER W. R., A rotary motion device suitable for use in high vacuum systems, *Vacuum*, **13,** 193 (1963).

613. HYMAN H. JR., and BILLINGS B. H., High transmission window for radiation of 3 to 14 μ wave length, *J. Opt. Soc. Amer.*, **37,** 113 (1947).

614. HYNES F. I. Alumina bodies for ceramic-to-metal seals, *Ceram. Ind.*, **67,** 87 (1956).

615. Imperial Chemical Industries Ltd., (London), *Metal Degreasing by Trichlorethylene*, 1957.

616. INANANANDA S., *High Vacua*, Van Nostrand, New York, 1947.

616a. Indium Corp. of America (1676 Lincoln Ave., Utica, N. Y.), *Indalloy Intermediate Solders*, 1957.

617. ITO G., SUZUKI H. *et al.*, Studies on electron-beam welding, *Trans. AVS Nat. Vac. Symp. 1961*, p. 693, Pergamon Press, Oxford, 1962.

618. IVANOV A. P., *Elektricheskie Istochniku Sveta* (Electrical Light Sources), Gosenergoizdat, Moscow, 1948.

619. JACKSON R., and QUARELL A. G., A simple, high-vacuum release valve, *J. Sci. Instr.*, **15,** 208 (1938).

619a. JACOBS T. A., GIEDT R. R., and DURRAN D. A., High-pressure, high vacuum valve for shock tube application, *Rev. Sci. Instr.*, **33,** 488 (1962).

620. JACOBS R. B., and ZUHR F. H., New development in vacuum engineering, *J. Appl. Phys.*, **18,** 34 (1947).

621. JACOBUS D. D., and POLK I. J., Vacuum System. Part I: Vacuum Chamber. *Rev. Sci. Instr.*, **24,** 821 (1953).

622. JACQUET P. A., Electrolytical and chemical polishing of metals, *Metallurgical Rev.*, **1,** 157 (1956).

623. JAECKEL R. *Kleinste Drucke, Ihre Erzeugung und Messung*, Springer, Berlin, 1950.

624. JAECKEL R., Entgasung, *Trans. AVS Nat. Vac. Symp. 1961*, p. 17, Pergamon Press, Oxford, 1962.

625. JAGER G., POCHERT., and ROGENBUCK W., Zwei neue Bauelemente für die Vakuumtechnik, *Die Technik*, **15,** 27 (1960).

626. JAMES D. H., and PHILLIPS C. S. G., A simple gas-flow control of high efficiency, *J. Sci. Instr.*, **29,** 362 (1952).

627. JAMES D. H., and PHILLIPS C. S. G., A simple pressure controller, *J. Sci. Instr.*, **31,** 193 (1954).

627a. JANECKE D., Neuzeitliche Metallentfettung, *Vacuum-Technik*, **7,** 52 (1958).

628. JAMIESON W. D., Simple refractory-metal and glass-metal seals, *J. Sci. Instr.*, **35,** 73 (1958).

628a. JEBSEN-MARWEDEL H., *Glastechnische Fabrikationsfehler*, Berlin, Heidelberg, 1959.

629. JEFFERSON T. B., *The Welding Encyclopedia*, McGraw-Hill, New York, 1955.

630. JENKINS D. E. P., Vakuumdichte Keramic-Metall-Verbindungen, *Vacuum-Technik*, **4,** 107 (1955) and *Electronic Engng.* **27,** 290 (1955).

631. JENKINS R. O., The construction of small vacuum leaks of constant value, *J. Sci. Instr.*, **35,** 428 (1958).

632. JENKINS W. A., A bubble counter for vacuum system, *Rev. Sci. Instr.*, **22,** 845 (1951).

633. JENKS G. H., A convenient leak for testing He leak detectors, *Rev. Sci. Instr.*, **21,** 674 (1950).

634. JENNINGS B., A large, straight-through vacuum valve for Van Der Graaf generators, *Rev. Sci. Instr.*, **20,** 366 (1949).

635. JENSEN N., Vapor pressure of plastic materials, *J. Appl. Phys.*, **27,** 1460 (1956).

636. JOHNSON D. P., Behaviour of refractory oxides and metals alone and in combination, *in vacuo* at high temperatures, *J. Am. Ceram. Soc.*, **33,** 168 (1950).

637. JOHNSON I. W., and GOOD W. M., Gas leak valve with improved characteristics, *Rev. Sci. Instr.*, **32,** 219 (1961).

638. Johnson, Matthey & Co. Ltd., (78 Hatton Garden, London, E. C. 1), *The Use of Pre-placed Brazing Inserts*, Cat. 162 (1959).

639. Johnson, Matthey & Co. Ltd., *Silver Brazing of Stainless Steel*, Cat. 187 (1961).
640. JOHNSON R. P., Construction of filament surfaces, *Phys. Rev.*, **54**, 459 (1938).
641. JOHNSON R. P., WHITE A. B., and NELSON R. B., An engine for smoothing small tungsten wires, *Rev. Sci. Instr.*, **9**, 253 (1938).
642. JOHNSTON W. H., (Lab.), *Basic Radiation Studies*, Quart. Rep. Nov. 1959 – Jan. 1960; TID-5612.
643. JONAS B., and SEITZ G., Demountable seals for glass high-vacuum equipment, *Philips, Techn. Rev.*, **22**, 53 (1960–61).
644. JONES B., and MEYER F. R., Ultrasonic welding of structural aluminium alloys, *Welding J.*, **37**, 81 s. (1958).
645. JONES G. O., *Glass*, John Wiley, New York, 1956.
646. JORDAN I. R., New developments in vacuum sealing, *Trans. AVS Nat. Vac. Symp. 1961*, p. 1302, Pergamon Press, Oxford, 1962.
647. JOSSEM E. L., Admission of pure gases to vacuum systems, *Rev. Sci. Instr.*, **11**, 164 (1940).
647a. JOUANIGOT M., Un poste industriel d'évaporation sous vide, *Le Vide*, **3**, 381 (1948).
648. JUENKER D. W., SWAAY M. VON, and BIRCHENALL C. E., The use of palladium diffusion membranes for the purification of hydrogen, *Rev. Sci. Instr.*, **26**, 888 (1955).
649. KALLWEIT H. (Telefunken GmbH.), *Joining Glass to Metals*, Germ. Pat. 950, 872 (1956).
650. KALMUS P. I. P., A simple flexible vacuum joint, *Vacuum*, **9**, 147 (1959).
651. KALPERS H., Warmfeste Stahle für Glasseinschmelzformung, *Glas-Email-Keramo Techn.*, 1, 8 (1950).
652. KALSING H., Glasses for glass-to-metal seals, *Sprechsaal*, **86**, 212—392 (1953).
652a. KAMMERER (Belgisch Gladbach) *Hochvakuumventile*, Cat. E. K.
652b. KAMPHAUSEN H. A., Method for making multiple glass–tungsten rod seals, *J. Sci. Instrum.*, **40**, 605 (1963).
653. KAPPEN G. M. VAN, A metal vacuum valve, *Appl. Sci. Res.*, **3B**, 141 (1953).
654. KAPTZOV N. A., *Electrical Phaenomena in Gases and Vacuum*, Editura Tehnica, Bucharest, 1955.
655. KARLSSON E., and SIEGBAHN K., A new design for beta-ray spectrograph for relative measurements, *Nucl. Instr. & Meth.*, **7**, 113 (1960).
656. KATAN I. L., Vacuum powder insulation for low temperatures, *Vacuum*, **1**, 191 (1951).
657. KATZ H., and STETTER W., (Siemens-Halske Akt. Ges.) *Vacuum-tight Sealing of Metals to Glass for Electric Discharge Tubes*, Germ. Pat. 954, 628 (1956).
658. KAUFMAN A. B., Selecting solders for low temperature service, *Mater. in Design. Engng.* **48** (6), 114 (1958).
659. KEITH J. K., *Leak Rate versus Absolute Pressure Completion*, Rep. Y-1347 (1961).
660. KELMAN L. R., WILKINSON W. D., *et al.*, The safe handling of radioactive-pyrophoric materials. Glovebox design and construction. *Nucleonics* **14**, 65 (Apr. 1956).
661. KENDALL J. M., Permeation of air through walls of platin tubing used in low pressure system, *Trans. AVS Nat. Vac. Symp. 1957*, p. 120, Pergamon Press, Oxford, 1959.
662. KENT P. I. C., Glass-refractory seals, *J. Sci. Instr.*, **30**, 485 (1953).
663. KENT T. B., A hydrogen Pirani leak detector using a charcoal trap., *J. Sci. Instr.*, **32**, 132 (1955).
664. KENTY C., and REUTER F. W., An apparatus for micro gas analysis, *Rev. Sci. Instr.*, **18**, 918 (1947).
665. KENTY C., McLeod gauge with an electrical contact, *Rev. Sci. Instr.*, **22**, 217 (1951).
666. KENTY C., New self-locking mercury cut-off, *Rev. Sci. Instr.*, **22**, 844 (1951).
667. KERSTEN H., A needle valve for gas X-ray tubes, *Rev. Sci. Instr.*, **6**, 175 (1936).
668. KERSTEN H., and Dwight C. H., Color changes in glasses due to X-rays, *J. Phys. Chem.*, **35**, 627 (1933).
669. KERWIN H., *Arc and Acetylene Welding*, McGraw-Hill, New York, 1944.

670. KESSELRING F., *Leakproof Glass-to-Metal Seals*, Swiss. Pat. 310, 625 (1956).

671. KIENEL G., and LORENZ A., A getter ion pump for technical purposes, Haereus Hanau (Description), *Vakuum-Technik*, **9,** 3 (1960).

672. KIENEL G., and LORENZ A., Eine doppelwandige Ultrahochvakuum-Anlage für Technische Anwendungszwecke, *Vakuum-Technik*, **9,** 217 (1960).

672a. KINDER E., Ein einfaches Mehrwege-Vakuumventil, *Vakuum-Technik*, **8,** 38 (1959).

673. KING L. D. P., High-vacuum valve and seal, *Rev. Sci. Instr.*, **19,** 83 (1948).

673a. KING G. W., Flap valve for long narrow apertures, *J. Sci. Instr.*, **37,** 439 (1960).

673b. KING J. G., Production leak testing of large pressure vessels, *Trans. AVS Nat. Vac. Symp. 1961*, p. 1106, Pergamon Press, Oxford, 1962.

674. KINGERY W. D., Role of surface energies and wetting in metal–ceramic sealing, *Bull. Am. Ceram. Soc.*, **35,** 108 (1956).

675. KINGSTON W. E., Recorder dillatometer, *Metal. Progr.*, **44,** 1115 (1954).

676. KIRBY P. L., Conductivity of glasses, *Brit. J. Appl. Phys.*, **1,** 193 (1950).

677. KIRLY R. K., Teflon, *J. Res. Nat. Bur. Stand.*, **57,** 91 (1956).

678. KIRCHNER F., A fast simple method for ultra-high vacuum, *Naturwiss.*, **48,** 548 (1961).

679. KIRSLIS S. S., and ASMANES C., Improved glass valve, *Rev. Sci. Instr.*, **26,** 615 (1955).

680. KISTEMAKER J., The influence of fractionizing and viscosity effects in mass-spectrometric gas handling systems, *Physica*, **18** (3) 163 (1952).

681. KITAIGORODSKY I. L., *Technologie des Glases* (Translation from Russian), Verlag Technik, Berlin, 1957.

682. KITTLE D. B., *Evaluation of Seal Weld Joint Designs*, Knolls Atomic Power Labs., Rep. KAPL-M-DBK-2 (1957).

683. KIYOSHI HISANO, Standardization, project relative to vacuum engineering in Japan, *Adv. Vac. Sci. Technol.*, p. 80, Pergamon Press, Oxford, 1960.

684. KLAPFER A., GARBE S., and SCHMIDT W., Residual gases in vacuum systems, *Trans. AVS Nat. Vac. Symp. 1959*, p. 27, Pergamon Press, Oxford, 1960.

685. KLEEN W., KATZ H., and KAHL P., (Siemens-Halske Akt., Ges.), *Vacuumtight Soldered Joint, especially for Electric Discharge Vessels*, Germ. Pat. 1,028,241 (1958).

686. KLEIN E. J., A method of cutting threads in glass and ceramic tubes, *Rev. Sci. Instr.* **31,** 571 (1960).

687. KLEINTEICH R., Pflege und Wartung von glasernen Schliffteilen, *Vakuum-Technik*, **4,** 44 (1955).

687a. KLEINTEICH R., Die Glass-Metall Verschmelzung, *Glas-Email-Keramo-Techn.* **7,** 313 (1956).

688. KLEMENC A., *Die Behandlung und Reindarstellung von Gasen*, Springer Verlag, Wien, 1948.

689. KLIPPING G., and MIRGEL K. H., Ein Hochvakuum-Verschlussventil, *Vakuum-Technik*, **10,** 21 (1961).

690. KLOPFER A., Das Erreichen und Messen von tiefen Drucken, *Ztschr. Angew. Phys.*, **13,** 480 (1961).

691. KLOPFER F. H., SEAGONDOLLAR L. W., and SMITH R. K., Leak localizer for Van de Graaf accelerator tubes, *Rev. Sci. Instr.*, **23,** 245 (1952).

692. KLUMB H., and SCHRÖTTER A. W., Durchlassigkeitsuntersuchungen an Metall- und Kunststoffolien bei verschiedenen Temp., *Vakuum-Technik*, **10,** 175 (1961).

693. KNAPP O., Zusammensetzung vakuumtechnischer Gläser, *Silikattechnik*, **6,** 99 (1955).

694. KNAPP O., Solderglass, *Silikattechnik*, **9,** 153 (1958).

695. KNAUSS H. P., A device for controlling the flow of gas into a vacuum system, *Rev. Sci. Instr.*, **2,** 750 (1931).

696. KNECHT W. L., Application of pressed powder techniques for production of metal-to-ceramic seals, *Ceram. Age*, **63,** 12 (1954).

696a. KNIGHT R. D., High voltage vacuum feedthrough, *Rev. Sci. Instr.*, **34,** 303 (1963).

697. KNOLL M., *Materials and Processes of Electron Devices*, Springer Verlag, Berlin, 1959.
698. KNOLL M. E., Active-metal soldering of crystalline quartz, *Rev. Sci. Instr.* **32**, 83 (1961).
699. KNOR Z., Mercury float valve, *Rev. Sci. Instr.*, **31**, 351, (1960).
700. KNUDSEN A. W., Metalic vacuum-tight gasket, *Rev. Sci. Instr.*, **23**, 566 (1952).
701. KNUDSEN A. W., Fast-closing vacuum valve, *Rev. Sci. Instr.*, **27**, 148 (1956).
702. KOBAYASHI S., and YADA K., A standard leak utilising gas permeation, *J. Jap. Vac. Soc.*, **7**, 189 (July, 1957).
703. KOBAYASHI S., and YADA K., Leak test of rubber gaskets, *Adv. Vac. Sci. Technol.*, p. 248, Vol. I., Pergamon Press, Oxford, 1960.
704. KOENIG J. H., New developments in ceramics, *Mater. in Design. Engng.*, **47** (5), 121 (1958).
704a. KOFOID M. J., and ZIESKE P., Vacuum lock for protecting oxide coated cathodes, *Rev. Sci. Instr.*, **33**, 1115 (1962).
705. KOHL W. H., Baustoffe für die Herstellung von Elektronenröhren, *Vakuum-Technik*, **4**, 157 (1955).
706. KOHL W. H., *Materials and Techniques for Electron Tubes*, Reinhold, New York, 1960.
707. KOHL W., Tube techniques, *Vacuum*, **11**, 53 (1961).
708. KOHLER W., and ROMPE R., *Die Elektrischen Leuchthröhren*, Braunschweig, 1933.
709. KOHLRAUSCH F., *Praktische Physik*, B. G. Teubner, Leipzig, 1953.
710. KOOPMAN K. H., and KITTLE D. B., *Evaluation of Seal Weld Joint Design*, Rep. KAPL-M-DBK-2 (Suppl. 1), 1957.
711. KOOPS R., and ZEISS C., *Device to protect Windows in Vacuum Evaporation Plant*, Germ. Pat. Z. 4921 VI/486 (1955).
711a. KORFF S. A., *Geiger Counters*, in Fluge: *Handbuch der Physik*, Vol. 45, p. 76, Springer Verlag, Berlin, 1958.
712. KOROLEV V. I., *Osnovi Vacuumnoi Techniki* (Foundations of Vacuum Techniques), Gosenergoisdad, Moscow, 1953.
713. KOTOWSKI J., Keramische Baustoffe für Keramikröhren, *Vakuum-Technik*, **10**, 7 (1961).
714. KOTOWSKI J., Aktiv-Metal-Keramik-Verbindung, *Vakuum-Technik*, **11**, 21 (1962).
715. KOTOWSKI J., and MUSCHAWECK I., Über die Haftfestigkeit, von Metallisierungen auf Keramiken, *Vakuum-Technik*, **10**, 141 (1961).
716. KRAMERS W. J., and DENNARD F., A high temperature resistance heated high vacuum furnace, *Vacuum*, **3**, 155 (1953).
717. KREMERS H. C., Optical silver chloride, *J. Opt. Soc. Am.* **37**, 337 (1947).
718. KRIEG O., Hochvakuum-Nadelfeinventil mit Federrohrabdichtung, *Ztschr. Techn. Phys.*, **23**, 314 (1942).
719. KRIEGER G. L., *High Vacuum Techniques*, Sandia Corp, SCTM-76-60-14 (1960).
720. KROHN R., 2-inch packless through-vacuum valve, *Rev. Sci. Instr.*, **18**, 682 (1947).
721. KRONBERGER H., Vacuum techniques in the atomic energy industry, *Proc. of Inst. Mech. Engrs.* **172**, 113 (1958); *Engineer*, **15** (11), 702 (1957).
721a. KUBASCHEWSKI O., and HOPKINS B. E., *Oxidation of Metals and Alloys*, Butterworth, London, 1953.
721b. KUHNER M., Les scellements céramique–métal, *Le Vide*, **2**, 194 (1947).
722. KUNZL V., and SLAVIK J. B., Gaseinlassventil, *Ztschr. Techn. Phys.*, **16**, 272 (1935).
723. KURIE F. N. D., Vacuum systems, seals and valves, *Rev. Sci. Instr.*, **19**, 485 (1948).
723a. KURIE F. N. D., The technique of high intensity bombardment with fast particles, *Rev. Sci. Instr.*, **10**, 199 (1939).
724. KUTZELNIGG A., and GRÜTTNER S., *Galvanische Polierbäder*, Germ. Pat. 936, 124 (1956).
725. LABEYRIE J., Vacuum-tight sealing of glass and mica, *J. Phys. Radium.*, **11**, 20 (1950).
726. LABEYRIE J., and LEGER P., Jointure mica–verre, *Le Vide*, **6**, 951 (1951).

727. Lab. des Basses Pressions, Raccord rapide pour petites canalisations, *Le Vide*, **13**, **296** (1958).
728. Laboratoires des Basses Pressions (87 rue A. G. Belin, Argenteuil, France), *Le Mécavide* (catalogue).
729. LACH E., and PERLOW G. J., Small vacuum valve, *Rev. Sci. Instr.*, **27**, 336 (1956).
729a. LAKE G. J., LINDLEY P. P., and THOMAS A. G., High Vacuum sliding seal, *J. Sci. Instr.*, **40**, 40 (1963).
729b. LA FORGE L. H., Application of ceramic sections in high power pulsed klystrons, *Bull. Am. Ceram. Soc.*, **35**, 117 (March 1956).
730. LAMBE J., Glass vacuum valve, *Rev. Sci. Instr.*, **20**, 831 (1949).
731. LANDECKER K., and GRAY A. J., Diffusion of gases through nickel and design of a convenient leak for hydrogen and deuterium, *Rev. Sci. Instr.*, **25**, 1151 (1954).
732. LANDER H. J., HESS W. T., and WHITE S. S., Electron-beam welding of high strength alloys, *Trans. AVS Nat. Vac. Symp.* **1961**, p. 685, Pergamon Press, Oxford, 1962.
733. LANE C. T., Kovar–glass seals at liquid helium temperature, *Rev. Sci. Instr.*, **20**, 140 (1949).
734. LANG R., A mechanical valve for high-vacuum systems, *Rev. Sci. Instr.*, **10**, 196 (1939).
735. LANGE N. A., *Handbook of Chemistry*, McGraw Hill, New York, 1961.
736. LANGE J., Large ultra high vacuum valve, *Rev. Sci. Instr.*, **30**, 602, (1959).
737. LANGE W. J. and ALPERT D., Step-type demountable metal vacuum joint, *Rev. Sci. Instr.*, **28** 726 (1957).
738. LANGE W. J., *Ultra-high Vacuum Techniques*, Westinghouse Progr. Rep. AECU-3889 (1957) and 3890 (1957).
739. LANGE W. J., and RIMERSINA H., *Use of Teflon in Ultra-High Vacuum System*, Westinghouse Rep., NYO-8704 (1959).
740. LANGMUIR I., Flames of atomic hydrogen, *Ind. & Eng. Chem.*, **19**, 667 (1927).
741. LAPORTE H., *Hochvacuum, seine Erzeugung, Messung und Anwendungen*, W. Knapp, Halle, 1957.
742. LASZLO H., DE, Copper-to-glass seals, *J. Sci. Instr.*, **10**, 296 (1933).
743. LAUFER M. K., Standard leaks and their calibration by constant pressure change in volume technique, *Trans. AVS Nat. Vac. Symp. 1961*, p. 1086, Pergamon Press, Oxford, 1962.
744. LANG E. P., Cleaning of glass, *Ind. & Eng. Chem. Anal.*, **6**, 111 (1934).
745. LAUGHNER V. H., and HARGAN A. D., *Handbook of Fastening and Joining of Metal Parts*, McGraw-Hill, New York, 1956.
746. LAURITSEN C., and LAURITSEN P., Vacuum Fittings, *Rev. Sci. Instr.*, **19**, 919 (1948).
746a. LAVROV O. V., *et al.*, A fast vacuum slide gate, *Instr. and Exp. Techn.*, November, 502 (1963).
747. LAW R. R., Brazing with gold by diffusion, *R.C.A. Rev.*, **9**, 646 (1948).
748. LAWRENCE R. B., A simplified method of calculating pressure drop in vacuum piping, *Trans. AVS Nat. Vac. Symp. 1954*, p. 55, Pergamon Press, Oxford, 1955.
749. LAWSON R. W., Degassing properties of nickel, *Brit. J. Appl. Phys.*, **13**, 115 (1962).
750. LEBLANC J. A., Vacuum gauge calibration system, *Trans. AVS Nat. Vac. Symp. 1956*, p. 79, Pergamon Press, Oxford, 1957.
751. LEBLANC M., *La Technique du Vide*, A. Colin, Paris, 1951.
752. LECK J. H., Pressure measurements in vacuum systems, *The Institute of Physics*, London, 1957.
753. LEE H., and NEVILLE K., *Epoxy Resins*, McGraw-Hill, New York, 1957.
754. LEE F. R. N., and YOUNG S. A., A new vacuum joint, *J. Sci. Instr.*, **31**, 70 (1954).
755. LEFFERT C. B., KASNER W. H., and DONAHUE T. M., Counterbalanced bellows in vacuum systems, *Rev. Sci. Instr.*, **27**, 1084 (1956).
756. LEISEGANG S., Elektronenmikroskope, in: *Handbuch der Physik*, Vol. 33, p. 416, 443, Springer Verlag, Berlin, 1956.

757. LEMPICKI A., A demountable vacuum system for secondary emission studies, *J. Sci. Instr.*, **32,** 221 (1955).

758. LESEK F., and SYTAR M., Vacuum sampling device, *Chem. Techn.*, **10,** 421 (1958).

759. LEUTNER K., "Araldite"—an artificial resin for preparing working models in ceramic industry, *Glass-Email*, **92,** 377 (1959).

760. LEVANTINE A. D., Space thermal environment simulation facilities, *Trans. AVS Nat. Vac. Symp. 1961*, p. 1270, Pergamon Press, Oxford, 1962.

761. LEWIN G., MASK R., Theory of dissimilar tubular seals of glass, ceramics and metals for critical application, *Trans. AVS Nat. Vac. Symp. 1958*, p. 44, Pergamon Press, Oxford, 1959.

762. LEYBOLD (Köln—Bayental, Germany), *Hochvakuumtechnik*, Catalogues HV-104, HV-110, HV-112 E (1962).

763. LEYBOLD (Köln—Bayental, Germany), *Directions for use of Gas Ballast Pumps*, (Catalogue).

764. LICHTMAN D., and HEBLING A., Ultra-high vacuum demountable system for mass spectrometer gas analysis, *Trans. AVS Nat. Vac. Symp. 1960*, p. 187, Pergamon Press, Oxford, 1961.

764a. LICHTMAN D., and OSWALD R., Short-length, in-line, all-metal valve, *Rev. Sci. Instr.*, **34,** 1145 (1963).

765. LIEBSON M., Sealing molybdenum leads to alumino-silicate glass, *Rev. Sci. Instr.*, **32,** 204 (1961).

765a. LILLIE H. R., Basic principles of glass annealing, *Glass. Ind.*, **31,** 355 (1950).

766. LILLIE H. R., Flow point of glass, *J. Am. Ceram. Soc.*, **35,** 149 (1952).

766a. LILLIE H. R., Viscosity–time–temperature relations in glass at annealing temperatures, *J. Am. Ceram. Soc.*, **16,** 619 (1933).

767. LILLIE H. R., New viscosity values for glasses at the annealing and strain points, *J. Am. Ceram. Soc.*, **37,** 111 (1954).

768. LINCOLN I. C., Welding with twin carbon arc, *J. Amer. Inst. Electr. Eng.*, **49,** 110 (1930).

769. LINDEMANN C. F., and LINDEMANN C. A., Für weiche Röntgenstrahlen durchlässiges Glas, *Z. Röntgenkunde*, **13,** 141 (1911).

770. LINDSAY W. B., Improved demountable high temperature mica vacuum window, *Rev. Sci. Instr.*, **32,** 748 (1961).

770a. LINEWEAVER J. L., The detection of leaks: ultra-sensitive method using the He leak detector, *Inst. Radio Eng. Trans.*, **E. D. 5,** 28 (1958).

770b. LIPSON H. G., and BOUTHILLETTE L. O., Epoxy vacuum seal for introduction of leads into cryogenic equipment, *Vacuum*, **13,** 145 (1963).

771. LISSBERGER P. H., Vacuum sluice valve of large aperture, *J. Sci. Instr.*, **38,** 132 (1961).

772. LITTLETON J. T., The effect of temperature treatment on glass-to-metal seals, *J. Am. Ceram. Soc.*, **18,** 239 (1935).

773. LITTLETON J. T., The measurement of the softening point of glasses, *J. Am. Ceram. Soc.*, **10,** 259 (1927); *Trans. Soc. Glass. Technol.* **24,** 176 (1940).

774. LITTLETON J. T., and ROBERTS E. H., Determination of the annealing point in glass, *J. Opt. Soc. Am.*, **4,** 224 (1920).

774a. LITTMANN C., and BERKOWITZ-MATTUCK J. B., Method for the introduction of small, measured gas samples into a flow system, *Rev. Sci. Instr.*, **32,** 1154 (1961).

775. LITTON C. V., *Metal-to-Ceramic Seals*, U.S. Pat. 2,647,070 (1953).

776. LLOYD H., A flexible attachment for use in vacuum apparatus, *J. Sci. Instr.*, **13,** 171 (1936).

777. LLOYD J. T., Thin glass counter windows, *J. Sci. Instr.*, **26,** 126 (1949).

778. LOCKENVITZ A. E., HUGHES D. C. *et al.*, A high vacuum valve, *Rev. Sci. Instr.*, **19,** 272 (1948).

779. LOEB L. B., *The Kinetic Theory of Gases*, McGraw-Hill, New York, 1934.

780. LOH L. T., NEILL H. W. *et al.*, Remote opening and sealing of metal tubes, *Rev. Sci. Instr.*, **23,** 339 (1952).

781. LONG R. A., and RUPPENDER R. R., High-temperature-alloy fusion brazing for titanium and titanium alloys, *Welding J.*, **33**, 1087 (1954).
782. LORENZ A., (HERAEUS W. C., GmbH), *Vacuum Tight, Conducting Sintered Cermets Melted into Glass, especially Silica Glass*, Germ. Pat. 963,012 (1957).
783. LORD R. C., and MCDONALD R. S., Vacuum-tight seal between LiF and Silver, *Rev. Sci. Instr.*, **23**, 442 (1952).
784. LOSSING F. P., and TICKNER A. W., Free radicals by mass spectrometry *J. Chem. Phys.*, **20**, 907 (1952).
785. LOUCKES F. J., Unique sliding seal for a vacuum chamber, *Rev. Sci. Instr.*, **28**, 468 (1957).
786. LOUGHRIDGE F. A., WONG W. S., Improved reliability of soft glass to metal vacuum-tight seals, *Trans. AVS Nat. Vac. Symp. 1959*, p. 283, Pergamon Press, Oxford, 1960.
786a. LOWDEN R. S., and HOH F. C., Fast gas valve for plasma research, *Rev. Sci. Instr.*, **33**, 1236 (1962).
787. LUCAS L. R., and HERNANDEZ H. P., *Inflatable Gasket for the 72-Inch Bubble Chamber*, Rep. UCRL-8526 (1959); *Rev. Sci. Instr.*, **30**, 941 (1959).
788. LUDWIG C., Hartmetallwerkzeugen für Glasbearbeitung, *Fertigungstechn.*, **3**, 383 (1953).
789. LUPINEK M., Soldering and sealing with HF, *Slabopr. Obzor.*, **14**, 28 (1953).
790. LYONS L. R., Epoxy resin seal, *Chem. Eng.*, 144 (Aug. 6, 1962).
791. LYUBIMOV M. L., SHAKHOV K. P., and YUKVIDIN Y. A., Experience in designing and manufacturing all-metal vacuum systems, *Adv. Vac. Sci. Techn.*, p. 146, Pergamon Press, Oxford, 1960.
791a. MACDONALD K. A. *et al.*, *Ceramic-to-Metal Seals*, U.S. Pat. 2,836,885 (1958).
792. MACHLETT R. R., and STEEN G. W., *Method for Making Metal–Glass Seals*, U.S. Pat. 2,462,205 (1949).
793. MACKINNON-LITTLE D., Silicone rubber as a sealing material *Rubber and Plastic Age*, **42**, (10), 1239 (1961).
794. MAGLIO M. M., Removal of silicone grease from glass surfaces, *Chem. Anal.*, **36**, 22 (1947).
794a. MAHLER W., FELMEY H. V., Vacuum tube opener, *Rev. Sci. Instr.*, **33**, 1127 (1962).
795. MAIRS K. H., Fe–Ni–Co alloys for glass-to-metal seals, *J. Metals*, **40**, 460 (1952).
796. MAKER F. L., *Design of a Large Vacuum Vessel*, A.E.C. Rep, LRL-157 (1954).
797. MALIC B., Making glass-to-metal seals for laboratory purposes by a HF heating method, *Rev. Trav. Inst. Res. Struct. Materie*, (Belgrade), **1**, 75 (1952); *Chem. Abstr.*, **47**, 6189e (1953).
798. MALTZ J., and PIERRE V. DE, Metallurgy of titanium, *Metal Progr.*, 189 (Aug. 1950).
798a. MANCEBO L., *A Cold High-Vacuum Seal without Gaskets*, UCRL-Rep.-6978 (1962).
799. MANFREDI R. E., and NOLTE H. J., Application of ceramics to vacuum tubes, *Bull. Am. Ceram. Soc.*, **35**, 105 (1956).
800. MANN J. B., Improved metal gasket vacuum systems, *Rev. Sci. Instr.*, **27**, 1083 (1956).
801. MANNERS M., Glasses for vacuum tubes, *Electr. Eng.*, **25**, 512 (1953).
802. MANTELL C. L., (Editor), *Engineering Materials Handbook*, McGraw-Hill, New York, 1958.
803. MARK J. T., and DREYER K., Ultra-high vacuum system developments for the model-C Stellarator, *Trans. AVS Nat. Vac. Symp. 1959*, p. 177, Pergamon Press, Oxford, 1960.
804. MARK J. T., and HENDERSON W. G., Problems of very large systems, *Trans. AVS Nat. Vac. Symp. 1961*, p. 31, Pergamon Press, Oxford, 1962.
805. MARKER R. C., Vacuum techniques and components used for a continuously pumped linear electron accelerator, *Adv. Vac. Sci. Techn.*, p. 494, Pergamon Press, Oxford, 1960.
806. MARKLEY F., ROMAN R., and VOSECEK R., Outgassing data for several epoxy resins

and rubbers for the Zero Gradient Synchrotron, *Trans. AVS Nat. Vac. Symp. 1961*, p. 78, Pergamon Press, Oxford, 1962.
807. MARKS R., Method for making very small holes in metals, *Rev. Sci. Instr.*, **28**, 381 (1957).
808. MARSDEN C. P., Eine Übersicht über die Anwendung von keramischen Werkstoffen als Hülle für Vakuum-Röhren, *Vakuum-Technik*, **9**, 43 (1960).
809. MARSHALL A., Wide temperature range pump seal, *Rev. Sci. Instr.*, **24**, 1151 (1953).
810. MARSHALL J., Acceleration of plasma into vacuum, *Proc. 2nd Int. Conf. on Peaceful Uses of Atomic Energy, Geneva 1958*, Vol. 31, p. 341.
811. MARTIN A. J., *Metal Seals in Vacuum Equipment*, AWRE. Rep. 0–40 (1954).
812. MARTIN A. J., *Degassing of Vacuum System by Ionisation Bombardment*, AWRE. - Rep. 0–11 (1955).
812a. MARTIN G., Vacuum properties of silver chloride, *Rev. Sci. Instr.*, **34**, 707 (1963).
813. MARTIN F. W., Stresses in glass-to-metal seals (the cylindrical seal), *J. Am. Ceram. Soc.*, **33**, 224 (1950).
813a. MARTIN J., Évolution des orifices de vidage des tubes électroniques, *Le Vide*, **2**, 308 (1947).
814. MARTIN M., *Use of Redox Adhesives for Bounding Metal*, Rep. MDDC-1166 (1947).
815. MARTIN S. L., Vacuum technique, *Chem. Proc. Engng.*, **34**, 276 (Sept. 1953).
816. MARTIN S. L., High vacuum technique, *Chem. Proc. Engng.*, **35**, 301 (Oct. 1954).
817. MARTIN J. H., Adjustable glass capillary gas leak, *Rev. Sci. Instr.*, **19**, 404 (1948).
818. MARTIN L. H., and HILL R. D., *A Manual of Vacuum Practice*, Univ. Press, Melbourne, 1949.
819. MARTIN S. T., Soldered metal to glass or porcelain seals, *Rev. Sci. Instr.*, **11**, 205 (1940).
820. MARTINSON E. N., Production of ultra-low pressures in glass devices by means of gas sorbtion by deposited titanium films, *Instr. & Exp. Techn.*,U.R.S.S. No. 5, 798 (1960).
820a. MARTON F., Ball valves, *Instr. & Control Systems*, **34**, 1845 (1961).
821. MARTON L., SIMPSON J. A., *et al.*, Ultra-high vacuum ultraviolet monochromator, *Trans. AVS Nat. Vac. Symp. 1961*, p. 633, Pergamon Press, Oxford, 1962.
822. MARTZ D. E., New method for sealing infrared window to glass, *Rev. Sci. Instr.*, **32**, 214 (1961).
823. MARX H., *Glass und Keramik*, Fachenbuchverlag, Leipzig. 1957.
823a. MASSEY B. J., Production of thin self-supporting carbon films, *Trans. AVS Nat. Vac. Symp.* **1961**, p. 992, Pergamon Press, Oxford, 1962.
824. *Material Selector Issue, Materials in Design. Engng.*, **54**, 5 (1961).
825. MATHESON W. G., Vacuum systems and techniques in lamps and electronic tube industries, *Trans. AVS Nat. Vac. Symp.* **1954**, p. 104, Pergamon Press Oxford, 1955
826. MATTHEWS J. C., A small vacuum induction furnace, *J. Sci. Instr.*, **34**, 62 (1957).
827. MCCARTHY H. J., *Glass-to-Metal Seal*, U.S. Pat. 2,422,628 (1947).
828. MCCARTHY H. J., *Glass-to-Metal Seals*, U.S. Pat., 2,717,475 (1955).
828a. MCDOWELL C. A., *Mass Spectrometry*, Mc Graw-Hill, New York, 1963.
829. MCFARLAND S. H., ANDERSON R., and WELLS, J., Effect of mercury on a metal high vacuum valve. *Rev. Sci. Instr.*, **29**, 529 (1958).
830. MCGUINESS E. W., How high temperatures affect epoxy adhesives, *Mater. & Methods*, **43** (3), 121 (1956).
831. MCCUIRE J. C., New method for soft soldering metals and ceramics, *Rev. Sci. Instr.*, **26**, 893 (1955).
832. MCINTOSH R. O., and COLTMAN J. W., Toggle-operated vacuum valve, *Rev. Sci. Instr.*, **20**, 135 (1949).
833. MCKNIGHT W. H., Pressed and sintered glass powder shapes, *Mater. & Methods*, **40** (10), 94 (1954).
834. MCLEAN D., and FARMER M. H., Measurement of force required to pull through a 1 inch diameter Wilson vacuum seal, *J. Sci. Instr.*, **30**, 293 (1953).
835. MCNALLY F. A., Motion within a vacuum system, *Rev. Sci. Instr.*, **21**, 813 (1950).

836. MEBS R. W., and ROESER W. F., *Solders and Soldering*, Nat. Bur. Stand. Circ. 492 (1950).
837. MEDICUS G., Rubber solution gaskets for high vacuum work, *Rev. Sci. Instr.*, **23,** 646 (1952).
838. MEDICUS G., Remountable, bakeable tubes by soldering joint technique, *Rev. Sci. Instr.*, **23,** 647 (1952).
839. MEDICUS G., Rubber ring gasket for conical vacuum joints, *Rev. Sci. Instr.*, **23,** 647 (1952).
839a. MEDICUS. G., and JEHN W., Bakeable UHV valve with long life, *Rev. Sci. Instr.*, **33,** 1265 (1962).
840. MEECHAN C. J., and SOSIN A., Electrically insulated, thermally conducting vacuum seal for low temperatures, *Rev. Sci. Instr.*, 29, 323 (1958).
841. MEHNERT K., Elastic and plastic properties of fluoropolymers, *Chem.Ing.-Technik,* **27,** 284 (1955).
842. MEIER J. W., High-power-density electron beam welding of several materials, *Trans. AVS Nat., Vac. Symp. 1961.* p. 670, Pergamon Press, Oxford, 1962.
843. MELLER K., *Elektrische Lichbogenschweissung*, Leipzig, 1932.
843a. MELMED A. J., Electrical lead into vacuum tubes, *Rev. Sci. Instr.*, **34,** 307 (1963).
844. MELNICK L. M., Glass for neutron absorbtion, *J. Am. Ceram. Soc.* **34,** 82 (1951).
845. MERKEL E., Teflon as sealing material, *Chemie Ing.-Technik,* **27,** 279 (1955); **31,** 10 (1959).
846. MERKEL M. K. G., (Asbest und Gummiwerke, Hamburg) *Teflon, Hostaflon TF, Fluon,* (Catalogue 1961).
847. MERRIAM J. C., Adhesive bonding, *Mater. in Design. Engng.*, **50** (3); 113 (1959).
848. *Metals Handbook,* 8th Ed., Am. Soc. for Metals, 1961.
849. METCALFE R. A., and TRABERT F. W., Performance of a double-walled ultra-high vacuum chamber, *Trans. AVS Nat. Vac. Symp. 1961*, p. 1211, Pergamon Press, Oxford, 1962.
850. METHFESSEL S., *Dunne Schichten (Ihre Herstellung und Messung)* W. Knapp, Halle (Saale), 1953.
851. Metro-Vickers Co., *Production of Substantially Vacuum-Tight Joints between Metallic Members*, Brit. Pat. 727, 909.
852. METZLER A. J., Bakable, high-vacuum isolation valve, *Rev. Sci. Instr.*,**33,**130(1962).
852a. MEUNIER R., and BONPAS M., Mica windows for GM tubes, *Le Vide,* **8,** 1342 (1953).
853. MEYER B., and STODIE K. W., New cloud-chamber valve, *Rev. Sci. Instr.*, **24,** 76 (1953).
853a. Micafil Ltd. (P.O.B. 48, Zürich, Switzerland), *Pompes à vide*, (Catalogue).
854. MICHAELSON H. B., High temperature materials for vacuum service, *Mater. & Methods*, **38** (6), 110 (1953).
855. MICHELSON C., *Welded Seal-Ring Vacuum Closures*, Oak Ridge Lab.-CF-59-8-80 (1959).
856. Midlands Silicones Ltd., (19 Upper Brook Street, London, W. 1 England). *Properties and Applications of "Silastomer" silicone rubber,* Catalogue K-1-5 (1958).
857. MIGNOLET I. C. P., Simple and multiple gas micro-inductors for adsorption studies, *J. Sci. Instr.*, **30,** 15 (1953).
858. MILLEN T. S., The design of industrial vacuum system, *J. Sci. Instr.*, **28** (Suppl. 1), 7 (1951).
859. MILLER S. A., and CHASE V. A., Epoxy resin is strong at 500 °F plus, *Mater. in Design. Engng.*, **51** (4), 17 (1960).
860. MILLER G. H., PRITCHARD G. O. *et al.*, Modified greaseless mercury cut-off, *Rev. Sci. Instr.*, **31,** 466 (1960).
861. MILLER J. L., and ROBINSON J. E. L., The three-dimensional adjustment of an electrode in vacuo, *J. Sci. Instr.*, **9,** 264 (1932).
862. MILLER M. A., and RUSSELL A. S., Vacuum-tightness of welded and brazed aluminium containers, *Welding J.*, **32,** 116 (1953).
863. MILLER C. F., and SHEPARD R. W., Evaluation of Pyroceram, Code 9606, as a

suitable envelope for vacuum devices operating at high ambient temperature, *Vacuum*, **11**, 58 (1961).
864. MILLERON N., Utilisation of the surface tension of liquid metals in making high vacuum seals, *Trans. AVS Nat. Vac. Symp. 1957*, p. 38, Pergamon Press, Oxford, (1958); *Rep. UCRL*-4938 (1957).
865. MILLERON N., Some component design permitting ultra-high vacuum with large oil diffusion pump. *Trans. Avs Nat. Vac. Symp. 1958.*, p. 143, Pergamon Press, Oxford, (1959).
865a. MILLERON N., *Vacuum Sealing Means for low Vacuum Pressures*, U.S. Pat. 3,038,731 (1962).
866. MILLICAN R., *Low Flow Variable Leak*, A.E.C. Rep. KY-166 (1956).
867. MILLER T., SZALKAY, *et al.*, *Vakuumtechnika*, Nehézipari Kiadó, Budapest, 1953.
868. Ministry of Agriculture, *Machine Milking*, Bill. No. 177, London, 1959.
869. MIRANDA F. J., and JONES W. M., Vacuum seals for continuously evacuated accelerator tubes operating in high ambient pressure, *Vacuum*, **2**, 259 (1952).
870. MITCHELL L., Ceramics, *Ind. Eng. Chem.*, **48** (2), 1702 (1956).
870a. MOENICH J., TREKA R., Ring Vacuum system zero gradient synchrotron, Argonne National Lab., *Trans. AVS Nat. Vac. Symp.* **1961**, p. 1133, Pergamon Press, Oxford, 1962.
871. MOLL J. TH., Coating technique in ultra-high vacuum, *Trans. AVS Nat. Vac. Symp.* **1961**, p. 995, Pergamon Press, Oxford, 1962.
872. MOLLER T. H. Factors influencing the calibration and application of standard leaks, *Trans. AVS Nat. Vac. Symp.* **1954.**, p. 121, Pergamon Press Oxford, 1955.
873. MOLNAR J. P., and HARTMAN C. D., Data on porcelain rod leak, *Rev. Sci. Instr.*, **21**, 394 (1950).
873a. MOLYNEUX F., Optimum (Material) Selection, in Hepner I. L. (Edit.)., *Materials of Constructions for Chemical Plant*, Leonard Hill, London, 1962.
874. MONACK A. J., *Method of Sealing Glass to Iron*, U.S. Pat. 2,345,278 (1944).
875. MONACK A. J., Theory and practice of glass–metal seals, *Glass. Ind.*, **27**, 389 (1946).
876. MONACK A. J., Metallizing of glass and ceramics, *Glass Ind.*, **28**, 21 (1947).
877. MONACK A. J., Glass-to-metal seals in electronic components and applications, *Electr. Mftg.* **39**, 96—180 (1947).
878. MÖNCH G., CH., *Hochvakuumtechnik*, R. A. Lang, Pössneck (1950).
879. MÖNCH G. CH., *Neues und Bewährtes aus der Hochvakuumtechnik* Verlag Technik, Berlin, 1961.
879a. MONGODIN G., *et al.*, Organisation et méthodes mises en oeuvre pour l'étanchéité de l'usine de séparation isotopique de Pierrelatte, *Le Vide*, **18**, 528 (1963).
880. MOORADIAN V. G., Selecting materials for electrical contacts, *Mater. & Meth.*, **44** (3), 121 (1956).
881. MOORE H. R., A tapered plug vacuum valve, *J. Sci. Instr.*, **33**, 282 (1956).
882. MOORE H. R., Demountable vacuum seal for attaching an end-plate to glass tube, *Rev. Sci. Instr.*, **29**, 737 (1958).
883. MOORE M., High-vacuum valve, *J. Sci. Instr.*, **21**, 162 (1944).
884. MOORE M. J., Insulated power-lead-in for vacuum system, *J. Sci. Instr.*, **27**, 231 (1950).
885. MOOZ W. E., The design and operation of high vacuum systems, *Chem. Eng. Progr.*, **49**, 197 (1953).
886. MORAND, M. (Edit.), *Traité Pratique de Technique du Vide*, Eyrolles, Gauthier Villars, Paris, 1961.
887. MOREAU J., Le probléme des vannes en technique du vide, *Le Vide*, **14**, 3 (1959).
888. MOREY G. W., *The Properties of Glass*, Reinhold, New York, 1954.
889. MORRIS D. D., The production of metal-to-silica joints and vacuum seals, *J. Sci. Instr.*, **11**, 232 (1934).
890. MORRISON J., A controlled gas leak, *Rev. Sci. Instr.*, **24**, 230 (1953).
891. MORRISON J., Leak control tube, *Rev. Sci. Instr.*, **24**, 546 (1953).

891a. MORRISON J., Gas collection and analysis employed in vacuum tube problems, *Trans. AVS Nat. Vac. Symp. 1957*, p. 100, Pergamon Press, Oxford, 1958.

892. MORRISON J., An ultra-high vacuum system for low energy electron diffraction studies, *Trans AVS Nat. Vac. Symp. 1961*, p. 1183, Pergamon Press, Oxford, 1962.

893. MORRISON S. J., The design of glass vacuum chambers of elliptic cross section, *Engineer*, **113**, 426 (1952).

894. MORSE R. S., High vacuum technology, *Ind. Eng. Chem.*, **19**, 1064 (1947).

895. MORTIMER J. W., Argonarc welding of Cu., *Welding J.*, **27**, 715 (1948).

896. MORTON A. H., and JEFFREY P. M., A positive action vacuum valve, *J. Sci. Instr.*, **30**, 58 (1953).

897. MOSS A. R., and Brooker H. R. *Some Aspects of German Soldering, Brazing and Welding Methods*, Bios-Final Rep. 1844 (1947).

898. MOSS C. J., Use of Araldite, *Plastics*, **16** (162), 3 (1951).

899. MOSS C. J., Araldite, *Electronic Engng.*, **21**, 389 (1949).

900. MUELLER M. H., Vacuum seal for thin metal windows, *Rev. Sci., Instr.*, **27**, 411 (1956).

901. MUELLER M. H., and BILINSKI R. B., Sensitive leak control, *Rev. Sci. Instr.*, **22**, 704 (1951).

901a. MUGGLETON A. H. F., and HOWE F. A., The preparation of thin O^{18} targets, *Nucl. Instr. & Meth.*, **12**, 192 (1961).

902. MUGGLETON A. H. F., and HOWE F. A., The preparation of thin selfsupporting boron films, *Nucl. Instr., & Meth.*, **13**, 211 (1961).

902a. MULLANEY D., Bakeable high conductance vacuum valve, *Rev. Sci. Instr.*, **33**, 1003 (1962).

903. MÜLLER A., Aircomatic welding, *Welding J.*, **29**, 458 (1950).

904. MÜLLER A., A vacuum valve using a platinum wire, *Stud. Cercetari Fiz.*, (Roumania), **12** (1), 153 (1961).

904a. MÜLLER K., Vakuumdichte Metall-Keramikverbindungen, *Elektrot. Ztschr. (ETZ)*, **B10**, 69 (1958).

905. MÜLLER K. G., *Vakuumtechnische Berechnungsgrundlagen*, Verlag Chemie, Weinheim, 1961.

906. MUNDAY G. L., Ultra-high vacuum technology, *Nucl. Instr. & Meth.*, **4**, 367 (1959).

906a. MUNDAY G., A totally enclosed gate valve, *J. Sci. Instr.*, **40**, 603 (1963).

907. MUNDAY G. L., and ZINSMEISTER G., Experience with the vacuum equipment for the CERN 25 GeV Proton-Synchrotron, *Trans. AVS Nat. Vac. Symp. 1961*, p. 1154, Pergamon Press, Oxford, 1962.

908. MUNSON R. J., Production of small holes in glass, *Rev. Sci. Instr.*, **26**, 236 (1955).

909. MURGATROYD J. B., Annealing of commercial glasses, *J. Soc. Glass Technol.*, **14**, 2 (1930).

910. MYLONAS C., and GREEK M., The optical system of polariscopes as used in photoelasticity *J. Sci. Instr*, **25**, 77, (1948).

911. National Bureau of Standards, CS-21-58, *Interchangeable Ground Glass Joints, Stopcocks and Stoppers*, U.S. Government Print. Office, 1958.

912. National Research Corp., (NRC, 160 Charlemont Str. Newton, Mass., U.S.A.), *High Vacuum Equipment* (Cat.).

913. NAVIAS L., and GREEN R. L., Dielectric properties of glasses at high frequencies function of their composition, *J. Am. Ceram. Soc.*, **29**, 267 (1946).

913a. NEHER H. V., Metal system for chemical reactions and for studying properties of gases and liquids, *Rev. Sci. Instr.*, **28**, 267 (1957).

914. NEHER H. V., and JOHNSTON A. R., Techniques useful in evacuating and pressurising metal chambers, *Rev. Sci. Instr.*, **25**, 517 (1954).

915. NELSON L. S., Windowed Dewar vessels for use at low temperatures, *Rev. Sci. Instr.*, **27**, 655 (1956).

916. NELSON R. S., Insulated terminals for high vacuum systems using demountable glass–metal seals, *J. Sci. Instr.*, **37**, 358 (1960).

917. Nelson E. H., Weald H. and Power E. B., *Quartz-to-Metal Seal*, U.S. Pat. 2,699,847 (1955).
918. Neppiras E. A., and Foskett R. D., Ultrasonic treatment of materials, *Philips Techn. Rdschau.*, **19**, 37 (1957).
919. Nerken A. A., Experiments on flow of gases through leaks, *Trans. AVS Nat. Vac. Symp. 1956*, p. 1, Pergamon Press, Oxford, 1957.
920. Nester R. G., Method for joining metal pipe to glass tubing *Rev. Sci. Instr.*, **25**, 95 (1954).
921. Nester R. G., High vacuum gas valve, *Rev. Sci. Instr.*, **27**, 874 (1956).
922. Nester R. G., High-speed flexible-blade stirrer and stirrer seal for high vacuum use, *Rev. Sci. Instr.*, **27**, 1080 (1956).
922a. Neubert J., Hochvakuum Gallium-ventil, *Exp. Techn. der Physik*, **10**, 146 (1962).
923. Neumann E., *Die Physikalischen Grundlagen der Leuchtstofflampen und Leuchtrohren*, Verlag Technik, Berlin, 1954.
924. Neville H. H., Non-sticking float valve for vacuum systems *J. Sci. Instr.*, **32**, 488 (1955).
925. Newitt J. H., Brazing in vacuum, *Steel*, **4** (6), 107 (1956).
926. N.G.N. Electrical Ltd., (Accrington, Lancs, England), *High Vacuum Equipment*, (Catalogue).
926a. Nicolaev L. V., Quartz–metal bonds, *Instr. & Experim. Techn.*, No. **5**, 1037 (Sept.–Oct. 1962).
927. Nicollian E. H., A laboratory leak detector using an omegatron, *Trans. AVS Nat. Vac. Symp. 1960*, p. 82, Pergamon Press, Oxford, 1961.
928. Nief G. A., molecular leak for the mass spectrometer, *J. Chim. Phys.*, **49**, 149 (1952).
929. Nier A. O., A mass spectrometer for isotope and gas analysis, *Rev. Sci. Instr.*, **18**, 398 (1947).
930. Nier A. O. C., Ney E. P. *et al.*, Adjustable gas leak, *Rev. Sci. Instr.*, **18**, 191 (1947).
931. Nier A. O. Abbott T. A., *et sal.*, Recording mass spectrometer for process analysis, *Analyt. Chem.*, **20**, 188 (1948).
932. Nikolsky S. A., Labzun P. K. *et al.*, *Lubricant for Vacuum Joints*, USSR Pat. 117,911 (1959).
933. Nitzsche S. *et al.*, Silicongummi, *Elektrot. Ztg. (ETZ)*, **A 74**, 71 (1953).
934. Niviere P., Une methode pour la détection des fuites tres minces dans les tuyaux, *Genie Chim.*, **75**, 122 (1956).
935. Noggle T. S., Blewitt T. H., *et al.*, Thermal e.m.f. free vacuum seal for electrical lead wires, *Rev. Sci. Instr.*, **28**, 464 (1957).
936. Nöller H., *Vakuum-Technik* in "Handbuch fur Hochfrequenz und Elektrotechniker", Funk, Techn. Verl., Berlin, 1957.
936a. Nolte H. J., and Spurck R. F., Metallizing ceramic Members, U.S. Pat. 2,667,427 (1954), U.S. Pat. 2,667,432 (1954), *Television Engng.*, **1**, 14 (1950).
937. Norbury I., An easily made flexible vacuum-tight joint, *J. Sci. Instr.*, **29**, 204 (1952).
938. Nordberg M. E., Uviolglasses for mercury lamps, *J. Am. Ceram. Soc.*, **30**, 174 (1947).
939. Normand C. E., *Gaseous Flow*, Rep. AECD-4084 (1955).
940. Normand C. E., Use of a standard orifice in the calibration of vacuum gauges, *Trans. AVS Nat. Vac. Symp. 1961*, p. 534, Pergamon Press, Oxford, 1962.
941. Normand C. E., and Rathkamp W. R., *Outgassing and Vaporization Characteristics of Vacuum Greases*, Rep. Y-642 (Rev), (1950).
942. Normand L. E., and Rathkamp W. R., Evaluation of vacuum gasket materials, *Vacuum*, **3**, 200 (1953).
943. Norton F. J., An "Opener-Upper" for gas samples, *Rev. Sci. Instr.*, **20**, 620 (1949).
944. Norton F. J., Permeation problems in high vacuum, *Trans. AVS Nat. Vac. Symp. 1954*, p. 47, Pergamon Press, Oxford, 1955.
945. Norton F. J., Vacuum tube opener, *Rev. Sci. Instr.*, **26**, 238 (1955).

946. Norton F. J., Permeation of gases through solids, *J. Appl. Phys.*, **28**, 34 (1957).
947. Norton F. J. Gas permeation through the vacuum envelope, *Trans. AVS Nat. Vac. Symp. 1961*, p. 8, Pergamon Press, Oxford, 1962.
948. Nowak P., Silicone in der Elektrotechnik, *Elektrot. Ztschr.*, **B. 7**, 354 (1955).
949. Nürnberger C. E., and Livingston R., Color changes of glass due to X rays, *J. Phys. Chem.*, **41**, 691 (1939).
950. Oatley C. W., The experimental determination of the speed of a vacuum pump and of components of a vacuum system, *Brit. J. Appl. Phys.*, **5**, 358 (1954).
951. Ochert N., and Steckelmacher W., Leak detection practice, with particular reference to the hydrogen palladium method, *Vacuum*, **2**, 125 (1952).
951a. Ochert N., and Steckelmacher W., The flow characteristics of capillary leaks, *Brit. J. Appl. Phys.*, **2**, 332 (1951).
952. Oehme F., and Frey H. P., Einbrenn — Versilberungsverfahren für Glasoberflächen, *Chem. Techn.*, **8**, 498 (1956).
953. Officine Galileo, (Via C. Bini 44, Firenze, Italy), High vacuum Equipment, (Çatalogue).
954. Oldfield L. F., Sealing glass to metal, *The Times Review of Industry*, Sept. 1958.
955. Oranje P. J., *Gasentladungslampen*, Philips, Techn. Bibl. Eindhoven, 1943.
956. Ortel Y., Vanne à clapet détachable, *Le Vide*, **12**, 444 (1957).
957. Osborne A. D., Device for admitting gas at constant rate to vacuum apparatus, *J. Sci. Instr.*, **36**, 370 (1959).
958. Osram — G.E.C. Glass Works (London), *Glass* (Catalogue).
959. Padmos A. A., The estimation of the thermal expansion coefficient of a glass by the measurement of stresses produced by fusion to a standard glass, *Glass Ind.*, **33** (4), 199 (1952).
960. Padmos A. A., and de Vries J., Stresses in glass and their measurement, *Philips. Techn. Rev.*, **9**, 277 (1947—48).
961. Pagano F., Pump as a gas source, *Trans. AVS Nat. Vac. Symp. 1961*, p. 86, Pergamon Press, Oxford, 1962.
962. Paige E., *A Metal High Vacuum Manifold*, AEC. Rep. MLM-810 (1953).
963. Pakswer S., and Dowd J., Bakeable indium seals, *Rev. Sci. Instr.*, **33**, 780 (1962).
964. Palmer F., A method of attaching a fluorite window to glass apparatus, *Phy. Rev.*, **45**, 556 (1934).
965. Palmour H., Review of high temperature metal–ceramic seals, *J. Electrochem. Soc.*, **102**, 160 C, (1955).
966. Palmour H., and Adam H., Ein Bericht über hitzbestandige Metall-Keramik-Verbindungen, *Vakuum-Technik*, **4**, 121 (1955).
966a. Papagosta J. P., and Murdock D. M., Brazing thin beryllium windows, *Rev. Sci. Instr.*, **33**, 1289 (1962).
966b. Papirov J. J., Ultra-high vacuum demountable seals and valves that can be heated, *Instr. Exp. Techn.*, **2**, 225 (Nov. 1962).
967. Parker W., A method for the separation of gold irradiated mercury isotopes, *Nucl. Instr. & Meth.*, **8**, 354 (1960).
968. Parker W. C., and Gunditz V., Rapid carrier free separation technique, *Nucl. Instr. & Meth.*, **14**, 71 (1961).
969. Parker W. B., and Mark J. T., A large ultra-high vacuum valve, *Trans. AVS Nat. Vac. Symp. 1960*, p. 21, Pergamon Press, Oxford, 1961.
970. Parks W. G., and Carritt D. E., An improved stop-cock substitute, *Rev. Sci. Instr.*, **10**, 148 (1939).
971. Parr L. M., and Hendley C. A., *Laboratory Glass Blowing*, G. Newnes, London, 1956.
972. Partridge J. H., *Glass-to-metal seals*, The Soc. of Glass Technol., Elmfield, Sheffield, 1949.
973. Pask J. A., New techniques in Glass-to-Metal Sealing, *Proc. IRE*, **36**, 286 (1948).
974. Pask J. A., Expansion curves of Corning glasses, *Product. Engng.*, 129 (Jan. 1950).

975. Pattee H. H. Jr., Demountable ultra-high vacuum joint, *Rev. Sci. Instr.*, **25**, 1132 (1954).
976. Pattee H. H., A metal vacuum joint suitable for field emitters, *Phys. Rev.*, **98**, 283 (1955).
977. Paty L., High-vacuum, gas-leak valves, *Vacuum*, **8**, 80 (1958).
978. Paty L., and Schürer P., New ultra-high vacuum valve, *Rev. Sci. Instr.*, **28**, 654 (1957).
979. Paul F. W., Convenient type of vacuum chamber for evaporation of optical films, *Rev. Sci. Instr.*, **15**, 152 (1944).
980. Paul W., Ein Massenspektrometer Zur Bestimmung von Isotopen Mischungsverhältnissen, *Ztschr. Physik*, **124**, 244 (1948).
981. Pearsal C. S., New brazing methods for joining non-metallic materials to metals, *Mater. & Meth.* **30**, 61 (July 1949).
981a. Pentland M., *et al.*, Heat sealing and sealing of metal electrodes into fragile glass bulbs under high vacuum, *J. Sci. Instr.*, **32**, 364 (1955).
982. Pequignot M. and Verna M., L'utilisation de scellements à la silice dans la fabrication de tubes spèciaux à multiples traversées de courant, *Le Vide*, **14**, 290 (1959).
982a. Perrine T. D., Needle valve stopcock, *Anal. Chem.*, **27**, 286 (1956).
983. Perott L. F., *Glass and Metal Seal*, U.S. Pat. 772, 150 (1950).
984. Perry H. A., and Wagner R. H., *Adhesive Bonding of Reinforced Plastics* McGraw-Hill, New York, 1959.
985. Perry J. T., Mechanically sealed beryllium windows for permanently evacuated X-ray tubes, *Rev. Sci. Instr.*, **27**, 759 (1956).
986. Pesterfield E. E., and Einstein P. A., Coaxial cable termination vacuum seal, *J. Sci. Instr.*, **32**, 77 (1955).
986a. Peters R., Metal gaskets for the Berkeley 88-inch cyclytron, *Nucl. Instr. & Meth.*, **18** (19), 549 (1962).
987. Pfeiffer A., GmbH (Wetzlar, Germany), *Hochvakuumtechnik* (Cat. No. 1000).
988. Piatti H., Contrôle d'hermeticité non destructeur des petites enveloppes scellées, *Le Vide*, **17**, 250 (1962).
989. Pierre J., Les joints metalloplastiques en technique du vide, *Le Vide*, **15**, 381 (1960).
990. Pike E. W., On rubber connections for mercury system, *Rev. Sci. Instr.*, **6**, 328 (1935).
991. Pincus A. G., Mechanism of ceramic-to-metal adherence, *Ceram. Age.*, **63**, 16 (1954).
992. Pirani M., and Yarwood J., *Principles of Vacuum Engineering*, Chapman & Hall, London, 1961.
993. Plastic & Rubber Prod. Co. (Los Angeles), *O-Ring and Custom Molded Rubber Products* (Catalogue, 1962).
994. Plastic & Rubber Prod. Co. (Los Angeles), *Teflon Back-up Rings* (Catalogue 1962).
995. Pockman L. T., New construction for lead gasket seals, *Rev. Sci. Instr.*, **10**, 105 (1939).
996. Pollard J., High vacuum pipe connectors, *J. Sci. Instr.*, **30**, 25 (1953).
997. Pollard J., Progress in vacuum technology, *Rep. Progr. Phys.*, **22**, 33 (1959).
998. Pollermann M., *Bauelemente der Physikalischen Technik*, Springer, Berlin, 1955.
999. Pompes D. J., and Meyer E., Construction of glass bellows, *Rev. Sci. Instr.*, **12**, 368 (1941).
1000. Pondy P. R., A high temperature ceramic–metal seal made with low vapour pressure materials, *Proc. 4th Nat. Conf. on Tube Technique*, p. 29, University Press, New York, 1959.
1000a. Porteous P., An apparatus for specimen changing in pumped liquid helium through an air lock, *J. Sci. Instr.*, **40**, 47 (1963).
1001. Porubszky J., *Thermal treatment of the glass*, in Millner, *Vakuumtechnika*, Nehézipari K., Budapest, 1953.
1002. Power E. B., and Nelson E. H., *Method of Making Quartz-to-Metal Seals*, U.S. Pat. 2,659,964 (1948).

1003. POWER B. D., and ROBSON F. C., Experiences with demountable ultra-high vacuum systems, *Trans. AVS Nat. Vac. Symp. 1961*, p. 1175, Pergamon Press, Oxford, 1962.
1004. Precision Rubber Prod. Corp. (3110 Oak Ridge Drive, Dayton, Ohio, U.S.A.), *O-Rings* (Cat. 1958).
1005. PREISWERK E., Recent developments in ethoxyline resins, *Light Metals*, 594 (Nov. 1951).
1006. PREISWERK E., Aethoxylinharze in der Elektrotechnik, *Elektrotechn. Ztschr.* (ETZ), **B5**, 1 (1953).
1007. PRESTON F. W., Mechanical properties of glass, *J. Appl. Phys.* **13**, 623 (1942).
1008. PRESTON F. W., A re-examination of the Adams and Williamson law of annealing, *J. Soc. Glass Technol.*, **36** (10), 287 (1952).
1009. PREVOT F., Les principles génèraux et la physique d'un bon vide, *Genie Chim.*, **73**, 1 (Jan. 1955).
1010. PRIEM M., Method for sealing thermocouple wire through fused silica for use at high temperatures, *J. Sci. Instr.*, **39**, 130 (1962).
1011. PRIOR A. C., Neoprene vacuum gasket for wires, *J. Sci. Instr.*. **35**, 382 (1958).
1012. PROBYN H. C., A liquid level regulator for replenishing cold trap, *J. Sci. Instr.*, **33**, 322 (1956).
1013. PROVAN A. G., SHAW B. A., and WILLIAMS C. I., A window material for high temperature X-ray cameras, *J. Sci. Instr.*, **33**, 445 (1956).
1014. PRUGNE P., Compressed air operated valve for use in straight pipelines of vacuum evaporation plant, *J. Phys. Radium*, **12**, 66A (1951).
1014a. PRUGNE M., and GARIN P., L'utilisation des membranes en technique du vide, *Le Vide*, **7**, 1197 (1952).
1015. PRYSLAK N. E., Metal-to-ceramic seals for magnetron waveguides, *Electronics*, **27**, 282 (Jan. 1954).
1016. PRYSLAK N. E., A sandwich-type metal-to-ceramic, vacuum-tight seal, *Ceram. Age*, **65** (3), 21 (1955).
1016a. PULFRICH H., *Glass-to-Metal Seals*, Germ. Pat. 659, 427 (1936).
1017. PULFRICH H., Stellungnahme für Veröffentlichung von D.E.P. Jenkins "Vakuumdichte Keramik-Metall-Verbindungen"., *Vakuum-Technik*, **4**, 113 (1955).
1018. Pulsometer Eng. Co., (Nine Elms, Reading, England), *Pulsometer Pumps and Unions* (Catalogue).
1019. PUMPHREY W. J., Developments in arc welding, *Metallurgia*, **65**, 159 (1962).
1020. QUARRINGTON J. E., Metal-to-glass seals for vacuum work at low temperatures, *J Sci. Instr.*, **31**, 387 (1954).
1021. Q.V.F. Ltd. (Stoke-on-Trent, Staffs., England), *Glass for Industry* (Catalogue).
1022. RAATS E. *et al.*, Simple greaseless valve, *J. Sci. Instr.*, **34**, 510 (1957).
1023. RADABAUGH W. D. JR., Improved small valve, *Rev. Sci. Instr.*, **24**, 1005 (1953).
1024. RAIN M., Chemical cleaning of oxidized Ni-Fe-Co alloys for glass-to-metal seals, *J. Sci. Instr.*, **36**, 479 (1959).
1025. RANDOLPH M. L., SLATER M., and PARRISH D. L., O-Ring mounting of bellows, *Rev. Sci. Instr.*, **27**, 878 (1956).
1026. RAPP R. A., New techniques with attainment of high vacua, *Rev. Sci. Instr.*, **30**, 839 (1959).
1027. RAMSPERGER H. C., A greaseless and chemically inert valve for high vacua, *Rev. Sci. Instr.*, **2**, 738 (1931).
1028. RASOR S., and NUDING J. M., *Radiation and Heat Resistant Vacuum Seals*, North Am. Av. Rep. NAA-SR-158 (1955).
1028a. RAUCH C. J., and KERNAN W. C., Small He II tight optical window *Rev. Sci. Instr.*, **33**, 496 (1962).
1029. RAWSON H., The theory of stress in two-compoment glass-to-metal tube seals, *J. Sci. Instr.*, **26**, 25 (1949).
1030. RAWSON H., A theory of stresses in glass butt seals, *Brit. J. Appl. Phys.*, **2**, 151 (1951)
1031. RAWSON H., A method of sealing sapphire to glass, *J. Sci. Instr.*, **28**, 208 (1951).

1032. RAWSON H., A note on oxidation-resistant coatings on molybdenum and their uses in glass seals for mercury lamps, *J. Soc. Glass. Technol.*, **39,** 211T, 1955.

1033. RAWSON H., and DENTON E. P., The glass sealing properties of titanium and zirconium, *Brit. J. Appl. Phys.*, **5,** 352 (1954).

1034. RAWSON H., and DENTON E. P., Propriétés du titane et du zirconium employés dans les scellements du verre, *Le Vide*, **10,** 202 (1955).

1035. REA R. F., and RIPPLE J. W., Cutting and grinding ceramics, *Am. Ceram. Soc. Bull.* **36,** 163 (1957).

1036. READ W. T., An optical method for measuring the stress in glass bulbs, *J. Appl. Phys.*, **21,** 250 (1950).

1036a. REDMAN J. D., RICHARDSON D. M. *et al.*, *Vacuum Systems techniques and Material Studies*, ORNL Rep. 3472.

1037. REDSTON G. D., Iron–Nickel–Cobalt alloy for sealing to glass, *J. Sci. Instr.*, **23,** 53 (1946).

1038. REDSTON G. D., and STANWORTH J. E., Glass-to-metal seals, *J. Soc. Glass Technol.*, **29,** 48 T (1945).

1039. REDSTON G. D., and STANWORTH J. E., The effect of various treatments on the stresses in glass-to-metal seals, *J. Sci. Instr.*, **25,** 138 (1948).

1040. REDSTON G. D., and STANWORTH J. E., The theoretical development of simplified annealing schedules, *J. Soc. Glass. Technol.*, **32,** (2), 32 (1948).

1041. REEBER M. D., Demountable high-vacuum seal for use at liquid helium temp., *Rev. Sci. Instr.*, **32,** 1150 (1961).

1042. REID G. C., Making thin films of polythene, *J. Sci. Instr.*, **30,** 210 (1953).

1043. REILLY E. G., Combination high vacuum and pressure valve, *Rev. Sci. Instr.*, **24,** 875 (1953).

1044. REILLY E. G., Insulated union for metal tubing, *Rev. Sci. Instr.*, **24,** 879 (1953).

1045. REIMANN A. L., Coppered tungsten seals through hard glass, *J. Sci. Instr.*, **23,** 121 (1946).

1046. REIMANN A. L., *Vacuum technique*, Chapman & Hall London, 1952.

1047. REINARTZ J. L., Industrial brazing by pulse technique, *Electronics*, **23,** 78 (Jan. 1950).

1048. RETZLOFF O., Flexible vacuum seal, *Rev. Sci. Instr.*, **20,** 324 (1949).

1049. REYNOLDS F. L. *The Use of Indium in High-Vacuum Equipment*, Rep. UCRL-2989 (1955).

1050. REYNOLDS F. L., All-metal vacuum valve using an indium seat, *Trans. AVS Nat. Vac. Symp. 1955*, p. 74, Pergamon Press, Oxford, 1956; *Vakuum-Technik*, **4,** 181 (1955).

1051. RICHARD E. E., Attachment of a heavy object to a high vacuum line by means of a ball joint and spring suspension, *Vacuum*, **12,** 170 (1962).

1051a. RICHARDS C. B., A demountable vacuum seal, *J. Sci. Instr.*, **31,** 431 (1954).

1052. RICHARDS L. A., Low vacuum pressure control apparatus, *Rev. Sci. Instr.*, **2,** 49 (1931).

1053. RICHARDS R. J., High vacuum seal-off valve, *Rev. Sci. Instr.*, **25,** 520 (1954).

1053a. RICHARDS J. R., Valve and adapter for grease-free connection of glass sample tubes to a mass spectrometer inlet system, *J. Sci. Instr.*, **37,** 69 (1960).

1054. RICHARDSON H. K., Gasburner for glass blowing, *Ceram. Ind.*, **52,** (3), 89 (1949).

1055. RICHARDSON H. K., Sulphur in gas for glass working, *Glass. Ind.* **35,** 319 (June 1954).

1056. RIDDIFORD L., and LILLEY H., The design of high vacuum valves, *J. Sci. Instr.*, **29,** 296 (1952).

1057. RIDENOUR L. N., and HARNWELL G. P., Vacuum packing for valves, *Rev. Sci. Instr.*, **12,** 157 (1941).

1057a. RIDYARD H. N., A metal-to-glass joint, *J. Am. Chem. Soc.*, **46,** 287 (1924).

1058. RIECK G. D., and VERBEEK L. H., Lamps with laque protecting coating *Philips Techn. Rev.*, **12,** 189 (1951).

1059. RIEMERSTMA H., and LANGE W. J., *Ultra-High Vacuum Techniques*, Westinghouse Progress Rep. NYO-2615 (1959).

1060. RIEMERSTMA H., and LANGE W. J., *Ultra-High Vacuum Techniques*, Westinghouse Progr. Rep. NYO-8705 (1959).
1061. RIGGS F. B., Vacuum sealing of gold wire leads to a differential termopile, *Rev. Sci. Instr.*, **32,** 366 (1961).
1061a. RIGGS F. B., Construction of a small valve for high vacuum, *Rev. Sci. Instr.*, **33,** 1114 (1962).
1062. RILEY M. W., How to select and specify glass, *Mater. & Meth.*, **44** (5), 139 (1956)
1063. RILEY M. W., Selection and design of fluocarbon plastics, *Mater. & Meth.* **45** (6), 129 (1957).
1064. RILEY M. W., How to design and specify rubber parts, *Mater. in Design Engng.*, **55** (1), 95 (1962).
1065. RINEHART D. W., and BONINO J. J., Dielectric losses of some simple ternary silicate glasses, *J. Am. Ceram. Soc.*, **42,** 107 (Mar. 1959).
1066. RISZ E., Dosieren von Hg in Hg-Hochdruckbrenner, *Feinwerktechnik*, **62,** 155 (1958).
1067. RIVERA M., and LE RICHE R., A diferentially pumped ultra-high vacuum system, *Trans. AVS Nat. Vac. Symp. 1959*, p. 56, Pergamon Press, Oxford, 1960.
1068. ROBERTS R. H., and WALSH J. V., Vacuum lock for a mass spectrometer solids entry system, *Rev. Sci. Instr.*, **26,** 890 (1955).
1069. ROBERTS R. W., Rotary vacuum seal, *Rev. Sci. Instr.*, **32,** 750 (1961).
1069a. ROBERTS R. W., and VANDERSLICE T. A., *Ultra-High Vacuum and Its Applications*, Prentice-Hall, Englewood Cliffs, 1963.
1070. ROBERTS V., A simple method of making vacuum-tight coolable window seals for low temperature optical transmission cells, *J. Sci. Instr.*, **31,** 251 (1954).
1071. ROBERTS V., Coolable vacuum-tight window seals for optical use, *J. Sci. Instr.*, **36,** 99 (1959).
1072. ROBERTSON A. J. B., FABIAN D. J., CROCKER A. J., and DEWING J., *Laboratory Glass-Working*, Butterworth, London, 1957.
1073. ROBINSON N. W., Bakeable high-vacuum seals, *J. Sci. Instr.*, **34,** 121 (1957).
1074. ROEBEN F. J., and MCNALLY J. F., Removal of silicone grease from glass surfaces, *Chemist. Anal.*, **37,** 86 (1948).
1074a. ROGOZINSKI A., and WEIL J., Passage isolant avec anneau de garde étanche au vide et a des pressions élevées, *Le Vide*, **6,** 985 (1951).
1074b. RÖLLINGER W., Die Verwendung von Klammer-flanschen in der Vakuumtechnik, *Vakuum-Technik*, **13,** 42 (1964).
1074c. RÖLLINGER M., Vannes à membrane dépliante, *Le Vide*, **19** (110), 54 (1964).
1075. RONCI V. L., *Improvements Relating to Glass-to-Metal Seals*, Brit. Pat. 474, 706 (1937).
1076. ROSE J. E., Two aid in high vacuum technique, *Rev. Sci. Instr.*, **8,** 130 (1937).
1077. ROSE K., Silicones, *Mater. & Meth.*, **41** (2), 109 (1955).
1078. ROSE K., Siliconrubber (Properties), *Mater. & Meth.*, **41,** 118 (Feb. 1955).
1079. ROSE W., Packless valve flow regulator, *Rev. Sci. Instr.*, **21,** 772 (1950).
1079a. ROSENBERG P., Method for diminishing the sticking of mercury in capillaries, *Rev. Sci. Instr.*, **9,** 258 (1938).
1080. ROSENBLUM S., and WALEN R., Making very thin vacuum-tight glass windows, *J. Sci. Instr.*, **22,** 196 (1945).
1080a. ROSTOKER W., MCCAUGHEY J. M., and MARKUS H., *Embrittlement by Liquid Metals*, Chapman-Hall, London, 1960.
1081. ROTH A., Glass-to-metal seals, *Electricitatea*, (Roumania), **1,** 147, (1950).
1082. ROTH A., The measurement of stresses in technical glass, *Metrologia Aplicata*, (Roumania), **1** (1), 23 (1954).
1083. ROTH A., *Tehnica Vidului*, (Vacuumtechnique), Edit. Energetica, Bucharest, 1955.
1084. ROTH A., Some physico-chemical aspects of the metal–glass seals, *Studii-Cercet. Fizica* (Roum.), **7,** 337 (1956).
1085. ROTH A., Calculation of "leaky" vacuum installations, *Brit. Chem. Eng.*, **6,** 40 (1961).

1086. ROTH T., and ROTH A., Nomographic design of vacuum installations, *Brit. Chem. Eng.*, **5**, 392 (1960).
1087. ROTHERMEL J. J., X-ray absorbing glasses, *J. Am. Ceram. Soc.*, **32**, 153 (1949).
1088. ROTHSTEIN J., Rate of exhaust through a tube or orifice, *Rev. Sci. Instr.*, **29**, 243 (1958).
1089. ROUBEAU P., GARIN J. M., and PRUGNE M., Fuite réglable entièrement métallique, *Le Vide*, **11**, 453 (1956).
1089a. ROUND V., Fast closing vacuum valve of large aperture, *Rev. Sci. Instr.*, **33**, 758 (1962).
1090. RUBENS S. M., and HENDERSON J. E., Spherical ground joints for vacuum systems, *Rev. Sci. Instr.*, **10**, 49 (1939).
1091. RUEGER L. J., Teflon insert makes needle valve vacuum-tight, *Rev. Sci. Instr.*, **24**, 551 (1953).
1092. RUTHBERG S., and CREEDON F. E., Aluminium foil high vacuum gasket, *Rev. Sci. Instr.*, **26**, 1208 (1955).
1092a. SAECHTLING H., and ZEBROWSKI W., *Kunstoff Taschenbuch*, München 1957.
1093. SAGER T. P., and SUDIER M., *Permeability of Neoprene to Gases*, Nat. Bur. Stand. Res. R. P. 1166.
1094. SALSING W. W., *Utilizing Plastic Impregnation in Equipment for Nuclear Research*, Univ. Calif. Berkeley Radiation Lab., Rep. UCRL-3195 (1955).
1095. SANCIER K. M., Vacuum tight, flexible metal-glass seal, *Rev. Sci. Instr.*, **20**, 958 (1949).
1096. SANTELER D. J., Outgassing characteristics of various materials, *Trans. AVS Nat. Vac. Symp. 1958*, p. 1, Pergamon Press, Oxford, 1959.
1097. SANTELER D. J., and MOLLER T. W., Fluid flow conversion in leaks and capillaries, *Trans. AVS Nat. Vac. Symp. 1956*, p. 29, Pergamon Press, Oxford, 1957.
1098. SAVINSKII K. A., High vacuum equipment (in Russian) *Zavodskaya Labor* **9**, 1955; Vacuum **4**, 338 (1954).
1098a. SAWYER G. A., Unbacked thin films, *Rev. Sci. Instr.*, **23**, 604 (1952).
1099. SAXON D., and RICHARDS J., Vacuum lock for continuously evacuated systems, *Rev. Sci. Instr.*, **20**, 745 (1949).
1100. SAYERS F. J., Epoxy-resin joints for sealed-off high vacuum tubes, *J. Sci. Instr.*, **37**, 203 (1960).
1101. SCHACHER G. P., Device for opening small sealed objects under vacuum, *Analyt. Chem.*, **31**, 161 (1959).
1102. SCHÄFER W., and JAHN H., Epoxydharzklebetechnik, *Fertigungstechnik*, **6**, 544 (1956).
1103. SCHEFLAN L., and JACOBS M. B., *The Handbook of Solvents*, D. Van Nostrand, New York, 1953.
1104. SCHMITT O. H., A high-vacuum cut-off, *Rev. Sci. Instr.*, **8**, 68 (1937).
1105. SCHNEIDEMESSER B., A new method of sealing-off quartz vessels in vacuo, *Vacuum*, **4**, 489 (1954).
1105a. SCHOONMAKER T. P., *Choice of Inert Gas for Welding*, in Cook G. A. (Editor): *Argon, Helium and the Rare Gases*, Interscience, London, 1961.
1106. SCHRACK R. A., High vacuum gimbals, *Rev. Sci. Instr.*, **27**, 172 (1956).
1107. SCHRADE J., *Les Resines Epoxy*, Dunod, Paris, 1957.
1107a. SCHRAM A., La désorption sous vide, *Le Vide*, **18**, 55 (1963).
1108. SCHRIEVER W. W., Double seal vacuum coupling, *Rev. Sci. Instr.*, **24**, 402 (1953).
1109. SCHRODT V. N., Leak-proof joint for high vacuum work, *Chem. Eng.*, **67**, 128 (1960).
1110. SCHULZ G., Hostaflon, *Chem. Ing.-Techn.*, **24**, 544 (1952).
1111. SCHULZ L. G., Design of a vacuum photographic plate chamber, *Rev. Sci. Instr.*, **15**, 78 (1944).
1112. SCHWARZ E., A permanent vacuum seal, *J. Sci. Instr.*, **32**, 445 (1955).
1112a. SCHWARZ H., Electron beam processes at different voltages, *Trans. AVS Nat. Vac. Symp. 1961*, p. 699 Pergamon Press, London, 1962.

1113. Schweitzer J., Reflexions sur l'ultra-vide, Rapport CEA-1476 (1960).

1114. Scott R. B., *Cryogenic Engineering*, Van Nostrand, New York, 1959.

1115. Scott W. J., Glass-to-metal seal design, *J. Sci. Instr.*, **23**, 193 (1946).

1116. Sears G. W., and Hopke E. R., Effective vacuum cut-off, *Rev. Sci. Instr.*, **21**, 570 (1950).

1116a. Seavom (30 rue Raspail, Argenteuil, France) *Vacuum* catalogues.

1117. Sedden E., Tippett E. J., and Turner W. E. S., The electrical conductivity of sodium-metasilicate-silica glassed, *J. Soc. Glass. Technol.*, **16**, 450 T. (1932).

1117a. Seehof J., *et al.*, Electron-permeable window for cathode ray tubes, *Rev. Sci. Instr.*, **29**, 776 (1958).

1118. Seki Hajime, Simple demountable indium O-ring seal tight to HeII., *Rev. Sci. Instr.*, **30**, 943, (1959).

1119. Seren L., Thin glass window to withstand high pressures, *Rev. Sci. Instr.*, **19**, 123 (1948).

1120. Serfass E., and Muraca R., High vacuum cut-off, *Rev. Sci. Instr.*, **16**, 225 (1945).

1121. Setapen A. M., Hard Soldering technique, *Trans. Electrochem. Soc.*, **86**, 273 (1945).

1122. Servranckx R., and Leyniers R., Vanne de sécurité pour les installations de vide, *Le Vide*, **14**, 295 (1959).

1122a. S.F.I.T.V., (Commission Technique de la Société Française des Ingénieurs et Techniciens du Vide), Glossaire des termes de technique du vide, *Le Vide*, **18**, 486 (1963).

1123. Shand E. B., Industrial Glass Piping, *Ind. Eng. Chem.*, **46**, 179 (1954).

1124. Shand E. B., *Glass Engineering Handbook*, BcGraw-Hill, New York, 1958.

1125. Shapiro P., A large differential pumping system, *Trans. AVS Nat. Vac. Symp. 1961*, p. 1186, Pergamon Press, Oxford, 1962.

1126. Shatford P. A., O-ring sealed vacuum valves, *J. Sci. Inst.*, **29**, 336, (1952).

1127. Shaw M. R., Resistance welding of hard glasses, *Trans. Am. Inst. Electr. Eng. II* **71** (1), 384 (1953).

1128. Shaw M. R., Electrical welding of glass, *Electr. Engng.*, **72** (1), 31 (1953).

1129. Shell Chemical Co. Ltd. (Marlborough House, London), *Aplezon Products* (Catalogue).

1130. Shell Chemical Co. Ltd., *Epikote Resins*, Technical Bull, Res. 59/12.

1131. Sheriff E. V., Conductance of bends and concentric tubes in a vacuum systems, *J. Sci. Instr.*, **26**, 43 (1949).

1132. Shields W. R., *Mass Spectrometer Leak*, U.S. Pat. 2,956,771 (1960).

1133. Shire E. S., A thermally-operated gas-valve, *J. Sci. Instr.*, **31**, 192 (1954).

1134. Shon F. J., Steidlitz M., and Feldman M. H., Thin window metal irradiation cells for high vacuum or high pressure experiments, *Rev. Sci. Instr.*, **24**, 550 (1953).

1135. Siemens Ediswan Ltd., (155 Charing Cross Road, London, W.C. 2.), *Glass–Metal Seals* (Catalogue).

1136. Sikorsky J., and Woods H. J., A rotary vacuum seal, *J. Sci. Instr.*, **30**, 432 (1953).

1137. Sill R. C., All-glass valve with no moving parts, *Rev. Sci. Instr.*, **27**, 657 (1956).

1138. Simard G., and Steger J., High temperature sodium chloride to glass vacuum seal for infrared cells, *Rev. Sci. Instr.*, **17**, 156 (1946).

1139. Simonds H. R., *Sourcebook of the New Plastics*, Reinhold, New York, Vol. I (1959), Vol. II. (1961).

1140. Simons J. C., Jr., An ultra-high vacuum chamber for space simulation, *Trans. AVS Nat. Vac. Symp. 1959*, p. 48, Pergamon Press, Oxford, 1960.

1140a. Sinclair F. D., Inexpensive all-metal vacuum valve, *Rev. Sci. Instr.*, **34**, 437 (1963).

1141. Sistare G. H., Halbig J. J., and Grenell L. H., Silver brazing alloys for corrosion resistance joints in stainless steel, *Welding J.*, **33**, 137 (1954).

1142. Sitney L. R., High speed rotary vacuum seal, *Rev. Sci. Instr.*, **23**, 505 (1952).

1143. Skeits J., *Epoxy Resins*, Reinhold, New York, 1958.

1143a. Skeits J. *Handbook of Adhesives*, Chapman & Hall, London, 1962.

1144. Skellett A. M., Limitations of tubular ground-glass joint, *Rev. Sci. Instr.*, **7**, 179 (1936).

1145. SKIDMORE J. R., Apparatus for stirring a reaction mixture through a vapour-tight seal, *Analyt. Chem.*, **26,** 789 (1954).
1146. SKINNER H. W. B., and BURROW J. A., The technique of copper–Pyrex tube seals, *J. Sci. Instr.*, **7,** 290 (1930).
1147. SLACK F. G., An arrangement for obtaining a steady flow of gas at constant pressure, *Rev. Sci. Instr.*, **1,** 33 (1930).
1148. SLAVINSKI M. P. *Proprietățile fizico-chimice ale elementelor* (Physico-Chemical properties of elements), Edit. Tehnica, Bucharest, 1955.
1149. SLOAN L. G., Glass resistor for use in vacuum systems, *Rev. Sci. Instr.*, **27,** 1019 (1956).
1150. SMELT J., Verres pour tubes à television, *Verres et Refract.*, **6,** 75 (1952).
1150a. SMILEY V. N., Window seal for gas lasers, *Rev. Sci. Instr.*, **34,** 820 (1963).
1151. SMITH H. R., The technology of large mercury-pumped vacuum system, *Trans AVS Nat. Vac. Symp. 1955*, p. 27, Pergamon Press, Oxford, 1956; *Vakuum-Technik*, **5,** 117 (1956); **5,** 124 (1956).
1152. SMITH H. A., POSEY J. C., and THOMAS C. O., Mercury–glass checkvalves, *Rev. Sci. Instr.*, **30,** 202 (1959).
1153. SMITH J. R. W., Demountable vacuum seals, *J. Sci. Instr.*, **29,** 131 (1952).
1154. SMITH W. R., A quick-change seal opener, *Rev. Sci. Instr.*, **24,** 886 (1953).
1155. SMITH R. K., and SEAGONDOLLAR L. V., A sliding vane vacuum valve, *Rev. Sci. Instr.*, **23,** 767 (1952).
1156. SMITH P. L., and TAYLOR N. W., Helium permeation through glasses, *J. Am. Ceram. Soc.*, **23,** 139 (1940).
1157. SMITHELLS C. J., *Metals Reference Book*, Interscience, New York, 1955.
1158. SMITHER R. K., Controlled capillary gas leak, *Rev. Sci. Instr.*, **27,** 964 (1956).
1159. SMOTRICH H., and FELDMAN L., Three-directional motion vacuum-sealed joint, *Rev. Sci. Instr.*, **27,** 970 (1956).
1160. SMYTHE W. R., An adjustable vacuum leak, *Rev. Sci. Instr.*, **7,** 435 (1936).
1161. SNOW A. J., Method for introducing sublimable material into a vacuum system, *Rev. Sci. Instr.*, **23,** 246 (1952).
1162. SNYDER TH., and LIPKIN D., *Topics on Vacuum Technique*, in Graves-Froman: *Miscellaneous Physical and Chemical Techniques of the Los Alamos Project*, McGraw-Hill, New York, 1952.
1162a. SNYDER J. A., and STEEL E. C., Design and construction of a glass envelope closely coaxial to a high power electron beam, *Rev. Sci. Instr.*, **33,** 945 (1962).
1162b. SOGEV — Société Générale du Vide (Voie du Four, Choisy-le-Roi, Seine, France), *Vacuum equipment.*
1162c. SOA A. E., Entwicklung eines Nadelventils, *Exper. Techn. der Physik*, **12,** 202 (1964).
1163. SOHN J. S., and KUGLER A. N., Aircomatic welding, *Welding* J., **27,** 913 (1948).
1164. SOLED I., *Fasteners Handbook*, Reinhold, New York, 1957.
1165. SOWTER A. B., Cold welding of metals as Al, Cu., *Mater. & Meth.*, **28,** 60 (Nov. 1948.)
1166. SPARTA Manuf. Co. (Dover, Ohio, U.S.A.), *Teflon* Nr. 8 Publication.
1167. SPEES A. H., REYNOLDS C. A., *et al.*, Vacuum gasket at low temperatures, *Rev. Sci. Instr.*, **28,** 1090 (1957).
1168. SPENCE D. C., Design and operating criteria for environmental vacuum chambers, *Trans. AVS Nat. Vac. Symp. 1961*, p. 1228, Pergamon Press, Oxford, 1962.
1169. SPIGHEL J. S., High speed electronic vacuum valve, *Pribory i. Tekn. Eksper.*, No. 1, 151 (1959).
1170. SPINDLER G. P., Sapphire window seal, *Proc. of Symp. on the Art of Glassblowing* p. 28, Am. Sci. Glasblowers Soc., (Wilmington, Del.), 1959.
1171. SPRINGER L., *Laboratoriumsbuch für die Glasindustrie*, W. Knapp Halle (Saale), 1950.
1171a. SPINKS W. S., *Vacuum Technology*, Chapman & Hall, London, 1963.
1172. STALLMANN F. W., and KRUGER P. G., Insensitive needle valve, *Rev. Sci. Instr.*, **10,** 242 (1939).

1173. STAMPER J. A., Replaceable insulating vacuum seal for small electrical leads, *Rev. Sci. Instr.*, **33,** 491 (1962).
1174. Standard Telephones and Cables Ltd., (Connaught House, Aldwych, London, W.C. 2) *Hermetic Seals*, Catalogue MK101 (Aprl. 1958).
1175. STANIER H. M., and BEYNON J. H., A greaseless mercury-sealed vacuum tap, *J. Sci. Instr.*, **29,** 165 (1952).
1175a. STANIER B. J., and MEE C. H. B., The production of fine holes by a spark erosion technique, *J. Sci. Instr.*, **41,** 51 (1964).
1176. STANWORTH J. E. *Physical Properties of Glass*, Clarendon Press, Oxford, 1950.
1177. STANWORTH J. E., A Nickel–Chromium–Iron Alloy for sealing to glass, *J. Sci. Instr.*, **27,** 282 (1950).
1178. STANWORTH J. E., RAWSON H., and KNOWLES M., *Surface Treatment of Molybdenum and Tungsten for Glass Sealing*, U.S. Pat. 2,836,935 (1958).
1179. STARK D. S., The conductance of a column of vapour for gases at low pressure, *Brit. J. Appl. Phys.*, **8,** 210 (1957).
1179a. STARK J. A., and LANGSDORF A., A high vacuum plug valve, *Rev. Sci. Instr.*, **25,** 188 (1954).
1180. STECKELMACHER W., Seals and gaskets for ultra-high vacuum systems, *Vacuum*, **12,** 109 (1962).
1180a. STECKELMACHER W., and TINSLEY D. M., Thermal conductivity leak detectors for testing equipment by overpressure or vacuum, *Vacuum*, **12,** 153 (1962).
1181. STEIN F. S., Three-way vacuum valve, *Rev. Sci. Instr.*, **25,** 515 (1954).
1181a. STEINHERZ H. A., *Handbook of High Vacuum Engineering*, Reinhold, New York, 1963.
1182. STEINHERZ H. A., and REDHEAD P. A., Ultra-high vacuum, *Sci. Amer.*, **206** (3), 78 (1962).
1183. STERN J., Vacuum valve for glass system, *Rev. Sci Instr.*, **22,** 703, (1951).
1184. STERZER F., Simple, high-temperature, vacuum-tight mica window, *Rev. Sci. Instr.*, **28,** 208 (1957).
1185. STEVELS J. M., *Progress in the Theory of the Physical Properties of Glass*, Elsevier, New York, 1948.
1186. STEVELS J. M., Experiences et théories sur le facteur de puissance des verres en fonction de leur composition, *Verre et Refractaires.* 83 (1950); *Philips Res. Rep.*, **5,** 13 (1950); **6,** 34 (1951); *Philips Techn. Rdsch.* **13,** 385 (1952).
1187. STEVELS J. M., Quelques nouveautes dans les recherches sur le verre *Verre et Refractaires*, **1,** 3 (Feb. 1950).
1188. STEVENS C. M., Vacuum lock for routine solid analyses with a mass spectrometer, *Rev. Sci. Instr.*, **24,** 148 (1953).
1189. STEYSKAL H., *Arbeitsverfahren und Stoffkunde der Hochvakuumtechnik*, Physik-Velag, Mosbach (Baden), 1955.
1190. STIFF G. E., A connection system for vacuum piping, *Trans. AVS Nat. Vac. Symp. 1954* p. 21, Pergamon Press, Oxford, 1955.
1191. STINNETT A. J., Vibrating needle variable gas leak, *Rev. Sci. Instr.*, **24,** 883 (1953).
1192. STIVALA S. S., DENNIGER V. L., Epoxy resin as sealant for high vacuum systems, *Ind. Eng. Chem.*, **49,** 1106 (1957).
1193. STOCKDALE G. F., and TOOLEY F. V., Transmission changes of glasses, *J. Am. Ceram. Soc.* **33,** 11 (1950).
1193a. STODDARD C. K., and MOOZ W. E., Observations on the design and operation of high-vacuum systems, *Chem. Eng. Progr.*, **49,** (4), 197 (1953).
1194. STOHR J. A., *Electronic Welding of Metals*, Fuel Elements Conf. TID-7546, p. 6 (1957).
1195. STOHR J. A., Soudure des métaux sous vide, *Le Vide*, **13,** 163 (1958).
1196. STOHR J. A., and BRIOLA J., Vacuum welding of metals, *Welding and Metal Fabrication* **26,** 3 66 (1958).
1197. STOHR M., Elements de technique du vide. (I) Les joints, *Le Vide*, **2,** 205 (1947).
1197a. STOHR M., Elements de technique du vide, II. Les robinets., *Le Vide* **2,** 238 (1947).

1198. STOHR M., Elements de technique du vide, III. La recherche des fuites, *Le Vide*, **2**, 269 (1947).

1199. STOECKERT A. J., A novel method of fabricating ceramic stems, *Proc. Electronic Comp. Conf.* p. 228, 1959.

1200. STOLL S. J., Origin and analysis of gas in electron tubes, *Brit. J. Appl. Phys.*, **7**, 94 (1956).

1201. STRAUSS W. A., and EDSE R., Vacuum type gas flow calibrator, *Rev. Sci. Instr.*, **30**, 258 (1959).

1202. STRAUSS S., and MADORSKY S. L., Thermal degradation of unvulcanized and vulcanized rubber in vacuum, *Ind. Eng. Chem.*, **48**, 1212 (1956).

1203. STRINGER J. E. C., Miniature torroidal seals, *Rev. Sci. Instr.*, **24**, 395, (1953).

1204. STRNAD A. R., Mica windows assembly for use at elevated bake-out temperatures, *Rev. Sci. Instr.*, **29**, 533 (1958).

1205. STROHMAIER K., Ein einfaches Verfahren zur Herstellung grossflächiger durchsichtiger Haute aus Aluminiumoxyd einer Dicke 50–300 mμ, *Ztschr. Naturforsch.*, **6A**, 508 (1951).

1205a. STRONG J., On splitting mica, *Rev. Sci. Instr.*, **6**, 243 (1935).

1206. STRONG J., Use if synthetic rubber-like compounds in vacuum technique, *Rev. Sci. Instr.*, **10**, 104 (1939).

1207. STRONG J., *Procedures in Experimental Physics*, Prentice-Hall, Englewood Cliffs, N. J., 1956.

1208. STROTHER F. P. and GRANBERRY H., Glass-to-metal seals for vacuum systems, *Rev. Sci. Instr.*, **22**, 432 (1951).

1208a. STROUD P. T., Cathodic etching in a Penning cold cathode discharge, *Vacuum*, **9**, 269 (1959).

1209. STRUM R. G., and O'BRIEN H. J., Computing strength of vessels subjected to external pressure, *Trans. Am. Soc. Mech. Eng.*, **69**, 353 (1947).

1209a. STURCKEN E., HELLER R., and WEBER A. H., Percent beta-ray transmission through thin windows, *Phys. Rev.*, **78**, 327 (1950).

1209b. SUNDHEIM B. R., An absolute leak, *Rev. Sci. Instr.*, **35**, 228 (1964).

1210. SUN L. L., and SUN K. H., Glasses for X-ray absorbtion and transmission, *Glass. Ind.*, **29**, 686 (1948).

1211. SUN K. H., and SUN L. L. Glasses for neutron absorbtion and transmission, *Glass. Ind.*, **31**, 507 (1950).

1211a. SZALMA G., Glass metal seals by sintered glass (Roumanian), *Studii Cercet. Fizica*, **14** (3), 305 (1963).

1212. SZYMANOWITZ R., Graphite coating on glass, *Glass. Ind.*, **26**, 125 (1945).

1213. TANK S., *Method of Forming a Vacuum-Tight Bond between Ceramics and Metals*, U.S. Pat. 2,564,738 (1951).

1213a. TAPLIN J. F., and PHILIPS J. J., *Diaphragm Seals*, in *The Seals Book*, Machine Design, Jan. 1961.

1214. TARBES P., Rapport de la commission technique concernant la normalisation des canalisations metalliques pour le vide poussé, *Le Vide*, **9**, 212 (1954).

1215. TARBES P., and HENRY R. P., La normalisation des canalisations pour vide poussé, *Le Vide*, **16**, 347 (1961).

1215a. TASMAN H. A., BOERBOOM A. J. H., and KISTEMAKER J., Vacuum technique in conjunction with mass spectrometry, *Vacuum*, **13**, 33 (1963).

1216. TAYLOR J. B., A convenient method for introducing oxygen into evacuated systems, *Rev. Sci. Instr.*, **6**, 243 (1936).

1217. TAYLOR R. C., and WARD C. W., Lead-in seals for kinetic high vacuum systems, using rubbers, *Rev. Sci. Instr.*, **20**, 457 (1949).

1218. TENNENT R. M., Photomultiplier window for a liquid scintillator, *J. Sci. Instr.*, **36**, 148 (1959).

1218a. Thermal Syndicate Ltd. (Wallsend, Northumberland, England), *Vitreosil* (Cat. T6).

1218b. THIELE K., Ein elektromagnetisch gesteuertes Allglasventil für Ultrahochvakuum, *Vakuum-Technik*, **8**, 223 (1959).

1219. THIELSCH H., Stainless steels welding summary, *Welding J.*, **34**, 22s., (1955).

1219a. THIEME G., Mass spectrometer investigation of gas emission from plastics, *Vacuum*, **13**, 137 (1963).

1220. THOMAS E., Installation de vide pour une source d'ions, *Mem. Soc. Roy. Belge des Ing. et Industriels*, **1**, 3 (1953).

1220a. THOMAS A. M., JOHNSON D. P., and LITTLE J. M., Design of an interferometric oil manometer for vacuum measurement, *Trans. AVS Nat. Vac. Symp. 1962*, p. 468 Macmillan, New York, 1963.

1221. THOMPSON F. W., A magnetic mercury cut-off, *J. Sci. Instr.*, **31**, 145 (1954).

1222. THOMPSON H. R., Casting of Araldite sheets for photoelastic stress analysis, *J. Sci. Instr.*, **39**, 133 (1962).

1223. THOMSON G. P., and COCHRANE W., *Theory and Practice of Electron Diffraction*, MacMillan, London, 1939.

1223a. THORNESS R. B. and NIER A. O., Device for remote opening of a vacuum system, *Rev. Sci. Instr.*, **33**, 1005 (1962).

1224. THORNESS R. B., and NIER A. O., All-metal valve for ultra-high vacuum use, *Rev. Sci. Instr.*, **32**, 807 (1961).

1224a. TIMOSHENKO S., *Theory of Elastic Stability*, McGraw-Hill, New York, 1936.

1225. TIPTON F. W., *Design Handbook for O-Rings and Similar Elastic Seals*, Boeing Airplane Co., WADC-TR-59-428.

1225a. TOBIN R. A., Simple, bakeable, thin vacuum window, *Rev. Sci. Instr.*, **34**, 1450 (1963).

1226. TOBY S., and KUTSCHKE K. O., Alloy-filled cut-off for high temperatures, *Rev. Sci. Instr.*, **28**, 470 (1957).

1227. TODD B. J., Outgassing of Glass, *J. Appl. Phys.*, **26**, 1238 (1955).

1228. TODD B. J., Equilibrium between glass and water vapour at bake-out temperatures, *J. Appl. Phys.*, **27**, 1209 (1956).

1229. TOFFER H., and AMREIN R. L., Plastic-to-metal vacuum seal for rotating targets, *Rev. Sci. Instr.*, **31**, 348 (1960).

1230. TOLLIVER B. R., GRADIN J. H., and COOK R. L., Experiments in glass-to-metal sealing, *Ceram. Age*, **77** (3), 44 (1961).

1231. TOMBOULIAN D. H., and HARTMAN P. L., Spectral and angular distribution of U.V. radiation from the 300 MeV Cornell synchrotron, *Phys. Rev.*, **102**, 1423 (1956).

1232. TOOL A. Q., Relaxation of stresses in annealing glass, *J. Res. Natl. Bur. Stand.*, **34** (2), 199 (1945).

1233. TOPANELIAN E., and COGGESHALL N. D., Metal packless vacuum valve, *Rev. Sci. Instr.*, **17**, 38 (1946).

1234. TOWNES C. H., Greaseless vacuum valve, *Rev. Sci. Instr.*, **9**, 428 (1938).

1235. TRÉBUCHON G., and KIEFFER J., L'aspect physique des soudures verre–metal dans l'industrie des tubes électroniques, *Ann. Radioélectr.*, **5**, 125 (1950); **5**, 243 (1950); **5**, 407 (1950).

1236. TRENDELENBURG E. A., L'ultra vide, *Ind. Atomique*, **5** (9/10), 76 (1961).

1236a. TRENDELENBURG E. A., *Ultrahochvakuum*, G. Braun, Karlsruhe, 1963.

1237. TREVOY D. J., and TORPEY W. A., Shaft seal for vacuum apparatus, *Analyt. Chem.*, **24**, 1382 (1952).

1238. TURNBULL A. H., Vacuum technique in atomic energy research, *Vacuum*, **5**, 131 (1955).

1239. TURNBULL J. C., Recorder dilatometer, *J. Am. Ceram. Soc.*, **33**, 54 (1950).

1240. TURNBULL J. C. High conductivity glass-to-metal seal, *R.C.A. Rev.*, **13**, 291 (1952).

1240a. TURNBULL A. H., BARTON R. S., and RIVIÉRE I. C., *An Introduction to Vacuum Technique*, G. Newnes, London, 1962.

1241. TURNER J. O., *Plastics in Nuclear Engineering*, Reinhold, New York, 1961.

1242. TURNER J. A., PICKARD R. M., and HOFFMAN C. R., Ultra-high vacuum system, *J. Sci. Instr.*, **39**, 26 (1962).

1243. TURK E. P. DE, and LOCKENVITZ A. E., Method for grinding the internal wall of small capillaries, *Rev. Sci. Instr.*, **13**, 295 (1942).

1244. TUTHILL R. W., Argonarc welding, *Welding J.*, **26**, 766 (1947).

1244a. TWEEDDALE J. W., TENTARELLI L. A., and WHITE J. M., A method for producing low temperature metal–ceramic seals, *1962 Advances in Electron Tube Techniques*, p. 47, Pergamon Press, Oxford, 1963.

1245. TZAREV B. M., *Rascet i Konstruirovanie Electromih Lamp*, in Russian (Design and Construction of Electronic Tubes), Gosenergoizdat, Moscow, 1952.

1246. UBISCH H. VON, Gas sample intake for mass spectrometer, *J. Sci. Instr.*, **29**, 412 (1952).

1247. UBISCH H. VON, Transfer of movement into a vacuum by means of twisting a copper tube, *J. Sci. Instr.*, **33**, 200 (1956).

1248. UDIN H., FUNK R., and WULFF J., *Welding for Engineers*, John Wiley, New York, 1954.

1249. ULLMAN I. R., Commercial seals as seats in a bakeable valve, *Trans. AVS Nat. Vac. Symp. 1961*, p. 1323, Pergamon Press, Oxford, 1962.

1250. ULLMAN R., A large metal system permitting low base pressures, *Trans. AVS Nat. Vac. Symp. 1957.*, p. 95, Pergamon Press, Oxford, 1958.

1251. ULMICHEK L. G., *Proisvostvo Elektricheskih Lamp Nakalivania*, in Russian (Production of Incandescent Electric lamps), Gesenergoizdat, Moscow, 1958.

1251a. Ultek Corp. (920 Commercial Street, Palo-Alto, Calif., U.S.A.) *Vacuum Equipment*, Bull. B4 (1961).

1252. UMBLIA E., The metallisation of glass, *Glass*, 343 (Oct. 1951).

1253. URRY G. and URRY W. H., Automatic Toepler pump of improved design, *Rev. Sci Instr.*, **27**, 819 (1956).

1253a. Vacuum Research Comp. (420 Market Street, San Francisco 11, Calif. U.S.A.). *High Vacuum Valves*, (Data Sheets), 1958.

1253b. Vactronic Ins. (East Northpost, Long Island, New York, U.S.A.), *Vari-vac Adjustable Leak Valve*, (Cat. VV 50).

1254. VÁMBÉRI L., *Materials Used in High Vacuum Technology*, in Millner, *Vakuumtechnika*, Nehézipari K., Budapest, 1953.

1254a. VANCE R. W., and DUKE W. M., *Applied Cryogenic Engineering*, John Wiley, New York, 1962.

1255. VANDERSLICE T. A., Ultra-high vacuum, *Chem. Engng. News*, **39** (51), 86 (1961).

1256. VÁRADI P. F., Effect of pretreatment on the degassing of materials, *Trans. AVS Nat. Vac. Symp. 1961*, p. 73, Pergamon Press, Oxford, 1962.

1257. VARIAN R. H., Mica window for klystrons, *Electronics*, **25**, 112 (April 1952).

1257a. Varian Associates (611 Hansen Way, Palo Alto, Calif. U.S.A.), *Data Sheets*, 1962.

1258. VATTER H., Pumpröhrchen mit Weichlot, *Glas und Hochvakuumtechnik*, **1**, 79 (1952).

1259. VATTER H., Zur Geschichte der Keramik-Metall-Harlot-Technik, *Vakuum-Technik*, **4**, 180 (1955).

1260. VAUGHAN W., Non-lubricated, all glass, high vacuum valve, *Rev. Sci. Instr.*, **16**, 254 (1946).

1261. VAUGHAN I. R. M., Some high-power window failures, *IRE. Trans. on Electron Devices*, ED-8, 302 (1961).

1262. Veeco Vacuum Corp. (New Hyde Park, Long Island, N.Y., USA) *High Vacuum Equipment* (Catalogue).

1263. VELTE A. J., Procédé scellement ceramique–titane, *Le Vide*, **88**, 330 (1960).

1264. VENEMA A., Vacuum technique in Europe, *Vacuum*, **11**, 94 (1961).

1265. VENEMA A., and BANDRINGA M., The production and measurement of ultra-high vacua, *Philips Techn. Rev.*, **20**, 145 (1958/59).

1266. VERBA J. and HAWRYLAK R., Continuous rotation scattering chamber of new design, *Rev. Sci. Instr.*, **32**, 1037 (1961).

1266a. Via-Vac Ltd., (Wishaw, Lancs., England) *Vacuum Pumps and Accessories* (Catalogue).

1266b. V.I.C., Vacuum Instrument Corp. (Stepar Place, Huddington Station, New York), *Tubing Couplings for High Vacuum.*

1267. VILLIERS I. W. L., DE, Method for making vacuum feed-through terminals, *Rev. Sci. Instr.*, **29**, 527 (1958).

1268. VILLIERS I. W. L. DE, Methode pour la confection de traverses étanches au vide, *Le Vide*, **15**, 366 (1960).

1269. VIOLET F., *et al.* Glass in the radiotube industry, *Ann. Radioelectr.*, **2**, 24 (1947).

1270. VLAARDINGERBROEK M. T., An experimental disc-seal triode for 6000 Mc/s., *Philips techn. Rev.*, **21**, 167 (1959/60).

1271. VOGL T. P., and EVANS H. D., Electromagnetically operated ultra-high vacuum valve, *Rev. Sci. Instr.*, **27**, 657 (1956).

1271a. VOGL T. P., and MCINTOSH R. O., *Vacuum-Tight Windows with Wide Band Transmission Characteristics*, Westinghouse, Res. Rep. 71F189-R2-X, and U.S. Pat. 2,966,592/1960.

1272. VOLPE M. L., FULRATH R. M., *et al.*, *Fundamentals of Glass-to-Metal Bonding*, Rep. AECU-3560 (1957).

1273. VRIES H. L., DE, and STUIVER M., A proportional counter for low level counting with high efficiency, *Nucl. Instr. & Meth.*, **5**, 111 (1959).

1274. DE VRIES J. Testing of materials for glass-to-metal seals by means of stress birefringence, *Philips. Techn. Rev.*, **22**, 337 (1960/61).

1275. VUCCINO S., Preparation et étude de quelques films minces de haut polymers utilisés en physique nucléaire, *J. Phys. et Radium*, **13**, 543 (1952).

1276. Wacker-Chemie (München), *Silicone Rubber*, Catalogue 2283 (1961).

1277. WADEY W. G., Standards for vacuum fittings, *Vacuum*, **4**, 53 (1954).

1278. WADEY W. G., Proposed system of standard vacuum flanges, *Rev. Sci. Instr.*, **27**, 322 (1958).

1279. WAGENER J. S., and MARTH P. T., Analysis of gases of very low pressures by using the omegatron spectrometer, *J. Appl. Phys.*, **28**, 1027 (1957).

1280. WAGNER G., *Erzengung u. Messung von Hochvakuum*, F. Deuticke, Wien, 1950.

1281. WAHL J. S., FORBES S. G. *et al.*, New, low impedance, high-vacuum valve, *Rev. Sci. Instr.*, **23**, 379 (1952).

1282. WALDEN L., Laboratory cements and waxes, *J. Sci. Instr.*, **13**, 345 (1936).

1283. WALDSCHMIDT E., Evolution of gas and permeability to gas of constructional metals for vacuum systems, *Metall*, **19**, 20 (Oct. 1954). *A.E.A. Rep.* IGRL-T/C 25 (1956).

1284. WALLACE R. A., and VANDERVEER W. R., Brazing vacuum tight joints in high nickel alloys, *Mater. & Meth.*, **36**, 117 (1952).

1285. WALTON G. N. (Editor), *Glove Boxes and Shielded Cells*, Butterworth, London, 1958.

1286. WARD J. W., A method for high vacuum bakeout of metal vacuum systems, *Vacuum*, **12**, 27 (1962).

1287. WARMOLTZ N., and BOUWMEESTER E., Metal vacuum equipment, *Philips Techn. Rev.*, **21**, 173 (1959/60).

1288. WARSCHAUER D. M., Infrared windows for helium cryostats, *Rev. Sci. Instr.*, **27**, 419 (1956).

1288a. WATTS T. L., and SNEIDER C. J., Thin films for accurate measurement of charged particle ranges, *Nucl. Instr. & Meth.*, **21**, 296 (1963).

1288b. WATSON K., and BAUMAN C. A. Apparatus to measure the thermal conductivity of powders in Vacuum from 120 to 350°K., *Rev. Sci. Instr.*, **34**, 1235 (1963).

1289. WEBER A. H., and BAZZONI C. B., Quartz to Pyrex joint, *Rev. Sci. Instr.*, **8**, 170 (1937).

1290. WEIGEL R. G., Trübglaser, *Licht*, **4**, 1–39 (1934).

1291. WEINMANN R. W., and LANGMUIR I., Arcatom welding, *Gen. Electr., Rev.*, **29**, 160 (1926).

1292. WEINTROUB S., A demountable vacuum joint with clamp, *J. Sci. Instr.*, **14**, 315 (1937).

1292a. WEISBECK R., Gasreiniger und Dosierventil für Wasserstoff beruhend auf der selektiven Diffusion durch Nickel, *Vakuum-Technik*, **10**, 54 (1961).

1293. WEITZEL D. H., ROBBINS R. F. *et al.*, Low temperatures static seals using elastomers and plastics, *Rev. Sci. Instr.*, **31**, 1350 (1960).

1294. WELCH W. M., Scientific Co., (1515 Sedgwick Street, Dept. VA, (Chicago 10, Ill., U.S.A.). *Duo-Seal Pumps* (Catalogue).

1295. WENT J. J. (N. V. Philips), *Articles Consisting of a Glass Part and a Metal Part Melted Together*, Dutch Pat. 72, 275 (1953).

1296. WERTZ I. E., An improved vacuum septum, *Rev. Sci. Instr.*, **23**, 448 (1952).

1297. Westinghouse Electric Corp. *Ultra-High Vacuum Techniques*, Progr. Rep. 2616 (1959).

1298. WEXLER A., CORAK W. S., and CUNNINGHAM G. T., Low temperature gaskets, *Rev. Sci. Instr.*, **21**, 259 (1950).

1299. WEXLER A., and JACKET H. S., Storage container for liquid helium, *Rev. Sci. Instr.*, **22**, 282 (1951).

1300. WEYL W. A., Solarization of glass, *J. Soc. Glass Techn.*, **30**, 90T (1946).

1301. WEYL W. A., Practical importance of surface cleaning of gases, *Glass Ind.* **28**, 231 (May 1947).

1302. WEYL W. A., Dielectric properties of glasses function of their composition, *J. Soc. Glass. Techn.*, **33**, 220 (1949).

1302a. WAALEN I. I., Select and apply gaskets effecting, *Chem. Eng.*, **69** (20), 1 (Oct. 1962).

1302b. WHEATLEY J. C., Epoxy resin seals to copper and Nylon for Cryogenic Applications, *Rev. Sci. Instr.*, **35**, 765 (1964).

1303. WHEELER E. L., *Scientific Glass Blowing*, Interscience, New York, 1958.

1304. WHEELER W. R., and CARLSON M. A., Ultra-high vacuum flanges, *Trans. AVS Nat. Vac. Symp. 1961*, p. 1309, Pergamon Press, Oxford, 1962.

1304a. WHEELER W. R., Theory and application of metal gasket seals, *1963 Trans. American Vacuum Soc.*, p. 159. Macmillan, New York, 1963.

1305. WHETTEN N. R., and YOUNG I. R., Use of a silver tube to admit oxygen to a vacuum system, *Rev. Sci. Instr.*, **30**, 472, (1959).

1305a. WHITE B. L. *et al.*, Gas flow regulator for an rf ion source, *Rev. Sci. Instr.*, **33**, 1111 (1962).

1306. WHITE G. K., *Experimental Techniques in Low Temperature Physics*, Clarendon Press, Oxford, 1959.

1306a. WHITE P. A. F., and SMITH S. E., *Inert Atmospheres*, Butterworth, London, 1962.

1307. WHITE J. F., and SILVERMAN W. B., Solarisation of glass, *J. Am. Ceram. Soc.*, **33**, 252 (1950).

1307a. WHITEHEAD E. C., and COLGATE S. O., A simple technique for glassing metal tubing, *Vacuum*, **12**, 307 (1962).

1308. WHITEHEAD S., *Dielectric Breakdown of Solids*, Clarendon Press, Oxford, 1951.

1309. WHITNEY R. G., A glass transistor enclosure, *Semiconductor Products*, **2**, 32 (1959).

1310. WICKERS E., and SAYLOR C. P., Sealing platinum to Pyrex glass, *Rev. Sci. Instr.*, **10**, 245 (1939).

1311. WICK M., Siliconkautschuk, *Elektrotechn. u. Maschinenbau*, **74**, 355 (1957).

1312. WIDMER H., and KIRSCH J., High vacuum evaporator for radioactive materials, *Rev. Sci. Instr.*, **31**, 791, (1960).

1313. WIEDER H., and SMITH A. W., Electrical lead for vacuum system, *Rev. Sci. Instr.*, **29**, 794 (1958).

1314. Wiggin H., & Co. Ltd., (Birmingham, England), *Wilco-Wiggin Thermometals; Basic Information* (Bulletin).

1315. WILDY P. C., *A Method for Sealing Lithium Fluoride Windows to Pyrex Vacuum System*, Res. Jan. 1960, U.K. AEA (AERE-R-3201),

1316. WILLENS R. H., Bearings for high vacuum applications, *Rev. Sci. Instr.*, **31**, 574 (1960).

1316a. WILLENS R. H., Vacuum X-ray diffractometer for high temperature studies of metals sensitive to contamination by oxygen or nitrogen, *Rev. Sci. Instr.*, **33**, 1069 (1962).

1317. WILLIAMS N. T., Metal–ceramic vacuum seal, *Rev. Sci. Instr.*, **18**, 394 (1947).

1318. Williams N. T., Grooveless O-ring vacuum seal for bell jars and other demountable components, *Rev. Sci. Instr.*, **26,** 1207 (1955).

1318a. Williams W. S., Induction heating in a metal ultra-high vacuum system, *Rev. Sci. Instr.*, **32,** 743 (1961).

1319. Williams I. C., and Nielsen I. W., Wetting of original and metallized high-alumina surfaces by molten brazing solders, *J. Am. Ceram. Soc.*, **42,** 229 (1959).

1320. Willis C. H., *Preliminary Design for the Stellarator. II. Vacuum Chamber*, Tech. Memo Princeton Univ., AECD–3747 (1957).

1321. Willis J., Low temperature optical window seal used at 800 °K, *Rev. Sci. Instr.*, **29,** 1053 (1958).

1322. Willot W. H., Grinding hardness of glasses, *J. Soc. Glass. Technol.*, **34,** 77 (1950).

1323. Wilson E. J., An automatic Toepler pump and a circulating pump, *Vacuum*, **3,** 287 (1953).

1324. Wilson E. J., Manipulation of radioactive gases in high vacuum apparatus, *Vacuum*, **4,** 303 (1954).

1325. Wilson R. R., A vacuum-tight sliding seal, *Rev. Sci. Instr.*, **12,** 91 (1941).

1326. Winter A., Glass formation, *J. Am. Ceram. Soc.*, **40,** 54 (Feb. 1957).

1327. Wishart I., and Bancroft G. H., A new design for ultra-high vacuum valves, *Trans. AVS Nat. Vac. Symp. 1960*, p. 13, Pergamon Press, Oxford, 1961.

1327a. Wittrick W. H., Elastic circular ring compressed between rigid plates, *Austral. J. Appl. Sci.*, **13** (2), 172 (1962).

1328. Wittwer H., Eine einfache Methode zur Zerstörungsfreier Bestimung der Wandstärke von Glassröhren, *Vakuum-Technik*, **6,** 111 (1957).

1329. Witty R., Flow of gases in vacuum systems, *J. Sci. Instr.*, **26,** 316 (1949).

1330. Work R. H., Silica-glass helium leaks as standards in leak detection, *Trans. AVS Nat. Vac. Symp. 1958*, p. 126, Pergamon Press, Oxford, 1959.

1331. Wright D. A., Rare metals in electron tubes, *Radio Convention*, (1951); *J. Brit. I.R.E.*, **11,** 389 (1951).

1332. Wright E., A remote vacuum operated pipette controller, *Atomics*, **7,** 30 (1956).

1333. Wright R. C., *A New Type of Adjustable leak for High Vacuum Systems*, Rep. AECD-3872 (1955).

1334. Wroughton D. M., and Gregory E. D., *Metal Mixture for Sealing to Glass*, U.S. Pat. 2,699,015 (1955).

1335. Wu C. S., Meaker C. L., and Glassford H. A., Thin window counter with special mica-to-glass seal. *Rev. Sci. Instr.*, **18,** 693 (1947).

1335a. Wyckoff R. W. G., and Lagsdin J. B., An improved self-rectifying gas X-ray tube, *Rev. Sci. Instr.*, **7,** 35 (1936)

1336. Wyllie H. A., A gas inductor for a vacuum system, *J. Sci. Instr.*, **33,** 360 (1956).

1337. Willie H. A., A mercury-in-glass gas burette, *J. Sci. Instr.*, **34,** 410 (1957).

1338. Wyman W. L., High vacuum electron-beam fusion welding, *Welding J.* **37,** 49 (1958).

1339. Yarwood J., *High Vacuum Technique*, Chapman & Hall, London, 1948.

1340. Yarwood J., Isolation valves for vacuum systems, *Vacuum*, **3,** 398 (1953).

1341. Yarwood J., Ultra-high vacua, *J. Sci. Instr.*, **34,** 297 (1957).

1342. Yates W. A., Griffiths L. B. and Hill R. J., Ein in Ultrahochvakuum Systeme verwendbares Ventil, *Vacuum-Technik*, **11,** 122 (1962).

1343. Yergin P. F., Modifications of steam valves for high vacuum use, *Rev. Sci. Instr.*, **23,** 310 (1952).

1344. Yonts O. C., and Morgan H. W., Versatile glass-to-metal vacuum seal, *Rev. Sci. Instr.*, **26,** 619 (1955).

1345. Young J. F., *Materials and Processes*, John Wiley, New York, 1954.

1346. Young Y. R., Penetration of electrons in aluminium oxide films, *Phys. Rev.*, **103,** 292 (1956).

1347. Young Y. R., Vacuum limitations of rubber O-ring joints, *Rev. Sci. Instr.*, **29,** 795 (1958).

1348. Young Y. R., Cleaning techniques for rubber O-rings used in vacuum systems, *Rev Sci. Instr.*, **30,** 291 (1959).

1349. Young J. R., Use of an ion pump as a leak detector, *Rev. Sci. Instr.*, **32,** 85 (1961).

1349a. Young L., *Anodic Oxide Films*, Academic Press, London, 1961.

1350. Young J. R., and Whetten N. R., Purity of helium permeating through quartz into a vacuum system, *Rev. Sci. Instr.*, **32,** 453 (1961).

1351. Young J. R., and Whetten N. R., Techniques for the admission of high purity gases to a vacuum system, *Trans. AVS Nat. Vac. Symp. 1961*, p. 625, Pergamon Press, Oxford, 1962.

1351a. Young J. R., Ultra-high vacuum sealing material, *Rev. Sci. Instr.*, **35,** 116 (1964).

1352. Young R. S., and Snaddon R., Safety trap for vacuum line, *Lab. Practice*, **2,** 27 (Jan. 1953).

1352a. Zabel R. M., Vapour pressure of vacuum cements, *Rev. Sci. Instr.*, **4,** 233 (1933).

1353. Zackay V. F., Mitchell D. W., *et al.*, Fundamentals of glass-to-metal bonding, *J. Am. Ceram. Soc.*, **36,** 84 (1953); **40,** 269 (1957); **42,** 102 (1959).

1354. Zincke A., and Marcic B., *Technologie der Glassverschmelzungen*, Akadem. Verlag, Leipzig 1961.

1354a. Zincke A., Glas-Probleme in der Elektrotechnik, *Vakuum-Technik*, **7,** 93 (1958).

1355. Ziock K., Experience gained with various leak detection methods and apparatus, *Glas & Hochvakuumtechnik*, **2,** 292 (1953).

1356. Zobac L., *Zaklady Vacuove Techniky*, S.N.T.L., Praha, 1954.

1357. Zollman J. A., Martin J. E., Powell J. A., Ceramic, sapphire and glass seals for the model-C stellarator, *Trans. AVS. Nat. Vac. Symp. 1959*, p. 278, Pergamon Press, Oxford, 1960.

1358. Zunick M. J., and Gosling I. B., Hard glasses for X-ray tubes, *Glass Ind.*, **32,** (3), 117 (1951).

ADDITIONAL REFERENCES

[These appeared during the printing of the book. The numbers in square brackets refer to the corresponding sections in the book.]

1359. ABENDROTH R. P., Oxide formation and adherence on iron–cobalt–nickel glass sealing alloys, *Materials Res. and Stand.*, **5** (9), 459 (1965) [Sec. 24.11].

1360. ADAMS R. O., *Indium Seals*, Rep. RFP - 449, TID - 4500, Dec. 1964 [Sec. 38.51].

1361. Advac (174 Richmond Hill Avenue, Stanford, Conn., USA), *High Temperature Ceramic to Metal Hermetic Seals* (catalogue) [Sec. 25.42].

1362. AMOIGNON J., and LÉVÊQUE M., Élements de raccordement pour canalisations, *Le Vide*, **20** (117), 222 (1965) [Sec. 38.47].

1363. Andar Corp. (185 E, Evelyn Avenue, Mountain View, Calif, 94041, USA), *High Vacuum Equipment* (catalogue) [Sec. 11.3; 38.52].

1364. BANNENBERG J. B., Operating characteristics of a fast gas valve, *Trans. 3rd International Vacuum Congress*, vol. 2, Pergamon Press, Oxford, 1966. [Sec. 61.34].

1365. BATZER T. H., and RYAN J. F., Some new techniques in ultra-high vacuum, *Trans. AVS Nat. Vac. Symp. 1963*, p. 166, MacMillan, New York, 1964 [Sec. 38.6].

1366. BECK A. H. (Ed.), *Handbook of Vacuum Physics*, Pergamon Press, Oxford, 1965 [Sec. 11.3].

1367. BELONOGOV A. V., Safety valve for vacuum systems with self-sealing, *Instr. Exp. Techn.* No. 1 (Jan.–Febr.), 236 (1964) [Sec. 61.33].

1368. BONY J., and ROMMEL G., Mesure de la permeabilité des feuilles de Mylar, *Le Vide* **19** (112), 224 (1964) [Sec. 12.2].

1369. BÜLTEMANN H. J., *et al.*, Uhv-Hochtemperaturventile mit Golddraht-Ventildichtung, *Vakuum–Techn.*, **14,** 132 (1965) [Sec. 61.33 f].

1370. CHIRLIAN P. M., New high vacuum rotary feedthrough for liquid nitrogen, *Rev Sci. Instr.* **35,** 1078 (1964) [Sec. 6.2].

1371. CLAUS F. J., Sliding electrical contacts for ultrahigh vacuum, *Rev. Sci. Instr.* **36,** 1251 (1965) [Sec. 53.3].

1372. Compagnie Générale d'Électro–Ceramique (12 rue de la Baume, Paris 8e), *Sintered Oxides* (catalogue) [Sec. 42.1].

1373. COMSA G., and SIMIONESCU C., Bakeable metal valve with copper gasket, *Trans. 3rd International Vacuum Congress*, Vol. 2, Pergamon Press, Oxford, 1966 [Sec. 61.35 f].

1374. DEMETSOPOULLOS I. C., *et al.*, A demountable, bakeable, mechanically sealed, vacuum-tight beryllium window, *J. Sci. Instr.* **42,** 39 (1965) [Sec. 72.9].

1375. Denton Vacuum Inc. (8 Fellowship Rd, Cherry Hill Ind. Center, Cherry Hill, N.J, USA) *Vacuum System Accessories* (catalogue) [Sec. 43.3].

1376. EICHENBAUM A. L., *et al.*, Use of silver chloride seals for cesium vapor devices, *Rev. Sci. Instr.* **35,** 1056 (1964) [Sec. 3.4].

1377. Electrovac (Wien XIX, Rampengasse 5, Austria), *Hermetic Seals* (Catalogue) [Sec. 43.3].

1378. General Electric (Schnectady, N.Y 12305, USA), *Vacuum Products* (Catalogue 1965) [Sec. 43.3; 61.35 f; 71.31].

1379. General Technology Corp (Torrance and Hayward, Calif. USA), *Vacuum Products and Accessories* (Catalogue) [Sec. 61.32 h].

1380. Goddard Valve Division (705 Plantation Street, Worcester, Mass., USA), *High Vacuum Valves* (Catalogue) [Sec. 61.32 h].

1381. GIFFORD F. E., and DOLENGA A., Direct glass–metal seals, *Rev. Sci. Instr.*, **35,** 591 (1964) [Sec. 24.11].
1382. GROSS F., Über Konzentrische Mehrfacheinschmelzungen nach dem Druckglasprinzip, *Vakuum Techn.*, **14,** 97 (1965) [Sec. 24.43].
1383. HARPER M. E., and CARTER R., Vacuum brazing as a production technique for vacuum equipment, *Trans. 3rd International Vacuum Congress*, Vol. 2, Pergamon Press, Oxford, 1966 [Sec. 22.21].
1384. HAUFF A., KIENEL G., and WAMSER A., Aufbau und Wirkungsweise eines hydraulisch betätigten Ultravakuum–Durchgangsventils, *Vacuum–Technik*, **14,** 139 (1965) [Sec. 61.34b; 61.35f].
1385. HAWRYLAK R. A., CLINE D., and GRUBE D. F., An improved strip sealed continuous rotating scattering chamber, *Nucl. Instr. and Methods*, **36,** 237 (1965) [Sec. 51.78]
1386. HEYMANN D. and KEUR E., An all–metal gas pipette for use in mass spectrometry, *J. Sci. Instr.* **42,** 121 (1965) [Sec. 61.32g].
1387. HORNBECK G. H., High temperature seal of sapphire windows to ceramics, *Rev. Sci. Instr.* **36,** 845 (1965) [Sec. 72.2].
1388. HOUWINK R., and SALOMON G. (Ed.), *Adhesion and Adhesives*, Elsvier Publ. Corp., Amsterdam, 1965 [Sec. 3.2; 3.3].
1389. JEPSON S. R., and THIEMANN J. W., Test methods for vacuum–tight ceramic–metal joints, *Bull. Amer. Ceram. Soc.*, **43** (4), 355 (1964) [Sec. 25.5].
1390. JOHNSON H., and DESLATTES R. D., Production of alumina windows, *Rev. Sci. Instr.*, **36,** 1381 (1965) [Sec. 71.36c].
1391. KIMBER G. M. *et al.*, Method of mounting silica windows for shock tubes, *J. Sci. Instr.*, **42,** 44 (1965) [Sec. 25.25].
1392. KINDIG N. B., and SPICER W. E., Vacuum system and cleaving mechanism for photoemission measurement of CdS single crystals in the vacuum ultraviolet, *Rev. Sci. Instr.*, **36,** 759 (1965) [Sec. 72.6].
1393. KIRCHMER K., A 19-inch diameter bakeable metal–ceramic seal, *Trans. AVS Nat. Symp. 1963*, p. 170, Macmillan, N.Y. 1964 [Sec. 25.42].
1394. KLEINTEICH R., Die Kombinationen der Glas-Metallverschmelzung, *G.I.T. Glas-Instrum. Techn.* **8,** 553, 705, 778 (1964) and **9,** 22 (1965) [Sec. 24.1].
1395. KONDRASHEV L. F., *et.al.*, Thin vacuum-tight walls, *Instr. Exp. Techn.* No. 4, 625 (1960) [Sec. 71.31b].
1396. LEWIN G., *Fundamentals of Vacuum Science and Technology*, McGraw-Hill, New York, 1965 [Sec. 11.3].
1397. LEWIN G., and MULLANEY D., A fast acting ultrahigh vacuum gas inlet valve and its use in the production of special non-stationary pressure distribution in the C Stellarator, *Trans. AVS Nat. Symp. 1963*, p. 176, Macmillan, New York, 1964 [Sec. 61.35f].
1398. MACK J. L., and WILMOT G. B., Vacuum-tight seal for low-temperature optical windows, *Rev. Sci. Instr.*, **46,** 1265 (1965) [Sec. 14.32; 72.8].
1399. Mason–Renshaw Ind. (Box 157, Roslyn, Long Island, N.Y. USA), *Flexible Stainless Steel Tubing* (Catalogue 1965) [Sec. 51.3].
1400. Materials Research Corp. (Orangeburg, N.Y. USA), *Vacuum Collars and Components* (Catalogue) [Sec. 43.3].
1401. MAUGIS H., Brassage d'élements déstinés á la technique de vide, *Le Vide*, **19,** (111) 141 (1964) [Sec. 22.2].
1402. MEYER H., and WILDE E., Ausheizbare Thermoelementdurchführung für Hochvakuumapparaturen, *Exper. Tech. der Physik*, **12,** 447 (1965) [Sec. 4.2].
1403. Microperv Inc. (Cedar Grove, N.Y. USA), Pinch-off tools (Catalogue 1965) [Sec. 26.31].
1404. MOORE B. C., Vacuum sealing rough surfaces, *J. Vac. Sci. Technol.*, **2,** (3), 160 (1965) [Sec. 38.33; 38.51].
1405. MOTOVILOV O. A., Device for transmitting rotation into a vacuum chamber, *Instr. Exp. Techn.*, No. 3 (May–June), 717 (1964) [Sec. 52.2].

1406. NEAL R. B., The Stanford Two-Mile linear electron accelerator, *J. Vac. Sci. Technol.*, **2** (3) 149 (1965) [Sec. 22.32].

1407. NOBES M., Improved method for the production of thin self-supporting carbon foils, *J. Sci. Instr.*, **42**, 753 (1965) [Sec. 71.36].

1408. NORTHROP D. A., Metal reaction tubes employing cold-weld seals, *Rev. Sci. Instr.*, **36**, 400 (1965) [Sec. 26.31].

1409. OWEN C. W., *et.al.*, Vacuum system for MURA 50–MeV electron accelerator VIII, *Rev. Sci. Instr.*, **35**, 1444 (1964) [Sec. 38. 32].

1410. PEACOCK R. N., High current feedthrough assembly for ultrahigh vacuum systems, *J. Vac. Sci. Technol.*, **2**, (3), 161 (1965) [Sec. 43.2].

1411. PERESSE I., Passage étanche pour grande vitesse de rotation, *J. Phys. Appl.*, **26**, (3), 121 (1965) [Sec. 51.8].

1412. PUPP W., *Vakuumtechnik*, K. Thiemig, München, 1964.

1413. Radio Corporation of America (Lancaster, Pa., 17604, USA) *Vacuum Components* (Product Guide 1965) [Sec. 38.5; 43.3; 61.35 f; 71.3].

1414. RATHBUN F. O., Metal-to-metal and metal-gasketed seals for extreme environment applications, *Proc. Aerospace Fluid Power Syst. Equipm. Conf.* Los Angeles, May 1965, p. 156, SAE, New York.

1415. REED L., and MCRAE R. C., Evaporated metallizing on ceramics, *Bull. Amer. Ceram. Soc.*, **44** (1), 12 (1965) [Sec. 25.25].

1416. ROACH W. R., and WHEATLEY J. C., Electrical lead seals for cryogenic applications, *Rev. Sci. Instr.* **35**, 634 (1964) [Sec. 14.32; 43.1].

1417. ROBBINS R. F., and LUDTKE P. R., Review of static seals for cryogenic systems, *J. Spacecraft*, **1** (3), 253 (1964) [Sec. 14.32].

1418. ROBERTS R. W., and BAHM E. L., Modified Hoke valve for ultrahigh vacuum use, *J. Vac. Sci. Technol.*, **2**, 89 (1965) [Sec. 61.35f].

1419. RODGERS R., A pumpable vacuum valve for sealed-off metal cryostats, *J. Sci. Instr.*, **42**, 724 (1965) [Sec. 61.35a].

1420. ROSEBURY F., *Handbook of Electron Tube and Vacuum Techniques*, Addison–Wesly, Reading, Mass. 1965 [Sec. 11.3].

1421. ROTH A., and AMILANI A., Sealing factors, their measurement and use in the design of vacuum gasket seals, *Trans. 3rd Internat. Vacuum Congress*, Pergamon Press, Oxford, 1966 [Sec. 38.1].

1422. ROTH A., Nomographic design of vacuum gasket seals, *Vacuum* **16** (1966) in Press [Sec. 38.1].

1423. SHEFFIELD J. C., Ultrahigh vacuum gate valve, *Rev. Sci. Instr*, **36**, 1269 (1965) [Sec. 61.35f].

1424. STEVENS R. H., Fluxless internal solder seal for metal tubing, *Rev. Sci. Instr.*, **36**, 1059 (1965) [Sec. 26.33].

1425. STRUNIN V. P., and FRANKEVICH E. L., Preparation of glass and quartz diaphragms for mass spectrometer filling system, *Instr. Exp. Techn.* No. 2 (Mar.–Apr.), 443 (1964) [Sec 61.42].

1426. TOWE K. M., Carbon films for electron microscopy; A reliable method for stripping from glass surface, *Rev. Sci. Instr.*, **36**, 1247 (1965) [Sec. 71.36].

1427. Ulvac (217 Idogaya, Shimo-cho, Miniami-Ku, Jokohama, Japan), *Vacuum Equipment* (Catalogue) [Sec. 11.3].

1428. Vacuum Engineering Comp. (49 Pollard Street, North Billerica, Mass. 01862, USA), *Vacuum Equipment* (Catalogue) [Sec. 61.35f].

1429. Vacuum Processes Inc., (P.O. Box 523, Richardson, Texas 75081. USA), *Vac-U-lok* (Catalogue) [Sec. 61.3].

1430. VEILLON C., and WINEFORDNER J. D., Inexpensive liquid–metal glass vacuum valve, *Rev. Sci. Instr.*, **36**, 229 (1965) [Sec. 61.32b].

1431. VOGL T. P., *et.al.*, Vacuum tight windows with wide band transmission characteristics, *Rev. Sci. Instr.*, **36**, 1439 (1965) [Sec. 71.33; 72.6].

1432. Whittaker Corp. (16217 Lindbergh Street, Van Nuys, Calif, 91406, USA) *Vacuum Products* (Catalogue) [Sec. 61.35f].

1433. WIESEMANN K., Ein hochheizbare Präzisions-Schiebedurchführung für zwei Koordinaten, *Vakuum-Techn.* **14**, 45 (1965) [Sec. 51.4].

1434. ZURMÜHLE R. W., A scattering chamber for charged particle work, *Nucl. Instr. and Methods.* **36**, 168 (1965) [Sec. 51.78].

INDEX